Groundwater Mechanics

Groundwater Mechanics

OTTO D.L. STRACK

Professor of Civil Engineering
Department of Civil and Mineral Engineering
University of Minnesota
Minneapolis, Minnesota 55455

PRENTICE HALL *Englewood Cliffs, New Jersey 07632*

Library of Congress Cataloging-in-Publication Data

Strack, Otto D. L.
  Groundwater Mechanics.

  Bibliography: p. 675
  Includes index.
  1. Groundwater Flow—Mathematical models. I. Title.
TC176.S77 1988 551.49 87-18714
ISBN 0-13-365412-5

Editorial/production supervision
and interior design: *Mary Jo Stanley, Carolyn Fellows, and Debbie Young*
Cover design: *Edsal Enterprises*
Manufacturing buyer: *Cindy Grant and Mary Noonan*

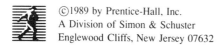

© 1989 by Prentice-Hall, Inc.
A Division of Simon & Schuster
Englewood Cliffs, New Jersey 07632

All rights reserved. No part of this book may be
reproduced, in any form or by any means,
without permission in writing from the publisher.

Printed in the United States of America
10 9 8 7 6 5 4 3 2 1

ISBN 0-13-365412-5

Prentice-Hall International (UK) Limited, *London*
Prentice-Hall of Australia Pty. Limited, *Sydney*
Prentice-Hall Canada Inc., *Toronto*
Prentice-Hall Hispanoamericana, S.A., *Mexico*
Prentice-Hall of India Private Limited, *New Delhi*
Prentice-Hall of Japan, Inc., *Tokyo*
Simon & Schuster Asia Pte. Ltd., *Singapore*
Editora Prentice-Hall do Brasil, Ltda., *Rio de Janeiro*

To Andrine

Contents

Preface   xvii

Introduction   1

    1   Overview   1

1   Basic Equations   5

    2   Basic Concepts   5
         The Specific Discharge, 6
         Pressure and Head, 7

    3   Darcy's Law   8
         Rectilinear Flow, 8
         Intrinsic Permeability, 10
         Range of Validity of Darcy's Law, 10
         General Form of Darcy's Law, 11
         Anisotropy, 13

    4   Continuity of Flow   15

    5   Fundamental Equations   16

2 Types of Flow — 19

6 Horizontal Confined Flow 20
One-dimensional Flow, 20
Basic Equations, 22
Radial Flow, 25
Two-dimensional Flow, 27
Partial Penetration, 33
Piezometric Contours, 34

7 Shallow Unconfined Flow 36
Basic Equations, 36
Differential Equation for the Discharge Potential, 40
One-dimensional Flow, 40
Two-dimensional Flow, 42

8 Combined Shallow Confined and Unconfined Flow 48
Method of Solution, 49
One-dimensional Flow, 51
Radial Flow, 52
Dewatering Problem, 53
*Implementation of Uniform Flow with Wells
  in the Fortran Program SLW, 56*

9 Shallow Unconfined Flow with Rainfall 74
Basic Equations, 74
One-dimensional Flow, 76
Radial Flow, 77
Superposition, 79
Radial Flow Toward a Well, 79
Rainfall on an Island with an Elliptical Boundary, 80
Local Infiltration, 82
Implementation of Infiltration in SLW, 89

10 Shallow Interface Flow 96
Condition Along the Interface, 96
Potentials, 98
Shallow Confined Interface Flow, 99
*Combined Shallow Confined Interface Flow
  and Confined Flow, 101*
Shallow Unconfined Interface Flow, 106
*Combined Shallow Unconfined Interface Flow
  and Unconfined Flow, 108*
Shallow Unconfined Interface Flow with Rainfall, 111

11 Aquifers with Vertically Varying Hydraulic Conductivity 112
Continuous Discharge Potentials, 113
Flow in an Aquifer with an Arbitrary Number of Layers, 115
Girinskii Potentials, 118
Implementation of Layering in SLW, 120

Contents

12 Shallow Flow in Aquifers with Clay Laminae   126
    Basic Equations and Method of Solution, 127
    One-dimensional Shallow Flow in Aquifers
      with Clay Laminae, 133
    Two-dimensional Shallow Flow in Aquifers
      with Clay Laminae, 139
    Upper and Lower Aquifers with Different Hydraulic
      Conductivities, 142

13 Shallow Interface Flow in Aquifers
    with Impermeable Laminae   143
    The Single-Aquifer Zone, 144
    The Upper Aquifer, 144
    The Lower Aquifer, 145
    The Conditions at the Edge of the Lamina, 148
    One-dimensional Flow, 152
    Validation, 157

14 Shallow Semiconfined Flow   160
    Basic Equations, 160
    General Solution for One-dimensional Flow, 162
    Leakage into an Aquifer Bounded by a Long River, 163
    Axi-Symmetric Flow, 165
    Flow Toward a Well in an Infinite Leaky Aquifer, 167
    Two-dimensional Flow, 170

15 Shallow Flow in Systems of Aquifers Separated
    by Leaky Layers   172
    Semiconfined Systems of Aquifers Separated
      by Leaky Layers, 172
    Example 15.1: Flow with a Well in Each Aquifer, 176
    Confined Systems of Aquifers Separated by Leaky Layers, 178
    Unconfined Systems of Aquifers Separated
      by Leaky Layers, 186

16 Transient Shallow Flow   192
    Transient Shallow Confined Flow, 192
    Transient Shallow Unconfined Flow, 195
    Transient Shallow Flow Toward Wells, 197

17 Two-dimensional Flow in the Vertical Plane   203
    Basic Equations, 203
    Horizontal Drains, 205
    The Graphical Method, 206
    Anisotropy, 212

18 Three-dimensional Flow   215
    The Point-Sink, 215
    The Method of Images, 216

3 Elementary Harmonic Solutions 219

19 Some Properties of Harmonic Functions 219
  The Maximum Modulus Theorem, 219
  Boundary Value Problems, 221
  The Stream Function Ψ, 221
  The Stream Function for Uniform Flow, 225
  The Stream Function for a Well, 226
  Streamlines for Uniform Flow with a Well, 228
  Flow Net for Flow from a Recharge Well Toward a Well, 231
  Example 19.1: Cooling Problem, 234
  Example 19.2: Intercepting Flow Between Wells, 236

20 The Method of Images for a Circle 241
  The Location of the Image, 241
  Example 20.1: An Eccentric Well in a Circular Island, 241
  Example 20.2: An Eccentric Well in a Circular Island with Rainfall, 243
  Example 20.3: A Circular Lake, 244
  A Lake in a Uniform Flow, 245
  The Potential and the Stream Function for a Dipole of Arbitrary Orientation, 248
  Streamlines for a Lake in a Uniform Flow, 250
  Flow Around a Cylindrical Impermeable Object, 253

21 The Vortex 255
  The Potential and the Stream Function for a Vortex, 255
  Superposition, 256
  The Method of Images for Vortices, 257

22 Algebra of Complex Numbers 262
  The Complex Plane, 263
  Elementary Algebraic Operations, 263
  The Complex Conjugate, the Modulus, and the Argument, 265
  Polar Coordinates, 266

23 Complex Functions 269
  A Complex Function, 269
  Examples of Analytic Functions, 271
  Some General Properties of Analytic Functions, 275

24 Complex Functions in Groundwater Flow 278
  The Complex Potential and the Discharge Function, 278
  The Uniform Flow, 278
  The Well, 279
  The Dipole, 279
  The Vortex, 279
  A Lake in a Uniform Flow, 280
  Unconfined Flow Toward a Drain, 281

25 Complex Line Integrals 283
  The Line-Sink, 283
  The Line-Dipole, 291
  The Line-Doublet, 297
  Singular Cauchy Integrals, 298
  The Computer Program SLWL, 302

4 Fluid Particle Paths and Solute Transport 311

26 Numerical Determination of Fluid Particle Paths 311
  A Procedure for Tracing Streamlines
    Using the Stream Function, 312
  A Procedure for Tracing Streamlines and Path Lines
    Using the Velocity Vector, 313
  Determination of Streamlines in SLWL, 318

27 Three-dimensional Streamlines in
  Dupuit-Forchheimer Models 319
  Steady Shallow Flow, 320
  Transient Shallow Flow, 337

28 Solute Transport 340
  Simplifying Assumptions, 340
  Solute Movement, 340
  Transport Equation, 341
  Integration Along Characteristics, 343
  The Distribution Factor, 345
  Contaminant Transport in a Layered Aquifer, 345

5 Applications of Conformal Mapping 349

29 The Principles of Conformal Mapping 349
  The Properties of a Conformal Map, 350
  Solving Boundary-Value Problems, 352

30 An Application of the Bilinear Transformation 352
  The Bilinear Transformation, 353
  A Lake in an Aquifer with Wells and a Uniform Far-Field, 355

31 An Elliptic Lake in an Infinite Aquifer with Wells
  and a Uniform Far-Field 358
  The Mapping Function, 358
  An Elliptic Lake in a Uniform Flow, 361
  An Elliptic Lake with a Well, 363
  A Finite Canal, 363
  An Elliptic Impermeable Object, 364
  Flow in Two Confined Aquifers Separated
    by an Impermeable Layer with an Elliptic Opening, 365

32 Applications of the Schwarz-Christoffel Transformation 369
   The Transformation, 371
   Determination of the Parameters, 373
   Mapping the Upper Half Plane onto the Exterior
      of a Polygon, 374
   Flow Toward a Well Between Two Long Parallel Rivers, 375
   Flow Toward a Well Between Two Intersecting Rivers, 377
   Flow Underneath an Impermeable Dam, 379
   Flow Toward a Narrow Partially Penetrating Ditch
      in a Confined Aquifer, 381

33 Transformations with Piecewise Constant Argument 389
   Polygons that Contain both the Origin and Infinity, 390
   Polygons that Contain Infinity and Not the Origin, 392
   Polygons that Contain the Origin and Not Infinity, 395

6 Regional Aquifer Modeling Using the Analytic Element Method   403

34 Modeling a Single Homogeneous Aquifer 404
   Functions with Given Coefficients, 406
   Functions with Unknown Coefficients, 408
   The System of Equations, 409
   Type of Flow, 410

35 Modeling Areal Inhomogeneities 411
   Boundary Conditions for an Inhomogeneity
      in the Hydraulic Conductivity, 411
   Boundary Conditions for an Inhomogeneity
      in the Hydraulic Conductivity, the Aquifer Thickness,
      and the Base Elevation, 412
   A Cylindrical Inhomogeneity, 414
   The First-Order Line-Doublet, 415
   The System of Equations, 418

36 Leaky or Draining Objects 420
   The Boundary Condition Along a Leaky Object, 421
   The Boundary Condition Along a Draining Object, 422
   Modeling Thin Leaky Objects by Using Line-Doublets, 423
   Modeling Thin Impermeable Walls Using Line-Doublets, 424
   Modeling Narrow Draining Objects Using Line-Dipoles, 424
   Modeling Narrow Ideal Drains, 425

37 Variable Infiltration 425
   Infiltration Through River- and Lake-Bottoms, 425
   The Potential for an Area-Sink, 426
   Applications of Area-Sinks, 442

38 Elements of Order Higher Than One 447
   The Second-Order Line-Sink, 449

Contents   xiii

          The Second-Order Line Doublet, 450
          The Second-Order Triangular Area-Sink, 453

39   Analytic Elements for Thin Isolated Features   460
          The Double-Root Dipole, 461
          The Double-Root Doublet, 465
          Applications of Double-Root Elements, 465

40   Tip Elements   468
          The Single-Root Line-Dipole, 469
          The Single-Root Line-Doublet, 473

41   Curvilinear Elements   473
          The Complex Potential for a General Line-Dipole, 473
          Mapping a Circular Element on a Straight Line, 475

42   Analytic Elements for Closed Boundaries   479
          Two Sided Analytic Elements for Closed Boundaries, 479
          An Analytic Element for an Elliptic Inhomogeneity, 486

43   Modeling Flow in Two Aquifers Separated
      by a Thin Layer   491
          Formulation of the Problem, 493
          Homogeneous Hydraulic Conductivity, 493
          Inhomogeneous Hydraulic Conductivity, 495
          Leakage Through the Separating Bed, 496
          The Regional Effect of the Tennessee-Tombigbee
            Waterway, 499

44   Transient Effects, Three-dimensional Effects,
      and Computer Programs   506
          Transient Effects, 506
          Three-dimensional Effects, 507
          Implementations in Computer Models, 508

7 Solving Problems with Free Boundaries by Conformal Mapping   511

45   Boundary Types   512

46   The Hodograph   516
          Basic Equations, 516
          A Straight Equipotential, 516
          A Straight Streamline, 517
          A Straight Seepage Face, 517
          A Straight Outflow Face, 517
          A Phreatic Surface Without Infiltration or Evaporation, 518
          A Phreatic Surface with an Infiltration Rate N, 519
          An Interface Between Fresh Groundwater
            and Salt Groundwater at Rest, 521
          Example 46.1: Unconfined Flow Toward a Drain, 521

47	The Reference Function 523	
	Straight Boundary Segments, 524	
	A Phreatic Surface, 524	
	An Interface, 525	
	Other Auxiliary Functions, 526	
48	Outflow from a Layered Slope 530	
	Boundary Conditions, 531	
	The Hodograph for the Unconfined Case, 532	
	The Confined Case, 533	
49	Flow Toward a Vertical Embankment 539	
	The Hodograph and the Function $W_1 = W_1(\zeta)$, 539	
	The Zhukovski Function, 540	
	The Function $z = z(\omega)$, 542	
50	Unconfined Flow with a Semi-infinite Impermeable Lamina 544	
	The Hodograph and the Function $W_1 = W_1(\zeta)$, 544	
	The Function $\Omega = \Omega(\zeta)$, 547	
	The Function $z = z(\zeta)$, 547	
	Determination of the Parameters in the Solution, 548	
51	Interface Flow from an Equipotential Toward Drains 551	
	Flow from a Single Horizontal Equipotential Toward a Drain, 551	
	Flow Toward the Sea, 558	
52	Superposition; Problems with Drains 561	
	Superposition in the Upper Half Plane $\Im\zeta \geq 0$, 561	
	Application, 563	
	Problems with One Drain, 565	
53	Flow from Recharge Drains Toward Drains in a Phreatic Aquifer 569	
	Flow from a Single Recharge Drain, 570	
	Flow with an Arbitrary Number of Drains and Recharge Drains, 572	
	Solution For a System of One Recharge Drain and One Drain, 574	
54	Anisotropic Hydraulic Conductivity 577	
	Transformation of the Domain, 577	
	Boundary Conditions Along a Phreatic Surface Without Infiltration, 579	
	Boundary Conditions Along a Phreatic Surface with an Infiltration Rate N, 580	
	Flow from One Recharge Drain Toward a Drain in an Anisotropic Aquifer, 582	

8 The Boundary Integral Equation Method — 587

- 55 Green's Second Identity — 587
- 56 Boundary Integral Formulation in Two Dimensions — 589
  - *Evaluation of the Potential at Boundary Points, 591*
- 57 Boundary Integral Equation Method in Terms of Complex Variables — 592
  - *Complex Variable Formulation of the Boundary Integral, 592*
  - *Relation with Singular Cauchy Integrals, 593*
  - *Evaluation of the Complex Potential at Boundary Points, 595*
  - *Application: Flow in a Strip, 595*
- 58 Solving Boundary-Value Problems — 600
  - *Division into Subdomains, 600*
- 59 Boundary Integral Formulation in Three Dimensions — 603

9 Finite Difference and Finite Elements Methods — 605

- 60 Finite Difference Methods — 605
  - *Steady Flow, 606*
  - *Transient Flow, 609*
- 61 The Finite Element Method — 613
  - *The Variational Principle, 613*
  - *The Finite Element Method for Steady Flow, 616*
  - *Example 61.1: Flow Through a Strip, 617*
  - *The Finite Element Method for Transient Flow, 620*
  - *Formulation in Terms of Potentials, 621*

10 Analogue Methods — 625

- 62 The Membrane Analogue — 626
- 63 The Electric Analogue — 628
- 64 The Hele Shaw Analogue — 629
- 65 Scale Models; Measuring Hydrodynamic Dispersion — 631

A Expansion of a Function Raised to a Power — 633

B The Dilogarithm of Complex Argument — 637

The Many-Valued Dilogarithm, 638
Separation into Real and Imaginary Parts, 639
The Boundary Condition Along the Phreatic Surface for the Problem Solved in Section 50, 640

| C | Computer Programs SLW and SLWL | 643 |
| D | Additional Reading | 671 |

General Reading Material, 671
Selected Topics, 672

References	675
List of Symbols	683
Author Index	719
Subject Index	723

Preface

The present text evolved from a series of lecture notes written over the past twelve years. Nearly all of the material was taught in one form or another at the University of Minnesota. The material covered in Chapter 1, about half of Chapter 2, most of Chapters 3 and 4, and parts of Chapters 6, 8, 9, and 10 are taught in two courses at the undergraduate level. The material covered in the remaining parts of this text is taught at the graduate level.

The reader is expected to be familiar with elementary algebra, and to have some insight into the geology relevant to groundwater flow problems. The degree of difficulty of the material increases through the text; the reader is educated gradually in the necessary mathematics, using groundwater flow, wherever possible, as a vehicle for explaining the theorems and equations. The mathematical developments are thus part of the text nearly everywhere; two isolated mathematical developments are covered in appendices.

Three computer programs, one in BASIC and two in FORTRAN, are presented in this text; the programs are contained on a diskette provided with the book. Instructions for using the diskette are given in Appendix C.

One chapter is devoted to the analytic element method, which consists simply of approximating solutions by superposing analytic functions suitable for modeling aquifer features. This method is closely related to indirect boundary element methods but differs from these in that the elements are not necessarily obtained by distributing singularities along boundary segments, and that the analytic elements need not be boundary elements. The analytic element method may be viewed as an adaptation of classical analytic techniques for use in digital computers. Work on the analytic element method started eleven years ago in an attempt to model the regional effect of the Tennessee-Tombigbee Waterway, under contract with the Nashville District of the United States Army Corps of

Engineers. The method has been further developed and implemented in several computer programs with support from the University of Minnesota, Battelle Pacific Northwest Laboratories, the United States Bureau of Mines, Barr Engineering Company, and the J. L. Shiely Company. Several engineering companies in Minnesota applied computer programs based on the analytic element method to problems of regional groundwater flow. An analytic element model for flow in systems of up to four aquifers is currently under development at the University of Minnesota, supported by a grant from the Legislative Commission for Minnesota Resources.

I would like to express my appreciation to the agencies mentioned above for their support. The U.S. Army Corps of Engineers, Nashville District, kindly granted permission to include the regional modeling of the effect of the Tennessee-Tombigbee Waterway as an example in the text. The J. L. Shiely Company gave permission to use the regional modeling of the effect of mining on Grey Cloud Island, Minnesota, as an example.

Interaction with engineers and geologists in the field is an important factor in the development and communication of methods of groundwater modeling. The opportunity given to me by Barr Engineering Company to remain in continuous contact with field problems has been of great assistance in obtaining examples from practice. Five major projects carried out with the highly valued assistance of four geologists and engineers have contributed, either directly or indirectly, to the material contained in this text. These are the modeling of the aquifer system in the neighborhood of the Tennessee-Tombigbee Waterway, which was done in close cooperation with Joe Melnyk from the Nashville District of the U.S. Army Corps of Engineers, the modeling of the aquifer system near Grey Cloud Island, which was done with Kelton Barr from Barr Engineering Company, the modeling of a site in the Twin Cities area, which was carried out by Jeff Stoner from the United States Geological Survey, Minnesota, the modeling of three-dimensional flow in the neighborhood of a projected WP cave, and the modeling of flow in a coastal aquifer near a nuclear repository in Sweden. The latter two projects were carried out with Charlie Axelsson from Geosystems AB, Sweden. Although only the first two projects are described in this text, all five projects have contributed to the material in this book.

I thank my former teachers, Professors G. de Josselin de Jong and Arnold Verruijt, for their continued support. I am indebted to Arnold Verruijt, Bruce Hunt, and Kelton Barr for their many constructive comments. I also thank Ed Veling for his comments.

I thank Willem Jan Zaadnoordijk, for screening the text and for writing several dedicated computer programs used to generate some of the drawings in Secs. 34, 53, and 54. I thank R. W. Macneal for preparing the BASIC program presented in Sec. 12, and for writing the computer program used to produce the computer-generated drawings in Sec. 13. LeeAnn Hammerbeck wrote a computer program used to produce the contour plots in Figures 6.8 and 6.27. The plot reproduced in Figure 7.29 was generated by a computer program written by Gordon Heitzman. The computer program used to produce the flow nets in Sec. 48 was written by T. Gray Curtis. Many students who attended my courses over the years made constructive comments.

I thank Charles Fairhurst, Head of the Department of Civil and Mineral Engineering of the University of Minnesota, for his continuous encouragement and support over the past twelve years. I am very grateful to Henk Haitjema, with whom I had the opportunity to discuss much of the material presented here, and who wrote the computer program SYLENS, used to model the effect of the Tennessee-Tombigbee Waterway. The first draft of this book was typed by my wife, Andrine, using a wordprocessor written especially for this purpose. Since then, major portions of the book have been rewritten, and Andrine has retyped it using the technical typesetting program TEX, written by D. Knuth, with the aid of the macro package $\mathcal{A}_\mathcal{M}\mathcal{S}$-TEX. The book is truly a family effort. My son Erik assisted in preparing the final drawings. My mother helped me with literature review, and screened the text and all of the equations, carefully comparing character for character the formulae as they appear in the various versions of the book. My wife assisted enormously in the production by careful typesetting, reading, screening, and commenting. I dedicate this work to her.

Otto D. L. Strack
Minneapolis, Minnesota

Introduction

1 OVERVIEW

The subject matter covered in this text is the mathematical description of fluid flow through porous media, with emphasis on the modeling of steady flow by the use of analytic techniques. There are two major objectives: The first is to instruct the reader in solving elementary groundwater flow problems analytically. Solving such problems requires the ability to approximate the actual flow problem by a simpler one that can be solved analytically, yet still provides insight into the essence of the real problem. Even if the problem cannot be handled adequately by simple means and recourse to a numerical solution is necessary, the determination and interpretation of relatively crude approximate solutions often provides an insight that can be used with advantage in selecting and setting up the numerical model that will ultimately be used to solve the problem. The second objective is to instruct the reader in the application and development of models capable of handling complex problems of groundwater flow.

In view of the above two objectives, emphasis is placed on a detailed coverage of methods for solving a variety of problems, rather than on providing a catalogue of existing solutions. A brief overview of related literature is supplied in Appendix D in order to provide the reader with information on approaches and solutions that are not covered in this text.

The subject matter is organized in sections, numbered from 1 through 65; equations are numbered per section. The sections are grouped in chapters, each covering one main topic.

Chapter 1 is concerned with the basic concepts and the derivation of the governing equations. Many groundwater flow problems, in particular problems of regional aquifer flow, are not solved exactly; rather, approximate governing equations are derived for the aquifer configuration considered. The mathematical formulations for the various aquifer configurations are unified by the use of discharge potentials, expressed differently for each type of flow corresponding to a particular aquifer configuration. The mathematical formulations thus become similar and solutions and techniques are often transferable from one type of flow to another.

The most common aquifer configurations and types of flow are covered in Chapter 2, including the basic equations and elementary methods of solution. The idea of formulating groundwater flow problems in terms of discharge potentials is not new, although rarely applied. Girinski [1946] presented discharge potentials for three types of flow; the author (Strack [1976, 1981(a)]) developed this idea further; in this text, from Chapter 2 onward, the mathematical formulation is carried out entirely in terms of discharge potentials.

An important class of groundwater flow problems is governed by a partial differential equation known as Laplace's equation. Solutions to this differential equation are called harmonic functions, and have specific properties. Elementary harmonic functions and their main properties are discussed in the context of groundwater flow problems in Chapter 3. This chapter ends with an introduction to complex variable approaches in groundwater flow.

Chapter 4 is devoted to determining streamlines and pathlines, and to modeling advective solute transport with retardation and decay.

A major simplification of the mathematical formulation of problems governed by Laplace's equation is achieved by the use of complex variable techniques. Complex variable methods were at one time commonly used in the field of groundwater mechanics. Authors of classical texts such as Muskat [1937], Polubarinova-Kochina [1952, 1962], and Aravin and Numerov [1953, 1965] have devoted major parts of their books to complex variable approaches. Although these methods are still covered in many contemporary textbooks on groundwater mechanics (e.g., Verruijt [1982], Bear [1972, 1979]), the emphasis has shifted in the direction of numerical methods. A development toward numerical models based on analytic formulations has recently become evident, however, (e.g., Kirkham and Powers [1971], Van der Veer [1978], Hunt and Isaacs [1981], Strack and Haitjema [1981a, 1981b], Strack [1982], Liggett and Liu [1983], and Sidiropoulos et al [1983]). For two-dimensional problems these techniques are, or can be, formulated with advantage in terms of complex variables. For this reason, a considerable part of the present text is devoted to complex variable techniques. Classical applications of complex variable methods to groundwater flow problems are covered in Chapters 3, 5 and 7. Chapter 6 is devoted to the analytic element method, which is formulated in terms of complex variables. The analytic element method is presented here in full detail for the first time, and is intended primarily for the modeling of regional groundwater flow. The principle of the method is very simple: the superposition of analytic functions, each being suitable for modeling a particular aquifer feature.

A related technique, the boundary integral equation method (Liggett and Liu [1983], Hunt and Isaacs [1981]), is covered in Chapter 8. Numerical methods based on discretization of the governing differential equation, the finite difference, and finite element methods, are discussed in Chapter 9. Textbooks focusing on the application of the latter three methods are available (e.g., Liggett and Liu [1983], Wang and Anderson [1982], Bear and Verruijt [1987]); therefore only the basic principles involved are discussed in this text. Analogue methods, finally, are covered in Chapter 10.

An attempt has been made to make the sections as independent from each other as possible, although cross references are made. The reader thus may skip certain sections and proceed to others without losing the ability to follow the discussion. This holds true particularly for many of the sections in Chapter 2. It is important to note, however, that an understanding of complex variable methods, as covered in Secs. 22 through 25, is a prerequisite for following the discussions in Chapters 5, 6, 7, and part of Chapter 8.

Chapter 1

Basic Equations

Flow of water in a body of soil occurs through interconnected openings between the soil particles. The path of a water particle through the soil is erratic, and its velocity changes continuously: the velocity is large in the small pores and small in the larger ones.

For most engineering groundwater flow problems, it is not necessary to determine the path that the water particles follow in their way through the soil. It usually is sufficient to determine average velocities, average flow paths, the discharge flowing through a given area of soil, or the pressure distribution in the soil. We will, throughout this text, work with averages and ignore the actual paths of flow. The term *rectilinear flow*, for example, will be used when the *average* flow is in one direction.

The theory of groundwater flow is based on a law discovered by the Frenchman Darcy in 1856. After the introduction of the basic concepts, we will discuss the experiment performed by Darcy and present his law in its simplest form. We will then present the generalized form of Darcy's law and the equation of continuity, and finish the chapter by combining these two equations into one governing equation for steady flow of a homogeneous fluid in a porous medium.

2 BASIC CONCEPTS

The basic quantities used to describe groundwater flow are velocity, discharge, pressure, and head. These quantities will be discussed in this section.

The Specific Discharge

The specific discharge is defined as the volume of water flowing through a unit area of soil per unit time. The units of specific discharge are $m^3/(m^2 s)$ or m/s and thus are the same as those of a velocity. Specific discharge sometimes is called discharge velocity, but we will use the term specific discharge to avoid confusion with a velocity. We will represent the specific discharge by the letter q.

The seepage velocity is the average velocity occurring at a certain point of the porous medium. The seepage velocity will be represented as v. It is obtained by dividing the specific discharge by the area of voids present in a unit area of porous medium. If the porosity of the medium is n, then the area of voids per unit area is n and therefore

$$v = \frac{q}{n} \qquad (2.1)$$

Since n is always less than 1, v is always larger than q.

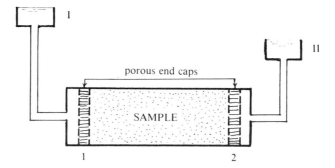

Figure 1.1 Flow through a cylinder.

The concepts of specific discharge and seepage velocity may be illustrated by considering water flowing through a cylindrical tube filled with sand which is contained in the space between the end caps 1 and 2, as is shown in Figure 1.1. The cylinder is filled with water and is connected to two reservoirs, I and II, with different water levels. The water will flow through the cylinder as a result of the difference in water levels. The level of reservoir II is controlled by overflow. Water must be poured into reservoir I in order to maintain its level. By measuring the rate at which water is poured into reservoir I, one determines the total amount, Q, of water flowing through the cylinder per unit time. The specific discharge is found by dividing Q by the cross-sectional area of the cylinder, A, i.e.,

$$q = \frac{Q}{A} \qquad (2.2)$$

The seepage velocity v is found from (2.1) to be

$$v = \frac{q}{n} = \frac{Q}{nA} \tag{2.3}$$

The seepage velocity may be used to find the average time t needed for a water particle to travel through the sample. If the length of the sample is L, the time t is obtained from the relation

$$vt = L \tag{2.4}$$

or

$$t = \frac{L}{v} = \frac{LnA}{Q} \tag{2.5}$$

see (2.3).

Pressure and Head

The hydraulic head at a certain point $\mathcal{P}$ in a soil body is defined as the level to which the water rises in an open standpipe with its lower end at point $\mathcal{P}$ (see Figure 1.2). The hydraulic head is defined as a level and is measured with respect to a reference level or datum. We will represent the hydraulic head by the letter ϕ, and refer to it often simply as *the head*. The units of ϕ are the units of length, m.

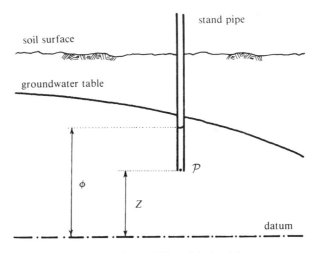

Figure 1.2 Definition of the head ϕ.

The pressure at point $\mathcal{P}$ is found from the weight of the water column above $\mathcal{P}$ in the standpipe. If the elevation of $\mathcal{P}$ above the reference level is Z [m], then the height of the water column above $\mathcal{P}$ is $\phi - Z$. Denoting the pressure as p [N/m^2], the density of water as ρ [kg/m^3] and the acceleration of gravity as g [m/s^2], the pressure at $\mathcal{P}$ is

$$p = \rho g(\phi - Z) \qquad (2.6)$$

The elevation Z of point $\mathcal{P}$ above the reference level is known as the elevation head of point $\mathcal{P}$. The head, ϕ, can be expressed in terms of pressure, p, and elevation head, Z, by the use of (2.6) as follows:

$$\boxed{\phi = \frac{p}{\rho g} + Z} \qquad (2.7)$$

The fraction $p/(\rho g)$, with the units of (N/m^2)/(N/m^3), or m, is known as the *pressure head*. Thus, (2.7) can be given in words as: "head equals pressure head plus elevation head."

3 DARCY'S LAW

Darcy's law (Darcy [1856]) is an empirical relation for the specific discharge in terms of the head. The original form of this law is applicable to rectilinear flow of a homogeneous liquid only. A general form of Darcy's law exists, and will be discussed after treating the case of rectilinear flow.

Rectilinear Flow

Darcy found that the amount of flow through a cylinder of sand of cross-sectional area A increases linearly with the difference in head at the ends of the sample (see Figure 1.3). The head at end cap 1 is ϕ_1. This is seen from Figure 1.3 by noticing that the pipe or hose connecting reservoir I to the sample can be viewed as a standpipe. Similarly, the head at end cap 2 is ϕ_2. Darcy's law for the experiment of Figure 1.3 is

$$Q = kA\frac{\phi_1 - \phi_2}{L} \qquad (3.1)$$

where Q is the discharge [m^3/s], A is the cross-sectional area of the sample [m^2] and L is the length of the sample [m]. The proportionality constant k is known as the hydraulic conductivity. It follows from (3.1) that the dimensions of k are those of a velocity, [m/s]. We will sometimes use the term *resistance to flow*, borrowing this concept from electrokinetics. Resistance to flow is the inverse of hydraulic conductivity ($1/k$).

Writing (3.1) in terms of the specific discharge q, we obtain, with $q = Q/A$,

$$q = k\frac{\phi_1 - \phi_2}{L} \qquad (3.2)$$

Sec. 3   Darcy's Law

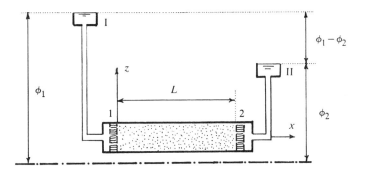

Figure 1.3 Darcy's experiment.

If the head is measured at various points inside the sample of Figure 1.3, it appears to vary linearly over the sample. If a coordinate system is chosen with the x axis running along the axis of the sample and the origin at the end cap 1, we may write the following expression for ϕ,

$$\phi = \phi_1 + \frac{\phi_2 - \phi_1}{L} x \tag{3.3}$$

This equation represents the straight line of Figure 1.4, where the head is plotted as a function of position over the sample. It follows from (3.3) that

$$\frac{d\phi}{dx} = \frac{\phi_2 - \phi_1}{L} \tag{3.4}$$

so that (3.2) may be written as

$$q_x = -k \frac{d\phi}{dx} \tag{3.5}$$

The index x in q_x is used to indicate that the specific discharge is in the x direction. The derivative $d\phi/dx$ is known as the hydraulic gradient for flow in the x direction.

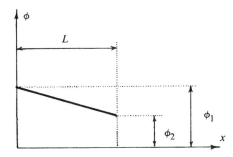

Figure 1.4 Linear variation of ϕ.

Intrinsic Permeability

The hydraulic conductivity is a material constant, which depends upon the properties of both the fluid and the soil. It is possible to define another constant, the intrinsic permeability κ, which depends only upon the soil properties and is related to the hydraulic conductivity as

$$k = \frac{\kappa \rho g}{\mu} \tag{3.6}$$

where μ is the dynamic viscosity [Ns/m^2]. The dimension of κ is m^2. The intrinsic permeability is used primarily when the density or the viscosity of the fluid varies with position. In this text, however, only fluids with homogeneous properties are considered and therefore the classical hydraulic conductivity k is used. Values for k and κ are listed in Table 1.1 for some natural soils.

Table 1.1	*Permeabilities for Some Natural Soils*	
	k [m/s]	κ [m^2]
Clays	$< 10^{-9}$	$< 10^{-17}$
Sandy clays	$10^{-9} - 10^{-8}$	$10^{-16} - 10^{-15}$
Peat	$10^{-9} - 10^{-7}$	$10^{-16} - 10^{-14}$
Silt	$10^{-8} - 10^{-7}$	$10^{-15} - 10^{-14}$
Very fine sands	$10^{-6} - 10^{-5}$	$10^{-13} - 10^{-12}$
Fine sands	$10^{-5} - 10^{-4}$	$10^{-12} - 10^{-11}$
Coarse sands	$10^{-4} - 10^{-3}$	$10^{-11} - 10^{-10}$
Sand with gravel	$10^{-3} - 10^{-2}$	$10^{-10} - 10^{-9}$
Gravels	$> 10^{-2}$	$> 10^{-9}$

(Source: A. Verruijt, *Theory of Groundwater Flow*, © 1970, p. 10. Reprinted by permission of the author.)

Range of Validity of Darcy's Law

Darcy's law is restricted to specific discharges less than a certain critical value. The critical specific discharge depends upon the grain size of the soil and the specific mass and the viscosity of the fluid. The criterion for assessing the validity of Darcy's law in a given case is often expressed in terms of the Reynolds number Re, defined for groundwater flow as follows:

$$Re = \frac{D\rho}{\mu} q \tag{3.7}$$

where D is the average grain diameter [m]. The Reynolds number is dimensionless. The range of validity of Darcy's law is defined by a relation obtained experimentally,

$$Re \leq 1 \tag{3.8}$$

If the Reynolds number is larger than 1, Darcy's law is not valid and other, more complex, equations of motion must be used.

Darcy's law is valid for most cases of flow through soils. This is seen by substituting some average values for q, D, ρ and μ in (3.7). The dynamic viscosity, μ, of water at a temperature of 10^0 C is about $1.3 * 10^{-3}$ N s/m^2 and ρ is about 10^3 kg/m^3. The average particle size of coarse sand is about $0.4 * 10^{-3}$ m in diameter. Substitution of these values in (3.7) yields

$$Re = (0.3 * 10^3)q \tag{3.9}$$

This number will be smaller than 1 if

$$q < 3.3 * 10^{-3} \text{ m/s} = 3.3 \text{ mm/s} \tag{3.10}$$

This is a large value for the flow of groundwater. The hydraulic conductivity ranges from less than 10^{-9} m/s for clays to about 10^{-3} m/s for coarse sands. Furthermore, k is equal to the specific discharge occurring when the hydraulic gradient is 1, a large value. The specific discharge is the product of the hydraulic gradient and k, and will thus usually be less than 10^{-3} m: Darcy's law indeed appears to be valid for most cases of flow through soils.

General Form of Darcy's Law

In practical problems, the flow is rarely rectilinear and neither the direction of flow nor the magnitude of the hydraulic gradient is known. The simple form (3.5) of Darcy's law is not suitable for solving problems; it is necessary to use a generalized form of Darcy's law which gives a relation between the specific discharge vector and the hydraulic gradient. The direction of the specific discharge vector will vary with position and is in the direction of flow. The magnitude of this vector represents the amount of water flowing per unit time through a plane of unit area normal to the direction of flow. In three dimensions, the specific discharge vector has three components. With reference to a Cartesian coordinate system x, y, z, the three components of the specific discharge vector are represented as q_x, q_y, and q_z. The form of Darcy's law for three-dimensional flow through an isotropic porous medium is

$$\begin{aligned} q_x &= -k\frac{\partial \phi}{\partial x} \\ q_y &= -k\frac{\partial \phi}{\partial y} \\ q_z &= -k\frac{\partial \phi}{\partial z} \end{aligned} \tag{3.11}$$

Because the three components of the specific discharge vector are proportional to the three components of the hydraulic gradient, with k as the proportionality factor, the specific discharge vector points in the direction of the hydraulic gradient. We will sometimes represent the specific discharge vector with components (q_x, q_y, q_z) briefly as q_i. The index i then stands for x, y, or z. The partial derivatives $\partial \phi/\partial x$, $\partial \phi/\partial y$, and $\partial \phi/\partial z$ represent the three components of the hydraulic gradient. We may write the components of this vector as $\partial_x \phi, \partial_y \phi$, and $\partial_z \phi$, where the ∂ with the index stands for differentiation with respect to the coordinate represented by the index. The hydraulic gradient can then be written as

$$\partial_i \phi = \left[\frac{\partial \phi}{\partial x}, \frac{\partial \phi}{\partial y}, \frac{\partial \phi}{\partial z}\right] \tag{3.12}$$

The notation with indices is known as the indicial notation or tensor notation, and has the advantage of compactness. The three equations (3.11), for example, can be written as one,

$$q_i = -k\partial_i \phi \tag{3.13}$$

Darcy's law (3.11) may be written in terms of pressure by the use of (2.6),

$$\begin{aligned} q_x &= -\frac{k}{\rho g}\frac{\partial p}{\partial x} \\ q_y &= -\frac{k}{\rho g}\frac{\partial p}{\partial y} \\ q_z &= -k - \frac{k}{\rho g}\frac{\partial p}{\partial z} \end{aligned} \tag{3.14}$$

where the z coordinate points vertically upward, so that $\partial Z/\partial z = 1$ (see Figure 1.2). Equations (3.11) and (3.14) are equivalent only if the density of the fluid is constant. Equation (3.11) is wrong for variable ρ, as may be demonstrated, following Verruijt [1970], by considering the case of groundwater of variable density at rest, so that $q_x = q_y = q_z = 0$. We integrate (3.11) and use (2.7) to express ϕ in terms of the pressure,

$$\phi = \text{constant} = \frac{p}{\rho g} + Z \tag{3.15}$$

Integration of (3.14) yields for this case

$$p = -\int_{Z_0}^{Z} \rho g dz \tag{3.16}$$

where Z_0 is a reference level. Since (3.15) is not applicable to water of variable density at rest, and is obtained from (3.11), it follows that the latter equation cannot be used for cases of variable density, at least not with the definition (2.6) for ϕ. Equation (3.16), however, is correct and (3.14) is indeed valid for variable density.

Cases where variable density must be considered are not covered in this text, and Darcy's law will be used in the form (3.11), with ϕ defined by (2.7).

Anisotropy

We assumed thus far that the hydraulic conductivity k is the same in all directions. In practice, however, the soil often is layered with the hydraulic conductivity being different in the directions parallel and normal to the layers. This is illustrated in Figure 1.5(a), where layers of sand are sandwiched between thin layers of clay. We consider the case that there is no flow normal to the plane of drawing, the x, y plane. We introduce Cartesian coordinates x^* and y^* such that the x^* axis is parallel to the layers. It follows from Figure 1.5(b) that

$$x = x^* \cos \alpha - y^* \sin \alpha$$
$$y = x^* \sin \alpha + y^* \cos \alpha$$
(3.17)

We write Darcy's law in terms of the x^*, y^* coordinate system, denoting the components of the specific discharge vector in the x^* and y^* directions as q_x^* and q_y^*:

$$q_x^* = -k_1 \frac{\partial \phi}{\partial x^*}$$
$$q_y^* = -k_2 \frac{\partial \phi}{\partial y^*}$$
(3.18)

where k_1 and k_2 represent the values of the hydraulic conductivity in the directions parallel and normal to the layers. These directions are called the principal directions, and k_1 and k_2 the principal values of the hydraulic conductivity.

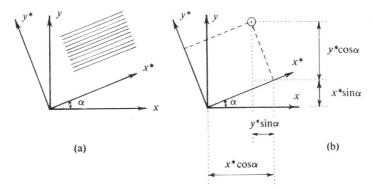

Figure 1.5 Anisotropic hydraulic conductivity.

It is possible to write Darcy's law in terms of vector components in the x and y directions. The expressions for q_x and q_y in terms of q_x^* and q_y^* are similar to (3.17):

$$q_x = q_x^* \cos \alpha - q_y^* \sin \alpha$$
$$q_y = q_x^* \sin \alpha + q_y^* \cos \alpha$$
(3.19)

Using (3.18), we obtain

$$q_x = -k_1 \frac{\partial \phi}{\partial x^*} \cos \alpha + k_2 \frac{\partial \phi}{\partial y^*} \sin \alpha$$
$$q_y = -k_1 \frac{\partial \phi}{\partial x^*} \sin \alpha - k_2 \frac{\partial \phi}{\partial y^*} \cos \alpha \quad (3.20)$$

By application of the chain rule we find

$$\frac{\partial \phi}{\partial x^*} = \frac{\partial \phi}{\partial x}\frac{\partial x}{\partial x^*} + \frac{\partial \phi}{\partial y}\frac{\partial y}{\partial x^*} = +\frac{\partial \phi}{\partial x} \cos \alpha + \frac{\partial \phi}{\partial y} \sin \alpha$$
$$\frac{\partial \phi}{\partial y^*} = \frac{\partial \phi}{\partial x}\frac{\partial x}{\partial y^*} + \frac{\partial \phi}{\partial y}\frac{\partial y}{\partial y^*} = -\frac{\partial \phi}{\partial x} \sin \alpha + \frac{\partial \phi}{\partial y} \cos \alpha \quad (3.21)$$

where the partial derivatives $\partial x/\partial x^*$, $\partial y/\partial x^*$, $\partial x/\partial y^*$, and $\partial y/\partial y^*$ are obtained by differentiating (3.17). Combining (3.20) and (3.21) we obtain Darcy's law for anisotropic hydraulic conductivity:

$$q_x = -k_{xx}\frac{\partial \phi}{\partial x} - k_{xy}\frac{\partial \phi}{\partial y}$$
$$q_y = -k_{yx}\frac{\partial \phi}{\partial x} - k_{yy}\frac{\partial \phi}{\partial y} \quad (3.22)$$

where

$$k_{xx} = k_1 \cos^2 \alpha + k_2 \sin^2 \alpha$$
$$k_{xy} = k_{yx} = (k_1 - k_2) \sin \alpha \cos \alpha \quad (3.23)$$
$$k_{yy} = k_1 \sin^2 \alpha + k_2 \cos^2 \alpha$$

It is of interest to note that each component of the specific discharge vector depends upon both components of the hydraulic gradient. These two vectors therefore are not collinear, as opposed to the isotropic case. This may also be seen from (3.18) considering the case that the hydraulic gradient is inclined at 45° to the layers ($\partial \phi/\partial x^* = \partial \phi/\partial y^*$). Because k_1 is larger than k_2 the flow will be inclined at an angle less than 45 degrees to the layers ($q_y^*/q_x^* < 1$).

For the general case of three-dimensional flow, Darcy's law (3.22) becomes

$$q_x = -k_{xx}\frac{\partial \phi}{\partial x} - k_{xy}\frac{\partial \phi}{\partial y} - k_{xz}\frac{\partial \phi}{\partial z}$$
$$q_y = -k_{yx}\frac{\partial \phi}{\partial x} - k_{yy}\frac{\partial \phi}{\partial y} - k_{yz}\frac{\partial \phi}{\partial z} \quad (3.24)$$
$$q_z = -k_{zx}\frac{\partial \phi}{\partial x} - k_{zy}\frac{\partial \phi}{\partial y} - k_{zz}\frac{\partial \phi}{\partial z}$$

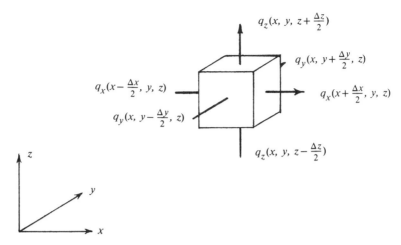

Figure 1.6 Continuity of flow.

The nine coefficients k_{ij} ($i = x, y, z$) are called the coefficients of a second rank tensor, the hydraulic conductivity tensor. Of these nine coefficients, only six are different; the tensor is symmetric:

$$k_{xy} = k_{yx} \qquad k_{xz} = k_{zx} \qquad k_{yz} = k_{zy} \qquad (3.25)$$

We will consider problems with anisotropic hydraulic conductivity in this text, but, unless explicitly stated, assume that the hydraulic conductivity is isotropic.

4 CONTINUITY OF FLOW

Darcy's law furnishes three equations of motion for the four unknowns q_x, q_y, q_z, and ϕ. A fourth equation is required for a complete description of groundwater flow. This equation is the mass balance equation, which for steady flow reduces to the equation of continuity. It results from the consideration that no water can be stored or produced in an elementary block of soil filled with water. Such a block is shown in Figure 1.6. The sides of the block are Δx, Δy, and Δz, and the center is at (x, y, z). The rectangle is so small that the variation of the magnitudes of the components of q_i along the sides can be neglected with respect to their value at the center of each side. The inflow through each side then is equal to the normal component of q_i multiplied by the area of the side. The net inflow ΔQ [m³/s] leaving the block is obtained from Figure 1.6, where only the components of q_i normal to the sides are drawn, and is set equal to zero:

$$\begin{aligned}\Delta Q = 0 = \ &\Delta y \Delta z [q_x(x + \tfrac{\Delta x}{2}, y, z) - q_x(x - \tfrac{\Delta x}{2}, y, z)] \\ &+ \Delta x \Delta z [q_y(x, y + \tfrac{\Delta y}{2}, z) - q_y(x, y - \tfrac{\Delta y}{2}, z)] \\ &+ \Delta x \Delta y [q_z(x, y, z + \tfrac{\Delta z}{2}) - q_z(x, y, z - \tfrac{\Delta z}{2})] \end{aligned} \qquad (4.1)$$

Division by $\Delta x \Delta y \Delta z$ yields

$$\frac{q_x(x + \frac{\Delta x}{2}, y, z) - q_x(x - \frac{\Delta x}{2}, y, z)}{\Delta x}$$
$$+ \frac{q_y(x, y + \frac{\Delta y}{2}, z) - q_y(x, y - \frac{\Delta y}{2}, z)}{\Delta y}$$
$$+ \frac{q_z(x, y, z + \frac{\Delta z}{2}) - q_z(x, y, z - \frac{\Delta z}{2})}{\Delta z} = 0 \tag{4.2}$$

Passing to the limit for $\Delta x \to 0$, $\Delta y \to 0$, and $\Delta z \to 0$, the three terms in (4.2) become partial derivatives and we obtain

$$\boxed{\frac{\partial q_x}{\partial x} + \frac{\partial q_y}{\partial y} + \frac{\partial q_z}{\partial z} = 0} \tag{4.3}$$

which is the equation of continuity for steady groundwater flow. This equation is applicable also to transient flow in an incompressible porous medium.

5 FUNDAMENTAL EQUATIONS

Darcy's law and the continuity equation together provide four equations for the four unknown quantities q_x, q_y, q_z, and ϕ. Three of these equations will be eliminated by substituting the derivatives $-k\partial_i\phi$ for q_i in the continuity equation. This yields

$$\frac{\partial}{\partial x}\left[k\frac{\partial \phi}{\partial x}\right] + \frac{\partial}{\partial y}\left[k\frac{\partial \phi}{\partial y}\right] + \frac{\partial}{\partial z}\left[k\frac{\partial \phi}{\partial z}\right] = 0 \tag{5.1}$$

which is the governing equation for steady groundwater flow in an isotropic porous medium. If the hydraulic conductivity k is a constant, (5.1) may be written as

$$\boxed{\frac{\partial^2 \phi}{\partial x^2} + \frac{\partial^2 \phi}{\partial y^2} + \frac{\partial^2 \phi}{\partial z^2} = 0} \tag{5.2}$$

which is Laplace's equation in three dimensions. The fundamental equations for steady flow through isotropic porous media consist of either (5.1) or (5.2) and Darcy's law,

$$\boxed{\begin{aligned} q_x &= -k\frac{\partial \phi}{\partial x} \\ q_y &= -k\frac{\partial \phi}{\partial y} \\ q_z &= -k\frac{\partial \phi}{\partial z} \end{aligned}} \tag{5.3}$$

Solving groundwater flow problems amounts to solving the differential equation of Laplace with the appropriate boundary conditions. Sometimes these conditions are given in terms of the head ϕ, as along the bank of a canal or river, and sometimes in terms of the discharge vector, as along a slurry wall where the component of q_i normal to the wall is zero.

Only few solutions to Laplace's equation in three dimensions exist. Fortunately, many practical flow problems either are two-dimensional or can be approximated by a two-dimensional analysis. Such problems will be discussed in the next chapter, along with approximate methods which make it possible to solve three-dimensional problems by a two-dimensional analysis.

Chapter 2

Types of Flow

Compared to other fields of applied mechanics, steady flow of groundwater is governed by a comparatively simple differential equation. The difficulty in solving engineering problems, however, is the diversity of natural conditions encountered. Groundwater flow problems may be categorized into types of flow according to the natural conditions under which the flow takes place. We will define and discuss these types of flow in this chapter, treating one type of flow in each section.

Many of the natural types of flow are three-dimensional but can be covered by good approximation by a two-dimensional analysis. For each type of flow we will discuss the approximations made, derive the basic equations, present some elementary methods of solution, and apply these to one or more practical problems. The scope of this chapter is limited to homogeneously permeable porous media. Where possible, we will introduce potentials and discharges for each type of flow in such a way that the various mathematical formulations become similar, if not identical.

The solution of practical problems may be viewed as a process that consists of three main steps:

1. Simplification of the real problem in such a way that it becomes tractable, while retaining the important characteristics of the problem
2. Formulation of the physical quantities in terms of abstract variables and functions used in the mathematical description of the problem
3. Determination of the mathematical solution and interpretation of the results.

This procedure will be followed when solving practical problems at the end of each section.

6 HORIZONTAL CONFINED FLOW

We will call a flow horizontal if it takes place only in the horizontal plane. Horizontal flow thus occurs if the vertical component of the specific discharge vector, q_z, is zero throughout the flow region, i.e., if

$$q_z = 0 \qquad (6.1)$$

Flow is called *confined* if the groundwater flows between two impermeable boundaries. These impermeable boundaries are known as the *confining boundaries*. If the confining boundaries are horizontal and the flow in the space between them is horizontal, we will call the flow *horizontal confined flow*. The permeable layer between the confining layers is called a *water bearing layer* or *aquifer*.

Horizontal confined flow either is one-dimensional or two-dimensional. In some cases, two-dimensional flow may be radial. We will discuss horizontal confined flow for cases of one-dimensional flow, radial flow, and two-dimensional flow below.

One-dimensional Flow

An example of a case of one-dimensional flow is given in Figure 2.1. The flow occurs in an aquifer between two long, parallel, straight rivers. An x, y, z coordinate system is selected as indicated in the figure. The x and y axes are normal and parallel to the rivers, respectively, and the z axis points vertically upwards. The origin is halfway between the rivers, the distance between the rivers is L, and the thickness of the aquifer is H. The heads in rivers $\mathcal{AB}$ ($x = -L/2$) and $\mathcal{CD}$ ($x = L/2$) are ϕ_1 and ϕ_2, respectively. It follows from symmetry that no flow will occur in the y direction, so that, with (6.1),

$$q_y = q_z = 0 \qquad (6.2)$$

It follows from Darcy's law, (3.11), that $\partial \phi / \partial y = \partial \phi / \partial z = 0$; ϕ does not vary in the y and z directions and is a function of x alone. Equation (5.2) therefore reduces to

$$\frac{d^2 \phi}{dx^2} = 0 \qquad (6.3)$$

and integration gives

$$\frac{d\phi}{dx} = A \qquad (6.4)$$

where A is a constant. Integration of (6.4) gives the general solution for one-dimensional flow:

$$\phi = Ax + B \qquad (6.5)$$

Sec. 6 Horizontal Confined Flow

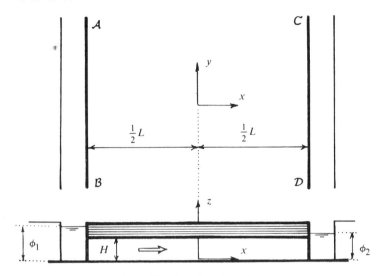

Figure 2.1 One-dimensional flow.

This solution covers all cases of one-dimensional flow. The particular solution that fits the flow case of Figure 2.1 is found by application of the boundary conditions. These boundary conditions are found from Figure 2.1 to be

$$\begin{aligned} x &= -\tfrac{1}{2}L & \phi &= \phi_1 \\ x &= \tfrac{1}{2}L & \phi &= \phi_2 \end{aligned} \tag{6.6}$$

The first boundary condition gives with (6.5)

$$\phi_1 = -\tfrac{1}{2}AL + B \tag{6.7}$$

and the second one yields

$$\phi_2 = \tfrac{1}{2}AL + B \tag{6.8}$$

Solving (6.7) and (6.8) for A and B, the following is obtained:

$$A = -\frac{\phi_1 - \phi_2}{L} \qquad B = \tfrac{1}{2}(\phi_1 + \phi_2) \tag{6.9}$$

Substitution of these values for A and B in (6.5) gives

$$\phi = -(\phi_1 - \phi_2)\frac{x}{L} + \tfrac{1}{2}(\phi_1 + \phi_2) \tag{6.10}$$

The specific discharge is found by the use of Darcy's law (3.11)

$$q_x = \frac{k(\phi_1 - \phi_2)}{L} \qquad q_y = 0 \qquad q_z = 0 \tag{6.11}$$

It follows that the specific discharge does not vary with position, and increases linearly with both the difference in head at the two rivers and the coefficient of hydraulic conductivity. It decreases linearly with the distance, L, between the two rivers.

The discharge flowing through the aquifer per unit length of river bank is found by multiplying the specific discharge q_x by the thickness of the aquifer, H. Denoting this discharge as Q_x, we find from (6.11) that

$$Q_x = q_x H = \frac{kH(\phi_1 - \phi_2)}{L} \tag{6.12}$$

Basic Equations

In problems of horizontal confined flow, we are interested primarily in the discharge occurring over the thickness of the aquifer, rather than in the specific discharge, and introduce the discharge vector Q_i. For one-dimensional flow this vector has one component, Q_x, defined by (6.12). For two-dimensional flow the discharge vector has two components, Q_x and Q_y, and is defined as $Q_i = Hq_i = H(q_x, q_y)$, or

$$\begin{aligned} Q_x &= Hq_x \\ Q_y &= Hq_y \end{aligned} \tag{6.13}$$

The two components of the discharge vector (Q_x, Q_y) can be expressed in terms of derivatives of the head ϕ by use of (6.13) and Darcy's law (3.11). This gives

$$\begin{aligned} Q_x &= Hq_x = H\left[-k\frac{\partial \phi}{\partial x}\right] \\ Q_y &= Hq_y = H\left[-k\frac{\partial \phi}{\partial y}\right] \end{aligned} \tag{6.14}$$

Since both H and k are constants, (6.14) can be written as

$$\begin{aligned} Q_x &= -\frac{\partial[kH\phi]}{\partial x} \\ Q_y &= -\frac{\partial[kH\phi]}{\partial y} \end{aligned} \tag{6.15}$$

These equations suggest the introduction of a new variable, Φ, defined as follows,

$$\boxed{\Phi = kH\phi + C_c} \tag{6.16}$$

where C_c is an arbitrary constant. Since $\partial C_c/\partial x = \partial C_c/\partial y = 0$, (6.15) can be simplified by the use of (6.16)

$$\boxed{\begin{aligned} Q_x &= -\frac{\partial \Phi}{\partial x} \\ Q_y &= -\frac{\partial \Phi}{\partial y} \end{aligned}} \tag{6.17}$$

Sec. 6　　Horizontal Confined Flow

We will refer to the function Φ as the *discharge potential for horizontal confined flow*, or briefly as the *potential* Φ.

Since $q_z = -k(\partial \phi/\partial z) = 0$ for all cases of horizontal confined flow, the governing equation becomes

$$\frac{\partial^2 \phi}{\partial x^2} + \frac{\partial^2 \phi}{\partial y^2} = 0 \qquad (6.18)$$

Furthermore, it follows from (6.16) and (6.18) that

$$\frac{\partial^2}{\partial x^2}\left[\frac{\Phi - C_c}{kH}\right] + \frac{\partial^2}{\partial y^2}\left[\frac{\Phi - C_c}{kH}\right] = \frac{1}{kH}\left[\frac{\partial^2 \Phi}{\partial x^2} + \frac{\partial^2 \Phi}{\partial y^2}\right] = 0 \qquad (6.19)$$

or

$$\boxed{\frac{\partial^2 \Phi}{\partial x^2} + \frac{\partial^2 \Phi}{\partial y^2} = 0} \qquad (6.20)$$

Problems of horizontal confined flow can be solved by determining a function Φ that fulfills Laplace's equation and satisfies the boundary conditions.

Equipotentials and Streamlines

Equipotentials are defined as curves of constant potential. A set of equipotentials can be obtained by the use of computer graphics, for example, and provides us with a visual tool for studying the solution to a flow problem. A plot of equipotentials, which are sometimes called potentiometric contours, shows the variation of the potential throughout a flow domain in much the same way as the contour lines on a topographic map show the ground elevations. In addition, both the direction and the magnitude of the discharge vector can be derived from a plot of equipotentials as is shown below. The discharge vector is directed normal to the equipotentials. This is seen as follows, by the use of a local s, n Cartesian coordinate system, where s is normal and n is tangent to the equipotential at a point $\mathcal{P}$, as is shown in Figure 2.2. Writing (6.17) in terms of the coordinates s and n, we obtain

$$\begin{aligned}Q_s &= -\frac{\partial \Phi}{\partial s} \\ Q_n &= -\frac{\partial \Phi}{\partial n}\end{aligned} \qquad (6.21)$$

The potential does not vary in the direction n, because n is tangent to the equipotential at $\mathcal{P}$. Hence, Q_n is zero so that the discharge vector points in the direction s, which is normal to the equipotential. It may be noted that this is true only if the magnitude of the discharge vector is unequal to zero. Point $\mathcal{P}$ is called a *stagnation point* if both components of flow are zero at $\mathcal{P}$; in that case the equipotential and the streamline are not mutually orthogonal at $\mathcal{P}$.

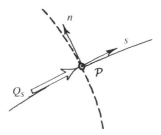

Figure 2.2 Equipotentials and streamlines.

Because the discharge vector is normal to the equipotentials, the flow occurs in the direction of the steepest slope of the plot of potentiometric contours. *Discharge streamlines* are curves whose tangent at any point coincides with the discharge vector; see the solid curve in Figure 2.2. The word discharge in "discharge streamlines" is added to indicate that these streamlines correspond to the discharge vector, rather than to the specific discharge vector. When confusion is unlikely, however, we will omit the word discharge, and refer briefly to streamlines. We cannot, at this stage, determine mathematical expressions for points of the streamlines; such expressions will be presented in Chapters 3 and 4. However, knowing that the streamlines are normal to the equipotentials, we can sketch them for any given plot of potentiometric contours. The magnitude of the discharge vector can be estimated from the distance between consecutive equipotentials, provided that the difference in values of Φ on such equipotentials is kept constant throughout the flow domain. If this difference is $\Delta\Phi$, and the distance between two consecutive equipotentials is Δs, then the first equation of (6.21) may be approximated as

$$Q_s \approx -\frac{\Delta\Phi}{\Delta s} \qquad (6.22)$$

Since $\Delta\Phi$ is constant, Q_s is inversely proportional to the distance between the equipotentials. It follows from the definition of the streamlines that no flow can occur across a streamline. Thus, the discharge flowing between two streamlines is constant, provided that no water enters the aquifer through the upper or lower boundaries. If the distance between two consecutive streamlines at a point $\mathcal{P}$ is Δn, then the discharge ΔQ flowing between these streamlines is approximately $\Delta Q \approx Q_s \Delta n \approx -\Delta\Phi \Delta n/\Delta s$. If we draw the streamlines such that the distances Δn and Δs are approximately equal, ΔQ is about equal to $-\Delta\Phi$. The amount of flow between any pair of streamlines is then constant throughout the mesh of equipotentials and streamlines, which is called a flow net.

For the simple case of one-dimensional flow shown in Figure 2.1, the equipotentials are straight lines that are parallel to the two rivers, and will all be the same distance apart.

Problem 6.1

Determine both $\Delta\Phi$ and Δs for the case of Figure 2.1, if there are to be 4 equipotentials, not counting the river banks. Draw the equipotentials, determine Q_s by the use of (6.22), and check this value against (6.12).

Radial Flow

Radial horizontal confined flow occurs when there is symmetry with respect to a vertical axis. An example of radial flow is the case of flow toward a well at the center of a circular island, as illustrated in Figure 2.3. We will determine the potential for this flow case by applying Darcy's law and the equation of continuity separately rather than solving the differential equation directly. Let the discharge of the well be Q [m^3/sec], the radius of the well r_w, the radius of the island R, the head at the boundary ϕ_0, and the thickness of the aquifer H. It follows from the symmetry with respect to the axis of the well that the flow is radial, i.e., the discharge vectors at all points are directed toward the center of the island. Since the radial coordinate r points away from the center, the discharge vector Q_r points in the negative r direction, i.e.,

$$Q_r < 0 \tag{6.23}$$

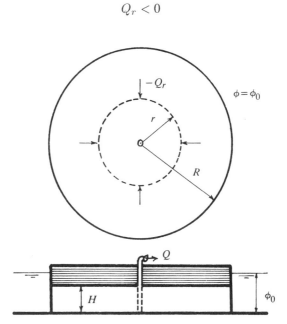

Figure 2.3 A well in a circular island.

It follows from continuity of flow that the same amount of water must pass through all cylinders with radius r ($r_w \leq r \leq R$) and height H. This amount equals the discharge Q of the well and may be expressed as follows, noting that $(-Q_r)$ is positive:

$$Q = 2\pi r(-Q_r) \tag{6.24}$$

It follows from Darcy's law (6.17) that

$$Q_r = -\frac{d\Phi}{dr} \qquad (6.25)$$

and substitution in (6.24) gives

$$Q = 2\pi r \frac{d\Phi}{dr} \qquad (6.26)$$

or

$$\frac{d\Phi}{dr} = \frac{Q}{2\pi r} \qquad (6.27)$$

Integration yields

$$\boxed{\Phi = \frac{Q}{2\pi} \ln r + C} \qquad (6.28)$$

where C is a constant of integration. This equation represents the general solution for flow toward a well in an infinite confined aquifer. The value of the constant C for flow toward a well in a circular island is found from the boundary condition at $r = R$,

$$r = R \qquad \phi = \phi_0 \qquad (6.29)$$

Taking the arbitrary constant C_c in (6.16) as zero, the relation between potential and head becomes

$$\Phi = kH\phi \qquad (6.30)$$

and the boundary condition may be expressed in terms of Φ as

$$r = R \qquad \Phi = \Phi_0 = kH\phi_0 \qquad (6.31)$$

The constant C in (6.28) is found upon application of (6.31) to (6.28) and equals $\Phi_0 - Q/(2\pi)\ln R$, so that (6.28) becomes

$$\Phi = kH\phi = \frac{Q}{2\pi} \ln \frac{r}{R} + \Phi_0 \qquad (6.32)$$

or, after division by kH, noticing that Φ_0 equals $kH\phi_0$

$$\phi = \frac{Q}{2\pi kH} \ln \frac{r}{R} + \phi_0 \qquad (6.33)$$

This equation is known as the Thiem equation (Thiem [1906]).

Sec. 6   Horizontal Confined Flow 27

The Head at the Well

It may be necessary to calculate the head that must be maintained at the well to obtain a certain discharge of Q. If the radius of the well is r_w, this head, ϕ_w, is found from (6.33) by substituting r_w for r, which gives

$$\phi_w = \phi_0 + \frac{Q}{2\pi k H} \ln \frac{r_w}{R} \tag{6.34}$$

Since the flow is confined, the water table at the well must be above the upper impervious boundary. Hence, ϕ_w must be greater than H:

$$\phi_w = \phi_0 + \frac{Q}{2\pi k H} \ln \frac{r_w}{R} \geq H \tag{6.35}$$

If this condition is not fulfilled, the flow near the well is not confined and (6.33) is not applicable.

Problem 6.2

Prove that the equipotentials for the case of Figure 2.3 are circles. It is given that $r_w = 0.01 * R$. Draw 9 equipotentials, not including the equipotentials $r = r_w$ and $r = R$, for a value of R of your choice. Do this by solving (6.35) for Q and substituting the resulting expression for Q in (6.32).

Two-dimensional Flow

The differential equation that describes two-dimensional horizontal confined flow in the x, y plane is Laplace's equation (6.20).

$$\frac{\partial^2 \Phi}{\partial x^2} + \frac{\partial^2 \Phi}{\partial y^2} = 0 \tag{6.36}$$

All solutions to problems of horizontal flow must fulfill this equation. The solution to a particular problem then is selected by application of the boundary conditions.

A relatively simple method of solution is known as the *method of images* and is based on the *superposition principle*. This method will be discussed below.

The Superposition Principle

The superposition principle states that if two different functions, say Φ_1 and Φ_2, are solutions of Laplace's equation, then the function,

$$\Phi(x, y) = C_1 \Phi_1(x, y) + C_2 \Phi_2(x, y) \tag{6.37}$$

also is a solution of Laplace's equation.

The proof of the superposition principle is as follows: Since $\Phi_1(x, y)$ and $\Phi_2(x, y)$ are solutions of the differential equation we have

$$C_1 \left[\frac{\partial^2 \Phi_1}{\partial x^2} + \frac{\partial^2 \Phi_1}{\partial y^2} \right] + C_2 \left[\frac{\partial^2 \Phi_2}{\partial x^2} + \frac{\partial^2 \Phi_2}{\partial y^2} \right] = C_1 * 0 + C_2 * 0 = 0 \tag{6.38}$$

Since C_1 and C_2 are constants, this may be written as

$$\frac{\partial^2 [C_1 \Phi_1]}{\partial x^2} + \frac{\partial^2 [C_2 \Phi_2]}{\partial x^2} + \frac{\partial^2 [C_1 \Phi_1]}{\partial y^2} + \frac{\partial^2 [C_2 \Phi_2]}{\partial y^2}$$

$$= \frac{\partial^2 [C_1 \Phi_1 + C_2 \Phi_2]}{\partial x^2} + \frac{\partial^2 [C_1 \Phi_1 + C_2 \Phi_2]}{\partial y^2} = 0 \qquad (6.39)$$

It follows from (6.37) and (6.39) that

$$\frac{\partial^2 \Phi}{\partial x^2} + \frac{\partial^2 \Phi}{\partial y^2} = 0 \qquad (6.40)$$

as asserted. It may be noted that superposition of solutions is possible without violating the differential equation if this equation is linear in Φ. Laplace's equation is only one of many linear differential equations for which superposition holds.

The Method of Images

The method of images consists of combined use of the superposition principle and symmetry. As an example, consider the cases of horizontal confined flow illustrated in Figures 2.4(a) and (b). The dotted curves in the plan views are equipotentials and the solid curves are streamlines. Figure 2.4(a) represents the case of a discharge well and a recharge well in a confined aquifer of infinite extent. A discharge well is one which withdraws water from the aquifer. A recharge well pumps water into the aquifer. The term "well" may be used with reference to either a discharge or a recharge well. Both discharge and recharge may be represented as Q, where Q is positive for a discharge well and negative for a recharge well. Figure 2.4(b) represents the case of two discharge wells. We will determine the potential as a function of position for each case, starting with Figure 2.4(a).

A Cartesian x, y, z coordinate system is chosen as indicated in the figure, with the origin halfway between the wells and the x axis passing through the recharge well. The distance between the wells is $2d$. The potential Φ for this flow problem is found by superposition of two potentials, Φ_1 and Φ_2.

The function Φ_1 is the potential for one well ($\mathcal{W}_\mathcal{W}$ in Figure 2.4[a]) in an infinite aquifer and is of the form (6.28),

$$\Phi_1 = \frac{Q_1}{2\pi} \ln r_1 + C_1 \qquad (6.41)$$

where Q_1 is the discharge of the well. The value of Φ_1 at any given point is found by measuring the distance r_1 from the well to that point and substituting this value for r_1 in (6.41). Thus, r_1 represents the distance from the well to the point where Φ_1 is to be evaluated. We will refer to r_1 as the local radial coordinate from $\mathcal{W}_\mathcal{W}$.

The function Φ_2 corresponds to one recharge well ($\mathcal{W}_\mathcal{R}$ in Figure 2.4[a]) in an infinite aquifer. An amount Q_2 is pumped into the aquifer at the recharge well. The

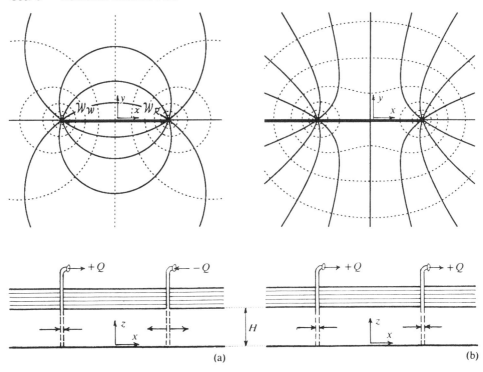

Figure 2.4 The method of images.

potential Φ_2 for a recharge well in an infinite aquifer is found from (6.28) by letting Q be negative, $Q = -Q_2$ ($Q_2 > 0$),

$$\Phi_2 = -\frac{Q_2}{2\pi} \ln r_2 + C_2 \tag{6.42}$$

where r_2 is the local radial coordinate emanating from $\mathcal{W}_\mathcal{R}$.

According to the superposition principle, the sum Φ of Φ_1 and Φ_2,

$$\Phi = \frac{Q_1}{2\pi} \ln r_1 - \frac{Q_2}{2\pi} \ln r_2 + (C_1 + C_2) \tag{6.43}$$

represents a solution of Laplace's equation; it is the potential for flow from a recharge well $\mathcal{W}_\mathcal{R}$ toward a well $\mathcal{W}_\mathcal{W}$ (see Figure 2.4).

For the special case that discharge and recharge are equal, i.e., if

$$Q_1 = Q_2 = Q \tag{6.44}$$

then (6.43) becomes

$$\Phi = \frac{Q}{2\pi} \ln \frac{r_1}{r_2} + C \tag{6.45}$$

where $C = C_1 + C_2$. The equipotentials shown in Figure 2.4(a) correspond to this case. This solution has the property that the head is constant along the y axis, which is the bisecting perpendicular of the line connecting the wells. Each point of the y axis is at the same distance from $\mathcal{W}_\mathcal{W}$ and $\mathcal{W}_\mathcal{R}$. Hence, for an arbitrary point of the y axis we have that $r_1 = r_2 = r_0$, where r_0 is the distance from that point to either $\mathcal{W}_\mathcal{R}$ or $\mathcal{W}_\mathcal{W}$. Substitution of r_0 for both r_1 and r_2 in (6.45) yields

$$\Phi = \frac{Q}{2\pi} \ln \frac{r_0}{r_0} + C = C \tag{6.46}$$

Hence, Φ is indeed constant along the y axis, as asserted. This is seen also from the potentiometric contours shown in Figure 2.4(a).

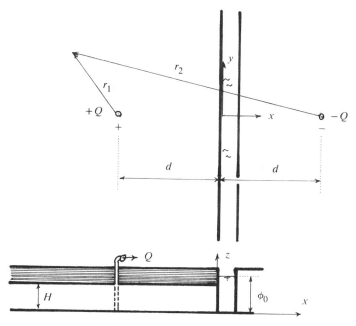

Figure 2.5 Flow toward a well near a river.

It follows that the potential for flow from a recharge well toward a well can be used to solve the problem of a well near a straight river where the head is constant and equal to ϕ_0. This is illustrated in Figure 2.5. The boundary condition along river bank $\mathcal{AB}$ is that ϕ is constant and equals ϕ_0. The distance from the well to the river bank is d. The function

$$\Phi = \frac{Q}{2\pi} \ln \frac{r_1}{r_2} + \Phi_0 \tag{6.47}$$

fulfills the boundary condition along the river bank, $r_1 = r_2$, and thus represents the discharge potential for the flow problem of Figure 2.5. The local radial coordinates r_1

Sec. 6   Horizontal Confined Flow

and r_2 can be expressed in terms of the Cartesian coordinates x and y. It may be seen from Figure 2.5 that

$$r_1 = \sqrt{(x+d)^2 + y^2} \tag{6.48}$$

$$r_2 = \sqrt{(x-d)^2 + y^2} \tag{6.49}$$

Substitution of these expressions for r_1 and r_2 in (6.47) yields

$$\Phi = \frac{Q}{2\pi} \ln \sqrt{\frac{(x+d)^2 + y^2}{(x-d)^2 + y^2}} + \Phi_0 \tag{6.50}$$

This potential describes the flow from a river along the y axis ($x = 0$) where the head is ϕ_0, to a well at $x = -d$, $y = 0$. The flow region is the half plane $x \leq 0$. Note that the solution was found by use of the *image well* with discharge $-Q$ (a recharge well) at $x = d$, $y = 0$. The image well is not real but is merely used as a means to obtain the solution.

We will next consider the case of two discharge wells shown in Figure 2.4(b). The potential for two wells of equal discharge Q and a distance $2d$ apart is obtained from (6.43) by replacing $-Q_2$ by $+Q$ and $+Q_1$ by $+Q$. This yields

$$\Phi = \frac{Q}{2\pi} \ln(r_1 r_2) + C \tag{6.51}$$

Substitution of (6.48) and (6.49) for r_1 and r_2 yields

$$\begin{aligned}\Phi &= \frac{Q}{2\pi} \ln \sqrt{[(x+d)^2 + y^2][(x-d)^2 + y^2]} + C \\ &= \frac{Q}{4\pi} \ln\{[(x+d)^2 + y^2][(x-d)^2 + y^2]\} + C\end{aligned} \tag{6.52}$$

This solution has the property that no flow occurs across the y axis, which is the perpendicular bisector of the line between the two wells in Figure 2.4(b). The y axis is a streamline, which means that the specific discharge vector is directed along that line, i.e., $Q_x(0, y) = 0$. This is seen as follows. The component Q_x of the discharge vector is found from (6.52) by use of Darcy's law, $Q_x = -\partial \Phi / \partial x$,

$$Q_x = -\frac{\partial \Phi}{\partial x} = -\frac{Q}{4\pi} \left[\frac{2(x+d)}{(x+d)^2 + y^2} + \frac{2(x-d)}{(x-d)^2 + y^2} \right] \tag{6.53}$$

which gives, for $x = 0$,

$$Q_x(0, y) = -\frac{Q}{4\pi} \left[\frac{2d}{d^2 + y^2} - \frac{2d}{d^2 + y^2} \right] = 0 \tag{6.54}$$

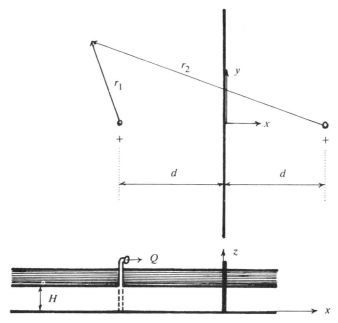

Figure 2.6 A well near impervious boundary.

as asserted. The equipotentials in Figure 2.4(b) intersect the y axis at right angles, because $Q_x(0, y)$ is zero and the flow is normal to the equipotentials.

The solution for flow toward two wells of equal discharge can be used to solve problems involving an impervious boundary and wells. An example is given in Figure 2.6. The potential (6.52) has the property that no flow occurs across the y axis (see Figure 2.6), and can be used to describe the problem of flow toward a well near an impervious boundary. The difference between the solutions to the problems of Figures 2.5 and 2.6 is the "sign" of the image well. The image well is a recharge well (indicated with a "$-$") in Figure 2.5 and a well (indicated with a "$+$") in Figure 2.6.

Examples of problems that can be solved by the use of the method of images are given in Figures 2.7(a) and 2.8(a). The image wells are indicated in Figures 2.7(b) and 2.8(b). The potentials are

$$\Phi = \frac{Q_1}{2\pi} \ln \frac{r_1}{r_3} - \frac{Q_2}{2\pi} \ln \frac{r_2}{r_4} + \Phi_0 \tag{6.55}$$

and

$$\Phi = \frac{Q_1}{2\pi} \ln \frac{r_1 r_4}{r_5 r_8} - \frac{Q_2}{2\pi} \ln \frac{r_2 r_3}{r_6 r_7} + \Phi_0 \tag{6.56}$$

for the flow cases shown in Figures 2.7(a) and 2.8(a), respectively, which contain equipotentials (the dotted curves) and streamlines (the solid ones). The image wells are shown in Figures 2.7(b) and 2.8(b). It is of interest to note that two equipotentials and two

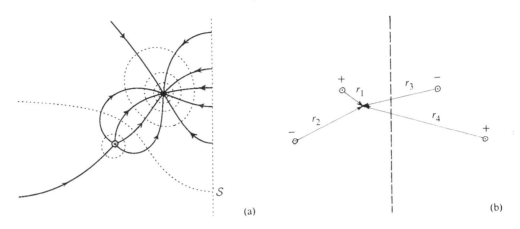

Figure 2.7 A well and a recharge well near a river.

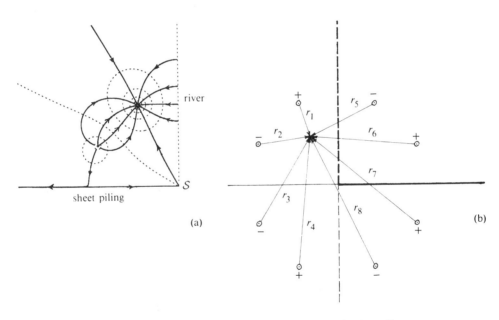

Figure 2.8 A well and a recharge well near a river and a sheet piling.

streamlines intersect at the point of intersection, S, of the river and the sheet piling: the equipotentials and streamlines are not orthogonal at this point. Point S is a stagnation point: the effects of all wells and image wells on the flow cancel there.

Partial Penetration

The boundaries discussed thus far, such as the rivers and sheet pilings, fully penetrate the aquifer; they are called *fully penetrating boundaries*. Boundaries that extend over part of the aquifer thickness are called *partially penetrating boundaries*. The flow in aquifers with partially penetrating boundaries exhibits components in the vertical direction. The vertical components of flow are large in the neighborhood of a partially penetrating boundary, but reduce rapidly with the distance from the boundary. As an illustration, equipotentials are sketched in Figure 2.9 for the case of flow between two partially penetrating rivers. The equipotentials are curved in the neighborhood of the river, where the components of flow in the vertical direction are large, and become nearly vertical at a distance of several times the aquifer thickness away from the boundaries. The application of an analysis of horizontal flow to such a case will produce errors, as a result of neglecting the resistance to flow in the vertical direction (recall that resistance to flow is the inverse of hydraulic conductivity ($1/k$); a zero resistance implies an infinite hydraulic conductivity). The relative error decreases as the total resistance to horizontal flow increases, so that the ratio of the neglected resistance to the total resistance decreases. Thus, the errors become less important as the rivers become further apart. The horizontal dimensions in real aquifers are often much larger than the vertical ones; for such cases the effect of partial penetration rarely need be considered. Wells also may be either fully penetrating or partially penetrating; a similar discussion as that given above applies to partially penetrating wells. We will assume for the time being that all boundaries and wells in problems of horizontal flow are fully penetrating.

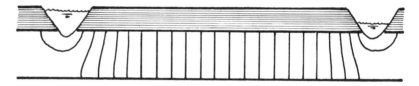

Figure 2.9 A partially penetrating river.

Piezometric Contours

Piezometric contours are curves along which the head is constant. The head may be expressed linearly in terms of the potential, see (6.16):

$$\phi = \frac{\Phi - C_c}{kH} \qquad (6.57)$$

The value of ϕ associated with each piezometric contour is obtained from the corresponding value of Φ by the use of (6.57). Because ϕ is a linear function of Φ, the difference in value of ϕ, $\Delta\phi$, on two neighboring equipotentials is constant throughout the flow domain, provided that $\Delta\Phi$ is constant.

Problem 6.3

A system of 4 wells, located at the 4 corners of a building pit, will be used to reduce the

Sec. 6   Horizontal Confined Flow

pressures directly below the confining layer that separates the pit from a confined aquifer, to a level that is safe with regard to stability of the bottom of the pit. This level is given as $0.5 * 10^5$ N/m².

The thickness of the aquifer is $H = 25$ m and the hydraulic conductivity is $k = 10^{-5}$ m/s. The aquifer is bounded by a long, straight river (see Figure 2.10). The heads are measured with respect to the base of the aquifer, and the head along the entire river bank is 40 m. The soil above the upper confining bed is dry. The sides of the building pit are 40 m and its center is 80 m from the river bank. There is no flow in the aquifer before the excavation.

Questions
(a) Calculate the discharge, Q, of each well necessary to maintain the pressure at or below the safe level everywhere underneath the pit, assuming that all wells have the same discharge.
(b) Calculate the two components of the discharge vector at the midpoint of the side of the pit closest to the canal.

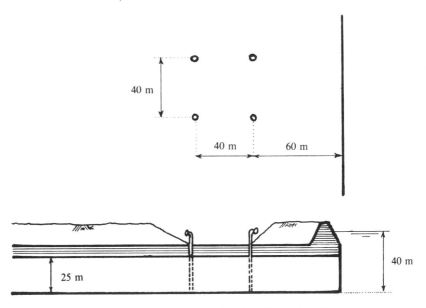

Figure 2.10   Dewatering of a building pit (Problem 6.3).

Problem 6.4

A system of a well and a recharge well in a confined aquifer of thickness H=30 m and $k = 10^{-4}$ m/s is used to produce groundwater for a cooling system. The wells are $2d = 100$ m apart, and the radius of each well is $r_w = 0.1$ m. The head in the aquifer is constant and equal to 35 m before the wells operate. The discharges of the well and the recharge well are $0.6 * 10^{-2}$ m³/s and $-0.6 * 10^{-2}$ m³/s, respectively.

Questions
(a) Calculate the largest distance from the midpoint between the wells at which the influence of the system on the head is 0.05 m.

(b) Find the maximum discharge circulated by the system such that the aquifer remains confined at the well (neglect r_w with respect to d).

7 SHALLOW UNCONFINED FLOW

An unconfined aquifer is one whose upper boundary is a water table, or phreatic surface. Problems with such a boundary are difficult to solve. Approximate solutions for such problems can be obtained when the flow is mainly horizontal so that the resistance to flow in the vertical direction may be neglected. A large class of groundwater flow problems fulfills this condition. We will consider such problems in this section, and make the additional assumption that the pores of the soil beneath the phreatic surface are completely filled with water, and the pores above it are completely filled with air. This assumption is justified for fairly permeable soils such as sand. The zone of the aquifer below the phreatic surface is sometimes called the *saturated zone*.

We will call a flow *shallow unconfined flow* when we neglect the vertical component of the specific discharge vector, q_z. The word shallow is added to avoid confusion; there are cases of unconfined flow where the resistance to flow in vertical direction may not be neglected.

Basic Equations

We derived the basic equations for horizontal confined flow directly from the fundamental equations presented in Chapter 1; the general equations applied because no approximations were made. For the case of shallow unconfined flow, however, approximations are being made, and we must take these into account when deriving the basic equations. The basic equations consist of an equation of motion and the continuity equation. We will introduce a discharge vector Q_i and a potential Φ for shallow unconfined flow in such a way that the basic equations become identical to those for horizontal confined flow.

The Dupuit-Forchheimer Assumption

Dupuit [1863] and Forchheimer [1886] independently proposed to neglect the variation of the head ϕ in the vertical direction. This assumption is usually identified with the assumption that the component q_z of the specific discharge vector is neglected. We will, however, state the Dupuit-Forchheimer assumption in a somewhat different way: the resistance to flow in the vertical direction is negligible (Strack [1984]). Thus, the magnitude of the vertical component q_z exists, but is not associated with a variation of ϕ:

$$q_z \neq 0 \qquad \frac{\partial \phi}{\partial z} = 0 \qquad (7.1)$$

The head thus is constant along vertical surfaces. The Dupuit-Forchheimer assumption is applicable to aquifers bounded below by an impervious base. This impervious base is taken as the reference level for the head (see Figure 2.11). If the elevation of the water

Sec. 7   Shallow Unconfined Flow

table above the impervious base is h, the head at a point $\mathcal{P}$ of the phreatic surface may be expressed in terms of the pressure as follows (see [2.7]):

$$\phi = \frac{p}{\rho g} + Z = \frac{p}{\rho g} + h \tag{7.2}$$

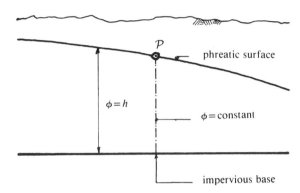

Figure 2.11   The Dupuit-Forchheimer assumption.

The pressure at the phreatic surface is equal to the atmospheric pressure. Taking the atmospheric pressure as zero, (7.2) becomes

$$\phi = h \tag{7.3}$$

Hence, the head along the phreatic surface is equal to the elevation head, h. Since ϕ is taken to be constant along vertical surfaces, ϕ equals h throughout the unconfined aquifer.

The Discharge Vector

We defined the discharge vector in Sec. 6 as the product of the specific discharge vector and the thickness of the aquifer H. For the case of unconfined flow, the aquifer thickness varies and equals h, and we obtain

$$\begin{aligned} Q_x &= h q_x \\ Q_y &= h q_y \end{aligned} \tag{7.4}$$

The physical interpretation for Q_i is the same for shallow unconfined flow as for horizontal confined flow; the magnitude of Q_i equals the amount of flow through a vertical section of unit width. Application of Darcy's law (3.11) to (7.4) yields

$$\begin{aligned} Q_x &= -kh\frac{\partial \phi}{\partial x} \\ Q_y &= -kh\frac{\partial \phi}{\partial y} \end{aligned} \tag{7.5}$$

Since $h = \phi$, this becomes

$$Q_x = -k\phi \frac{\partial \phi}{\partial x}$$
$$Q_y = -k\phi \frac{\partial \phi}{\partial y} \qquad (7.6)$$

or, since k is a constant,

$$Q_x = -\frac{\partial \left[\frac{1}{2}k\phi^2\right]}{\partial x}$$
$$Q_y = -\frac{\partial \left[\frac{1}{2}k\phi^2\right]}{\partial y} \qquad (7.7)$$

The Discharge Potential

Introducing the discharge potential for shallow unconfined flow as

$$\boxed{\Phi = \tfrac{1}{2}k\phi^2 + C_u} \qquad (7.8)$$

where C_u is an arbitrary constant, (7.7) may be written as

$$\boxed{\begin{aligned} Q_x &= -\frac{\partial \Phi}{\partial x} \\ Q_y &= -\frac{\partial \Phi}{\partial y} \end{aligned}} \qquad (7.9)$$

It is noted that these equations are the same as those derived for horizontal confined flow (see [6.17]).

Equipotentials and Piezometric Contours

Equipotentials are defined in the same way as for horizontal confined flow; the concept of an equipotential is universal and is applicable to any potential function. Information regarding the discharge vector may be obtained from a plot of potentiometric contours as outlined in Sec. 6. Piezometric contours are obtained by letting the head ϕ be constant. The head ϕ is expressed in terms of Φ by the use of (7.8):

$$\boxed{\phi = \sqrt{\frac{2(\Phi - C_u)}{k}}} \qquad (7.10)$$

Because ϕ is constant whenever Φ is constant, each equipotential is a piezometric contour. A plot of potentiometric contours obtained for constant $\Delta\Phi$ may be viewed as a plot of piezometric contours, with the value of ϕ on each contour obtained from the value of Φ

Sec. 7 Shallow Unconfined Flow

by the use of (7.10). However, ϕ is not a linear function of Φ, and therefore $\Delta\phi$, the difference in value of ϕ on two consecutive equipotentials, is not constant. A plot of piezometric contours with constant $\Delta\phi$ therefore must be constructed separately from a plot of equipotentials. Equipotentials serve to gain insight in the magnitude and direction of the discharge vector. Piezometric contours are used as a graphical representation of the head ϕ.

Problem 7.1

Prove that the magnitude of the specific discharge vector is approximately inversely proportional to the distance between two neighboring piezometric contours (compare [6.22]).

The Continuity Equation

We will express the continuity equation for shallow unconfined flow in terms of the components of the discharge vector. Unlike the case of confined flow, inflow or outflow may occur along the upper boundary, as a result of rainfall or evaporation. We will, for the time being, make restriction to cases where neither evaporation nor rainfall occurs. Continuity of flow then can be applied to a column of height h and cross-sectional area $\Delta x \Delta y$ in much the same way as was done in Sec. 4. A plan view of the aquifer is given in Figure 2.12. The flow through the sides of the column is expressed in terms of the components of the discharge vector Q_i. We obtain, by an analysis similar to the one outlined in Sec. 4:

$$\frac{\partial Q_x}{\partial x} + \frac{\partial Q_y}{\partial y} = 0 \qquad (7.11)$$

This equation is the continuity equation for shallow unconfined flow without rainfall or evaporation.

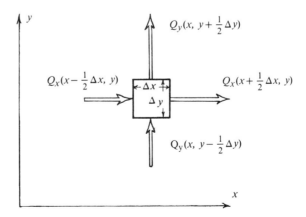

Figure 2.12 Continuity of flow in an unconfined aquifer.

Differential Equation for the Discharge Potential

Substitution of expressions (7.9) for Q_x and Q_y in (7.11) yields

$$\frac{\partial^2 \Phi}{\partial x^2} + \frac{\partial^2 \Phi}{\partial y^2} = 0 \qquad (7.12)$$

It is seen by comparing (7.12) with (6.20) that the governing equations for horizontal confined and shallow unconfined flow are the same. These flows therefore can be treated mathematically in the same manner. It is noted that this result was achieved by the use of the discharge potential Φ, which was introduced by Girinski [1946] in a somewhat different form.

The difference between horizontal confined flow and shallow unconfined flow lies in the expressions for Φ. We defined Φ for horizontal confined flow as a linear function of ϕ

$$\Phi = kH\phi + C_c \qquad (7.13)$$

and for shallow unconfined flow Φ is given by a quadratic function,

$$\Phi = \tfrac{1}{2} k \phi^2 + C_u \qquad (7.14)$$

Plots of equipotentials can be viewed as plots of piezometric contours for cases of horizontal confined flow. For cases of shallow unconfined flow, however, the plots of equipotentials and piezometric contours differ in the spacing between the contours.

We will solve problems of unconfined flow in three steps: first we express the boundary conditions in terms of the potential, second we solve the differential equation with these boundary conditions, and third we obtain the head from the potential by the use of (7.10). Examples of such solutions are given in the following.

One-dimensional Flow

Consider an unconfined aquifer bounded by two long, parallel rivers. A cross-section through the aquifer is given in Figure 2.13. The heads along the river banks are ϕ_1 and ϕ_2 and the distance between the banks is L. An x, y, z coordinate system is chosen as indicated in the figure. The y coordinate is parallel to the rivers and the origin is midway between the two river banks. The boundary conditions in terms of the head ϕ are

$$\begin{aligned} x &= -\tfrac{1}{2}L & \phi &= \phi_1 \\ x &= \tfrac{1}{2}L & \phi &= \phi_2 \end{aligned} \qquad (7.15)$$

and may be expressed in terms of the potential Φ by the use of (7.14). The constant C_u can be chosen arbitrarily and we will, unless stated otherwise, take C_u as zero. Then,

$$\begin{aligned} x &= -\tfrac{1}{2}L & \Phi &= \Phi_1 = \tfrac{1}{2} k \phi_1^2 \\ x &= \tfrac{1}{2}L & \Phi &= \Phi_2 = \tfrac{1}{2} k \phi_2^2 \end{aligned} \qquad (7.16)$$

Sec. 7 Shallow Unconfined Flow

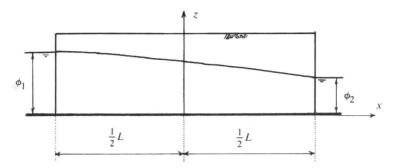

Figure 2.13 One-dimensional flow in an unconfined aquifer.

For one-dimensional flow, Φ is a function of x alone, and the differential equation (7.12) reduces to $d^2\Phi/dx^2 = 0$, with the general solution:

$$\Phi = Ax + B \tag{7.17}$$

Application of the boundary conditions (7.16) yields

$$\begin{aligned}\Phi_1 &= \tfrac{1}{2}k\phi_1^2 = -\tfrac{1}{2}AL + B \\ \Phi_2 &= \tfrac{1}{2}k\phi_2^2 = \tfrac{1}{2}AL + B\end{aligned} \tag{7.18}$$

Hence,

$$A = \frac{\Phi_2 - \Phi_1}{L} \qquad B = \tfrac{1}{2}(\Phi_1 + \Phi_2) \tag{7.19}$$

and the expression (7.17) for the potential becomes

$$\Phi = -(\Phi_1 - \Phi_2)\frac{x}{L} + \tfrac{1}{2}(\Phi_1 + \Phi_2) \tag{7.20}$$

The discharge Q_x is found from $Q_x = -\partial\Phi/\partial x$,

$$Q_x = \frac{\Phi_1 - \Phi_2}{L} \tag{7.21}$$

Substitution of the expressions (7.16) for Φ_1 and Φ_2 gives

$$Q_x = \frac{k(\phi_1^2 - \phi_2^2)}{2L} \tag{7.22}$$

This formula for the leakage through an unconfined aquifer, which could be called a "dam with vertical faces," is due to Dupuit. It was shown by Charny [1951] that the Dupuit formula is exact, notwithstanding the simplifying assumptions used to derive it.

An expression for the head ϕ is found from (7.10) and (7.20)

$$\phi = \sqrt{\tfrac{1}{2}(\phi_1^2 + \phi_2^2) - \frac{\phi_1^2 - \phi_2^2}{L}x} \qquad (C_u = 0) \tag{7.23}$$

This equation represents a parabola, known as the Dupuit parabola.

Problem 7.2

Draw 4 piezometric contours for the case of Figure 2.13 if $\phi_1 = 20$ m, $\phi_2 = 10$ m, and $L = 1000$ m.

Two-dimensional Flow

The results obtained in Sec. 6 for confined flow can be applied directly to unconfined flow, because the governing equations for confined and unconfined flow in terms of the discharge potentials are identical. The potential for flow toward a well in a circular island, illustrated in Figure 2.14, is given by (6.32),

$$\Phi = \frac{Q}{2\pi} \ln \frac{r}{R} + \Phi_0 \tag{7.24}$$

where the potential is now defined in terms of ϕ by (7.14) rather than (7.13), so that the expression for ϕ becomes

$$\tfrac{1}{2} k \phi^2 = \frac{Q}{2\pi} \ln \frac{r}{R} + \tfrac{1}{2} k \phi_0^2 \tag{7.25}$$

or

$$\phi = \sqrt{\frac{Q}{\pi k} \ln \frac{r}{R} + \phi_0^2} \tag{7.26}$$

Note that although the potentials for radial confined and radial unconfined flow are the same, the expressions for ϕ, (6.33) and (7.26), are quite different.

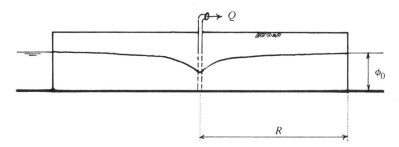

Figure 2.14 Unconfined flow toward a well.

It is important to note that, although the superposition principle applies to the potential, it does not apply to the head; the differential equation is linear in terms of Φ, but not in terms of ϕ. Superposition of two solutions of the form (7.24) therefore is allowed, but superposition of two expressions of the form (7.26) is wrong.

Sec. 7   Shallow Unconfined Flow

Example 7.1.   As an example of an application of the method of images to a case of shallow unconfined flow, consider the case indicated in Figure 2.15. A well is to be drilled in an unconfined aquifer near a long straight river. Prior to the operation of the well, uniform flow occurs in the direction normal to the river and the head along vertical planes parallel to the river is constant. The head at a distance L from the river bank prior to operation of the well is ϕ_1. The well is to be located at a distance d from the river bank. The problem is to determine the maximum discharge of the well for which no river water enters the aquifer after steady flow conditions have been reached.

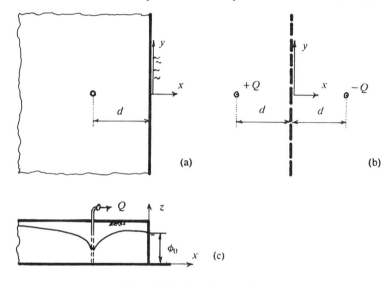

Figure 2.15   A well near a river.

A coordinate system is chosen, as indicated in Figure 2.15, with the x axis pointing normal to the river away from the aquifer, and the y axis along the river bank. First we determine the potential Φ for the uniform flow in the x direction occurring before the well operates. The boundary conditions are

$$\begin{aligned} x &= 0 & \phi &= \phi_0 & \Phi &= \Phi_0 = \tfrac{1}{2}k\phi_0^2 \\ x &= -L & \phi &= \phi_1 & \Phi &= \Phi_1 = \tfrac{1}{2}k\phi_1^2 \end{aligned} \qquad (7.27)$$

For uniform flow in the x direction, Φ is a function of x alone, i.e.,

$$\Phi = Ax + B \qquad (7.28)$$

see (7.17). Application of the boundary conditions (7.27) to (7.28) yields the following expression for Φ:

$$\Phi = -\frac{\Phi_1 - \Phi_0}{L}x + \Phi_0 \qquad (7.29)$$

The discharge Q_{xo} flowing toward the river before the well operates is found from $Q_x = -\partial\Phi/\partial x$ and (7.29):

$$Q_{x0} = \frac{\Phi_1 - \Phi_0}{L} \tag{7.30}$$

Equation (7.29) thus can be written as

$$\Phi = -Q_{x0}x + \Phi_0 \tag{7.31}$$

Second, we consider the case that the well draws an amount Q from the aquifer and that $Q_{x0} = 0$. We determine the potential for the situation occurring after the well has been operating over a sufficiently long period to obtain steady-state flow conditions. The potential Φ is obtained by the method of images. The image well is a recharge well located at a distance d from the river bank (see Figure 2.15[b]). The corresponding potential is

$$\Phi = \frac{Q}{2\pi} \ln \frac{r_1}{r_2} + C \tag{7.32}$$

where r_1 and r_2 are local radial coordinates with their respective origins at the well ($\mathcal{W}_\mathcal{W}$) and recharge well ($\mathcal{W}_\mathcal{R}$).

The potential for the combined flow, i.e., for flow toward the river with the well operating, is found by superposition. Adding the potentials (7.31) and (7.32), we obtain

$$\Phi = -Q_{x0}x + \Phi_0 + \frac{Q}{2\pi} \ln \frac{r_1}{r_2} + C \tag{7.33}$$

This potential must fulfill the boundary condition along the river, i.e., Φ must be equal to Φ_0 for $x = 0$, $r_1 = r_2 = r_0$. It follows from (7.33) that

$$x = 0 \qquad \Phi = \Phi_0 + C \tag{7.34}$$

In order that Φ be Φ_0 for $x = 0$, the constant C must vanish

$$C = 0 \tag{7.35}$$

so that (7.33) becomes

$$\Phi = -Q_{x0}x + \frac{Q}{2\pi} \ln \frac{r_1}{r_2} + \Phi_0 \tag{7.36}$$

It is noted that the ratio r_1/r_2 approaches unity far away from the well. The potential (7.36) then reduces to the one for the flow occurring prior to operation of the well. The influence of the well thus is noticeable only in a limited area. A plot of equipotentials (the dotted curves) and streamlines (the solid curves) is given in Figure 2.16 as an illustration. The highest value of Φ, $\Phi_{\max}$, occurs along the two short equipotentials drawn at the extreme left of the plot. The curves are obtained by letting Φ be $\Phi_{\max} - j\Delta\Phi$, $j = 0, 1, 2, \ldots, n-1$, where n is the total number of equipotentials. An exception is

Sec. 7   Shallow Unconfined Flow   45

the special equipotential adjacent to the river, which intersects itself at point S in Figure 2.16. The flow stagnates at this point of intersection, where both Q_x and Q_y are equal to zero: this point is a stagnation point. Along the x axis, the flow to the right of S is toward the river, and the flow to the left of S is toward the well, as may be seen by inspecting the streamlines. The streamline through the stagnation point is called the *dividing streamline*, because the flow to the one side of the dividing streamline is toward the well whereas the flow to the other side is toward the river.

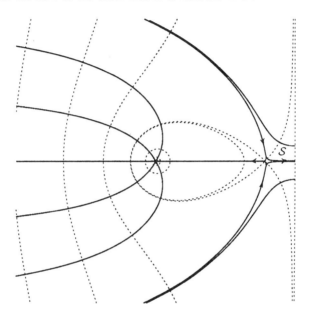

Figure 2.16 Equipotentials for flow toward a well near a river.

The question asked in stating the problem was to determine the maximum discharge of the well such that no river water enters the aquifer. To answer this question, we will examine the sign of the component Q_x of the discharge vector along the river bank. No water will enter the aquifer from the river if the flow everywhere along the river bank is out of the aquifer, i.e., if Q_x is positive. Hence, the condition is

$$x = 0 \qquad Q_x \geq 0 \qquad (7.37)$$

With Darcy's law, $Q_x = -\partial \Phi / \partial x$, this becomes

$$x = 0 \qquad \frac{\partial \Phi}{\partial x} \leq 0 \qquad (7.38)$$

We express r_1 and r_2 in terms of x and y in order to differentiate (7.36) with respect to x. This gives

$$r_1^2 = (x + d)^2 + y^2 \qquad r_2^2 = (x - d)^2 + y^2 \qquad (7.39)$$

Substitution in (7.36) gives, with $\ln(r_1/r_2) = \frac{1}{2}\ln(r_1/r_2)^2$,

$$\Phi = -Q_{x0}x + \frac{Q}{4\pi}\ln\frac{(x+d)^2 + y^2}{(x-d)^2 + y^2} + \Phi_0$$

$$= -Q_{x0}x + \frac{Q}{4\pi}\left\{\ln[(x+d)^2 + y^2] - \ln[(x-d)^2 + y^2]\right\} + \Phi_0 \tag{7.40}$$

Differentiation with respect to x yields

$$\frac{\partial\Phi}{\partial x} = -Q_{x0} + \frac{Q}{4\pi}\left[\frac{2(x+d)}{(x+d)^2 + y^2} - \frac{2(x-d)}{(x-d)^2 + y^2}\right] \tag{7.41}$$

The condition (7.38) can now be written as follows:

$$\frac{\partial\Phi}{\partial x} = -Q_{x0} + \frac{Q}{4\pi}\frac{4d}{d^2 + y^2} \leq 0 \tag{7.42}$$

or

$$\frac{Q}{4\pi}\frac{4d}{y^2 + d^2} \leq Q_{x0} \tag{7.43}$$

The expression at the left of the $\leq$ sign has its maximum value for $y = 0$. Thus, condition (7.43) will be fulfilled along the entire river bank, i.e., for all values of y, if

$$\frac{Q}{\pi}\frac{1}{d} \leq Q_{x0} \tag{7.44}$$

or

$$Q \leq \pi d Q_{x0} \tag{7.45}$$

This equation gives the maximum discharge of the well [m³/s] in terms of Q_{x0} [m²/s] and d [m]. This maximum discharge increases linearly with the distance between well and river bank, as is seen from (7.45).

For the critical case that $Q = \pi d Q_{x0}$, Q_x is positive along the entire river bank, except at the origin, where $Q_x = 0$. The river bank is an equipotential and therefore Q_y is zero there. Thus, both Q_x and Q_y are zero at $x = 0$, $y = 0$ for the critical case; the origin is a stagnation point. This is illustrated in Figure 2.17(a), where the equipotentials and streamlines are reproduced for the critical case. If the ratio $\pi d Q_{x0}/Q$ is greater than unity, the stagnation point lies to the left of the river bank and outflow occurs along the entire bank (see [7.44]). The equipotentials shown in Figure 2.16 apply to this case; $\pi d Q_{x0}/Q = 1.05$. If the ratio $\pi d Q_{x0}/Q$ is less than unity, inflow occurs along a portion of the river bank, centered at the origin. This is shown in Figure 2.17(b), which applies to the case that $\pi d Q_{x0}/Q = 0.95$. The section of inflow is between the two stagnation points labeled S_1 and S_2.

Problem 7.3

Determine the y coordinates of the two stagnation points in Figure 2.17(b).

Sec. 7 Shallow Unconfined Flow

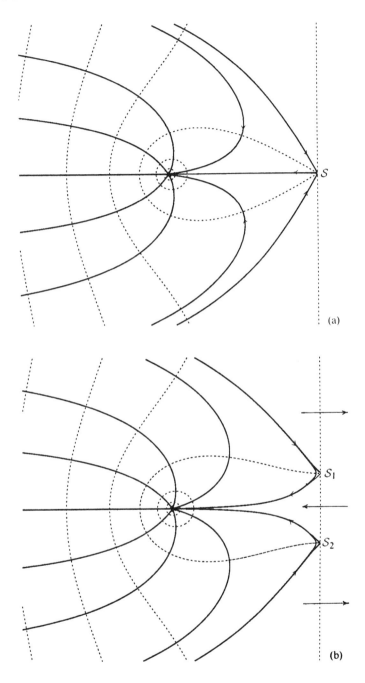

Figure 2.17 Streamlines and equipotentials for (a) the critical case ($Q = \pi d Q_{x0}$) and (b) the case that $\pi d Q_{x0}/Q = 0.95$.

Problem 7.4

Consider the problem of one-dimensional shallow unconfined flow through a dam with vertical faces (see Figure 2.18). The dam is divided into two zones, consisting of different materials which have hydraulic conductivities k_1 and k_2. The lengths of the zones are L_1 and L_2, and the heads along the faces are ϕ_1 and ϕ_2.

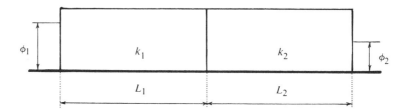

Figure 2.18 Flow through an inhomogeneous dam (Problem 7.4).

Questions

(a) Express Q_x in terms of the parameters of the problem.
(b) Calculate Q_x if $k_1 = 10^{-5}$ m/s, $k_2 = 10^{-12}$ m/s, $L_1 = 100$ m, $L_2 = 100$ m, $\phi_1 = 20$ m, and $\phi_2 = 10$ m.

Problem 7.5

A dewatering system consisting of 4 wells will be used to obtain dry working conditions in a building pit in an unconfined aquifer of hydraulic conductivity $k = 10^{-5}$ m/s. The aquifer is bounded by a long straight river and there is no flow before the wells operate. The head along the river bank is constant, and is $\phi_0 = 25$ m above the base of the aquifer. The building pit is square and the bottom of the pit will be 20 m above the aquifer base. The sides of the pit are 50 m in length, and the center of the pit is 100 m away from the river bank.

Questions

(a) Calculate the discharge Q per well that is necessary to obtain dry working conditions.
(b) Answer the question stated under (a) if it is given that there is a uniform flow of discharge $Q_{x0} = 1.1 * 10^{-6}$ m^2/s in the x direction before the wells operate.
(c) Verify that the wells can pump the calculated discharges by checking that the heads at the wells are above the base of the aquifer. The radii of all wells are .25 m.

8 COMBINED SHALLOW CONFINED AND UNCONFINED FLOW

Combined shallow confined and unconfined flow occurs if the head is less than the aquifer thickness H in some zones and greater than H in others. Such unconfined zones may occur around wells or near boundaries. Problems of combined shallow confined and unconfined flow can be solved in a way suggested by Girinski [1946], as is explained in the following.

Method of Solution

We will refer to the boundaries between confined and unconfined zones as the interzonal boundaries and measure ϕ with respect to the base of the aquifer. Along the interzonal boundaries the head, then, equals the thickness, H, of the aquifer.

A single potential Φ will be used throughout the flow region. In the confined zones, this potential is given by (7.13),

$$\Phi = kH\phi + C_c \qquad (\phi \geq H) \tag{8.1}$$

and in the unconfined zones, Φ is defined by (7.14),

$$\Phi = \tfrac{1}{2}k\phi^2 + C_u \qquad (\phi < H) \tag{8.2}$$

We will choose the constants C_c and C_u such that the potential is continuous across the interzonal boundary. Thus, (8.1) and (8.2) must yield the same value when $\phi = H$,

$$kH^2 + C_c = \tfrac{1}{2}kH^2 + C_u \tag{8.3}$$

This condition can be met by choosing C_c and C_u such that

$$C_c = C_u - \tfrac{1}{2}kH^2 \tag{8.4}$$

One of the two constants can be set arbitrarily. We will let C_u be zero, so that (8.4) becomes

$$C_c = -\tfrac{1}{2}kH^2 \qquad C_u = 0 \tag{8.5}$$

The single potential Φ is thus defined, depending on the magnitude of ϕ, as follows:

$$\boxed{\Phi = kH\phi - \tfrac{1}{2}kH^2 \qquad (\phi \geq H)} \tag{8.6}$$

$$\boxed{\Phi = \tfrac{1}{2}k\phi^2 \qquad (\phi < H)} \tag{8.7}$$

The discharge vector has the same physical meaning for horizontal confined flow as it has for shallow unconfined flow: it points in the direction of flow and its magnitude equals the amount of water flowing through a cross section of unit width that extends over the saturated thickness of the aquifer. By continuity of flow, the component of the discharge vector normal to the interzonal boundary, Q_n, must be the same on either side of that boundary (see Figure 2.19). Furthermore, the interzonal boundary is an equipotential $\Phi = \tfrac{1}{2}kH^2$, and therefore the tangential component of the discharge vector along the interzonal boundary is zero. The discharge vector is thus continuous across the interzonal boundary; there exists a single discharge vector throughout the aquifer,

$$\boxed{Q_x = -\frac{\partial \Phi}{\partial x}} \tag{8.8}$$

$$\boxed{Q_y = -\frac{\partial \Phi}{\partial y}} \tag{8.9}$$

(compare [6.17] and [7.9]). The continuity equation (7.11) must be satisfied in the entire flow domain, including the interzonal boundary,

$$\frac{\partial Q_x}{\partial x} + \frac{\partial Q_y}{\partial y} = 0 \tag{8.10}$$

and (8.8) and (8.9) yield Laplace's equation for the potential Φ,

$$\frac{\partial^2 \Phi}{\partial x^2} + \frac{\partial^2 \Phi}{\partial y^2} = 0 \tag{8.11}$$

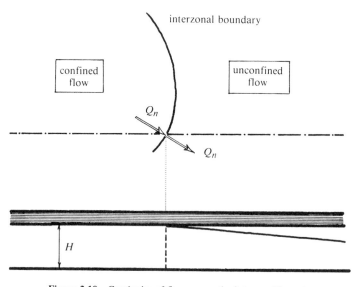

Figure 2.19 Continuity of flow across the interzonal boundary.

Sec. 8 Combined Shallow Confined and Unconfined Flow

It follows that problems of combined confined-unconfined flow can be solved without taking the location of the interzonal boundary into account; if so desired, it can be calculated after determining Φ as a function of position. A choice must be made between (8.6) and (8.7) when writing the boundary conditions in terms of Φ; (8.6) applies along boundary segments where the flow is confined, and (8.7) must be used where the flow is unconfined.

When calculating the head, for example, in order to obtain piezometric contours, either (8.6) or (8.7) must be chosen according to the value of Φ:

$$\phi = \frac{\Phi + \frac{1}{2}kH^2}{kH} \qquad (\Phi \geq \tfrac{1}{2}kH^2) \tag{8.12}$$

$$\phi = \sqrt{\frac{2\Phi}{k}} \qquad (\Phi < \tfrac{1}{2}kH^2) \tag{8.13}$$

Some examples of the application of this approach follow.

One-dimensional Flow

Consider the problem of one-dimensional flow illustrated in Figure 2.20. The aquifer is confined, of length L, and of thickness H. An x, y, z coordinate system is chosen as shown in the figure. The boundary conditions are

$$x = -\tfrac{1}{2}L \qquad \phi = \phi_1 > H \tag{8.14}$$

and

$$x = \tfrac{1}{2}L \qquad \phi = \phi_2 < H \tag{8.15}$$

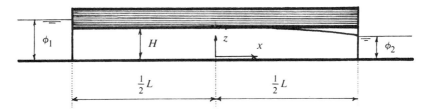

Figure 2.20 One-dimensional combined confined-unconfined flow.

The aquifer is confined at $x = -L/2$ and unconfined at $x = L/2$, and the boundary values of Φ are obtained by the use of (8.6) and (8.7), respectively,

$$x = -\tfrac{1}{2}L \qquad \Phi = \Phi_1 = kH\phi_1 - \tfrac{1}{2}kH^2 \tag{8.16}$$

and

$$x = \tfrac{1}{2}L \qquad \Phi = \Phi_2 = \tfrac{1}{2}k\phi_2^2 \qquad (8.17)$$

The expression for the potential is identical to that obtained for the problem of Figure 2.13 (see [7.20]),

$$\Phi = -(\Phi_1 - \Phi_2)\frac{x}{L} + \tfrac{1}{2}(\Phi_1 + \Phi_2) \qquad (8.18)$$

but the expressions for Φ_1 and Φ_2 in terms of the head are different from those applying to (7.20). The expression for Q_x becomes

$$Q_x = \frac{\Phi_1 - \Phi_2}{L} = \frac{kH\phi_1 - \tfrac{1}{2}kH^2 - \tfrac{1}{2}k\phi_2^2}{L} \qquad (8.19)$$

The location of the interzonal boundary, $x = x_b$, is obtained from (8.18) by letting $\Phi = \tfrac{1}{2}kH^2$, which gives with (8.16) and (8.17),

$$\frac{x_b}{L} = \frac{\tfrac{1}{2}(kH\phi_1 - \tfrac{3}{2}kH^2 + \tfrac{1}{2}k\phi_2^2)}{kH\phi_1 - \tfrac{1}{2}kH^2 - \tfrac{1}{2}k\phi_2^2} \qquad (8.20)$$

Note that $x_b/L = -\tfrac{1}{2}$ when $\phi_1 = H$, and $x_b/L = \tfrac{1}{2}$ when $\phi_2 = H$.

Problem 8.1

Determine the potential as a function of x for the confined and unconfined zones in Figure 2.20 separately. Join the two solutions by requiring that there be continuity of flow across the interzonal boundary. Express both the discharge vector and x_b in terms of the parameters of the problem and compare these results with (8.19) and (8.20).

Radial Flow

As a second application, we will consider the case of radial flow toward a well of radius r_w in a confined aquifer (see Figure 2.21). The flow near the well is unconfined; the head at the well, ϕ_w, is less than the aquifer thickness, H. The expression for the potential is identical to those obtained for confined flow and for unconfined flow in Secs. 6 and 7, and is

$$\Phi = \frac{Q}{2\pi} \ln \frac{r}{R} + \Phi_0 \qquad (8.21)$$

where Φ_0 is the potential at $r = R$. If the head at $r = R$ is ϕ_0, and if the flow is confined there, Φ_0 is given by

$$\Phi_0 = kH\phi_0 - \tfrac{1}{2}kH^2 \qquad (\phi_0 > H) \qquad (8.22)$$

Sec. 8 Combined Shallow Confined and Unconfined Flow

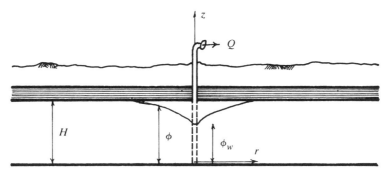

Figure 2.21 Combined confined-unconfined flow toward a well.

The head at the well is less than H, so that

$$\Phi_w = \tfrac{1}{2}k\phi_w^2 \qquad (\phi_w < H) \tag{8.23}$$

Application of (8.23) to (8.21) yields, with (8.22),

$$\tfrac{1}{2}k\phi_w^2 = \frac{Q}{2\pi}\ln\frac{r_w}{R} + kH\phi_0 - \tfrac{1}{2}kH^2 \tag{8.24}$$

and solving for Q, we obtain

$$Q = 2\pi\frac{\tfrac{1}{2}k\phi_w^2 - kH\phi_0 + \tfrac{1}{2}kH^2}{\ln\frac{r_w}{R}} \tag{8.25}$$

The interzonal boundary is a circle. The radius of this circle will be denoted as r_b and is obtained from (8.21) by setting $\Phi = \tfrac{1}{2}kH^2$,

$$\tfrac{1}{2}kH^2 = \frac{Q}{2\pi}\ln\frac{r_b}{R} + \Phi_0 \tag{8.26}$$

or, solving for r_b and using (8.22)

$$r_b = Re^{2\pi kH(H-\phi_0)/Q} \tag{8.27}$$

Problem 8.2

Draw 9 piezometric contours between the concentric circles $r = r_w$ and $r = R$, if $r_w = 0.01R$, $\phi_0 = 1.2H$ and $\phi_w = 0.2H$ for the case of radial flow of the preceding example.

Dewatering Problem

The technique explained in this section can be applied to dewatering problems where a building pit is dug through the upper confining bed of an aquifer. A case of a circular building pit that is to be dewatered by means of six wells is illustrated in Figure 2.22. An x, y, z Cartesian coordinate system is chosen with the origin at the center of the pit

54 Types of Flow   Chap. 2

as shown in the figure. The wells are at a distance B from the center of the pit, and are spaced uniformly along the circle $x^2 + y^2 = B^2$. The bottom of the pit is at an elevation H_0 above the base of the aquifer, which has thickness H. It is assumed for the sake of simplicity that there is no flow before the wells operate. The wells all have a discharge Q, which must be determined such that the head inside a cylinder of radius B around the center of the pit is less than or equal to H_0:

$$0 \leq x^2 + y^2 \leq B^2 \qquad \phi < H_0 \tag{8.28}$$

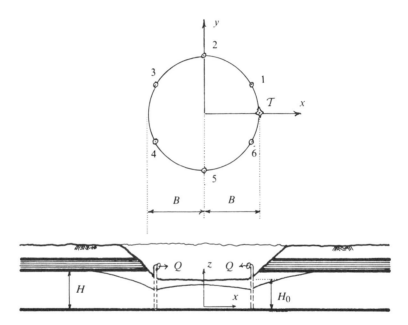

Figure 2.22 A circular building pit dug into a confined aquifer.

The head at a point $\mathcal{P}$ with coordinates $x_p = R$, $y_p = 0$ is assumed to be unaffected by the wells and is ϕ_0, where $\phi_0 > H$. The potential for this flow problem has the form

$$\Phi = \frac{Q}{2\pi} \sum_{j=1}^{6} \ln[r_j(x,y)] + C \tag{8.29}$$

where $r_j(x,y)$ is the local radial coordinate emanating from well j, which is centered at (x_j, y_j), i.e.,

$$r_j^2(x,y) = (x - x_j)^2 + (y - y_j)^2 \tag{8.30}$$

Sec. 8 Combined Shallow Confined and Unconfined Flow

The coordinates of the six wells are obtained from Figure 2.22:

$$x_1 = x_6 = -x_3 = -x_4 = B\cos\frac{\pi}{6} \tag{8.31}$$

$$y_1 = y_3 = -y_4 = -y_6 = \tfrac{1}{2}B \tag{8.32}$$

$$x_2 = x_5 = 0 \tag{8.33}$$

$$y_2 = -y_5 = B \tag{8.34}$$

The condition at point $\mathcal{P}$ is

$$x = R \quad y = 0 \quad \phi = \phi_0 > H \quad \Phi = \Phi_0 = kH\phi_0 - \tfrac{1}{2}kH^2 \tag{8.35}$$

and application to (8.29) yields

$$C = \Phi_0 - \frac{Q}{2\pi}\sum_{j=1}^{6}\ln r_j(R,0) \tag{8.36}$$

so that (8.29) becomes

$$\Phi = \frac{Q}{2\pi}\sum_{j=1}^{6}\ln\frac{r_j(x,y)}{r_j(R,0)} + \Phi_0 \tag{8.37}$$

The groundwater table along the boundary of the pit will be highest at points midway between the wells, such as point $\mathcal{T}$ in Figure 2.22. Water will flow from these points inward to the center of the pit and from there to the wells. The head will therefore be higher at point $\mathcal{T}$ than anywhere inside the cylinder of radius B. Condition (8.28) will thus be met if the head at point $\mathcal{T}$ is equal to H_0. The flow is unconfined at $\mathcal{T}$ so that the potential there equals $\tfrac{1}{2}kH_0^2$. Substitution of B for x, zero for y and $\tfrac{1}{2}kH_0^2$ for Φ in (8.37) yields, with (8.35),

$$\tfrac{1}{2}kH_0^2 = \frac{Q}{2\pi}\sum_{j=1}^{6}\ln\frac{r_j(B,0)}{r_j(R,0)} + kH\phi_0 - \tfrac{1}{2}kH^2 \tag{8.38}$$

and the discharge Q is obtained as

$$Q = 2\pi\frac{\tfrac{1}{2}kH_0^2 - kH\phi_0 + \tfrac{1}{2}kH^2}{\sum_{j=1}^{6}\ln\frac{r_j(B,0)}{r_j(R,0)}} \tag{8.39}$$

As an illustration, the interzonal boundary $\Phi = \tfrac{1}{2}kH^2$ and the equipotential $\Phi = \tfrac{1}{2}kH_0^2$ are shown in Figure 2.23(a) for the case that $B/R = 0.1$, $B/H = 10$, $H_0/H = 0.7$, and

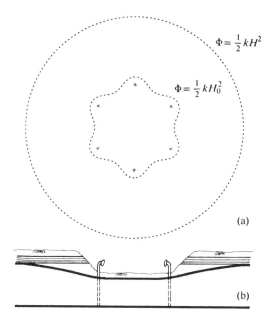

Figure 2.23 The equipotentials $\Phi = \frac{1}{2}kH^2$ and $\Phi = \frac{1}{2}kH_0^2$ (a) and sectional view (b).

$\phi_0/H = 1.4$. The corresponding value for $Q/(kH^2)$ obtained from (8.39) is 0.31. A cross-sectional view along the x axis is shown in Figure 2.23(b).

Implementation of Uniform Flow with Wells in the FORTRAN Program SLW

The solutions to Laplace's equation discussed so far become more useful when implemented in a computer program. In the following, we will examine a program written in ANSI FORTRAN 77, with the capability of free-format data input and graphics support. We will refer to this program as SLW, which stands for Single Layer with Wells. The program is written for the IBM PC, but can be easily adapted to other machines, which should only require rewriting of some of the subroutines for graphics.

The primary purpose of the program is to demonstrate applications of a selection of the material presented in this text: it is kept as simple as possible. For example, no provision is made to protect against entering more data (e.g., for wells) than can be stored, and more weight is given to readability of the program than to computational efficiency.

Use is made of three packages of service subroutines: one for free-format data input (written by P. A. Cundall), one for producing piezometric contours and equipotentials, and one for plotting. These packages will not be discussed in the text; the source code of all routines is contained on the diskette supplied with this book; for information on how to use the diskette, see Appendix C.

Sec. 8   Combined Shallow Confined and Unconfined Flow

We will next discuss the implementation in this computer program of a combination of uniform flow in an arbitrary direction with wells. The potential for uniform flow presented so far is valid only for flow in the x direction, and has the form (7.31):

$$\Phi = -Q_{x0}x + C \tag{8.40}$$

If the flow is in the y direction rather than in the x direction, then the potential is, by analogy,

$$\Phi = -Q_{y0}y + C \tag{8.41}$$

where Q_{y0} is the discharge flowing in the y direction. Thus, if there is a uniform flow with a discharge vector with components Q_{x0} and Q_{y0}, then the potential is

$$\Phi = -Q_{x0}x - Q_{y0}y + C \tag{8.42}$$

If the angle between the uniform flow and the x axis is α and the discharge of the uniform flow is Q_0, then the components Q_{x0} and Q_{y0} may be expressed as

$$\begin{aligned} Q_{x0} &= Q_0 \cos \alpha \\ Q_{y0} &= Q_0 \sin \alpha \end{aligned} \tag{8.43}$$

so that (8.42) may be written as

$$\Phi = -Q_0(x \cos \alpha + y \sin \alpha) \tag{8.44}$$

If there are n wells of discharge Q_j located at (x_j, y_j) $(j = 1, 2, \ldots, n)$ then the general potential for uniform flow and wells becomes

$$\Phi = -Q_0(x \cos \alpha + y \sin \alpha) + \sum_{j=1}^{n} \frac{Q_j}{4\pi} \ln[r_j^2(x,y)] + C \tag{8.45}$$

where r_j is given by (8.30).

To implement (8.45) in a computer program, we first provide the facility for input of the data (in a routine called SETUP), and then write function subroutines for the various parts of the potential. Prior to doing this, however, we will establish some rules for notation, intended to reduce the possibility for making errors. These rules are as follows:

1. All variables of type CHARACTER*1 have A as their first letter. The corresponding declaration is IMPLICIT CHARACTER*1 (A).

2. All integer values have I, J, K, N, or M as their first letter. A declaration is not necessary, as variables with these characters as their first letter are integer by default.

3. All logical variables have L as their first letter. The corresponding declaration is IMPLICIT LOGICAL (L).

4. All real variables have R as their first letter. Again, a declaration is not necessary, as all variables with R as their first letter are real by default.

5. Most function subroutine names have F as their second letter; the first letter is chosen according to the type of function following rules 1 though 4.
6. The second and third letters of the names of variables in common blocks represent a code corresponding to the common block name.

The Main Program

The main program is very short, and merely serves to specify the dimensions of a matrix for storage of values to be contoured (RMAT), and of a scratch array used by the contouring routine (RSCRT). The listing of the main program is as follows:

```
      PROGRAM SL
      DIMENSION RMAT(50,50),RSCRT(81)
C     RMAT   : matrix used for contouring: RMAT(IMXSZE,IMXSZE)
C     RSCRTC : array used for contouring , minimum size
C              is 1+(IMXSZE*IMXSZE-1)/31
      IMXSZE=50
      CALL SETUP(IMXSZE,RMAT,RSCRT)
      STOP
      END
```

The Routine Setup

The routine SETUP, called by the main program, serves for entering data and for issuing various commands. The listing contains explanations of all statements of importance; it will suffice to make some general remarks in the text. SETUP contains a list of command words, each of which can be entered by the user of the program. The routine MATCH is called when an input line is entered, and determines whether it contains one of the command words. If so, the program jumps to the appropriate label and begins execution of the command.

The only computation contained in this routine is the determination of the constant C in the solution. This constant is computed from the condition that the head be equal to a given value ϕ_0 at a point with coordinates x_0 and y_0, referred to in the program as the reference point. The corresponding value of the potential is obtained by the use of a function $\Phi = \Phi(\phi)$, which is given either by (8.6) or (8.7), depending on the type of flow. Hence,

$$\Phi_0 = -Q_0(x_0 \cos \alpha + y_0 \sin \alpha) + \sum_{j=1}^{n} \frac{Q_j}{4\pi} \ln[r_j^2(x_0, y_0)] + C \qquad (8.46)$$

where

$$\Phi_0 = \Phi(\phi_0) \qquad (8.47)$$

Sec. 8 Combined Shallow Confined and Unconfined Flow

Solving (8.46) for the constant, we obtain:

$$C = \Phi_0 + Q_0(x_0 \cos \alpha + y_0 \sin \alpha) - \sum_{j=1}^{n} \frac{Q_j}{4\pi} \ln[r_j^2(x_0, y_0)] \qquad (8.48)$$

The listing of the routine SETUP is as follows:

```
c       Routine for processing input.
c       RMAT    =  matrix of maximum size IMXSZE*IMXSZE.
c       RSCRT   =  scratch array.
        SUBROUTINE SETUP(IMXSZE,RMAT,RSCRT)
c       All variables beginning with L are logical, all variables
c       beginning with A are character*1.
        IMPLICIT LOGICAL(L),CHARACTER*1(A)
        DIMENSION RMAT(*),RSCRT(*)
c       AWORD and AWOGR contain key-words.
        DIMENSION AWORD(55),AWOGR(10)
        CHARACTER*32 AFILE
c       Common block used for free-format input. ALINE=input line,
c       LERROR,LMISS are set true if an error is detected.
        COMMON /CBMATC/ ILPNT(40),LMISS,LERROR,NPBAD,NCHAR
        COMMON /CBMATA/ ALINE(80),AFILE
c       Common block with grid data.
        COMMON /CBGRID/ RGRX1,RGRY1,RGRX2,RGRY2,RGRMAX,RGRMIN,
     .                  NGRX,NGRY,RGRXG(150),RGRYG(150),RGRDX,RGRDY,
     .                  IGRX1,IGRY1,IGRX2,IGRY2,RGRFAX,RGRFAY,NGRDTX,
     .                  NGRDTY,RGREPS
c       Common block with aquifer data.
        COMMON /CBAQ/   RAQPRM,RAQTHK,RAQBAS
c       Common block with reference point data.
        COMMON /CBRF/   RRFX,RRFY,RRFFI,RRFCON
c       Common block with data for given functions.
        COMMON /CBGV/   RGVUQ0,RGVUAL,RGVUU(2)
c       Common block with well data.
        COMMON /CBWL/   NWL,RWLXC(200),RWLYC(200),RWLQ(200),RWLRAD(200)
c       Functions that can be contoured: RFPOT=potential,
c       RFHEAD=piezometric head.
        EXTERNAL RFPOT,RFHEAD
c       List of main keywords. The routine MATCH matches the word entered
c       in the input line to the words below. It returns the position in
c       the list of the keyword for which a match is found. Thus, if you
c       enter WELL then MATCH will return position number 4. The words
c       are separated by blanks and an exclamation mark denotes the end
c       of the list.
```

```
      DATA AWORD   /'A','Q','U','I',' ',
     .              'R','E','F','E',' ',
     .              'U','N','I','F',' ',
     .              'W','E','L','L',' ',
     .              'S','O','L','V',' ',
     .              'W','I','N','D',' ',
     .              'G','R','I','D',' ',
     .              'P','L','O','T',' ',
     .              'L','A','Y','O',' ',
     .              'C','H','E','C',' ',
     .              'P','A','U','S',' ',
     .              'S','W','I','T',' ',
     .              'S','T','O','P','!'/
c     Keywords defining which function should be contoured.
      DATA AWOGR   /'H','E','A','D',' ',
     .              'P','O','T','E','!'/
c     Reading is from unit 5, writing of all messages is to the
c     default unit (*), and writing of numerical output is to unit 7.
c     Open unit 7 as the keyboard (default).
      OPEN(UNIT=7,FILE='CON')
c     Initialization of the logicals LMISS and LERROR.
      LERROR=.FALSE.
      LMISS=.FALSE.
 10   IF(LMISS) WRITE(*,9001)
      IF(LERROR.AND..NOT.LMISS) WRITE(*,9002)
      LERROR=.FALSE.
      LMISS=.FALSE.
      WRITE(*,9005)
c     Read an input line.
      READ(5,9000) ALINE
c     Tidy up the input line.
      CALL TIDY
c     Find the number (JUMP) of the word in AWORD that matches the
c     first keyword entered in the input line.
      CALL MATCH(AWORD,1,JUMP)
c     Return to label 10 if no match was found, an error was detected,
c     or the input line is empty.
      IF(LMISS.OR.LERROR.OR.ILPNT(2).EQ.0) GOTO 10
c     Go to the appropriate label, depending on the keyword entered by
c     the user.
      GOTO( 100, 200, 300, 400, 500, 600, 700, 800, 900,1000,
     .     1100,1200,8900),JUMP
c     Aquifer data; read hydraulic conductivity, aquifer thickness,
c     and, if entered, aquifer base elevation. The function RVAR(IPAR)
```

Sec. 8   Combined Shallow Confined and Unconfined Flow   61

```
c        is part of the MATCH package, interprets the IPAR'th word of the
c        input line as a real variable, and returns it as a real variable.
c        The array ILPNT initially is filled with zeros. When the input
c        line is cleaned up, non-zero numbers are entered in the positions
c        1 through NPAR+1, where NPAR is the number of parameters in the
c        input line. Thus, if ILPNT(NPAR+1) is non-zero, then there are at
c        least NPAR parameters in the input line.
  100    RAQPRM=RVAR(2)
         RAQTHK=RVAR(3)
         IF(ILPNT(5).NE.0) RAQBAS=RVAR(4)
         GOTO 10
c        Reference point; read coordinates of reference point (a point in
c        the aquifer where the head is given), and the head at the
c        reference point.
  200    RRFX=RVAR(2)
         RRFY=RVAR(3)
         RRFFI=RVAR(4)
         GOTO 10
c        Uniform flow; read the discharge of uniform flow and the angle
c        (in degrees) that the direction of uniform flow makes with the
c        x axis.
  300    RGVUQ0=RVAR(2)
         RGVUAL=RVAR(3)
c        Convert the angle from degrees to radians.
         RARG=RGVUAL*3.1415926/180.
c        Compute the cosine and sine of the above angle, and store them
c        as components of a unit vector.
         RGVUU(1)=COS(RARG)
         RGVUU(2)=SIN(RARG)
         GOTO 10
c        Wells; add one to the number of wells, then store the coordinates
c        of the well, its discharge, and its radius.
  400    NWL=NWL+1
         RWLXC(NWL)=RVAR(2)
         RWLYC(NWL)=RVAR(3)
         RWLQ(NWL)=RVAR(4)
         RWLRAD(NWL)=RVAR(5)
c        If an error occurred, reduce the number of wells by one, so that
c        the user can re-enter the data for this well.
         IF (LERROR) NWL=NWL-1
         GOTO 10
c        Solve; compute the constant, RRFCON, such that the head is equal
c        to the specified value at the reference point. The potential due
c        to all functions in the solution is the sum of RFPGVN and
```

```
c       RRFCON. The value of the head at the reference point, RRFFI, is
c       converted to a potential by a function called RFPOTH(RHEAD),
c       which resides in the module AQUIFER. Thus, in order that the
c       potential function (RFPOT) at the reference point be equal to
c       RFPOTH(RRFFI), the constant RRFCON must be computed from:
  500   RRFCON=RFPOTH(RRFFI)-RFPGVN(RRFX,RRFY)
        GOTO 10
c       Window; set the coordinates of the lower left corner and the
c       upper right corner to the four values entered:
  600   RGRX1=RVAR(2)
        RGRY1=RVAR(3)
        RGRX2=RVAR(4)
        RGRY2=RVAR(5)
        GOTO 10
c       Grid; set the number of grid points, NGRX, to one plus the number
c       of intervals. The function IVAR(IPAR) is similar to RVAR(IPAR),
c       but interprets the IPAR'th word as an integer.
  700   NGRX=IVAR(2)+1
        IF(LERROR) GOTO 10
c       Prevent the number of grid points to exceed the maximum (assuming
c       a square grid).
        IF(NGRX.GE.IMXSZE) NGRX=IMXSZE-1
c       If the user did not specify a second keyword (HEAD or POTENTIAL),
c       use the default (HEAD) this situation occurs if only 2 entries
c       exist on the input line, i.e., if ILPNT(3) is the last
c       non-zero item:
        IF(ILPNT(4).EQ.0) THEN
           JUMP=1
c       otherwise call MATCH to find out which keyword was entered.
        ELSE
           CALL MATCH(AWOGR,3,JUMP)
        ENDIF
c       Call the routine that initializes the grid: it fills the arrays
c       RGRXG AND RGRYG with coordinates of points in the grid.
        CALL GRINIT
c       Go to the appropriate label according to the function to be
c       contoured.
        GOTO(710,720),JUMP
c       Piezometric heads: fill the matrix with values of the head at
c       the grid points.
  710   CALL FILLXY(RMAT,NGRX,IMXSZE,RFHEAD,LWRONG)
c       if the total number of grid points exceeded imxsze*imxsze give
c       error message.
        IF(LWRONG) WRITE(*,9004)
```

Sec. 8 Combined Shallow Confined and Unconfined Flow

```
             GOTO 10
c      Equipotentials; fill the matrix with values of the potential at
c      the grid points.
 720         CALL FILLXY(RMAT,NGRX,IMXSZE,RFPOT,LWRONG)
c      if the total number of grid points exceeded imxsze*imxsze give
c      error message.
             IF(LWRONG) WRITE(*,9004)
             GOTO 10
c      Contouring; first check whether the number of grid points, NGRX,
c      is set.
 800         IF(NGRX.EQ.0.OR.NGRY.EQ.0) THEN
               WRITE(*,9003)
               GOTO 10
             ENDIF
c      Display the maximum and minimum values encountered in the grid;
c      these values are stored by FILLXY in the common block CBGRID.
c      Ask the user to enter the minimum value to be contoured and the
c      contour interval
             WRITE (*,9010) RGRMIN,RGRMAX
c      Read ALINE, containing the desired values.
             READ(5,9000) ALINE
c      Call TIDY to tidy up the input line.
             CALL TIDY
c      Next, obtain the two above values.
             RLEVEL=RVAR(1)
             RINCRM=RVAR(2)
             IF(LERROR) GOTO 10
c      Set the screen in graphics mode.
             CALL DRAWIN
c      Produce a layout of all entered wells, plotted as circles on the
c      screen.
             CALL LAYOUT
c      Call the contouring routine to produce the contour plot on the
c      screen.
             ISIZE=NGRX
 850         CALL CONTUR
     .           (RMAT,ISIZE,NGRX,NGRY,IGRX1,IGRY1,RGRDX,RGRDY,RLEVEL,RSCRT)
c      Increment the contour level and check whether it exceeds the
c      maximum. If not, call CONTUR with the next level.
             RLEVEL=RLEVEL+RINCRM
             IF (RLEVEL.LE.RGRMAX) GOTO 850
c      Wait until the user presses the enter key prior to leaving
c      graphics mode.
             READ(5,*)
```

```
c       Re-set the screen to text mode.
        CALL DRAWOUT
        GOTO 10
c       Layout: Produce a layout on the screen.
 900    CALL DRAWIN
        CALL LAYOUT
        READ(5,*)
        CALL DRAWOUT
        GOTO 10
c       Check; call the routine for data retrieval.
1000    CALL CHECK
        GOTO 10
c       Pause; causes the program to pause, giving access to DOS
c       commands. Press enter to return to program.
1100    PAUSE
        GOTO 10
c       The following statements make it possible to switch input
c       and output devices. The first argument of the command
c       SWITCH denotes the new input device (e.g., the name of a
c       file with input data prepared in advance). The command
c       SWITCH CON will cause the input to return to the keyboard.
c       The latter command must be the last command in an input file.
c       The second argument (e.g., a filename or the name of a device
c       such as PRN) will cause the output generated by the routine
c       CHECK to be directed to this file or device. The second
c       argument is optional. The function GETFN(IPAR) contained in
c       the match package returns in the variable AFILE the filename
c       entered on position IPAR.
1200    CALL GETFN(2)
        IF(LMISS.OR.LERROR) GOTO 10
        CLOSE(5)
        OPEN(UNIT=5,FILE=AFILE)
        IF(ILPNT(4).NE.0) THEN
          CALL GETFN(3)
          CLOSE(7)
          OPEN(UNIT=7,FILE=AFILE)
        ENDIF
        GOTO 10
c       Leave this routine and stop the program.
8900    RETURN
c       Format for reading an input line.
9000    FORMAT(80A1)
c       Error messages.
9001    FORMAT('++++ILLEGAL OR MISSING PARAMETERS IN MAIN+++'/)
```

Sec. 8   Combined Shallow Confined and Unconfined Flow

```
 9002 FORMAT('++++ILLEGAL COMMAND IN MAIN+++'/)
 9003 FORMAT('++++GRID NOT FILLED+++'/)
 9004 FORMAT(
     .'++++GRID TOO LARGE, PLEASE REDUCE NUMBER OF GRIDPOINTS+++')
c     Command summary.
 9005 FORMAT(
     .'+<AQUIFER>(K,H)[BASE]'/
     .' <UNIFLOW>(Q,ANGLE)'/
     .' <WELL>(X,Y,Q,R)'/
     .' <REFERENCE>(X,Y,FI)'/
     .' <SOLVE>'/
     .' <WINDOW>(X1,Y1 X2,Y2)'/
     .' <LAYOUT>'/
     .' <GRID>(N)[<HEAD>/<POTENTIAL>]'/
     .' <PLOT>'/
     .' <CHECK>'/
     .' <SWITCH>(FILEIN)[FILEOUT]'/
     .' <PAUSE>'/
     .' <STOP>'/)
c     Message with maximum and minimum levels of contours.
 9010 FORMAT('+MIN. LEVEL= ',E14.6,' MAX. LEVEL=',E14.6/
     .       '  (MIN. LEVEL, INCREMENT)'/)
      END
c
c     The subroutine LAYOUT, when called, causes a layout to be
c     displayed. At present this layout contains only wells. It
c     calls subroutines that reside in the modules GIVEN and WELL and
c     provide their part of the layout. These modules are groups of
c     subroutines that have to do with the uniform flow and the
c     wells, respectively.
c
      SUBROUTINE LAYOUT
      CALL GVLAY
      CALL WLLAY
      RETURN
      END
c     RMAT   = matrix of maximum size IMXSZE*IMXSZE.
c     RSCRT  = scratch array.
```

The Main Functions

The functions grouped together under the heading "Main Functions" are separated into functions whose coefficients are known a priori, referred to in the program as "given functions," and functions that contain the contributions of all functions implemented in the program. This is done with applications in mind that will be discussed in Sec. 25; at present the only function with a coefficient that is not known a priori is the constant C. The main functions have as task only to add together the contributions of the uniform flow and the wells, which are programmed separately.

The main functions return the value of the potential or the head at any given point (x, y) in the aquifer. Besides these functions, there are subroutines which are called with x, y and the two components of the discharge vector; they add their contributions to these components and return the result.

```
c      Potential function
       REAL FUNCTION RFPOT(RX,RY)
c      The potential equals the sum of RFPGVN and the constant,
c      returned by RFRFC
       RFPOT=RFPGVN(RX,RY)+RFRFC(RX,RY)
       RETURN
       END
c      Components of the discharge function. At present, the only
c      contribution is from DVGVN, which is called with a pair of
c      coordinates and a discharge vector. It adds its contribution
c      to the two components of the discharge vector.
       SUBROUTINE DISV(RX,RY,RVQ)
       DIMENSION RVQ(2)
       RVQ(1)=.0
       RVQ(2)=.0
       CALL DVGVN(RX,RY,RVQ)
       RETURN
       END
c      Piezometric head.
       REAL FUNCTION RFHEAD(RX,RY)
       RPOT=RFPOT(RX,RY)
c      The function RFHEDP(RPOT) converts potentials to heads, and
c      resides in the aquifer module.
       RFHEAD=RFHEDP(RPOT)
       RETURN
       END
c      Contributions of all functions with given coefficients to the
c      potential: the function RFGVG in the module GIVEN and the
c      the function RFWLG in the module WELL.
       REAL FUNCTION RFPGVN(RX,RY)
       RFPGVN=RFGVG(RX,RY)+RFWLG(RX,RY)
       RETURN
       END
c      Contributions of all functions with given coefficients to the
c      components of the discharge vector.
```

Sec. 8   Combined Shallow Confined and Unconfined Flow

```
      SUBROUTINE DVGVN(RX,RY,RVQ)
      DIMENSION RVQ(2)
      CALL DVGVG(RX,RY,RVQ)
      CALL DVWLG(RX,RY,RVQ)
      RETURN
      END
```

The Module Aquifer

The module AQUIFER consists of a BLOCK DATA SUBPROGRAM where the default values of hydraulic conductivity, aquifer thickness and base elevation are set, a routine to convert heads to potentials, and a routine to convert potentials to heads. The conversion is valid for combined confined-unconfined flow. The flow may be forced to be unconfined everywhere by setting the aquifer thickness to a value larger than the head at any point in the aquifer.

```
c     Initialization of common block with aquifer data.
c     Set aquifer data to default values.
c     RAQPRM = hydraulic conductivity.
c     RAQTHK = aquifer thickness.
c     RAQBAS = aquifer base elevation.
      BLOCK DATA BDAQ
      COMMON /CBAQ/ RAQPRM,RAQTHK,RAQBAS
      DATA RAQPRM,RAQTHK,RAQBAS /1.,1.,.0/
      END
c     Function to convert head RFI to potential RFPOTH.
      REAL FUNCTION RFPOTH(RFI)
      COMMON /CBAQ/ RAQPRM,RAQTHK,RAQBAS
      RFIL=RFI-RAQBAS
      IF (RFIL.LE..0) THEN
c        Set the potential to zero if the head is below the base
c        elevation:
         RFPOTH=.0
      ELSEIF (RFIL.GT.RAQTHK) THEN
c        Confined flow:
         RFPOTH=RAQPRM*RAQTHK*RFIL-0.5*RAQPRM*RAQTHK*RAQTHK
      ELSE
c        Unconfined flow:
         RFPOTH=0.5*RAQPRM*RFIL*RFIL
      ENDIF
      RETURN
      END
c     Function to convert potential RPHI to head RFHEDP.
      REAL FUNCTION RFHEDP(RPHI)
```

```
      COMMON /CBAQ/ RAQPRM,RAQTHK,RAQBAS
      RKHH2=.5*RAQPRM*RAQTHK*RAQTHK
      IF(RPHI.LE.0.0) THEN
C       If the potential is negative set the head equal to the base
C       elevation:
        RFHEDP=RAQBAS
      ELSEIF(RPHI.GT.RKHH2) THEN
C       Confined flow:
        RFHEDP=(RPHI+RKHH2)/(RAQPRM*RAQTHK)+RAQBAS
      ELSE
C       Unconfined flow:
        RFHEDP=SQRT(2.*RPHI/RAQPRM)+RAQBAS
      ENDIF
      RETURN
      END
```

The Module Given

The module GIVEN here contains only the contributions of the uniform flow to the potential and discharge vector components. It also contains a routine GVLAY which will later be used to provide the layout of given functions.

```
C     Initialization of common block with data of uniform flow.
C     RGVUQ0   = discharge of uniform flow [L2/T].
C     RGVUAL   = angle between direction of uniform flow and x axis.
C     RGVUU(I) = components of unit vector in the direction of uniform
C                flow.
      BLOCK DATA BDGV
      COMMON /CBGV/    RGVUQ0,RGVUAL,RGVUU(2)
      DATA RGVUQ0,RGVUAL,RGVUU /0.,0.,1.,0./
      END
C     Contribution of given functions to the potential.
      REAL FUNCTION RFGVG(RX,RY)
      COMMON /CBGV/    RGVUQ0,RGVUAL,RGVUU(2)
      RFGVG=-RGVUQ0*(RGVUU(1)*RX+RGVUU(2)*RY)
      RETURN
      END
C     Contribution of given functions to the discharge vector.
      SUBROUTINE DVGVG(RX,RY,RVQ)
      DIMENSION RVQ(2)
      COMMON /CBGV/    RGVUQ0,RGVUAL,RGVUU(2)
      RQX=RGVUQ0*RGVUU(1)
      RQY=RGVUQ0*RGVUU(2)
      RVQ(1)=RVQ(1)+RQX
```

Sec. 8 Combined Shallow Confined and Unconfined Flow

```
      RVQ(2)=RVQ(2)+RQY
      RETURN
      END
c     Layout: currently no contribution.
      SUBROUTINE GVLAY
      COMMON /CBGV/   RGVUQ0,RGVUAL,RGVUU(2)
      RETURN
      END
```

The Module Reference

The module REFERENCE consists of a BLOCK DATA SUBPROGRAM where the default values of x_0, y_0, and ϕ_0 are set, and a function that returns the value of the constant in the expression for the potential:

```
c     Initialization of common block for reference point
c     RRFX  = x coordinate of reference point.
c     RRFY  = y coordinate of reference point.
c     RRFFI = head at reference point.
c     RRFCON= constant in the potential function.
      BLOCK DATA BDRF
      COMMON /CBRF/ RRFX,RRFY,RRFFI,RRFCON
      DATA RRFFI,RRFX,RRFY,RRFCON /4*.0/
      END
c     Function returning the constant in the potential.
      REAL FUNCTION RFRFC(RX,RY)
      COMMON /CBRF/ RRFX,RRFY,RRFFI,RRFCON
      RFRFC=RRFCON
      RETURN
      END
```

The Module Well

The module WELL consists of a BLOCK DATA SUBPROGRAM where the number of wells is initialized to zero, a function that adds the contributions of all wells to the potential, and a function that computes the potential due to a single well of unit discharge. It further contains a subroutine that adds the contributions of all wells to the components of the discharge vector, a subroutine that adds the contribution of a single well to the components of the discharge vector, and a subroutine that draws a circle with the radius of the well around the well center. The latter subroutine makes use of a subroutine RDRAWC(XC,YC,R) that draws a circle of radius R centered at (XC,YC).

```
c     Initialization of common block with well data.
c     Set the number of wells to zero.
c     NWL        = number of wells.
```

```
c      RWLXC(IWL) = x coordinate of the center of well number IWL.
c      RWLYC(IWL) = y coordinate of the center of well number IWL.
c      RWLQ(IWL)  = discharge of well number IWL.
c      RWLRAD(IWL) = radius of well number IWL.
       BLOCK DATA BDWL
       COMMON /CBWL/ NWL,RWLXC(200),RWLYC(200),RWLQ(200),RWLRAD(200)
       DATA NWL /0/
       END
c      Contribution of all wells to the potential.
       REAL FUNCTION RFWLG(RX,RY)
       COMMON /CBWL/ NWL,RWLXC(200),RWLYC(200),RWLQ(200),RWLRAD(200)
       RFWLG=.0
       DO 100 IWL=1,NWL
          RFWLG=RFWLG+RWLQ(IWL)*RFWELL(RX,RY,IWL)
 100   CONTINUE
       RETURN
       END
c      Contribution of a single well of unit discharge to the potential.
       REAL FUNCTION RFWELL(RX,RY,IWELL)
       COMMON /CBWL/ NWL,RWLXC(200),RWLYC(200),RWLQ(200),RWLRAD(200)
       DATA RPI /3.1415926/
       RADSQ =(RX-RWLXC(IWELL))**2+(RY-RWLYC(IWELL))**2
       RADWSQ=RWLRAD(IWELL)**2
       IF(RADSQ.GE.RADWSQ) THEN
          RFWELL=.25/RPI*ALOG(RADSQ)
       ELSE
c         For points inside the well, return the value at the well
c         screen.
          RFWELL=.25/RPI*ALOG(RADWSQ)
       ENDIF
       RETURN
       END
c      Contribution of all wells to the discharge vector.
       SUBROUTINE DVWLG(RX,RY,RVQ)
       DIMENSION RVQ(2)
       COMMON /CBWL/ NWL,RWLXC(200),RWLYC(200),RWLQ(200),RWLRAD(200)
       DO 100 IWL=1,NWL
          CALL DVWELL(RX,RY,RVQ,IWL)
 100   CONTINUE
       RETURN
       END
c      Contribution of a single well to the discharge vector.
       SUBROUTINE DVWELL(RX,RY,RVQ,IWELL)
       DIMENSION RVQ(2)
```

```
      COMMON /CBWL/ NWL,RWLXC(200),RWLYC(200),RWLQ(200),RWLRAD(200)
      DATA RPI /3.1415926/
      RADSQ =(RX-RWLXC(IWELL))**2+(RY-RWLYC(IWELL))**2
      RADWSQ=RWLRAD(IWELL)**2
      IF(RADSQ.GE.RADWSQ) THEN
        RO2PRS=1./(2.*RPI*RADSQ)
        RQX=-RWLQ(IWELL)*(RX-RWLXC(IWELL))*RO2PRS
        RQY=-RWLQ(IWELL)*(RY-RWLYC(IWELL))*RO2PRS
      ELSE
        RQX=-RWLQ(IWELL)/(2.*RPI*RWLRAD(IWELL))
        RQY=.0
        IF(RX.LT.RWLXC(IWELL)) RQX=-RQX
      ENDIF
      RVQ(1)=RVQ(1)+RQX
      RVQ(2)=RVQ(2)+RQY
      RETURN
      END
c     layout of all wells
      SUBROUTINE WLLAY
      COMMON /CBWL/ NWL,RWLXC(200),RWLYC(200),RWLQ(200),RWLRAD(200)
      DO 10 IWL=1,NWL
        CALL RDRAWC(RWLXC(IWL),RWLYC(IWL),RWLRAD(IWL))
 10   CONTINUE
      RETURN
      END
```

The Module Check

All the basic functions have now been introduced. Compilation of the source code presented in this subsection together with the routines provided on the diskette, followed by linking, will result in a program that is capable of producing piezometric contours for any combination of wells and uniform flow. However, it often is desirable to generate numerical output besides graphical output. The routine CHECK provides this facility. It can be used to display the data entered into the program, and to generate the values of the piezometric head, the potential, and the two components of the discharge vector at an arbitrary point. The routine makes use of free-format input similarly as SETUP does. The listing of this routine is as follows:

```
c     Routine for data retrieval.
      SUBROUTINE CHECK
      IMPLICIT LOGICAL(L),CHARACTER*1(A)
      DIMENSION AWORD(45),RVQ(2)
      CHARACTER*32 AFILE
      COMMON /CBMATC/ ILPNT(40),LMISS,LERROR,NPBAD,NCHAR
      COMMON /CBMATA/ ALINE(80),AFILE
```

```
      COMMON /CBAQ/    RAQPRM,RAQTHK,RAQBAS
      COMMON /CBRF/    RRFX,RRFY,RRFFI,RRFCON
      COMMON /CBGV/    RGVUQ0,RGVUAL,RGVUU(2)
      COMMON /CBWL/    NWL,RWLXC(200),RWLYC(200),RWLQ(200),RWLRAD(200)
c     list of keywords in check.
      DATA AWORD /'A','Q','U','I',' ',
     .            'R','E','F','E',' ',
     .            'U','N','I','F',' ',
     .            'W','E','L','L',' ',
     .            'C','O','N','T',' ',
     .            'H','E','A','D',' ',
     .            'P','O','T','E',' ',
     .            'D','I','S','C',' ',
     .            'R','E','T','U','!'/
      LERROR=.FALSE.
      LMISS=.FALSE.
   10 IF(LMISS) WRITE(*,9001)
      IF(LERROR.AND..NOT.LMISS) WRITE(*,9002)
      LERROR=.FALSE.
      LMISS=.FALSE.
      WRITE(*,9005)
      READ(5,9000) ALINE
      CALL TIDY
      CALL MATCH(AWORD,1,JUMP)
      IF(LMISS.OR.LERROR.OR.ILPNT(2).EQ.0) GOTO 10
      GOTO( 100, 200, 300, 400, 500, 600, 700, 800,8900),JUMP
c     Aquifer data.
  100 WRITE(7,9010) RAQPRM,RAQTHK,RAQBAS
      GOTO 10
c     Reference point.
  200 WRITE(7,9020) RRFX,RRFY,RRFFI,RRFCON
      GOTO 10
c     Uniform flow.
  300 WRITE(7,9030) RGVUQ0,RGVUAL,RGVUU(1),RGVUU(2)
      GOTO 10
c     Wells.
  400 IF(NWL.GT.0) THEN
         WRITE(7,9040)
         WRITE(7,9041) (I,RWLXC(I),RWLYC(I),RWLQ(I),RWLRAD(I),I=1,NWL)
      ENDIF
      GOTO 10
c     Head at reference point: specified value versus computed value.
  500 WRITE(7,9050)  RRFFI,RFHEAD(RRFX,RRFY)
      GOTO 10
```

Sec. 8 Combined Shallow Confined and Unconfined Flow

```
c       Head.
 600    RX=RVAR(2)
        RY=RVAR(3)
        IF(LERROR) GOTO 10
        WRITE(7,9060) RX,RY,RFHEAD(RX,RY)
        GOTO 10
c       Potential.
 700    RX=RVAR(2)
        RY=RVAR(3)
        IF(LERROR) GOTO 10
        WRITE(7,9070) RX,RY,RFPOT(RX,RY)
        GOTO 10
c       Discharge vector.
 800    RX=RVAR(2)
        RY=RVAR(3)
        IF(LERROR) GOTO 10
        CALL DISV(RX,RY,RVQ)
        WRITE(7,9080) RX,RY,RVQ(1),RVQ(2)
        GOTO 10
c       Return to SETUP.
 8900   RETURN
 9000   FORMAT(80A1)
 9001   FORMAT('++++ILLEGAL OR MISSING PARAMETERS IN CHECK+++'/)
 9002   FORMAT('++++ILLEGAL COMMAND IN CHECK+++'/)
 9005   FORMAT(
       .'+<AQUIFER><REFERENCE><UNIFLO><WELL><CONTROL><HEAD>(X,Y)'/
       .'  <DISCHARGE>(X,Y)<POTENTIAL>(X,Y)<RETURN>'/)
 9010   FORMAT(
       .'AQUIFER:    PERMEABILITY   THICKNESS     ELEVATION OF BASE'/
       .'         ',3E14.6)
 9020   FORMAT(
       .'REFERENCE:       X              Y            HEAD         CONSTANT'/
       .'         ',4E14.6)
 9030   FORMAT('  UNIFORM FLOW  ANGLE        COS(ANGLE)     SIN(ANGLE)'/
       .    4E14.6)
 9040   FORMAT('WELLS: GIVEN: NO XCENTER       YCENTER ',
       .'       DISCHARGE      RADIUS')
 9041   FORMAT('             ',I4,4E14.6)
 9050   FORMAT('                      SPECIFIED HEAD    CALCULATED HEAD'/
       .       ' REFERENCE POINT  :',E14.6,4X,E14.6)
 9060   FORMAT('X,Y,HEAD    ',3E14.6)
 9070   FORMAT('X,Y,PHI     ',3E14.6)
 9080   FORMAT('X,Y,QX,QY   ',4E14.6)
        END
```

Expansion of the Program

The program will be gradually expanded. As we progress through this chapter, we will add infiltration from rainfall, infiltration through the bottoms of circular ponds, and the capability to deal with layering in the aquifer. In the next chapter, we will convert the program to deal with complex numbers, and include the effects of creeks and fully penetrating rivers and lakes. In Chapter 4, we will add the capability to plot streamlines.

Problem 8.3

Use the computer program to verify your answer to Problem 6.3. Enter the wells with the computed discharges, and set the head at a point of the river bank equal to the given boundary value. Let the program reproduce the piezometric contour with the maximum value allowed inside the building pit: this contour must be outside the pit and touch one of its sides at one point. Use the routine CHECK to verify your answer to question (b) of Problem 6.3.

Problem 8.4

Use the program to reproduce the piezometric contours drawn for Problems 6.1, 6.2, 7.2, and 8.2. Note: For the cases of one-dimensional flow, choose the reference point at one of the two rivers, and enter the discharge Q_{x0}, computed as the difference in value of the potential along the river bank.

Problem 8.5

Enter the data corresponding to Figure 2.17(b) in SLW. Use the program to determine the heads at the two stagnation points computed in Problem 7.3. Reproduce the piezometric contours through these points.

Problem 8.6

Verify your answer to problem 7.5, questions (b) and (c), by the use of SLW (compare Problem 8.3).

9 SHALLOW UNCONFINED FLOW WITH RAINFALL

Water may percolate downward through the soil above the phreatic surface of an unconfined aquifer. This water may infiltrate, for example, as the result of rainfall or artificial infiltration, or may infiltrate through the bottom of a creek or pond that is located above the phreatic surface. In the first case the infiltration occurs over the entire aquifer or a large part of it. In the latter two cases the infiltration is local. We will in this section first derive the basic equations for shallow unconfined flow with rainfall, and discuss both uniform and local infiltration for one-dimensional, radial, and two-dimensional shallow unconfined flow along with some examples.

Basic Equations

The infiltration rate will be represented as N [m/s] and is equal to the amount of water per unit area that enters the aquifer per unit time. The resistance to flow in the vertical

Sec. 9   Shallow Unconfined Flow with Rainfall

direction is neglected in accordance with the Dupuit-Forchheimer assumption, and the infiltration is modeled as taking place without any change in head over the height of the aquifer. The process by which the rainwater moves down toward and into the aquifer is not considered in this section.

With this assumption, the condition that the inflow into a volume with sides Δx and Δy in the x and y directions be equal to the outflow becomes (see Figure 2.24):

$$\left[Q_x(x - \tfrac{\Delta x}{2}, y)\right]\Delta y + \left[Q_y(x, y - \tfrac{\Delta y}{2})\right]\Delta x + N\Delta x \Delta y$$
$$= \left[Q_x(x + \tfrac{\Delta x}{2}, y)\right]\Delta y + \left[Q_y(x, y + \tfrac{\Delta y}{2})\right]\Delta x \qquad (9.1)$$

Dividing by $\Delta x \Delta y$ and passing to the limit for $\Delta x \to 0, \Delta y \to 0$, the following is obtained:

$$\boxed{\frac{\partial Q_x}{\partial x} + \frac{\partial Q_y}{\partial y} = N} \qquad (9.2)$$

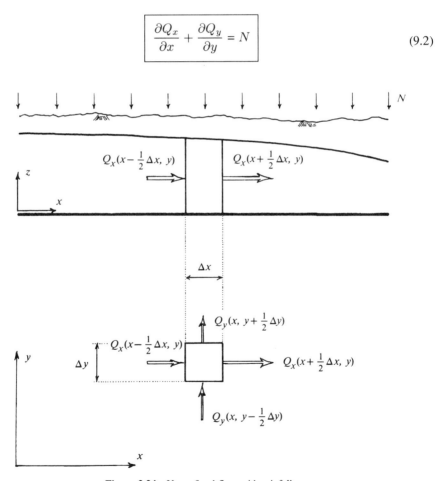

Figure 2.24 Unconfined flow with rainfall.

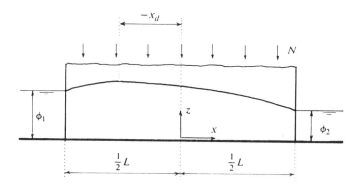

Figure 2.25 One-dimensional unconfined flow with rainfall.

The expression for the potential in terms of the head, (7.14) with $C_u = 0$, remains unchanged and the discharge vector equals minus the gradient of Φ. Hence, the differential equation for the potential becomes the Poisson equation:

$$\boxed{\frac{\partial^2 \Phi}{\partial x^2} + \frac{\partial^2 \Phi}{\partial y^2} = \nabla^2 \Phi = -N} \qquad (9.3)$$

where the operator ∇^2 stands for $\partial^2/\partial x^2 + \partial^2/\partial y^2$, and is pronounced "Nabla second." The infiltration rate N may be a function of position. For the time being, however, we will assume that N is a constant.

One-dimensional Flow

For one-dimensional flow, (9.3) becomes

$$\frac{d^2 \Phi}{dx^2} = -N \qquad (9.4)$$

with the general solution

$$\Phi = -\tfrac{1}{2} N x^2 + Ax + B \qquad (9.5)$$

As an example, consider the case of flow between two long and straight parallel rivers, as illustrated in Figure 2.25. The distance between the rivers is L. The origin of an x, y, z coordinate system is chosen to be halfway between the rivers, with the x axis normal to the river banks. The boundary conditions become

$$x = -\tfrac{1}{2} L \qquad \phi = \phi_1 \qquad \Phi = \Phi_1 = \tfrac{1}{2} k \phi_1^2 \qquad (9.6)$$

and

$$x = \tfrac{1}{2} L \qquad \phi = \phi_2 \qquad \Phi = \Phi_2 = \tfrac{1}{2} k \phi_2^2 \qquad (9.7)$$

Application of the boundary conditions (9.6) and (9.7) to the general solution (9.5) yields

$$\Phi = -\tfrac{1}{2}N\left[x^2 - (\tfrac{1}{2}L)^2\right] - \frac{\Phi_1 - \Phi_2}{L}x + \tfrac{1}{2}(\Phi_1 + \Phi_2) \tag{9.8}$$

The expression for the discharge vector component Q_x is found from (9.8) upon differentiation,

$$Q_x = -\frac{d\Phi}{dx} = Nx + \frac{\Phi_1 - \Phi_2}{L} \tag{9.9}$$

The location of the divide, $x = x_d$, the point where Φ has a maximum and $d\Phi/dx = 0$, is found by setting Q_x equal to zero,

$$Nx_d = -\frac{\Phi_1 - \Phi_2}{L} \qquad (-\tfrac{1}{2}L \le x_d \le \tfrac{1}{2}L) \tag{9.10}$$

The divide is at the center of the aquifer when $\Phi_1 = \Phi_2$, as would be expected. There will not always be a divide: if the value obtained for x_d from (9.10) is larger than $\tfrac{1}{2}L$ or less than $-\tfrac{1}{2}L$, there is no divide and the flow occurs in one direction throughout the aquifer. If, for example, x_d is greater than $\tfrac{1}{2}L$, all flow is in the negative x direction and all rainwater flows off to the boundary at $x = -\tfrac{1}{2}L$. If $-\tfrac{1}{2}L < x_d < \tfrac{1}{2}L$, there is a divide and the amount of rainwater flowing per unit width of the aquifer to the boundary at $x = \tfrac{1}{2}L$ is $(\tfrac{1}{2}L - x_d)N$ whereas the remaining amount, $(\tfrac{1}{2}L + x_d)N$ flows toward the boundary at $x = -\tfrac{1}{2}L$.

Radial Flow

The potential for radial flow with rainfall is obtained by application of Darcy's law and the condition that there be continuity of flow (see Figure 2.26). Consider the case of rain falling on a circular island of radius R. There is no well in the island and the infiltrated rainwater flows off to the boundary. Use is made of the radial coordinate r and the head at the boundary $r = R$ is ϕ_0. Hence,

$$r = R \qquad \phi = \phi_0 \qquad \Phi = \Phi_0 = \tfrac{1}{2}k\phi_0^2 \tag{9.11}$$

The amount of rainwater infiltrating inside a cylinder of radius r equals $N\pi r^2$. By continuity of flow, this amount of water must flow through the wall of the cylinder. Since the flow is radial, the component Q_r of the discharge vector is constant along the cylinder wall; the total flow through this wall equals $2\pi r Q_r$. The continuity equation thus may be written as follows:

$$2\pi r Q_r = N\pi r^2 \tag{9.12}$$

or

$$Q_r = \tfrac{1}{2}Nr \tag{9.13}$$

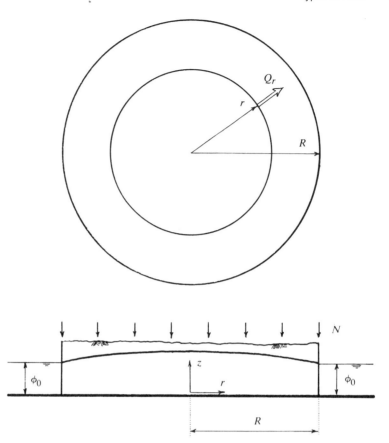

Figure 2.26 Radial shallow unconfined flow with rainfall.

Application of Darcy's law yields

$$Q_r = -\frac{d\Phi}{dr} = \tfrac{1}{2}Nr \qquad (9.14)$$

so that

$$\boxed{\Phi = -\tfrac{1}{4}Nr^2 + C} \qquad (9.15)$$

where C is a constant of integration. Application of the boundary condition (9.11) to (9.15) yields $C = \tfrac{1}{4}NR^2 + \Phi_0$ so that (9.15) becomes

$$\Phi = -\tfrac{1}{4}N(r^2 - R^2) + \Phi_0 \qquad (9.16)$$

Superposition

The differential equation for shallow unconfined flow with rainfall is the Poisson equation (9.3),

$$\nabla^2 \Phi = -N \qquad (9.17)$$

Because the Poisson equation is linear, superposition of solutions is possible without violating the differential equation. Let Φ_1 be a solution to the Poisson equation,

$$\nabla^2 \Phi_1 = -N \qquad (9.18)$$

whereas Φ_2 is a solution to the Laplace equation,

$$\nabla^2 \Phi_2 = 0 \qquad (9.19)$$

Then the sum, $\Phi = \Phi_1 + \Phi_2$, fulfills the Poisson equation,

$$\nabla^2 \Phi = \nabla^2(\Phi_1 + \Phi_2) = \nabla^2 \Phi_1 + \nabla^2 \Phi_2 = -N \qquad (9.20)$$

Superposition is useful for solving problems of unconfined flow with rainfall. These problems, then, are solved in two steps: first a particular solution to the Poisson equation is selected and, second, a solution to Laplace's equation is determined such that the sum of the two solutions meets the boundary conditions. As an example, the problem of unconfined flow with rainfall and a well in the center of a circular island will be solved by superposition.

Radial Flow Toward a Well

The problem of radial unconfined flow with a well at the center of a circular island is illustrated in Figure 2.27. The discharge of the well is Q, the radius of the island is R and the head at $r = R$ is ϕ_0. Hence, the boundary condition at $r = R$ is given by (9.11). The particular solution of the Poisson equation is (9.16). Addition of the potential for unconfined flow toward a well, (7.24), to (9.16) yields

$$\Phi = -\tfrac{1}{4}N(r^2 - R^2) + \frac{Q}{2\pi} \ln \frac{r}{R} + C \qquad (9.21)$$

It is noted that a new constant C is introduced and the constants Φ_0 in (7.24) and (9.16) are omitted; it is advisable to introduce a new constant after superimposing individual solutions and to determine this constant from a boundary condition. Application of the boundary condition (9.11) to (9.21) yields $C = \Phi_0$ and (9.21) becomes

$$\Phi = -\tfrac{1}{4}N(r^2 - R^2) + \frac{Q}{2\pi} \ln \frac{r}{R} + \Phi_0 \qquad (9.22)$$

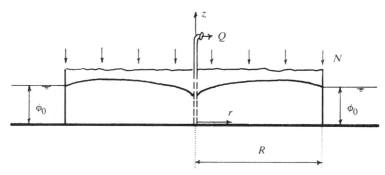

Figure 2.27 Radial flow toward a well.

The expression for the component Q_r of the discharge vector is obtained from (9.22) upon differentiation,

$$Q_r = -\frac{d\Phi}{dr} = \tfrac{1}{2}Nr - \frac{Q}{2\pi r} \qquad (9.23)$$

The divide occurs at $r = r_d$, where $Q_r = 0$, i.e.,

$$\tfrac{1}{2}Nr_d = \frac{Q}{2\pi r_d} \qquad (9.24)$$

or

$$r_d = \sqrt{\frac{Q}{\pi N}} \qquad (r_d \le R) \qquad (9.25)$$

Similarly, as in the case of one-dimensional flow, there may not be a divide in the aquifer; there only is a divide when $r_d < R$.

Rainfall on an Island with an Elliptical Boundary

Circular islands rarely occur in reality; a solution whereby the boundary of the island is an ellipse would be more useful. Consider the following potential in this context:

$$\Phi = -\tfrac{1}{2}\frac{N}{\frac{1}{a^2} + \frac{1}{b^2}}\left[\left(\frac{x}{a}\right)^2 + \left(\frac{y}{b}\right)^2 - 1\right] + \Phi_0 \qquad (9.26)$$

This potential fulfills the differential equation

$$\nabla^2 \Phi = -N \qquad (9.27)$$

and, in addition, has the value Φ_0 along the boundary of an ellipse with principal axes of lengths a and b:

$$\left(\frac{x}{a}\right)^2 + \left(\frac{y}{b}\right)^2 = 1 \qquad \Phi = \Phi_0 \qquad (9.28)$$

Sec. 9   Shallow Unconfined Flow with Rainfall

In view of these properties, the potential (9.26) describes the flow in an elliptic island.

The potential (9.26) may be generalized for the case that the principal axis of length a makes an angle α with the x axis. We introduce an x^*, y^* Cartesian coordinate system with its origin at the center of the ellipse; the potential then may be written in terms of this coordinate system as follows:

$$\Phi = -\frac{1}{2}\frac{N}{\frac{1}{a^2}+\frac{1}{b^2}}\left[\left(\frac{x^*}{a}\right)^2 + \left(\frac{y^*}{b}\right)^2 - 1\right] + \Phi_0 \qquad (9.29)$$

The coordinates x^* and y^* may be expressed in terms of x and y as follows, using Figure 1.5 and noticing that the origin of the x^*, y^* coordinate system is at $x = x_c$, $y = y_c$:

$$\begin{aligned} x^* &= (x - x_c)\cos\alpha + (y - y_c)\sin\alpha \\ y^* &= -(x - x_c)\sin\alpha + (y - y_c)\cos\alpha \end{aligned} \qquad (9.30)$$

The potential (9.29) may be written in a form that applies to the case that either a or b is zero:

$$\Phi = -\frac{1}{2}\frac{N}{a^2+b^2}\left[b^2(x^*)^2 + a^2(y^*)^2 - a^2b^2\right] + \Phi_0 \qquad (9.31)$$

This may be written in terms of x and y by the use of (9.30) as follows:

$$\begin{aligned}\Phi = -\frac{1}{2}\frac{N}{a^2+b^2}\Big\{&b^2\big[(x-x_c)^2\cos^2\alpha + 2(x-x_c)(y-y_c)\sin\alpha\cos\alpha \\ &+ (y-y_c)^2\sin^2\alpha\big] \\ +&a^2\big[(x-x_c)^2\sin^2\alpha - 2(x-x_c)(y-y_c)\sin\alpha\cos\alpha \\ &+ (y-y_c)^2\cos^2\alpha\big] - a^2b^2\Big\} + \Phi_0 \end{aligned}$$
$$\qquad (9.32)$$

Collecting terms, this may be simplified to

$$\begin{aligned}\Phi = -\frac{1}{2}\frac{N}{a^2+b^2}\big[&(a^2\sin^2\alpha + b^2\cos^2\alpha)(x-x_c)^2 \\ &-2(a^2-b^2)(x-x_c)(y-y_c)\sin\alpha\cos\alpha \\ &+(a^2\cos^2\alpha + b^2\sin^2\alpha)(y-y_c)^2 - a^2b^2\big] + \Phi_0\end{aligned} \qquad (9.33)$$

This function covers all cases of uniform infiltration that we have discussed so far: if $a = 0$ and $\alpha = 0$, then (9.33) reduces to

$$\Phi = -\tfrac{1}{2}N(x-x_c)^2 + \Phi_0 = -\tfrac{1}{2}Nx^2 + Nx_cx - \tfrac{1}{2}Nx_c^2 + \Phi_0 \qquad (9.34)$$

which is equivalent to (9.5) with Nx_c replacing A and $-\tfrac{1}{2}Nx_c^2 + \Phi_0$ replacing B. If $a = b = R$, then (9.33) reduces to the potential (9.16) for radial flow with rainfall.

Expressions for the components Q_x and Q_y of the discharge vector are obtained from (9.33) by the use of Darcy's law:

$$Q_x = -\frac{\partial \Phi}{\partial x} = \frac{N}{a^2 + b^2}\Big[\ (a^2 \sin^2 \alpha + b^2 \cos^2 \alpha)(x - x_c)$$
$$-(a^2 - b^2)(y - y_c)\sin \alpha \cos \alpha\Big]$$
$$Q_y = -\frac{\partial \Phi}{\partial y} = \frac{N}{a^2 + b^2}\Big[-(a^2 - b^2)(x - x_c)\sin \alpha \cos \alpha$$
$$+(a^2 \cos^2 \alpha + b^2 \sin^2 \alpha)(y - y_c)\Big]$$
(9.35)

We will implement equations (9.33) and (9.35) in the computer program SLW at the end of this section.

Local Infiltration

When the bottom of a ditch or a pond is not in contact with the phreatic surface, water will leak through the bottom, filter down to the phreatic surface and finally join the flow in the aquifer. We will not attempt to model the unsaturated flow in the zone between the bottom of the ditch or pond and the phreatic surface. For the sake of simplicity, we assume that the infiltrating water percolates vertically downward through the unsaturated zone. We will first discuss infiltration through the bottom of a long ditch for cases of one-dimensional flow, and second the infiltration through the bottom of a circular pond.

One-dimensional Flow

Consider the case of infiltration through the bottom of a long ditch of width b, which is parallel to the two infinite boundaries of an unconfined aquifer. There is no rainfall, a Cartesian coordinate system is chosen as shown in Figure 2.28, the width of the aquifer is L, and the heads along the two boundaries are ϕ_1 and ϕ_2:

$$\begin{aligned} x &= -\tfrac{1}{2}L & \phi &= \phi_1 & \Phi &= \Phi_1 = \tfrac{1}{2}k\phi_1^2 \\ x &= \tfrac{1}{2}L & \phi &= \phi_2 & \Phi &= \Phi_2 = \tfrac{1}{2}k\phi_2^2 \end{aligned}$$
(9.36)

The coordinates of the boundaries of the ditch are ξ_1 and ξ_2 (see Figure 2.28), so that

$$\xi_2 - \xi_1 = b$$
(9.37)

The rate of infiltration through the bottom of the ditch is N_1 [m/s]. The differential equations for the potential Φ are

$$\begin{aligned} -\tfrac{1}{2}L \leq x \leq \xi_1 & \quad \frac{d^2\Phi}{dx^2} = 0 \\ \xi_1 \leq x \leq \xi_2 & \quad \frac{d^2\Phi}{dx^2} = -N_1 \\ \xi_2 \leq x \leq \tfrac{1}{2}L & \quad \frac{d^2\Phi}{dx^2} = 0 \end{aligned}$$
(9.38)

Sec. 9 Shallow Unconfined Flow with Rainfall

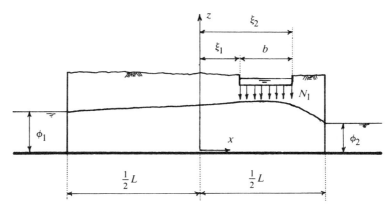

Figure 2.28 Local infiltration for one-dimensional flow.

We will solve problems with ditches by making use of special solutions, which we will refer to as influence functions. Two of these influence functions are as follows:

$$F_1 = -\frac{x - x_2}{L} \qquad F_2 = \frac{x - x_1}{L} \qquad (9.39)$$

where x_2 and x_1 represent the coordinates of the right- and left-hand boundaries, respectively, and L is the width of the aquifer. For the case of Figure 2.28, x_1 and x_2 are as follows:

$$x_1 = -\tfrac{1}{2}L \qquad x_2 = \tfrac{1}{2}L \qquad (9.40)$$

The functions (9.39) represent straight lines, and

$$\frac{d^2 F_1}{dx^2} = \frac{d^2 F_2}{dx^2} = 0 \qquad (9.41)$$

The function F_1 is zero at $x = x_2$ and unity at $x = x_1$, whereas F_2 is zero at $x = x_1$ and unity at $x = x_2$; this is seen from (9.39), with the aid of the relation:

$$x_2 - x_1 = L \qquad (9.42)$$

We can make use of the functions (9.39) to solve the general one-dimensional problem. If, for example, the potential at $x = x_2$ is Φ_2 and at $x = x_1$ equals Φ_1, then the solution is

$$\Phi = \Phi_1 F_1 + \Phi_2 F_2 = -\Phi_1 \frac{x - x_2}{L} + \Phi_2 \frac{x - x_1}{L} \qquad (9.43)$$

Another influence function serves to implement rainfall, and has the form:

$$F_r = -\tfrac{1}{2}(x - x_2)(x - x_1) \qquad (9.44)$$

It vanishes at $x = x_2$ and $x = x_1$, and fulfills the following differential equation:

$$\nabla^2 F_r = -1 \tag{9.45}$$

If there is infiltration of a rate N on the aquifer with boundary values Φ_1 and Φ_2, then the potential Φ is obtained by adding a term NF_r to (9.43), which yields

$$\begin{aligned}\Phi &= \Phi_1 F_1 + \Phi_2 F_2 + N F_r \\ &= -\Phi_1 \frac{x - x_2}{L} + \Phi_2 \frac{x - x_1}{L} - \tfrac{1}{2} N(x - x_2)(x - x_1)\end{aligned} \tag{9.46}$$

It is noted that the influence function F_r does not affect the boundary values of Φ; its addition to the potential serves only to include rainfall.

We will use the above influence functions to construct yet another influence function that serves to include infiltration over an area extending between $x = \xi_1$ and $x = \xi_2$, and vanishes at the boundaries $x = x_1$ and $x = x_2$. This influence function, which we will write as $G_d(x, \xi_1, \xi_2)$, where the index d stands for ditch, is depicted in Figure 2.29(a): it is obtained by addition of the functions G_0 and aF_2, $(a \geq 0)$, which are plotted versus x in Figures 2.29(b) and (c). We require that the function G_0 meets the following conditions:

$$x_1 \leq x \leq \xi_1 \qquad G_0 = 0 \tag{9.47}$$

$$\xi_1 \leq x \leq \xi_2 \qquad \frac{d^2 G_0}{dx^2} = -1 \tag{9.48}$$

$$\xi_2 \leq x \leq x_2 \qquad \frac{d^2 G_0}{dx^2} = 0 \tag{9.49}$$

The function F_2 is defined by (9.39) and addition of aF_2 to G_0, where a is the value of G_0 at $x = x_2$, yields G_d. Because G_0 is zero for $x_1 < x \leq \xi_1$, the derivative dG_0/dx at ξ_1 is zero. We thus obtain the following two boundary conditions for the determination of G_0 along the interval $\xi_1 \leq x \leq \xi_2$:

$$x = \xi_1 \qquad G_0 = 0 \qquad \frac{dG_0}{dx} = 0 \tag{9.50}$$

It follows from (9.48) that $dG_0/dx = -x + C$, where C is a constant of integration, and application of the second condition in (9.50) yields

$$\xi_1 \leq x \leq \xi_2 \qquad \frac{dG_0}{dx} = -(x - \xi_1) \tag{9.51}$$

Integration and subsequent use of the first condition in (9.48) gives

$$\xi_1 \leq x \leq \xi_2 \qquad G_0 = -\tfrac{1}{2}(x - \xi_1)^2 \tag{9.52}$$

The values of G_0 and dG_0/dx at $x = \xi_2$ are used as boundary values for G_0 on the interval $\xi_2 \leq x \leq x_2$; these values are

$$x = \xi_2 \qquad G_0 = -\tfrac{1}{2}(\xi_2 - \xi_1)^2 \qquad \frac{dG_0}{dx} = -(\xi_2 - \xi_1) \tag{9.53}$$

Sec. 9 Shallow Unconfined Flow with Rainfall

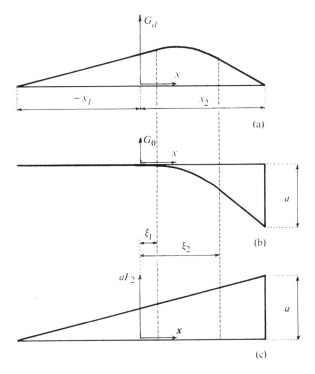

Figure 2.29 Determination of the influence function G_d.

The function G_0 is linear for $\xi_2 \leq x \leq x_2$ and application of the two boundary conditions (9.53) to this linear function gives the following result:

$$\xi_2 \leq x \leq x_2 \qquad G_0 = -(\xi_2 - \xi_1)(x - \xi_2) - \tfrac{1}{2}(\xi_2 - \xi_1)^2 \qquad (9.54)$$

The constant a equals $-G_0$ at $x = x_2$ (see Figure 2.29),

$$a = (\xi_2 - \xi_1)(x_2 - \xi_2) + \tfrac{1}{2}(\xi_2 - \xi_1)^2 \qquad (9.55)$$

The function $G_d(x, \xi_1, \xi_2)$ is obtained by adding aF_2 to G_0:

$$x_1 \leq x \leq \xi_1 \qquad G_d(x, \xi_1, \xi_2) = aF_2 = \left[(\xi_2 - \xi_1)(x_2 - \xi_2) + \tfrac{1}{2}(\xi_2 - \xi_1)^2\right] \frac{x - x_1}{L} \qquad (9.56)$$

$$\xi_1 \leq x \leq \xi_2 \qquad G_d(x, \xi_1, \xi_2) = aF_2 - \tfrac{1}{2}(x - \xi_1)^2 \qquad (9.57)$$

$$\xi_2 \leq x \leq x_2 \qquad G_d(x, \xi_1, \xi_2) = aF_2 - (\xi_2 - \xi_1)(x - \xi_2) - \tfrac{1}{2}(\xi_2 - \xi_1)^2 \qquad (9.58)$$

The function $G_d(x, \xi_1, \xi_2)$ models an infiltration of rate 1 along the interval $\xi_1 \leq x \leq \xi_2$ and vanishes at the two boundaries. It can be applied to the case of an infiltration rate

N_1 simply by multiplication by a factor N_1. The potential for the flow problem of Figure 2.28 is now readily obtained by superposition as follows, using (9.36), (9.39), and (9.40):

$$\Phi = -\Phi_1 \frac{x - \frac{1}{2}L}{L} + \Phi_2 \frac{x + \frac{1}{2}L}{L} + N_1 G_d(x, \xi_1, \xi_2) \qquad (9.59)$$

Using (9.37) and (9.40), equations (9.56) through (9.58) may be simplified as follows:

$$-\tfrac{1}{2}L \leq x \leq \xi_1 \qquad G_d(x, \xi_1, \xi_2) = aF_2 = \left[(\tfrac{1}{2}L - \xi_2)b + \tfrac{1}{2}b^2\right] \frac{x + \tfrac{1}{2}L}{L} \qquad (9.60)$$

$$\xi_1 \leq x \leq \xi_2 \qquad G_d(x, \xi_1, \xi_2) = aF_2 - \tfrac{1}{2}(x - \xi_1)^2 \qquad (9.61)$$

$$\xi_2 \leq x \leq \tfrac{1}{2}L \qquad G_d(x, \xi_1, \xi_2) = aF_2 - b(x - \xi_2) - \tfrac{1}{2}b^2 \qquad (9.62)$$

The influence function G_d may be used to model infiltration through n ditches, located at $\xi_{2j-1} \leq x \leq \xi_{2j}, j = 1, 2, \ldots, n$, where the infiltration rate through the bottom of the jth ditch is N_j. The corresponding potential is obtained by superposition:

$$\Phi = -\Phi_1 \frac{x - \tfrac{1}{2}L}{L} + \Phi_2 \frac{x + \tfrac{1}{2}L}{L} + \sum_{j=1}^{n} N_j G_d(x, \xi_{2j-1}, \xi_{2j}) \qquad (9.63)$$

It is recalled that we introduced the influence function $G_d(x, \xi_1, \xi_2)$ in such a way that

1. In each zone of the aquifer it fulfills the appropriate differential equation.
2. The conditions of both continuity of potential and continuity of flow are fulfilled at the boundaries between the zones.
3. The influence function vanishes at the boundaries of the aquifer.

It is a consequence of these properties of G that superposition of an arbitrary number of functions of this kind is possible by violating neither the differential equations nor the boundary conditions.

Problem 9.1

Establish, for the flow problem of Figure 2.28, the conditions for which:
1. All water infiltrating through the bottom of the ditch leaves the aquifer at $x = -\tfrac{1}{2}L$.
2. All water infiltrating through the bottom of the ditch leaves the aquifer at $x = \tfrac{1}{2}L$.
3. Some of the water infiltrating through the bottom of the ditch leaves the aquifer at $x = -\tfrac{1}{2}L$, and some at $x = \tfrac{1}{2}L$. Determine the fraction of the total amount that leaves the aquifer at $x = -\tfrac{1}{2}L$.

Problem 9.2

Two ditches of width b are used for infiltration into an unconfined aquifer. The ditches are long and parallel to the two boundaries of the aquifer; the flow is one-dimensional. The distance from the center of each ditch to the nearest boundary is $d + b/2$, the width of the aquifer is L, and the heads at the boundaries are ϕ_1 and ϕ_2. The infiltration rate through the bottom of the one ditch is N_1 and of the other is N_2. The hydraulic conductivity is k.

Questions
(a) Determine expressions for the ratios N_1/k and N_2/k so that the phreatic surface between the ditches is horizontal and a distance H above the aquifer base.
(b) Determine the conditions that H, ϕ_1, ϕ_2, b, and d must satisfy in order that the infiltration rates N_1 and N_2 are less than k. (Infiltration at a rate larger than k is physically impossible.)

Radial Flow

An influence function useful for modeling infiltration through circular ponds can be determined in much the same fashion as was done above for one-dimensional flow. This function will be used for infinite aquifers and we will choose it such that it vanishes at the boundary of the pond. We will represent this function as $G_p(x, y, x_1, y_1, R)$, where the index p stands for pond, and where x_1 and y_1 are the coordinates of the center of the pond and R is its radius. The function will apply to a unit infiltration rate through the bottom of the pond. Hence, G_p must fulfill the following differential equations:

$$0 \leq r_1 \leq R \quad \nabla^2 G_p = -1 \qquad (9.64)$$

$$R \leq r_1 < \infty \quad \nabla^2 G_p = 0 \qquad (9.65)$$

where

$$r_1 = \sqrt{(x - x_1)^2 + (y - y_1)^2} \qquad (9.66)$$

The expression for G_p that applies to the area below the pond is obtained from (9.16) with $N = 1$ and $\Phi_0 = 0$, and with r_1 replacing r:

$$0 \leq r_1 \leq R \quad G_p = -\tfrac{1}{4}(r_1^2 - R^2) \qquad (9.67)$$

The discharge at $r_1 = R$ is directed normal to the boundary of the pond because the flow is radial, and the total discharge flowing out of the pond of unit infiltration rate is Q, with

$$Q = \pi R^2 \qquad (9.68)$$

Continuity of flow through the wall of the cylinder of radius R about the center of the pond requires that the discharge enters the area outside this cylinder radially. Furthermore, G_p must fulfill Laplace's equation outside the pond, and the following function

$$R \leq r_1 < \infty \quad G_p = -\frac{\pi R^2}{2\pi} \ln \frac{r_1}{R} = -\frac{R^2}{2} \ln \frac{r_1}{R} \qquad (9.69)$$

meets these conditions. Expressions (9.67) and (9.69) both vanish at $r_1 = R$, so that the potential is continuous at $r_1 = R$. The function G_p is thus defined as follows:

$$0 \leq r_1 \leq R \quad G_p(x, y, x_1, y_1, R) = -\tfrac{1}{4}[(x - x_1)^2 + (y - y_1)^2 - R^2]$$
$$(9.70)$$

$$R \leq r_1 < \infty \quad G_p(x, y, x_1, y_1, R) = -\frac{R^2}{4} \ln \frac{(x - x_1)^2 + (y - y_1)^2}{R^2}$$
$$(9.71)$$

As an example, we will solve the problem of flow from a circular pond of radius R and infiltration rate N_1 with its center at a distance d from the bank of a long river (see Figure 2.30). The head along the river bank is ϕ_0. An x, y, z coordinate system is chosen as indicated in the figure, with the y axis along the river bank, the x axis pointing away from the aquifer, and the negative x axis passing through the center of the pond. Far away from the pond, the flow is uniform in the x direction and has a discharge Q_{x0}. We will henceforth refer to flow fields near infinity as *far-fields*.

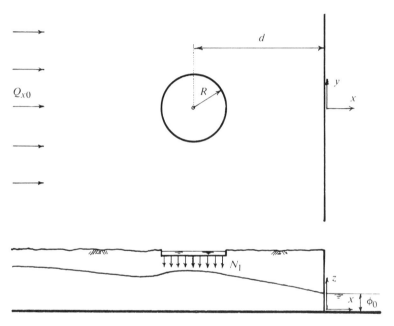

Figure 2.30 Flow from a circular pond toward a river.

The function $G_p(x, y, x_1, y_1, R)$ for this case becomes

$$0 \leq r_1 \leq R \qquad G_p(x, y, -d, 0, R) = -\tfrac{1}{4}\left[(x+d)^2 + y^2 - R^2\right] \qquad (9.72)$$

$$R \leq r_1 < \infty \qquad G_p(x, y, -d, 0, R) = -\frac{R^2}{4} \ln \frac{(x+d)^2 + y^2}{R^2} \qquad (9.73)$$

where

$$r_1^2 = (x+d)^2 + y^2 \qquad (9.74)$$

The potential due to the flow from the pond and valid in the area outside the pond equals that of a recharge well of strength $\pi N_1 R^2$. The boundary condition along the river bank,

$$x = 0 \qquad \phi = \phi_0 \qquad \Phi = \Phi_0 = \tfrac{1}{2} k \phi_0^2 \qquad (9.75)$$

Sec. 9   Shallow Unconfined Flow with Rainfall

thus can be met by placing an image well of discharge $\pi N_1 R^2$ at $x = d, y = 0$. The complete potential for this problem thus becomes

$$\Phi = -Q_{x0}x + N_1 G_p(x, y, -d, 0, R) + \frac{N_1 R^2}{4} \ln\left[(x - d)^2 + y^2\right] + C \qquad (9.76)$$

The constant C is found by applying the boundary condition (9.75) to (9.76). Since $r_1 > R$ along the river bank, (9.73) represents G_p there and we obtain

$$\Phi_0 = \frac{-N_1 R^2}{4} \ln \frac{d^2 + y^2}{R^2} + \frac{N_1 R^2}{4} \ln(d^2 + y^2) + C \qquad (9.77)$$

so that

$$C = \Phi_0 - \frac{N_1 R^2}{4} \ln R^2 \qquad (9.78)$$

Substitution of this expression for C in (9.76) yields

$$\Phi = -Q_{x0}x + N_1 G_p(x, y, -d, 0, R) + \frac{N_1 R^2}{4} \ln \frac{(x - d)^2 + y^2}{R^2} + \Phi_0 \qquad (9.79)$$

where G_p is defined by (9.72) and (9.73).

Problem 9.3
Infiltration occurs at a rate N_1 through the bottom of a circular pond of radius R in an infinite aquifer. If the pond were not present, there would be no flow. A well is drilled through the bottom of the pond and screened in the aquifer. Determine the location and the discharge of this well such that exactly all infiltrated water is captured. Determine an expression for the potential as a function of position.

Implementation of Infiltration in SLW

In the following we will implement in the FORTRAN program SLW the functions for uniform infiltration from rainfall and for infiltration through the bottoms of circular ponds. We will choose the function given by (9.33) to model the uniform infiltration; this function has the advantage that its parameters can be chosen so as to meet the condition that the head is constant either along straight equipotentials at an arbitrary angle, or along a circle, or along an ellipse at any orientation and of any aspect ratio.

Implementation of these functions in the program requires the modification of the routine SETUP so that the parameters associated with the new functions can be entered, the expansion of the common block CBGV to communicate these parameters between the various subroutines, and, finally, the programming of the functions; these will be placed in the module GIVEN. We will further add the facility to retrieve data associated with these functions by the use of the module CHECK. In order to save space, new listings will not be given of the subroutines SETUP and CHECK; it will suffice to list only the statements that must either be changed or added.

The Module Given

The function that adds the contributions of all ponds to the potential is a direct implementation of (9.72) and (9.73); an IF-THEN-ELSE construction is used to distinguish between points inside and outside the ponds. The contributions of all ponds to the discharge vector components are obtained by differentiation of (9.72) and (9.73). The function that adds the contribution of the uniform infiltration is a direct implementation of (9.33); the contribution to the components of the discharge vector is programmed by implementing (9.35). The listing of the modified module GIVEN is as follows:

```
c       Initialization of common block with data of rain, uniform flow
c       and ponds.
c       RGVUQ0          = discharge of uniform flow [L2/T].
c       RGVUAL          = angle in degrees between direction of uniform
c                         flow and x axis.
c       RGVUU(I)        = components of unit vector in the direction of
c                         uniform flow.
c       RGVRN           = infiltration from rainfall [L/T].
c       RGVRX,RGVRY     = x and y coordinates of the center of the
c                         elliptical equipotentials generated by rainfall
c                         alone.
c       RGVRAS          = square of the a axis of the ellipse.
c       RGVRBS          = square of the b axis of the ellipse.
c       RGVRAL          = angle between the a axis of ellipse and the
c                         x axis
c       NGVP            = number of ponds.
c       RGVPXC(IPOND)   = x coordinate of the center of pond number IPOND.
c       RGVPYC(IPOND)   = y coordinate of the center of pond number IPOND.
c       RGVPRD(IPOND)   = radius of pond number IPOND.
c       RGVPSG(IPOND)   = infiltration rate of pond number IPOND.
c
        BLOCK DATA BDGV
        COMMON /CBGV/   RGVUQ0,RGVUAL,RGVUU(2),
       .                RGVRN,RGVRX,RGVRY,RGVRAS,RGVRBS,RGVRAL,
       .                NGVP,RGVPXC(20),RGVPYC(20),RGVPRD(20),RGVPSG(20)
        DATA RGVUQ0,RGVUAL,RGVUU /0.,0.,1.,0./
        DATA RGVRN,RGVRX,RGVRY,RGVRAS,RGVRBS,RGVRAL
       .    /.0   ,.0   ,.0   ,1.   ,1.    ,.0   /
        DATA NGVP /0/
        END
c       Contribution of given functions to the potential.
        REAL FUNCTION RFGVG(RX,RY)
        COMMON /CBGV/   RGVUQ0,RGVUAL,RGVUU(2),
       .                RGVRN,RGVRX,RGVRY,RGVRAS,RGVRBS,RGVRAL,
       .                NGVP,RGVPXC(20),RGVPYC(20),RGVPRD(20),RGVPSG(20)
```

Sec. 9   Shallow Unconfined Flow with Rainfall

```
c     Contribution of the uniform flow.
      RFGVG=-RGVUQ0*(RGVUU(1)*RX+RGVUU(2)*RY)
c     Contribution of the rainfall.
      IF(RGVRN.NE.0.) THEN
        RCOS  =COS(RGVRAL)
        RSIN  =SIN(RGVRAL)
        RCOSSQ=RCOS**2
        RSINSQ=RSIN**2
        RCROSS=2.*(RGVRAS-RGVRBS)*RSIN*RCOS
        RCOF  =-.5*RGVRN/(RGVRAS+RGVRBS)
        RFGVG =RFGVG
     .         +RCOF*((RGVRAS*RSINSQ+RGVRBS*RCOSSQ)*(RX-RGVRX)**2
     .         -      RCROSS*(RX-RGVRX)*(RY-RGVRY)
     .         +      (RGVRAS*RCOSSQ+RGVRBS*RSINSQ)*(RY-RGVRY)**2)
      ENDIF
c     Contribution of the ponds.
      RFGVG=RFGVG+RFGVGP(RX,RY)
      RETURN
      END
c     Potential due to all ponds.
      REAL FUNCTION RFGVGP(RX,RY)
      COMMON /CBGV/    RGVUQ0,RGVUAL,RGVUU(2),
     .                 RGVRN,RGVRX,RGVRY,RGVRAS,RGVRBS,RGVRAL,
     .                 NGVP,RGVPXC(20),RGVPYC(20),RGVPRD(20),RGVPSG(20)
      RFGVGP=.0
      IF(NGVP.EQ.0) RETURN
      DO 100 IPOND=1,NGVP
        RADSQ =(RX-RGVPXC(IPOND))**2+(RY-RGVPYC(IPOND))**2
        RADPSQ=RGVPRD(IPOND)
        IF(RADSQ.GE.RADPSQ) THEN
c          Outside the pond:
           RFPPD=-.25*RADPSQ*ALOG(RADSQ/RADPSQ)
        ELSE
c          Inside the pond:
           RFPPD=-.25*(RADSQ-RADPSQ)
        ENDIF
        RFGVGP=RFGVGP+RGVPSG(IPOND)*RFPPD
  100 CONTINUE
      RETURN
      END
c     Contribution of the given functions to the discharge vector.
      SUBROUTINE DVGVG(RX,RY,RVQ)
      DIMENSION RVQ(2)
      COMMON /CBGV/    RGVUQ0,RGVUAL,RGVUU(2),
```

```
                          RGVRN,RGVRX,RGVRY,RGVRAS,RGVRBS,RGVRAL,
                          NGVP,RGVPXC(20),RGVPYC(20),RGVPRD(20),RGVPSG(20)
c     Contribution of uniform flow.
      RQX=RGVUQ0*RGVUU(1)
      RQY=RGVUQ0*RGVUU(2)
c     Contribution of rainfall.
      RCOS  =COS(RGVRAL)
      RSIN  =SIN(RGVRAL)
      RCOSSQ=RCOS**2
      RSINSQ=RSIN**2
      RCROSS=(RGVRAS-RGVRBS)*RSIN*RCOS
      RCOF  =RGVRN/(RGVRAS+RGVRBS)
      RQX=RQX+RCOF*((RGVRAS*RSINSQ+RGVRBS*RCOSSQ)*(RX-RGVRX)-
     .             RCROSS*(RY-RGVRY))
      RQY=RQY+RCOF*(-RCROSS*(RX-RGVRX)+(RGVRAS*RCOSSQ+RGVRBS*RSINSQ)*
     .             (RY-RGVRY))
      RVQ(1)=RVQ(1)+RQX
      RVQ(2)=RVQ(2)+RQY
c     Contribution of ponds.
      CALL DVGVGP(RX,RY,RVQ)
      RETURN
      END
c     Contribution of all ponds to the discharge vector.
      SUBROUTINE DVGVGP(RX,RY,RVQ)
      DIMENSION RVQ(2)
      COMMON /CBGV/    RGVUQ0,RGVUAL,RGVUU(2),
     .                 RGVRN,RGVRX,RGVRY,RGVRAS,RGVRBS,RGVRAL,
     .                 NGVP,RGVPXC(20),RGVPYC(20),RGVPRD(20),RGVPSG(20)
      IF(NGVP.EQ.0) RETURN
      DO 100 IPOND=1,NGVP
        RADSQ =(RX-RGVPXC(IPOND))**2+(RY-RGVPYC(IPOND))**2
        RADPSQ=RGVPRD(IPOND)
        IF(RADSQ.GE.RADPSQ) THEN
c         Outside the pond:
          RRO2RS=RADPSQ/(2.*RADSQ)
          RQX=RGVPSG(IPOND)*(RX-RGVPXC(IPOND))*RRO2RS
          RQY=RGVPSG(IPOND)*(RY-RGVPYC(IPOND))*RRO2RS
        ELSE
c         Inside the pond:
          RQX=.5*RGVPSG(IPOND)*(RX-RGVPXC(IPOND))
          RQY=.5*RGVPSG(IPOND)*(RY-RGVPYC(IPOND))
        ENDIF
        RVQ(1)=RVQ(1)+RQX
        RVQ(2)=RVQ(2)+RQY
```

Sec. 9 Shallow Unconfined Flow with Rainfall

```
 100   CONTINUE
       RETURN
       END
c      Layout; the ponds are plotted as circles.
       SUBROUTINE GVLAY
       COMMON /CBGV/    RGVUQ0,RGVUAL,RGVUU(2),
     .                  RGVRN,RGVRX,RGVRY,RGVRAS,RGVRBS,RGVRAL,
     .                  NGVP,RGVPXC(20),RGVPYC(20),RGVPRD(20),RGVPSG(20)
       DO 100 IPOND=1,NGVP
         CALL RDRAWC(RGVPXC(IPOND),RGVPYC(IPOND),RGVPRD(IPOND))
 100   CONTINUE
       RETURN
       END
```

The listing contains the modified common block; this common block must replace the former ones in SETUP and in CHECK.

The Routine Setup

We add the command words RAIN and POND to the data statement of AWORD in SETUP, and change the dimension of AWORD accordingly; the following statements replace the old ones:

```
       DIMENSION AWORD(75),AWOGR(10)
       DATA AWORD   /'A','Q','U','I',' ',
     .               'R','E','F','E',' ',
     .               'U','N','I','F',' ',
     .               'R','A','I','N',' ',
     .               'P','O','N','D',' ',
     .               'W','E','L','L',' ',
     .               'S','O','L','V',' ',
     .               'W','I','N','D',' ',
     .               'G','R','I','D',' ',
     .               'P','L','O','T',' ',
     .               'L','A','Y','O',' ',
     .               'C','H','E','C',' ',
     .               'P','A','U','S',' ',
     .               'S','W','I','T',' ',
     .               'S','T','O','P','!'/
```

We must add two new labels to the computed GOTO; we choose labels 320 and 340 to correspond to the command words RAIN and POND, which are the fourth and fifth command words in the list, respectively. The new computed GOTO is

```
       GOTO( 100, 200, 300, 320, 340, 400, 500, 600, 700, 800, 900,1000,
     .       8900),JUMP
```

The statements associated with these labels store the data corresponding to rainfall and ponds and are as follows:

```
c       Infiltration from rainfall.
 320    RGVRN=RVAR(2)
        RGVRX=RVAR(3)
        RGVRY=RVAR(4)
c       Principal axes of the ellipse; store
```
a^2 and b^2.
```
        RGVRAS=RVAR(5)*RVAR(5)
        RGVRBS=RVAR(6)*RVAR(6)
c       Read the angle between the a axis and the x axis;
c       convert it to radians.
        RGVRAL=RVAR(7)*3.1415926/180.
        GOTO 10
c       Ponds; add 1 to the number of ponds, store the coordinates or
c       the center and the infiltration rate.
 340    NGVP=NGVP+1
        RGVPXC(NGVP)=RVAR(2)
        RGVPYC(NGVP)=RVAR(3)
        RGVPRD(NGVP)=RVAR(4)
        RGVPSG(NGVP)=RVAR(5)
c       If an error occurred, reduce the number of ponds by 1 so that
c       the user can enter the data again.
        IF(LERROR) NGVP=NGVP-1
        GOTO 10
```

The final modification of SETUP is the change of the format statement which controls the content of the command prompt:

```
9005 FORMAT(
    .'+<AQUIFER>(K,H)[BASE]'/
    .' <UNIFLOW>(Q,ANGLE)'/
    .' <RAIN>(INF.RATE,XCTR,YCTR,A AXIS,B AXIS,ANGLE)'/
    .' <POND>(XCTR,YCTR,RAD,INF.RATE)'/
    .' <WELL>(X,Y,Q,R)'/
    .' <REFERENCE>(X,Y,FI)'/
    .' <SOLVE>'/
    .' <WINDOW>(X1,Y1 X2,Y2)'/
    .' <LAYOUT>'/
    .' <GRID>(N)[<HEAD>/<POTENTIAL>]'/
    .' <PLOT>'/
    .' <CHECK>'/
    .' <SWITCH>(FILEIN)[FILEOUT]'/
    .' <PAUSE>'/
    .' <STOP>'/)
```

Sec. 9   Shallow Unconfined Flow with Rainfall

The routine SETUP is now prepared for the new functions.

The Routine Check

Only minor modifications are necessary in CHECK: besides replacing the common block BDGV with the new one, we add the command words RAIN and POND to the list of command words and modify the dimension statement of AWORD:

```
      DIMENSION AWORD(55),RVQ(2)
      DATA AWORD /'A','Q','U','I',' ',
     .            'R','E','F','E',' ',
     .            'U','N','I','F',' ',
     .            'R','A','I','N',' ',
     .            'P','O','N','D',' ',
     .            'W','E','L','L',' ',
     .            'C','O','N','T',' ',
     .            'H','E','A','D',' ',
     .            'P','O','T','E',' ',
     .            'D','I','S','C',' ',
     .            'R','E','T','U','!'/
```

The computed GOTO must be changed, and two new labels added with the statements necessary to print data regarding the new functions:

```
      GOTO( 100, 200, 300, 320, 340, 400, 500, 600, 700, 800,
     .      8900),JUMP
c     Infiltration from rainfall.
  320 WRITE(7,9032) RGVRN,RGVRX,RGVRY,SQRT(RGVRAS),SQRT(RGVRBS),
     .              RGVRAL*180./3.1415926
      GOTO 10
c     ponds.
  340 IF(NGVP.GT.0) THEN
         WRITE(7,9034)
         WRITE(7,9035)
     .        (I,RGVPXC(I),RGVPYC(I),RGVPRD(I),RGVPSG(I),I=1,NGVP)
         GOTO 10
      ENDIF
```

The final changes and additions are the modification of the format statement for the command line, and the addition of three new format statements:

```
 9005 FORMAT(
     .'+<AQUIFER><REFERENCE><UNIFLO><RAIN><POND><WELL><CONTROL>'/
     .' <HEAD>(X,Y)<DISCHARGE>(X,Y)<POTENTIAL>(X,Y)<RETURN>'/)
 9030 FORMAT('   UNIFORM FLOW   ANGLE         COS(ANGLE)     SIN(ANGLE)'/
     .       4E14.6)
 9032 FORMAT('   RAIN           XCENTER        YCENTER'/
```

```
       .        ,3E14.6/
       .      '                    A-AXIS         B-AXIS          ANGLE'/
       .      '              ',3E14.6)
 9034 FORMAT('PONDS:  NO XCENTER        YCENTER         RADIUS ',
      .'           INFILTRATION')
 9035 FORMAT('     ',I4,4E14.6)
```

Problem 9.4

Implement the new functions in your version of SLW. Use the computer model to simulate the flow case governed by (9.8) and verify that the flow rates through the boundaries at $x = -\frac{1}{2}L$ and at $x = \frac{1}{2}L$ are as given in the text. Verify further that the program reproduces the divide in the proper place by using the command DISCHARGE in the routine CHECK. Note that you cannot enter the heads along the two boundaries directly; however, you can reproduce them by centering the rainfall midway between the river banks, by setting the head at a point of one of the river banks to the given value, and by entering uniform flow with a discharge computed such that the head along the second river bank matches the boundary value.

Problem 9.5

Use the computer model to simulate flow toward a well at the center of a circular island with rainfall, and verify that the program produces the divide at the location computed by the use of (9.25).

Problem 9.6

Use the computer model to verify your answer to Problem 9.3.

10 SHALLOW INTERFACE FLOW

We will refer to a flow as interface flow when the aquifer contains two fluids that are separated by an interface rather than a transition zone, and when one of the fluids is at rest. Coastal aquifers are often modeled with good approximation by assuming interface flow. The fresh water, then, is assumed to flow over salt water at rest. If the aquifer is sufficiently shallow with respect to its extent so that the resistance to flow in the vertical direction may be neglected, we will call the flow shallow interface flow.

Condition Along the Interface

A cross section through an aquifer with fresh water flowing over salt water at rest is shown in Figure 2.31. The aquifer is bounded above by a phreatic surface and below by an interface that separates the fresh water from salt water at rest. The distances between these surfaces and sea level are denoted as h_f and h_s, where the indices f and s stand for fresh and salt. All heads are measured from a reference level, which lies a distance H_s below sea level. The head in the fresh water, ϕ, is expressed in terms of the pressure p_f by the use of (2.7) as

$$\phi = \frac{p_f}{g\rho_f} + Z \qquad (10.1)$$

Sec. 10 Shallow Interface Flow

where ρ_f is the density of the fresh water and the elevation head Z is measured from the reference level. The head ϕ_s in the salt water is written similarly in terms of the pressure p_s in the salt water,

$$\phi_s = \frac{p_s}{g\rho_s} + Z \qquad (10.2)$$

where ρ_s is the density of the salt water. The pressure must be single-valued at all points of the interface, such as point $\mathcal{P}$ in Figure 2.31. This gives, with (10.1) and (10.2), and with $Z = H_s - h_s$ at $\mathcal{P}$:

$$g\rho_f(\phi - H_s + h_s) = g\rho_s(\phi_s - H_s + h_s) \qquad (10.3)$$

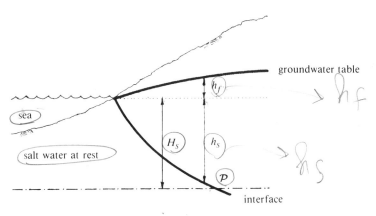

Figure 2.31 The Ghyben-Herzberg equation.

We neglect the resistance to flow in the vertical direction, the Dupuit-Forchheimer assumption, so that the equipotentials are vertical: $\phi = H_s + h_f$. We further limit the analysis to cases where the salt water is at rest so that the head in the salt water equals the elevation of the sea level above the reference level, i.e., $\phi_s = H_s$. This yields with (10.3), after rearrangement of terms and division by $g\rho_f$:

$$\boxed{h_f = h_s \frac{\rho_s - \rho_f}{\rho_f}} \qquad (10.4)$$

This equation is known as the Ghyben-Herzberg equation and was derived independently by Badon Ghyben [1888] and Herzberg [1901]. It is of interest to note that Du Commun [1828] should really be credited with the discovery of the Ghyben-Herzberg equation (Carlston [1963]). It may be noted that the Ghyben-Herzberg equation is equally valid when the upper boundary of the aquifer is a horizontal impermeable boundary rather than a phreatic surface. In that case, h_f represents the piezometric head with respect to sea level.

Potentials

In describing shallow interface flow, we will make use of discharge potentials. These potentials were introduced by Strack [1976], and are obtained similarly as the ones given in Sec. 8 for horizontal confined and shallow unconfined flow. We will denote the saturated thickness of the aquifer as h and make restriction to problems of flow where h is a linear function of the head ϕ, i.e.,

$$\boxed{h = \alpha\phi + \beta} \tag{10.5}$$

where α and β are constants that depend upon the type of flow. The components of the discharge vector, Q_x and Q_y, are equal to h times the components q_x and q_y of the specific discharge vector, compare (7.4), and we obtain with Darcy's law:

$$Q_x = hq_x = -kh\frac{\partial \phi}{\partial x}$$
$$Q_y = hq_y = -kh\frac{\partial \phi}{\partial y} \tag{10.6}$$

We will first consider the case that α in (10.5) is unequal to zero. Substituting (10.5) for h in (10.6) and integrating, we obtain:

$$Q_x = -k(\alpha\phi + \beta)\frac{\partial \phi}{\partial x} = -\frac{\partial}{\partial x}\left[\tfrac{1}{2}k\alpha\left(\phi + \frac{\beta}{\alpha}\right)^2 + C\right] \quad (\alpha \neq 0)$$
$$Q_y = -k(\alpha\phi + \beta)\frac{\partial \phi}{\partial y} = -\frac{\partial}{\partial y}\left[\tfrac{1}{2}k\alpha\left(\phi + \frac{\beta}{\alpha}\right)^2 + C\right] \quad (\alpha \neq 0) \tag{10.7}$$

where C is a constant of integration and the hydraulic conductivity k is constant. The potential Φ for $\alpha \neq 0$ thus is given by

$$\boxed{\Phi = \tfrac{1}{2}k\alpha\left(\phi + \frac{\beta}{\alpha}\right)^2 + C \quad (\alpha \neq 0)} \tag{10.8}$$

It may be noted that the potential (8.2) for shallow unconfined flow is retrieved from (10.8) by noticing that for that case $\phi = h$, or $\alpha = 1$, $\beta = 0$, which yields $\Phi = \tfrac{1}{2}k\phi^2 + C$. For the case that $\alpha = 0$, (10.5) and (10.6) yield

$$Q_x = \beta q_x = -k\beta\frac{\partial \phi}{\partial x} = -\frac{\partial}{\partial x}(k\beta\phi + C)$$
$$Q_y = \beta q_y = -k\beta\frac{\partial \phi}{\partial y} = -\frac{\partial}{\partial y}(k\beta\phi + C) \tag{10.9}$$

Sec. 10 Shallow Interface Flow

which gives the following expression for the potential Φ,

$$\boxed{\Phi = k\beta\phi + C \qquad (\alpha = 0)} \qquad (10.10)$$

The potential for horizontal confined flow is retrieved from (10.10) by setting $\beta = H$. For all cases, the discharge vector equals the gradient of the potential,

$$Q_x = -\frac{\partial \Phi}{\partial x}$$
$$Q_y = -\frac{\partial \Phi}{\partial y} \qquad (10.11)$$

Shallow Confined Interface Flow

We will refer to the flow as shallow confined interface flow when the aquifer is bounded above by a horizontal impermeable boundary and below by an interface (see Figure 2.32). If the upper boundary of the aquifer lies at a distance H above the reference level, the aquifer thickness h may be expressed as follows:

$$h = h_s - (H_s - H) \qquad (10.12)$$

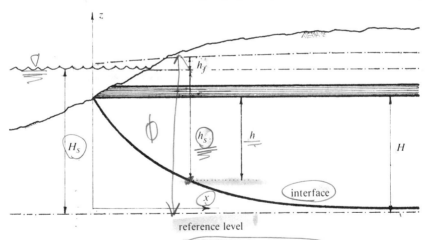

Figure 2.32 Shallow confined interface flow.

The head ϕ is given by

$$\phi = h_f + H_s \qquad (10.13)$$

Expressing h_s in terms of h_f by the use of the Ghyben-Herzberg equation (10.4) and using (10.12) and (10.13), the aquifer thickness may be written as

$$h = \frac{\rho_f}{\rho_s - \rho_f} h_f - (H_s - H) = \frac{\rho_f}{\rho_s - \rho_f}\phi - \frac{\rho_s}{\rho_s - \rho_f}H_s + H \qquad (10.14)$$

The constants α and β in (10.5) become

$$\alpha = \frac{\rho_f}{\rho_s - \rho_f} \qquad \beta = -\frac{\rho_s}{\rho_s - \rho_f} H_s + H \qquad (10.15)$$

The potential for shallow confined interface flow is obtained from (10.8) and (10.15) as

$$\Phi = \tfrac{1}{2} k \frac{\rho_f}{\rho_s - \rho_f} \left[\phi - \frac{\rho_s}{\rho_f} H_s + \frac{\rho_s - \rho_f}{\rho_f} H \right]^2 + C_{ci} \qquad (10.16)$$

where C_{ci} replaces the constant C; the index ci stands for confined interface flow.

The condition of continuity of flow may be applied to the discharge vector (Q_x, Q_y) in the same way as was done for shallow unconfined flow in Sec. 7, compare (7.11). This gives

$$\frac{\partial Q_x}{\partial x} + \frac{\partial Q_y}{\partial y} = 0 \qquad (10.17)$$

and the differential equation for the potential is obtained from (10.11) and (10.17)

$$\nabla^2 \Phi = 0 \qquad (10.18)$$

It is important to note that the basic equations in terms of the potential are the same as those treated in Sections 6 through 8; the difference between the various types of flows appears only in the definition of the potentials in terms of head.

Elevation of the Interface

The elevation Z of the interface above the reference level equals the difference between H_s and h_s (see Figure 2.31):

$$Z = H_s - h_s \qquad (10.19)$$

By using the Ghyben-Herzberg equation (10.4) to express h_s in terms of h_f and (10.13) to express h_f in terms of ϕ, we obtain

$$\boxed{Z = \frac{\rho_s}{\rho_s - \rho_f} H_s - \frac{\rho_f}{\rho_s - \rho_f} \phi} \qquad (10.20)$$

Example 10.1: Uniform Flow Toward the Coast

As an example, we will consider the case of uniform flow toward the coast. An x, y, z coordinate system is chosen as shown in Figure 2.32, with the x axis pointing inland from the coast, the y axis along the coast, and the z axis vertically upward. The head at a distance L from the coast is ϕ_1. The flow is one-dimensional in the x direction and the corresponding solution to (10.18) may be written as

$$\Phi = Ax + B \qquad (10.21)$$

Sec. 10 Shallow Interface Flow

The boundary condition along the coast is obtained in terms of ϕ as follows. The pressure p_f at a point on the coast and just below the confining layer equals the weight of the column of *salt* water above that point, which equals $g\rho_s(H_s - H)$ (see Figure 2.32). The elevation of the confining layer above the reference level is H, and application of (10.1) gives, representing the head along the coast as ϕ_0,

$$x = 0 \qquad \phi = \phi_0 = \frac{p_f}{g\rho_f} + Z = \frac{\rho_s}{\rho_f}(H_s - H) + H = \frac{\rho_s}{\rho_f}H_s - \frac{\rho_s - \rho_f}{\rho_f}H \qquad (10.22)$$

The corresponding value for the potential is found from (10.16) to be C_{ci},

$$x = 0 \qquad \Phi = \Phi_0 = C_{ci} \qquad (10.23)$$

The condition $\phi = \phi_1$ at $x = L$ may be expressed in terms of Φ,

$$x = L \qquad \Phi = \Phi_1 = \tfrac{1}{2}k\frac{\rho_f}{\rho_s - \rho_f}\left[\phi_1 - \frac{\rho_s}{\rho_f}H_s + \frac{\rho_s - \rho_f}{\rho_f}H\right]^2 + C_{ci} \qquad (10.24)$$

Application of these boundary conditions to (10.21) gives

$$\Phi = \frac{\Phi_1 - \Phi_0}{L}x + \Phi_0 \qquad (10.25)$$

The head ϕ as a function of x is obtained from (10.25) using the definition (10.16) for Φ,

$$\phi = \frac{\rho_s}{\rho_f}H_s - \frac{\rho_s - \rho_f}{\rho_f}H + \sqrt{\frac{2}{k}\frac{\rho_s - \rho_f}{\rho_f}\left[\frac{\Phi_1 - \Phi_0}{L}x + \Phi_0 - C_{ci}\right]} \qquad (10.26)$$

Problem 10.1
1. Demonstrate that the values of ϕ computed by the use of expression (10.26) are not affected by the choice of C_{ci}.
2. Plot 9 points of the interface between $x = L = 10H$ and $x = 0$ if $\rho_s = 1025$ kg/m^3, $\rho_f = 1000$ kg/m^3, $H_s = 1.1H$, and $\phi_1 = 1.3H$.
3. Why is it not necessary to know the value of k for the latter exercise?
4. What is h along the coast? Do you believe that this value is realistic? If not, why would the model predict an unrealistic value?

Problem 10.2
Adapt the program SLW to shallow confined interface flow by modifying the module AQUIFER.

Combined Shallow Confined Interface Flow and Confined Flow

We next consider the case that a horizontal impervious base bounds the aquifer from below, and choose this base as the reference level (see Figure 2.33). Proper choice of the constant C_{ci} in the definition of the potential for shallow confined interface flow enables us to make the transition from confined interface flow to confined flow. This is done in much the same way as in Sec. 8 where we combined unconfined flow with confined flow in one aquifer. We will represent the zones of confined interface flow and of confined flow as zones 1 and 2, respectively (see Figure 2.33). The potential in the confined zone will be represented by (8.6):

$$\text{zone 2:} \quad \Phi = kH\phi - \tfrac{1}{2}kH^2 \tag{10.27}$$

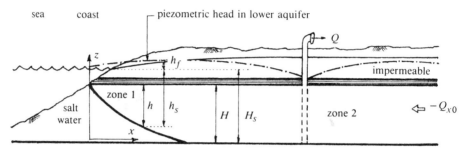

Figure 2.33 Combined confined flow-confined interface flow. (Source: Adapted from Strack [1976].)

The interzonal boundary lies at the tip of the saltwater tongue, where $h = H$. The corresponding value for ϕ is obtained from (10.14),

$$\phi_t = \frac{\rho_s}{\rho_f} H_s \tag{10.28}$$

where the subscript t refers to tip. We will require that the potential, which is defined by (10.16) in zone 1 and by (10.27) in zone 2, is single-valued across the tip of the saltwater tongue. That is, substitution of ϕ_t for ϕ in both (10.16) and (10.27) must give the same value:

$$\tfrac{1}{2}k\frac{\rho_f}{\rho_s - \rho_f}\left[\frac{\rho_s}{\rho_f}H_s - \frac{\rho_s}{\rho_f}H_s + \frac{\rho_s - \rho_f}{\rho_f}H\right]^2 + C_{ci} = kH\frac{\rho_s}{\rho_f}H_s - \tfrac{1}{2}kH^2 \tag{10.29}$$

so that

$$C_{ci} = k\frac{\rho_s}{\rho_f}HH_s - \tfrac{1}{2}k\frac{\rho_s}{\rho_f}H^2 \tag{10.30}$$

and the potential in zone 1 becomes, with (10.16)

$$\text{zone 1:} \quad \Phi = \tfrac{1}{2}k\frac{\rho_f}{\rho_s - \rho_f}\left[\phi - \frac{\rho_s}{\rho_f}H_s + \frac{\rho_s - \rho_f}{\rho_f}H\right]^2 + k\frac{\rho_s}{\rho_f}HH_s - \tfrac{1}{2}k\frac{\rho_s}{\rho_f}H^2 \tag{10.31}$$

Problem 10.3
Consider the problem solved in Example 10.1, but now with an impermeable base at a distance H_s below sea level. Determine the position of the tip of the saltwater tongue in terms of x.

Flow Toward a Well in a Coastal Aquifer

We will next consider the case of flow toward a well in the confined coastal aquifer depicted in Figure 2.33. The well is located at a distance d from the coast, and the problem is to determine the maximum discharge of the well for which no salt water is pumped. A cross section through the well and a plan view of the aquifer are shown in Figure 2.34. It is given that the flow without the well is described by (10.25). Defining Q_{x0} as the discharge flowing in the x direction when the well is not present, (10.25) may be written as

$$\Phi = -Q_{x0}x + \Phi_0 \tag{10.32}$$

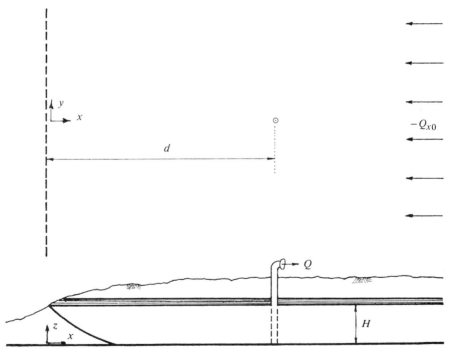

Figure 2.34 Confined flow toward a well in a coastal aquifer.

It is noted that Q_{x0} is negative in this case because the flow occurs in the negative x direction toward the coast. The value for Φ_0 is given by (10.23) with C_{ci} given by (10.30).

$$\Phi_0 = k\frac{\rho_s}{\rho_f}HH_s - \tfrac{1}{2}k\frac{\rho_s}{\rho_f}H^2 \tag{10.33}$$

The potential for flow with a well is obtained by application of the method of images: The potential for a well of discharge Q at $x = d$, $y = 0$ and an image recharge well at $x = -d$, $y = 0$ gives $\Phi = 0$ along the coast. Addition of the latter potential to (10.32) will therefore yield the function that meets the boundary conditions:

$$\Phi = -Q_{x0}x + \frac{Q}{4\pi}\ln\frac{(x-d)^2 + y^2}{(x+d)^2 + y^2} + \Phi_0 \tag{10.34}$$

The equation for the tip of the tongue of salt water is found from (10.34) by setting Φ equal to the constant value that the potential assumes along the tip. The latter value is obtained by substituting expression (10.28), ϕ_t, for ϕ in either (10.27) or (10.31) which yields

$$\Phi_t = k\frac{\rho_s}{\rho_f}HH_s - \tfrac{1}{2}kH^2 \tag{10.35}$$

and the equation for the tip of the tongue becomes, using (10.33) for Φ_0,

$$k\frac{\rho_s}{\rho_f}HH_s - \tfrac{1}{2}kH^2 = -Q_{x0}x + \frac{Q}{4\pi}\ln\frac{(x-d)^2 + y^2}{(x+d)^2 + y^2} + k\frac{\rho_s}{\rho_f}HH_s - \tfrac{1}{2}k\frac{\rho_s}{\rho_f}H^2 \tag{10.36}$$

which may be written in terms of the dimensionless variables λ_c and μ by dividing both sides by $-\tfrac{1}{2}Q_{x0}d$ as follows:

$$\lambda_c = 2\frac{x}{d} + \frac{\mu}{2\pi}\ln\frac{(\frac{x}{d}-1)^2 + (\frac{y}{d})^2}{(\frac{x}{d}+1)^2 + (\frac{y}{d})^2} \tag{10.37}$$

where

$$\lambda_c = -\frac{kH^2}{dQ_{x0}}\frac{\rho_s - \rho_f}{\rho_f} \qquad \mu = -\frac{Q}{dQ_{x0}} \tag{10.38}$$

It may be noted that Q_{x0} is negative because the uniform flow occurs in the negative x direction, so that μ is positive. If Q is gradually increased while keeping all other flow parameters constant, the tip will tend to move inland toward the well. This is illustrated in Figure 2.35 where the tip of the tongue is represented in the dimensionless $x/d, y/d$ plane for the case that $\lambda_c = 0.5$; only the part of the tip for $y/d \geq 0$ is shown. The curve for $\mu = 2$ intersects the x axis inland from the well; the well is pumping salt water, and the assumption that the salt water is at rest is violated. The dashed-dotted curve for $\mu = 1.63$ applies to the unstable case that the salt water is on the verge of starting to flow. This is seen with the aid of Figure 2.36 where the equipotentials are shown that apply to the unstable case. The dashed-dotted curve represents the tip of the saltwater tongue. Although the values of Φ along the equipotentials are greater than Φ_t in most

Sec. 10   Shallow Interface Flow

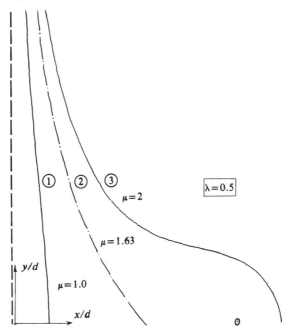

Figure 2.35 Locations of the tip for various values of μ.

of the aquifer inland from the tip, there exists an area, shaded in the figure, where the values of Φ are less than Φ_t. Point S is a stagnation point, and although the head at S is just high enough to keep the tip at point S, any reduction in head at S will cause the salt water to enter the shaded area of low heads: the situation is unstable.

Unstable conditions thus correspond to the stagnation point S lying on the tip of the saltwater tongue. These conditions are described mathematically by requiring that the coordinates (x_s, y_s) of the stagnation point fulfill equation (10.37) for the tip. Equations for the coordinates of the stagnation point are obtained by setting the expressions for Q_x and Q_y obtained by differentiation of (10.34) equal to zero:

$$Q_x = -\frac{\partial \Phi}{\partial x} = Q_{x0} - \frac{Q}{2\pi}\left[\frac{x_s - d}{(x_s - d)^2 + y_s^2} - \frac{x_s + d}{(x_s + d)^2 + y_s^2}\right] = 0 \qquad (10.39)$$

and

$$Q_y = -\frac{\partial \Phi}{\partial y} = -\frac{Q}{2\pi}\left[\frac{y_s}{(x_s - d)^2 + y_s^2} - \frac{y_s}{(x_s + d)^2 + y_s^2}\right] = 0 \qquad (10.40)$$

The only roots of (10.40) occur for either $y_s = 0$ or $x_s = 0$. The case $x_s = 0$ can be disregarded, however, because for $x_s = 0$ the stagnation point(s) are on the coast, which means that salt water is entering the aquifer: this violates the assumption that salt water is at rest. Thus, $y_s = 0$, and (10.39) becomes

$$Q_{x0} - \frac{Q}{2\pi}\left[\frac{1}{x_s - d} - \frac{1}{x_s + d}\right] = Q_{x0} - \frac{Q}{2\pi}\left[\frac{2d}{x_s^2 - d^2}\right] = 0 \qquad (y_s = 0) \qquad (10.41)$$

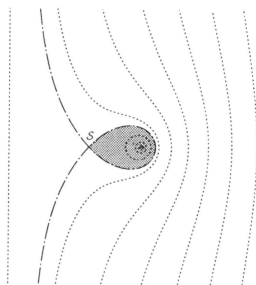

Figure 2.36 Equipotentials for the unstable case.

Solving this equation for x_s, we obtain, writing $-Q/(Q_{x0}d)$ as μ,

$$\frac{x_s}{d} = +\sqrt{1 - \frac{\mu}{\pi}} \qquad y_s = 0 \tag{10.42}$$

where the + sign is selected because x_s is known to be positive. Substitution of this for x_s and y_s in (10.37) gives the following condition for instability to occur:

$$\lambda_c = 2\sqrt{1 - \frac{\mu}{\pi}} + \frac{\mu}{\pi} \ln \frac{1 - \sqrt{(1 - \mu/\pi)}}{1 + \sqrt{(1 - \mu/\pi)}} \tag{10.43}$$

This equation represents a curve, which is shown in Figure 2.37(a), where λ_c is plotted versus μ. This curve divides the plane into two domains: points plotting above the curve correspond to two-fluid flow, and points plotting below it to one-fluid flow. The points in the plot labeled as (1), (2), and (3) correspond to the curves labeled equally in Figure 2.35. Equation (10.42) is represented graphically in Figure 2.37(b).

Shallow Unconfined Interface Flow

A shallow coastal unconfined aquifer is shown in Figure 2.38. The aquifer thickness h now may be expressed in terms of h_s and h_f as

$$h = h_f + h_s \tag{10.44}$$

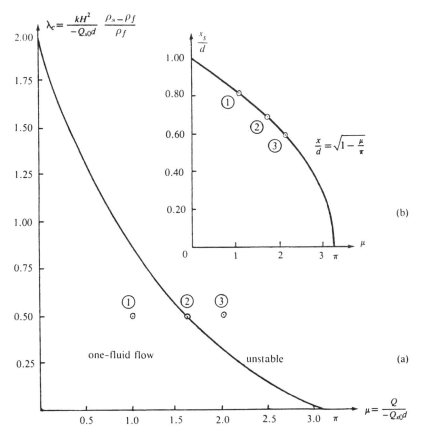

Figure 2.37 Flow parameters for an unstable situation. (Source: Adapted from Strack [1976].)

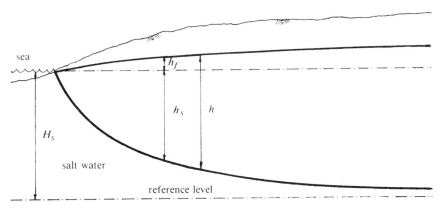

Figure 2.38 Shallow unconfined interface flow.

Using the Ghyben-Herzberg equation (10.4) to express h_s in terms of h_f and solving (10.44) for h_f, we obtain

$$h_f = \frac{\rho_s - \rho_f}{\rho_s} h \tag{10.45}$$

It is seen from the figure that ϕ may be expressed as

$$\phi = H_s + h_f \tag{10.46}$$

The general form (10.8) of the potential Φ may be applied to shallow unconfined interface flow by expressing h linearly in terms of ϕ according to (10.5). Using (10.45) and (10.46), we obtain

$$h = \frac{\rho_s}{\rho_s - \rho_f} \phi - \frac{\rho_s}{\rho_s - \rho_f} H_s \tag{10.47}$$

so that the constants α and β in (10.5) become

$$\alpha = \frac{\rho_s}{\rho_s - \rho_f} \qquad \beta = -\frac{\rho_s}{\rho_s - \rho_f} H_s \tag{10.48}$$

The potential Φ is obtained from (10.8) as

$$\Phi = \tfrac{1}{2} k \frac{\rho_s}{\rho_s - \rho_f} (\phi - H_s)^2 + C_{ui} \tag{10.49}$$

Elevation of the Interface

The expression (10.20) for the elevation Z of the interface was obtained from the Ghyben-Herzberg equation and the identity $\phi = H_s + h_f$, which are both equally valid for unconfined interface flow as for confined interface flow. The elevation of the interface for unconfined interface flow thus is given by (10.20):

$$Z = \frac{\rho_s}{\rho_s - \rho_f} H_s - \frac{\rho_f}{\rho_s - \rho_f} \phi \tag{10.50}$$

Problem 10.4
1. Determine the head as a function of position for the problem described in Example 10.1, but with a phreatic surface replacing the upper confining boundary.
2. Plot 9 points of the interface between $x = L = 10 H_s$ and $x = 0$ if $\rho_s = 1025$ kg/m^3, $\rho_f = 1000$ kg/m^3, and $\phi_1 = 1.1 H_s$.

Combined Shallow Unconfined Interface Flow and Unconfined Flow

Two types of flow may occur in the aquifer if the lower boundary is an impermeable base: the flow will be shallow unconfined interface flow near the coast and shallow

Sec. 10 Shallow Interface Flow

unconfined flow at some distance away from the coast. Such problems can be solved by the use of a single potential Φ in the same fashion as was done in the foregoing. The constants C_u and C_{ui} involved in (8.2) and (10.49) must be chosen such that Φ is continuous across the interzonal boundary, which occurs at the tip of the tongue, where $\phi = \rho_s H_s / \rho_f$. Application of this condition of continuity to (8.2) and (10.49) yields

$$\tfrac{1}{2} k \left[\frac{\rho_s}{\rho_f} H_s \right]^2 + C_u = \tfrac{1}{2} k \frac{\rho_s}{\rho_s - \rho_f} \left[\frac{\rho_s - \rho_f}{\rho_f} H_s \right]^2 + C_{ui} \qquad (10.51)$$

or

$$C_{ui} = \tfrac{1}{2} k \frac{\rho_s}{\rho_f} H_s^2 \qquad C_u = 0 \qquad (10.52)$$

where the choice $C_u = 0$ is arbitrary. The expressions for the potential thus become

$$\Phi = \tfrac{1}{2} k \frac{\rho_s}{\rho_s - \rho_f} (\phi - H_s)^2 + \tfrac{1}{2} k \frac{\rho_s}{\rho_f} H_s^2 \qquad (\phi \leq \frac{\rho_s}{\rho_f} H_s) \qquad (10.53)$$

and

$$\Phi = \tfrac{1}{2} k \phi^2 \qquad (\phi \geq \frac{\rho_s}{\rho_f} H_s) \qquad (10.54)$$

As an example, the problem illustrated in Figure 2.34 will be solved for the case that the upper boundary of the aquifer is a phreatic surface. The expression for the potential as a function of position will be as before, i.e., (10.34). However, the expression for the potential along the coast, Φ_0, will be different. It is seen from Figure 2.38 that the head along the coast equals H_s, so that the potential Φ_0 is obtained from (10.53) as

$$\Phi_0 = \tfrac{1}{2} k \frac{\rho_s}{\rho_f} H_s^2 \qquad (10.55)$$

The potential along the tip of the tongue is found by setting ϕ equal to $\rho_s H_s / \rho_f$ in either (10.53) or (10.54):

$$\Phi_t = \tfrac{1}{2} k \left[\frac{\rho_s}{\rho_f} H_s \right]^2 \qquad (10.56)$$

The equation for the tip of the tongue is obtained by setting Φ equal to Φ_t in (10.34), which yields, with (10.55):

$$\tfrac{1}{2} k \left[\frac{\rho_s}{\rho_f} H_s \right]^2 = -Q_{x0} x + \frac{Q}{4\pi} \ln \frac{(x-d)^2 + y^2}{(x+d)^2 + y^2} + \tfrac{1}{2} k \frac{\rho_s}{\rho_f} H_s^2 \qquad (10.57)$$

This equation is written in terms of dimensionless variables by dividing both sides by $-\tfrac{1}{2} Q_{x0} d$, yielding

$$\lambda_u = 2 \frac{x}{d} + \frac{\mu}{2\pi} \ln \frac{(\frac{x}{d} - 1)^2 + (\frac{y}{d})^2}{(\frac{x}{d} + 1)^2 + (\frac{y}{d})^2} \qquad (10.58)$$

where μ is given in (10.38) and λ_u is defined as

$$\lambda_u = -\frac{kH_s^2}{dQ_{x0}}\frac{\rho_s}{\rho_f}\frac{\rho_s - \rho_f}{\rho_f} \qquad (10.59)$$

It appears that the present case of an aquifer bounded above by a phreatic surface differs from the confined case only in the expressions for Φ_0 and λ_u or λ_c; compare (10.58) and (10.37). The results represented in Figure 2.37 may therefore be applied to the present case by replacing λ_c by λ_u.

We will next examine the location of the most inland point of the tongue as a function of the parameters μ and λ_u or λ_c. It appears from Figure 2.35 that the most inland point lies on the x axis. We will denote the x coordinate of this point as L_c and L_u for the confined and unconfined cases, respectively. Expressions for L_c and L_u are obtained by setting y equal to zero and x equal to L_c in (10.37), and to L_u in (10.58). The resulting equations are represented graphically in Figure 2.39.

The case of uniform combined shallow unconfined interface flow and unconfined flow toward the coast without a well was solved by Henry [1959] without making use of discharge potentials. The discharge Q_{x0} for that case may be expressed as

$$Q_{x0} = -\frac{\Phi_1 - \Phi_0}{L} \qquad (10.60)$$

where Φ_1 is the potential at some distance away from the coast and Φ_0 the potential along the coast. If $\phi_1 > \rho_s H_s/\rho_f$ and $\phi_0 = H_s$ represent the corresponding values of the head, then (10.60) may be written as follows (see [10.53] and [10.54]):

$$Q_{x0} = -\frac{\frac{1}{2}k\left[\phi_1^2 - \frac{\rho_s}{\rho_f}H_s^2\right]}{L} \qquad (10.61)$$

Youngs [1971] showed that (10.61) gives the exact value of the discharge (i.e., the discharge obtained without making the Dupuit-Forchheimer assumption) if the boundary at $x = L$ is an equipotential and if the coast is vertical.

Problem 10.5

Consider the problem shown in Figure 2.34, but now with the well inactive, the upper impermeable layer elevated above sea level ($H > H_s$), and assuming that $\phi_1 > H$.
1. List three types of flow that occur in the aquifer if $H > \rho_s H_s/\rho_f$.
2. Select the constants in the various expressions for the potential such that it is continuous throughout the aquifer for the case that $H > \rho_s H_s/\rho_f$. Determine the location of the tip of the saltwater tongue if $L = 10H$, $\rho_s = 1025$ kg/m^3, $\rho_f = 1000$ kg/m^3, $H_s = 0.8H$, $\phi_1 = 1.1H$. Plot 9 points of the interface between the coast and the tip of the tongue.
3. List the three types of flow that occur in the aquifer if $H < \rho_s H_s/\rho_f$. Repeat part 2 for $H < \rho_s H_s/\rho_f$, and $H_s = 0.98H$ rather than $0.8H$.

Sec. 10   Shallow Interface Flow   111

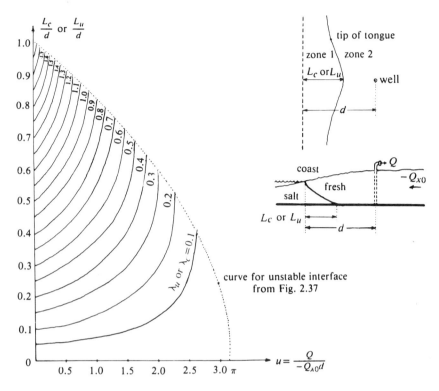

Figure 2.39 L_c/d and L_u/d as functions of λ_c, λ_u, and μ. (Source: Adapted from Strack [1976].)

Shallow Unconfined Interface Flow with Rainfall

Shallow unconfined interface flow with rainfall may be treated in the same way as unconfined flow without an interface. This is possible because the basic equations in terms of the discharge vectors and the potentials are identical for these types of flow.

The Interface in an Island

Consider the problem of shallow unconfined interface flow combined with unconfined flow in the circular island of radius R illustrated in Figure 2.40. The potential may be written in the form

$$\Phi = -\tfrac{1}{4}N(r^2 - R^2) + \Phi_0 \tag{10.62}$$

$$\Phi = \tfrac{1}{2}k\frac{\rho_s}{\rho_s - \rho_f}[\phi - H_s]^2 + \tfrac{1}{2}k\frac{\rho_s}{\rho_f}H_s^2 \qquad (\phi \leq \frac{\rho_s}{\rho_f}H_s)$$
$$\Phi = \tfrac{1}{2}k\phi^2 \qquad (\phi \geq \frac{\rho_s}{\rho_f}H_s) \qquad (10.63)$$

see (10.53) and (10.54). The value for Φ_0 is given by (10.55). The tip of the saltwater tongue occurs when $\Phi = \Phi_t$, see (10.56), and the corresponding radius r_t is found from (10.62) as

$$r_t = R\sqrt{1 - \frac{2k}{N}\left[\frac{H_s}{R}\right]^2 \frac{\rho_s}{\rho_f}\frac{\rho_s - \rho_f}{\rho_f}} \qquad (10.64)$$

Note that the interface lies entirely above the base of the aquifer when the root is imaginary.

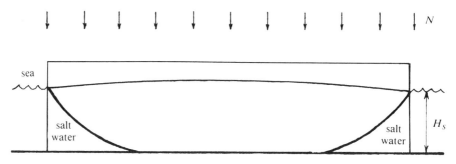

Figure 2.40 A circular island in the sea.

Problem 10.6

Determine the potential for the problem of Figure 2.40, but with a well of discharge Q at the center of the island. Find the equation that holds for the unstable condition of the interface.

Problem 10.7

Determine an expression for the location of the tip of the saltwater tongue for rainfall on an island with an elliptic boundary, using (9.26).

11 AQUIFERS WITH VERTICALLY VARYING HYDRAULIC CONDUCTIVITY

Aquifer systems often consist of a stack of aquifers of different hydraulic conductivities, sometimes separated by layers of low hydraulic conductivity. We call the separating layers *aquicludes* if they are impermeable, and *leaky layers* or *aquitards* if their hydraulic conductivity is not zero, but much less than the hydraulic conductivities of the aquifers. If the vertical variations in the hydraulic conductivity are relatively small, and the resistance to flow in the vertical direction may be neglected, the head will vary little in the vertical

Sec. 11   Aquifers with Vertically Varying Hydraulic Conductivity

direction. Such problems may be solved by the use of potentials first introduced by Girinski [1946]. Problems where different heads must be assigned to each aquifer in the system will be discussed in Sections 12 through 16.

We will first present discharge potentials in a form somewhat different from that introduced by Girinski, emphasizing continuity of the potentials across interzonal boundaries. We will then present the potential as introduced by Girinski, along with a generalization due to Youngs [1971]. It is assumed, throughout this section, that the hydraulic conductivity k is a function of the coordinate in vertical direction, z:

$$k = k(z) \tag{11.1}$$

Continuous Discharge Potentials

The two components of the discharge vector are defined by the relation

$$\begin{aligned} Q_x &= \int_{Z_b}^{Z_t} q_x(x,y,z)dz \\ Q_y &= \int_{Z_b}^{Z_t} q_y(x,y,z)dz \end{aligned} \tag{11.2}$$

where Z_b and Z_t represent the lower and upper boundaries of the saturated zone of flowing groundwater in the aquifer. Note that equations (11.2) are valid for three-dimensional flow. Using Darcy's law, (11.2) may be written as

$$\begin{aligned} Q_x &= -\int_{Z_b}^{Z_t} k(z)\frac{\partial \phi}{\partial x}dz \\ Q_y &= -\int_{Z_b}^{Z_t} k(z)\frac{\partial \phi}{\partial y}dz \end{aligned} \tag{11.3}$$

According to the Dupuit-Forchheimer assumption, ϕ is independent of z, so that (11.3) may be written as

$$\begin{aligned} Q_x &= -\frac{\partial \phi}{\partial x}\int_{Z_b}^{Z_t} k(z)dz \\ Q_y &= -\frac{\partial \phi}{\partial y}\int_{Z_b}^{Z_t} k(z)dz \end{aligned} \tag{11.4}$$

We make the assumption that Z_t and Z_b may be expressed as functions of ϕ alone:

$$\begin{aligned} Z_t &= Z_t(\phi) \\ Z_b &= Z_b(\phi) \end{aligned} \tag{11.5}$$

It may be noted that all types of flow discussed so far satisfy this condition. The transmissivity, T, of the aquifer is defined as

$$T = \int_{Z_b}^{Z_t} k(z)dz = T(\phi) \tag{11.6}$$

which is a function of ϕ alone, as follows from (11.5). The expressions (11.4) for the components of the discharge vector may now be written as:

$$Q_x = -T(\phi)\frac{\partial \phi}{\partial x} = -\frac{\partial \Phi}{\partial x}$$
$$Q_y = -T(\phi)\frac{\partial \phi}{\partial y} = -\frac{\partial \Phi}{\partial y} \tag{11.7}$$

where the potential Φ is defined as

$$\Phi = \int T(\phi)d\phi + C \tag{11.8}$$

The constant of integration, C, must be chosen such that the potential is continuous across interzonal boundaries.

The components of the specific discharge vector are now functions of z because the hydraulic conductivity is a function of z:

$$q_x = -k(z)\frac{\partial \phi}{\partial x}$$
$$q_y = -k(z)\frac{\partial \phi}{\partial y} \tag{11.9}$$

These components may be expressed in terms of the components of the discharge vector by the use of (11.7):

$$q_x = \frac{k(z)}{T}Q_x$$
$$q_y = \frac{k(z)}{T}Q_y \tag{11.10}$$

All discharge potentials presented so far may be obtained by the use of (11.8). For example, if the flow is unconfined with a base at $Z_b = 0$ and the hydraulic conductivity is uniform, then

$$T(\phi) = \int_0^\phi k d\phi = k\phi \tag{11.11}$$

so that (11.8) yields:

$$\Phi = \int k\phi d\phi + C = \tfrac{1}{2}k\phi^2 + C \qquad (11.12)$$

Flow in an Aquifer with an Arbitrary Number of Layers

Aquifers often have a layered structure, and the approach presented in this section makes it possible to include the effect of these layers on the flow in an approximate fashion. It should be kept in mind, however, that the analysis is valid only if the differences in the hydraulic conductivities are small enough that variations of the head in the vertical direction can be neglected. We will consider the case that the aquifer is bounded between two parallel impervious horizontal boundaries, but allow for combined confined-unconfined flow.

We number the layers as $1, 2, \ldots, n$, where layer 1 is the bottom layer, represent the base of layer j as b_j, and denote the base of the upper confining bed as b_{n+1}. The thickness, H_n, of layer j may thus be written as

$$H_j = b_{j+1} - b_j \qquad (j = 1, 2, \ldots, n) \qquad (11.13)$$

The hydraulic conductivity is uniform within each layer, and is represented for layer j as k_j. If layer j is fully saturated, then its transmissivity is T_j, with

$$T_j = k_j H_j = k_j(b_{j+1} - b_j) \qquad (11.14)$$

The head at any point of the aquifer may vary between the base elevation (b_1) and values larger than the elevation of the upper confining bed (b_{n+1}). If ϕ is equal to some value between b_m and b_{m+1} ($m = 1, 2, \ldots, n$), then the layers $m+1, m+2, \ldots, n$ do not contribute to the transmissivity. Hence, if T represents the total transmissivity of the aquifer, then

$$T = \sum_{j=1}^{m-1} T_j + k_m(\phi - b_m) \qquad (b_m \leq \phi \leq b_{m+1}) \qquad (11.15)$$

We will adopt the convention that summations are not to be carried out (i.e., are set to zero) if the upper bound ($m - 1$ in this case) is less than the lower one. The potential Φ may be expressed in terms of ϕ by the use of (11.8) and (11.15) as follows:

$$\Phi = \int \left[\sum_{j=1}^{m-1} T_j + k_m(\phi - b_m) \right] d\phi + C_m \qquad (b_m \leq \phi \leq b_{m+1}) \qquad (11.16)$$

Integration yields

$$\Phi = \sum_{j=1}^{m-1} T_j \phi + \tfrac{1}{2} k_m (\phi - b_m)^2 + C_m \qquad (b_m \leq \phi \leq b_{m+1}) \qquad (11.17)$$

where the index m in the constant C_m denotes that the aquifer is saturated up to a level somewhere in layer m.

We must now choose the constants C_m ($m = 1, 2, \ldots, n$) in such a way that the potential varies continuously if the head varies continuously between all possible levels in the aquifer. We will see that this is accomplished if we choose C_m such that (11.17) becomes:

$$\Phi = \sum_{j=1}^{m-1} \left[T_j(\phi - b_j) - \tfrac{1}{2}k_j(b_{j+1} - b_j)^2 \right] + \tfrac{1}{2}k_m(\phi - b_m)^2 \quad (b_m \leq \phi \leq b_{m+1}) \quad (11.18)$$

For the special case that $m = 1$, (11.18) reduces to

$$\Phi = \tfrac{1}{2}k_1(\phi - b_1)^2 \quad (b_1 \leq \phi \leq b_2) \quad (11.19)$$

We will demonstrate that the potential (11.18) indeed varies continuously across the interzonal boundary where the groundwater table intersects the base of layer m. We determine the potential for the case that $b_{m-1} \leq \phi \leq b_m$ ($m > 1$), and then show that the resulting expression for the potential is equal to (11.18) for the limiting case that $\phi = b_m$. If $b_{m-1} \leq \phi \leq b_m$ then (11.18) becomes

$$\Phi = \sum_{j=1}^{m-2} \left[T_j(\phi - b_j) - \tfrac{1}{2}k_j(b_{j+1} - b_j)^2 \right] + \tfrac{1}{2}k_{m-1}(\phi - b_{m-1})^2$$

$$(b_{m-1} \leq \phi \leq b_m \quad m > 1) \quad (11.20)$$

If $\phi = b_m$, then (11.18) and (11.19) become

$$\Phi_m = \sum_{j=1}^{m-1} \left[T_j(b_m - b_j) - \tfrac{1}{2}k_j(b_{j+1} - b_j)^2 \right] \quad (m > 1)$$

$$\Phi_1 = 0 \quad (11.21)$$

where Φ_m represents the value of the potential for $\phi = b_m$. With $\phi = b_m$, (11.20) becomes

$$\Phi_m = \sum_{j=1}^{m-2} \left[T_j(b_m - b_j) - \tfrac{1}{2}k_j(b_{j+1} - b_j)^2 \right] + \tfrac{1}{2}k_{m-1}(b_m - b_{m-1})^2 \quad (m > 1) \quad (11.22)$$

It follows from (11.14) that

$$\tfrac{1}{2}k_{m-1}(b_m - b_{m-1})^2 = T_{m-1}(b_m - b_{m-1}) - \tfrac{1}{2}k_{m-1}(b_m - b_{m-1})^2 \quad (11.23)$$

Thus, (11.22) is indeed identical to (11.21): the potential varies continuously across the interzonal boundary where the phreatic surface passes through the base of layer m, as asserted. Since m may be any number between 1 and n, the potential (11.18) is valid for any position of the phreatic surface below the upper boundary.

Sec. 11   Aquifers with Vertically Varying Hydraulic Conductivity

If the aquifer is entirely confined, then the potential becomes

$$\Phi = \sum_{j=1}^{n} \left[T_j(\phi - b_j) - \tfrac{1}{2} k_j (b_{j+1} - b_j)^2 \right] \qquad (b_{n+1} \leq \phi) \qquad (11.24)$$

We can use the potential given by either (11.18) or (11.24) to model flow in a layered system; the potential obeys either the Laplace or the Poisson equation, depending upon whether or not there is infiltration. It can be obtained in the usual manner, provided that we use the appropriate expression to transform boundary conditions given in terms of heads into boundary conditions in terms of the potential. Once the potential is known as a function of position, we must invert either (11.18) or (11.24) to compute the head ϕ. If $\Phi > \Phi_{n+1}$, where Φ_{n+1} is obtained from (11.24) as,

$$\Phi_{n+1} = \sum_{j=1}^{n} \left[T_j(b_{n+1} - b_j) - \tfrac{1}{2} k_j (b_{j+1} - b_j)^2 \right] \qquad (11.25)$$

then the flow is confined in all layers. The head, then, may be expressed in terms of the potential by the use of (11.24) as follows:

$$\phi = \frac{\Phi + \sum_{j=1}^{n} \left[T_j b_j + \tfrac{1}{2} k_j (b_{j+1} - b_j)^2 \right]}{\overset{n}{T}} \qquad (\Phi > \Phi_{n+1}) \qquad (11.26)$$

where

$$\overset{n}{T} = \sum_{j=1}^{n} T_j \qquad (11.27)$$

Equation (11.26) may be written in a slightly different form as

$$\phi = \frac{\Phi - \sum_{j=1}^{n} \left[T_j(b_{n+1} - b_j) - \tfrac{1}{2} k_j (b_{j+1} - b_j)^2 \right]}{\overset{n}{T}} + \frac{\sum_{j=1}^{n} T_j b_{n+1}}{\overset{n}{T}} \qquad (\Phi > \Phi_{n+1}) \qquad (11.28)$$

or

$$\phi = b_{n+1} + \frac{\Phi - \Phi_{n+1}}{\overset{n}{T}} \qquad (\Phi > \Phi_{n+1}) \qquad (11.29)$$

where use is made of (11.25) and (11.27).

If the potential has some value between Φ_m and Φ_{m+1}, then (11.18) applies. We write this equation in a different form:

$$\Phi = \sum_{j=1}^{m-1} \left\{ T_j[(\phi - b_m) + (b_m - b_j)] - \tfrac{1}{2} k_j (b_{j+1} - b_j)^2 \right\} + \tfrac{1}{2} k_m (\phi - b_m)^2$$

$$(\Phi_m \leq \Phi \leq \Phi_{m+1}) \qquad (11.30)$$

In view of (11.21) this may be written as

$$\Phi = \sum_{j=1}^{m-1} T_j(\phi - b_m) + \Phi_m + \tfrac{1}{2} k_m(\phi - b_m)^2 \qquad (\Phi_m \leq \Phi \leq \Phi_{m+1}) \qquad (11.31)$$

or

$$\tfrac{1}{2} k_m(\phi - b_m)^2 + \sum_{j=1}^{m-1} T_j(\phi - b_m) - (\Phi - \Phi_m) = 0 \qquad (\Phi_m \leq \Phi \leq \Phi_{m+1}) \qquad (11.32)$$

The two roots of this quadratic equation in terms of $(\phi - b_m)$ are

$$(\phi - b_m)_{1,2} = \frac{-\overset{m-1}{T} \pm \sqrt{\left(\overset{m-1}{T}\right)^2 + 2 k_m(\Phi - \Phi_m)}}{k_m} \qquad (\Phi_m \leq \Phi \leq \Phi_{m+1}) \qquad (11.33)$$

where $\overset{m-1}{T}$ is defined by (11.27) (with $m-1$ replacing n). Since $\phi - b_m$ is positive, the + sign applies, so that

$$\phi = b_m - \frac{\overset{m-1}{T}}{k_m} + \sqrt{\left(\frac{\overset{m-1}{T}}{k_m}\right)^2 + \frac{2(\Phi - \Phi_m)}{k_m}} \qquad (\Phi_m \leq \Phi \leq \Phi_{m+1}) \qquad (11.34)$$

The analysis is now complete. The equations are suitable for evaluation by hand only if the number of layers is small. The equations are, however, suitable for implementation in a computer program: at the end of this section we will modify SLW for layered aquifers.

Girinski Potentials

The potentials introduced by Girinski are rather similar to the potential (11.8); they differ in form and do not contain the additive constant, but otherwise render the same result.

For the derivation of Girinski potentials use is made of Leibnitz's rule for differentiation of a definite integral (see, for example, Abramowitz and Stegun [1965]):

$$\frac{\partial}{\partial x} \int_{Z_b}^{Z_t} F(x,y,z)dz = \int_{Z_b}^{Z_t} \frac{\partial}{\partial x} F(x,y,z)dz + F(x,y,Z_t)\frac{\partial Z_t}{\partial x} - F(x,y,Z_b)\frac{\partial Z_b}{\partial x} \qquad (11.35)$$

Since $k(z)$ is independent of x and y, the integrands in (11.3) may be interpreted as the derivatives $\partial F/\partial x$ and $\partial F/\partial y$, and the application of Leibnitz's rule to (11.3) yields

$$Q_x = -\frac{\partial}{\partial x} \int_{Z_b}^{Z_t} k(z)\phi\, dz + k(Z_t)\phi_t \frac{\partial Z_t}{\partial x} - k(Z_b)\phi_b \frac{\partial Z_b}{\partial x}$$
$$Q_y = -\frac{\partial}{\partial y} \int_{Z_b}^{Z_t} k(z)\phi\, dz + k(Z_t)\phi_t \frac{\partial Z_t}{\partial y} - k(Z_b)\phi_b \frac{\partial Z_b}{\partial y} \qquad (11.36)$$

Sec. 11   Aquifers with Vertically Varying Hydraulic Conductivity

where

$$\phi_t = \phi(x, y, Z_t) \qquad \phi_b = \phi(x, y, Z_b) \tag{11.37}$$

Note that these expressions are valid for three-dimensional flow. Girinski potentials are obtained by adopting the Dupuit-Forchheimer assumption. If the flow is unconfined then $Z_t = h$, $Z_b = 0$, and $\phi = \phi_t = h$ so that (11.36) may be written as

$$\begin{aligned}
Q_x &= -\frac{\partial}{\partial x} \int_0^\phi k(z)\phi\, dz + k(\phi)\phi \frac{\partial \phi}{\partial x} = -\frac{\partial}{\partial x}\int_0^\phi k(z)(\phi - z)\, dz \\
Q_y &= -\frac{\partial}{\partial y}\int_0^\phi k(z)\phi\, dz + k(\phi)\phi\frac{\partial \phi}{\partial y} = -\frac{\partial}{\partial y}\int_0^\phi k(z)(\phi - z)\, dz
\end{aligned} \tag{11.38}$$

Note that Leibnitz's rule is applied in reverse order to write the second term in each equation as the derivative of an integral. The Girinski potential thus becomes

$$\Phi = \int_0^h k(z)(\phi - z)\, dz \tag{11.39}$$

If k is constant the familiar potential for unconfined flow is obtained:

$$\Phi = k\phi^2 - \tfrac{1}{2}k\phi^2 = \tfrac{1}{2}k\phi^2 \tag{11.40}$$

Equation (11.39) is applicable to cases of confined flow provided that the upper limit in the integral is set equal to the aquifer thickness H. That this is true does not follow from (11.36) directly: setting $Z_t = H$ and $Z_b = 0$ yields only the first term of the integrand in (11.39). However, if $h = H$ the second term in this integrand yields a constant upon integration and thus may be included. This constant, fortuitously, makes the potential continuous with (11.39) across the interzonal boundary. Setting $h = H$ in (11.39) we obtain for $k(z) = k = $ constant:

$$\Phi = kH\phi - \tfrac{1}{2}kH^2 \tag{11.41}$$

Girinski also presented a potential for interface flow (Girinski, 1947) but this function is not continuous with (11.40) across the interzonal boundary with an unconfined zone.

Youngs [1971] did not adopt the Dupuit-Forchheimer assumption, and expressed ϕ in (11.36) in terms of the pressure p by the use of the relation $\phi = p/(\rho g) + z$. This yields the following for Q_x:

$$\begin{aligned}
Q_x = &-\frac{\partial}{\partial x}\int_{Z_b}^{Z_t} k(z)\frac{p}{\rho g}\, dz - \frac{\partial}{\partial x}\int_{Z_b}^{Z_t} z k(z)\, dz \\
&+ k(Z_t)\left(\frac{p_t}{\rho g} + Z_t\right)\frac{\partial Z_t}{\partial x} - k(Z_b)\left(\frac{p_b}{\rho g} + Z_b\right)\frac{\partial Z_b}{\partial x}
\end{aligned} \tag{11.42}$$

where

$$p_t = p(x, y, Z_t) \qquad p_b = p(x, y, Z_b) \tag{11.43}$$

Application of Leibnitz's rule to the second integral in (11.42) yields after rearrangement:

$$Q_x = -\frac{\partial}{\partial x}\int_{Z_b}^{Z_t} k(z)\frac{p}{\rho g}dz + k(Z_t)\frac{p_t}{\rho g}\frac{\partial Z_t}{\partial x} - k(Z_b)\frac{p_b}{\rho g}\frac{\partial Z_b}{\partial x} \qquad (11.44)$$

If the flow is unconfined, $p_t = 0$ and we obtain

$$Q_x = -\frac{\partial}{\partial x}\int_{Z_b}^{Z_t} k(z)\frac{p}{\rho g}dz - k(Z_b)\frac{p_b}{\rho g}\frac{\partial Z_b}{\partial x} \qquad (11.45)$$

A similar expression results for Q_y. Youngs applied (11.45) together with the condition of continuity of flow to establish an approximate method for modeling flow in aquifers with lower boundaries of any shape.

Of particular interest is that (11.45) is exact: no approximations have been made. For the case of interface flow, p_b may be expressed in terms of Z_b, and the right-hand side of (11.45) may be written as the derivative of a potential. For one-dimensional flow, Q_x is a constant, equal to the difference in values of the potential at the boundaries. If these boundaries are vertical and the pressures along them are known, then the exact value of Q_x can be computed. In this way Youngs demonstrated that (10.61) is exact.

Problem 11.1

Solve the problem of one-dimensional horizontal confined flow in an aquifer system consisting of two layers, each of thickness $\frac{1}{2}H$. The boundary conditions are

$$x = -\tfrac{1}{2}L \quad \phi = \phi_1 \qquad x = \tfrac{1}{2}L \quad \phi = \phi_2$$

The hydraulic conductivity in the upper layer is k_1 and in the lower layer is k_2. The conditions along any interface between two layers of different hydraulic conductivities are that there is continuity both of flow and of head across the interface. Solve the problem in each layer separately and demonstrate that the conditions across the interface are met. Determine the discharge through each layer and the total discharge through the system. Compare the latter answer by solving the problem by the use of the potential obtained from (11.8). Is the answer approximate or exact?

Implementation of Layering in SLW

Making SLW suitable for the simulation of flow in a layered aquifer requires the implementation of the expressions that convert potentials to heads and vice versa in the module AQUIFER. It further requires that the common block CBAQ be modified to communicate the new aquifer parameters between the various functions and subroutines, and that the input of these parameters be accommodated in SETUP.

The Module Aquifer

The modified module AQUIFER follows exactly the analysis presented in this section; some of the comment statements in the listing that follows refer to the equations in the text by number.

Sec. 11 Aquifers with Vertically Varying Hydraulic Conductivity

```
c      Initialization of common block with aquifer data.
c      Set aquifer data to default values.
c      NAQLAY   = number of layers in the aquifer.
c      RAQK(I)  = hydraulic conductivity of layer I.
c      RAQT(I)  = transmissivity of layer I.
c      RAQB(I)  = elevation of the base of layer I.
c      RAQPHI(I) = value of the potential corresponding to a head equal
c                 to the elevation of the base of aquifer I.
c      RAQTSM   = transmissivity of the aquifer if it is entirely
c                 saturated.
c      NAQLMX   = maximum number of layers, given by the dimension
c                 statement in the common block.
       BLOCK DATA BDAQ
       COMMON /CBAQ/ NAQLAY,RAQK(20),RAQB(21),RAQT(20),RAQPHI(21),
      .              RAQTSM,NAQLMX
       DATA NAQLAY /1/
       DATA RAQK(1),RAQB(1),RAQB(2),RAQT(1),RAQPHI(1),RAQPHI(2),RAQTSM
      . / 1.     ,.0    ,1.    ,1.    ,.0     ,.5      ,1.   /
       DATA NAQLMX /20/
       END
c      Computation of RAQTSM, and (for all layers) of RAQT(I) and
c      of  RAQPHI(I).
       SUBROUTINE AQINIT
       COMMON /CBAQ/ NAQLAY,RAQK(20),RAQB(21),RAQT(20),RAQPHI(21),
      .              RAQTSM,NAQLMX
       RAQTSM=.0
       DO 100 ILAYER=1,NAQLAY
          RAQT(ILAYER)=RAQK(ILAYER)*(RAQB(ILAYER+1)-RAQB(ILAYER))
          RAQTSM=RAQTSM+RAQT(ILAYER)
  100  CONTINUE
       DO 200 ILAYER=1,NAQLAY+1
          RAQPHI(ILAYER)=RFAQPH(ILAYER)
  200  CONTINUE
       RETURN
       END
c      Function for computing the value of RAQPHI(M) (see [11.21]).
       REAL FUNCTION RFAQPH(M)
       COMMON /CBAQ/ NAQLAY,RAQK(20),RAQB(21),RAQT(20),RAQPHI(21),
      .              RAQTSM,NAQLMX
       RFAQPH=.0
       IF(M.EQ.1) RETURN
       DO 100 J=1,M-1
          RFAQPH=RFAQPH+RAQT(J)*(RAQB(M)-RAQB(J))
      .         -.5*RAQK(J)*(RAQB(J+1)-RAQB(J))**2
```

```
  100   CONTINUE
        RETURN
        END
c       Function to convert head RFI to potential RFPOTH.
        REAL FUNCTION RFPOTH(RFI)
        COMMON /CBAQ/ NAQLAY,RAQK(20),RAQB(21),RAQT(20),RAQPHI(21),
       .              RAQTSM,NAQLMX
        RFPOTH=.0
c       Return 0 if the head is below the base of the aquifer.
        IF(RFI.LE.RAQB(1)) RETURN
c       Confined flow in all layers, apply (11.24).
        IF(RFI.GE.RAQB(NAQLAY+1)) THEN
          RFPOTH=RFAQPC(RFI,NAQLAY)
          RETURN
        ENDIF
c       Unconfined conditions occur; find the layer that contains the
c       water table, then compute the potential, using (11.31).
        DO 100 ILAYER=1,NAQLAY
          IF(RFI.GE.RAQB(ILAYER).AND.RFI.LT.RAQB(ILAYER+1)) THEN
            RFPOTH=.5*RAQK(ILAYER)*(RFI-RAQB(ILAYER))**2
            IF(ILAYER.GT.1) RFPOTH=RFPOTH+RFAQPC(RFI,ILAYER-1)
            RETURN
          ENDIF
  100   CONTINUE
        RETURN
        END
c       Function for computing the contributions of all fully saturated
c       layers to the potential.
        REAL FUNCTION RFAQPC(RFI,ILAYER)
        COMMON /CBAQ/ NAQLAY,RAQK(20),RAQB(21),RAQT(20),RAQPHI(21),
       .              RAQTSM,NAQLMX
c       First determine the sum of the transmissivities of all fully
c       saturated layers.
        RTJ=.0
        DO 100 J-1,ILAYER
          RTJ=RTJ+RAQT(J)
  100   CONTINUE
c       Next compute the contribution of all fully saturated layers
c       (the first 2 terms in [11.31]).
        RFAQPC=RTJ*(RFI-RAQB(ILAYER+1))+RAQPHI(ILAYER+1)
        RETURN
        END
c       Function to convert potential RPHI to head RFHEDP.
        REAL FUNCTION RFHEDP(RPOT)
```

```
      COMMON /CBAQ/ NAQLAY,RAQK(20),RAQB(21),RAQT(20),RAQPHI(21),
     .              RAQTSM,NAQLMX
      RFHEDP=RAQB(1)
c     Return elevation of base if the potential is <= 0.
      IF(RPOT.LE..0) RETURN
c     Confined flow in all layers; use (11.29)
      IF(RPOT.GE.RAQPHI(NAQLAY+1)) THEN
        RFHEDP=RAQB(NAQLAY+1)+(RPOT-RAQPHI(NAQLAY+1))/RAQTSM
        RETURN
      ENDIF
c     Unconfined conditions occur; first determine the sum of
c     the transmissivities of all fully saturated layers.
      DO 100 ILAYER=1,NAQLAY
c        find the layer that contains the water table, then compute the
c        head, using (11.34).
         IF(RPOT.GE.RAQPHI(ILAYER).AND.RPOT.LT.RAQPHI(ILAYER+1)) THEN
           RTILM1=.0
           DO 50 J=1,ILAYER-1
             RTILM1=RTILM1+RAQT(J)
  50       CONTINUE
           RFHEDP=RAQB(ILAYER)-RTILM1/RAQK(ILAYER)
     .           +SQRT((RTILM1/RAQK(ILAYER))**2
     .           +2.*(RPOT-RAQPHI(ILAYER))/RAQK(ILAYER))
           RETURN
         ENDIF
 100  CONTINUE
      RETURN
      END
```

The Routine Setup

The modifications in SETUP are as follows: The command word LAYER is added to provide for modification of the hydraulic conductivity and thickness of any layer. The command AQUIFER remains in place; however, as arguments we now require only the number of layers for which parameters are to be entered and the elevation of the aquifer base. We then add statements that cause the program to prompt the user to enter the hydraulic conductivities and thicknesses of the layers in sequence. The user can later modify the parameters of any of the layers by entering the command LAYER followed by the layer number and a new hydraulic conductivity and thickness. After any data regarding the layers have been entered, the routine AQINIT is called to ensure that the various constants necessary for the computations are set. Both the modified and the new statements in SETUP follow.

The line containing the dimension statement of AWORD, the list of command words including the new command LAYER, and the modified computed GOTO statement are

```
      DIMENSION AWORD(80),AWOGR(10)
      DATA AWORD   /'A','Q','U','I',' ',
     .              'L','A','Y','E',' ',
     .              'R','E','F','E',' ',
     .              'U','N','I','F',' ',
     .              'R','A','I','N',' ',
     .              'P','O','N','D',' ',
     .              'W','E','L','L',' ',
     .              'S','O','L','V',' ',
     .              'W','I','N','D',' ',
     .              'G','R','I','D',' ',
     .              'P','L','O','T',' ',
     .              'L','A','Y','O',' ',
     .              'C','H','E','C',' ',
     .              'P','A','U','S',' ',
     .              'S','W','I','T',' ',
     .              'S','T','O','P','!'/
      GOTO( 100, 160, 200, 300, 320, 340, 400, 500, 600, 700, 800, 900,
     .      1000,1100,1200,8900),JUMP
```

The statements for entering aquifer data become

```
c     Aquifer data; First read the number of layers and the elevation
c     of the base of the aquifer.
 100  NAQLAY=IVAR(2)
      IF(NAQLAY.GT.NAQLMX) THEN
        WRITE(*,9006) NAQLMX
        GOTO 10
      ENDIF
      RAQB(1)=RVAR(3)
      IF(LMISS.OR.LERROR) GOTO 10
c     Prompt the user to enter the hydraulic conductivity and thickness
c     of each layer.
      ILAYER=1
 110  WRITE(*,9007) ILAYER
      READ(5,9000) ALINE
      CALL TIDY
      RAQK(ILAYER)=RVAR(1)
      RAQB(ILAYER+1)=RAQB(ILAYER)+RVAR(2)
      IF(LMISS.OR.LERROR) THEN
c     If an error occurred, prompt the user to enter the data for this
c     layer again.
        WRITE(*,9008) ILAYER
      ELSE
c     Otherwise, add 1 to the number of layers.
```

Sec. 11   Aquifers with Vertically Varying Hydraulic Conductivity   **125**

```
              ILAYER=ILAYER+1
          ENDIF
c     If the data set for the layers is not yet complete, read the
c     data for the next layer.
          IF(ILAYER.LE.NAQLAY) GOTO 110
c     Initialize the aquifer parameters.
          CALL AQINIT
          GOTO 10
c     Overwrite the data of layer nr. ILAYER.
  160     ILAYER=IVAR(2)
          RAQK(ILAYER)=RVAR(3)
          RAQB(ILAYER+1)=RAQB(ILAYER)+RVAR(4)
c     Initialize the aquifer parameters.
          CALL AQINIT
          GOTO 10
```

The following format statements have either been modified or added:

```
c     command line:
 9005 FORMAT(
     .'+<AQUIFER>(NLAYER,BASE)'/
     .' <LAYER>(NLAYER,PERM,THICKNESS)'/
     .' <POND>(XCTR,YCTR,RAD,INF.RATE)'/
     .' <UNIFLOW>(Q,ANGLE)'/
     .' <RAIN>(INF.RATE,XCTR,YCTR,A AXIS,B AXIS,ANGLE)'/
     .' <POND>(XCTR,YCTR,RAD,INF.RATE)'/
     .' <WELL>(X,Y,Q,R)'/
     .' <REFERENCE>(X,Y,FI)'/
     .' <SOLVE>'/
     .' <WINDOW>(X1,Y1 X2,Y2)'/
     .' <LAYOUT>'/
     .' <GRID>(N)[<HEAD>/<POTENTIAL>]'/
     .' <PLOT>'/
     .' <CHECK>'/
     .' <SWITCH>(FILEIN)[FILEOUT]'/
     .' <PAUSE>'/
     .' <STOP>'/)
c     Messages associated with layering.
 9006 FORMAT('++++MAXIMUM NR. OF LAYERS (= ',I4,') EXCEEDED+++'/)
 9007 FORMAT(
     .'+ENTER HYDRAULIC CONDUCTIVITY AND THICKNESS FOR LAYER',I4/)
 9008 FORMAT(
     .'++++ILLEGAL OR MISSING PARAMETERS; RE-ENTER DATA FOR LAYER',I4/)
```

The Routine Check

The modifications in CHECK are limited to WRITE and FORMAT statements:

```
c      Aquifer data.
 100   WRITE(7,9010)
       DO 190 ILAYER=1,NAQLAY
         RTHICK=RAQB(ILAYER+1)-RAQB(ILAYER)
         WRITE(7,9011) ILAYER,RAQK(ILAYER),RAQB(ILAYER),RTHICK,
      .               RAQT(ILAYER)
 190   CONTINUE
       GOTO 10
 9010  FORMAT(
      .'   LAYER   HYDR.CONDUCT.  BASE           THICKNESS       ',
      .'TRANSMISSIVITY')
 9011  FORMAT(I3,3X,4E14.6)
```

Problem 11.2

Save the version of SLW created in Sec. 9, then implement the layering. Check the new routines for accuracy in steps as follows:

1. Run the former version of the program with $k = 1$ and $H = 1$, with a uniform flow of discharge rate 1 in the x direction, and with the head set to 0 at $(1, 0)$. Create a plot of piezometric contours for a window defined by the points $(-1, -1)$ and $(1, 1)$. Use CHECK to compute the head at $(-0.5, 0)$, $(0, 0)$, and $(0.5, 0)$. Repeat this for the new version of the program, subdividing the aquifer in 10 layers, each of hydraulic conductivity 1 and thickness .1. Compare your results; they should be the same.

2. Run the program again with 10 layers, each of thickness .1, but now with hydraulic conductivities equal to $1, 2, \ldots, 10$ (10 corresponds to the uppermost layer). Enter a uniform flow with a discharge of 1 in the x direction and a well of discharge 2.5 at $(0, 0)$. Produce piezometric contours, and verify that there are no discontinuities. Check, at least for one layer, that the head is continuous where the phreatic surface intersects the top of the layer by using the command HEAD.

3. Solve the problem of uniform unconfined flow in the x direction for an aquifer consisting of a bottom layer of thickness $\frac{1}{2}$ and hydraulic conductivity 1 and an upper layer of thickness $\frac{1}{2}$ and hydraulic conductivity 5. Set the heads at the planes $x = -10$ and $x = 10$ equal to 1 and 0, respectively. Determine the head at 19 points distributed uniformly along the x axis. Run the computer program and compare your results.

12 SHALLOW FLOW IN AQUIFERS WITH CLAY LAMINAE

Aquifers often contain thin clay laminae. These laminae have little effect on the flow if they are limited in extent and if the flow is predominantly horizontal. However, we should take the effect of laminae into account if they cover large areas, so that the flow in the aquifers above and below the laminae may be quite different. We will at first assume that the laminae have negligible thickness, that they are impermeable, and horizontal. We will further assume that the base of the aquifer is horizontal, that hydraulic conductivities above and below the laminae are both equal to k, and that the laminae do not overlap. An example of such an aquifer is given in Figure 2.41.

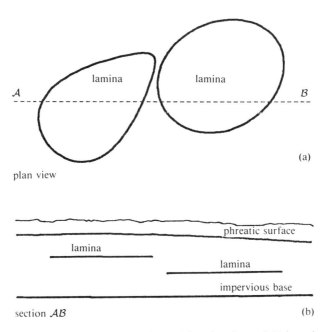

Figure 2.41 An aquifer with clay laminae; (a) is a plan view and (b) is section AB.

The approach presented here is based upon the Dupuit-Forchheimer assumption that the resistance to flow in the vertical direction can be neglected and follows Strack [1981(a)]. In this approach, use is made of a formulation whereby the aquifer system is treated as a single unit, with its own discharge vector and potential.

We will refer to a region of the aquifer where there is no lamina as a single-aquifer zone, and to a region where a lamina is present as a dual-aquifer zone. All variables associated with the single-aquifer zone will be denoted by the superscript s, and those associated with the upper and lower aquifers in a dual-aquifer zone will be denoted by the superscripts u and l, respectively. The head in any aquifer or layer is measured with respect to the base of the aquifer system.

Basic Equations and Method of Solution

We will begin by writing the basic equations first for the dual-aquifer zone, considering the upper and lower aquifers separately, and review the equations for the single-aquifer zone. These formulations then are combined to cover the flow in the entire aquifer system.

The Upper Aquifer

The upper aquifer in the dual-aquifer zone is unconfined, with the lamina serving as its base. Denoting the elevation of the lamina above the base of the aquifer system as H (see Figure 2.42), the saturated thickness of the upper aquifer as $\overset{u}{h}$, and the head as $\overset{u}{\phi}$ we may write,

$$\overset{u}{h} = \overset{u}{\phi} - H \tag{12.1}$$

The potential $\overset{u}{\Phi}$ for the upper aquifer is found from (7.8), where we replace ϕ by $\overset{u}{h}$, choose C_u as zero, and use (12.1):

$$\boxed{\overset{u}{\Phi} = \tfrac{1}{2}k(\overset{u}{\phi} - H)^2} \tag{12.2}$$

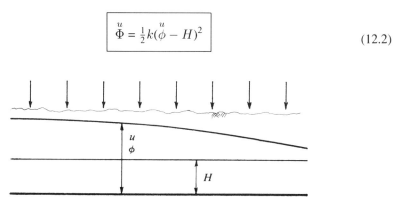

Figure 2.42 The upper aquifer.

The discharge vector in the upper layer is represented as $(\overset{u}{Q}_x, \overset{u}{Q}_y)$, and equals minus the gradient of $\overset{u}{\Phi}$:

$$\begin{aligned}\overset{u}{Q}_x &= -\frac{\partial \overset{u}{\Phi}}{\partial x} \\ \overset{u}{Q}_y &= -\frac{\partial \overset{u}{\Phi}}{\partial y}\end{aligned} \tag{12.3}$$

If there is infiltration of rate N, the differential equation for $\overset{u}{\Phi}$ is given by (9.3):

$$\nabla^2 \overset{u}{\Phi} = -N \tag{12.4}$$

The boundary values to be applied to this differential equation are the heads $\overset{u}{\phi}$ along the perimeter of the lamina; at present these heads are unknown, and we will proceed with describing the flow in the other aquifers of the system before attempting to specify these boundary values.

Sec. 12   Shallow Flow in Aquifers with Clay Laminae

The Lower Aquifer

We denote the head and potential in the lower aquifer in the dual-aquifer zone as $\overset{l}{\phi}$ and $\overset{l}{\Phi}$, respectively. Allowing for unconfined conditions to occur ($\overset{l}{\phi} < H$), we adopt the definitions (8.6) and (8.7):

$$\boxed{\begin{aligned} \overset{l}{\Phi} &= kH\overset{l}{\phi} - \tfrac{1}{2}kH^2 & (\overset{l}{\phi} \geq H) \\ \overset{l}{\Phi} &= \tfrac{1}{2}k\overset{l}{\phi}{}^2 & (\overset{l}{\phi} < H) \end{aligned}} \qquad (12.5)$$

The components $\overset{l}{Q}_x$ and $\overset{l}{Q}_y$ of the discharge vector in the lower aquifer are equal to minus the gradient of $\overset{l}{\Phi}$ (compare [8.8] and [8.9]),

$$\begin{aligned} \overset{l}{Q}_x &= -\frac{\partial \overset{l}{\Phi}}{\partial x} \\ \overset{l}{Q}_y &= -\frac{\partial \overset{l}{\Phi}}{\partial y} \end{aligned} \qquad (12.6)$$

and the potential fulfills the differential equation (8.11),

$$\nabla^2 \overset{l}{\Phi} = 0 \qquad (12.7)$$

The Single-Aquifer Zone

The flow in the single-aquifer zone is shallow unconfined flow with infiltration; the potential $\overset{s}{\Phi}$ equals

$$\boxed{\overset{s}{\Phi} = \tfrac{1}{2}k\phi^2} \qquad (12.8)$$

and the components of the discharge vector $(\overset{s}{Q}_x, \overset{s}{Q}_y)$ may be represented as

$$\begin{aligned} \overset{s}{Q}_x &= -\frac{\partial \overset{s}{\Phi}}{\partial x} \\ \overset{s}{Q}_y &= -\frac{\partial \overset{s}{\Phi}}{\partial y} \end{aligned} \qquad (12.9)$$

The potential $\overset{s}{\Phi}$ fulfills the differential equation (9.3) for unconfined flow with rainfall:

$$\nabla^2 \overset{s}{\Phi} = -N \qquad (12.10)$$

The Comprehensive Discharge Vector and the Comprehensive Potential

We introduce the comprehensive discharge vector $(\overset{d}{Q}_x, \overset{d}{Q}_y)$ in the dual-aquifer zone as the vectorial sum of the discharge vectors in the upper and lower aquifers:

$$\overset{d}{Q}_x = \overset{u}{Q}_x + \overset{l}{Q}_x$$
$$\overset{d}{Q}_y = \overset{u}{Q}_y + \overset{l}{Q}_y \tag{12.11}$$

It follows from (12.3) and (12.6) that

$$\overset{d}{Q}_x = -\frac{\partial(\overset{u}{\Phi} + \overset{l}{\Phi})}{\partial x} = -\frac{\partial \overset{d}{\Phi}}{\partial x}$$
$$\overset{d}{Q}_y = -\frac{\partial(\overset{u}{\Phi} + \overset{l}{\Phi})}{\partial y} = -\frac{\partial \overset{d}{\Phi}}{\partial y} \tag{12.12}$$

We refer to $\overset{d}{\Phi}$ as the comprehensive potential in the dual-aquifer zone. It follows from (12.2) and (12.5) that

$$\overset{d}{\Phi} = \tfrac{1}{2}k(\overset{u}{\phi} - H)^2 + kH\overset{l}{\phi} - \tfrac{1}{2}kH^2 \qquad (\overset{l}{\phi} \geq H)$$
$$\overset{d}{\Phi} = \tfrac{1}{2}k(\overset{u}{\phi} - H)^2 + \tfrac{1}{2}k\overset{l}{\phi}^2 \qquad (\overset{l}{\phi} < H) \tag{12.13}$$

The Conditions at the Edge of the Lamina

Thus far, we considered the dual- and single-aquifer zones separately. In solving the problem of combined flow, we must consider the conditions to be applied along the perimeter of the lamina. In the first place, we observe that there must be continuity of flow across this perimeter: the components $\overset{s}{Q}_n$ and $\overset{d}{Q}_n$ normal to the edge of the lamina must be equal,

$$\overset{s}{Q}_n = \overset{d}{Q}_n \tag{12.14}$$

In the second place, we make the assumption that the heads in the upper and lower aquifers along the lamina edge are equal to that in the single-aquifer zone if the head ϕ is greater than H, as in Figure 2.41. This assumption is in keeping with the Dupuit-Forchheimer assumption that the resistance to flow in a vertical direction may be neglected. Labeling heads and potentials along the edge of the lamina by the subscript e, the following equation may be written:

$$\overset{u}{\phi}_e = \overset{l}{\phi}_e = \phi_e \qquad (\phi \geq H) \tag{12.15}$$

Sec. 12   Shallow Flow in Aquifers with Clay Laminae

If the conditions along the edge of the lamina are unconfined, the head in the upper aquifer is set to its lowest possible value (H) and the head in the lower aquifer is set equal to that in the single aquifer zone:

$$\overset{u}{\underset{e}{\phi}} = H \qquad \overset{l}{\underset{e}{\phi}} = \underset{e}{\phi} \qquad (\underset{e}{\phi} < H) \tag{12.16}$$

This condition is an approximation of the actual situation shown in Figure 2.43(a): water flowing over the lamina edge percolates down through an unsaturated zone below the lamina. In writing (12.16), however, we assume that the phreatic surface in the lower aquifer separates dry soil with atmospheric pressure from saturated soil as illustrated in Figure 2.43(b). In this assumption we follow investigators analyzing flow in aquifers with semipermeable clay laminae (Mualem and Bear [1974], Brock [1976]).

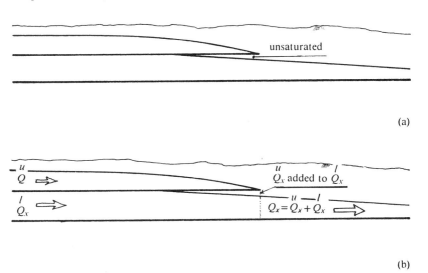

Figure 2.43 Unconfined flow underneath the lamina. (Source: Adapted from Strack [1981(a)].)

Continuity of the Comprehensive Potential

We next investigate the behavior of the potentials $\overset{s}{\Phi}$ and $\overset{d}{\Phi}$ in the single- and dual-aquifer zones along the perimeter of the lamina. The potential $\overset{d}{\underset{e}{\Phi}}$ along the edge of the lamina may be expressed in terms of the head $\underset{e}{\phi}$ by the use of (12.13), (12.15), and (12.16) as follows:

$$\begin{aligned}\overset{d}{\underset{e}{\Phi}} &= \tfrac{1}{2}k(\underset{e}{\phi} - H)^2 + kH\underset{e}{\phi} - \tfrac{1}{2}kH^2 = \tfrac{1}{2}k\underset{e}{\phi}^2 \qquad (\underset{e}{\phi} \geq H) \\ \overset{d}{\underset{e}{\Phi}} &= \tfrac{1}{2}k(H - H)^2 + \tfrac{1}{2}k\underset{e}{\phi}^2 = \tfrac{1}{2}k\underset{e}{\phi}^2 \qquad (\underset{e}{\phi} < H)\end{aligned} \tag{12.17}$$

These expressions are identical to the one for the single-aquifer zone, obtained from (12.8):

$$\overset{s}{\underset{e}{\Phi}} = \tfrac{1}{2}k\phi_e^2 \tag{12.18}$$

Because the potentials $\overset{s}{\underset{e}{\Phi}}$ and $\overset{d}{\underset{e}{\Phi}}$ are equal,

$$\overset{s}{\underset{e}{\Phi}} = \overset{d}{\underset{e}{\Phi}} \tag{12.19}$$

the components of the gradients of $\overset{d}{\Phi}$ and $\overset{s}{\Phi}$ tangential to the edge are equal to each other. The normal components of these gradients are equal across the edge as follows from the continuity constraint (12.14) combined with (12.9) and (12.12). Hence, the gradients of $\overset{s}{\Phi}$ and $\overset{d}{\Phi}$ are equal to each other in terms of both components. We may therefore define a single comprehensive potential Φ, equal to $\overset{s}{\Phi}$ in the single-aquifer zone and to $\overset{d}{\Phi}$ in the dual-aquifer zone:

$$\boxed{\begin{aligned}\Phi &= \overset{s}{\Phi} = \tfrac{1}{2}k\phi^2 && \text{(single-aquifer zone)}\\ \Phi &= \overset{d}{\Phi} = \overset{u}{\Phi} + \overset{l}{\Phi} && \text{(dual-aquifer zone)}\end{aligned}} \tag{12.20}$$

This potential, along with its partial derivatives, is continuous throughout the aquifer system. The comprehensive discharge vector is defined accordingly:

$$\boxed{\begin{aligned}Q_i &= \overset{s}{Q}_i && \text{(single-aquifer zone)}\\ Q_i &= \overset{d}{Q}_i = \overset{u}{Q}_i + \overset{l}{Q}_i && \text{(dual-aquifer zone)}\end{aligned}} \tag{12.21}$$

where i represents either x or y. It follows that the comprehensive discharge vector equals minus the gradient of the potential:

$$\begin{aligned}Q_x &= -\frac{\partial \Phi}{\partial x}\\ Q_y &= -\frac{\partial \Phi}{\partial y}\end{aligned} \tag{12.22}$$

It follows from (12.4), (12.7), (12.10), and (12.20) that the comprehensive potential fulfills the Poisson equation

$$\nabla^2 \Phi = -N \tag{12.23}$$

throughout the aquifer system.

Method of Solution

For a general case of unconfined flow with an arbitrary number of nonoverlapping laminae, the flow domain will consist of a single-aquifer zone interspersed with dual-aquifer zones. We will assume in this section that the heads in the aquifer(s) are known along the boundaries, so that the boundary values of the comprehensive potential Φ can be computed by means of (12.20) along with (12.2) and (12.5). The comprehensive potential may be determined as a function of position by the use of the methods discussed earlier. Once the comprehensive potential is known, the head ϕ_e along the edge of each lamina may be computed by the use of (12.17), and the corresponding values of the potential in the lower aquifer are obtained from (12.5). The potential $\overset{l}{\Phi}$ may then be determined by applying the boundary values to Laplace's equation in the area below the lamina. The potential in the upper aquifer, finally, is determined from the second equation in (12.20), in which both Φ and $\overset{l}{\Phi}$ are known.

Determining $\overset{l}{\Phi}$ requires solving Laplace's equation in a closed domain; closed-form solutions can only rarely be obtained. The tools required to determine $\overset{l}{\Phi}$ for the general case of a lamina of arbitrary shape are presented in Chapter 5, and a general approach for solving problems with clay laminae will be presented in Chapter 6. Some special cases, however, will be discussed in this chapter, where we deal mainly with one-dimensional flow, and in Chapter 5.

One-Dimensional Shallow Flow in Aquifers with Clay Laminae

We next apply the approach presented above to the case of one-dimensional shallow flow illustrated in Figure 2.44. The aquifer system is unconfined, and there is infiltration of rate N. A clay lamina lies at a distance H above the base. The aquifer is bounded by two long parallel rivers with heads ϕ_1 and ϕ_2, and the origin of an x, z coordinate system is chosen at the intersection of the base with the left-hand boundary. The width of the aquifer is L and the edges of the lamina are at $x = x_1$ and $x = x_2$. The heads at the boundaries are ϕ_1 (at $x = 0$) and ϕ_2 (at $x = L$), and the corresponding boundary values for Φ are Φ_1 and Φ_2:

$$\begin{aligned} x = x_1 & \quad \Phi = \Phi_1 = \tfrac{1}{2} k \phi_1^2 \\ x = x_2 & \quad \Phi = \Phi_2 = \tfrac{1}{2} k \phi_2^2 \end{aligned} \quad (12.24)$$

The comprehensive potential is obtained as a function of x by solving the differential equation

$$\frac{d^2 \Phi}{dx^2} = -N \qquad (12.25)$$

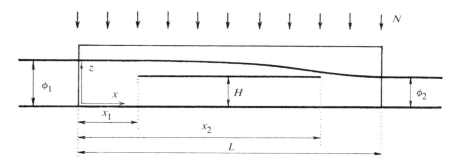

Figure 2.44 Shallow one-dimensional flow with a lamina.

with the boundary conditions (12.24). The solution to this equation may be written by the use of the influence functions F_1, F_2, and F_r defined in Sec. 9 (see [9.39] and [9.44]):

$$\begin{aligned}\Phi = \Phi(x) &= \Phi_1 F_1(x) + \Phi_2 F_2(x) + N F_r \\ &= -\Phi_1 \frac{x-L}{L} + \Phi_2 \frac{x}{L} - \tfrac{1}{2} N x(x-L) \end{aligned} \quad (12.26)$$

The values of Φ along the edges at $x = x_1$ and $x = x_2$ are obtained from (12.26) by substituting x_1 and x_2 for x:

$$\Phi_1^c = \Phi(x_1) \qquad \Phi_2^c = \Phi(x_2) \quad (12.27)$$

with the function $\Phi(x)$ defined by (12.26). The heads along the edges, ϕ_1^c and ϕ_2^c are found by the use of (12.18) as follows:

$$\phi_1^c = \sqrt{\frac{2\Phi_1^c}{k}} \qquad \phi_2^c = \sqrt{\frac{2\Phi_2^c}{k}} \quad (12.28)$$

The boundary values of $\overset{l}{\Phi}$ are obtained from (12.5):

$$\begin{aligned} \overset{l}{\Phi}_1^c &= kH\phi_1^c - \tfrac{1}{2} kH^2 & (\phi_1^c \geq H) \\ \overset{l}{\Phi}_1^c &= \tfrac{1}{2} k {\phi_1^c}^2 & (\phi_1^c < H) \end{aligned} \quad (12.29)$$

and

$$\begin{aligned} \overset{l}{\Phi}_2^c &= kH\phi_2^c - \tfrac{1}{2} kH^2 & (\phi_2^c \geq H) \\ \overset{l}{\Phi}_2^c &= \tfrac{1}{2} k {\phi_2^c}^2 & (\phi_2^c < H) \end{aligned} \quad (12.30)$$

Sec. 12 Shallow Flow in Aquifers with Clay Laminae   135

The potential $\overset{l}{\Phi}$ fulfills Laplace's equation

$$\frac{d^2\overset{l}{\Phi}}{dx^2} = 0 \qquad (12.31)$$

and the solution may be written by the use of the influence functions F_1 and F_2 defined by (9.39) as follows:

$$\overset{l}{\Phi} = \overset{l}{\Phi}_{1_e}\frac{x - x_2}{x_1 - x_2} + \overset{l}{\Phi}_{2_e}\frac{x - x_1}{x_2 - x_1} \qquad (12.32)$$

The potential $\overset{u}{\Phi}$ is found from the second equation in (12.20):

$$\overset{u}{\Phi} = \Phi - \overset{l}{\Phi} = -\Phi_1\frac{x-L}{L} + \Phi_2\frac{x}{L} - \tfrac{1}{2}Nx(x-L) - \overset{l}{\Phi}_{1_e}\frac{x - x_2}{x_1 - x_2} - \overset{l}{\Phi}_{2_e}\frac{x - x_1}{x_2 - x_1} \qquad (12.33)$$

The heads in the various zones and layers may now be determined. It follows from (12.20) that in the single-aquifer zone,

$$\phi = \sqrt{\frac{2\Phi}{k}} \qquad (0 \le x \le x_1 \quad x_2 \le x \le L) \qquad (12.34)$$

The head $\overset{u}{\phi}$ is found from (12.2):

$$\overset{u}{\phi} = H + \sqrt{\frac{2\overset{u}{\Phi}}{k}} \qquad (x_1 \le x \le x_2) \qquad (12.35)$$

and (12.5) yields the head in the lower aquifer,

$$\overset{l}{\phi} = \frac{\overset{l}{\Phi} + \tfrac{1}{2}kH^2}{kH} \qquad (\overset{l}{\Phi} \ge \tfrac{1}{2}kH^2)$$

$$\overset{l}{\phi} = \sqrt{\frac{2\overset{l}{\Phi}}{k}} \qquad (\overset{l}{\Phi} < \tfrac{1}{2}kH^2) \qquad (12.36)$$

where it is noted that $\overset{l}{\Phi}$ equals $\tfrac{1}{2}kH^2$ when $\overset{l}{\phi}$ equals H; this value marks the transition from confined to unconfined flow.

The phreatic surfaces are shown for two cases in Figures 2.45 and 2.46; for both cases $N/k = .01$, $L/H = 10$, $x_1/H = 2$, $x_2/H = 8$, and $\phi_2/H = 0.5$. For the case of Figure 2.45 $\phi_1/H = 1.5$, and for the case of Figure 2.46 $\phi_1/H = 0.9$. It may be noted that these values for the aquifer parameters are chosen for illustrative purposes only; for real aquifers L/H is usually much larger and N/k much smaller than the values chosen here. The plots were generated by the use of an elementary program written in BASIC, which runs on the IBM PC with a color graphics adapter. The listing is reproduced in the following text; note that statements that are too long to be displayed on one line have been broken into two lines. These lines should be joined prior to running the program.

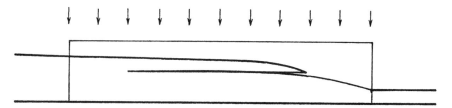

Figure 2.45 Phreatic surfaces for $N/k = 0.01$, $L/H = 10$, $x_1/H = 2$, $x_2/H = 8$, $\phi_1/H = 1.5$, and $\phi_2/H = 0.5$.

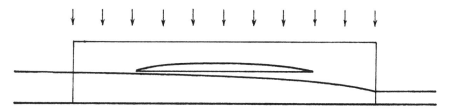

Figure 2.46 Phreatic surfaces for $N/k = 0.01$, $L/H = 10$, $x_1/H = 2$, $x_2/H = 8$, $\phi_1/H = 0.9$, and $\phi_2/H = 0.5$.

```
10  KEY OFF 20   CLS
30  INPUT"L,H,k,N > ",L,H,K,N
     : IF L<0 OR H>L OR H<0 OR K<0 OR N<0 THEN GOTO 30
40  INPUT"fi1,fi2 > ",FI1,FI2
     : IF FI1>2*H OR FI2>2*H OR FI1<0 OR FI2<0 THEN GOTO 40
50  INPUT"x1,x2   > ",X1,X2
     : IF X1>L OR X2>L OR X1>X2 OR X1<0 OR X2<0 THEN GOTO 50
60  INPUT"Nr. of Increments for heads    > ",NDELX
     : IF NDELX<0 THEN GOTO 60
70  INPUT"Output to F(ile or P(rinter (F/P) > ",A$
80  IF ((A$="F") OR (A$="f")) THEN OPEN "SFL.OUT" FOR OUTPUT AS #1
90  KH=K*H
100 HKHS=.5*KH*H
110 ' Potential as a function of head for unconfined flow.
120 DEF FN PHIHU(HEAD)=.5*K*HEAD*HEAD
130 ' Potential as a function of head for confined flow.
140 DEF FN PHIHC(HEAD)=KH*HEAD-HKHS
150 ' Head as a function of potential for unconfined flow.
160 DEF FN FIPOTU(POT)=SQR(2*POT/K)
170 ' Head as a function of potential for confined flow.
180 DEF FN FIPOTC(POT)=(POT+HKHS)/KH
190 ' Boundary values of the potential.
200 PHI1=FN PHIHU(FI1) : PHI2=FN PHIHU(FI2)
```

Sec. 12 Shallow Flow in Aquifers with Clay Laminae

```
210 ' Comprehensive potential as a function of position.
220 DEF FN PHI(X)=-(PHI1*(X-L)/L)+(PHI2*X/L)-(.5*N*X*(X-L))
230 ' Values of the comprehensive potential at the edges of the lamina.
240 PHIE1=FN PHI(X1) : PHIE2=FN PHI(X2)
250 ' Heads at the edges of the lamina.
260 FIE1=FN FIPOTU(PHIE1) : FIE2=FN FIPOTU(PHIE2)
270 ' Values of the potential in the lower aquifer at the edges of the
280 ' lamina.
290 IF FIE1<=H THEN PHILE1=FN PHIHU(FIE1) : ELSE PHILE1=FN PHIHC(FIE1)
300 IF FIE2<=H THEN PHILE2=FN PHIHU(FIE2) : ELSE PHILE2=FN PHIHC(FIE2)
310 ' Potential in the lower aquifer as a function of position.
320 DEF FN PHIL(X)=PHILE1*(X-X2)/(X1-X2)+PHILE2*(X-X1)/(X2-X1)
330 ' Potential in the upper aquifer as a function of position.
340 DEF FN PHIU(X)=FN PHI(X)-FN PHIL(X)
350 ' Initialization of the graphics screen.
360 GOSUB 1040
370 ' Plotting of the layout.
380 WINDOW (0,0)-(1.25*L,.78125*L) : LINE (0,0)-(1.25*L,0)
390 XL=.125*L
400 LINE (0,0)-(XL,FI1),1,BF : LINE (1.125*L,0)-(1.25*L,FI2),1,BF
410 LINE (1.125*L,0)-(1.125*L,2*H) : LINE(XL,0)-(XL,2*H)
420 LINE (XL-.125*H,2.125*H)-(XL,2*H),1
430 LINE -(XL,2.5*H),1
440 LINE (XL+.125*H,2.125*H)-(XL,2*H),1
450 IF XL<1.125*L THEN LINE -(XL+.125*L,2*H)
460 XL=XL+.125*L
470 IF XL<=1.125*L THEN GOTO 420
480 LINE (.125*L+X1,H)-(.125*L+X2,H),2
490 ' find heads
500 DELT=L/NDELX
510 '------ Plot heads in the single-aquifer zone (0<=x<=x1)------------
520 IF ((A$="F") OR (A$="f")) THEN PRINT #1,"x,fi"
        : ELSE LPRINT"x,fi"
530 X=0
540 WHILE X<=X1
550   PHI=FN PHI(X) : IF PHI<0 THEN PHI=0
560   FI=FN FIPOTU(PHI)
570   IF ((A$="F") OR (A$="f")) THEN PRINT #1, X,FI
        : ELSE LPRINT X,FI
580   IF X>0 THEN LINE -(X+.125*L,FI),1
        : ELSE LINE (X+.125*L,FI)-(X+.125*L,FI),1
590   X=X+DELT
600 WEND
610 '------ Plot heads in the upper aquifer---------------------------
```

```
620 IF ((A$="F") OR (A$="f"))` THEN PRINT #1,"x,fiu"
       : ELSE LPRINT"x,fiu"
630 X=X1
640 WHILE X<=X2
650   PHIU=FN PHIU(X) : IF PHIU<0 THEN PHIU=0
660   FIU=H+FN FIPOTU(PHIU)
670   IF (((A$="F") OR (A$="f")) AND (FIU<=2*H)) THEN PRINT #1,X,FIU
680   IF (((A$="F") OR (A$="f")) AND (FIU >2*H)) THEN PRINT #1,X,">2H"
690   IF (((A$="P") OR (A$="p")) AND (FIU<=2*H)) THEN LPRINT    X,FIU
700   IF (((A$="P") OR (A$="p")) AND (FIU >2*H)) THEN LPRINT    X,">2H"
710   IF ((FIU<=2*H) AND (X>X1)) THEN LINE -(X+.125*L,FIU),1
720   IF ((FIU >2*H) AND (X>X1)) THEN LINE -(X+.125*L,2*H)
730   IF ((FIU<=2*H) AND (X = X1)) THEN
          LINE (X+.125*L,FIU)-(X+.125*L,FIU),1
740   IF ((FIU >2*H) AND (X = X1)) THEN
          LINE (X+.125*L,2*H)-(X+.125*L,2*H)
750   X=X+DELT
760 WEND
770 '------ Plot heads in the lower aquifer ---------------------------
780 IF ((A$="F") OR (A$="f")) THEN PRINT #1,"x,fil"
       : ELSE LPRINT"x,fil"
790 X=X1-DELT
800 WHILE X<=X2
810   X=X+DELT
820   IF X>X2 THEN GOTO 920
830   PHIL=FN PHIL(X)
840   IF (((A$="F") OR (A$="f")) AND (PHIL>HKHS)) THEN PRINT #1,X,">2H"
850   IF (((A$="P") OR (A$="p")) AND (PHIL>HKHS)) THEN LPRINT    X,">2H"
860   IF PHIL>=HKHS THEN LINE (X+.125*L-DELT,H)-(X+.125*L,H),2
870   IF PHIL>=HKHS THEN GOTO 810 : ELSE FIL=FN FIPOTU(PHIL)
880   IF ((A$="F") OR (A$="f")) THEN PRINT #1, X,FIL
          : ELSE LPRINT X,FIL
890   IF X>X1 THEN LINE -(X+.125*L,FIL),1
          : ELSE LINE (X+.125*L,FIL)-(X+.125*L,FIL),1
900 WEND
910 '------ PLot heads in the single-aquifer zone (x2<=x<=L)------------
920 IF ((A$="F") OR (A$="f")) THEN PRINT #1,"x,fi" : ELSE LPRINT"x,fi"
930 X=X2
940 WHILE X<=L
950   PHI=FN PHI(X) : IF PHI<0 THEN PHI=0
960   FI=FN FIPOTU(PHI)
970   IF ((A$="F") OR (A$="f")) THEN PRINT #1, X,FI : ELSE LPRINT X,FI
980   IF X>X2 THEN LINE -(X+.125*L,FI),1
          : ELSE LINE (X+.125*L,FI)-(X+.125*L,FI),1
```

Sec. 12   Shallow Flow in Aquifers with Clay Laminae   **139**

```
990  X=X+DELT
1000 WEND
1010 ' Leave graphics mode
1020 GOSUB 1100
1030 END
1040 DEF SEG=0    '------ Switch to graphics monitor --------------------
1050 POKE &H410, (PEEK(&H410) AND &HCF) OR &H10
1060 SCREEN 1,0,0,0 : SCREEN 0 : WIDTH 40 : LOCATE ,,1,6,7
1070 SCREEN 1,0
1080 KEY OFF
1090 RETURN
1100 DEF SEG=0    '------ Switch to monochrome monitor ------------------
1110 POKE &H410, (PEEK(&H410) OR &H30)
1120 SCREEN 0 : WIDTH 80 : LOCATE ,,1,12,13
1130 CLS
1140 KEY ON
1150 RETURN
```

Problem 12.1
1. Determine the flow into the lower aquifer at $x = x_1$ for the case of Figure 2.45.
2. Compute $\overset{u}{Q}_x, \overset{l}{Q}_x$, and Q_x at $x = x_2$ for that same case, and verify that the condition of continuity of flow is met.

Problem 12.2
Modify the BASIC program given above to include in the upper aquifer a ditch of width b, centered at $x = x_d$, and with an infiltration rate N_d.

Two-dimensional Shallow Flow in Aquifers with Clay Laminae

As was noted above, most two-dimensional problems in aquifers with clay laminae require more advanced methods of solution than those presented in this chapter. We will, therefore, limit ourselves here to a simple problem: uniform flow in a system of two aquifers separated by a thin impermeable layer of infinite extent. The sole connection between the aquifers consists of an abandoned well, i.e., a well that is capped off and is no longer being pumped.

This problem differs from those discussed in the foregoing in that the lamina is infinite in extent rather than finite. The approach, however, remains applicable, but the conditions at infinity (the far-field conditions) now must be specified for both the upper and lower aquifers. We choose the origin of an x, y coordinate system at the center of the well, and assume that the far-fields in the two aquifers consist of uniform flows of discharges $\overset{u}{Q}_{x0}$ and $\overset{l}{Q}_{x0}$ in the x direction (see Figures 2.47[a] and [b]). There are no other features in the aquifer system, and no water is withdrawn from the well (see Figure

2.47[c]). The comprehensive potential therefore is the potential for uniform flow with a discharge Q_{x0}

$$\Phi = -Q_{x0}x + C \qquad (12.37)$$

where Q_{x0} is the sum of the discharges in the two aquifers:

$$Q_{x0} = \overset{u}{Q}_{x0} + \overset{l}{Q}_{x0} \qquad (12.38)$$

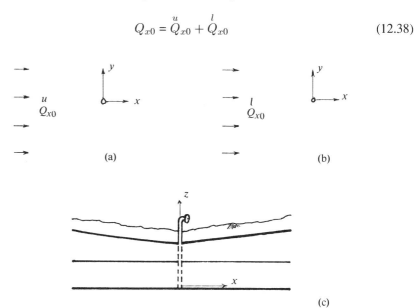

Figure 2.47 An abandoned well.

If the heads at a point $(x_0, 0)$ in the two aquifers are $\overset{u}{\phi}_0$ and $\overset{l}{\phi}_0$,

$$x = x_0 \qquad y = 0 \qquad \left\{ \begin{array}{ll} \overset{u}{\phi} = \overset{u}{\phi}_0 & (\overset{u}{\phi}_0 > H) \\ \overset{l}{\phi} = \overset{l}{\phi}_0 & (\overset{l}{\phi}_0 > H) \end{array} \right\} \qquad (12.39)$$

then the values of Φ, $\overset{u}{\Phi}$, and $\overset{l}{\Phi}$ at $(x_0, 0)$ are (compare [12.2], [12.5], and [12.20]):

$$x = x_0 \qquad y = 0 \qquad \left\{ \begin{array}{l} \overset{u}{\Phi} = \overset{u}{\Phi}_0 = \tfrac{1}{2}k(\overset{u}{\phi}_0 - H)^2 \\ \overset{l}{\Phi} = \overset{l}{\Phi}_0 = kH\overset{l}{\phi}_0 - \tfrac{1}{2}kH^2 \\ \Phi_0 = \overset{u}{\Phi}_0 + \overset{l}{\Phi}_0 \end{array} \right\} \qquad (12.40)$$

Using this condition with (12.37) to solve for the constant C, we obtain

$$\Phi = -Q_{x0}(x - x_0) + \Phi_0 \qquad (12.41)$$

Sec. 12 Shallow Flow in Aquifers with Clay Laminae

The head at the well is found as follows. Assuming that the well radius is small, we neglect variations of the comprehensive potential along the well boundary, and select the point $x = r_w$, $y = 0$ to represent the well. This gives the following expression for the comprehensive potential, Φ_w, at the well boundary:

$$\Phi_w = Q_{x0}(x_0 - r_w) + \Phi_0 \tag{12.42}$$

Since the heads in the two aquifers along the circumference of the well are equal, the well boundary acts as a single aquifer zone so that the comprehensive potential may be represented there by the first equation in (12.20) and we obtain:

$$\phi_w = \sqrt{\frac{2\Phi_w}{k}} = \sqrt{\frac{2}{k}[Q_{x0}(x_0 - r_w) + \Phi_0]} \tag{12.43}$$

We may now solve the flow problem in the lower aquifer. The potential $\overset{l}{\Phi}$ is that for uniform flow with a well at $(0,0)$ and may be written as

$$\overset{l}{\Phi} = -\overset{l}{Q}_{x0}x + \frac{\overset{l}{Q}}{4\pi}\ln(x^2 + y^2) + \overset{l}{C} \tag{12.44}$$

where $\overset{l}{Q}$ represents the discharge leaving the lower aquifer through the well. The constant $\overset{l}{C}$ may be solved for by applying the condition (12.40) to (12.44) which yields

$$\overset{l}{C} = \overset{l}{\Phi}_0 + \overset{l}{Q}_{x0}x_0 - \frac{\overset{l}{Q}}{4\pi}\ln(x_0^2) \tag{12.45}$$

so that (12.44) becomes

$$\overset{l}{\Phi} = -\overset{l}{Q}_{x0}(x - x_0) + \frac{\overset{l}{Q}}{4\pi}\ln\frac{x^2 + y^2}{x_0^2} + \overset{l}{\Phi}_0 \tag{12.46}$$

The discharge $\overset{l}{Q}$, finally, may be expressed in terms of the parameters of the problem by the use of (12.43). Depending upon the magnitude of ϕ_w, the potential $\overset{l}{\Phi}_w$ is given by

$$\begin{aligned}\overset{l}{\Phi}_w &= \tfrac{1}{2}k\phi_w^2 & (\phi_w < H) \\ \overset{l}{\Phi}_w &= kH\phi_w - \tfrac{1}{2}kH^2 & (\phi_w \geq H)\end{aligned} \tag{12.47}$$

and we obtain, upon substitution of r_w for x, 0 for y, and $\overset{l}{\Phi}_w$ for $\overset{l}{\Phi}$ in (12.46):

$$\overset{l}{\Phi}_w = -\overset{l}{Q}_{x0}(r_w - x_0) + \frac{\overset{l}{Q}}{4\pi}\ln\frac{r_w^2}{x_0^2} + \overset{l}{\Phi}_0 \tag{12.48}$$

Solving this for the unknown discharge $\overset{l}{Q}$, we find

$$\overset{l}{Q} = 4\pi \frac{\overset{l}{\Phi}_0 - \overset{l}{\Phi}_w + \overset{l}{Q}_{x0}(x_0 - r_w)}{\ln \frac{x_0^2}{r_w^2}} \qquad (12.49)$$

An expression for the potential in the upper aquifer is obtained from the relation $\overset{u}{\Phi} = \Phi - \overset{l}{\Phi}$: subtraction of (12.46) from (12.41) yields, with (12.38),

$$\overset{u}{\Phi} = -(\overset{u}{Q}_{x0} + \overset{l}{Q}_{x0})(x - x_0) + \Phi_0 + \overset{l}{Q}_{x0}(x - x_0) - \frac{\overset{l}{Q}}{4\pi} \ln \frac{x^2 + y^2}{x_0^2} - \overset{l}{\Phi}_0 \qquad (12.50)$$

Simplifying this, and writing Φ_0 as $\overset{u}{\Phi}_0 + \overset{l}{\Phi}_0$, we get

$$\overset{u}{\Phi} = -\overset{u}{Q}_{x0}(x - x_0) - \frac{\overset{l}{Q}}{4\pi} \ln \frac{x^2 + y^2}{x_0^2} + \overset{u}{\Phi}_0 \qquad (12.51)$$

We observe that the water drawn from the well by the lower aquifer ($\overset{l}{Q}$) is supplied by the upper aquifer; if the well is a discharge well in the lower aquifer then it is a recharge well in the upper aquifer.

Abandoned wells may be vehicles for spreading of contaminants in aquifers; if a contaminant travels with the groundwater in the upper aquifer and reaches the abandoned well, it will spread into the lower aquifer. An extreme example of such a spreading is the case that $\overset{l}{Q}_{x0} = 0$, and $\overset{l}{Q} < 0$. The well then acts as a discharge well in the upper aquifer and as a recharge well in the lower aquifer, spreading the contaminant radially away from the well.

Upper and Lower Aquifers with Different Hydraulic Conductivities

Often the two aquifers are separated by a single impermeable bed with openings. The technique outlined above may be adapted to account for different hydraulic conductivities in the upper and lower layers. If these conductivities are $\overset{u}{k}$ and $\overset{l}{k}$ then

$$\begin{aligned} H < z & \qquad k = \overset{u}{k} \\ 0 \leq z \leq H & \qquad k = \overset{l}{k} \end{aligned} \qquad (12.52)$$

The expressions for $\overset{u}{\Phi}$ and $\overset{l}{\Phi}$, given by (12.2) and (12.5), now become

$$\boxed{\overset{u}{\Phi} = \tfrac{1}{2}\overset{u}{k}(\overset{u}{\phi} - H)^2} \qquad (12.53)$$

and

$$\overset{l}{\Phi} = \overset{l}{k}H\overset{l}{\phi} - \tfrac{1}{2}\overset{l}{k}H^2 \qquad (\overset{l}{\phi} \geq H)$$
$$\overset{l}{\Phi} = \tfrac{1}{2}\overset{l}{k}\overset{l}{\phi}^2 \qquad (\overset{l}{\phi} < H)$$
(12.54)

The comprehensive potential in the single aquifer zone may be expressed as the sum of $\overset{u}{\Phi}$ and $\overset{l}{\Phi}$, with $\phi = \overset{u}{\phi} = \overset{l}{\phi} > H$:

$$\overset{s}{\Phi} = \tfrac{1}{2}\overset{u}{k}(\phi - H)^2 + \overset{l}{k}H\phi - \tfrac{1}{2}\overset{l}{k}H^2 \qquad (\phi \geq H)$$
(12.55)

If the head is less than H, $\overset{s}{\Phi}$ becomes

$$\overset{s}{\Phi} = \tfrac{1}{2}\overset{l}{k}\overset{l}{\phi}^2 \qquad (\phi < H)$$
(12.56)

The method of solution remains the same as that outlined in the foregoing; the changes are limited to the definitions of the potentials in terms of head.

Problem 12.3

Solve the problem of an abandoned well illustrated in Figure 12.47 with $\overset{u}{k}$ and $\overset{l}{k}$ representing the hydraulic conductivities in the upper and lower aquifers. Determine an expression in terms of the aquifer parameters for the ratio of $\overset{l}{Q}$ for the case that $\overset{u}{k} = \overset{l}{k} = k$ to $\overset{l}{Q}$ for the case that $\overset{u}{k} = 2k$, $\overset{l}{k} = k$.

13 SHALLOW INTERFACE FLOW IN AQUIFERS WITH IMPERMEABLE LAMINAE

We have seen in the preceding two sections how shallow interface flow in coastal aquifers may be handled by the use of suitable potentials, and how problems in aquifers with thin impermeable laminae may be described by means of a comprehensive potential. These two techniques may be combined to cover problems in coastal aquifers with thin clay laminae. We will adhere to the notation introduced in the previous sections, and begin our analysis by developing expressions for the potentials in the single and dual aquifer zones, considering the various types of flow possible in coastal aquifers, with the restriction that the upper boundary is a phreatic surface. We will present the equations for the general case of two-dimensional shallow interface flow with laminae, and discuss an example of one-dimensional flow in a coastal aquifer.

The Single-Aquifer Zone

The upper boundary in the single-aquifer zone is a phreatic surface. Unconfined interface flow occurs wherever the lower boundary is an interface, and unconfined flow occurs where the impervious base forms the lower boundary. The two types of flow are shown in Figure 2.48, where the abbreviations u and ui stand for unconfined flow and unconfined interface flow. The expressions for the potential are given by (10.49) for unconfined interface flow and by (8.2) for unconfined flow:

$$ui: \quad \Phi = \tfrac{1}{2}k\frac{\rho_s}{\rho_s - \rho_f}(\phi - H_s)^2 + C_{ui} \qquad (13.1)$$

$$u: \quad \Phi = \tfrac{1}{2}k\phi^2 + C_u \qquad (13.2)$$

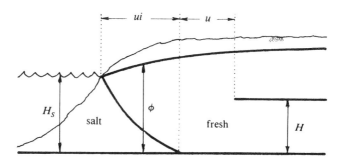

Figure 2.48 Flow conditions in the single-aquifer zone.

The condition that Φ is continuous across the interzonal boundary yields (compare [10.51] and [10.52]):

$$C_{ui} - C_u = \tfrac{1}{2}k\frac{\rho_s}{\rho_f}H_s^2 \qquad (13.3)$$

The Upper Aquifer

The flow in the upper aquifer may be either unconfined interface flow or unconfined flow (see Figure 2.49). The head in the upper aquifer is represented as $\overset{u}{\phi}$ and is measured from the base of the aquifer system. The expressions (13.1) and (13.2) correspond to the case that the aquifer base is at a distance H_s below sea level. We observe from Figure 2.49 that interface flow occurs in the upper aquifer only if $H_s > H$. The base of the upper aquifer then is a distance $H_s - H$ below sea level and the expressions for

Sec. 13   Shallow Interface Flow in Aquifers with Impermeable Laminae   145

$\overset{u}{\Phi}$ are obtained from (13.1) and (13.2) by replacing ϕ by $\overset{u}{\phi} - H$ and H_s by $H_s - H$. Introducing the constants $\overset{u}{C}_{ui}$ and $\overset{u}{C}_u$, the expressions for $\overset{u}{\Phi}$ become

$$ui: \quad \overset{u}{\Phi} = \tfrac{1}{2}k\frac{\rho_s}{\rho_s - \rho_f}(\overset{u}{\phi} - H_s)^2 + \overset{u}{C}_{ui} \qquad (H_s > H) \qquad (13.4)$$

$$u: \quad \overset{u}{\Phi} = \tfrac{1}{2}k(\overset{u}{\phi} - H)^2 + \overset{u}{C}_u \qquad (13.5)$$

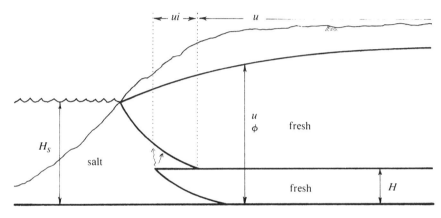

Figure 2.49 Flow conditions in the upper aquifer.

The conditions that the constants $\overset{u}{C}_{ui}$ and $\overset{u}{C}_u$ must fulfill in order that $\overset{u}{\Phi}$ is continuous throughout the upper aquifer are obtained from (13.3) by replacing C_{ui} by $\overset{u}{C}_{ui}$, C_u by $\overset{u}{C}_u$, and H_s by $H_s - H$. Selecting $\overset{u}{C}_u$ as zero, the constants become

$$\overset{u}{C}_u = 0 \qquad \overset{u}{C}_{ui} = \tfrac{1}{2}k\frac{\rho_s}{\rho_f}(H_s - H)^2 \qquad (13.6)$$

The Lower Aquifer

Four different types of flow may occur in the lower aquifer. These are confined interface flow (*ci*), unconfined interface flow (*ui*), confined flow (*c*), and unconfined flow (*u*). These four types of flow are represented in the lower aquifers of Figure 2.50(a) and Figure 2.50(b). The expressions for the potential in the zones of unconfined interface flow and unconfined flow of the lower aquifer are the same as the expressions (13.1) and (13.2) for the comprehensive potential. Introducing the constants $\overset{l}{C}_{ui}$ and $\overset{l}{C}_{u}$, the expressions for $\overset{l}{\Phi}$ become

$$ui: \quad \overset{l}{\Phi} = \tfrac{1}{2}k\frac{\rho_s}{\rho_s - \rho_f}(\overset{l}{\phi} - H_s)^2 + \overset{l}{C}_{ui} \tag{13.7}$$

$$u: \quad \overset{l}{\Phi} = \tfrac{1}{2}k\overset{l}{\phi}^2 + \overset{l}{C}_{u} \tag{13.8}$$

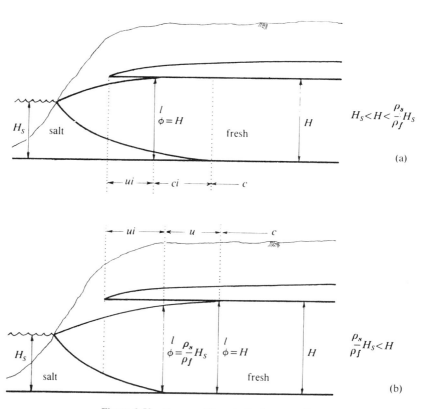

Figure 2.50 Flow conditions in the lower aquifer.

The expressions for $\overset{l}{\Phi}$ in the zones of confined interface flow and confined flow are given by (10.16) and (8.1) as

$$ci: \quad \overset{l}{\Phi} = \tfrac{1}{2}k\frac{\rho_f}{\rho_s - \rho_f}\left[\overset{l}{\phi} - \frac{\rho_s}{\rho_f}H_s + \frac{\rho_s - \rho_f}{\rho_f}H\right]^2 + \overset{l}{C}_{ci} \tag{13.9}$$

$$c: \quad \overset{l}{\Phi} = kH\overset{l}{\phi} + \overset{l}{C}_{c} \tag{13.10}$$

Sec. 13   Shallow Interface Flow in Aquifers with Impermeable Laminae   147

The head ϕ_t at the tip of the saltwater tongue equals $\rho_s H_s/\rho_f$ (see [10.28]). Furthermore, $\overset{l}{\phi}$ equals H at the boundary between the zones of confined and unconfined flow. The heads at the interzonal boundaries are listed below (see Figures 2.50[a] and [b]):

$$ci - ui: \quad \overset{l}{\phi} = H \qquad\qquad ci - c: \quad \overset{l}{\phi} = \frac{\rho_s}{\rho_f} H_s \qquad (13.11)$$

$$u - ui: \quad \overset{l}{\phi} = \frac{\rho_s}{\rho_f} H_s \qquad u - c: \quad \overset{l}{\phi} = H \qquad (13.12)$$

For the case of Figure 2.50(a) the phreatic surface intersects the clay lamina ($\overset{l}{\phi} = H$) downstream from the location of the tip of the saltwater tongue ($\overset{l}{\phi} = \rho_s H/\rho_f$) because H is less than $\rho_s H/\rho_f$. For the case of Figure 2.50(b) H is greater than $\rho_s H/\rho_f$ so that the tip lies downstream from the point where the phreatic surface intersects the lamina. The values that the head ϕ may assume in each of the four zones are

$$ui: \quad H_s < H < \frac{\rho_s}{\rho_f} H_s \qquad H_s < \overset{l}{\phi} < H \qquad (13.13)$$

$$ui: \quad \frac{\rho_s}{\rho_f} H_s < H \qquad H_s < \overset{l}{\phi} < \frac{\rho_s}{\rho_f} H_s \qquad (13.14)$$

$$ci: \quad H < \frac{\rho_s}{\rho_f} H_s \qquad H < \overset{l}{\phi} < \frac{\rho_s}{\rho_f} H_s \qquad (13.15)$$

$$u: \quad \frac{\rho_s}{\rho_f} H_s < H \qquad \frac{\rho_s}{\rho_f} H_s < \overset{l}{\phi} < H \qquad (13.16)$$

$$c: \quad H < \overset{l}{\phi} \qquad \frac{\rho_s}{\rho_f} H_s < \overset{l}{\phi} \qquad (13.17)$$

The above definitions of the four zones are obtained by inspection of Figure 2.50 as follows. Unconfined interface flow (ui) may occur in the lower aquifer only if $H_s < H$ since the head in an interface zone cannot drop below sea level and $\overset{l}{\phi}$ is less than H in a zone of unconfined flow. Moving with increasing head through the zone of unconfined interface flow, the interzonal boundary will be reached when $\overset{l}{\phi} = H$ if $H < \rho_s H_s/\rho_f$ (see Figure 2.50[a]). If, however, $\rho_s H/\rho_f < H$, the interzonal boundary will be reached when $\overset{l}{\phi} = \rho_s H_s/\rho_f$ (see Figure 2.50[b]). These considerations lead to (13.13) and (13.14). The zone of confined interface flow is defined by (13.15) and it follows that confined interface flow can exist only if $H < \rho_s H/\rho_f$.

The four constants in (13.7) through (13.10) are determined by requiring that the potential $\overset{l}{\Phi}$ be continuous throughout the lower aquifer for any sequence of zones that is physically possible. Using the conditions (13.11') and (13.12) at the interzonal boundaries and the four expressions (13.7) through (13.10) for $\overset{l}{\Phi}$ we obtain, choosing $\overset{l}{C}_{ci} = 0$,

$$\overset{l}{C}_{ci} = 0 \tag{13.18}$$

$$\overset{l}{C}_{ui} = \tfrac{1}{2}k\frac{\rho_s}{\rho_f}(H_s - H)^2 \tag{13.19}$$

$$\overset{l}{C}_{u} = \tfrac{1}{2}k\frac{\rho_s}{\rho_f}(H^2 - 2HH_s) \tag{13.20}$$

$$\overset{l}{C}_{c} = \tfrac{1}{2}k\frac{\rho_s - \rho_f}{\rho_f}H^2 - kHH_s\frac{\rho_s}{\rho_f} \tag{13.21}$$

The Conditions at the Edge of the Lamina

We will examine the conditions along the edge of the lamina for the three cases labeled as cases 1, 2, and 3 in Figures 2.51(a) through (c).

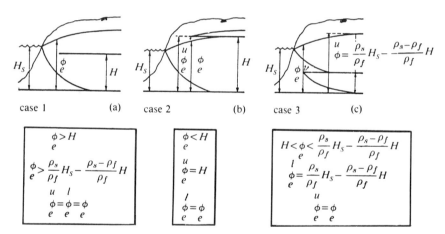

Figure 2.51 Conditions at the edge of the lamina.

<u>Case 1</u>

For case 1, fresh groundwater flows above and below the edge of the lamina, and there is no interface in the upper aquifer. Since the phreatic surface is above the edge, the head $\underset{e}{\phi}$ must be greater than H,

$$H < \underset{e}{\phi} \tag{13.22}$$

There is no interface in the upper aquifer so that $\overset{u}{\underset{e}{\phi}}$ must be greater than the value of $\overset{u}{\phi}$ that corresponds to the tip of a saltwater tongue in the upper aquifer, i.e., $\overset{u}{\phi}_t < \underset{e}{\phi}$. An

Sec. 13   Shallow Interface Flow in Aquifers with Impermeable Laminae   149

expression for $\overset{u}{\phi}_t$ is gotten by replacing in (10.28) ϕ_t by $\overset{u}{\phi}_t - H$, and H_s by $H_s - H$, thus taking into account that the base of the upper aquifer lies a distance H above the base of the system:

$$\overset{u}{\phi}_t = H + \frac{\rho_s}{\rho_f}(H_s - H) = \frac{\rho_s}{\rho_f}H_s - \frac{\rho_s - \rho_f}{\rho_f}H \tag{13.23}$$

The final condition for case 1 to occur is that the heads in the upper, lower, and single aquifers along the edge of the lamina are all equal to $\underset{c}{\phi}$. The above conditions associated with case 1 may be summarized as follows:

$$\text{case 1}: \quad H < \underset{e}{\phi} \quad \frac{\rho_s}{\rho_f}H_s - \frac{\rho_s - \rho_f}{\rho_f}H < \underset{c}{\phi} \quad \underset{e}{\overset{u}{\phi}} = \underset{e}{\phi} = \underset{e}{\overset{l}{\phi}} \tag{13.24}$$

Case 2

Case 2 applies if $\underset{c}{\phi}$ is less than H; the conditions at the edge of the lamina are that $\underset{e}{\overset{l}{\phi}} = \underset{e}{\phi}$ and $\underset{e}{\overset{u}{\phi}} = H$:

$$\text{case 2}: \quad \underset{e}{\phi} < H \quad \underset{e}{\overset{u}{\phi}} = H \quad \underset{e}{\phi} = \underset{e}{\overset{l}{\phi}} \tag{13.25}$$

Case 3

For case 3, the fresh water flowing from underneath the lamina mixes with the salt water above. Some simplifying assumptions will be adopted rather than to attempt to give an exact mathematical description of the complex physical phenomena associated with the flow for this case. Mualem and Bear [1974] considered a case similar to that of Figure 2.51(c) but with a leaky layer. These authors performed Hele Shaw experiments and observed that a mixing zone takes the place of the interface that occurs in other areas of the flow domain. A clear limit was observed to this mixing zone, which is ignored in the analysis; the interface is taken to continue through the mixing zone. As was done by Mualem and Bear, we estimate the shape of the interface, its approximate position, and the rate of seaward flow by assuming that the fresh water spilling from underneath the lamina will join the fresh water flow above the lamina. As a second assumption, the pressure p_f in the fresh water at the edge of the lamina is taken to be equal to the pressure p_s in the stationary salt water,

$$p_f = \rho_f g(\underset{e}{\overset{l}{\phi}} - H) = p_s = \rho_s g(H_s - H) \tag{13.26}$$

so that

$$\underset{e}{\overset{l}{\phi}} = \frac{\rho_s}{\rho_f}H_s - \frac{\rho_s - \rho_f}{\rho_f}H \tag{13.27}$$

The heads in the single and upper aquifers are equal at the edge:

$$\underset{e}{\overset{u}{\phi}} = \underset{c}{\phi} \tag{13.28}$$

An interface exists above the lamina, which implies that $\overset{u}{\underset{e}{\phi}}$ is greater than H and less than ϕ_t. The conditions for case 3 to occur are thus:

$$\text{case 3}: \quad H < \underset{e}{\phi} < \frac{\rho_s}{\rho_f}H_s - \frac{\rho_s - \rho_f}{\rho_f}H \quad \underset{e}{\overset{l}{\phi}} = \frac{\rho_s}{\rho_f}H_s - \frac{\rho_s - \rho_f}{\rho_f}H \quad \underset{e}{\overset{u}{\phi}} = \underset{e}{\phi} \quad (13.29)$$

The Comprehensive Potential

The constants in the expressions for the various potentials are given above except for the constants C_{ui} and C_u in (13.1) and (13.2), which must fulfill (13.3). These remaining constants are found by requiring that the potential Φ is continuous across the edge of the lamina.

Five different types of flow may occur at the edge of the lamina and are illustrated in Figures 2.52(a) through (e). The values for $\underset{e}{\phi}$, $\underset{e}{\overset{u}{\phi}}$, and $\underset{e}{\overset{l}{\phi}}$ are indicated in the figures. For the case of Figure 2.52(a) the flows in the single aquifer and in the upper aquifer above the edge of the lamina are unconfined interface flows. The heads at the edge of the lamina are given by (13.29). The expressions for the potentials at the edge, $\underset{e}{\Phi}$, $\underset{e}{\overset{u}{\Phi}}$, and $\underset{e}{\overset{l}{\Phi}}$, are found by use of (13.29), (13.1), (13.4), and (13.9), which gives

$$\underset{e}{\Phi} = \tfrac{1}{2}k\frac{\rho_s}{\rho_s - \rho_f}(\underset{e}{\phi} - H_s)^2 + C_{ui}$$

$$\underset{e}{\overset{u}{\Phi}} = \tfrac{1}{2}k\frac{\rho_s}{\rho_s - \rho_f}(\underset{e}{\overset{u}{\phi}} - H_s)^2 + \overset{u}{C}_{ui} \quad (13.30)$$

$$\underset{e}{\overset{l}{\Phi}} = \overset{l}{C}_{ci}$$

where $\overset{u}{C}_{ui}$ and $\overset{l}{C}_{ci}$ are given by (13.6) and (13.18)

$$\overset{u}{C}_{ui} = \tfrac{1}{2}k\frac{\rho_s}{\rho_f}(H_s - H)^2 \quad \overset{l}{C}_{ci} = 0 \quad (13.31)$$

The condition that Φ be continuous across the edge gives, using (13.30) and (13.31) and representing Φ as $\overset{u}{\Phi} + \overset{l}{\Phi}$ in the dual aquifer zone,

$$C_{ui} = \overset{u}{C}_{ui} = \tfrac{1}{2}k\frac{\rho_s}{\rho_f}(H_s - H)^2 \quad (13.32)$$

The remaining constant, C_u, is found from (13.3) and (13.32) to be

$$C_u = \tfrac{1}{2}k\frac{\rho_s}{\rho_f}(H^2 - 2HH_s) \quad (13.33)$$

Sec. 13   Shallow Interface Flow in Aquifers with Impermeable Laminae   151

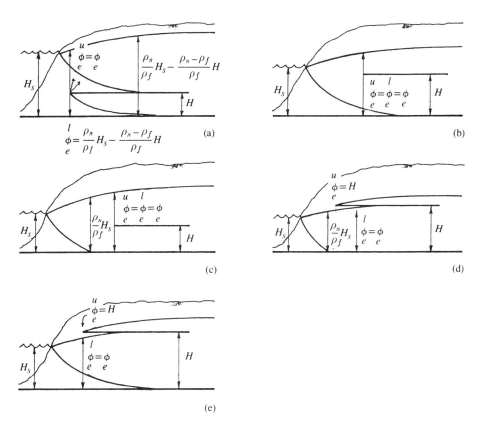

Figure 2.52   The five possible flow types at the edge of the lamina.

It may be verified by the reader that Φ is indeed continuous across the edge of the lamina for the remaining four flow cases of Figure 2.52.

As a summary, the expressions for the potentials and constants are given below. These are obtained from (13.1), (13.2), (13.4) through (13.10), (13.18) through (13.21), (13.32), and (13.33):

$$ui: \quad \Phi = \tfrac{1}{2}k\frac{\rho_s}{\rho_s - \rho_f}(\phi - H_s)^2 + C_{ui} \tag{13.34}$$

$$u: \quad \Phi = \tfrac{1}{2}k\phi^2 + C_u \tag{13.35}$$

$$ui: \quad \overset{u}{\Phi} = \tfrac{1}{2}k\frac{\rho_s}{\rho_s - \rho_f}(\overset{u}{\phi} - H_s)^2 + C_{ui} \tag{13.36}$$

$$u: \quad \overset{u}{\Phi} = \tfrac{1}{2}k(\overset{u}{\phi} - H)^2 \tag{13.37}$$

$$ui: \quad \overset{l}{\Phi} = \tfrac{1}{2}k\frac{\rho_s}{\rho_s - \rho_f}(\overset{l}{\phi} - H_s)^2 + C_{ui} \tag{13.38}$$

$$ci: \overset{l}{\Phi} = \tfrac{1}{2}k\frac{\rho_f}{\rho_s - \rho_f}\left[\overset{l}{\phi} - \frac{\rho_s}{\rho_f}H_s + \frac{\rho_s - \rho_f}{\rho_f}H\right]^2 \tag{13.39}$$

$$u: \overset{l}{\Phi} = \tfrac{1}{2}k\overset{l}{\phi}^2 + C_u \tag{13.40}$$

$$c: \overset{l}{\Phi} = kH\overset{l}{\phi} + \overset{l}{C}_c \tag{13.41}$$

where

$$C_{ui} = \tfrac{1}{2}k\frac{\rho_s}{\rho_f}(H_s - H)^2 \tag{13.42}$$

$$C_u = \tfrac{1}{2}k\frac{\rho_s}{\rho_f}(H^2 - 2HH_s) \tag{13.43}$$

$$\overset{l}{C}_c = C_u - \tfrac{1}{2}kH^2 \tag{13.44}$$

The method of solution for two-dimensional problems is similar to that described in Sec. 12. First the boundary conditions are expressed in terms of Φ, and this function is determined. Second, the values of ϕ along the edges of all laminae are computed from the values of Φ. Third, the values of either $\overset{u}{\Phi}$ or $\overset{l}{\Phi}$ are established along the boundary of each lamina, so that this function can be found by solving the differential equation. The latter aspect of the approach requires methods that we have not discussed as yet; we will therefore restrict the present analysis to one-dimensional flow.

One-dimensional Flow

We will consider a case of one-dimensional flow in a coastal aquifer, both as an illustration and a validation of the approach. A solution to a special case of this problem is found in Bear [1979]. The problem is similar to that studied by Mualem and Bear [1974], who present the results of a series of Hele Shaw experiments which will be used for a comparison with predictions from the analysis. Although the Hele Shaw experiments involve a layer that is leaky rather than impermeable, the ratio of the hydraulic conductivity of the leaky layer to that of the flow domain was small, $3.97 * 10^{-3}$. Neglecting the leakage appears not to cause a significant deviation from the test results for most of the experiments.

The problem of one-dimensional flow is illustrated in Figure 2.53. The origin of a Cartesian x, y, z coordinate system is taken at the base of the aquifer system directly below the coastline. The z axis points vertically upward, the y axis is parallel to the coast, and the x axis points toward the land. The flow is seaward and uniform. The separating layer extends from $x = x_1$, $z = H$ to $x = x_2$, $z = H$. The head is given to be ϕ_1 at $x = L$ in the single aquifer zone. The boundary conditions in terms of heads are

$$x = 0 \quad \phi = H_s \qquad x = L \quad \phi = \phi_1 \tag{13.45}$$

Sec. 13 Shallow Interface Flow in Aquifers with Impermeable Laminae

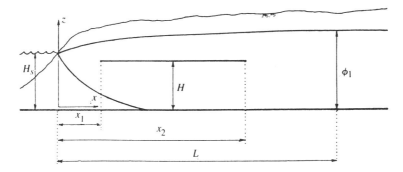

Figure 2.53 Uniform flow in a coastal aquifer.

For the sake of simplicity, restriction is made to cases where the tip of the saltwater tongue does not extend beyond the plane $x = x_2$. The boundary conditions in terms of the comprehensive potential are found from (13.34), (13.35), and (13.45), noticing that unconfined interface flow occurs at $x = 0$ and unconfined flow at $x = L$,

$$x = 0 \quad \Phi = \Phi_0 = C_{ui} \qquad x = L \quad \Phi = \Phi_1 = \tfrac{1}{2}k\phi_1^2 + C_u \tag{13.46}$$

with C_{ui} and C_u given by (13.42) and (13.43). There is no rainfall and the differential equation for the comprehensive potential reduces to $d^2\Phi/dx^2 = 0$ with the general solution

$$\Phi = -Q_{x0}x + B \tag{13.47}$$

where the magnitude of Q_{x0} ($Q_{x0} < 0$) represents the total discharge flowing toward the coast. Application of the boundary conditions (13.46) to (13.47) yields

$$\Phi = -Q_{x0}x + C_{ui} \tag{13.48}$$

where

$$Q_{x0} = -\frac{\Phi_1 - \Phi_0}{L} = -\tfrac{1}{2}k\frac{\phi_1^2 - \frac{\rho_s}{\rho_f}H_s^2}{L} \tag{13.49}$$

see (13.46), (13.42), and (13.43). It may be noted that the magnitude of Q_{x0} is not affected by the clay lamina.

The Head in the Single-Aquifer Zone

Interface flow occurs in the single-aquifer zone wherever the head ϕ is less than $\phi_t = \rho_s H_s/\rho_f$. The value for the comprehensive potential at the tip is found from (13.34) to be

$$\Phi_t = \tfrac{1}{2}k\frac{\rho_s}{\rho_f}\frac{\rho_s - \rho_f}{\rho_f}H_s^2 + C_{ui} \tag{13.50}$$

154   Types of Flow   Chap. 2

There is unconfined interface flow if $\Phi < \Phi_t$ and unconfined flow if $\Phi > \Phi_t$. The head ϕ is expressed in the single-aquifer zone in terms of Φ by use of (13.34) and (13.35) as follows (see Figure 2.54),

$$\phi = \sqrt{\frac{2}{k}\frac{\rho_s - \rho_f}{\rho_s}[\Phi(x) - C_{ui}]} + H_s \qquad (\Phi < \Phi_t) \qquad (13.51)$$

$$\phi = \sqrt{\frac{2}{k}[\Phi(x) - C_u]} \qquad (\Phi \geq \Phi_t) \qquad (13.52)$$

with $\Phi(x)$ given by (13.48). Equations (13.51) and (13.52) make it possible to determine the head as a function of position in the single-aquifer zone. The elevation Z of the interface above the base is given by (10.50):

$$Z = \frac{\rho_s}{\rho_s - \rho_f} H_s - \frac{\rho_f}{\rho_s - \rho_f} \phi \qquad (13.53)$$

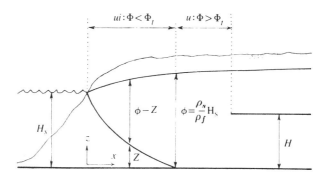

Figure 2.54 The elevation of the interface.

The Conditions at the Edge of the Lamina

The head ϕ in the single-aquifer zone at the edge of the lamina at $x = x_1$ is found from (13.51) and (13.52) by substituting x_1 for x. The head in the lower aquifer at $x = x_1$ is obtained from (13.24), (13.25), or (13.29). If case 3 applies, the value of $\overset{l}{\underset{e1}{\Phi}}$ is zero (compare [13.39]):

$$\overset{l}{\underset{e1}{\phi}} = \frac{\rho_s}{\rho_f} H_s - \frac{\rho_s - \rho_f}{\rho_f} H \qquad \overset{l}{\underset{e1}{\Phi}} = 0 \qquad \left(H < \overset{}{\underset{e1}{\phi}} < \frac{\rho_s}{\rho_f} H_s - \frac{\rho_s - \rho_f}{\rho_f} H \right) \qquad (13.54)$$

If either case 1 or case 2 applies, the flow type in the lower aquifer at $x = x_1$ must be determined from the value of $\underset{e1}{\phi}$ and the flow type definitions (13.13) through (13.17).

Sec. 13   Shallow Interface Flow in Aquifers with Impermeable Laminae   155

With the flow type known, the proper expression may be selected from the four definitions (13.38) through (13.41) for $\overset{l}{\Phi}_{e1}$, and the value of $\overset{l}{\Phi}_{e1}$ determined by substituting ϕ for $\overset{l}{\phi}_{e1}$.

We will assume that the flow in the lower aquifer is confined at the second edge of the lamina, $x = x_2$. The head $\overset{}{\phi}_{e2}$ at this edge, then, is found from (13.52) with $x = x_2$, and the corresponding expression for $\overset{l}{\Phi}, \overset{l}{\Phi}_{e2}$, is found from (13.41):

$$\overset{l}{\Phi}_{e2} = kH\phi_{e2} + \overset{l}{C}_c \tag{13.55}$$

The Potential $\overset{l}{\Phi}(x)$

If $\overset{l}{Q}_{x0}$ represents the discharge flowing in the lower aquifer ($\overset{l}{Q}_{x0} < 0$) then

$$\overset{l}{\Phi} = -\overset{l}{Q}_{x0}x + D \tag{13.56}$$

Application of the conditions at the edges of the lamina,

$$x = x_1 \quad \overset{l}{\Phi} = \overset{l}{\Phi}_{e1} \qquad x = x_2 \quad \overset{l}{\Phi} = \overset{l}{\Phi}_{e2} \tag{13.57}$$

to (13.56) yields

$$\overset{l}{\Phi} = -\overset{l}{Q}_{x0}(x - x_2) + \overset{l}{\Phi}_{e2} \tag{13.58}$$

where

$$\overset{l}{Q}_{x0} = -\frac{\overset{l}{\Phi}_{e2} - \overset{l}{\Phi}_{e1}}{x_2 - x_1} \tag{13.59}$$

The Heads in the Lower Aquifer

Prior to converting potentials to heads in the lower aquifer we establish relations for determining the flow type from the value of the potential. Case 3, illustrated in Figure 2.51(c) and defined by (13.29), is not included in the flow type definitions (13.13) through (13.17) and is treated separately. This special case may occur only if $H < H_s$, and if $\overset{l}{\phi}$ ranges between $\overset{l}{\phi}_{e1}$ (see [13.54]) and $\overset{l}{\phi}_t = \rho_s H_s/\rho_f$ in the zone of confined interface flow

$$ci: \quad H < H_s \quad \frac{\rho_s}{\rho_f}H_s - \frac{\rho_s - \rho_f}{\rho_f}H \leq \overset{l}{\phi} \leq \frac{\rho_s}{\rho_f}H_s \tag{13.60}$$

The flow type definitions (13.13) through (13.17) and (13.60) may be written in terms of potentials by the use of (13.38) through (13.44). Using the relation $(\rho_s/\rho_f)\rho_s/(\rho_s-\rho_f) = \rho_s/(\rho_s-\rho_f) + \rho_s/\rho_f$, we obtain

$ui:\ H_s < H < \dfrac{\rho_s}{\rho_f}H_s$

$$\tfrac{1}{2}k\dfrac{\rho_s}{\rho_f}(H_s - H)^2 < \overset{l}{\Phi} < \tfrac{1}{2}k\dfrac{\rho_s}{\rho_s-\rho_f}\dfrac{\rho_s}{\rho_f}(H_s - H)^2 \qquad (13.61)$$

$ui:\ \dfrac{\rho_s}{\rho_f}H_s < H$

$$\tfrac{1}{2}k\dfrac{\rho_s}{\rho_f}(H_s - H)^2 < \overset{l}{\Phi} < \tfrac{1}{2}k\dfrac{\rho_s-\rho_f}{\rho_f}\dfrac{\rho_s}{\rho_f}H_s^2 + \tfrac{1}{2}k\dfrac{\rho_s}{\rho_f}(H_s - H)^2 \qquad (13.62)$$

$ci:\ H_s < H < \dfrac{\rho_s}{\rho_f}H_s$

$$\tfrac{1}{2}k\dfrac{\rho_s}{\rho_s-\rho_f}\dfrac{\rho_s}{\rho_f}(H_s - H)^2 < \overset{l}{\Phi} < \tfrac{1}{2}k\dfrac{\rho_s-\rho_f}{\rho_f}H^2 \qquad (13.63)$$

$ci:\ H < H_s \qquad 0 < \overset{l}{\Phi} < \tfrac{1}{2}k\dfrac{\rho_s-\rho_f}{\rho_f}H^2 \qquad (13.64)$

$u\ :\ \dfrac{\rho_s}{\rho_f}H_s < H$

$$\tfrac{1}{2}k\dfrac{\rho_s^2}{\rho_f^2}H_s^2 + \tfrac{1}{2}k\dfrac{\rho_s}{\rho_f}(H^2 - 2HH_s) < \overset{l}{\Phi} < \tfrac{1}{2}kH^2 + \tfrac{1}{2}k\dfrac{\rho_s}{\rho_f}(H^2 - 2HH_s) \qquad (13.65)$$

$c\ :\ \tfrac{1}{2}kH^2 + \tfrac{1}{2}k\dfrac{\rho_s}{\rho_f}(H^2 - 2HH_s) < \overset{l}{\Phi} \qquad \tfrac{1}{2}k\dfrac{\rho_s-\rho_f}{\rho_f}H^2 < \overset{l}{\Phi} \qquad (13.66)$

The above six conditional expressions make it possible to determine the flow type uniquely from the value of $\overset{l}{\Phi}$ computed from (13.58) for any value of x. Knowing the flow type, the proper equation may be selected from (13.38) through (13.41) and the head determined. The elevation of the interface is found by the use of (13.53).

The Heads in the Upper Aquifer

The potential $\overset{u}{\Phi}(x)$ is found from the relation $\overset{u}{\Phi} = \Phi - \overset{l}{\Phi}$ with $\Phi(x)$ and $\overset{l}{\Phi}(x)$ given by (13.48) and (13.58):

$$\overset{u}{\Phi} = -(Q_{x0} - \overset{l}{Q}_{x0})x + C_{ui} - \overset{l}{Q}_{x0}x_2 - \underset{e2}{\overset{l}{\Phi}} \qquad (13.67)$$

Interface flow occurs in the upper aquifer if $H_s > H$ and $H < \underset{e1}{\phi} < \overset{u}{\phi}_t$ (see [13.23]);

Sec. 13   Shallow Interface Flow in Aquifers with Impermeable Laminae   157

the interface extends from the edge at $x = x_1$ to the interzonal boundary where $\overset{u}{\phi} = \overset{u}{\phi}_t$. The value for $\overset{u}{\Phi}$ at the tip, $\overset{u}{\Phi}_t$, is found from (13.23) and (13.36) as

$$\overset{u}{\Phi}_t = \tfrac{1}{2}k\frac{\rho_s - \rho_f}{\rho_f}\frac{\rho_s}{\rho_f}(H_s - H)^2 + C_{ui} \qquad (H_s > H) \qquad (13.68)$$

There is unconfined interface flow if $\overset{u}{\Phi} < \overset{u}{\Phi}_t$, and unconfined flow if $\overset{u}{\Phi} > \overset{u}{\Phi}_t$ and the head $\overset{u}{\phi}$ is obtained from $\overset{u}{\Phi}$ by use of (13.36) and (13.37) as follows:

$$\overset{u}{\phi} = \sqrt{\frac{2}{k}\frac{\rho_s - \rho_f}{\rho_s}\left[\overset{u}{\Phi}(x) - C_{ui}\right]} + H_s \qquad (H_s > H \quad \overset{u}{\Phi} < \overset{u}{\Phi}_t) \qquad (13.69)$$

$$\overset{u}{\phi} = \sqrt{\frac{2}{k}\overset{u}{\Phi}(x)} + H \qquad\qquad (\overset{u}{\Phi} > \overset{u}{\Phi}_t) \qquad (13.70)$$

The position of the interface is found by the use of (13.53).

Plots of the Phreatic Surfaces and Interfaces

Plots of phreatic surfaces and interfaces are given in Figures 2.55(a) through (e) and correspond to the flow cases illustrated in Figures 2.52(a) through (e). The plots are obtained from a simple computer program. All plots were made for $\rho_s/\rho_f = 1.053$. The values of the other aquifer parameters are listed in Table 2.1.

Table 2.1	Aquifer Parameters for Figure 2.55				
Case	x_1/H	x_2/H	L/H	H_s/H	ϕ_1/H
a	0.5636	5.0485	5.0545	1.9697	2.0739
b	1.0000	5.0000	7.0000	1.2000	1.2900
c	1.0000	5.0000	7.0000	1.2000	1.6000
d	0.5000	2.5000	3.5000	0.9000	1.1000
e	0.5000	2.5000	3.5000	0.9250	1.0500

Validation

The technique presented in this section has not been published before and needs to be validated. We will discuss a comparison with physical experiments as a validation. The physical experiments are carried out by means of a viscous flow analogue, known as a Hele Shaw model, consisting of two stiff parallel transparent plates with a viscous fluid flowing between them. Hele Shaw experiments are discussed in some detail in Chapter 10; for the present comparison the details of the experimental procedure are not

158 Types of Flow Chap. 2

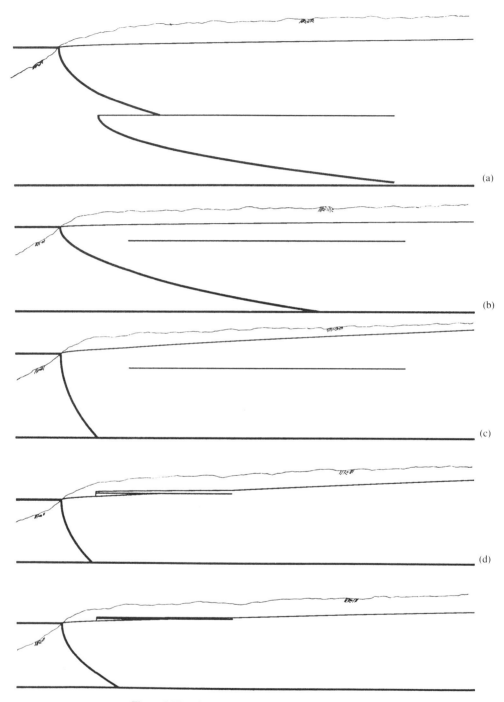

Figure 2.55 Five plots of phreatic surfaces and interfaces.

Sec. 13   Shallow Interface Flow in Aquifers with Impermeable Laminae   159

important. The results of four Hele Shaw experiments done by Mualem and Bear [1974] are compared with those obtained analytically and presented in Figures 2.56(a) through (d). The values for k and ρ_s/ρ_f are 39.33 cm/s and 1.053; the values of the other flow parameters are listed in Table 2.2.

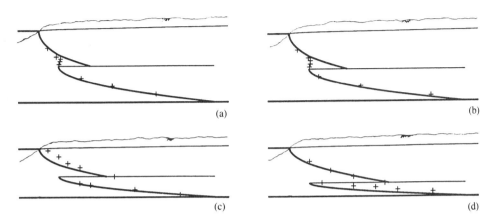

Figure 2.56   Comparison of analytical results with Mualem and Bear's experiments.

Table 2.2	Aquifer Parameters (in cm) for Figure 2.56					
Case	x_1	x_2	L	H	H_s	ϕ_1
a	9.3	83.3	83.4	16.5	32.5	34.22
a	9.3	83.3	83.4	14.5	30.5	32.12
a	9.3	83.3	83.4	6.1	22.1	23.27
a	9.3	83.3	83.4	6.1	22.1	23.27

It may be noted that the values for k and ρ_s/ρ_f were inferred from the discharges and the heights of the phreatic surfaces, measured from the figures presented by Mualem and Bear, and may be somewhat in error. The dots in the figures represent some of the points obtained experimentally; the solid curves are obtained analytically. The agreement appears to be good, except for the mixing zone that occurs just above the edge of the lamina in Figures 2.56(a) and (b) and except for the interface in the lower aquifer of Figure 2.56(d) where the omission of the leakage may be the explanation for the deviation of the analytical solution from the experiment. The lower interface intersects the semipermeable layer in the experiment of Figure 2.56(d), whereas the layer is taken to be impervious in the present analysis where the lower interface must reach the edge of the lamina.

Mualem and Bear analyzed the flow problem, taking the leaky layer into account and linearizing the governing differential equation. Discharges obtained experimentally and analytically by Mualem and Bear and from the present analysis are listed in Table 2.3.

Table 2.3		\multicolumn{3}{c}{Comparison of Discharge Rates}		
		Experiment	M and B	Present Analysis
H_s (cm)	H (cm)	\multicolumn{3}{c}{Q_{x0} (cm^2/s)}		
35.6	19.67	15.91	15.91	16.68
32.5	16.50	13.66	13.33	13.94
30.5	14.50	12.26	11.61	12.24
25.4	9.40	8.82	8.39	8.49
23.7	7.70	7.85	7.20	7.39
22.1	6.10	7.20	6.45	6.43
18.5	2.50	5.16	4.52	4.50

It may be noted that the resistance of the leaky layer used in the experiment is high, which explains the relatively good agreement between computed and observed interfaces. The leakage does not, however, affect the total discharge in the aquifer system in analyses where the Dupuit-Forchheimer approximation is adopted and continuity of total flow is assumed. The discrepancies between the discharges obtained analytically by Mualem and Bear and those computed here probably are a result of the linearization of some of Mualem and Bear's equations.

14 SHALLOW SEMICONFINED FLOW

The confining beds of a shallow confined aquifer are never truly impermeable; we indicate by the term "confined" that the leakage through the confining beds is negligibly small. If the leakage cannot be neglected the aquifer is referred to as a semiconfined aquifer and the leaking confining bed as a leaky layer or a semipermeable layer. We will use the term *shallow semiconfined flow* whenever the aquifer is sufficiently shallow that the resistance to flow in the vertical direction may be neglected.

The simplest case of shallow semiconfined flow occurs if the head is constant above the upper semipermeable layer, and the lower boundary of the aquifer is impermeable. Such a case would apply in reality if there is little or no flow in the aquifer above the semipermeable layer. We will derive the basic equations for this type of flow and consider several examples. In the next section, we will discuss methods of modeling flow in systems of several aquifers separated by several leaky layers.

Basic Equations

A cross section through a semiconfined aquifer is shown in Figure 2.57; the hydraulic conductivity and thickness of the aquifer are k amd H. We will label all quantities associated with the semipermeable layer with an asterisk: the hydraulic conductivity and

Sec. 14 Shallow Semiconfined Flow

thickness of the semipermeable layer are k^* and H^*. The constant head in the aquifer above the leaky layer is ϕ^*. All heads are measured with respect to the impermeable base of the aquifer.

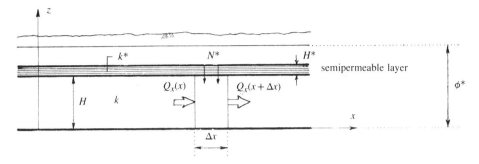

Figure 2.57 A semiconfined aquifer.

We assume that the hydraulic conductivity of the leaky layer is much less (at least a factor of 10) than the hydraulic conductivity k of the aquifer,

$$k^* \ll k \qquad (14.1)$$

The pressure, and therefore the head, is continuous across the interface between the aquifer and the leaky layer; it follows that the horizontal components of the gradient of ϕ ($\partial\phi/\partial x$, $\partial\phi/\partial y$) are continuous across the interface. This implies, with Darcy's law, that the horizontal component of flow in the leaky layer is a factor k^*/k less than that in the aquifer. Conversely, we expect a considerable difference between the heads ϕ and ϕ^* and therefore a vertical component of flow in the leaky layer that is much larger than the horizontal one. This is the basis for neglecting in the leaky layer the horizontal component of flow with respect to the vertical one.

We may view a semiconfined aquifer as a confined one with an infiltration rate N^* through the upper confining bed, as shown in Figure 2.57, and use the potential for shallow confined flow introduced in Sec. 6,

$$\Phi = kH\phi + C_{sc} \qquad (14.2)$$

where C_{sc} is a constant. The discharge vector equals minus the gradient of the potential, as usual,

$$Q_x = -\frac{\partial \Phi}{\partial x} \qquad Q_y = -\frac{\partial \Phi}{\partial y} \qquad (14.3)$$

The differential equation for this potential will be the same as that for unconfined flow with infiltration rate N^*, cf. (9.3),

$$\nabla^2 \Phi = -N^* \qquad (14.4)$$

By Darcy's law, the leakage N^* (which is positive downward) is equal to

$$N^* = k^* \frac{\phi^* - \phi}{H^*} \qquad (14.5)$$

Multiplying both the numerator and denominator by kH, we obtain

$$N^* = -\frac{kH(\phi - \phi^*)}{kHH^*/k^*} \qquad (14.6)$$

We introduce the leakage factor λ, which has the dimension of length and is defined as

$$\lambda = \sqrt{kHH^*/k^*} = \sqrt{kHc} \qquad (14.7)$$

where c is known as the resistance of the leaky layer,

$$c = H^*/k^* \qquad (14.8)$$

and has the dimension of time.

If we choose the constant C_{sc} in (14.2) as $-kH\phi^*$, i.e.,

$$\boxed{\Phi = kH(\phi - \phi^*)} \qquad (14.9)$$

then (14.6) becomes, using (14.7)

$$N^* = -\frac{\Phi}{\lambda^2} \qquad (14.10)$$

Substitution in the differential equation (14.4) yields

$$\boxed{\nabla^2 \Phi = \frac{\Phi}{\lambda^2}} \qquad (14.11)$$

General Solution for One-dimensional Flow

For the case of one-dimensional flow the differential equation (14.11) reduces to

$$\frac{d^2\Phi}{dx^2} = \frac{\Phi}{\lambda^2} \qquad (14.12)$$

The general solution to this equation is obtained by trying a function of the form

$$\Phi = Ce^{\alpha x} \qquad (14.13)$$

which yields

$$C\alpha^2 e^{\alpha x} = \frac{C}{\lambda^2} e^{\alpha x} \qquad (14.14)$$

Sec. 14 Shallow Semiconfined Flow

so that
$$\alpha = \pm 1/\lambda \tag{14.15}$$

The general solution to (14.12) therefore is
$$\Phi = Ae^{x/\lambda} + Be^{-x/\lambda} \tag{14.16}$$

where the two constants A and B are to be determined from the boundary conditions.

Leakage into an Aquifer Bounded by a Long River

As an example, we consider the problem of leakage into a semiconfined aquifer bounded by a long river, sketched in Figure 2.58. The boundary conditions are that the head approaches ϕ^* far away from the river and equals ϕ_0 along the river bank. Choosing the y axis along the river bank and the x axis as shown in the figure, the boundary conditions become

$$x = 0 \quad \phi = \phi_0 \qquad x \to \infty \quad \phi \to \phi^* \tag{14.17}$$

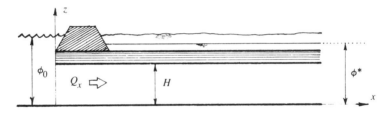

Figure 2.58 Leakage into an aquifer bounded by a long river.

Using the expression (14.9) for the potential, these conditions may be written as
$$x = 0 \quad \Phi = \Phi_0 = kH(\phi_0 - \phi^*) \qquad x \to \infty \quad \Phi \to 0 \tag{14.18}$$

The general solution for one-dimensional flow is given by (14.16),
$$\Phi = Ae^{x/\lambda} + Be^{-x/\lambda} \tag{14.19}$$

The second condition in (14.18) can be satisfied only if A is zero,
$$A = 0 \tag{14.20}$$

and application of the first one gives
$$\Phi_0 = B \tag{14.21}$$

so that the solution becomes

$$\Phi = \Phi_0 e^{-x/\lambda} = kH(\phi_0 - \phi^*)e^{-x/\lambda} \qquad (14.22)$$

The total amount of leakage into the aquifer is, by continuity of flow, equal to the discharge leaving the aquifer at the river bank. It follows from Darcy's law (14.3) and (14.22) that

$$Q_x = -\frac{\partial \Phi}{\partial x} = \frac{\Phi_0}{\lambda} e^{-x/\lambda} \qquad (14.23)$$

which becomes at $x = 0$:

$$Q_x(0) = \frac{\Phi_0}{\lambda} = \frac{kH(\phi_0 - \phi^*)}{\lambda} \qquad (14.24)$$

We observe that the leakage is proportional to the difference in head and inversely proportional to the leakage factor. Thus, the leakage factor appears to be a good measure of the amount of leakage.

Problem 14.1

A semiconfined aquifer is bounded by two infinite rivers that are a distance L apart. The y axis is midway between the rivers, and the heads along the river banks at $x = -L/2$ and $x = L/2$ are ϕ_1 and ϕ_2, respectively. The head above the leaky layer is ϕ^*. Using the symbols for the aquifer parameters introduced in the text above:
1. Determine the head as a function of position.
2. Determine the total leakage into the aquifer in terms of the parameters of the problem.

Problem 14.2

A long straight dam separates two lakes as shown in Figure 2.59; the levels in the lakes are ϕ_1^* and ϕ_2^*. A leaky layer forms the bottoms of the lakes, which may be approximated as being semi-infinite, and separates the lakes from an aquifer of thickness H. The hydraulic conductivity and thickness of the leaky layer are k^* and H^*. The centerline of the dam falls along the y axis, and the width of the dam may be neglected.

Questions
(a) Find the head in the aquifer as a function of position.
(b) Derive an expression for the flow underneath the dam per unit length of the dam.

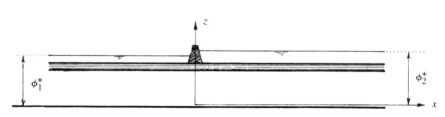

Figure 2.59 Leakage underneath a dam separating two lakes (Problem 14.2).

Sec. 14   Shallow Semiconfined Flow   165

Problem 14.3
   Determine the flow underneath the dam of Figure 2.60 if an impervious sheet is placed over the interval $0 \leq x \leq L$ as shown in the figure.

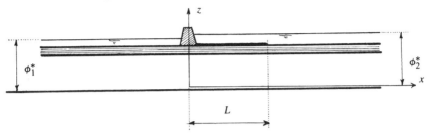

Figure 2.60   Leakage underneath a dam with an impervious sheet (Problem 14.3).

Axisymmetric Flow

We will derive the differential equation for axisymmetric semiconfined flow by considering continuity of flow in the volume between the cylinders of radii r and $(r + \Delta r)$ (see Figure 2.61). Defining two functions $f(r)$ and $g(r)$ as

$$f(r) = 2\pi r Q_r \qquad g(r) = \pi r^2 \qquad (14.25)$$

the condition of continuity may be written as

$$f(r + \Delta r) - f(r) = N^*[g(r + \Delta r) - g(r)] \qquad (14.26)$$

where N^* is the leakage, expressed earlier by (14.10),

$$N^* = -\frac{\Phi}{\lambda^2} \qquad (14.27)$$

Dividing both sides of (14.26) by Δr and passing to the limit for $\Delta r \to 0$, we obtain

$$\frac{df}{dr} = -\frac{\Phi}{\lambda^2}\frac{dg}{dr} \qquad (14.28)$$

or, with (14.25),

$$2\pi \frac{d}{dr}(rQ_r) = -\frac{\Phi}{\lambda^2} 2\pi r \qquad (14.29)$$

Using Darcy's law, $Q_r = -d\Phi/dr$, we obtain the following differential equation:

$$\frac{1}{r}\frac{d}{dr}\left[r\frac{d\Phi}{dr}\right] = \frac{\Phi}{\lambda^2} \qquad (14.30)$$

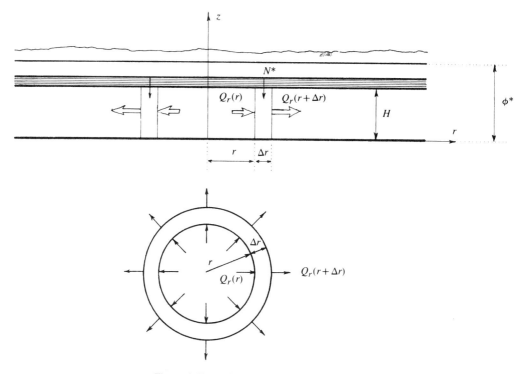

Figure 2.61 Axisymmetric semiconfined flow.

It is of interest to note that (14.30) is a special case of (14.11): the Laplacian $\nabla^2 \Phi$ for radial flow reduces to the following expression:

$$\nabla^2 \Phi = \frac{1}{r} \frac{d}{dr}\left[r \frac{d\Phi}{dr} \right] = \frac{d^2\Phi}{dr^2} + \frac{1}{r}\frac{d\Phi}{dr} \tag{14.31}$$

and (14.30) may be written as

$$\boxed{\frac{d^2\Phi}{dr^2} + \frac{1}{r}\frac{d\Phi}{dr} = \frac{\Phi}{\lambda^2}} \tag{14.32}$$

This equation often is transformed by making the substitution

$$\xi = r/\lambda \tag{14.33}$$

so that, with

$$\frac{d\Phi}{dr} = \frac{d\Phi}{d\xi}\frac{d\xi}{dr} = \frac{1}{\lambda}\frac{d\Phi}{d\xi} \qquad \frac{d^2\Phi}{dr^2} = \frac{1}{\lambda^2}\frac{d^2\Phi}{d\xi^2} \tag{14.34}$$

Sec. 14 Shallow Semiconfined Flow

the differential equation (14.32) becomes

$$\frac{d^2\Phi}{d\xi^2} + \frac{1}{\xi}\frac{d\Phi}{d\xi} - \Phi = 0 \qquad (14.35)$$

This equation is known as Bessel's modified differential equation of order zero. The general solution to this equation is written as

$$\Phi = AI_0(\xi) + BK_0(\xi) = AI_0(r/\lambda) + BK_0(r/\lambda) \qquad (14.36)$$

where $I_0(\xi)$ and $K_0(\xi)$ are called modified Bessel functions of order zero and of the first and second kind, respectively. Series expansions for these functions are available, as well as tables and subroutines for numerical evaluation. These are given in mathematical handbooks and will not be presented here. The derivatives of the above functions are modified Bessel functions of order one:

$$\frac{dI_0(\xi)}{d\xi} = I_1(\xi) \qquad \frac{dK_0(\xi)}{d\xi} = -K_1(\xi) \qquad (14.37)$$

An expression for Q_r is obtained upon differentiation of (14.36) with respect to r:

$$Q_r = -\frac{A}{\lambda}I_1(r/\lambda) + \frac{B}{\lambda}K_1(r/\lambda) \qquad (14.38)$$

Graphs of the above four Bessel functions are presented in Figure 2.62. The constants A and B are determined by application of the boundary conditions, as shown for an example in the following section.

Flow Toward a Well in an Infinite Leaky Aquifer

The case of radial semiconfined flow toward a well in an infinite leaky aquifer is illustrated in Figure 2.63. The discharge of the well is Q, and the origin of the radial coordinate system r, z is at the center of the well; the symbols for the various aquifer parameters are as introduced earlier. It is given that the head in the aquifer approaches ϕ^* at large distances from the well, i.e.,

$$r \to \infty \qquad \Phi \to 0 \qquad (14.39)$$

Furthermore, the well pumps a discharge Q and the boundary condition at the boundary of the well, $r = r_w$, is

$$r = r_w \qquad Q = 2\pi r_w[-Q_r(r_w)] \qquad (14.40)$$

where it is noted that Q_r is negative for flow toward the well.

It is seen from the graph of I_0 in Figure 2.62 that this function approaches infinity for $r \to \infty$: the condition (14.39) will be met only if the constant A in (14.36) is zero,

$$A = 0 \qquad (14.41)$$

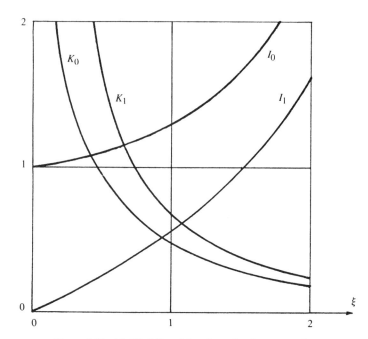

Figure 2.62 Modified Bessel functions of orders zero and one.

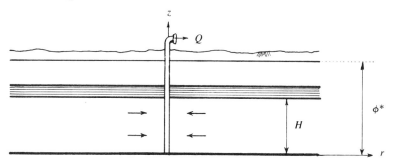

Figure 2.63 Semiconfined flow toward a well.

Application of (14.40) to the expression (14.38) for Q_r gives

$$Q = -2\pi r_w \frac{B}{\lambda} K_1(r_w/\lambda) \tag{14.42}$$

or

$$B = -\frac{Q\lambda}{2\pi r_w} \frac{1}{K_1(r_w/\lambda)} \tag{14.43}$$

Sec. 14 Shallow Semiconfined Flow

The expression for the potential is obtained by using (14.41) and (14.43) in the general solution (14.36), which yields

$$\Phi = -\frac{Q\lambda}{2\pi r_w} \frac{K_0(r/\lambda)}{K_1(r_w/\lambda)} \qquad (14.44)$$

A comparison of this solution with the one obtained for confined flow may be made using the following approximations (Abramowitz and Stegun [1965]):

$$\begin{aligned} K_0(r/\lambda) &\approx -\ln\frac{r}{1.123\lambda} & (r/\lambda \ll 1) \\ K_1(r/\lambda) &\approx \lambda/r & (r/\lambda \ll 1) \end{aligned} \qquad (14.45)$$

These approximations are sufficiently accurate for practical purposes for $r/\lambda < 0.2$. For many cases, λ is larger than 100; for example, if $k/k^* = 10^3$, and $HH^* = 10$ m^2, then $\lambda = \sqrt{10^4}$ m $=100$ m so that the approximations (14.45) are accurate for $r < 20$ m. The potential (14.44) thus becomes in the neighborhood of the well:

$$\Phi \approx -\frac{Q\lambda}{2\pi r_w} \frac{-\ln\frac{r}{1.123\lambda}}{\lambda/r_w} = \frac{Q}{2\pi}\ln r - \frac{Q}{2\pi}\ln(1.123\lambda) \qquad (14.46)$$

and we observe that the potential behaves near the well as in a confined aquifer. Apparently, the effect of the leakage through the leaky layer inside a cylinder of radius $r/\lambda = 0.2$ is not important, and the effect of leakage outside that cylinder is accounted for by the constant in (14.46).

It is to be noted that the solution (14.44) is valid only if the head at the well is larger than H. The potential at the well is obtained from (14.46) by setting r equal to r_w,

$$\Phi_w = kH(\phi_w - \phi^*) = \frac{Q}{2\pi}\ln\frac{r_w}{1.123\lambda} \qquad (14.47)$$

and ϕ_w must fulfill the condition

$$\phi_w = \phi^* + \frac{Q}{2\pi kH}\left[\ln\frac{r_w}{1.123\lambda}\right] \geq H \qquad (14.48)$$

Problem 14.4

Determine, in terms of the aquifer parameters, the leakage Q_0 through the leaky layer in a circular semiconfined aquifer of radius R in a lake with head $\phi = \phi_0$. The head above the leaky layer is ϕ^*. Define Q_0 as being positive for leakage into the aquifer.

Problem 14.5

Answer the question stated in problem 14.4 if a well is located at the center of the island and pumps a discharge Q.

Two-dimensional Flow

The differential equation (14.11) for two-dimensional semiconfined flow

$$\nabla^2 \Phi = \frac{\Phi}{\lambda^2} \qquad (14.49)$$

is linear so that the superposition principle holds. A number of two-dimensional problems may be solved by the method of images; as an example, consider the problem of a well in a leaky aquifer near a river, illustrated in Figure 2.64. The origin of an x, y, z coordinate system is chosen at a point of the riverbank, with the z axis pointing vertically upward, the y axis along the riverbank, and the x axis into the aquifer. The well is located at $x = d$, $y = 0$; the discharge and radius of the well are Q and r_w. There are three boundary conditions. The first one is that the head is ϕ_0 along the river bank:

$$x = 0 \qquad \phi = \phi_0 \qquad (14.50)$$

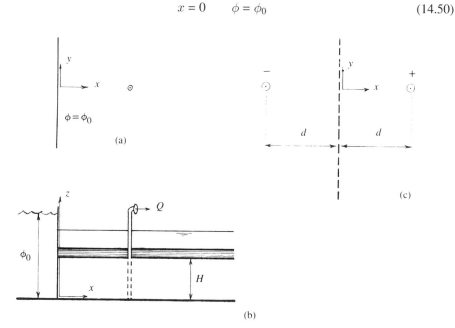

Figure 2.64 A well in a leaky aquifer near a river.

The corresponding value of the potential is obtained from (14.9):

$$x = 0 \qquad \Phi = \Phi_0 = kH(\phi_0 - \phi^*) \qquad (14.51)$$

The second boundary condition is that the head approaches ϕ^* for $x \to \infty$:

Sec. 14 Shallow Semiconfined Flow

$$x \to \infty \qquad \phi \to \phi^* \qquad \Phi \to 0 \qquad (14.52)$$

The third boundary condition is that the well draws an amount Q from the aquifer. Denoting the local radial coordinate emanating from the well as r_1, this boundary condition becomes

$$r_1 = r_w \qquad Q = -\int_0^{2\pi} Q_{r1}(r_w, \theta) r_w d\theta \qquad (14.53)$$

where $Q_{r1}(r_1, \theta)$ represents the component of the discharge vector in the direction r_1. Note that this condition reduces to (14.40) for the case that Q_{r1} is independent of θ.

We solve the problem in two steps: first we solve the problem illustrated in Figure 2.64(c) for flow from an image recharge well at $x = -d$, $y = 0$ toward the well at $x = d$, $y = 0$, and second we determine an additional solution to meet the boundary condition along the river bank. Denoting the local radial coordinate emanating from the image recharge well as r_2, we obtain the potential for flow with the two wells by superposition of two solutions of the form (14.44):

$$\Phi = -\frac{Q\lambda}{2\pi r_w} \left[\frac{K_0(r_1/\lambda) - K_0(r_2/\lambda)}{K_1(r_w/\lambda)} \right] \qquad (14.54)$$

This potential vanishes along the river bank, as is seen by noticing that r_1 equals r_2 for $x = 0$. As opposed to solutions of the Laplace's equation, a constant potential cannot be added to (14.54) to satisfy the boundary condition (14.51): a constant does not fulfill the differential equation (14.49).

We may, however, add a potential for uniform flow chosen such that it equals Φ_0 at $x = 0$. The boundary conditions (14.51) and (14.52) are identical to those for the problem of Figure 2.58 with the solution given by (14.22),

$$\Phi = \Phi_0 e^{-x/\lambda} \qquad (14.55)$$

Addition of the potentials (14.54) and (14.55) yields the complete solution to the problem:

$$\Phi = \Phi_0 e^{-x/\lambda} - \frac{Q\lambda}{2\pi r_w} \left[\frac{K_0(r_1/\lambda) - K_0(r_2/\lambda)}{K_1(r_w/\lambda)} \right] \qquad (14.56)$$

This potential fulfills the boundary conditions (14.51) and (14.52). It satisfies condition (14.53) in an approximate fashion. The term in (14.56) due to the well at $x = d$, $y = 0$ contributes an amount Q to the integral in (14.53). The contributions of the other two terms are small; by continuity of flow each of these terms contributes to the integral an amount equal to their portion of the leakage term integrated over the area of the well. For most cases this amount is small with respect to the discharge of the well and may be neglected.

15 SHALLOW FLOW IN SYSTEMS OF AQUIFERS SEPARATED BY LEAKY LAYERS

We will consider problems of flow in systems of aquifers separated by leaky layers in this section. The aquifers need be considered together in formulating the problem since they interact through the separating layers. If there is no variation in the thicknesses and hydraulic conductivities of the aquifers and the leaky layers, then the mathematical formulation is relatively straightforward.

The first to consider flow in systems of an arbitrary number of aquifers separated by leaky layers was Mjatiev [1947]; Huisman and Kemperman [1951] independently solved a problem of flow in a system of two aquifers with a well in each layer. These analyses are applicable to semiconfined systems, i.e., the uppermost boundary is a leaky layer with a constant head ϕ^* above it. We will discuss the method of solution for problems of semiconfined flow with an arbitrary number, n, of aquifers and treat the problem of two wells in a semiconfined system of two aquifers as an example.

The technique introduced by Mjatiev, and discussed for a system of two aquifers by Polubarinova-Kochina [1962], may be modified to account for the case that the upper boundary is impermeable, i.e., for horizontal confined flow in a system of aquifers separated by leaky layers. We will outline the approach for solving problems in such systems with an arbitrary number of aquifers, and treat the problem of flow in a system with two aquifers as an example.

In practice the uppermost aquifer of the system is often unconfined, leading to a nonlinearity in the system of differential equations. Following the generally adopted practice of linearization, the system may be described in an approximate fashion in much the same way as for the confined case. We will adapt the technique applied to the latter case to systems with an upper unconfined aquifer, and solve a problem in a system of two aquifers, as an example.

Semiconfined Systems of Aquifers Separated by Leaky Layers

We will examine the case of a system of n aquifers separated by leaky layers, illustrated in Figure 2.65 for the case that n=3. The aquifers are labeled as $1, 2, \ldots, n$ with aquifer 1 being the uppermost one; the separating layers are numbered $1, 2, \ldots, n$, also from the top down. The hydraulic conductivity and thickness of layer i are k_i and H_i, the resistance of the jth separating layer is c_j and the head in aquifer i is $\overset{i}{\phi}$. Expressions for the specific discharge $\overset{i}{q}_z^*$ through separating layers $i = 1, 2, \ldots, n$ are obtained from Figure 2.65 as (compare [14.5])

$$\overset{1}{q}_z^* = \frac{\overset{1}{\phi} - \phi^*}{c_1}$$

$$\overset{i}{q}_z^* = \frac{\overset{i}{\phi} - \overset{i-1}{\phi}}{c_i} = \frac{\overset{i}{\phi} - \phi^*}{c_i} - \frac{\overset{i-1}{\phi} - \phi^*}{c_i} \qquad (i = 2, 3 \ldots, n)$$

(15.1)

Sec. 14   Shallow Flow in Systems of Aquifers Separated by Leaky Layers   173

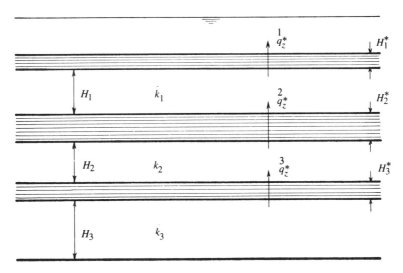

Figure 2.65 A semiconfined system of three aquifers.

Defining potentials for layers $i = 1, 2, \ldots, n$ as

$$\overset{i}{\Phi} = k_i H_i(\overset{i}{\phi} - \phi^*) \tag{15.2}$$

equations (15.1) may be expressed as,

$$\overset{1}{q_z^*} = a_{1,1}\overset{1}{\Phi}$$
$$\overset{i}{q_z^*} = -a_{i-1,i}\overset{i-1}{\Phi} + a_{i,i}\overset{i}{\Phi} \tag{15.3}$$

where

$$a_{i,j} = \frac{1}{k_i H_i c_j} \tag{15.4}$$

The leakage $\overset{i}{N^*}$ into aquifer i may be expressed in terms of the potentials by the use of (15.3) and Figure 2.65 as follows:

$$\overset{1}{N^*} = -\overset{1}{q_z^*} + \overset{2}{q_z^*} = -(a_{1,1} + a_{1,2})\overset{1}{\Phi} + a_{2,2}\overset{2}{\Phi} \tag{15.5}$$

$$\overset{i}{N^*} = -\overset{i}{q_z^*} + \overset{i+1}{q_z^*} = a_{i-1,i}\overset{i-1}{\Phi} - (a_{i,i} + a_{i,i+1})\overset{i}{\Phi} + a_{i+1,i+1}\overset{i+1}{\Phi} \quad (i = 2, 3, \ldots, n-1) \tag{15.6}$$

$$\overset{n}{N^*} = -\overset{n}{q_z^*} = a_{n-1,n}\overset{n-1}{\Phi} - a_{n,n}\overset{n}{\Phi} \tag{15.7}$$

The differential equation for the potential in layer i is given by (14.4),

$$\nabla^2 \overset{i}{\Phi} = -\overset{i}{N^*} \tag{15.8}$$

Using (15.5) through (15.7) we obtain the following system of n coupled differential equations:

$$\nabla^2 \overset{1}{\Phi} = (a_{1,1} + a_{1,2})\overset{1}{\Phi} - a_{2,2}\overset{2}{\Phi} \tag{15.9}$$

$$\nabla^2 \overset{i}{\Phi} = -a_{i-1,i}\overset{i-1}{\Phi} + (a_{i,i} + a_{i,i+1})\overset{i}{\Phi} - a_{i+1,i+1}\overset{i+1}{\Phi} \quad (i = 2, 3, \ldots, n-1) \tag{15.10}$$

$$\nabla^2 \overset{n}{\Phi} = -a_{n-1,n}\overset{n-1}{\Phi} + a_{n,n}\overset{n}{\Phi} \tag{15.11}$$

We have thus far used several particular solutions, both one-dimensional and axi-symmetric, to the differential equation

$$\nabla^2 W(x, y, \omega) = \omega^2 W(x, y, \omega) \tag{15.12}$$

It appears that solutions to the above systems of n differential equations may be obtained by setting

$$\overset{i}{\Phi} = \overset{i}{A} W(x, y, \omega) \tag{15.13}$$

Substitution of this expression for $\overset{i}{\Phi}$ in (15.9) through (15.11) yields, using (15.12),

$$\overset{1}{A}\omega^2 = (a_{1,1} + a_{1,2})\overset{1}{A} - a_{2,2}\overset{2}{A} \tag{15.14}$$

$$\overset{i}{A}\omega^2 = -a_{i-1,1}\overset{i-1}{A} + (a_{i,i} + a_{i,i+1})\overset{i}{A} - a_{i+1,i+1}\overset{i+1}{A} \quad (i = 2, 3, \ldots, n-1) \tag{15.15}$$

$$\overset{n}{A}\omega^2 = -a_{n-1,n}\overset{n-1}{A} + a_{n,n}\overset{n}{A} \tag{15.16}$$

This system of equations is homogeneous in terms of the constants $\overset{i}{A}$ and has nontrivial solutions only if the determinant vanishes. The latter condition yields an equation of order n in terms of ω^2, which has n different roots, denoted as ω_j^2 ($j = 1, 2, \ldots, n$). For each of these roots there exists a set of constants $\overset{i}{A}_j$. All equations may be divided on both sides by $\overset{1}{A}_j$, and the ratios $\overset{i}{A}_j/\overset{1}{A}_j$ determined, beginning with (15.14) to find $\overset{2}{A}_j/\overset{1}{A}_j$, then proceeding to $i = 2$ and solving the second equation for $\overset{3}{A}_j/\overset{1}{a}_j$, using the value for $\overset{2}{A}_j/\overset{1}{A}_j$, and so on until all constants are found. This yields $n - 1$ constants $\overset{i}{\alpha}_j$ ($i = 2, 3, \ldots, n$) for each ω_j:

$$\overset{i}{\alpha}_j = \overset{i}{A}_j/\overset{1}{A}_j \tag{15.17}$$

Sec. 15   Shallow Flow in Systems of Aquifers Separated by Leaky Layers

There are n different functions $W = W(x, y, \omega_j)$, each for one value of ω_j. The particular solution of the system of differential equations in terms of the functions W may now be written as a sum of all solutions that satisfy the system:

$$\overset{i}{\Phi} = \sum_{j=1}^{n} \overset{i}{\alpha}_j \mu_j W(x, y, \omega_j) \qquad (i = 1, 2, \ldots, n) \tag{15.18}$$

where

$$\mu_j = \overset{1}{A}_j \tag{15.19}$$

represents n unknown parameters. For the case of one-dimensional flow there are two particular solutions W_1 and W_2, given by (compare [14.13])

$$W_1(x, y, \omega_j) = e^{\omega_j x} \tag{15.20}$$

and

$$W_2(x, y, \omega_j) = e^{-\omega_j x} \tag{15.21}$$

Writing $\mu_j W$ in (15.18) as a linear combination $\overset{1}{\mu}_j W_1 + \overset{2}{\mu}_j W_2$ we obtain

$$\overset{i}{\Phi} = \sum_{j=1}^{n} \overset{i}{\alpha}_j \left[\overset{1}{\mu}_j e^{\omega_j x} + \overset{2}{\mu}_j e^{-\omega_j x} \right] \qquad (i = 1, 2, \ldots, n) \tag{15.22}$$

There are $2n$ coefficients $\overset{1}{\mu}_j$ and $\overset{2}{\mu}_j$, which may be determined from $2n$ boundary conditions, two in each layer. A similar approach may be followed for other forms of W, for example, the Bessel functions K_0 and I_0.

We will apply the technique outlined above to a system with two aquifers and solve an elementary problem as an example. The interested reader is referred to Heitzman [1977] for a detailed treatment of systems of two and three layers.

Two Aquifers

For the case of two aquifers ($n = 2$) the equations (15.14) and (15.16) reduce to

$$\overset{1}{A}_j \omega_j^2 = (a_{1,1} + a_{1,2})\overset{1}{A}_j - a_{2,2}\overset{2}{A}_j \tag{15.23}$$

$$\overset{2}{A}_j \omega_j^2 = -a_{1,2}\overset{1}{A}_j + a_{2,2}\overset{2}{A}_j \tag{15.24}$$

Division of each equation by $\overset{1}{A}_j = \mu_j$ gives, writing $\overset{2}{A}_j/\overset{1}{A}_j$ as $\overset{2}{\alpha}_j$,

$$\omega_j^2 = (a_{1,1} + a_{1,2}) - a_{2,2}\overset{2}{\alpha}_j \tag{15.25}$$

$$\overset{2}{\alpha}_j \omega_j^2 = -a_{1,2} + a_{2,2}\overset{2}{\alpha}_j \tag{15.26}$$

Solving the first equation for $\overset{2}{\alpha}_j$ and substituting the result in the second one we obtain the following equation for ω_j^2:

$$\omega_j^4 - (a_{1,1} + a_{1,2} + a_{2,2})\omega_j^2 + a_{1,1}a_{2,2} = 0 \tag{15.27}$$

with the solutions

$$\omega_j^2 = \tfrac{1}{2}(a_{1,1} + a_{1,2} + a_{2,2}) \pm \tfrac{1}{2}\sqrt{[(a_{1,1} + a_{2,2}) + a_{1,2}]^2 - 4a_{1,1}a_{2,2}} \tag{15.28}$$

Working out the expression under the square root sign this becomes

$$\omega_j^2 = \tfrac{1}{2}(a_{1,1} + a_{1,2} + a_{2,2}) \pm \tfrac{1}{2}\sqrt{(a_{1,1} - a_{2,2})^2 + 2(a_{1,1} + a_{2,2})a_{1,2} + a_{1,2}^2} \tag{15.29}$$

The expression under the square root sign is always positive since all constants $a_{i,j}$ are positive for physical reasons. The root obtained by using the + sign is positive and the product of the roots of (15.27) equals $a_{1,1}a_{2,2}$; both roots are therefore always positive. With the roots ω_1^2 and ω_2^2 obtained from (15.29), the constant $\overset{2}{\alpha}_j$ may be expressed in terms of $a_{i,j}$ by the use of (15.25):

$$\overset{2}{\alpha}_j = \frac{a_{1,1} + a_{1,2} - \omega_j^2}{a_{2,2}} \tag{15.30}$$

The expression for the potential $\overset{i}{\Phi}$ is obtained from (15.18) in terms of a particular solution W as follows, using that $\overset{1}{\alpha}_j = 1$ (see [15.17])

$$\overset{1}{\Phi} = \mu_1 W(x, y, \omega_1) + \mu_2 W(x, y, \omega_2) \tag{15.31}$$

$$\overset{2}{\Phi} = \overset{2}{\alpha}_1 \mu_1 W(x, y, \omega_1) + \overset{2}{\alpha}_2 \mu_2 W(x, y, \omega_2) \tag{15.32}$$

where μ_1 and μ_2 are two unknown constants, to be determined from the boundary conditions.

Example 15.1: Flow with a Well in Each Aquifer

As an application, we consider the case that a discharge $\overset{1}{Q}$ is pumped from a well in the upper aquifer. The aquifer system is infinite and we require that the potential vanishes at infinity, so that only the Bessel function $K_0(\omega_j r)$ need be considered. The expressions for the potentials (15.31) and (15.32) thus become

$$\overset{1}{\Phi} = \overset{1}{\mu}_1 K_0(\omega_1 r) + \overset{1}{\mu}_2 K_0(\omega_2 r) \tag{15.33}$$

$$\overset{2}{\Phi} = \overset{2}{\alpha}_1 \overset{1}{\mu}_1 K_0(\omega_1 r) + \overset{2}{\alpha}_2 \overset{1}{\mu}_2 K_0(\omega_2 r) \tag{15.34}$$

Sec. 15 Shallow Flow in Systems of Aquifers Separated by Leaky Layers

where the superscript 1 is used in $\overset{1}{\mu}_j$ to indicate that the well is located in the upper aquifer. The potential in the lower aquifer, $\overset{2}{\Phi}$, must be finite below the well since the well extends only into aquifer 1. For small values of $\omega_j r$, the Bessel function $K_0(\omega_j r)$ approaches $-\ln r + C_j$, where C_j is a constant (see [14.45]); in order that $\overset{2}{\Phi}$ is finite at $r = 0$, the coefficients of the two Bessel functions in (15.34) must be equal in magnitude and opposite in sign:

$$\overset{1}{\mu}_2 = -\frac{\overset{2}{\alpha}_1 \overset{1}{\mu}_1}{\overset{2}{\alpha}_2} \tag{15.35}$$

The well pumps a discharge $\overset{1}{Q}$, so that (14.40) must be satisfied. It follows upon differentiation of $\overset{1}{\Phi}$ that

$$\overset{1}{Q}_r = -\frac{d\overset{1}{\Phi}}{dr} = \overset{1}{\mu}_1 \omega_1 K_1(\omega_1 r) + \overset{1}{\mu}_2 \omega_2 K_1(\omega_2 r) \tag{15.36}$$

and application of a condition of the form (14.40) yields

$$\overset{1}{Q} = -2\pi r_w [\overset{1}{\mu}_1 \omega_1 K_1(\omega_1 r_w) + \overset{1}{\mu}_2 \omega_2 K_1(\omega_2 r_w)] \tag{15.37}$$

Using (15.35) the following expression for $\overset{1}{\mu}_1$ is obtained:

$$\overset{1}{\mu}_1 = \frac{-\overset{1}{Q}\overset{2}{\alpha}_2}{2\pi r_w [\overset{2}{\alpha}_2 \omega_1 K_1(\omega_1 r_w) - \overset{2}{\alpha}_1 \omega_2 K_1(\omega_2 r_w)]} \tag{15.38}$$

All coefficients are now expressed in terms of known quantities, and the head may be computed in each of the two aquifers.

A similar approach may be followed for the case that the well is screened only in the lower aquifer. We replace $\overset{1}{\mu}_j$ in (15.33) and (15.34) by $\overset{2}{\mu}_j$. The potential $\overset{1}{\Phi}$ now must be finite at $r = 0$, and we obtain from (15.33), replacing $\overset{1}{\mu}_j$ by $\overset{2}{\mu}_j$,

$$\overset{2}{\mu}_1 = -\overset{2}{\mu}_2 \tag{15.39}$$

The discharge component $\overset{2}{Q}_r$ is obtained by differentiating (15.34) and is used in a condition of the form (14.40). This gives, denoting the discharge of the well as $\overset{2}{Q}$ and using (15.39):

$$\overset{2}{Q} = -2\pi r_w \overset{2}{\mu}_1 [\overset{2}{\alpha}_1 \omega_1 K_1(\omega_1 r_w) - \overset{2}{\alpha}_2 \omega_2 K_1(\omega_2 r)] \tag{15.40}$$

Solving this for $\overset{2}{\mu}_1$ we find

$$\overset{2}{\mu}_1 = \frac{-\overset{2}{Q}}{2\pi r_w[\overset{2}{\alpha}_1 \omega_1 K_1(\omega_1 r_w) - \overset{2}{\alpha}_2 \omega_2 K_1(\omega_2 r_w)]} \tag{15.41}$$

As a final example, we will discuss the case that a well operates in the upper aquifer at $x = \frac{1}{2}d$, $y = 0$ and a recharge well in the lower one at $x = -\frac{1}{2}d$, $y = 0$. The discharge and recharge are equal to Q, and the local radial coordinates r_1 and r_2 emanate from the well and the recharge well, respectively. The expressions for the potentials for this case are obtained by superposition of the solutions obtained above:

$$\overset{1}{\Phi} = \overset{1}{\mu}_1 K_0(\omega_1 r_1) + \overset{1}{\mu}_2 K_0(\omega_2 r_1) + \overset{2}{\mu}_1 K_0(\omega_1 r_2) + \overset{2}{\mu}_2 K_0(\omega_2 r_2) \tag{15.42}$$

$$\overset{2}{\Phi} = \overset{2}{\alpha}_1 \overset{1}{\mu}_1 K_0(\omega_1 r_1) + \overset{2}{\alpha}_2 \overset{1}{\mu}_2 K_0(\omega_2 r_1) + \overset{2}{\alpha}_1 \overset{2}{\mu}_1 K_0(\omega_1 r_2) + \overset{2}{\alpha}_2 \overset{2}{\mu}_2 K_0(\omega_2 r_2) \tag{15.43}$$

where $\overset{1}{Q} = Q$ and $\overset{2}{Q} = -Q$ in (15.38) and (15.41).

Problem 15.1

Determine the head at the well for the case of flow from a recharge well in the upper aquifer toward a well in the lower one discussed in the foregoing. Express the value of Q in terms of the aquifer parameters for the case that $\overset{2}{\phi}_w = H_2$.

Problem 15.2

Solve the problem for flow in a system of two aquifers toward a well that is screened in both aquifers. The total discharge of the well is Q, and the radius of the well is r_w. Use the symbols introduced in this section to denote the various aquifer parameters. Derive expressions for the discharges $\overset{1}{Q}$ and $\overset{2}{Q}$ withdrawn through the segments of the well in layers 1 and 2 by requiring that the heads along the well boundaries in the two aquifers be the same.

Confined Systems of Aquifers Separated by Leaky Layers

We will next treat systems of aquifers separated by leaky layers with an impermeable upper boundary, and refer to these as confined leaky systems. The system of differential equations for a confined system of n layers (see Figure 2.66) is derived in a similar fashion as for the above semiconfined system. A difference comes about as a result of the absence of leakage through the upper boundary,

$$\overset{1}{q}_z^* = 0 \tag{15.44}$$

Sec. 14   Shallow Flow in Systems of Aquifers Separated by Leaky Layers   **179**

The head ϕ^* has no meaning for confined leaky systems, and we set ϕ^* equal to zero in the definition (15.2) of the potential in terms of head. The leakage $\overset{1}{N^*}$ into the upper aquifer is obtained from (15.5) with $\overset{1}{q_z^*} = 0$,

$$\overset{1}{N^*} = -a_{1,2}\overset{1}{\Phi} + a_{2,2}\overset{2}{\Phi} \tag{15.45}$$

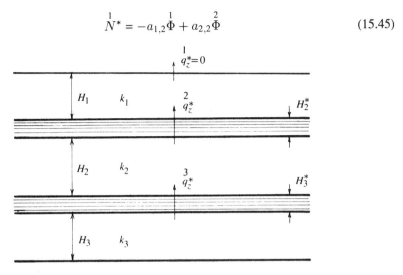

Figure 2.66 A confined system of three aquifers.

The leakage into all other layers is given by (15.6) and (15.7), and the differential equation (15.9) for $\overset{1}{\Phi}$ changes to

$$\nabla^2 \overset{1}{\Phi} = a_{1,2}\overset{1}{\Phi} - a_{2,2}\overset{2}{\Phi} \tag{15.46}$$

but (15.10) and (15.11) are valid for the present case:

$$\nabla^2 \overset{i}{\Phi} = -a_{i-1,i}\overset{i-1}{\Phi} + (a_{i,i} + a_{i,i+1})\overset{i}{\Phi} - a_{i+1,i+1}\overset{i+1}{\Phi} \quad (i = 2, 3, \ldots, n-1) \tag{15.47}$$

$$\nabla^2 \overset{n}{\Phi} = -a_{n-1,n}\overset{n-1}{\Phi} + a_{n,n}\overset{n}{\Phi} \tag{15.48}$$

There is no net inflow into the system through the upper and lower boundaries and therefore the comprehensive potential Φ, defined as the sum of all potentials $\overset{i}{\Phi}$,

$$\Phi = \sum_{i=1}^{n} \overset{i}{\Phi} = \sum_{i=1}^{n} k_i H_i \overset{i}{\phi} \tag{15.49}$$

fulfills Laplace's equation

$$\nabla^2 \Phi = 0 \tag{15.50}$$

This may be verified from (15.46) through (15.48), observing that all terms to the right of the equal sign cancel upon addition. We separate the potential $\overset{i}{\Phi}$ in two functions $\overset{i}{F}$ and $\overset{i}{G}$,

$$\overset{i}{\Phi} = \overset{i}{F} + \overset{i}{G} \tag{15.51}$$

where the functions $\overset{i}{F}$ are chosen such that

$$\sum_{i=1}^{n} \overset{i}{F} = \Phi \tag{15.52}$$

so that, by (15.49),

$$\sum_{i=1}^{n} \overset{i}{G} = 0 \tag{15.53}$$

Furthermore, we require that all $\overset{i}{F}$ are harmonic

$$\nabla^2 \overset{i}{F} = 0 \tag{15.54}$$

and that these functions fulfill the set of equations (15.46) through (15.48):

$$a_{1,2}\overset{1}{F} - a_{2,2}\overset{2}{F} = 0 \tag{15.55}$$

$$-a_{i-1,i}\overset{i-1}{F} + (a_{i,i} + a_{i,i+1})\overset{i}{F} - a_{i+1,i+1}\overset{i+1}{F} = 0 \quad (i = 2, 3, \ldots, n-1) \tag{15.56}$$

$$-a_{n-1,n}\overset{n-1}{F} + a_{n,n}\overset{n}{F} = 0 \tag{15.57}$$

We may solve this system of equations by expressing all functions $\overset{i}{F}$ in terms of $\overset{1}{F}$ starting with the equation for $i = 1$ and proceeding until $i = n - 1$. It may be noted that (15.57) is redundant: any equation may be replaced by a linear combination of the system of equations. Choosing this combination as the sum of all equations we obtain the identity $0 = 0$, which is always fulfilled so that there are only $n - 1$ independent equations. The above procedure yields $n - 1$ expressions of the form

$$\overset{i}{F} = \overset{i}{\gamma}\overset{1}{F} \tag{15.58}$$

where

$$\overset{1}{\gamma} = 1 \tag{15.59}$$

Sec. 15 Shallow Flow in Systems of Aquifers Separated by Leaky Layers

Using (15.52) as the nth equation we obtain

$$\sum_{i=1}^{n} \overset{i}{F} = \sum_{i=1}^{n} \overset{i}{\gamma}\overset{1}{F} = \Phi \tag{15.60}$$

so that

$$\overset{1}{F} = \frac{\Phi}{\sum_{i=1}^{n} \overset{i}{\gamma}} \tag{15.61}$$

and (15.58) yields

$$\overset{i}{F} = \frac{\overset{i}{\gamma}\Phi}{\sum_{j=1}^{n} \overset{j}{\gamma}} \tag{15.62}$$

Because the functions $\overset{i}{F}$ all fulfill the differential equations (15.46) through (15.48), the functions $\overset{i}{G}$ must be solutions to this system of equations as well:

$$\nabla^2 \overset{1}{G} = a_{1,2}\overset{1}{G} - a_{2,2}\overset{2}{G} \tag{15.63}$$

$$\nabla^2 \overset{i}{G} = -a_{i-1,i}\overset{i-1}{G} + (a_{i,i} + a_{i,i+1})\overset{i}{G} - a_{i+1,i+1}\overset{i+1}{G} \quad (i = 2, 3, \ldots, n-1)$$
$$\tag{15.64}$$

$$\nabla^2 \overset{n}{G} = -a_{n-1,n}\overset{n-1}{G} + a_{n,n}\overset{n}{G} \tag{15.65}$$

Utilizing (15.53), we may express $\overset{n}{G}$ as

$$\overset{n}{G} = -\sum_{i=1}^{n-1} \overset{i}{G} \tag{15.66}$$

and substitute this expression for $\overset{n}{G}$ in (15.64) for $i = n-1$, thus reducing the system to $n-1$ equations in terms of the $n-1$ unknown functions $\overset{i}{G}$, $i = 1, 2, \ldots, n-1$.

The comprehensive potential may be determined similarly as was done in Sec. 12 for problems of shallow flow with laminae, and the functions $\overset{i}{G}$ are obtained as summations of particular solutions to Bessel's equation. Each potential contains a harmonic part and a part that fulfills Bessel's equation. By the separation presented above we have split the problem into two parts: the determination of a harmonic potential and the selection of appropriate solutions to Bessel's equation.

Flow in a Confined System of Two Aquifers

As an example, we will consider the case of two aquifers, $n = 2$. The equations (15.55) through (15.57) now reduce to

$$a_{1,2}\overset{1}{F} - a_{2,2}\overset{2}{F} = 0 \tag{15.67}$$

$$-a_{1,2}\overset{1}{F} + a_{2,2}\overset{2}{F} = 0 \tag{15.68}$$

As was noted before, the last equation is redundant, and we obtain

$$\overset{2}{F} = \frac{a_{1,2}}{a_{2,2}}\overset{1}{F} = \overset{2}{\gamma}\overset{1}{F} \tag{15.69}$$

(compare [15.58]), where

$$\overset{2}{\gamma} = \frac{a_{1,2}}{a_{2,2}} \tag{15.70}$$

Using (15.52) we obtain

$$\overset{1}{F} = \frac{1}{1 + \overset{2}{\gamma}}\Phi \qquad \overset{2}{F} = \frac{\overset{2}{\gamma}}{1 + \overset{2}{\gamma}}\Phi \tag{15.71}$$

The differential equations (15.63) through (15.65) for $\overset{i}{G}$ reduce to

$$\nabla^2 \overset{1}{G} = a_{1,2}\overset{1}{G} - a_{2,2}\overset{2}{G} \tag{15.72}$$

$$\nabla^2 \overset{2}{G} = -a_{1,2}\overset{1}{G} + a_{2,2}\overset{2}{G} \tag{15.73}$$

Using (15.66) these equations become

$$\overset{2}{G} = -\overset{1}{G} \tag{15.74}$$

$$\nabla^2 \overset{1}{G} = (a_{1,2} + a_{2,2})\overset{1}{G} \tag{15.75}$$

The function $\overset{1}{G}$ may be written as a sum of particular solutions W to the differential equation

$$\nabla^2 W = \omega^2 W \tag{15.76}$$

where

$$\omega^2 = a_{1,2} + a_{2,2} \tag{15.77}$$

We will consider some elementary cases of flow as applications below.

Sec. 15 Shallow Flow in Systems of Aquifers Separated by Leaky Layers

Uniform Flow. Consider a case of uniform flow with a comprehensive discharge Q_{x0} in the x direction. The corresponding expression for the comprehensive potential is

$$\Phi = -Q_{x0}x + C \tag{15.78}$$

where C is a constant. It follows from (15.71) that

$$\overset{1}{F} = \frac{1}{1+\overset{2}{\gamma}}(-Q_{x0}x + C) \tag{15.79}$$

and

$$\overset{2}{F} = \frac{\overset{2}{\gamma}}{1+\overset{2}{\gamma}}(-Q_{x0}x + C) \tag{15.80}$$

The flow is one-dimensional and the general solution of (15.75) is

$$\overset{1}{G} = Ae^{\omega x} + Be^{-\omega x} \tag{15.81}$$

The aquifer system is infinite, and we expect on physical grounds that the leakage will remain finite at infinity. Since the leakage is proportional to $\overset{1}{G}$, this means that

$$A = B = 0 \tag{15.82}$$

and

$$\overset{1}{\Phi} = \overset{1}{F} \qquad \overset{2}{\Phi} = \overset{2}{F} \tag{15.83}$$

Thus, $\nabla^2 \overset{1}{\Phi} = \nabla^2 \overset{2}{\Phi} = 0$ so that there is no leakage anywhere between the layers. It appears that the uniform flows in the two aquifers of a confined leaky system adjust themselves under steady-state conditions such that the heads in the two aquifers are equal and there is no leakage between them: the system behaves as a single aquifer. The heads in the two aquifers will be different only locally, such as in the case of a well screened in one of the aquifers.

Flow Toward a Well in the Upper Aquifer. The case of a well screened in the upper aquifer will serve as a second example. If the well is located at the origin, if its discharge is Q, and if r is the local radial coordinate, then the comprehensive potential becomes

$$\Phi = \frac{Q}{2\pi}\ln r + C \tag{15.84}$$

where C is a constant. The functions $\overset{1}{F}$ and $\overset{2}{F}$ are obtained from (15.71) and (15.84), giving

$$\overset{1}{F} = \frac{1}{1+\overset{2}{\gamma}}\left[\frac{Q}{2\pi}\ln r + C\right] \tag{15.85}$$

and

$$\overset{2}{F} = \frac{\overset{2}{\gamma}}{1+\overset{2}{\gamma}} \left[\frac{Q}{2\pi} \ln r + C \right] \tag{15.86}$$

The function $\overset{1}{G}$ will be of the form

$$\overset{1}{G} = AK_0(\omega r) \tag{15.87}$$

The constant A is determined from the condition that the head is finite in aquifer 2 for $r = 0$. With $\overset{2}{\Phi} = \overset{2}{F} + \overset{2}{G} = \overset{2}{F} - \overset{1}{G}$ we obtain

$$\lim_{r \to 0} \left[\frac{\overset{2}{\gamma}}{1+\overset{2}{\gamma}} \frac{Q}{2\pi} \ln r - AK_0(\omega r) \right] + \frac{\overset{2}{\gamma}}{1+\overset{2}{\gamma}} C = M \tag{15.88}$$

where M is some finite number. The function $K_0(\omega r)$ may be approximated for small values of r by (14.45)

$$K_0(\omega r) \approx -\ln r + C_0 \qquad r \ll 1 \tag{15.89}$$

where C_0 equals $-\ln(\omega/1.123)$. The two singular terms in (15.88) will cancel in the limit for $r \to 0$ if

$$A = -\frac{\overset{2}{\gamma}}{1+\overset{2}{\gamma}} \frac{Q}{2\pi} \tag{15.90}$$

so that (15.87) becomes

$$\overset{1}{G} = -\frac{\overset{2}{\gamma}}{1+\overset{2}{\gamma}} \frac{Q}{2\pi} K_0(\omega r) \tag{15.91}$$

The potentials $\overset{1}{\Phi}$ and $\overset{2}{\Phi}$ may now be determined as a function of r by the use of the relations $\overset{i}{\Phi} = \overset{i}{F} + \overset{i}{G}$, $\overset{2}{G} = -\overset{1}{G}$, and (15.85), (15.86), and (15.91):

$$\overset{1}{\Phi} = \frac{1}{1+\overset{2}{\gamma}} \left[\frac{Q}{2\pi} \ln r + C \right] - \frac{\overset{2}{\gamma}}{1+\overset{2}{\gamma}} \frac{Q}{2\pi} K_0(\omega r) \tag{15.92}$$

and

$$\overset{2}{\Phi} = \frac{\overset{2}{\gamma}}{1+\overset{2}{\gamma}} \left[\frac{Q}{2\pi} \ln r + C \right] + \frac{\overset{2}{\gamma}}{1+\overset{2}{\gamma}} \frac{Q}{2\pi} K_0(\omega r) \tag{15.93}$$

Sec. 15   Shallow Flow in Systems of Aquifers Separated by Leaky Layers

As a check on the solution, we verify that the discharge of the well in aquifer 1 is indeed Q. Approximating $K_0(\omega r)$ for small values of ωr by (14.45) we obtain

$$\overset{1}{\Phi} \approx \frac{1}{1+\overset{2}{\gamma}}\left[\frac{Q}{2\pi}\ln r + C\right] + \frac{\overset{2}{\gamma}}{1+\overset{2}{\gamma}}\frac{Q}{2\pi}\left[\ln r + \ln \frac{\omega}{1.123}\right] \qquad (\omega r \ll 1) \qquad (15.94)$$

Hence, in the neighborhood of the well, $\overset{1}{\Phi}$ approaches the potential for a well, $Q/(2\pi)\ln r$. It is important to note, however, that the solution is approximate: as the well radius increases, the approximation (15.94) becomes slightly inaccurate. This inaccuracy is caused by the leakage through the dividing layer in the area inside the well, which is incorrectly entered into the solution.

The case of a well in the upper aquifer in a field of uniform flow in the x direction may be solved by adding the potentials given by (15.83), (15.79), and (15.80) to (15.92) and (15.93), combining the two constants C into a new one:

$$\overset{1}{\Phi} = \frac{1}{1+\overset{2}{\gamma}}\left[-Q_{x0}x + \frac{Q}{2\pi}\ln r + C\right] - \frac{\overset{2}{\gamma}}{1+\overset{2}{\gamma}}\frac{Q}{2\pi}K_0(\omega r) \qquad (15.95)$$

and

$$\overset{2}{\Phi} = \frac{\overset{2}{\gamma}}{1+\overset{2}{\gamma}}\left[-Q_{x0}x + \frac{Q}{2\pi}\ln r + C\right] + \frac{\overset{2}{\gamma}}{1+\overset{2}{\gamma}}\frac{Q}{2\pi}K_0(\omega r) \qquad (15.96)$$

The value of the constant C may be determined by specifying the head at some reference point in one of the two aquifers. It may be noted that the two aquifers are connected to the extent that the head at the reference point in the second aquifer cannot be set independently from that in the first.

Problem 15.3

Determine the solution for radial flow toward a single well of discharge Q screened only in the lower aquifer.

Problem 15.4

Determine the solution for radial flow toward a single well of discharge Q screened in both aquifers.

Unconfined Systems of Aquifers Separated by Leaky Layers

We refer to an unconfined leaky aquifer system if the upper aquifer in a stack of aquifers separated by leaky layers is unconfined. The approach for solving problems in such systems is similar to that for the confined systems treated in the foregoing. The two main differences are:

1. There may be infiltration from rainfall so that the comprehensive potential is not harmonic but fulfills the Poisson equation.

2. In order to keep the system of differential equations linear, the potential $\overset{1}{\Phi}$ for the upper aquifer is linearized in the expression for the leakage $\overset{2}{q}{}_z^*$.

The aquifer system is illustrated in Figure 2.67. Denoting the infiltration from rainfall as N, we obtain for $\overset{1}{q}{}_z^*$

$$\overset{1}{q}{}_z^* = -N \tag{15.97}$$

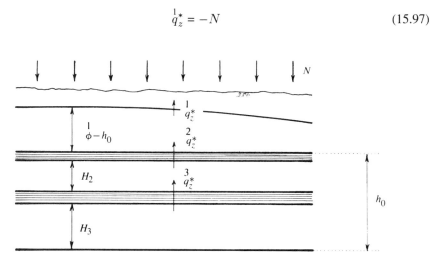

Figure 2.67 An unconfined system of three aquifers separated by leaky layers.

We represent by h_0 the elevation of the base of the uppermost aquifer above the base of the aquifer system. The expression of potential $\overset{1}{\Phi}$ for the upper aquifer in terms of the head $\overset{1}{\phi}$ then becomes

$$\overset{1}{\Phi} = \tfrac{1}{2} k_1 (\overset{1}{\phi} - h_0)^2 \tag{15.98}$$

The discharge $\overset{2}{q}{}_z^*$ is expressed as

$$\overset{2}{q}{}_z^* = \frac{\overset{2}{\phi} - \overset{1}{\phi}}{c_2} \tag{15.99}$$

The head $\overset{1}{\phi}$ is a function of the square root of $\overset{1}{\Phi}$, and we avoid the resulting nonlinearity as follows. Let H_1 represent the estimated average saturated thickness of the upper aquifer, so that

$$\overset{1}{\phi} - h_0 = H_1 + \varepsilon \tag{15.100}$$

Sec. 15   Shallow Flow in Systems of Aquifers Separated by Leaky Layers

where ε is the deviation from the average, which is assumed to be small. Substitution of $H_1 + \varepsilon$ for $\overset{1}{\phi} - h_0$ in (15.98) yields

$$\overset{1}{\Phi} = \tfrac{1}{2}k_1(H_1 + \varepsilon)^2 = \tfrac{1}{2}k_1 H_1^2 + k_1 H_1 \varepsilon + \tfrac{1}{2}k_1 \varepsilon^2 \tag{15.101}$$

Neglecting the term $\tfrac{1}{2}k_1 \varepsilon^2$ with respect to the others we obtain

$$\overset{1}{\Phi} \approx k_1 H_1(\varepsilon + H_1) - \tfrac{1}{2}k_1 H_1^2 = k_1 H_1(\overset{1}{\phi} - h_0) - \tfrac{1}{2}k_1 H_1^2 \tag{15.102}$$

and (15.99) is replaced by

$$\overset{2}{q_z^*} = -\frac{k_1 H_1(\overset{1}{\phi} - h_0) - \tfrac{1}{2}k_1 H_1^2}{k_1 H_1 c_2} + \frac{k_2 H_2 \overset{2}{\phi}}{k_2 H_2 c_2} - \frac{h_0 + \tfrac{1}{2} H_1}{c_2} \tag{15.103}$$

or

$$\overset{2}{q_z^*} \approx -a_{1,2}\overset{1}{\Phi} + a_{2,2}\overset{2}{\Phi} - \beta \tag{15.104}$$

where

$$\beta = \frac{h_0 + \tfrac{1}{2}H_1}{c_2} \tag{15.105}$$

The infiltration into the upper aquifer is

$$\overset{1}{N^*} = -\overset{1}{q_z^*} + \overset{2}{q_z^*} = -a_{1,2}\overset{1}{\Phi} + a_{2,2}\overset{2}{\Phi} - \beta + N \tag{15.106}$$

The leakage into the second aquifer is obtained from (15.6) with $i = 2$ and $\overset{2}{q_z^*}$ given by (15.104),

$$\overset{2}{N^*} = -\overset{2}{q_z^*} + \overset{3}{q_z^*} = a_{1,2}\overset{1}{\Phi} - (a_{2,2} + a_{2,3})\overset{2}{\Phi} + a_{3,3}\overset{3}{\Phi} + \beta \tag{15.107}$$

The other values of $\overset{i}{N^*}$ are identical to those given by (15.6) and (15.7). The differential equation for the potentials $\overset{i}{\Phi}$, $\nabla^2 \overset{i}{\Phi} = -\overset{i}{N^*}$, become (compare [15.46] through [15.48]):

$$\nabla^2 \overset{1}{\Phi} = a_{1,2}\overset{1}{\Phi} - a_{2,2}\overset{2}{\Phi} + \beta - N \tag{15.108}$$

$$\nabla^2 \overset{2}{\Phi} = -a_{1,2}\overset{1}{\Phi} + (a_{2,2} + a_{2,3})\overset{2}{\Phi} - a_{3,3}\overset{3}{\Phi} - \beta \tag{15.109}$$

$$\nabla^2 \overset{i}{\Phi} = -a_{i-1,i}\overset{i-1}{\Phi} + (a_{i,i} + a_{i,i+1})\overset{i}{\Phi} - a_{i+1,i+1}\overset{i+1}{\Phi} \quad (i = 3, 4, \ldots n-1) \tag{15.110}$$

$$\nabla^2 \overset{n}{\Phi} = -a_{n-1,n}\overset{n-1}{\Phi} + a_{n,n}\overset{n}{\Phi} \tag{15.111}$$

The infiltration into the system through the upper boundary is N so that the comprehensive potential fulfills the Poisson equation

$$\nabla^2 \Phi = -N \tag{15.112}$$

where

$$\Phi = \sum_{i=1}^{n} \overset{i}{\Phi} = \tfrac{1}{2} k_1 (\overset{1}{\phi} - h_0)^2 + \sum_{i=2}^{n} k_i H_i \overset{i}{\phi} \tag{15.113}$$

It is important to notice that the potential $\overset{1}{\Phi}$ is given by the quadratic form (15.98); the linearization was carried out only for the purpose of expressing the leakage term as a linear function of the potentials.

We decompose the potentials $\overset{i}{\Phi}$ into two functions $\overset{i}{F}$ and $\overset{i}{G}$, as in the case of a confined system, but add to each potential a constant $\overset{i}{C}$ and drop the requirements that $\nabla^2 \overset{i}{F} = 0$:

$$\overset{i}{\Phi} = \overset{i}{F} + \overset{i}{G} + \overset{i}{C} \tag{15.114}$$

where

$$\sum_{i=1}^{n} \overset{i}{F} = \Phi \tag{15.115}$$

and

$$\sum_{i=1}^{n} \overset{i}{G} = 0 \tag{15.116}$$

and

$$\sum_{i=1}^{n} \overset{i}{C} = 0 \tag{15.117}$$

We impose the conditions (15.55) through (15.57) to the functions $\overset{i}{F}$, which therefore may be expressed in terms of the comprehensive potential as

$$\overset{i}{F} = \frac{\overset{i}{\gamma}}{\sum_{j=1}^{n} \overset{j}{\gamma}} \Phi = \overset{i}{\gamma}{}^* \Phi \tag{15.118}$$

where the $\overset{i}{\gamma}$ are defined by (15.58), and

$$\overset{i}{\gamma}{}^* = \frac{\overset{i}{\gamma}}{\sum_{j=1}^{n} \overset{j}{\gamma}} \tag{15.119}$$

Sec. 15   Shallow Flow in Systems of Aquifers Separated by Leaky Layers

Substitution of (15.114) for $\overset{i}{\Phi}$ in the system of differential equations (15.108) through (15.111) yields, combining (15.112) and (15.118) to express $\nabla^2 \overset{i}{F}$ as $-\overset{i}{\gamma}{}^* N$, and using (15.55) through (15.57), (15.119), and (15.112),

$$\nabla^2 \overset{1}{G} = a_{1,2}\overset{1}{G} - a_{2,2}\overset{2}{G} + a_{1,2}\overset{1}{C} - a_{2,2}\overset{2}{C} + \overset{1}{\gamma}{}^* N + \beta - N \tag{15.120}$$

$$\nabla^2 \overset{2}{G} = -a_{1,2}\overset{1}{G} + (a_{2,2} + a_{2,3})\overset{2}{G} - a_{3,3}\overset{3}{G}$$
$$\qquad - a_{1,2}\overset{1}{C} + (a_{2,2} + a_{2,3})\overset{2}{C} - a_{3,3}\overset{3}{C} + \overset{2}{\gamma}{}^* N - \beta \tag{15.121}$$

$$\nabla^2 \overset{i}{G} = -a_{i-1,i}\overset{i-1}{G} + (a_{i,i} + a_{i,i+1})\overset{i}{G} - a_{i+1,i+1}\overset{i+1}{G}$$
$$\qquad - a_{i-1,i}\overset{i-1}{C} + (a_{i,i} + a_{i,i+1})\overset{i}{C} - a_{i+1,i+1}\overset{i+1}{C} + \overset{i}{\gamma}{}^* N \qquad (i = 3, 4, \ldots, n-1) \tag{15.122}$$

$$\nabla^2 \overset{n}{G} = -a_{n-1,n}\overset{n-1}{G} + a_{n,n}\overset{n}{G} - a_{n-1,n}\overset{n-1}{C} + a_{n,n}\overset{n}{C} + \overset{n}{\gamma}{}^* N \tag{15.123}$$

We choose the constants $\overset{i}{C}$ such that

$$a_{1,2}\overset{1}{C} - a_{2,2}\overset{2}{C} + \overset{1}{\gamma}{}^* N + \beta - N = 0 \tag{15.124}$$

$$-a_{1,2}\overset{1}{C} + (a_{2,2} + a_{2,3})\overset{2}{C} - a_{3,3}\overset{3}{C} + \overset{2}{\gamma}{}^* N - \beta = 0 \tag{15.125}$$

$$-a_{i-1,i}\overset{i-1}{C} + (a_{i,i} + a_{i,i+1})\overset{i}{C} - a_{i+1,i+1}\overset{i+1}{C} + \overset{i}{\gamma}{}^* N = 0 \qquad (i = 3, 4, \ldots, n-1) \tag{15.126}$$

$$-a_{n-1,n}\overset{n-1}{C} + a_{n,n}\overset{n}{C} + \overset{n}{\gamma}{}^* N = 0 \tag{15.127}$$

One of these equations may be replaced by another, obtained by any linear combination of the equations. Thus, (15.127) may be replaced by the sum of all equations, which yields the identity $0 = 0$, as is seen by noticing that the sum of all constants $\overset{i}{\gamma}{}^*$ is 1. We will use (15.117) to replace the equation eliminated above. All of the constants may then be expressed in terms of $\overset{1}{C}$, beginning with (15.124) which yields $\overset{2}{C}$ in terms of $\overset{1}{C}$, then using this expression for $\overset{2}{C}$ in (15.125), which yields $\overset{3}{C}$ in terms of $\overset{1}{C}$, and so on. Continuing this process until $i = n - 1$ we obtain expression of the form

$$\overset{i}{C} = \overset{i}{a}\overset{1}{C} + \overset{i}{b} \qquad (i = 1, 2, \ldots, n) \tag{15.128}$$

Using (15.117), we obtain

$$\sum_{i=1}^{n} \overset{i}{a}\overset{1}{C} + \sum_{i=1}^{n} \overset{i}{b} = 0 \tag{15.129}$$

so that

$$\overset{1}{C} = -\frac{\sum_{i=1}^{n} \overset{i}{b}}{\sum_{i=1}^{n} \overset{i}{a}} \qquad (15.130)$$

The differential equations (15.120) through (15.123) for the functions $\overset{i}{G}$ now may be simplified, using (15.124) through (15.127),

$$\nabla^2 \overset{1}{G} = a_{1,2}\overset{1}{G} - a_{2,2}\overset{2}{G} \qquad (15.131)$$

$$\nabla^2 \overset{i}{G} = -a_{i-1,i}\overset{i-1}{G} + (a_{i,i} + a_{i,i+1})\overset{i}{G} - a_{i+1,i+1}\overset{i+1}{G} \qquad (i = 2, 3, \ldots, n-1) \qquad (15.132)$$

$$\nabla^2 \overset{n}{G} = -a_{n-1,n}\overset{n-1}{G} + a_{n,n}\overset{n}{G} \qquad (15.133)$$

These differential equations are identical to those valid for the confined system. The method of solution differs from that for the confined case only in that rainfall may be included and that the constants $\overset{i}{C}$ must be added to each potential.

Flow in an Unconfined System of Two Aquifers

We will consider the case of two aquifers as an example, and follow the steps outlined for the case of confined flow. The expressions for the functions $\overset{1}{F}$ and $\overset{2}{F}$ are given by (15.71)

$$\overset{1}{F} = \frac{1}{1+\overset{2}{\gamma}}\Phi \qquad \overset{2}{F} = \frac{\overset{2}{\gamma}}{1+\overset{2}{\gamma}}\Phi \qquad (15.134)$$

where $\overset{2}{\gamma}$ is given by (15.70)

$$\overset{2}{\gamma} = \frac{a_{1,2}}{a_{2,2}} \qquad (15.135)$$

Expressions for the constants $\overset{1}{C}$ and $\overset{2}{C}$ are obtained from (15.124) and (15.117), which yields

$$\overset{1}{C} = -\overset{2}{C} = \frac{N(1-\overset{1}{\gamma}{}^*) - \beta}{a_{1,2} + a_{2,2}} \qquad (15.136)$$

Using (15.59), (15.119), and (15.135) this may be written as

$$\overset{1}{C} = -\overset{2}{C} = \frac{1}{a_{1,2} + a_{2,2}}\left[N\frac{a_{1,2}}{a_{1,2} + a_{2,2}} - \beta\right] \qquad (15.137)$$

Sec. 15 Shallow Flow in Systems of Aquifers Separated by Leaky Layers

and the potentials $\overset{1}{\Phi}$ and $\overset{2}{\Phi}$ may be expressed by the use of (15.114), (15.116), and (15.134) as

$$\overset{1}{\Phi} = \frac{1}{1+\overset{2}{\gamma}}\Phi + \overset{1}{C} + \overset{1}{G} \qquad (15.138)$$

and

$$\overset{2}{\Phi} = \frac{\overset{2}{\gamma}}{1+\overset{2}{\gamma}}\Phi - \overset{1}{C} - \overset{1}{G} \qquad (15.139)$$

where $\overset{1}{C}$ is given by (15.137). The functions Φ and $\overset{1}{G}$ are determined similarly as for the confined system.

Uniform Flow. The potentials for uniform flow in the x direction are obtained using (15.79) and (15.80) with (15.138) and (15.139), noticing that $\overset{1}{G}$ must be zero in order that the leakage at infinity remains finite:

$$\overset{1}{\Phi} = \frac{1}{1+\overset{2}{\gamma}}(-Q_{x0}x + C) + \overset{1}{C} \qquad (15.140)$$

$$\overset{2}{\Phi} = \frac{\overset{2}{\gamma}}{1+\overset{2}{\gamma}}(-Q_{x0}x + C) + \overset{2}{C} \qquad (15.141)$$

It is important to note that the leakage is zero. This result is inaccurate: the leakage is computed using a linearized expression for the potential $\overset{1}{\Phi}$. Because the potential $\overset{1}{\Phi}$ is quadratic in terms of the head $\overset{1}{\phi}$, the two heads $\overset{1}{\phi}$ and $\overset{2}{\phi}$ obtained from (15.140) and (15.141) are different even though the leakage is predicted to be zero. Keil [1982] compared the results of the solution for a system of two aquifers with finite boundaries to observations made on a Hele-Shaw model, and found that the discrepancies were well within the accuracy of the experiment.

Flow Toward a Well Screened in the Upper Aquifer. The functions $\overset{1}{F}$, $\overset{2}{F}$, and $\overset{1}{G}$ derived for the case of flow in a confined system toward a well screened in the upper aquifer are applicable to the case that the system is unconfined. These functions are given by (15.85), (15.86), and (15.91) and we obtain

$$\overset{i}{\Phi} = \overset{i}{F} + \overset{i}{G} + \overset{i}{C} \qquad (i=1,2) \qquad (15.142)$$

where the constants $\overset{i}{C}$ are given by (15.137).

Flow in Unconfined Systems of Three Aquifers

For the sake of brevity, we will not discuss cases of more than two aquifers. The interested reader is referred to Keil [1982] for a detailed treatment of unconfined systems of three aquifers, including problems with wells screened in any number of aquifers, rainfall, uniform flow, and advanced functions that make it possible to implement conditions along finite boundaries. Keil presents the listing of a FORTRAN program applicable to unconfined leaky aquifer systems.

16 TRANSIENT SHALLOW FLOW

As mentioned in the introduction, the emphasis is placed in this text on the application of analytic techniques to steady-state groundwater flow. For the sake of completeness, however, the basic equations for shallow transient confined and unconfined flows will be derived in this section. These equations will be used only in Chapter 4 in connection with contaminant transport and in Chapter 9 in connection with approximate methods.

Transient Shallow Confined Flow

Transient effects in aquifer systems come about when boundary values or infiltration rates vary with time. A common example is a well that is switched on; upon starting the pump the heads and pressures in the aquifer system change gradually until, for all practical purposes, steady-state conditions are reached. In a confined aquifer, the transient effects are caused by the compression of the grain skeleton as a result of decreasing pressures; if both the aquifer material and the fluid were incompressible, steady-state conditions would be reached instantaneously.

The problem of transient flow in a confined aquifer is a coupled one; the deformation of the grain skeleton is coupled to the groundwater flow. The problem is very complex, as the constitutive equations for soil are highly nonlinear, even under dry conditions, and the coupling with the groundwater flow increases the complexity further still. Biot [1941] formulated the coupled problem mathematically, approximating the grain skeleton as a linear elastic material. Fortunately, the pressure changes due to groundwater flow are usually small as compared to the overburden stresses in the grain skeleton, which allows approximations to be made that simplify the problem considerably. We will formulate the problem in terms of two equations: the equation describing storage, and a simplified equation for the deformation of the grain skeleton. All strains and stresses are taken positive for extension and tension, respectively.

The Storage Equation

The storage equation expresses that water may be stored in an elementary volume V of porous material due both to an increase of V and a decrease in the volume V_w of the water. We assume that the aquifer is shallow so that all properties may be taken constant over the aquifer thickness H. A decrease in storage of an elementary volume

Sec. 16 Transient Shallow Flow

will force water to flow out of V (see Figure 2.68) and has a similar effect upon the aquifer as infiltrating water, which also forces water to leave the elementary volume; by continuity of flow the rate of outflow equals the infiltration rate N^*. By this analogy we make use of the differential equation $\nabla^2 \Phi = -N^*$, interpreting N^* as the rate of decrease in storage. If $e_0 = \Delta V/V$ and $ne_w = \Delta V_w/V$ (n is the porosity) are the volume strains of soil and water, both taken with respect to the same volume and assumed to be constant over the thickness of the aquifer, then the rate of decrease N^* in storage equals the aquifer thickness H multiplied by the sum of the rate of decrease $-\partial e_0/\partial t$ of the volume strain and the rate of increase $n\partial e_w/\partial t$ of the volume strain of the water, so that

$$N^* = -H\left[\frac{\partial e_0}{\partial t} - n\frac{\partial e_w}{\partial t}\right] \tag{16.1}$$

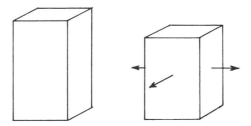

Figure 2.68 Decrease in storage of an elementary volume of porous material results in outflow.

With $\nabla^2 \Phi = -N^*$, where $\Phi = kH\phi$, we obtain

$$\nabla^2 \Phi = H\left[\frac{\partial e_0}{\partial t} - n\frac{\partial e_w}{\partial t}\right] \tag{16.2}$$

Adopting a linear law for the compression of water,

$$\frac{\partial e_w}{\partial t} = -\beta\frac{\partial p}{\partial t} \tag{16.3}$$

(16.2) becomes

$$\nabla^2 \Phi = H\left[\frac{\partial e_0}{\partial t} + n\beta\frac{\partial p}{\partial t}\right] \tag{16.4}$$

This differential equation contains the volume strain e_0, which is dependent upon the effective isotropic stress σ_0, which is generally not a function of the pressure p alone. We will adopt the approximate approach introduced by Jacob [1940] and Terzaghi [1943] discussed in the following.

Simplified Stress-Strain Relation

The simplified stress-strain relation is based upon the following three approximations:

1. The vertical total stress σ_{zz} does not change with time; this approximation is based on the notion that the effect of pore pressure changes is small with respect to the overburden, and implies that

$$\frac{\partial \sigma_{zz}}{\partial t} = 0 \qquad (16.5)$$

2. The total stress can be written, following Terzaghi, as the difference between an effective stress σ'_{zz} and the pore pressure p

$$\sigma_{zz} = \sigma'_{zz} - p \qquad (16.6)$$

3. The horizontal deformations in the soil are neglected.

It follows from the third approximation that

$$\frac{\partial e_0}{\partial t} = \frac{\partial e_{zz}}{\partial t} = m_v \frac{\partial \sigma'_{zz}}{\partial t} \qquad (16.7)$$

where m_v is, somewhat ambiguously, called the coefficient of volume compressibility. It follows from (16.5) and (16.6) that

$$\frac{\partial \sigma'_{zz}}{\partial t} = \frac{\partial p}{\partial t} \qquad (16.8)$$

so that (16.7) becomes

$$\frac{\partial e_0}{\partial t} = m_v \frac{\partial p}{\partial t} \qquad (16.9)$$

Combining (16.4) and (16.9) we may write:

$$\nabla^2 \Phi = H(m_v + n\beta) \frac{\partial p}{\partial t} \qquad (16.10)$$

Neglecting the variations of k, H, and ρ with time we obtain

$$\frac{\partial \Phi}{\partial t} = kH \frac{\partial}{\partial t} \left[\frac{p}{\rho g} + Z \right] = \frac{kH}{\rho g} \frac{\partial p}{\partial t} \qquad (16.11)$$

so that (16.10) may be written in terms of Φ as

$$\nabla^2 \Phi = \frac{S_s}{k} \frac{\partial \Phi}{\partial t} \qquad (16.12)$$

where S_s [1/m] is the coefficient of specific storage:

$$\boxed{S_s = \rho g (m_v + n\beta)} \qquad (16.13)$$

Sec. 16   Transient Shallow Flow

Table 2.4 *Typical Values of m_v*

	m_v [m²/N]
Clay	$10^{-6} - 10^{-8}$
Sand	$10^{-7} - 10^{-9}$
Gravel	$10^{-8} - 10^{-10}$
Jointed rock	$10^{-8} - 10^{-10}$
Sound rock	$10^{-9} - 10^{-11}$
Water (β)	$4.4 * 10^{-10}$

(Source: R. Allan Freeze/John Cherry, *Groundwater*, © 1979, p. 55.
Reprinted by permission of Prentice Hall, Englewood Cliffs, New Jersey.)

Some typical values of m_v are given in Table 2.4.

Transient Shallow Unconfined Flow

The elastic storage is several orders of magnitude less than the storage due to vertical movement of a phreatic surface, and therefore the elastic storage is usually neglected for transient shallow unconfined flow. The rate of decrease in storage, N^*, due to a rate of lowering $-\partial\phi/\partial t$ of a phreatic surface (see Figure 2.69) is equal to $-S_p\partial\phi/\partial t$, where S_p is the coefficient of phreatic storage, representing the fraction of a unit volume available for storage, i.e.,

$$N^* = -S_p \frac{\partial \phi}{\partial t} \tag{16.14}$$

The differential equation for transient flow becomes

$$\nabla^2 \Phi = S_p \frac{\partial \phi}{\partial t} - N \tag{16.15}$$

where N represents infiltration from rainfall. This differential equation, known as the Boussinesq equation, is nonlinear and only a few exact solutions of limited practical interest exist (see Boussinesq [1904] and Aravin and Numerov [1965]).

Equation (16.15) may be linearized as follows. Writing ϕ as $\overline{\phi} + \varepsilon$, where $\overline{\phi}$ is some average value, we may approximate the expression for the potential as follows:

$$\Phi = \tfrac{1}{2}k(\overline{\phi} + \varepsilon)^2 \approx \tfrac{1}{2}k\overline{\phi}^2 + k\overline{\phi}\varepsilon \tag{16.16}$$

so that

$$\frac{\partial \Phi}{\partial t} \approx k\overline{\phi}\frac{\partial \varepsilon}{\partial t} = k\overline{\phi}\frac{\partial \phi}{\partial t} \tag{16.17}$$

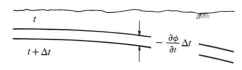

Figure 2.69 Decrease in storage due to lowering of the phreatic surface.

and (16.15) becomes

$$\nabla^2 \Phi = \frac{S_p}{k\bar{\phi}} \frac{\partial \Phi}{\partial t} - N \qquad (16.18)$$

Defining the coefficient of the specific storage S_s as

$$S_s = \frac{S_p}{\bar{\phi}} \qquad (16.19)$$

(16.18) becomes

$$\boxed{\nabla^2 \Phi = \frac{S_s}{k} \frac{\partial \Phi}{\partial t} - N} \qquad (16.20)$$

It is of interest to note that, in the absence of rainfall, the linearized form of the differential equation for transient shallow unconfined flow is the same as the one for transient shallow confined flow (compare [16.12]). The solution to (16.12) given below therefore is applicable to both types of flow.

Transient Shallow Flow Toward Wells

The solution for flow toward a well in an aquifer of infinite extent is one of the most commonly used solutions for transient groundwater flow, and is due to Theis [1935]. The problem is axisymmetric, and we write the Laplacian $\nabla^2 \Phi$ in (16.12) by the use of (14.31) in terms of the radial coordinate r:

$$\frac{\partial^2 \Phi}{\partial r^2} + \frac{1}{r} \frac{\partial \Phi}{\partial r} = \frac{S_s}{k} \frac{\partial \Phi}{\partial t} \qquad (16.21)$$

The Theis solution satisfies the following boundary conditions:

$$r \to \infty \qquad \phi = \phi_0 \qquad \Phi = \Phi_0 \qquad (16.22)$$

$$\lim_{r \to 0} r \frac{\partial \Phi}{\partial r} = \frac{Q}{2\pi} \qquad (t > t_0) \qquad (16.23)$$

Sec. 16   Transient Shallow Flow

where, for simplicity, the radius of the well is taken to be infinitely small. The solution to the above problem may be determined by the application of Laplace transforms. We will omit the derivation, present the solution, and demonstrate that it fulfills both the differential equation and the boundary conditions:

$$\boxed{\Phi = -\frac{Q}{4\pi} E_1(u) + \Phi_0} \tag{16.24}$$

where E_1 is the exponential integral and the dimensionless variable u is given by

$$u = \frac{S_s r^2}{4k(t - t_0)} \tag{16.25}$$

The exponential integral (see Abramowitz and Stegun [1965]) is defined as

$$E_1(u) = \int_u^\infty \frac{e^{-\xi}}{\xi} d\xi \tag{16.26}$$

In conjunction with the Theis solution, the exponential integral is often referred to as the well function W,

$$W(u) = E_1(u) \tag{16.27}$$

To demonstrate that (16.24) is a solution to the differential equation we differentiate the exponential integral twice with respect to r and once with respect to t. We write (16.26) as

$$E_1(u) = \int_u^\infty f(\xi) d\xi = F(\infty) - F(u) \tag{16.28}$$

where $f(\xi) = e^{-\xi}/\xi = dF/d\xi$, and differentiate $E_1(u)$ with respect to u

$$\frac{dE_1(u)}{du} = -f(u) = -\frac{e^{-u}}{u} \tag{16.29}$$

The derivatives with respect to r and t are obtained by the use of the chain rule and (16.25) as

$$\frac{\partial E_1}{\partial r} = -\frac{e^{-u}}{u} \frac{2S_s r}{4k(t - t_0)} = -2\frac{e^{-u}}{r} \tag{16.30}$$

and

$$\frac{\partial^2 E_1}{\partial r^2} = 2\frac{e^{-u}}{r} \frac{2S_s r}{4k(t - t_0)} + 2\frac{e^{-u}}{r^2} = 2\frac{e^{-u}}{r^2}(2u + 1) \tag{16.31}$$

and

$$\frac{\partial E_1}{\partial t} = \frac{e^{-u}}{u} \frac{S_s r^2}{4k(t - t_0)^2} = \frac{e^{-u}}{t - t_0} \tag{16.32}$$

Using (16.30) through (16.32) we substitute the expression (16.24) for the potential in the differential equation (16.21):

$$-\frac{Q}{2\pi}\frac{e^{-u}}{r^2}(2u+1) + \frac{Q}{2\pi}\frac{e^{-u}}{r^2} = -\frac{Q}{4\pi}\frac{S_s}{k}\frac{e^{-u}}{(t-t_0)} \qquad (16.33)$$

Multiplication of both sides by $(\pi/Q)r^2 e^u$ gives, using (16.25), the identity $-u = -u$ so that the differential equation is fulfilled. For $r \to \infty$ the exponential integral vanishes so that (16.22) is satisfied. It follows from (16.24) and (16.30) that

$$r\frac{\partial \Phi}{\partial r} = \frac{Q}{2\pi}e^{-u} \qquad (16.34)$$

In the limit for $r \to 0$ and $t > t_0$, u equals zero and (16.34) reduces to (16.23) so that the latter boundary condition is fulfilled as well.

The exponential integral may be conveniently approximated for computational purposes by various formulae listed in Abramowitz and Stegun [1965] as follows. For u in the range $0 \le u \le 1$ the formula by Allen [1954] yields results with a maximum error of $2 * 10^{-7}$:

$$E_1(u) = -\ln u + \left[\{[(a_5 u + a_4)u + a_3]u + a_2\}u + a_1\right]u + a_0 \qquad (16.35)$$

where

$$\begin{aligned}
&a_0 = -.57721566 \qquad a_1 = .99999193 \qquad a_2 = -.24991055 \\
&a_3 = .05519968 \qquad a_4 = -.00976004 \qquad a_5 = .00107857
\end{aligned} \qquad (16.36)$$

The second term in (16.35) is written such as to optimize numerical accuracy; this procedure is called Horner's rule. For u in the range $1 \le u < \infty$, Hastings [1955] presents a choice of two approximations, the more accurate one being more complex and therefore computationally slower. The first approximation is

$$ue^u E_1(u) = \frac{(u+b_1)u + b_0}{(u+c_1)u + c_0} + \varepsilon(u) \qquad (|\varepsilon(u)| < 5 * 10^{-5}) \qquad (16.37)$$

where

$$\begin{aligned}
&b_0 = .250621 \qquad c_0 = 1.681534 \\
&b_1 = 2.334733 \qquad c_1 = 3.330657
\end{aligned} \qquad (16.38)$$

The second approximation is very accurate, and is given by

$$ue^u E_1(u) = \frac{\{[(u+b_3)u + b_2]u + b_1\}u + b_0}{\{[(u+c_3)u + c_2]u + c_1\}u + c_0} + \varepsilon(u) \qquad (|\varepsilon(u)| < 2 * 10^{-8}) \qquad (16.39)$$

where

$$b_0 = .2677737343 \quad c_0 = 3.9584969228$$
$$b_1 = 8.6347608925 \quad c_1 = 21.0996530827$$
$$b_2 = 18.0590169730 \quad c_2 = 25.6329561486 \quad (16.40)$$
$$b_3 = 8.5733287401 \quad c_3 = 9.5733223454$$

The above approximations are readily implemented in a computer program as a function subroutine. The constant a_0 in (16.35) is the opposite of Euler's constant, γ,

$$a_0 = -\gamma = -.57721566 \qquad (16.41)$$

and for small values of u, $E_1(u)$ may be approximated as

$$E_1(u) = -\ln u - \gamma \qquad (u \ll 1) \qquad (16.42)$$

and the potential (16.24) may be approximated as

$$\Phi = \frac{Q}{4\pi}\left[\ln\frac{S_s r^2}{4k(t-t_0)} + \gamma\right] + \Phi_0 = \frac{Q}{2\pi}\ln\frac{r}{R_i} + \Phi_0 \qquad (u \ll 1) \qquad (16.43)$$

where R_i is called the radius of influence,

$$R_i = \sqrt{\frac{4k(t-t_0)}{S_s e^\gamma}} \qquad (16.44)$$

The radius of influence is often used as an approximate measure of the distance over which the well has a noticeable effect: for $r = R_i$, Φ equals the initial value Φ_0. It is seen from (16.44) that the radius of influence increases with time. Note, however, that the above approximation is valid only for small values of u.

Cooper and Jacob [1946] suggested a graphical method for determining the transmissivity T,

$$T = kH \qquad (16.45)$$

and the specific storage coefficient S_s from a pumping test, using the approximation (16.43). We consider the case of a confined aquifer and rewrite (16.43) as follows (compare Lohman [1972]):

$$\phi_0 - \phi = s = \frac{Q}{4\pi T}\left[\ln e^{-\gamma} + \ln\frac{4T(t-t_0)}{Sr^2}\right] \qquad (16.46)$$

where $e^{-\gamma} \approx .562$, s is the draw-down and S the storativity, defined as

$$S = S_s H \qquad (16.47)$$

We write (16.46) in dimensionless form as follows (compare Verruijt [1982]):

$$\frac{\phi_0 - \phi}{\phi_1} = \frac{s}{\phi_1} = \frac{Q}{2\pi T \phi_1} \left[\ln \frac{\sqrt{4T_1(t-t_0)e^{-\gamma}/S_1}}{r} + \ln \sqrt{\frac{S_1 T}{ST_1}} \right] \quad (16.48)$$

where ϕ_1, S_1, and T_1 are arbitrary reference quantities. It follows from (16.48) that the dimensionless draw-down $(\phi_0 - \phi)/\phi_1$ when plotted versus $\ln\{[4T_1(t-t_0)/(S_1 e^\gamma)]^{1/2}/r\} = \ln \tau$ gives a straight line as shown in Figure 2.70. By measuring the draw-down at some fixed observation point at different times and plotting these data versus $\ln \tau$ a straight line is obtained. The slope of the line yields T, and the ordinate of the point of intersection with the draw-down axis gives the quantity $S_1 T/(ST_1)$ so that S may be computed as well.

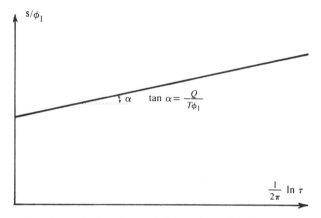

Figure 2.70 Determination of transmissivity and storativity from pumping tests.

Various other graphical methods have been developed for determining S and T, both for confined and unconfined aquifers, and using the exponential integral $E_1(u)$, rather than its approximation (16.42). These methods for pumping test data analysis are treated extensively in the literature and the interested reader is referred to texts such as Freeze and Cherry [1979] and Davis and DeWiest [1966]. An alternative way of determining aquifer parameters from pumping test data is to use a computer program based on the Theis solution.

The Theis solution is ideally suited for pumping test analysis. It is important in view of groundwater modeling applications to understand fully the implications of the boundary condition (16.22). This condition implies that the head at infinity is maintained at a constant level at all times. The well therefore draws all of its water from removal from storage, and none from other sources. In the steady-state solution for a single well in an infinite aquifer these other sources are lumped together at infinity. As a result, the Theis solution does not converge to the steady-state solution; it will eventually draw the water levels throughout the aquifer below the aquifer base. Assuming that $u \ll 1$,

Sec. 16   Transient Shallow Flow

the time at which the head at the well screen reaches the aquifer base ($\Phi_w = 0$) may be computed by the use of (16.43) as follows

$$\frac{Q}{2\pi} \ln \frac{r_w}{R_i} = -\Phi_0 \qquad (16.49)$$

or

$$R_i = r_w e^{2\pi \Phi_0 / Q} \qquad (16.50)$$

or, with (16.44)

$$(t - t_0)_{\max} = \frac{r_w^2 S_s e^\gamma}{4k} e^{4\pi \Phi_0 / Q} \qquad (16.51)$$

The governing differential equation as used here is linear, and therefore solutions may be superimposed. Since any steady-state solution of $\nabla^2 \Phi = 0$ satisfies (16.12), steady-state and transient solutions may be combined. As an example, consider the case of a well, pumping under steady-state a discharge Q. The well is switched off from $t = t_0$ until $t = t_1$. The potential that applies to this case may be written as

$$\Phi = \frac{Q}{2\pi} \ln \frac{r}{R} + \Phi_0 + \frac{Q}{4\pi}[E_1(u_0) - E_1(u_1)] \qquad (16.52)$$

where

$$u_0 = \frac{S_s r^2}{4k(t - t_0)} \qquad u_1 = \frac{S_s r^2}{4k(t - t_1)} \qquad (16.53)$$

The first term inside the brackets of (16.52) causes a recharge well to be switched on at time t_0, canceling the discharge of the well. The second term inside the brackets represents a well switched on at time t_1, from that time onward causing the original discharge to be restored. For large times, the two exponential integrals cancel. For sufficiently large times, u_0 and u_1 become small so that the approximation (16.43) applies:

$$\Phi = \frac{Q}{2\pi} \ln \frac{r}{R} + \Phi_0 + \frac{Q}{4\pi} \left[\ln \frac{S_s r^2}{4k(t - t_0)} - \ln \frac{S_s r^2}{4k(t - t_1)} \right] \qquad (16.54)$$

or

$$\Phi = \frac{Q}{2\pi} \ln \frac{r}{R} + \Phi_0 + \frac{Q}{4\pi} \ln \frac{t - t_1}{t - t_0} \qquad (16.55)$$

The second logarithm vanishes for $t \to \infty$, so that the solution reduces to the one for steady-state flow toward a well.

The method of images may be applied to transient solutions. As an example, consider the problem of flow toward a well near a river with $\Phi = \Phi_0$ along the y axis, shown in Figure 2.71. The well is located at $x = d$, $y = 0$ and starts pumping a discharge Q at time t_0. The solution is obtained by placing an image recharge well at $x = -d$, $y = 0$,

$$\Phi = -\frac{Q}{4\pi}[E_1(u_1) - E_1(u_2)] + \Phi_0 \tag{16.56}$$

where

$$u_1 = \frac{S_s[(x-d)^2 + y^2]}{4k(t-t_0)} \qquad u_2 = \frac{S_s[(x+d)^2 + y^2]}{4k(t-t_0)} \tag{16.57}$$

It is left as an exercise to the reader to determine the steady-state form of (16.56) by taking the limit for $t \to \infty$.

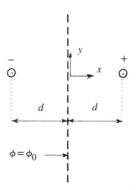

Figure 2.71 Transient flow toward a well near a river.

Problem 16.1
Determine the steady-state form of (16.56) by taking the limit for $t \to \infty$.

Problem 16.2
Write a function subroutine in FORTRAN for the exponential integral using the approximations (16.35) and (16.37), and implement this function in SLWL, to be used only for uniform hydraulic conductivity (i.e., no layering). In order to pass the value of $(t - t_0)$ to the function subroutine, it is suggested that a common block CBTIME be created, and a command TIME be added to SETUP to set the value of t. The coefficient of specific storage may be added to the common block CBAQ. A new common block will be necessary to accommodate the data for the transient wells.

Numerous solutions to transient problems involving wells have been determined by the application of Laplace transforms (see, for example, Hantush [1964] and Glover [1974]). The governing differential equation, known as the diffusion equation, occurs in various other fields, for example, that of conduction of heat in solids. Many of the solutions determined in the latter field (see, for example, Carslaw and Jaeger [1959]) are directly applicable to transient groundwater problems; in fact, the Theis solution is an example of such an application.

17 TWO-DIMENSIONAL FLOW IN THE VERTICAL PLANE

Two-dimensional flow in the vertical plane occurs if there exists a horizontal direction in which there is no flow. As an example, a case of flow underneath an impermeable dam is depicted in Figure 2.72. The flow occurs in the aquifer underneath the dam and is driven by the difference in head along the bottoms of the two lakes which are separated by the dam. Few cases of true two-dimensional flow in the vertical plane occur in reality, but a two-dimensional analysis often gives a good approximation. If the dam in Figure 2.72 is long with respect to its width, then the results obtained from a two-dimensional analysis will be accurate near the central portion of the dam and erroneous near the ends.

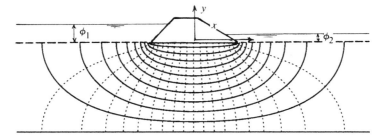

Figure 2.72 Flow underneath a dam.

The boundary of the flow domain often consists of several segments such as the horizontal streamlines and equipotentials in Figure 2.72. The elementary methods of solution discussed thus far are applicable only to problems with at the most two orthogonal infinite impermeable boundaries or equipotentials as boundaries. More sophisticated techniques that can be used to determine exact solutions of problems of two-dimensional flow in the vertical plane are covered in Chapters 5 and 7, numerical methods are treated in Chapters 8 and 9. We will focus our attention in this section on deriving the basic equations and give an application of the method of images. We will further discuss a particular approximate method, known as the graphical method, and apply it to a few problems involving dams. For all problems of two-dimensional flow in the vertical plane, we will use x, y, z Cartesian coordinates with the z axis normal to the plane of flow, the x and y axes in the plane of flow, and the y axis pointing vertically upward (see Figure 2.72).

Basic Equations

The basic equations are obtained directly from Darcy's law and the continuity equation presented in Chapter 1. There is no flow in the z direction, so that

$$q_z = 0 \qquad (17.1)$$

All discharges are calculated over a slice of aquifer of width B normal to the plane of flow. The components Q_x and Q_y of the discharge vector are introduced accordingly,

$$\boxed{\begin{aligned} Q_x &= Bq_x \\ Q_y &= Bq_y \end{aligned}} \qquad (17.2)$$

Darcy's law (5.3) gives with (17.2)

$$\begin{aligned} Q_x &= -\frac{\partial(kB\phi)}{\partial x} \\ Q_y &= -\frac{\partial(kB\phi)}{\partial y} \end{aligned} \qquad (17.3)$$

and we define the discharge potential Φ for two-dimensional flow in the vertical plane as

$$\Phi = kB\phi \qquad (17.4)$$

The components Q_x and Q_y may be written in terms of Φ as

$$\boxed{\begin{aligned} Q_x &= -\frac{\partial \Phi}{\partial x} \\ Q_y &= -\frac{\partial \Phi}{\partial y} \end{aligned}} \qquad (17.5)$$

With (17.1) the continuity equation (4.3) becomes

$$\frac{\partial q_x}{\partial x} + \frac{\partial q_y}{\partial y} = 0 \qquad (17.6)$$

or, after multiplication by the constant B,

$$\frac{\partial Q_x}{\partial x} + \frac{\partial Q_y}{\partial y} = 0 \qquad (17.7)$$

Combination of (17.5) and (17.7) yields Laplace's equation for the potential Φ,

$$\boxed{\nabla^2 \Phi = 0} \qquad (17.8)$$

The basic equations (17.5) and (17.8) are formally equal to those for the various types of shallow flow without leakage and infiltration. The elementary solutions and techniques discussed earlier in this chapter for the latter types of flow can therefore be applied to two-dimensional flow in the vertical plane.

Horizontal Drains

A horizontal drain is a drainpipe with a horizontal axis. If the head inside the drain is kept at a fixed level by pumping, flow will occur toward the drain. A case of flow toward a horizontal drain is illustrated in Figure 2.73. The drain lies in a deep aquifer, which is bounded above by an impermeable layer.

Flow toward a single drain of infinite length in an infinite aquifer is radial, and the potential that fulfills Laplace's equation for radial flow is

$$\Phi = \frac{Q}{2\pi} \ln r + C \tag{17.9}$$

where Q is the discharge flowing into the drain per length B. The aquifer in Figure 2.73 is semi-infinite; there is an impermeable boundary along $y = 0$. If the center of the drain is on the y axis at a depth d below the origin, then an image drain of discharge Q must be placed at $x = 0$, $y = d$ in order to meet the boundary condition that the x axis is impermeable. The potential is

$$\Phi = \frac{Q}{4\pi} \{\ln[x^2 + (y+d)^2] + \ln[x^2 + (y-d)^2]\} + C \tag{17.10}$$

where C is a constant. If, in addition, there is a uniform flow of discharge Q_{x0} parallel to the boundary, then a term $-Q_{x0}x$ must be added to (17.10),

$$\Phi = -Q_{x0}x + \frac{Q}{4\pi} \{\ln[x^2 + (y+d)^2] + \ln[x^2 + (y-d)^2]\} + C \tag{17.11}$$

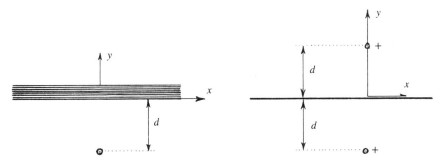

Figure 2.73 Flow toward a horizontal drain.

The constant C may be evaluated by requiring that the head be equal to some given value ϕ_0 at a point $\mathcal{P}$ with coordinates $x = x_0$, $y = y_0$:

$$x = x_0 \quad y = y_0 \quad \begin{Bmatrix} \phi = \phi_0 \\ \Phi = \Phi_0 = kB\phi_0 \end{Bmatrix} \tag{17.12}$$

Substitution in (17.11) of the expression obtained for C by applying (17.12) to (17.11) yields

$$\Phi = -Q_{x0}(x - x_0) + \frac{Q}{4\pi} \ln \frac{[x^2 + (y+d)^2][x^2 + (y-d)^2]}{[x_0^2 + (y_0+d)^2][x_0^2 + (y_0-d)^2]} + \Phi_0 \qquad (17.13)$$

An expression for the head in the aquifer is obtained by division of (17.13) by a factor kB.

The Graphical Method

The graphical method is simple, and requires little time and equipment to obtain a good approximation for many problems. It is used mostly to determine pressure distributions in embankments and dams, behind earth retaining structures, or for estimating seepage through dams and underneath hydraulic structures. The method is based upon the properties of equipotentials and streamlines (see Sec. 6). Equipotentials are curves along which the potential is constant. Streamlines are curves defined by requiring that the discharge vector is everywhere tangent to the streamline. Since the flow is normal to the equipotentials, as shown in Sec. 6, the streamlines and equipotentials are mutually orthogonal.

Flow Nets

A flow net is a graphical representation of equipotentials and streamlines in a flow region. An example of a flow net is shown in Figure 2.74. The intervals between consecutive streamlines and equipotentials are arbitrary but should be taken small enough so that the sides of the rectangles of the net are approximately straight. Consider the rectangle $\mathcal{ABFE}$ with sides Δs and Δn in Figure 2.74. The directions s and n are along the streamlines and equipotentials as shown in the figure. The magnitude of s at an arbitrary point $\mathcal{P}$ of a streamline equals the arc length of the section of that streamline between $\mathcal{P}$ and the equipotential $\mathcal{AD}$. The arc lengths of sections of the equipotentials are expressed similarly in terms of n. The flow through side $\mathcal{AB}$ will be denoted as ΔQ. Assuming that Q_s varies only slightly along $\mathcal{AB}$, ΔQ will be equal to the value of Q_s at a point halfway between $\mathcal{A}$ and $\mathcal{B}$, multiplied by Δn, i.e.,

$$\Delta Q \approx Q_s \Delta n \qquad (17.14)$$

Furthermore, $Q_s = -d\Phi/ds$ by Darcy's law or, approximately,

$$\Delta Q \approx -\frac{\Delta \Phi}{\Delta s} \Delta n \qquad (17.15)$$

where $\Delta \Phi$ is the difference in value of Φ on the equipotentials $\mathcal{EF}$ and $\mathcal{AB}$, i.e.,

$$\Delta \Phi = \Phi_{\mathcal{EF}} - \Phi_{\mathcal{AB}} \qquad (17.16)$$

Sec. 17   Two-dimensional Flow in the Vertical Plane

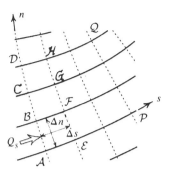

Figure 2.74   A flow net.

Since Φ decreases in the direction of flow, $\Phi_{\mathcal{EF}}$ is less than $\Phi_{\mathcal{AB}}$ so that $\Delta\Phi$ is negative and (17.15) may be written as

$$\Delta Q \approx \frac{|\Delta\Phi|}{\Delta s}\Delta n \tag{17.17}$$

We draw the flow net such that the rectangles are approximately square: $\Delta s = \Delta n$, and (17.17) becomes

$$\Delta Q \approx |\Delta\Phi| \tag{17.18}$$

No water can cross the streamlines: ΔQ is a constant inside the channel between the streamlines $\mathcal{BF}$ and $\mathcal{AE}$ by continuity,

$$\Delta Q = |\Delta\Phi| = \text{constant} \tag{17.19}$$

Hence, the drop in potential, $|\Delta\Phi|$, is the same for each pair of consecutive equipotentials: if the potential is known at the bordering equipotentials of the flow net, $|\Delta\Phi|$ can be found. If Φ along the equipotentials $\mathcal{AD}$ and $\mathcal{PQ}$ in Figure 2.74 is known, then

$$|\Delta\Phi| = \frac{\Phi_{\mathcal{AD}} - \Phi_{\mathcal{PQ}}}{m_e} \tag{17.20}$$

where m_e, the number of drops in potential, is 3 in the case of Figure 2.74. Since $|\Delta\Phi|$ is a constant throughout the flow net, ΔQ is the same for each pair of consecutive streamlines, by (17.18). The total discharge through the flow net is found by multiplying ΔQ by the number of channels between the boundary streamlines, m_s. Thus, by (17.18) and (17.20),

$$Q = m_s \Delta Q \approx m_s |\Delta\Phi| = \frac{m_s}{m_e}\left(\Phi_{\mathcal{AD}} - \Phi_{\mathcal{PQ}}\right) \tag{17.21}$$

The channels between the streamlines are called flow channels. The channels between the equipotentials are called equipotential channels.

We now may draw the following conclusion: if a flow net is a mesh of approximate squares, then the discharge flowing through the net may be approximated by (17.21). The numbers m_s and m_e represent the numbers of flow channels and equipotential channels, respectively. The value of Φ on any equipotential can be determined, and the values of Φ at points in between equipotentials may be computed either by interpolation, or by locally refining the mesh. The pressure is found from the head using the relation $p = \rho g[\phi - Z]$ where $\phi = \Phi/(kB)$.

A solution obtained by drawing a flow net will become more accurate as the squares are made smaller. Some applications of the graphical technique follow.

Applications of the Graphical Method

A good flow net fulfills the following conditions:

1. Equipotentials and streamlines are everywhere orthogonal.
2. Equipotentials and streamlines form approximate squares.
3. The flow net satisfies the boundary conditions for the problem.

The flow net for flow underneath a dam with a cutoff wall is shown in Figure 2.75. The boundary conditions are that $\mathcal{SPT}$ and $\mathcal{VW}$ are streamlines and that $\mathcal{RS}$ and $\mathcal{TU}$ are equipotentials; the streamlines of the net must intersect the boundaries $\mathcal{RS}$ and $\mathcal{TU}$ at right angles and the equipotentials must be perpendicular to $\mathcal{VW}$ and $\mathcal{SPT}$. The flow net is obtained by a trial-and-error procedure. One approach is to draw the streamlines first, for the case of Figure 2.75 from $\mathcal{TU}$ to $\mathcal{RS}$. The equipotentials are drawn next so as to obtain a pattern of approximate squares. The first estimate of the streamlines usually appears inaccurate after drawing the equipotentials; redrawing of the streamlines and equipotentials is necessary until the flow net is good.

The number of flow channels need not be an integer. The lowest curved streamline in Figure 2.75 was drawn in order to improve the flow net. The width of the channel enclosed between that streamline and the boundary streamline $\mathcal{VW}$ is about one tenth of the width of a flow channel. The total number of flow channels between $\mathcal{RS}$ and $\mathcal{VW}$ is therefore 3.1.

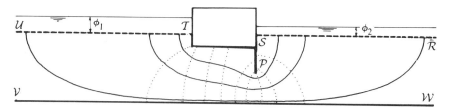

Figure 2.75 Flow underneath a dam with a cutoff wall. (Source: A. Verruijt, *Theory of Groundwater Flow*, © 1970, p. 147. Reprinted by permission of the author.)

The discharge flowing underneath the dam from $\mathcal{TU}$ to $\mathcal{RS}$ is found as follows. The difference in potential between $\mathcal{TU}$ and $\mathcal{RS}$ is

$$\Phi_1 - \Phi_2 = kB(\phi_1 - \phi_2) \tag{17.22}$$

Sec. 17   Two-dimensional Flow in the Vertical Plane

There are $m_e = 10$ equipotential channels between $\mathcal{TU}$ and $\mathcal{RS}$, so that

$$|\Delta \Phi| = \frac{\Phi_1 - \Phi_2}{m_e} = \frac{k(\phi_1 - \phi_2)B}{10} \tag{17.23}$$

Since $\Delta Q \approx |\Delta \Phi|$ and since there are $m_s = 3.1$ flow channels, the total discharge flowing underneath the dam is approximately given by

$$Q \approx m_s \Delta Q \approx m_s |\Delta \Phi| = 3.1 \frac{kB(\phi_1 - \phi_2)}{10} \tag{17.24}$$

If the difference in the heads between $\mathcal{TU}$ and $\mathcal{RS}$ is

$$\phi_1 - \phi_2 = 2.0 \text{ m} \tag{17.25}$$

and if the hydraulic conductivity is 10^{-4} m/sec,

$$k = 10^{-4} \text{ m/s} = 8.64 \text{ m/day} \tag{17.26}$$

then the seepage underneath the dam per B meters of length is found from (17.24) to be

$$Q = 3.1 \frac{10^{-4} * 2}{10} B = 6.2 * 10^{-5} * B \text{ m}^3/\text{sec} = 5.4 * B \text{ m}^3/\text{day} \tag{17.27}$$

It may be noted that the scale of the drawing has no influence on the discharge, which solely depends upon the geometry of the flow domain and the difference $\Phi_1 - \Phi_2$. A lengthening of the sheet piling $\mathcal{SP}$, however, will result in an increase of the number of equipotential channels; the discharge will decrease accordingly.

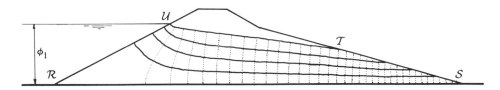

Figure 2.76   Flow through a dam. (Source: A. Verruijt, *Theory of Groundwater Flow*, © 1970, p. 148. Reprinted by permission of the author.)

Another flow net is shown in Figure 2.76. It represents the flow through a dam. The upper boundary $\mathcal{TU}$ of the flow net is a phreatic surface. It is assumed that the pores of the soil beneath the phreatic surface are completely filled with water whereas the pores above it are filled with air; the phreatic surface is a streamline. The phreatic surface represents a special kind of boundary condition: neither its form nor its position is known beforehand; it is called a free surface or a free boundary. The boundary conditions along this free surface are that (1) it is a streamline and (2) the pressure along it is equal to the atmospheric pressure, which usually is taken as zero. As follows from (2.7) the head ϕ along the free surface equals its elevation,

$$\phi = \frac{p}{\rho g} + Y = Y \tag{17.28}$$

where Y is now used to represent the elevation head rather than Z. Since $\Phi = kB\phi$, it follows that

$$\Phi = kBY \tag{17.29}$$

along the free surface. This implies that the vertical distance, ΔY, between the points of intersection of two consecutive equipotentials with the free surface is the same along the entire free surface, as is indicated in Figure 2.77. The free surface is drawn by trial and error: Its form and position are estimated, the flow net is drawn with the free surface as the boundary streamline, and it is determined whether the boundary condition (17.29) is met. If this condition is not met, the free surface is modified and the flow net redrawn. This procedure is repeated until the flow net is good.

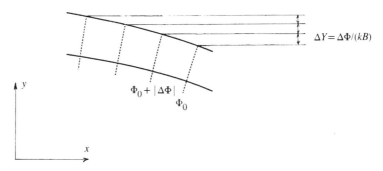

Figure 2.77 Detail of a phreatic surface.

The boundary segment $\mathcal{ST}$ of Figure 2.76 is called a seepage face; the water there leaves the dam, and the pressure is zero. Hence, the boundary condition along the seepage face is given by (17.29); the intersection points of consecutive equipotentials with the seepage face must be the same vertical distance apart along the entire seepage face.

Point $\mathcal{T}$ in Figure 2.76 is the intersection point of the free surface with the seepage face. The free surface is tangent to the seepage face at point $\mathcal{T}$; the proof of this statement is given by the use of the hodograph method in Chapter 7. The other boundary conditions for the case of Figure 2.76 are that $\mathcal{UR}$ is an equipotential ($\Phi = kB\phi_1 = \Phi_1$) and that $\mathcal{RS}$ is a streamline.

It requires several trials to obtain a good flow net involving a free surface such as the one represented in Figure 2.76. However, when the flow net is drawn carefully, the result will be quite accurate.

The total discharge flowing per unit width through the dam of Figure 2.76 is found as follows. The number of equipotential channels is 35. The potential decreases from $kB\phi_1$ along $\mathcal{RS}$ to 0 at point $\mathcal{S}$, so that

$$|\Delta\Phi| = \frac{kB\phi_1}{35} \tag{17.30}$$

Since there are four flow channels, the total discharge, Q, per unit length of dam is

$$Q = m_s \Delta Q = 4\Delta Q \approx 4|\Delta \Phi| = \frac{4kB\phi_1}{35} \qquad (17.31)$$

If ϕ_1 is 10 meters and k is again 10^{-4} m/sec or 8.64 m/day, and if $B = 1$ m,

$$\phi_1 = 10 \text{ m} \qquad k = 10^{-4} \text{ m/s} = 8.64 \text{ m/day} \qquad B = 1 \text{ m} \qquad (17.32)$$

then Q becomes:

$$Q = \frac{4}{35} 10^{-3} \text{ m}^3/\text{s} = 1.14 * 10^{-4} \text{ m}^3/\text{s} \approx 10 \text{ m}^3/\text{day} \qquad (17.33)$$

Problem 17.1

A long waterway will be dug into a thick horizontal impervious clay layer resting on a layer of fine sand of thickness H (see Figure 2.78). The layer of fine sand is bounded below by coarse sand. The head in the coarse sand is constant and equal to ϕ_1 ($\phi_1 > H$). A long sand drain of width d runs along the center of the waterway and is used for pressure relief during construction. The head underneath the trench at the interface between clay and fine sand must be kept below level H_0. The flow is two-dimensional.

Determine the maximum allowable head that must be maintained at the bottom of the sand drain. (Note that the resistance to flow in the sand drain can be neglected and that it is assumed that any head between ϕ_1 and H can be maintained at the bottom of the sand drain.) Copy the figure and draw your flow net in the figure using the graphical method.

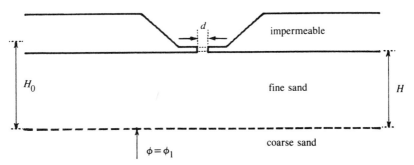

Figure 2.78 Flow toward a sand drain (Problem 17.1).

Anisotropy

The graphical method is not directly applicable to problems with anisotropic hydraulic conductivity, because there does not exist a potential: If x^* and y^* are Cartesian coordinates with the x^* axis parallel to the major principal direction of hydraulic conductivity tensor, then Darcy's law is given by (3.18):

$$q_x^* = -k_1 \frac{\partial \phi}{\partial x^*}$$
$$q_y^* = -k_2 \frac{\partial \phi}{\partial y^*} \qquad (17.34)$$

Indeed, we cannot define a single potential Φ since the coefficients in the two preceding equations are different.

It is possible, however, to transform the flow domain to another domain with coordinates X, Y such that a potential Φ may be defined in that domain. Let the transformation be given by

$$X = x^*$$
$$Y = \beta y^* \qquad (17.35)$$

We choose the constant β such that the head ϕ satisfies Laplace's equation in the transformed domain. Substitution of (17.34) in the continuity equation (4.3), written in terms of the x^*, y^* coordinate system, gives

$$\frac{\partial q_x^*}{\partial x^*} + \frac{\partial q_y^*}{\partial y^*} = -k_1 \frac{\partial^2 \phi}{\partial (x^*)^2} - k_2 \frac{\partial^2 \phi}{\partial (y^*)^2} = 0 \qquad (17.36)$$

With (17.35) this becomes

$$-k_1 \frac{\partial^2 \phi}{\partial X^2} - k_2 \beta^2 \frac{\partial^2 \phi}{\partial Y^2} = 0 \qquad (17.37)$$

In order to obtain Laplace's equation the constant β must be equal to

$$\beta = \sqrt{\frac{k_1}{k_2}} \qquad (17.38)$$

Since the head fulfills Laplace's equation in terms of X and Y, we can solve the problem in the transformed domain, for example, by the graphical method. We cannot compute flow rates from such a solution without first determining the isotropic hydraulic conductivity k_i in the transformed domain. This may be done by considering a case of one-dimensional flow in the x^* direction as follows (see Figure 2.79). Darcy's law in the transformed domain is

$$q_X = -k_i \frac{\partial \phi}{\partial X}$$
$$q_Y = -k_i \frac{\partial \phi}{\partial Y} \qquad (17.39)$$

Sec. 17 Two-dimensional Flow in the Vertical Plane

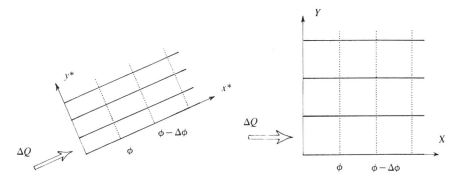

Figure 2.79 Transformation of a flow net.

The flow net in the x^*, y^* plane is such that a mesh of squares is obtained upon transformation to the X, Y plane. Denoting the difference in head between consecutive potentials in either plane as $\Delta\phi$, the discharge ΔQ flowing between two streamlines may be expressed as

$$\Delta Q = q_x^* \Delta y^* = -k_1 \frac{\Delta\phi}{\Delta x^*} \Delta y^* = -k_1 \frac{\Delta\phi}{\Delta X} \frac{\Delta Y}{\beta} \tag{17.40}$$

where use is made of (17.35). It follows from (17.39) that

$$\Delta Q = q_X \Delta Y = -\underset{i}{k} \frac{\Delta\phi}{\Delta X} \Delta Y \tag{17.41}$$

The above two expressions for ΔQ must be equal, so that

$$\underset{i}{k} = \frac{k_1}{\beta} = \sqrt{k_1 k_2} \tag{17.42}$$

We may now write Darcy's law for the discharge vector components Q_X and Q_Y in terms of a potential Φ defined in the isotropic domain as follows:

$$\begin{aligned} Q_X &= -\underset{i}{k} B \frac{\partial\phi}{\partial X} = -\frac{\partial\Phi}{\partial X} \\ Q_Y &= -\underset{i}{k} B \frac{\partial\phi}{\partial Y} = -\frac{\partial\Phi}{\partial Y} \end{aligned} \tag{17.43}$$

where

$$\Phi = \underset{i}{k} B \phi = B \phi \sqrt{k_1 k_2} \tag{17.44}$$

*Flow Through an Anisotropic Aquifer Underneath a Dam
with a Cutoff Wall*

As an example, we will reconsider the case of flow underneath a dam with a cutoff wall. The thickness of the aquifer and the depth of the cutoff wall are half of what they are for the case of Figure 2.75. The principal directions of the hydraulic conductivity are horizontal and vertical; $k_1 = 10^{-4}$ m/s, and $k_2 = k_1/4$:

$$k_2 = k_1/4 = 0.25 * 10^{-4} \text{ m/s} \tag{17.45}$$

The difference in head across the dam is $\phi_1 - \phi_2$, with

$$\phi_1 - \phi_2 = 2 \text{ m} \tag{17.46}$$

The transformation formulas become, with $x^* = x$ and $y^* = y$:

$$X = x$$
$$Y = \beta y = \sqrt{\frac{k_1}{k_2}} y = 2y \tag{17.47}$$

Thus, the boundary of the transformed flow domain, shown in Figure 2.80(b), is obtained from the flow domain in Figure 2.80(a) by multiplying all vertical coordinates by 2. The flow net may now be drawn in the isotropic domain, is identical to that shown in Figure 2.75, and is reproduced in Figure 2.80(b). This flow net is transformed by the use of (17.47), which results in the flow net in the anisotropic domain shown in Figure 2.80(a).

The potential Φ may be expressed in terms of ϕ by the use of (17.44) and (17.45)

$$\Phi = \underset{i}{k} B\phi = B\sqrt{\tfrac{1}{4}k_1^2}\phi = \tfrac{1}{2}Bk_1\phi = .5 * 10^{-4} B\phi \text{ m}^3/\text{s} \tag{17.48}$$

The difference in value of Φ across the flow domain is obtained by the use of (17.46) and (17.48):

$$\Phi_1 - \Phi_2 = 10^{-4} B \text{ m}^3/\text{s} \tag{17.49}$$

The number of equipotential channels (m_e) is 10 so that

$$|\Delta\Phi| = \frac{\Phi_1 - \Phi_2}{m_e} B = 10^{-5} B \text{ m}^3/\text{s} \tag{17.50}$$

There are 3.1 flow channels (compare [17.24]); the total discharge Q underneath the dam equals

$$Q = 3.1 \Delta Q = 3.1 |\Delta\Phi| = 3.1 * 10^{-5} B \text{ m}^3/\text{s} \tag{17.51}$$

It is usually difficult to estimate the effect of anisotropy; this effect depends upon the direction of flow. The value of k_2 will, for example, have little effect if the flow is primarily in the x^* direction.

Sec. 18 Three-dimensional Flow   215

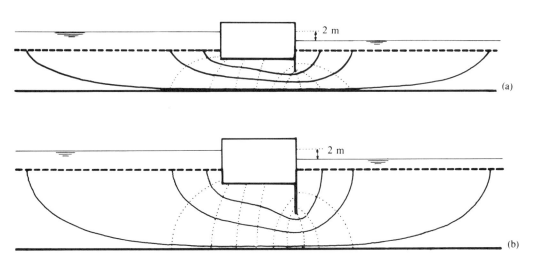

Figure 2.80 Flow in an anisotropic aquifer underneath a dam.

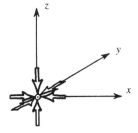

Figure 2.81 Radial flow toward a point-sink.

18 THREE-DIMENSIONAL FLOW

Only few exact solutions to problems of three-dimensional flow exist. We will derive the potential for flow in an infinite aquifer toward a point-sink, and apply the method of images to solve the problem of flow toward a point-sink in a deep aquifer bounded above by a horizontal impermeable layer.

The Point-Sink

We will derive the potential for flow toward a point-sink in an infinite aquifer by the use of the continuity equation and Darcy's law as follows (see Figure 2.81). The flow is radial, and we denote the component of the discharge vector in the radial (r) direction as q_r, where q_r is positive in the positive r direction, i.e., if pointing away from the point sink. By continuity of flow, the discharge flowing through a sphere of radius r and area $4\pi r^2$ around the sink must be equal to the discharge Q of the point-sink,

$$Q = -4\pi r^2 q_r \tag{18.1}$$

Defining the potential Φ [m²/s] for three-dimensional flow as $k\phi$,

$$\Phi = k\phi \qquad (18.2)$$

Darcy's law may be written as

$$q_r = -k\frac{d\phi}{dr} = -\frac{d\Phi}{dr} \qquad (18.3)$$

where it is noted that, by symmetry, Φ is a function of r alone. Combining (18.1) and (18.3) we get

$$\frac{d\Phi}{dr} = \frac{Q}{4\pi r^2} \qquad (18.4)$$

Integration yields

$$\Phi = -\frac{Q}{4\pi}\frac{1}{r} + \Phi_0 \qquad (18.5)$$

where r may be expressed as

$$r = \sqrt{(x-x_1)^2 + (y-y_1)^2 + (z-z_1)^2} \qquad (18.6)$$

with x_1, y_1, and z_1 representing the coordinates of the center of the sink in a Cartesian x, y, z coordinate system. It is of interest to note that, as opposed to the two-dimensional case of flow toward a well, the potential equals a constant at infinity. It follows from (18.5) that this constant equals Φ_0,

$$r \to \infty \quad \Phi \to \Phi_0 \qquad (18.7)$$

The reason why the potential is unbounded at infinity for two-dimensional flow toward a well follows. Viewed in infinite space, the two-dimensional solution for a well represents radial flow toward an infinite line (called a line-sink) along the z axis; for any value of z the radial component q_r is the same and lies in a plane parallel to the x, y plane. The discharge of this infinite line-sink is infinite, as opposed to the discharge of a point-sink which is finite. Drawing an infinite discharge appears to require that the potential increases logarithmically with the distance from the well. We always apply the two-dimensional solution to horizontal flow in an aquifer of finite thickness (in which the discharge is indeed finite), not realizing perhaps that viewed in three-dimensional space this discharge is infinite.

The Method of Images

The method of images can be used to solve a limited number of problems of three-dimensional flow. Consider the problem of flow toward a point-sink in a semi-infinite

Sec. 18 Three-dimensional Flow

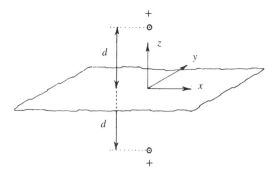

Figure 2.82 Flow toward a point-sink below an impermeable layer.

aquifer bounded above by a horizontal impermeable layer. We choose an x, y, z coordinate system such that the plane $z = 0$ coincides with the upper boundary of the aquifer (see Figure 2.82). The point-sink has a discharge Q and is located at $x_1 = y_1 = 0$, $z_1 = -d$.

In order to meet the boundary condition that the plane $z = 0$ is impermeable we place an image sink at $x_2 = y_2 = 0$, $z_2 = d$. The potential thus becomes

$$\Phi = -\frac{Q}{4\pi}\left[\frac{1}{\sqrt{x^2 + y^2 + (z+d)^2}} + \frac{1}{\sqrt{x^2 + y^2 + (z-d)^2}}\right] + \Phi_0 \qquad (18.8)$$

where Φ_0 represents the potential at infinity.

The case that there is a uniform flow of discharge q_{x0} in the x direction in addition to the well may be covered by adding the potential for uniform flow, $\Phi = -q_{x0}x$, to (18.8),

$$\Phi = -q_{x0}x - \frac{Q}{4\pi}\left[\frac{1}{\sqrt{x^2 + y^2 + (z+d)^2}} + \frac{1}{\sqrt{x^2 + y^2 + (z-d)^2}}\right] + C \qquad (18.9)$$

The constant C may be determined by requiring that the head be ϕ_0 at some reference point $x = x_0$, $y = z = 0$:

$$\Phi_0 = k\phi_0 = -q_{x0}x_0 - \frac{Q}{2\pi}\frac{1}{\sqrt{x_0^2 + d^2}} + C \qquad (18.10)$$

Solving this equation for C and substituting the result in (18.9) we obtain

$$\Phi = -q_{x0}(x - x_0) - \frac{Q}{4\pi}\left[\frac{1}{\sqrt{x^2 + y^2 + (z+d)^2}}\right.$$
$$\left. + \frac{1}{\sqrt{x^2 + y^2 + (z-d)^2}} - \frac{2}{\sqrt{x_0^2 + d^2}}\right] + \Phi_0$$

(18.11)

It is possible to solve the problem of flow toward a point-sink in a confined aquifer by the use of the method of images. In that case, the imaging need be continued indefinitely: an infinite number of images is necessary to meet the boundary conditions along both the upper and lower impermeable boundaries of the aquifer exactly. The procedure is somewhat involved and will not be discussed here. The reader is referred to Muskat [1937] and Haitjema [1982, 1985] for three-dimensional modeling in confined aquifers.

Problem 18.1

Prove that the potential (18.11) meets the boundary condition that $q_z = 0$ at $z = 0$.

Chapter 3

Elementary Harmonic Solutions

Most of the types of flow discussed in Chapter 2 are governed by either the Laplace equation $\nabla^2 \Phi = 0$ or the Poisson equation $\nabla^2 \Phi = -N$. Functions that fulfill Laplace's equation are called *harmonic functions* and play an important role in modeling groundwater flow. These harmonic functions are useful not only in solving Laplace's equation. They play, for example, a role in solving the Poisson equation: the solution to $\nabla^2 \Phi = -N$ for a given problem may be written as the sum of a harmonic function and a particular solution of $\nabla^2 \Phi = -N$.

The present chapter is devoted primarily to harmonic functions. We will first discuss some of the properties of these functions, then consider elementary harmonic functions such as the potential for a well, and finally discuss complex variable methods for solving boundary value problems.

19 SOME PROPERTIES OF HARMONIC FUNCTIONS

Harmonic functions have distinct mathematical properties. We will present some of these properties here without formal mathematical proof, but try to explain them from the viewpoint of groundwater mechanics. The interested reader is referred to the handbooks on potential theory for a more complete treatment of harmonic functions.

The Maximum Modulus Theorem

The maximum modulus theorem states that the values of a harmonic function Φ inside a domain $\mathcal{D}$ bounded by a boundary $\mathcal{B}$ are in between the maximum and minimum values of Φ on $\mathcal{B}$. We have already applied this theorem when designing dewatering systems of building pits: It is sufficient to check the values of Φ along the boundary of the pit; if Φ is less than the maximum allowable value everywhere on the boundary, then Φ is less than this value everywhere inside. We arrived at this property by reasoning that the potential cannot have a maximum inside a domain if there is no accretion; flow would occur away from this maximum in all directions, which violates continuity of flow.

We will express this reasoning in mathematical terms as follows. Consider a portion of a confined aquifer, bounded by two concentric circles $\mathcal{C}_1$ and $\mathcal{C}_2$ of radii R_1 and R_2, centered at the origin of an r, θ polar coordinate system (see Figure 3.1). There are no sources of water inside the circles, and therefore the net outflow through any circle of radius r ($r \leq R_2$) is zero:

$$\int_0^{2\pi} Q_r r d\theta = -\int_0^{2\pi} \frac{\partial \Phi}{\partial r} r d\theta = 0 \qquad (r \leq R_2) \tag{19.1}$$

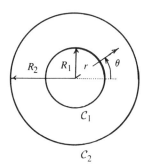

Figure 3.1 The mean value theorem.

Division of (19.1) by r and subsequent integration from R_1 to R_2 yields

$$\int_{R_1}^{R_2} \int_0^{2\pi} \frac{\partial \Phi}{\partial r} d\theta dr = \int_0^{2\pi} \int_{R_1}^{R_2} \frac{\partial \Phi}{\partial r} dr d\theta = \int_0^{2\pi} [\Phi(R_2, \theta) - \Phi(R_1, \theta)] d\theta = 0 \tag{19.2}$$

where the change in the sequence of integration is allowed if Φ and its derivatives are continuous inside $\mathcal{C}_2$. It follows from (19.2) that

$$\int_0^{2\pi} \Phi(R_1, \theta) d\theta = \int_0^{2\pi} \Phi(R_2, \theta) d\theta \tag{19.3}$$

Contracting the circle $\mathcal{C}_1$ to a point, (19.3) becomes, for $R_1 \to 0$:

$$\Phi_0 = \frac{1}{2\pi R_2} \int_0^{2\pi} \Phi(R_2, \theta) R_2 d\theta \tag{19.4}$$

where Φ_0 is the value of the potential at the center of $\mathcal{C}_2$. This result is known as the mean value theorem, and states that the value of a harmonic function at the center of

Sec. 19 Some Properties of Harmonic Functions

a circle equals the mean value along that circle. This property might explain the term "harmonic function." The maximum modulus theorem can be deduced from the mean value theorem: Φ cannot have a maximum at any point $\mathcal{P}$ inside a domain, because it is the average of the values along a small circle of radius ε around $\mathcal{P}$: there will always be both a higher and a lower value of Φ on the circle than at its center.

Boundary Value Problems

There exist infinitely many harmonic functions. The function that applies to any given problem is determined by the boundary conditions. There are various kinds of boundaries: some boundaries are finite, some contain infinity, and in other cases the boundary consists of separate closed contours. Domains bounded by the latter types of boundaries are called multiply connected. If only one boundary exists, the domain is simply connected.

Conditions along the boundaries may be specified in many different ways. The three most common kinds of boundary value problems will be listed here:

1. The Dirichlet problem: all boundary conditions are given in terms of the potential Φ.
2. The Neumann problem: the derivative $d\Phi/dn$ in the direction normal to the boundary is given everywhere along the boundary.
3. The mixed boundary value problem: Φ is given along certain segments of the boundary and $d\Phi/dn$ along all the others.

In general, we may state that along each boundary segment one, and only one, condition must be specified in order to define a harmonic function fully.

We will see in the following chapters that many other types of boundary conditions occur in groundwater flow problems. Closed boundaries are found mainly in two-dimensional problems in the vertical plane, as discussed in Sec. 16. Regional aquifers rarely have such boundaries and are often best modeled as infinite domains with internal boundaries consisting of rivers, creeks, or lake boundaries. If the aquifer is infinite, a condition must be specified that controls the behavior of the solution near infinity. We will term such a condition the far-field condition, whereby far-field refers to the flow pattern near infinity. Sometimes this far-field condition specifies a uniform far-field; it always determines whether or not inflow or outflow occurs from or to infinity. We have, in Chapter 2, specified the latter condition usually by prescribing the head at some fixed point $\mathcal{P}$, which we called the reference point.

The Stream Function Ψ

We defined a discharge streamline in Sec. 6 as a curve that at each point is tangent to the discharge vector. Expressed in mathematical terms, this definition becomes (see Figure 3.2)

$$\frac{dy/ds}{dx/ds} = \frac{Q_y}{Q_x} \tag{19.5}$$

where s is the arc length measured along the streamline, which is represented symbolically as

$$x = x(s) \qquad y = y(s) \qquad (19.6)$$

It follows from (19.5) and Darcy's law that

$$Q_y \frac{dx}{ds} - Q_x \frac{dy}{ds} = -\frac{\partial \Phi}{\partial y} \frac{dx}{ds} + \frac{\partial \Phi}{\partial x} \frac{dy}{ds} = 0 \qquad (19.7)$$

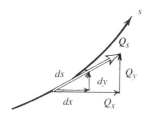

Figure 3.2 A discharge streamline.

We obtained plots of equipotentials in the previous chapter simply by setting Φ equal to a constant in the equation $\Phi = \Phi(x, y)$. Plots of streamlines may be obtained in the same fashion, provided that there exists a function that is constant along streamlines. We will see that such a function, called the stream function and represented as Ψ, indeed exists for problems governed by Laplace's equation ($\nabla^2 \Phi = 0$), e.g., for shallow aquifers, modeled using the Dupuit-Forchheimer assumption, where fluid cannot enter the aquifer through its upper or lower boundaries. If Ψ is to be constant along streamlines, then

$$\frac{d\Psi}{ds} = \frac{\partial \Psi}{\partial x} \frac{dx}{ds} + \frac{\partial \Psi}{\partial y} \frac{dy}{ds} = 0 \qquad (19.8)$$

It follows from (19.7) and (19.8) that this function must fulfill the following condition:

$$\boxed{\frac{\partial \Psi}{\partial x} = -\frac{\partial \Phi}{\partial y} \qquad \frac{\partial \Psi}{\partial y} = \frac{\partial \Phi}{\partial x}} \qquad (19.9)$$

These equations are known as the Cauchy-Riemann equations. It follows from Darcy's law that

$$Q_x = -\frac{\partial \Phi}{\partial x} = -\frac{\partial \Psi}{\partial y} \qquad Q_y = -\frac{\partial \Phi}{\partial y} = \frac{\partial \Psi}{\partial x} \qquad (19.10)$$

The stream function can be used to calculate the discharge ΔQ flowing in the channel between the streamlines $\mathcal{AB}$ and $\mathcal{DE}$ in Figure 3.3. This discharge is constant in the channel since flow can neither enter it through the top or bottom boundaries, nor across the streamlines. Consider the plane $\mathcal{BC}$, parallel to the y axis, and the plane $\mathcal{CD}$ parallel

Sec. 19   Some Properties of Harmonic Functions   223

to the x axis, where $\mathcal{B}$ and $\mathcal{D}$ lie on the same equipotential. By continuity of flow, the discharge ΔQ must pass through $\mathcal{BCD}$, so that

$$\Delta Q = \int_{\mathcal{D}}^{\mathcal{C}} Q_y dx + \int_{\mathcal{B}}^{\mathcal{C}} Q_x dy = \int_{\mathcal{D}}^{\mathcal{C}} \frac{\partial \Psi}{\partial x} dx - \int_{\mathcal{B}}^{\mathcal{C}} \frac{\partial \Psi}{\partial y} dy \qquad (19.11)$$

where use is made of (19.10). The integrations can be performed and

$$\Delta Q = \Psi_{\mathcal{C}} - \Psi_{\mathcal{D}} - \Psi_{\mathcal{C}} + \Psi_{\mathcal{B}} = \Psi_{\mathcal{B}} - \Psi_{\mathcal{D}} = -\int_{\mathcal{B}}^{\mathcal{D}} d\Psi \qquad (19.12)$$

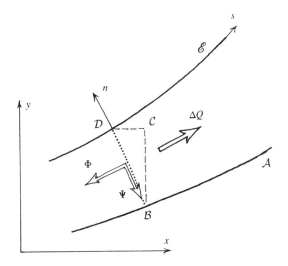

Figure 3.3 The discharge flowing between two streamlines.

It thus follows that the discharge flowing between two streamlines equals the difference in value of Ψ on these streamlines. ΔQ is positive, by definition, so that $\Psi_{\mathcal{B}} > \Psi_{\mathcal{D}}$: Ψ increases along the equipotential $\mathcal{BD}$ as shown in Figure 3.3. Φ increases in the direction against the flow; Φ and Ψ form a right-hand coordinate system.

The Cauchy-Riemann equations (19.9) represent two differential equations from which Ψ can be found if Φ is known. These two equations might be in conflict with each other; we must prove that the function Ψ obtained by integration is single-valued. We have seen that the difference in values of Ψ at two points equals the flow occurring between these points; if we consider the closed contour $\mathcal{C}$ in Figure 3.4, then the difference in value of Ψ at two points $\mathcal{P}^+$ and $\mathcal{P}^-$ located at infinitesimal distances on either side of a point $\mathcal{P}$ on $\mathcal{C}$ equals the net flow through $\mathcal{C}$. The stream function will therefore be single-valued only if this net inflow is zero:

$$\Delta Q = -\oint_{\mathcal{C}} d\Psi = \oint_{\mathcal{C}} Q_n d\tau = 0 \qquad (19.13)$$

where Q_n is the flow normal to the boundary C and τ is the arc length along C. Condition (19.13) is met if there is no inflow through the top and bottom boundaries of the aquifer inside C, i.e., if $\nabla^2 \Phi = 0$. The stream function Ψ is therefore single-valued if Φ is harmonic. It may be noted that the condition for single-valuedness may be expressed as

$$\frac{\partial^2 \Psi}{\partial x \partial y} = \frac{\partial^2 \Psi}{\partial y \partial x} \qquad (19.14)$$

and we see from (19.9) upon differentiation that this equation is fulfilled only if $\nabla^2 \Phi = 0$. We know for physical reasons that the potential Φ is single-valued so that

$$\frac{\partial^2 \Phi}{\partial x \partial y} = \frac{\partial^2 \Phi}{\partial y \partial x} \qquad (19.15)$$

and application to (19.9) shows that Ψ is harmonic:

$$\boxed{\nabla^2 \Psi = 0} \qquad (19.16)$$

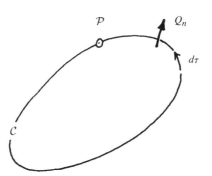

Figure 3.4 Single-valuedness of Ψ.

The potential and stream functions are called conjugate harmonic functions. The differential equation for the stream function is linear, so that stream functions may be superposed.

We may use (19.12) to express the rate of change of Φ along a streamline, $d\Phi/ds$, in terms of the rate of change of Ψ along an equipotential, $d\Psi/dn$. The arc lengths s and n are measured along a streamline and an equipotential in the directions shown in Figure 3.3. (The derivatives $d\Phi/ds$ and $d\Psi/dn$ are directional derivatives of the functions Φ and Ψ; they are not partial derivatives because s and n are lengths measured along curves, rather than coordinates.) The discharge ΔQ equals the integral along the equipotential BD of the discharge component $Q_s = -d\Phi/ds$ in the direction of flow. Thus, (19.12) may be written as

$$\Delta Q = \int_B^D Q_s \, dn = -\int_B^D \frac{d\Phi}{ds} dn = \Psi_B - \Psi_D \qquad (19.17)$$

Sec. 19   Some Properties of Harmonic Functions

Differentiation with respect to n gives, after multiplication by -1,

$$\frac{d\Phi}{ds} = \frac{d\Psi}{dn} \tag{19.18}$$

The Cauchy-Riemann conditions make it possible to determine the stream functions for the two elementary harmonic functions treated in Chapter 2: the potentials for uniform flow and for a well. We will do this in the following sections, and use the stream functions to determine streamlines for various combinations of wells and uniform flow.

The Stream Function for Uniform Flow

The two components Q_x and Q_y of the discharge vector are constants for rectilinear flow. Representing these constants as Q_{x0} and Q_{y0}, the following is obtained from Darcy's law:

$$Q_x = -\frac{\partial \Phi}{\partial x} = Q_{x0} \qquad Q_y = -\frac{\partial \Phi}{\partial y} = Q_{y0} \tag{19.19}$$

Integration of (19.19) yields the potential Φ as a function of x and y,

$$\Phi = -Q_{x0} x - Q_{y0} y + \Phi_0 \tag{19.20}$$

where Φ_0 is a constant.

Application of (19.19) to the Cauchy-Riemann equations (19.9) gives

$$\frac{\partial \Psi}{\partial y} = -Q_{x0} \qquad \frac{\partial \Psi}{\partial x} = Q_{y0} \tag{19.21}$$

Integration gives the following expression for the stream function:

$$\boxed{\Psi = -Q_{x0} y + Q_{y0} x} \tag{19.22}$$

where the constant of integration is taken to be zero: $\Psi = 0$ along the line $y = Q_{y0}/Q_{x0} x$.

The Flow Net

The flow net is obtained as follows. Equations for the equipotentials are obtained from (19.20) by setting Φ equal to constant values $\Phi_j = \Phi_0 + j\Delta\Phi$, where j assumes negative and positive integer values:

$$j\Delta\Phi = -Q_{x0} x - Q_{y0} y \tag{19.23}$$

Equations for the streamlines are obtained similarly, by setting Ψ in (19.22) equal to $\Psi_j = j\Delta\Psi$, where $\Delta\Psi = \Delta\Phi$,

$$j\Delta\Psi = Q_{y0} x - Q_{x0} y \tag{19.24}$$

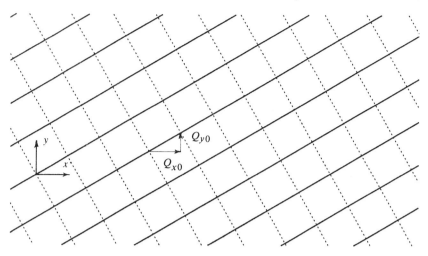

Figure 3.5 Flow net for uniform flow.

The two families of straight lines given by (19.23) and (19.24) are reproduced in Figure 3.5.

The Stream Function for a Well

The potential for a well may be written as follows:

$$\Phi = \frac{Q}{2\pi} \ln \frac{r}{R} + \Phi_0 \qquad (19.25)$$

where r is the radial coordinate emanating from the well and Φ_0 is the potential at $r = R$. We will obtain an expression for Ψ by using radial coordinates r, θ. By symmetry, the streamlines are straight lines through the well and the equipotentials are circles. Thus $ds = dr$ and $dn = r d\theta$ so that (19.18) becomes

$$\frac{\partial \Phi}{\partial r} = \frac{\partial \Psi}{r \partial \theta} \qquad (19.26)$$

and we obtain with (19.25)

$$\frac{\partial \Psi}{r \partial \theta} = \frac{Q}{2\pi r} \qquad (19.27)$$

so that

$$\boxed{\Psi = \frac{Q}{2\pi} \theta} \qquad (19.28)$$

It may be noted that we could add an arbitrary function $f(r)$ to (19.28) without violating (19.27). This function must reduce to a constant, however, in order that $d\Psi/ds =$

Sec. 19 Some Properties of Harmonic Functions

$\partial \Psi / \partial r = 0$. For convenience, we set this constant to zero. Ψ cannot be single-valued because the condition (19.13) is not fulfilled for any contour C around the well; this integral equals $-Q$. Indeed, the angle θ is discontinuous by an amount 2π along a line emanating from the well. We may choose this line, called the branch cut, as the negative x axis (see Figure 3.6). We denote θ as θ^+ just above this cut and as θ^- just below the cut, and

$$\theta^+ - \theta^- = 2\pi \tag{19.29}$$

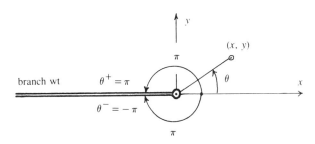

Figure 3.6 Branch cut for a well.

Introducing Ψ^+ and Ψ^- in a similar way, we see from (19.28) and (19.29) that,

$$\Psi^+ - \Psi^- = Q \tag{19.30}$$

The jump in the stream function results from the extraction of a discharge Q at the well; the potential is not harmonic at $r = 0$. Because the potential is not harmonic only at an isolated point, known as a singular point, the many-valuedness of Ψ may be restricted to a single line. The solution is said to have a *logarithmic singularity* at $r = 0$. It is possible to give a physical interpretation of the branch cut, which may help to understand the principle involved. In this interpretation, the branch cut is viewed as a narrow slot in which water can flow without resistance and without being governed by the values of Φ along the slot. Since Ψ jumps by an amount Q across the cut, the discharge flowing inside the slot is Q; since $\Psi^+ > \Psi^-$, the flow is to the left with respect to an arrow pointing from the $-$ side to the $+$ side of the slot, i.e., in the negative x direction for the case of Figure 3.6. Thus, the slot carries the water pumped by the well off to infinity, from where it flows back toward the well via the aquifer. Thus, the branch cut is the vehicle by which the mathematical model maintains continuity of flow. It is seen from this interpretation, that the branch cut can be chosen to have any shape, as long as it connects the well to infinity.

Branch cuts are a central feature in groundwater modeling by analytic functions; they provide a vehicle for using harmonic functions, in this case $\theta = \arctan(y/x)$, to model aquifer features that are associated with jumps.

Problem 19.1

Derive equation (19.30) by applying (19.13) to the problem of flow toward a well, taking for C a circle of radius R and centered at the well.

The Flow Net

The streamlines are straight lines emanating from the well. Setting Ψ equal to the constant values Ψ_j, the corresponding values for θ_j are found from (19.28):

$$\Psi_j = \frac{Q}{2\pi}\theta_j \tag{19.31}$$

Choosing 8 flow channels as shown in Figure 3.7, an amount of $\Delta\Psi = Q/8$ will flow through each channel. If we take $\Psi_1 = -\frac{1}{2}Q$, then

$$\Psi_j = \Psi_1 + (j-1)\Delta\Psi = -\frac{1}{2}Q + (j-1)\frac{1}{8}Q = \frac{Q}{2\pi}\theta_j \qquad (j = 1, 2, \ldots 9) \tag{19.32}$$

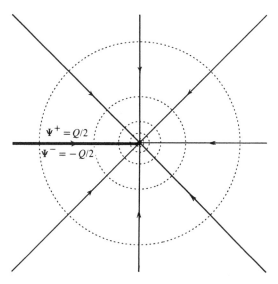

Figure 3.7 Flow net for a well.

The radii r_j of the corresponding circular equipotentials are found from (19.25) with $\Phi_j = \Phi_0 - j\Delta\Psi$ (note that $\Phi < \Phi_0$ for $r < R$, and $\Delta\Phi = \Delta\Psi$):

$$r_j = Re^{-2\pi j \Delta\Psi/Q} \tag{19.33}$$

Streamlines for Uniform Flow with a Well

We will next determine streamlines for uniform flow with a well in an infinite aquifer. The uniform flow is in the x direction and the corresponding potential is obtained from (19.20) by setting Q_{y0} equal to zero. Superposition of this potential with that for a well yields

$$\Phi = -Q_{x0}(x - L) + \frac{Q}{2\pi}\ln\frac{r}{L} + \Phi_0 \tag{19.34}$$

Sec. 19 Some Properties of Harmonic Functions

where the origin of the coordinate system is chosen at the well and the constant Φ_0 in (19.34) represents the potential at some reference point with coordinates $x = L$, $y = 0$.

The stream function is obtained by superposition of the expressions derived for uniform flow, (19.22), and for a well, (19.28). This yields, with $Q_{y0} = 0$,

$$\Psi = -Q_{x0}y + \frac{Q}{2\pi}\theta \qquad (19.35)$$

The angle θ is taken to be in the range from $-\pi$ to π,

$$-\pi \leq \theta \leq \pi \qquad (19.36)$$

and the stream function will be discontinuous along the negative x axis.

Along the positive x axis, θ and y are zero so that Ψ is 0 by (19.35): the positive x axis is a streamline (see Figure 3.8). Along the negative x axis, y is zero and θ equals either π or $-\pi$, depending upon whether the x axis is approached from above or below. It follows from (19.35) that Ψ has the dual values $+Q/2$ and $-Q/2$ along the negative x axis.

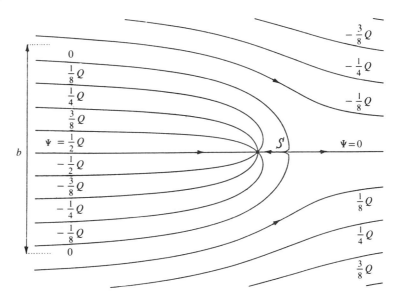

Figure 3.8 Streamlines for uniform flow with a well.

Far to the left of the well, for x approaching $-\infty$, the angle θ approaches either π or $-\pi$, depending upon whether y is positive or negative:

$$\Psi \to -Q_{x0}y + \frac{Q}{2} \qquad y > 0 \qquad x \to -\infty \qquad (19.37)$$

and

$$\Psi \to -Q_{x0}y - \frac{Q}{2} \qquad y < 0 \qquad x \to -\infty \qquad (19.38)$$

The first equation yields values of $\Psi \leq Q/2$ and the second one values of $\Psi \geq -Q/2$. For constant Ψ, the straight lines

$$y = \frac{-\Psi + \frac{Q}{2}}{Q_{x0}} \qquad \left(\Psi \leq \frac{Q}{2}\right) \qquad (19.39)$$

and

$$y = \frac{-\Psi - \frac{Q}{2}}{Q_{x0}} \qquad \left(\Psi \geq -\frac{Q}{2}\right) \qquad (19.40)$$

are asymptotes of the streamlines for $x \to -\infty$. It may be noted that values of Ψ between $+Q/2$ and $-Q/2$ correspond to two different streamlines, as a consequence of the many-valuedness of Ψ.

Far to the right of the well, for $x \to \infty$, the angle θ approaches 0 and the equations for the asymptotes of the streamlines on the downstream side of the well are found from (19.35) with $\theta = 0$,

$$y = \frac{-\Psi}{Q_{x0}} \qquad (19.41)$$

It is seen from (19.39) through (19.41) that the asymptotes of the streamlines on the downstream side of the well are a distance $Q/(2Q_{x0})$ closer to the x axis than on the upstream side. This is the result of the removal of an amount Q by the well, which flows over a width $b = Q/Q_{x0}$ in the far-field on the upstream side (see Figure 3.8).

The equations for the streamlines are obtained by setting Ψ equal to constant values Ψ_j in (19.35). Solving for θ, we obtain

$$\theta = \frac{2\pi}{Q}(\Psi_j + Q_{x0}y) = 2\pi \left[\frac{\Psi_j}{Q} + \frac{Q_{x0}}{Q}y\right] = 2\pi \left[\frac{\Psi_j}{Q} + \frac{y}{b}\right] \qquad (19.42)$$

Values for θ are obtained for various values of y, and the corresponding values of x are found from $\tan \theta = y/x$, or

$$x = y \cot \theta \qquad (19.43)$$

The streamlines reproduced in Figure 3.8 are obtained by setting Ψ_j equal to $\pm Q/2$, $\pm 3Q/8$, $\pm Q/4$, $\pm Q/8$, and zero. The points on the curves are found by substituting in (19.42) values for y within the asymptotes (19.39) and (19.40). The streamlines $\Psi = 0$ meet at the stagnation point S downstream from the well on the x axis. The location of point S is found by setting Q_x and Q_y equal to zero. Expressions for Q_x and Q_y are found from (19.34) upon differentiation:

$$Q_x = -\frac{\partial \Phi}{\partial x} = Q_{x0} - \frac{Q}{2\pi} \frac{x}{x^2 + y^2} \qquad (19.44)$$

and

Sec. 19 Some Properties of Harmonic Functions

$$Q_y = \frac{\partial \Phi}{\partial y} = -\frac{Q}{2\pi}\frac{y}{x^2+y^2} \qquad (19.45)$$

It follows from (19.45) that Q_y is zero only along the x axis, $y = 0$. The x coordinate of S thus is found by setting Q_x equal to zero with $y = 0$, which yields:

$$x_s = \frac{Q}{2\pi Q_{x0}} = \frac{b}{2\pi} \qquad y_s = 0 \qquad (19.46)$$

The streamlines may be obtained alternatively by contouring the stream function. A disadvantage of this approach is, however, that most numerical contouring routines will cause a black heavy line to be printed along the branch cut, due to the discontinuity in Ψ.

Problem 19.2

A strip of width B in a horizontal confined aquifer of thickness H and hydraulic conductivity k is contaminated. The flow is uniform of a discharge Q_{x0} in the x direction (see Figure 3.9). A well that penetrates the entire aquifer (a fully penetrating well) will be placed a distance d downstream from the front of the plume of contaminated water. Determine the smallest discharge of the well in terms of B, d, and Q_{x0}, such that all contaminated water will be captured. We assume that the contaminated water has the same physical properties as the groundwater and that the aquifer is incompressible, so that steady-state conditions are achieved immediately after the well starts pumping. It is given that the flow in the aquifer remains confined everywhere.

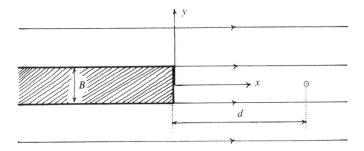

Figure 3.9 Capturing contaminated groundwater with a well (Problem 19.2).

Flow Net for Flow From a Recharge Well Toward a Well

Another example of a simple flow net is that for flow from a recharge well toward a well. The aquifer is infinite and the far-field is a constant potential. If the recharge well is located at $x = d$, $y = 0$, and the well at $x = -d$, $y = 0$, and if both recharge and discharge are Q, the potential becomes:

$$\Phi = \frac{Q}{4\pi}\ln\frac{(x+d)^2+y^2}{(x-d)^2+y^2} + \Phi_0 \qquad (19.47)$$

where Φ_0 is the constant potential at infinity. The equations for the equipotentials are obtained by setting Φ equal to $\Phi_j = \Phi_0 + j\Delta\Phi$ in (19.47):

$$\Phi_j = \Phi_0 + j\Delta\Phi = \frac{Q}{4\pi}\ln\frac{(x+d)^2 + y^2}{(x-d)^2 + y^2} + \Phi_0 \tag{19.48}$$

This yields

$$\frac{(x+d)^2 + y^2}{(x-d)^2 + y^2} = e^a \tag{19.49}$$

where

$$a = \frac{4\pi}{Q}j\Delta\Phi \tag{19.50}$$

The expression (19.49) for the family of equipotentials may be written in the following form:

$$x^2 - 2dx\frac{e^a + 1}{e^a - 1} + y^2 + d^2 = 0 \tag{19.51}$$

With $(e^a + 1)/(e^a - 1) = \coth\frac{a}{2}$, this may be simplified to

$$(x - d\coth\frac{a}{2})^2 + y^2 = \left[\frac{d}{\sinh(\frac{a}{2})}\right]^2 \tag{19.52}$$

which represents a circle, centered at x_c, y_c with

$$x_c = d\coth(\frac{a}{2}) \qquad y_c = 0 \tag{19.53}$$

and of radius R, with

$$R = \frac{d}{|\sinh(\frac{a}{2})|} \tag{19.54}$$

We determine the streamlines by the use of the stream function. This function is obtained by superposition:

$$\Psi = \frac{Q}{2\pi}(\theta_1 - \theta_2) \tag{19.55}$$

The angles θ_1 and θ_2 are defined in Figure 3.10(a); $\mathcal{W}_1$ and $\mathcal{W}_2$ mark the locations of the well and the recharge well, respectively. The streamlines are seen from (19.55) to correspond to constant values of $\theta_1 - \theta_2$. This difference is negative for points x, y with $y \geq 0$, and positive for points x, y where $y \leq 0$ (see Figure 3.10[a]), and equals the upper angle of the triangle $\mathcal{W}_1\mathcal{P}\mathcal{W}_2$ with $\mathcal{W}_1\mathcal{W}_2$ as the fixed base. The loci of points where $\theta_1 - \theta_2$ is constant therefore are circles through $\mathcal{W}_1$ and $\mathcal{W}_2$ as shown in Figure 3.10(b). The stream function is many-valued: approaching the line $\mathcal{W}_1\mathcal{W}_2$ from above, the angle $-(\theta_1 - \theta_2)$ tends to $+\pi$, $\theta_1 - \theta_2$ to $-\pi$, and the corresponding value of Ψ to

Sec. 19 Some Properties of Harmonic Functions

$(Q/2\pi)(-\pi) = -Q/2$, see (19.55) and Figure 3.11. Approaching the line $\mathcal{W}_1\mathcal{W}_2$ from below, $\theta_1 - \theta_2$ tends to π and Ψ approaches $Q/2$. The angle $\theta_1 - \theta_2$ vanishes along the remaining portion of the x axis: Ψ is zero there. The values listed for $\theta_1 - \theta_2$ in Table 3.1 correspond to $\Delta\Psi = Q/8$. The y coordinates, y_c, of the centers of the circles, and the radii R listed in the table are obtained from the geometry in Figure 3.10(b) as follows. Denoting the angle $-(\theta_1 - \theta_2)$ as α, the angle β enclosed between $\mathcal{W}_1\mathcal{W}_2$ and the line $\mathcal{W}_2\mathcal{C}$ from well $\mathcal{W}_2$ to the center $\mathcal{C}$ of the circle is seen to be $\pi/2 - \alpha$ (the angle between $\mathcal{W}_1\mathcal{W}_2$ and the tangent to the circle spans the same arc as the angle $\mathcal{W}_1\mathcal{P}\mathcal{W}_2$). The values for y_c are found from $y_c = d \tan \beta$ and those for R from $R^2 = y_c^2 + d^2$. The equations for the circular equipotentials shown in Figure 3.11 are obtained by the use of (19.52) with $\Delta\Phi = \Delta\Psi = Q/8$.

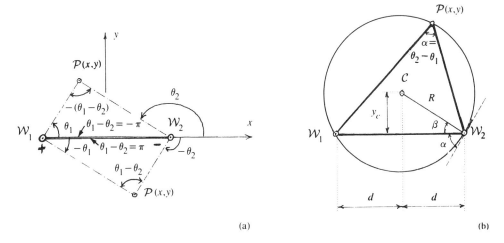

(a) (b)

Figure 3.10 Definition of the angles θ_1 and θ_2.

Table 3.1			Values of Ψ, $\theta_1 - \theta_2$, R, and y_c					
Ψ	$-\frac{1}{2}Q$	$-\frac{3}{8}Q$	$-\frac{1}{4}Q$	$-\frac{1}{8}Q$	0	$\frac{1}{8}Q$	$\frac{1}{4}Q$	$\frac{3}{8}Q$	$\frac{1}{2}Q$
$\theta_1 - \theta_2$	$-\pi$	$-\frac{3}{4}\pi$	$-\frac{1}{2}\pi$	$-\frac{1}{4}\pi$	0	$\frac{1}{4}\pi$	$\frac{1}{2}\pi$	$\frac{3}{4}\pi$	π
y_c	∞	$-d$	0	d	∞	$-d$	0	d	$-\infty$
R	∞	$\sqrt{2}\,d$	d	$\sqrt{2}\,d$	∞	$\sqrt{2}\,d$	d	$\sqrt{2}\,d$	∞

It may be observed from the flow net that the majority of the flow occurs in a comparatively narrow band; the two circular streamlines of radii d enclose a domain that carries half of the total discharge. The flow rates appear to decrease rapidly with the radii of the circular streamlines.

Problem 19.3

Determine $\int d\Psi$ along three contours enclosing: (1) both wells, (2) only the discharge well, and (3) only the recharge well.

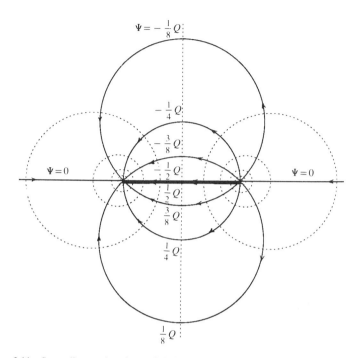

Figure 3.11 Streamlines and equipotentials for flow from a recharge well toward a well.

Example 19.1: Cooling Problem

A system of a well and a recharge well, each of strength Q, are used for cooling in an aquifer with an undisturbed field of uniform flow of discharge Q_{x0} in the x direction ($Q_{x0} > 0$). The purpose of the system is to pump cool water from the aquifer and inject it after use as hot water back into the aquifer. The wells are located on the x axis (see Figure 3.12). The problem is to determine the distance $2d$ between the wells such that none of the hot water is pumped by the well. We will neglect heat exchange between the water and the aquifer, as well as the effect of the temperature on the density.

The recharge well must be placed downstream from the well in order that the well may capture cold water. Choosing the origin of the x, y coordinate system midway between the wells, the potential becomes:

$$\Phi = -Q_{x0}x + \frac{Q}{4\pi} \ln \frac{(x+d)^2 + y^2}{(x-d)^2 + y^2} + \Phi_0 \tag{19.56}$$

where Φ_0 is the potential at the origin. In order that no water flows from the recharge well to the well, the flow must stagnate somewhere between the wells. The locations of the stagnation points are found by setting Q_x and Q_y equal to zero:

$$Q_x = -\frac{\partial \Phi}{\partial x} = Q_{x0} - \frac{Q}{2\pi}\left[\frac{x_s + d}{(x_s+d)^2 + y_s^2} - \frac{x_s - d}{(x_s-d)^2 + y_s^2}\right] = 0 \tag{19.57}$$

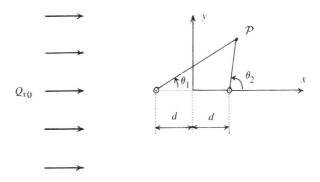

Figure 3.12 Cooling problem.

and

$$Q_y = -\frac{\partial \Phi}{\partial y} = -\frac{Q}{2\pi}\left[\frac{y_s}{(x_s+d)^2 + y_s^2} - \frac{y_s}{(x_s-d)^2 + y_s^2}\right] = 0 \quad (19.58)$$

where x_s and y_s are the coordinates of the stagnation points. It follows from the second equation that $Q_y = 0$ for $y_s = 0$, and for $x_s = 0$. In order that $Q_x = 0$ for $-d < x_s < d$, $y_s = 0$, (19.57) must have a real root for $y_s = 0$:

$$Q_{x0} - \frac{Q}{2\pi}\left[\frac{1}{x_s + d} - \frac{1}{x_s - d}\right] = 0 \quad (19.59)$$

which yields

$$\left[\frac{x_s}{d}\right]_{1,2} = \pm\sqrt{1 - \frac{Q}{\pi Q_{x0} d}} \quad (19.60)$$

A critical situation, whereby exchange of flow between the wells is on the verge of occurring, applies when there exists a single stagnation point:

$$\frac{Q}{\pi Q_{x0} d} = 1 \quad (19.61)$$

The well will capture injected hot water when $Q/(Q_{x0}d) \geq \pi$: the roots in (19.60) then are imaginary, the stagnation points lie on the y axis and hot water flows from the recharge well between these stagnation points toward the well. Streamlines are shown in Figure 3.13 for three cases: case (a) $Q/(Q_{x0}d) = 2.5$, (b) $Q/(Q_{x0}d) = \pi$, and (c) $Q/(Q_{x0}d) = 4$. The figure illustrates how the flows to the well and from the recharge well are completely separate in case (a), touch in case (b), and have a common domain in case (c), where the stagnation points lie on the y axis. The streamlines through the stagnation points for case (c) are not part of the contour plot (the set of values contoured do not contain the particular values of Ψ on the stagnation points). These streamlines are added for the purpose of illustration.

The streamlines were obtained by contouring the stream function using a computer program. The stream function is found by superposition of (19.22) with $Q_{y0} = 0$, and (19.28):

$$\Psi = -Q_{x0}y + \frac{Q}{2\pi}(\theta_1 - \theta_2) \tag{19.62}$$

where θ_1 and θ_2 are local coordinates as shown in Figure 3.12,

$$\begin{aligned} \theta_1 &= \arctan \frac{y}{x+d} & -\pi < \theta_1 \le \pi \\ \theta_2 &= \arctan \frac{y}{x-d} & -\pi < \theta_2 \le \pi \end{aligned} \tag{19.63}$$

It would be possible to obtain an expression of the form $x = x(y)$ from (19.62), which could be used to determine points of the streamlines. The procedure for determining such an expression is limited to a few elementary problems, however; we often resort to numerical techniques, contouring for example, to obtain plots of streamlines such as those in Figure 3.13. An alternative method for determining streamline plots consists of tracing the curves from point to point, and has the advantage over contouring that the jumps in values of the stream function across the line connecting the wells (the branch cut) may be taken into account without affecting the plot. We will discuss this method in Chapter 4. The latter technique is especially preferable when the branch cuts do not coincide with the streamlines, as will be the case if the uniform flow is inclined to the line connecting the wells.

Example 19.2: Intercepting Flow Between Wells

We will consider the problem of intercepting flow between two wells of discharge Q as a second example. The line connecting the two discharge wells is at right angles to a uniform flow of discharge Q_{x0}. The wells are a distance $2d$ apart as shown in Figure 3.14 and the problem is to determine d such that all flow between the wells is captured. If the origin is midway between the wells, the potential becomes

$$\Phi = -Q_{x0}x + \frac{Q}{4\pi} \ln\left\{[x^2 + (y-d)^2][x^2 + (y+d)^2]\right\} + C \tag{19.64}$$

where C is a constant. The water flowing between the wells will eventually enter the wells if the flow stagnates at least at one point on the x axis. The locations of the stagnation points are found by setting Q_x and Q_y equal to zero at (x_s, y_s):

$$Q_x = Q_{x0} - \frac{Q}{2\pi}\left[\frac{x_s}{x_s^2 + (y_s-d)^2} + \frac{x_s}{x_s^2 + (y_s+d)^2}\right] = 0 \tag{19.65}$$

Sec. 19 Some Properties of Harmonic Functions

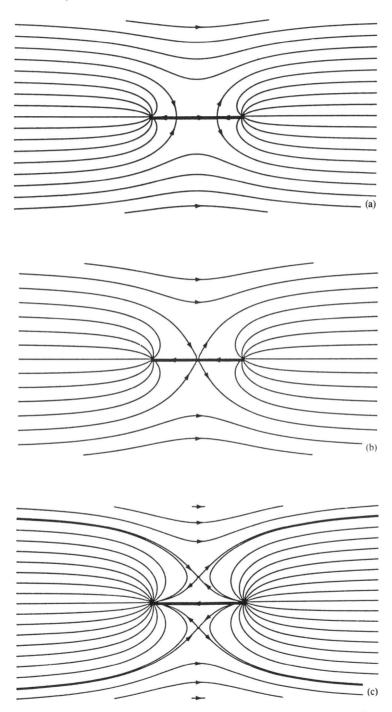

Figure 3.13 Streamlines for three cases of flow with a well and a recharge well.

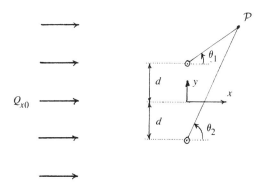

Figure 3.14 Two wells intercepting flow.

and

$$Q_y = -\frac{Q}{2\pi}\left[\frac{(y_s - d)}{x_s^2 + (y_s - d)^2} + \frac{(y_s + d)}{x_s^2 + (y_s + d)^2}\right] = 0 \qquad (19.66)$$

Multiplication of both sides of the second equation by a factor equal to the product of the denominators of the fractions inside the brackets yields, after some rearrangement:

$$2y_s\left(x_s^2 + y_s^2 - d^2\right) = 0 \qquad (19.67)$$

This means that the stagnation points either lie on the x axis, or on a circle of radius d centered at the origin. The stagnation points lie on the x axis if (19.65) has at least one real root for $y_s = 0$. This root is found from

$$Q_{x0} - \frac{Q}{\pi}\frac{x_s}{x_s^2 + d^2} = 0 \qquad (19.68)$$

or

$$\left[\frac{x_s}{d}\right]_{1,2} = \frac{Q}{2\pi Q_{x0}d} \pm \sqrt{\left[\frac{Q}{2\pi Q_{x0}d}\right]^2 - 1} \qquad \frac{y_s}{d} = 0 \qquad (19.69)$$

There exists at least one real root if

$$\frac{Q}{Q_{x0}d} \geq 2\pi \qquad (19.70)$$

A single stagnation point exists if the = sign applies; there are two stagnation points on the x axis if the > sign applies. That indeed all water flowing between the wells is eventually captured when (19.68) is fulfilled, is apparent from Figure 3.15. The plots (a), (b), and (c) in this figure correspond to $Q/(Q_{x0}d) = 9$, 2π, and 3, respectively. The figures illustrate how water passes through between the wells if the stagnation points do not lie on the x axis (plot [c]), and how no water passes through for plots (a) and (b). Note that case (b) is critical: the flow is on the verge of breaking through.

Sec. 19 Some Properties of Harmonic Functions

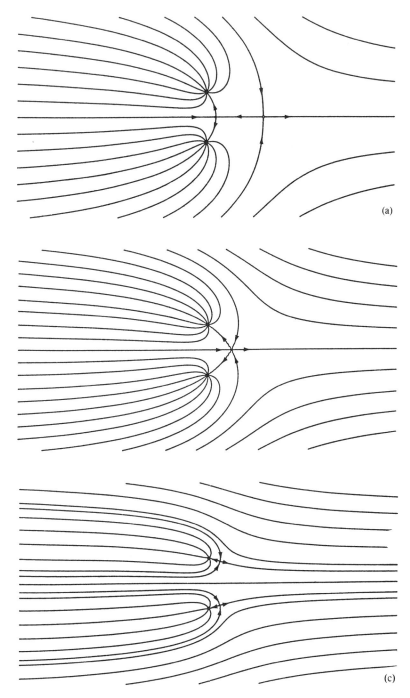

Figure 3.15 Streamlines for two wells intercepting flow: $Q/(Q_{x0}d) = 9$ (a), $Q/(Q_{x0}d) = 2\pi$ (b), and $Q/(Q_{x0}d) = 3$ (c).

Problem 19.4

Determine the coordinates of the stagnation points for the case of Figure 3.15(c).

The expression for the family of streamlines, used to generate the plots in Figure 3.15, are obtained by superposition of the stream functions derived earlier for uniform flow and for a well:

$$\Psi = -Q_{x0}y + \frac{Q}{2\pi}(\theta_1 + \theta_2) \tag{19.71}$$

It may be noted that the existence of jumps in the stream function is particularly disturbing for this case, since the branch cuts do not coincide with the streamlines. The plots above were obtained by tracing the streamlines from point to point, thereby eliminating the dark lines corresponding to the branch cuts. Figure 3.16 illustrates the effect of the jumps on the contour plots; it was produced by contouring the stream function for the case shown in Figure 3.15(a).

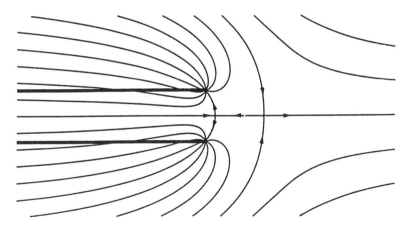

Figure 3.16 Contour plot of Ψ showing the branch cuts.

Problem 19.5

A well is used to recover water injected by a recharge well. Without the system operating, there is a uniform flow of discharge Q_{x0} in the x direction. Design the system such that all of the injected water is recovered under steady-state conditions.

Problem 19.6

Implement the stream function Ψ for wells and uniform flow in SLWL and reproduce the streamlines shown in Figures 3.8 and 3.16.

20 THE METHOD OF IMAGES FOR A CIRCLE

Certain boundary value problems involve circular boundaries; some of these problems may be solved by the method of images for a circle. This method is useful for problems involving islands that may be approximated as circular, and for problems in aquifers that contain a circular, or nearly circular, lake. We will generalize the method to handle more general boundaries through the application of conformal mapping techniques in Chapter 5.

The Location of the Image

Consider the problem of an eccentric well of discharge Q in a circular island, bounded by the equipotential $\Phi = \Phi_0$. The radius of the island is R, and the well is a distance p off-center (see Figure 3.17). The problem is to locate an image recharge well such that the system produces the desired boundary equipotential. We will solve this problem by using the flow net for a well and a recharge well shown in Figure 3.11. The well in this flow net lies eccentrically with respect to all circular equipotentials that surround it. The problem of an eccentric well inside a circular equipotential may therefore be solved by placing an image recharge well outside the circular boundary. Similarly, the problem of flow toward a well from a fully penetrating circular lake may be solved by placing an image recharge well inside the lake. The positions of the image wells may be determined from the position of the well with respect to the center of the circular equipotential, and from the radius R of the circle as follows. It is seen from Figure 3.11 that the well and the recharge well lie on a straight line through the center of the circle. The distances from this center to the well and the recharge well, p and p^*, are equal to $-x_c - d$ and $-x_c + d$, respectively, where x_c and d are defined in Figure 3.17. The product pp^* may be expressed in terms of x_c and d as follows:

$$pp^* = x_c^2 - d^2 \qquad (20.1)$$

Using expression (19.53) for x_c, we obtain

$$pp^* = d^2 \left(\coth^2 \tfrac{a}{2} - 1\right) = \frac{d^2}{\sinh^2 \tfrac{a}{2}} = R^2 \qquad (20.2)$$

where the radius for the circle, R, is introduced by the use of (19.54). The location of the image well is found from

$$\boxed{p^* = \frac{R^2}{p}} \qquad (20.3)$$

Example 20.1: An Eccentric Well in a Circular Island

The problem of unconfined flow toward an eccentric well in a circular island will be solved as an example. A cross-sectional view through the diameter of the circle that

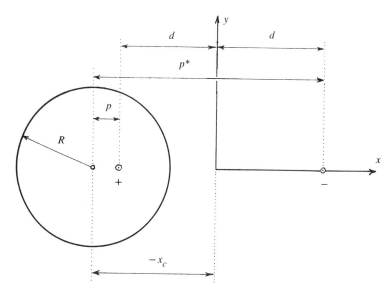

Figure 3.17 The method of images for a circle.

contains the well is shown in Figure 3.18. The origin of an x, y, z Cartesian coordinate system is at the center of the island, the well has a discharge Q and is located at $x = p$, $y = 0$, and the radius of the island is R. The head along the boundary is ϕ_0 and the boundary condition is

$$x^2 + y^2 = R^2 \qquad \phi = \phi_0 \qquad \Phi = \Phi_0 = \tfrac{1}{2}k\phi_0^2 \qquad (20.4)$$

where $\Phi = \tfrac{1}{2}k\phi^2$ for unconfined flow (see Sec. 7).

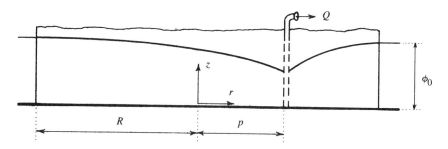

Figure 3.18 An eccentric well in a circular island.

The boundary condition may be met by placing an image recharge well at $x = R^2/p$, $y = 0$ and the potential becomes

$$\Phi = \frac{Q}{4\pi} \ln \frac{(x-p)^2 + y^2}{\left(x - \frac{R^2}{p}\right)^2 + y^2} + C \qquad (20.5)$$

Sec. 20 The Method of Images for a Circle

The constant C is obtained by application of the boundary condition (20.4). With $y^2 = R^2 - x^2$ along the circle, (20.5) becomes

$$\Phi_0 = \frac{Q}{4\pi} \ln \frac{x^2 - 2px + p^2 + R^2 - x^2}{x^2 - 2\frac{R^2}{p}x + \frac{R^4}{p^2} + R^2 - x^2} + C \tag{20.6}$$

or

$$\Phi_0 = \frac{Q}{4\pi} \ln \frac{p^2 - 2px + R^2}{(p^2 - 2px + R^2)\frac{R^2}{p^2}} + C = \frac{Q}{4\pi} \ln \frac{p^2}{R^2} + C \tag{20.7}$$

Hence, the constant C is given by

$$C = \Phi_0 - \frac{Q}{4\pi} \ln \frac{p^2}{R^2} \tag{20.8}$$

so that (20.5) becomes

$$\Phi = \frac{Q}{4\pi} \ln \left[\frac{(x-p)^2 + y^2}{\left(x - \frac{R^2}{p}\right)^2 + y^2} \frac{R^2}{p^2} \right] + \Phi_0 \tag{20.9}$$

It may be noted that the potential for an arbitrary number of eccentric wells inside a circular equipotential may be obtained by superposition, provided that the corresponding image wells are included.

Example 20.2: An Eccentric Well in a Circular Island with Rainfall

The problem of an eccentric well in a circular island with rainfall may be solved by superposing the solution (9.15) of $\nabla^2 \Phi = -N$ for radial flow to the potential (20.9). The constant is determined such that the boundary condition (20.4) is met, and the result is

$$\Phi = -\frac{N}{4}(x^2 + y^2 - R^2) + \frac{Q}{4\pi} \ln \left[\frac{(x-p)^2 + y^2}{\left(x - \frac{R^2}{p}\right)^2 + y^2} \frac{R^2}{p^2} \right] + \Phi_0 \tag{20.10}$$

We will use this solution to determine the maximum discharge of the well such that no inflow from the boundary occurs (see Figure 3.19). The inflow per unit length of boundary is equal to the radial component Q_r of the discharge vector. We determine Q_r by writing the potential (20.10) in terms of polar coordinates r and θ, with

$$x = r\cos\theta \qquad y = r\sin\theta \tag{20.11}$$

followed by differentiation with respect to r:

$$Q_r(r,\theta) = -\frac{\partial \Phi}{\partial r} = \frac{N}{2}r - \frac{Q}{2\pi} \left[\frac{r - p\cos\theta}{r^2 - 2pr\cos\theta + p^2} - \frac{r - \frac{R^2}{p}\cos\theta}{r^2 - 2\frac{R^2}{p}r\cos\theta + \frac{R^4}{p^2}} \right] \tag{20.12}$$

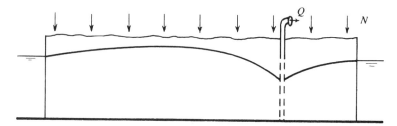

Figure 3.19 An eccentric well in a circular island with rainfall.

The radial component $Q_r(R,\theta)$ along the boundary is obtained by replacing r by R,

$$Q_r(R,\theta) = \frac{N}{2}R - \frac{Q}{2\pi}\left[\frac{R-p\cos\theta}{R^2-2pR\cos\theta+p^2} - \frac{R-\frac{R^2}{p}\cos\theta}{\frac{R^2}{p^2}(R^2-2pR\cos\theta+p^2)}\right] \quad (20.13)$$

or

$$Q_r(R,\theta) = \frac{N}{2}R - \frac{Q}{2\pi}\frac{1}{R}\frac{R^2-p^2}{R^2-2pR\cos\theta+p^2} \quad (20.14)$$

No inflow from the boundary will occur if Q_r is positive for all θ. Since the second term in (20.14) reaches a maximum for $\theta = 0$, Q_r will be positive for all θ if $Q_r(R,0) \geq 0$, i.e., if

$$\frac{N}{2} \geq \frac{Q}{2\pi}\frac{1}{R^2}\frac{R^2-p^2}{(R-p)^2} \quad (20.15)$$

Hence, the maximum value for Q is

$$Q_{\max} = \pi N R^2 \frac{(R-p)^2}{(R-p)(R+p)} = \pi N R^2 \frac{R-p}{R+p} \quad (20.16)$$

It follows that $Q_{\max} = 0$ when $p = R$ and $Q_{\max} = \pi N R^2$ when $p = 0$. In the latter case, the well captures all infiltrated rainwater that falls on the island.

Example 20.3: A Circular Lake

As a third example, we consider the problem of flow to or from a circular lake of radius R in an infinite aquifer, assuming that the lake penetrates the aquifer sufficiently far that the boundary of the lake may be approximated as an equipotential. Cartesian and polar coordinates are chosen with the origin at the center of the lake as shown in Figure 3.20. We first consider the case of radial flow from infinity toward the lake. The lake acts as a large well of radius R and discharge Q_0:

$$\Phi = \frac{Q_0}{2\pi}\ln\frac{r}{R} + \Phi_0 \quad (20.17)$$

Sec. 20   The Method of Images for a Circle

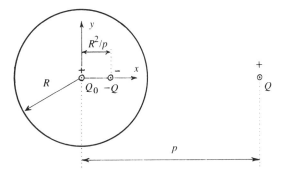

Figure 3.20   A well near a circular lake.

The discharge of the lake may be calculated if it is known that the head at a reference point remains at some fixed value. If a well of discharge Q is placed at a point $x = p$, $y = 0$ then an image recharge well must be introduced inside the lake at $x = R^2/p$, $y = 0$ (note that $R/p < 1$). The potential for the system is obtained by superposition:

$$\Phi = \frac{Q_0}{2\pi} \ln \frac{r}{R} + \frac{Q}{4\pi} \ln \left[\frac{(x-p)^2 + y^2}{\left(x - \frac{R^2}{p}\right)^2 + y^2} \frac{R^2}{p^2} \right] + \Phi_0 \qquad (20.18)$$

compare (20.10). Note that the discharge Q_0 is zero if there is no flow from or toward infinity. The above solution is applicable only to aquifers where the far-field is either radial or constant. In real aquifers, however, the far-field often may be approximated as a combination of radial flow and uniform flow. We will generalize the above solution to such cases by determining the solution for a lake in a field of uniform flow, and then adding this potential to (20.18).

Problem 20.1

For example 20.3, determine the total amount of flow into the lake if the potential at a point $x = L$, $y = 0$ ($L \gg R$) is kept at Φ_1.

A Lake in a Uniform Flow

The problem of a lake in a uniform flow is illustrated in Figure 3.21(a), and will be solved as a limiting case of the flow problem shown in Figure 3.21(b). This is done under the assumption that the potential for uniform flow may be obtained from the potential for flow from a recharge well toward a well, for the limiting case that the distance between the wells as well as the discharge and recharge of the wells approach infinity. Image wells are added to satisfy the condition along the lake boundary. Thus, the potential becomes

$$\Phi = \frac{Q}{4\pi} \ln \left[\frac{(x-p)^2 + y^2}{\left(x - \frac{R^2}{p}\right)^2 + y^2} \frac{R^2}{p^2} \right] - \frac{Q}{4\pi} \ln \left[\frac{(x+p)^2 + y^2}{\left(x + \frac{R^2}{p}\right)^2 + y^2} \frac{R^2}{p^2} \right] + \Phi_0 \qquad (20.19)$$

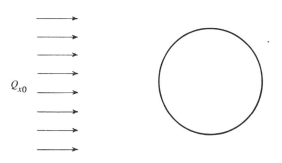

(a)

(b)

Figure 3.21 A well in a field of uniform flow.

We combine the terms due to the wells and those due to the image wells and rewrite the resulting expression:

$$\Phi = \frac{Q}{4\pi} \left[\ln \frac{1 + \frac{R^2}{p^2} \frac{R^2 + 2px}{x^2 + y^2}}{1 + \frac{R^2}{p^2} \frac{R^2 - 2px}{x^2 + y^2}} - \ln \frac{1 + \frac{x^2 + 2px + y^2}{p^2}}{1 + \frac{x^2 - 2px + y^2}{p^2}} \right] + \Phi_0 \qquad (20.20)$$

This may be written as follows, using that $\ln(1 + \varepsilon) \approx \varepsilon$ for small ε,

$$\Phi = \frac{Q}{4\pi} \left[\frac{R^2}{p^2} \frac{4px}{x^2 + y^2} - \frac{4px}{p^2} \right] + \Phi_0 + O(p^{-2}) \qquad (20.21)$$

Sec. 20 The Method of Images for a Circle

where the term $O(p^{-2})$ represents all remaining terms, which are of order p^{-2} or less. If we take the limit for $p \to \infty$ and $Q \to \infty$ such that

$$\lim_{\substack{Q \to \infty \\ p \to \infty}} \frac{Q}{\pi p} = Q_{x0} \qquad (20.22)$$

the potential becomes

$$\Phi = -Q_{x0}\left[x - \frac{xR^2}{x^2 + y^2}\right] + \Phi_0 \qquad (20.23)$$

Indeed, this potential approaches that for uniform flow near infinity: the second term in (20.23) vanishes and the first equals that for uniform flow of discharge Q_{x0} in the x direction. Along the circle, $x^2 + y^2 = R^2$, the potential becomes

$$\Phi = -Q_{x0}\left[x - \frac{xR^2}{R^2}\right] + \Phi_0 = \Phi_0 \qquad (20.24)$$

so that the boundary condition is met.

It may be noted that the term

$$\Phi = Q_{x0}\frac{xR^2}{x^2 + y^2} \qquad (20.25)$$

results from the well and recharge well inside the lake. These wells both have approached the origin: they were located at $(R^2/p, 0)$ and $(-R^2/p, 0)$ originally, and p has approached infinity. The limiting case of two wells of opposite and infinite charges at a single point is known as a *dipole* and (20.25) represents the potential for such a dipole.

The potential (20.23) lacks a radial component in its far-field, so that the net inflow into the lake is zero. The complete solution for a lake which extracts a net amount Q_0 from the aquifer is obtained by adding the potential (20.17) for radial flow toward a lake:

$$\Phi = -Q_{x0}\left[x - \frac{xR^2}{x^2 + y^2}\right] + \frac{Q_0}{4\pi}\ln\frac{x^2 + y^2}{R^2} + \Phi_0 \qquad (20.26)$$

The unknown quantity Q_0 is determined by requiring that the potential be Φ_1 at some given point $x = x_1$, $y = y_1$:

$$\Phi_1 = -Q_{x0}\left[x_1 - \frac{x_1 R^2}{x_1^2 + y_1^2}\right] + \frac{Q_0}{4\pi}\ln\frac{x_1^2 + y_1^2}{R^2} + \Phi_0 \qquad (20.27)$$

or

$$Q_0 = 4\pi\frac{\Phi_1 - \Phi_0 + Q_{x0}\left[x_1 - \frac{x_1 R^2}{x_1^2 + y_1^2}\right]}{\ln\frac{x_1^2 + y_1^2}{R^2}} \qquad (20.28)$$

The potential for the general case that the uniform flow is directed at an arbitrary angle β with the x axis may be obtained from the above result as follows. We modify

(20.23) such that the flow is in the y direction rather than in the x direction, by replacing Q_{x0} by Q_{y0}, and x by y:

$$\Phi = -Q_{y0}\left[y - \frac{yR^2}{x^2+y^2}\right] + \Phi_0 \qquad (20.29)$$

The uniform flow of discharge Q_0 in a direction β may be decomposed into two components:

$$Q_{x0} = Q_0 \cos\beta \qquad Q_{y0} = Q_0 \sin\beta \qquad (20.30)$$

The corresponding potential is found by superposition of (20.23) and (20.29), while modifying the constants to meet the boundary condition at $r = R$:

$$\Phi = -Q_{x0}x - Q_{y0}y + Q_{x0}\frac{xR^2}{x^2+y^2} + Q_{y0}\frac{yR^2}{x^2+y^2} + \Phi_0 \qquad (20.31)$$

The complete solution to the problem of flow in an infinite aquifer with a lake and a well is obtained by adding the potential (20.31) to (20.18) in order to account for uniform flow. This yields, again modifying the constants to match the boundary condition:

$$\Phi = \frac{Q_0}{4\pi}\ln\frac{x^2+y^2}{R^2} + \frac{Q}{4\pi}\ln\left[\frac{(x-p)^2+y^2}{\left(x-\frac{R^2}{p}\right)^2+y^2}\frac{R^2}{p^2}\right]$$
$$-Q_{x0}x - Q_{y0}y + Q_{x0}\frac{xR^2}{x^2+y^2} + Q_{y0}\frac{yR^2}{x^2+y^2} + \Phi_0 \qquad (20.32)$$

The dipole at the center of the lake may be considered as the image of a uniform far-field, whereby the strength and orientation of the dipole are determined in such a way that the condition along the lake boundary is met. The potential for a dipole of arbitrary orientation will be derived below, along with an expression for the stream function Ψ.

The Potential and the Stream Function for a Dipole of Arbitrary Orientation

The dipole obtained inside the circle of Figure 3.21 for the limiting case that the well and the recharge well coincide possesses an orientation. The x axis is a streamline by symmetry and the flow along this streamline is in the positive x direction because the recharge well lies on the + side of the origin. We identify the direction of flow along this axis of symmetry with the orientation of the dipole. The potential for a dipole oriented at an angle β to the x axis is obtained by superposition of (20.25) and the second term in (20.29). Introducing the components s_x and s_y of the strength s of the dipole as $s_x = 2\pi R^2 Q_{x0}$, $s_y = 2\pi R^2 Q_{y0}$, the potential becomes

$$\Phi = \frac{1}{2\pi}\frac{s_x x + s_y y}{x^2+y^2} \qquad (20.33)$$

Sec. 20   The Method of Images for a Circle   **249**

If the dipole is located at a point (x_0, y_0), the potential becomes

$$\Phi = \frac{1}{2\pi} \frac{s_x(x - x_0) + s_y(y - y_0)}{(x - x_0)^2 + (y - y_0)^2} \tag{20.34}$$

This potential may be written in terms of the local polar coordinates r and θ, with

$$x - x_0 = r\cos\theta \qquad y - y_0 = r\sin\theta \tag{20.35}$$

as follows:

$$\Phi = \frac{1}{2\pi} \frac{s_x \cos\theta + s_y \sin\theta}{r} \tag{20.36}$$

Expressing the components of the strength in terms of s and β as

$$s_x = s\cos\beta \qquad s_y = s\sin\beta \tag{20.37}$$

the potential becomes

$$\boxed{\Phi = \frac{s}{2\pi} \frac{\cos\theta \cos\beta + \sin\theta \sin\beta}{r} = \frac{s}{2\pi} \frac{\cos(\theta - \beta)}{r}} \tag{20.38}$$

We use this equation to determine the stream function. The Cauchy-Riemann conditions in terms of polar coordinates are (compare [19.26]):

$$\frac{\partial \Phi}{\partial r} = \frac{\partial \Psi}{r \partial \theta} \qquad \frac{\partial \Phi}{r \partial \theta} = -\frac{\partial \Psi}{\partial r} \tag{20.39}$$

Application of these conditions to (20.38) gives

$$\frac{\partial \Psi}{\partial \theta} = -\frac{s}{2\pi} \frac{\cos(\theta - \beta)}{r} \qquad \frac{\partial \Psi}{\partial r} = \frac{s}{2\pi} \frac{\sin(\theta - \beta)}{r^2} \tag{20.40}$$

and we obtain by integration of either equation:

$$\boxed{\Psi = -\frac{s}{2\pi} \frac{\sin(\theta - \beta)}{r}} \tag{20.41}$$

which represents the stream function for the dipole. We may write this stream function alternatively in terms of Cartesian coordinates:

$$\Psi = -\frac{s}{2\pi} \frac{r\sin\theta \cos\beta - r\cos\theta \sin\beta}{r^2} = -\frac{1}{2\pi} \frac{s_x(y - y_0) - s_y(x - x_0)}{(x - x_0)^2 + (y - y_0)^2} \tag{20.42}$$

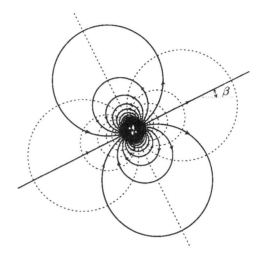

Figure 3.22 Flow net for a dipole.

Expressions for the equipotentials and streamlines are obtained from (20.34) and (20.42). These curves are families of circles through the dipole, with their centers on mutually orthogonal lines, as shown in Figure 3.22.

Problem 20.2

Determine the radii and coordinates of the centers of the families of streamlines and equipotentials for a dipole in terms of x_0, y_0, β, and s.

Streamlines for a Lake in a Uniform Flow

The potential for a lake in an infinite aquifer with a uniform flow in the x direction is given by (20.26),

$$\Phi = -Q_{x0}\left[x - \frac{xR^2}{x^2+y^2}\right] + \frac{Q_0}{4\pi}\ln\frac{x^2+y^2}{R^2} + \Phi_0 \qquad (20.43)$$

where the discharge Q_0 is the net inflow into the lake. This potential is composed of three terms: the first represents uniform flow, the second a dipole at the origin with strength components s_x and s_y with

$$s_x = 2\pi Q_{x0} R^2 \qquad s_y = 0 \qquad (20.44)$$

and the third term corresponds to a well of discharge Q_0 at the center of the lake. It may be noted that the dipole is oriented in the same direction as the uniform flow. The stream function is obtained by superposition of the three expressions for Ψ that correspond to the three terms in (20.43):

$$\Psi = -Q_{x0}\left[y + \frac{yR^2}{x^2+y^2}\right] + \frac{Q_0}{2\pi}\arctan\frac{y}{x} \qquad (20.45)$$

Sec. 20   The Method of Images for a Circle   251

The magnitude of Q_0 depends upon the fixed value of the potential at some point in the aquifer, and is given by (20.28).

A special case occurs when $Q_0 = 0$ and there is no net inflow into the lake; inflow occurs along the upstream face of the lake and outflow along the downstream face. The streamlines for this case are shown in Figure 3.23, and we note that this case applies only if the potential in the lake is equal to the value Φ_0 that would occur along the y axis if the lake were not present. This is illustrated in Figure 3.24 where the potential is plotted along the x axis. If the potential along the lake boundary is either higher or lower than this value, net inflow or outflow will occur.

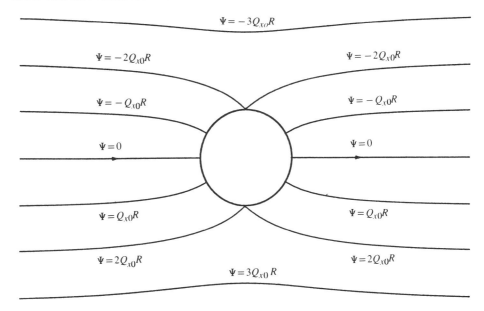

Figure 3.23   Streamlines for the case that $Q_0 = 0$.

Problem 20.3

Determine an expression for Q_r along the lake boundary if $Q_0 = 0$, and identify the inflow sections and the outflow sections of the boundary. Find an expression for the streamlines $\Psi = \Psi_j$ in the form $x = x(y)$ and determine the range of values that y may take along any given streamline outside the lake. Check your expression by calculating the coordinates of a few points on the streamlines in Figure 3.23.

Three different cases may apply when Q_0 is not zero:
Case 1: There is inflow into the lake along the entire circumference,

$$Q_r \leq 0 \qquad r = R \qquad -\pi < \theta \leq \pi \qquad (20.46)$$

where r and θ are polar coordinates emanating from the center of the lake.

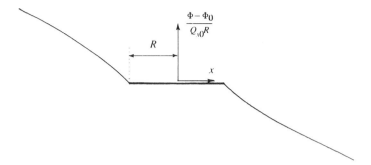

Figure 3.24 The potential along the x axis.

Case 2: There is inflow along part of the lake boundary and outflow along the remaining part.

Case 3: There is outflow from the lake along the entire circumference,

$$Q_r \geq 0 \qquad r = R \qquad -\pi < \theta < \pi \tag{20.47}$$

Case 1 applies when Q_r is negative along the entire boundary of the lake. In order to determine Q_r, we first write the potential (20.43) in terms of polar coordinates as follows:

$$\Phi = -Q_{x0}\left[r - \frac{R^2}{r}\right]\cos\theta + \frac{Q_0}{2\pi}\ln\frac{r}{R} + \Phi_0 \tag{20.48}$$

Partial differentiation with respect to r yields Q_r:

$$Q_r = Q_{x0}\left[1 + \frac{R^2}{r^2}\right]\cos\theta - \frac{Q_0}{2\pi r} \tag{20.49}$$

and we obtain for $r = R$:

$$Q_r(R, \theta) = 2Q_{x0}\cos\theta - \frac{Q_0}{2\pi R} \tag{20.50}$$

The maximum value occurs when $\theta = 0$, so that (20.46) applies when

$$\text{case 1:} \quad Q_0 \geq 4\pi R Q_{x0} \tag{20.51}$$

Case 3 applies when Q_r is positive along the entire lake boundary. The minimum value of Q_r occurs at $\theta = \pm\pi$, so that

$$\text{case 3:} \quad Q_0 \leq -4\pi R Q_{x0} \tag{20.52}$$

Case 2 applies for the remaining values of Q_0:

$$\text{case 2:} \quad -4\pi R Q_{x0} < Q_0 < 4\pi R Q_{x0} \tag{20.53}$$

Sec. 20 The Method of Images for a Circle

We will next determine the locations of the stagnation points for the various cases, and tie our findings in with the above observations. We do this by setting the polar components Q_r and Q_θ equal to zero; Q_r is given by (20.49) and Q_θ is found from

$$Q_\theta = -\frac{\partial \Phi}{r \partial \theta} = -Q_{x0}\left[1 - \frac{R^2}{r^2}\right]\sin\theta \qquad (20.54)$$

The latter expression is zero at $\theta = 0$, at $\theta = \pm\pi$ and at $r = R$. Substituting first zero and then π for θ in (20.49), and solving for $r = r_s$, the radial distance to the stagnation point, we obtain

$$\left[\frac{r_s}{R}\right]_{1,2} = \frac{Q_0}{4\pi Q_{x0} R} \pm \sqrt{\left[\frac{Q_0}{4\pi Q_{x0} R}\right]^2 - 1} \qquad (\theta_s = 0)$$

$$\left[\frac{r_s}{R}\right]_{1,2} = -\frac{Q_0}{4\pi Q_{x0} R} \pm \sqrt{\left[\frac{Q_0}{4\pi Q_{x0} R}\right]^2 - 1} \qquad (\theta_s = \pm\pi) \qquad (20.55)$$

The first equation gives only possible (i.e., positive) values for r_s if $Q_0/Q_{x0} > 0$, and the second one only if $Q_0/Q_{x0} < 0$.

When $r = R$, Q_θ vanishes and the second set of possible locations of the stagnation points are found from (20.49) to be

$$r_s = R \qquad \cos\theta_s = \frac{Q_0}{4\pi Q_{x0} R} \qquad (20.56)$$

We see that the stagnation points lie on the boundary of the lake only when $-1 \le Q_0/(4\pi Q_{x0} R) \le 1$ (otherwise θ_s does not exist). In this case (20.55) does not yield real roots, and case 2 applies (compare [20.53]). For the special case that $Q_0 = 0$, the stagnation points lie on $r_s = R$, $\theta_s = \pm\pi/2$. If either case 1 or case 3 applies, the roots in (20.55) are real and there exist two values for r_s/R. We observe from (20.55) that the product of these two roots equals 1; one stagnation point lies outside the lake and is real, the other one lies inside the lake and is discarded. As an illustration of case 1, the streamlines for the case that $Q_0/(Q_{x0}R) = 14 > 4\pi$ are shown in Figure 3.25. Note that the streamlines for the corresponding problem of case 3 have the same shape; if $Q_{x0} < 0$ and $Q_0 < 0$ the ratio $Q_0/(Q_{x0}R)$ does not change sign, but the direction of flow changes and the lake recharges. The flow pattern for that case is obtained simply by reversing the directions of the arrows in Figure 3.25 that point in the direction of flow.

Flow Around a Cylindrical Impermeable Object

We obtained the potential for a circular lake in a uniform flow by superposition of the potential for uniform flow and that for a dipole which is oriented in the direction of the flow ($\beta = 0$). We obtain the solution for flow around a cylindrical impermeable object, illustrated in Figure 3.26, if we orient the dipole against the uniform flow. The corresponding potential and stream functions are obtained from the solution for a lake

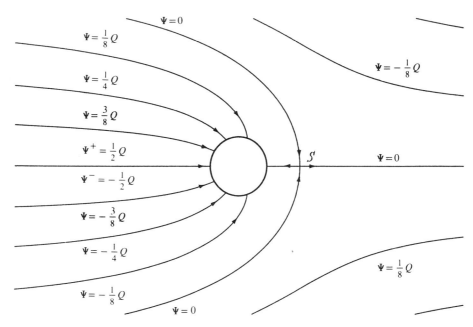

Figure 3.25 Streamlines for the case that $Q_0/(RQ_{x0}) = 14$.

in a uniform flow (with $Q_0 = 0$) by changing the signs of the second terms, see (20.43) and (20.45):

$$\Phi = -Q_{x0}\left[x + \frac{xR^2}{x^2 + y^2}\right] + C \tag{20.57}$$

and

$$\Psi = -Q_{x0}\left[y - \frac{yR^2}{x^2 + y^2}\right] \tag{20.58}$$

We see that indeed the stream function is constant along the perimeter of the cylinder, $x^2 + y^2 = R^2$.

The streamlines obtained from (20.58) are shown in Figure 3.26, and have the same shape as the equipotentials for a lake in a uniform flow with zero net inflow ($Q_0 = 0$).

Problem 20.4

Determine the potential and the stream function for the case that a well of discharge Q is located at $x = p$, $y = 0$ downstream from the impermeable object in Figure 3.26. Note that you will need two image wells, one at the origin and one at $x = R^2/p$, $y = 0$. Explain why you need two image wells, and prove that your solution meets all boundary conditions.

Sec. 21   The Vortex   255

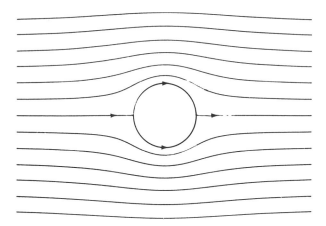

Figure 3.26 Streamlines for flow around a cylindrical obstacle.

21 THE VORTEX

We will study the potential and the stream function for a vortex as a final example of an elementary solution to Laplace's equation. A vortex is a point around which the water flows; the flow is rotational at the vortex. Vortices cannot occur in the interior of an aquifer, but exist at points of the boundary where the potential is discontinuous.

The Potential and the Stream Function for a Vortex

We will derive the potential and the stream function for a vortex for the case of unconfined flow around a lock illustrated in Figure 3.27. An infinitely long river along the x axis forms the boundary of a large unconfined aquifer. Flow in the aquifer results from a difference in head in two river sections that are separated by a lock. We neglect the width of the lock, which is located at $x = 0$, $y = 0$, denote the heads in the two river sections by ϕ_1 and ϕ_2, and write the boundary conditions as follows:

$$x < 0 \qquad y = 0 \qquad \phi = \phi_1 \qquad \Phi = \Phi_1 = \tfrac{1}{2}k\phi_1^2 \qquad (21.1)$$

and

$$x > 0 \qquad y = 0 \qquad \phi = \phi_2 \qquad \Phi = \Phi_2 = \tfrac{1}{2}k\phi_2^2 \qquad (21.2)$$

We seek the harmonic function that meets the preceding conditions and exhibits a discontinuity at the origin. The stream function for flow toward a well, $\Psi = Q\theta/(2\pi)$, possesses such a jump; we attempt to write the potential as a linear function of the polar coordinate θ,

$$\boxed{\Phi = \frac{A}{\pi}\theta + B} \qquad (21.3)$$

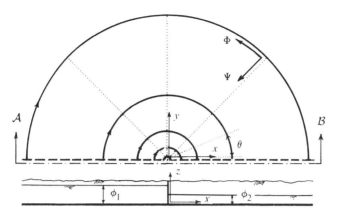

Figure 3.27 Unconfined flow around a lock.

where θ is measured from the x axis as shown in Figure 3.27, and where A and B are constants. Application of the boundary conditions (21.1) and (21.2) yields

$$x > 0 \quad y = 0 \quad \theta = 0 \quad \Phi_2 = B \tag{21.4}$$

$$x < 0 \quad y = 0 \quad \theta = \pi \quad \Phi_1 = A + B \tag{21.5}$$

Hence, $A = (\Phi_1 - \Phi_2)$ and $B = \Phi_2$ so that (21.3) becomes

$$\Phi = \frac{\Phi_1 - \Phi_2}{\pi}\theta + \Phi_2 \quad (0 \le \theta \le \pi) \tag{21.6}$$

It appears that the equipotentials are straight lines θ=constant, which emanate from the lock. The expression for Ψ is obtained from (21.3) by integration of the Cauchy-Riemann conditions (20.39), which yields

$$\boxed{\Psi = -\frac{A}{\pi}\ln r + C = -\frac{\Phi_1 - \Phi_2}{\pi}\ln\frac{r}{R}} \tag{21.7}$$

The value of the constant C in (21.7) is immaterial; only the difference in values of Ψ are of interest. The constant is expressed in terms of the arbitrary length R merely in order to obtain a dimensionless expression as the argument of the natural logarithm. It follows from (21.7) that the streamlines Ψ = constant are circles around the origin. The flow net reproduced in Figure 3.27 is obtained from the flow net for a well, Figure 3.7, simply by interchanging equipotentials and streamlines.

Superposition

Potentials for vortices can be superimposed just as those for wells and dipoles. For example, the presence of a second lock in the river may be accounted for by adding the corresponding potential and stream function to (21.6) and (21.7).

Sec. 21   The Vortex   257

We may model the case that there is a far-field of uniform flow directed toward the river by adding the appropriate terms to (21.6) and (21.7),

$$\Phi = -Q_{y0}y + \frac{\Phi_1 - \Phi_2}{\pi}\theta + \Phi_2 \qquad (21.8)$$

and

$$\Psi = Q_{y0}x - \frac{\Phi_1 - \Phi_2}{\pi}\ln\frac{r}{R} \qquad (21.9)$$

where Q_{y0} is negative if the flow is toward the river, i.e., in the negative y direction.

Problem 21.1

A building pit is planned to the north of a long, straight river containing a lock (see Figure 3.28). The head in the river to the west of the lock is $\phi_1 = 40$ m and the head to the east is $\phi_2 = 20$ m; the far-field does not contain uniform flow. The aquifer to the north of the river is unconfined. The dimensions of the pit are indicated in Figure 3.28. The bottom of the pit is 22 m above the base of the aquifer.

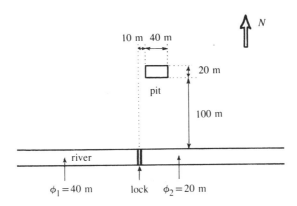

Figure 3.28 A building pit in an aquifer bounded by a river with a lock (Problem 21.1).

Questions
(a) Determine whether it is necessary to use wells to keep the bottom of the building pit dry.
(b) If necessary, design a system of no more than two wells that will keep the bottom of the building pit dry.

The Method of Images for Vortices

Certain boundary value problems involving vortices may be solved by the use of images. As an example, consider the flow problem illustrated in Figure 3.29 where an aquifer is bounded by two long straight orthogonal rivers. The origin of an x, y coordinate system is at the intersection of the rivers and a lock is located at $x = l$, $y = 0$. The boundary conditions in terms of the potential are as follows:

$$x = 0 \qquad y > 0 \qquad \Phi = \Phi_1 \qquad (21.10)$$

$$0 \leq x \leq 1 \qquad y = 0 \qquad \Phi = \Phi_1 \qquad (21.11)$$

$$1 < x < \infty \qquad y = 0 \qquad \Phi = \Phi_2 \qquad (21.12)$$

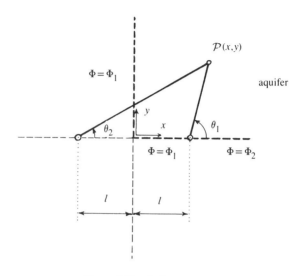

Figure 3.29 An image vortex.

The problem is solved by placing a vortex at $x = l$, $y = 0$ and an image vortex at $x = -l$, $y = 0$. If r_1, θ_1 and r_2, θ_2 are the local polar coordinates centered at the vortex and the image vortex, respectively, the expression for the potential becomes

$$\Phi = \frac{A_1}{\pi}\theta_1 + \frac{A_2}{\pi}\theta_2 + B \qquad (21.13)$$

The constants A_1, A_2, and B must be determined from the boundary conditions (21.10) through (21.12). It is seen from Figure 3.29 that $\theta_2 = \pi - \theta_1$ along the y axis, and application of the boundary condition (21.10) yields

$$\Phi_1 = \frac{A_1}{\pi}\theta_1 + \frac{A_2}{\pi}(\pi - \theta_1) + B \qquad (21.14)$$

This condition can be met only if the coefficient of the variable angle θ_1 vanishes, i.e., if

$$A_1 = A_2 = A \qquad (21.15)$$

Sec. 21   The Vortex   259

The angle θ_2 is zero and θ_1 is π along the interval $0 < x < l$ of the x axis, and application of condition (21.11) to (21.13) yields, using (21.15)

$$\Phi_1 = A + B \tag{21.16}$$

Application of the third boundary condition gives, with $\theta_1 = \theta_2 = 0$ for $y = 0, l < x < \infty$,

$$\Phi_2 = B \tag{21.17}$$

It follows from (21.16) and (21.17) that $A = \Phi_1 - \Phi_2$ and $B = \Phi_2$, so that the potential (21.13) becomes

$$\Phi = \frac{\Phi_1 - \Phi_2}{\pi}(\theta_1 + \theta_2) + \Phi_2 \tag{21.18}$$

It is of interest to note that the rotational flows around the two vortices are oriented in the same way; the vortices have identical strengths. Image vortices have the same strengths when the boundary between them is an equipotential.

Problem 21.2

Determine the strength of the image vortex for the flow problem sketched in Figure 3.29, if the y axis is a streamline, rather than an equipotential.

Problem 21.3

Consider the problem of two-dimensional flow in the vertical plane sketched in Figure 3.30. Two long parallel impervious dams separate three lakes with water levels as indicated in the figure. The widths of the dams may be neglected. The aquifer underneath the horizontal lake bottoms is infinite in extent.

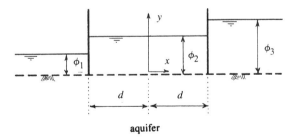

Figure 3.30 Flow underneath two dams (Problem 21.3).

Questions

(a) Determine the head in the aquifer as a function of position.
(b) Determine the location of the stagnation point.
(c) For the case that $\phi_2 = \frac{1}{2}(\phi_1 + \phi_3)$, determine an expression for the streamline through the stagnation point. Calculate the coordinates of points on this streamline and sketch the streamline.

Example 21.1: Flow near a Circular Lake with a Dam

The case of flow near a circular lake with a dam will be considered as an application of the method of images for vortices involving a circular boundary equipotential. The dam $\mathcal{V}_1\mathcal{V}_2$ in Figure 3.31 divides the lake into two parts with different water levels. The problem is to determine the flow in the surrounding aquifer. It will be assumed at first that the far-field is that of a constant potential.

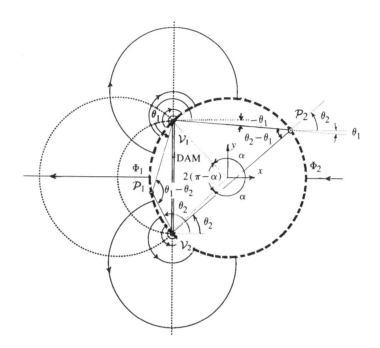

Figure 3.31 Flow near a lake with a dam.

The values of the potential along the two segments of the lake boundary are denoted as Φ_1 and Φ_2. The origin of a Cartesian x, y and of a polar r, θ coordinate system is at the center of the lake, which has a radius R. The boundary conditions may be written as follows, using the angle α defined in Figure 3.31:

$$-\alpha \leq \theta \leq \alpha \qquad r = R \qquad \Phi = \Phi_2 \qquad (21.19)$$

$$\alpha < \theta < \pi \qquad r = R \qquad \Phi = \Phi_1 \qquad (21.20)$$

$$-\pi < \theta < -\alpha \qquad r = R \qquad \Phi = \Phi_1 \qquad (21.21)$$

Jumps in the boundary values of the potential occur at points $\mathcal{V}_1$, $\theta = \alpha$, $r = R$ and $\mathcal{V}_2$, $\theta = -\alpha$, $r = R$: there are vortices at points $\mathcal{V}_1$ and $\mathcal{V}_2$. It is seen from the boundary values of the potential Φ that the flows around these vortices are in opposite directions.

Sec. 21 The Vortex

The strengths of the vortices therefore are opposite in sign but equal in magnitude, so that the potential will involve the difference of the angles θ_1 and θ_2, measured at points $\mathcal{V}_1$ and $\mathcal{V}_2$ as shown in Figure 3.31:

$$\Phi = \frac{A}{\pi}(\theta_1 - \theta_2) + B \tag{21.22}$$

where A and B are constants.

Care must be taken in defining the angle $\theta_1 - \theta_2$; this angle must be chosen such that no jumps in Φ occur inside the aquifer. We choose the branch cuts of θ_1 and θ_2 along the dam in the negative y direction, so that no discontinuity occurs outside the lake boundary (see Figure 3.31). The two jumps cancel each other along the branch cuts outside the dam, whereas the jump along the dam is due to vortex $\mathcal{V}_1$.

It remains to be shown that the constants A and B in (21.22) can be determined such that the potential meets the boundary conditions. The dam divides the circle into two arcs, and $\mathcal{P}_1$ and $\mathcal{P}_2$ represent arbitrary points on each of these arcs. The angles $\mathcal{V}_2\mathcal{P}_1\mathcal{V}_1$ and $\mathcal{V}_1\mathcal{P}_2\mathcal{V}_2$ are constant when $\mathcal{P}_1$ and $\mathcal{P}_2$ trace their respective arcs. The angles $\mathcal{V}_2\mathcal{P}_1\mathcal{V}_1$ and $\mathcal{V}_1\mathcal{P}_2\mathcal{V}_2$ span the same arcs as do the center angles of magnitudes 2α and $2(\pi - \alpha)$, respectively, so that

$$-\alpha \leq \theta \leq \alpha \qquad r = R \qquad \theta_2 - \theta_1 = \pi - \alpha \tag{21.23}$$

and

$$\left\{\begin{array}{c} \alpha < \theta < \pi \\ -\pi < \theta < -\alpha \end{array}\right\} \qquad r = R \qquad \theta_1 - \theta_2 = \alpha \tag{21.24}$$

Application of the boundary conditions (21.19) through (21.21) to the potential (21.22) yields, using (21.23) and (21.24),

$$\Phi_1 = \frac{A}{\pi}\alpha + B \tag{21.25}$$

and

$$\Phi_2 = \frac{A}{\pi}(\alpha - \pi) + B \tag{21.26}$$

so that

$$A = \Phi_1 - \Phi_2 \qquad B = \Phi_1 - (\Phi_1 - \Phi_2)\frac{\alpha}{\pi} \tag{21.27}$$

The potential thus becomes

$$\Phi = \frac{\Phi_1 - \Phi_2}{\pi}[(\theta_1 - \theta_2) - \alpha] + \Phi_1 \tag{21.28}$$

The case that the lake withdraws an amount Q_0 from the aquifer and that the far-field is one of uniform flow in the x direction is covered by adding the appropriate terms to (21.28), compare (20.43),

$$\Phi = \frac{\Phi_1 - \Phi_2}{\pi}[(\theta_1 - \theta_2) - \alpha] + \frac{Q_0}{2\pi} \ln \frac{r}{R} - Q_{x0}\left[x - \frac{xR^2}{r^2}\right] + \Phi_1 \qquad (21.29)$$

The unknown constant Q_0 may be obtained from a given value of the potential at some point away from the lake. Sometimes, however, the withdrawal Q_0 of the lake is known, from a water balance for the lake, for example.

The above treatment of a vortex concludes the discussion of elementary harmonic functions. The solutions for wells, dipoles, and vortices are called *singular solutions*, because the potential approaches infinity upon approaching the location of a well, a dipole, or a vortex. We will see later that continuous distributions of these singularities render functions that are useful in modeling regional aquifers. The analysis associated with both the derivation and the use of these functions becomes rather involved when carried out in terms of real analysis. Because both the analysis and formulae simplify considerably when carried out in terms of complex variables, we will introduce the use of complex variable techniques before treating distributed singularities.

Problem 21.4

A circular building pit in a river is to be dewatered by means of four wells, as shown in Figure 3.32. The building pit is surrounded by an impervious dam. An aquifer of thickness H is located underneath the river (see Figure 3.32[b]). Two dams, $\mathcal{AB}$ and $\mathcal{CD}$, connect the dam surrounding the pit with the river banks such that the line $\mathcal{ABCD}$ passes through the center of the pit. The following approximations are to be made:
1. The widths of the dams $\mathcal{AB}$ and $\mathcal{CD}$ are neglected.
2. The heads in the aquifer along the exterior of the dam surrounding the pit are ϕ_1 along the upstream arc $\mathcal{BC}$ and $\phi_2 = \phi_1 - \Delta\phi$ along the downstream arc $\mathcal{BC}$. The four wells are spaced uniformly along the inner circumference of the dam. The discharges of the pairs of upstream and downstream wells are Q_1 and Q_2, respectively. The radii of the wells are r_w and the radius of the circular pit is R. The width of the dam is b.

Question

Determine Q_1/k and Q_2/k in terms of the other parameters such that dry working conditions are obtained throughout the pit. It is given that $\phi_1 = 30$ m, $\phi_2 = \phi_1 - \Delta\phi = 25.7$ m, $H = 20$ m, $b = 15$ m, $R = 50$ m, and $r_w = 0.25$ m.

22 ALGEBRA OF COMPLEX NUMBERS

What follows in this section is a brief overview of the algebra of complex numbers. Readers familiar with complex analysis may skip Secs. 22 and 23, and continue reading with Sec. 24, where the elementary solutions determined previously are written in terms of complex functions. The reader interested in a more thorough treatment of complex analysis is referred to the handbooks on complex variable techniques (e.g., Churchill [1960], Von Koppenfels and Stallmann [1959], and Nehari [1952]).

Sec. 22   Algebra of Complex Numbers   263

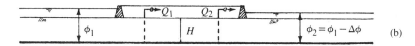

Figure 3.32 Flow in the aquifer below a circular building pit in a river (Problem 21.4).

The Complex Plane

We define a Cartesian x, y coordinate system as shown in Figure 3.33 and associate with each point (x, y) the complex number

$$z = x + iy \qquad (22.1)$$

The x, y plane is called the complex z plane, and the x and y axes are known as the real and the imaginary axis, respectively. We call the x and y coordinates of point z the real and imaginary parts of z, and write this as follows:

$$x = \Re(z) \qquad y = \Im(z) \qquad (22.2)$$

The symbol i in (22.1) separates the imaginary part from the real part.

Elementary Algebraic Operations

The addition or subtraction of complex numbers is achieved by adding or subtracting the real and imaginary parts. For example, if z_1 and z_2 are two complex numbers, then their sum z_3 is obtained as

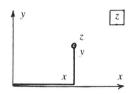

Figure 3.33 The complex plane.

$$z_3 = z_1 + z_2 = (x_1 + x_2) + i(y_1 + y_2) \tag{22.3}$$

Multiplication of complex numbers is defined in such a way that a clockwise rotation of the ξ, η coordinate system through an angle α into the x, y coordinate system corresponds to the complex multiplication:

$$z = \zeta z_0 \tag{22.4}$$

where

$$z = x + iy \qquad \zeta = \xi + i\eta \qquad z_0 = \cos\alpha + i\sin\alpha \tag{22.5}$$

This operation is illustrated in Figure 3.34. The complex number ζ is defined with reference to the real axis $\eta = 0$ and the imaginary axis $\xi = 0$; the complex number z_0 defines the rotation of the ξ, η coordinate system relative to the x, y coordinate system. The coordinate transformation is represented by the following set of equations:

$$\begin{aligned} x &= \xi \cos\alpha - \eta \sin\alpha = \xi x_0 - \eta y_0 \\ y &= \xi \sin\alpha + \eta \cos\alpha = \xi y_0 + \eta x_0 \end{aligned} \tag{22.6}$$

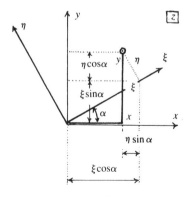

Figure 3.34 Complex multiplication.

Sec. 22 Algebra of Complex Numbers

The multiplication (22.4) may be worked out in terms of real algebra as follows:

$$x + iy = (\xi + i\eta)(x_0 + iy_0) = \xi x_0 + i^2 \eta y_0 + i(\xi y_0 + \eta x_0) \tag{22.7}$$

We see that (22.6) and (22.7) will correspond to each other, provided that we assign the value -1 to i^2

$$\boxed{i^2 = -1} \tag{22.8}$$

Note that it follows from this equation that $1/i = -i$:

$$\frac{1}{i} = \frac{i}{i^2} = -i \tag{22.9}$$

We note that the purely imaginary number i serves to separate the real and imaginary parts, and that multiplication by i is equivalent to rotating the coordinate system clockwise through $\pi/2$.

Following the above discussion, we expect that complex division by z_0 will yield the inverse of the coordinate transformation (22.6). We obtain from (22.4) that

$$\zeta = \frac{z}{z_0} = \frac{x + iy}{x_0 + iy_0} = \frac{(x + iy)(x_0 - iy_0)}{(x_0 + iy_0)(x_0 - iy_0)}$$

$$= \frac{xx_0 - i^2 yy_0 + i(x_0 y - xy_0)}{x_0^2 + y_0^2} \tag{22.10}$$

Using (22.5) and (22.8), we simplify this as follows:

$$\zeta = \xi + i\eta = x\cos\alpha + y\sin\alpha + i(-x\sin\alpha + y\cos\alpha) \tag{22.11}$$

which indeed corresponds to the inverse of (22.6), as asserted. Separation of the real and imaginary parts, i.e., isolation of terms containing a factor i from the others, was done in (22.10) by multiplying the numerator and denominator by $(x_0 - iy_0)$, which produced a real denominator.

The Complex Conjugate, the Modulus, and the Argument

The complex number

$$\boxed{\bar{z} = x - iy} \tag{22.12}$$

is called the complex conjugate of z, and is denoted by the superscribed bar. The point in the z plane that corresponds to $\bar{z}$ is the mirror image of the point z (see Figure 3.35). The expression

$$\boxed{|z| = \sqrt{z\bar{z}} = \sqrt{x^2 + y^2}} \tag{22.13}$$

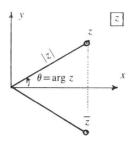

Figure 3.35 The complex conjugate, the modulus, and the argument of a complex number.

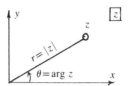

Figure 3.36 Polar coordinates.

is the modulus of z and equals the length of the vector pointing from the origin to point z. The angle θ enclosed between this vector and the x axis is called the argument of z,

$$\boxed{arg(z) = \theta = \arctan \frac{y}{x}} \tag{22.14}$$

The operations carried out in (22.10) may be written somewhat more elegantly by the use of the complex conjugate:

$$\zeta = \frac{z}{z_0} = \frac{z\bar{z}_0}{z_0\bar{z}_0} = \frac{(x+iy)(x_0-iy_0)}{|z_0|^2} = \frac{xx_0 + yy_0 + i(x_0y - y_0x)}{x_0^2 + y_0^2} \tag{22.15}$$

Polar Coordinates

Complex numbers can be represented in terms of polar coordinates r and θ (see Figure 3.36), as follows:

$$z = x + iy = r(\cos\theta + i\sin\theta) \tag{22.16}$$

An interesting result is obtained if we represent both $\sin\theta$ and $\cos\theta$ by their Taylor series expansions:

$$z = r\sum_{n=0}^{\infty}\left[(-1)^n\frac{\theta^{2n}}{(2n)!} + i(-1)^n\frac{\theta^{2n+1}}{(2n+1)!}\right] \tag{22.17}$$

Sec. 22   Algebra of Complex Numbers

Writing (-1) as i^2 and combining terms, we obtain the Taylor series of $e^{i\theta}$:

$$z = r \sum_{n=0}^{\infty} \left[\frac{(i\theta)^{2n}}{(2n)!} + \frac{(i\theta)^{2n+1}}{(2n+1)!} \right] = r \sum_{n=0}^{\infty} \left[\frac{(i\theta)^n}{n!} \right] = re^{i\theta} \tag{22.18}$$

The compact equation

$$\boxed{z = re^{i\theta} = |z|e^{i\,arg(z)}} \tag{22.19}$$

is the representation of a complex number in polar form, which involves both the modulus and the argument of z. The formula

$$\boxed{e^{i\theta} = \cos\theta + i\sin\theta} \tag{22.20}$$

is known as Euler's formula. Note that the modulus of $e^{i\theta}$ is 1,

$$|e^{i\theta}| = \sqrt{\cos^2\theta + \sin^2\theta} = 1 \tag{22.21}$$

If we replace θ by $n\theta$ in (22.20), we obtain de Moivre's formula:

$$e^{in\theta} = (e^{i\theta})^n = \cos(n\theta) + i\sin(n\theta) \tag{22.22}$$

If we choose $n = -1$ in (22.22), we obtain

$$e^{-i\theta} = \cos\theta - i\sin\theta \tag{22.23}$$

and addition and subtraction of (22.20) and (22.23) yields the representations of $\cos\theta$ and $\sin\theta$ in terms of exponential functions:

$$\cos\theta = \frac{1}{2}(e^{i\theta} + e^{-i\theta}) \qquad \sin\theta = \frac{1}{2i}(e^{i\theta} - e^{-i\theta}) \tag{22.24}$$

Of special interest are the complex numbers with modulus 1 and arguments 0, $\pm\pi/2$, $\pm\pi$, and 2π:

$$e^0 = 1 \quad e^{i\pi/2} = i \quad e^{-i\pi/2} = -i \quad e^{i\pi} = e^{-i\pi} = -1 \quad e^{2\pi i} = 1 \tag{22.25}$$

which are all obtained from Euler's formula (22.20). It further follows from (22.20) that

$$e^{i\theta + 2\pi k i} = e^{i\theta} \qquad (k = 0, \pm 1, \pm 2, \ldots) \tag{22.26}$$

Thus, addition of $2\pi k$ to the argument of a complex number does not change its value. A unique definition for $\theta = \arg(z)$ may be obtained, for example, by assigning to θ positive values between 0 and π for $y \geq 0$, and negative values between 0 and $-\pi$ for $y < 0$,

$$\begin{aligned} 0 \leq \theta = \arg(z) \leq \pi & \quad y = \Im z \geq 0 \\ -\pi < \theta = \arg(z) < 0 & \quad y = \Im z < 0 \end{aligned} \tag{22.27}$$

Note that this definition, or any other that might be chosen, must be taken into account when θ is computed by the use of (22.14).

The complex conjugate $\bar{z}$ may be written in polar form as

$$\bar{z} = x - iy = r(\cos\theta - i\sin\theta) = re^{-i\theta} \quad (22.28)$$

so that

$$z\bar{z} = re^{i\theta}re^{-i\theta} = r^2 = |z|^2 \quad (22.29)$$

which corresponds to (22.13).

The polar form is particularly useful for the operations of multiplication and division. If z_1 and z_2 are two complex numbers with moduli r_1 and r_2 and arguments θ_1 and θ_2, then we obtain the following expressions for $z_1 z_2$ and z_1/z_2:

$$z_1 z_2 = r_1 r_2 e^{i(\theta_1 + \theta_2)} = r_1 r_2 [\cos(\theta_1 + \theta_2) + i\sin(\theta_1 + \theta_2)] \quad (22.30)$$

and

$$\frac{z_1}{z_2} = \frac{r_1}{r_2} e^{i(\theta_1 - \theta_2)} = \frac{r_1}{r_2} [\cos(\theta_1 - \theta_2) + i\sin(\theta_1 - \theta_2)] \quad (22.31)$$

Thus, multiplication and division of two complex numbers results in addition and subtraction of their arguments, respectively.

We now return to the coordinate transformation (22.4) which consists of a rotation of axes through an angle α. It follows from (22.5) that

$$z_0 = \cos\alpha + i\sin\alpha = e^{i\alpha} \quad (22.32)$$

so that (22.4) may be written as

$$z = \zeta e^{i\alpha} \quad (22.33)$$

It follows that the arguments of z and ζ are related as follows:

$$\arg(z) = \arg(\zeta) + \alpha \quad (22.34)$$

an identity that is observed from the geometry in Figure 3.34. Since the x axis is obtained from the ξ axis by a clockwise rotation through α, we conclude from (22.33) that multiplication of a complex variable ζ by a factor $e^{i\alpha}$ corresponds to a clockwise rotation through α of the coordinate system.

Problem 22.1

Determine the real and imaginary parts of:

$(3 + 4i) + (1 - 6i)$ $(12 + i) - (3 + 0.5i)$

$(2 + 4i)(3 + i)$ $1/(2 + 8i)$ $(1 + i)/(1 - i)$

Determine the moduli and arguments of:

$1 + i$ $-1 + i$ $-1 - i$ $1 - i$ $(1 + i)(1 - i)$

23 COMPLEX FUNCTIONS

Complex functions are functional relationships between two complex variables. In this section, we will introduce complex functions, determine conditions for their continuity and differentiability, and give some examples. Our main objective is the application of complex functions to groundwater flow problems, and therefore we restrict our discussions of the mathematical foundations to a minimum. The interested reader is referred to the handbooks on complex analysis for a more rigorous discussion of complex functions.

A Complex Function

A complex function is a relationship between two complex variables, for example,

$$\Omega = \Omega(z) \tag{23.1}$$

where

$$\Omega = \Phi + i\Psi \qquad z = x + iy \tag{23.2}$$

We may write (23.1) in terms of a pair of real functions,

$$\Omega = \Omega(z) = \Phi(x,y) + i\Psi(x,y) \tag{23.3}$$

Although (23.1) has the appearance of a functional relationship between two variables, just as the real function $f = f(x)$, it really involves a total of four variables: the two independent variables x and y and the two dependent variables Φ and Ψ. A graphical representation of a complex function would require four-dimensional space, and we resort to a representation by the use of maps, as explained below. Consider the real function $f = x^2$ in the interval $\mathcal{AB}$: $0 \le x \le 2$. We can illustrate this relationship by the use of two one-dimensional axes, the x axis and the f axis, by marking the interval $\mathcal{AB}$ on each axis (see Figure 3.37). This graphical representation does not give the complete information of the functional relationship; it only illustrates where the interval is mapped on the f-line, and not how points inside this interval are mapped. Applying this same graphical representation to complex functions, we draw two planes, the Ω plane and the z plane (see Figure 3.38), and label corresponding points in these planes. If each point in a domain $\mathcal{D}_z$ in the z plane corresponds to one single point in the corresponding domain $\mathcal{D}_\Omega$ in the Ω plane, then the function $\Omega = \Omega(z)$ is called single-valued.

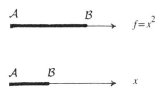

Figure 3.37 The map of a real function.

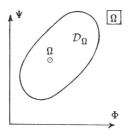 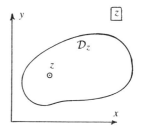

Figure 3.38 The complex Ω and z planes.

Continuity

The function $\Omega = \Omega(z)$ is continuous at a point z_0 if Ω approaches the same value $\Omega_0 = \Omega(z_0)$ for any path along which z tends to z_0. Hence, $\Omega(z)$ is continuous if

$$\lim_{z \to z_0} \Omega(z) = \Omega(z_0) \tag{23.4}$$

Differentiability; the Analytic Function

The derivative $d\Omega/dz = \Omega'(z)$ of the complex function $\Omega = \Omega(z)$ at a point z_0 is defined as

$$\left[\frac{d\Omega}{dz}\right]_{z=z_0} = \Omega'(z_0) = \lim_{z \to z_0} \frac{\Omega(z) - \Omega(z_0)}{z - z_0} \tag{23.5}$$

A complex function is differentiable at z_0 only if the limit in (23.5) is independent of the path along which z approaches z_0. A complex function is analytic at z_0 if it is differentiable throughout a neighborhood of z_0.

The definition (23.5) is rather restrictive, and we will see that Φ and Ψ must fulfill special conditions for Ω to be differentiable. We will establish these conditions by choosing two different directions of differentiation, and require that the expressions obtained for $d\Omega/dz$ be the same. First, we let z approach z_0 in a direction parallel to the x axis (see Figure 3.39): $z = x + iy_0$, and $z - z_0 = x + iy_0 - x_0 - iy_0 = x - x_0$, so that (23.5) becomes

$$\Omega'(z_0) = \lim_{x \to x_0} \left[\frac{\Phi(x, y_0) - \Phi(x_0, y_0)}{x - x_0} + i\frac{\Psi(x, y_0) - \Psi(x_0, y_0)}{x - x_0} \right]$$
$$= \frac{\partial \Phi}{\partial x} + i\frac{\partial \Psi}{\partial x} \tag{23.6}$$

where use is made of the definition of the partial derivative. Second, we let z approach z_0 in a direction parallel to the y axis: $z = x_0 + iy$, and $z - z_0 = i(y - y_0)$, so that we obtain

$$\Omega'(z_0) = \lim_{y \to y_0} \left[\frac{\Phi(x_0, y) - \Phi(x_0, y_0)}{i(y - y_0)} + i\frac{\Psi(x_0, y) - \Psi(x_0, y_0)}{i(y - y_0)} \right]$$
$$= \frac{\partial \Psi}{\partial y} - i\frac{\partial \Phi}{\partial y} \tag{23.7}$$

Sec. 23 Complex Functions

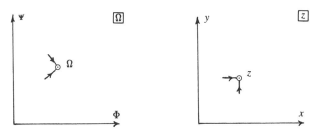

Figure 3.39 Complex differentiation.

We observe that (23.6) and (23.7) will be equal only if

$$\boxed{\frac{\partial \Phi}{\partial x} = \frac{\partial \Psi}{\partial y} \qquad \frac{\partial \Phi}{\partial y} = -\frac{\partial \Psi}{\partial x}} \tag{23.8}$$

which are the Cauchy-Riemann equations. Since we will obtain the same result for any orientation of the coordinate system, $d\Omega/dz$ in (23.5) will be single-valued for any direction $(z - z_0)$, provided that Φ and Ψ fulfill the Cauchy-Riemann conditions.

The potential and the stream function introduced earlier in this chapter must fulfill the Cauchy-Riemann conditions in order to be conjugate harmonic functions. We therefore conclude that the potential and stream function for any groundwater flow problem governed by Laplace's equation can be represented as the real and imaginary parts of a complex potential.

Examples of Analytic Functions

If a complex function is analytic at all points of a domain $\mathcal{D}$, then this function is said to be analytic in $\mathcal{D}$. It appears that most functions of interest in groundwater mechanics are analytic in the entire plane with the exception of isolated points, called singular points. If we deal with a complex function of which we know only that it is analytic in a domain $\mathcal{D}$, then we will state this explicitly. Otherwise, we will use the term "analytic function" somewhat loosely: we will call a function analytic if it is analytic in the entire plane with the possible exception of isolated singular points. Almost all differentiable real functions are analytic when written as complex functions of a complex variable z. Whether a complex function is analytic can be determined by verifying that the Cauchy-Riemann equations are satisfied. We will do this as an illustration only for the first example, and restrict our discussion of the other functions to determining their real and imaginary parts.

The Function $\Omega = z^2$

The real and imaginary parts of $\Omega = z^2$ are obtained as follows:

$$\Omega = \Phi + i\Psi = (x + iy)^2 = x^2 - y^2 + 2ixy \tag{23.9}$$

Hence,

$$\frac{\partial \Phi}{\partial x} = 2x \qquad \frac{\partial \Psi}{\partial y} = 2x \qquad \frac{\partial \Phi}{\partial y} = -2y \qquad \frac{\partial \Psi}{\partial x} = 2y \qquad (23.10)$$

so that the Cauchy-Riemann conditions are fulfilled, and the derivative $d\Omega/dz$ becomes

$$\frac{d\Omega}{dz} = \frac{\partial \Phi}{\partial x} + i\frac{\partial \Psi}{\partial x} = 2x + 2iy = 2z \qquad (23.11)$$

We see that the same result for Ω' is obtained by applying the rule for real differentiation to the complex function z^2. We state here without proof that this holds true for all analytic functions: the rules for real differentiation are directly transferable to complex functions.

The Taylor Series

It may be verified by the use of the Cauchy-Riemann conditions that sums and products of analytic functions are analytic. Thus, the series

$$\Omega(z) = \sum_{n=0}^{\infty} a_n (z - z_0)^n \qquad (23.12)$$

where z_0 is a constant, is analytic for all values of z for which it converges. We may obtain expressions for the coefficients a_n as follows. It appears from (23.12) that

$$\Omega(z_0) = a_0 \qquad (23.13)$$

The mth derivative of (23.12) is obtained by repeated differentiation as

$$\Omega^{(m)}(z) = \sum_{n=m}^{\infty} a_n [n - (m-1)][n - (m-2)] \ldots [n-1]n(z - z_0)^{n-m} \qquad (23.14)$$

so that

$$\Omega^{(m)}(z_0) = a_m [1 \cdot 2 \cdot 3 \ldots (m-1) \cdot m] \qquad (23.15)$$

which yields the following expression for a_m:

$$a_m = \frac{\Omega^{(m)}(z_0)}{m!} \qquad (23.16)$$

The series (23.12) thus may be written as

$$\Omega(z) = \sum_{n=0}^{\infty} \frac{\Omega^{(n)}(z_0)}{n!} (z - z_0)^n \qquad (23.17)$$

Sec. 23   Complex Functions   273

This series is known as the Taylor expansion of the analytic function $\Omega(z)$. We state here, without proof, that any function that is analytic within a circle of radius R may be represented by a Taylor series. Conversely, any function that may be represented in the form (23.17) in a domain $\mathcal{D}$ is analytic there. We see from (23.17) that all derivatives of an analytic function exist. Furthermore, the derivative of an analytic function is, itself, an analytic function.

With the aid of the Taylor series we introduce the concept *analytic at infinity*. A function $\Omega(z)$ is said to be analytic at infinity if it can be expanded in a Taylor series in terms of the variable $z^* = 1/z$ about $z^* = 0$. Thus, $\Omega = \Omega(z)$ is analytic at infinity if it can be represented by a series of the form:

$$\Omega = \sum_{n=0}^{\infty} b_n z^{-n} \qquad (z \to \infty) \tag{23.18}$$

near $1/z = 0$. We see, for example, that the function $\Omega = z^2$ is analytic in the entire z plane except infinity: $z = \infty$ is a singular point of this function.

It may be of interest to note that there exists only one function that is analytic *throughout* the z plane, i.e., including infinity. This function is the constant (Liouville's theorem).

The Function e^z

The real and imaginary parts of e^z are obtained from

$$e^z = e^{x+iy} = e^x e^{iy} = e^x(\cos y + i \sin y) \tag{23.19}$$

and the derivative of e^z is e^z:

$$\frac{d\Omega}{dz} = \frac{de^z}{dz} = e^z \tag{23.20}$$

The Function $\ln z$

The real and imaginary parts of $\ln z$ are obtained by writing z in polar form:

$$\ln z = \ln\left[|z|e^{i(\theta+2k\pi)}\right] = \ln|z| + i\theta + i2k\pi \qquad (k = 0, \pm 1, \pm 2, \ldots) \tag{23.21}$$

Note that this function is many-valued, as a consequence of the indeterminacy of $\arg(z)$. The principal values of $\ln z$ are usually associated with arguments of z within the range $-\pi < \arg(z) \leq \pi$.

The Function $\sin z$

The real and imaginary parts of $\sin z$ may be obtained by the use of the exponential representation (22.24):

$$\begin{aligned}\Omega = \sin z &= \tfrac{1}{2i}\left[e^{i(x+iy)} - e^{-i(x+iy)}\right] \\ &= \tfrac{1}{2i}\left[e^{-y}(\cos x + i\sin x) - e^{y}(\cos x - i\sin x)\right] \\ &= \tfrac{1}{2i}\left[-(e^y - e^{-y})\cos x + i(e^y + e^{-y})\sin x\right]\end{aligned} \tag{23.22}$$

which gives, using the definitions of the hyperbolic functions,

$$\sin z = \sin x \cosh y + i \cos x \sinh y \qquad (23.23)$$

Note that the modulus of the sine of a complex variable is not necessarily less than 1.

The Function $\cos z$

The real and imaginary parts of $\cos z$ may be determined similarly as for the sine:

$$\begin{aligned}\cos z &= \tfrac{1}{2}\left[e^{i(x+iy)} + e^{-i(x+iy)}\right] \\ &= \tfrac{1}{2}\left[e^{-y}(\cos x + i \sin x) + e^{y}(\cos x - i \sin x)\right]\end{aligned} \qquad (23.24)$$

or

$$\cos z = \cos x \cosh y - i \sin x \sinh y \qquad (23.25)$$

The modulus of $\cos z$ is not necessarily less than 1.

The Functions $\sinh z$ *and* $\cosh z$

The functions $\sinh z$ and $\cosh z$ may be expressed in terms of the sine and the cosine as follows. By definition, we have

$$\sinh z = \tfrac{1}{2}(e^{z} - e^{-z}) = \tfrac{i}{2i}\left[e^{i(-iz)} - e^{-i(-iz)}\right] = -i\sin(iz) \qquad (23.26)$$

and

$$\cosh z = \tfrac{1}{2}(e^{z} + e^{-z}) = \tfrac{1}{2}\left[e^{i(-iz)} + e^{-i(-iz)}\right] = \cos(iz) \qquad (23.27)$$

The real and imaginary parts may be obtained by the use of (23.23) and (23.25):

$$\sinh z = -i\sin(-y + ix) = \sinh x \cos y + i \cosh x \sin y \qquad (23.28)$$

and

$$\cosh z = \cos(-y + ix) = \cosh x \cos y + i \sinh x \sin y \qquad (23.29)$$

The Function $\sqrt{z}$

The real and imaginary parts of $\Omega = \sqrt{z}$ are readily obtained by writing z in polar form:

$$\Omega = \sqrt{z} = \sqrt{re^{i\theta + 2k\pi i}} = \sqrt{r}e^{i\theta/2}e^{k\pi i} = e^{k\pi i}\sqrt{r}\left(\cos\tfrac{\theta}{2} + i\sin\tfrac{\theta}{2}\right) \qquad (23.30)$$

We observe that two different roots are obtained, one for $k = 0$, and one for $k = 1$:

$$\sqrt{z} = \pm\sqrt{r}e^{i\theta/2} \qquad (23.31)$$

Sec. 23   Complex Functions   275

For any other choice of k, one of the two roots in (23.31) is obtained. In order to decide which root applies in any given problem, the range of values of $\arg(z)$ must be known, i.e., the branch cut must be chosen. Note that the function $\Omega = \sqrt{z}$ is not analytic at $z = 0$, because the derivatives of $\sqrt{z}$ do not exist at $z = 0$.

The Function z^α

Another many-valued function is the power z^α, where $|\alpha| < 1$. This function may be written as

$$\Omega = z^\alpha = r^\alpha e^{i\alpha\theta} e^{i\alpha 2k\pi} \tag{23.32}$$

The number of different roots depends upon the value of α. If α may be written as

$$\alpha = \frac{p}{n} \qquad p = 1, 2, \ldots \qquad n = 1, 2, \ldots \tag{23.33}$$

then there exist n different values for k that yield different roots. If α cannot be written in the form (23.33), then there exists an infinite number of different roots.

Examples of Functions That Are Not Analytic

The function $\Omega = \bar{z}$ is not an analytic function of z. This follows by application of the Cauchy-Riemann conditions:

$$\Omega = \Phi + i\Psi = \bar{z} = x - iy \tag{23.34}$$

so that

$$\frac{\partial \Phi}{\partial x} = 1 \qquad \frac{\partial \Psi}{\partial y} = -1 \qquad \frac{\partial \Phi}{\partial y} = 0 \qquad \frac{\partial \Psi}{\partial y} = 0 \tag{23.35}$$

and the Cauchy-Riemann equations are not fulfilled, as asserted. Any complex function of $\bar{z}$ is not an analytic function of z.

Another example of a function that is not analytic is $|z|$. Since Ψ is constant and Φ is not, the Cauchy-Riemann conditions are not be satisfied.

Some General Properties of Analytic Functions

We will briefly discuss some properties of analytic functions, without proof, that are used in what follows.

Inversion of Analytic Functions

The inverse $z = z(\Omega)$, of the analytic function $\Omega = \Omega(z)$, is an analytic function with the exception of points where $d\Omega/dz = 0$.

Analytic Functions of Analytic Functions

If both $\Omega = \Omega(\zeta)$ and $z = z(\zeta)$ are analytic functions of the complex variable ζ, then the function $\Omega = \Omega(\zeta(z)) = \Omega^*(z)$ is an analytic function of z. This important property makes it possible to use a parametric representation of an analytic function: the intermediate complex variable ζ in the above example acts as a complex reference parameter. For example, the function defined by the following pair of expressions

$$\Omega = a\zeta + b\ln\zeta \tag{23.36}$$

$$z = c\zeta + d\tan\zeta \tag{23.37}$$

where a, b, c, and d are constants, cannot be written without using the reference parameter ζ; ζ cannot be eliminated.

Functions Involving Complex Conjugates

If the complex function

$$\Omega = \Omega(z) \tag{23.38}$$

is an analytic function of z, then the function

$$\Omega^*(z) = \overline{\Omega(\overline{z})} \tag{23.39}$$

is also an analytic function of z.

Complex Differentiation

Even though complex differentiation involves four variables, it obeys the same rules as differentiation of a real function of a single real variable. The derivative of any of the analytic functions listed above is obtained in the same fashion as the real derivative of its real counterpart. For example: $d(\cos z)/dz = -\sin z$. Furthermore, the chain-rule applies just as for real functions:

$$\frac{d\Omega}{dz} = \frac{d\Omega}{d\zeta}\frac{d\zeta}{dz} \tag{23.40}$$

where $\Omega = \Omega(\zeta)$ and $z = z(\zeta)$. In addition, the property

$$\frac{d\zeta}{dz} = \frac{1}{\frac{dz}{d\zeta}} \tag{23.41}$$

holds for analytic functions, so that (23.40) may be written as

Sec. 23 Complex Functions

$$\frac{d\Omega}{dz} = \frac{\frac{d\Omega}{d\zeta}}{\frac{dz}{d\zeta}} \qquad (23.42)$$

This expression is particularly useful for differentiation of functions of the form (23.36) and (23.37), where the reference parameter cannot be eliminated. We obtain, for that case,

$$\frac{d\Omega}{dz} = \frac{a + \frac{b}{\zeta}}{c + \frac{d}{\cos^2 \zeta}} \qquad (23.43)$$

Complex Integration

Just like complex differentiation, complex integration is carried out formally in the same fashion as integration of a function of a single real variable. For example, the integral

$$\Omega = \int_{z_1}^{z_2} \frac{d\delta}{\delta - z} \qquad (23.44)$$

becomes

$$\Omega = \ln \frac{z_2 - z}{z_1 - z} \qquad (23.45)$$

It may be apparent from the above brief summary of complex analysis, that the use of complex variables greatly simplifies analytic work involved in deriving solutions for groundwater flow problems. We will demonstrate this in the next sections, where the solutions obtained thus far are written in terms of complex variables, and where we will discuss techniques for solving boundary-value problems.

Problem 23.1

Determine the real and imaginary parts of the following functions:

$$\tan z \qquad \coth z \qquad e^{\sin z} \qquad z^{1/3}$$

Determine the moduli and arguments of

$$\sinh z \qquad \cos z$$

Integrate the following:

$$\int \frac{1}{(z^2 + 1)} dz \qquad 1/(2i) \int \left[\frac{1}{(z - i)} - \frac{1}{(z + i)} \right] dz$$

Differentiate

$$\ln[\tan z] \qquad \exp(\sqrt{z})$$

24 COMPLEX FUNCTIONS IN GROUNDWATER FLOW

In this section we introduce the complex potential and the complex discharge function for groundwater flow problems. As an example, we write the basic solutions discussed earlier in terms of the complex potential: the well, the vortex, and the dipole. An additional example shows a solution that was originally obtained in terms of complex variables: the complex potential for flow in the vertical plane over a horizontal impervious base toward a horizontal drain.

The Complex Potential and the Discharge Function

The potential Φ and the stream function Ψ fulfill the Cauchy-Riemann conditions (23.8). Therefore, Φ and Ψ may be represented as the real and imaginary parts of an analytic function $\Omega = \Omega(z)$ of the complex variable $z = x + iy$, defined in the flow domain. This function is known as the *complex potential* Ω, with

$$\Omega = \Phi + i\Psi \qquad (24.1)$$

The components Q_x and Q_y of the discharge vector may be obtained by complex differentiation as follows. The *discharge function* W is defined as minus the derivative of the complex potential with respect to z,

$$W = -\frac{d\Omega}{dz} = -\frac{\partial \Phi}{\partial x} - i\frac{\partial \Psi}{\partial x} \qquad (24.2)$$

Using the Cauchy-Riemann equations (23.8) and Darcy's law, this becomes

$$W = -\frac{\partial \Phi}{\partial x} + i\frac{\partial \Phi}{\partial y} = Q_x - iQ_y \qquad (24.3)$$

Hence, the components Q_x and Q_y are obtained directly from $W = U + iV$,

$$W = U + iV \qquad U = Q_x \qquad V = -Q_y \qquad (24.4)$$

The Uniform Flow

The complex potential for uniform flow with components Q_{x0} and Q_{y0} is obtained by combining the expressions (19.20) and (19.22) for Φ and Ψ. This gives, discarding the constant Φ_0,

$$\Omega = -(Q_{x0} - iQ_{y0})z \qquad (24.5)$$

If the direction of the uniform flow is α, and if the discharge is Q_0, the complex number $Q_{x0} - iQ_{y0}$ may be written as

$$Q_{x0} - iQ_{y0} = Q_0(\cos\alpha - i\sin\alpha) = Q_0 e^{-i\alpha} \qquad (24.6)$$

Sec. 24   Complex Functions in Groundwater Flow

so that (24.5) becomes

$$\Omega = -Q_0 z e^{-i\alpha} \qquad (24.7)$$

The Well

As an example of the application of complex variables, consider the complex potential

$$\Omega = \frac{Q}{2\pi} \ln(z - z_w) + C \qquad (24.8)$$

where $z_w = x_w + iy_w$ and C is a real constant. It follows from (24.8) that

$$\Phi = \Re\Omega = \frac{Q}{4\pi} \ln|z - z_w|^2 + C = \frac{Q}{4\pi} \ln[(x - x_w)^2 + (y - y_w)^2] + C \qquad (24.9)$$

and

$$\Psi = \Im\Omega = \frac{Q}{2\pi} \arctan \frac{y - y_w}{x - x_w} = \frac{Q}{2\pi} \theta \qquad (24.10)$$

It appears that (24.8) represents the complex potential for a well with discharge Q at $z = z_w = x_w + iy_w$.

The Dipole

The complex potential for a dipole at $z = z_0 = x_0 + iy_0$ is the complex function with (20.38) as its real and (20.41) as its imaginary part:

$$\Omega = \Phi + i\Psi = \frac{s}{2\pi} \frac{\cos(\theta - \beta) - i\sin(\theta - \beta)}{r} = \frac{s}{2\pi} \frac{e^{i\beta}}{re^{i\theta}} \qquad (24.11)$$

The complex number $re^{i\theta}$ represents the vector pointing from z_0 to z, so that $re^{i\theta} = z - z_0$, and (24.11) becomes

$$\Omega = \frac{s}{2\pi} \frac{e^{i\beta}}{z - z_0} \qquad (24.12)$$

The Vortex

The complex potential for a vortex at $z = z_v$, finally, is given by

$$\Omega = \frac{A}{\pi i} \ln(z - z_v) + C' \qquad (24.13)$$

where A and C are a real and a complex constant, respectively. The reader may verify that the real and imaginary parts of (24.13) indeed represent the potential and stream functions for a vortex at $z = z_v$, cf. (21.3) and (21.7).

A Lake in a Uniform Flow

An example of the use of the discharge function will be given by determining the stagnation points for the flow problem of a lake in a uniform flow, discussed in Sec. 20, Figure 3.25. Combining (20.43) and (20.45) in a complex potential, we obtain

$$\Omega = -Q_{x0}\left[z - \frac{R^2}{z}\right] + \frac{Q_0}{2\pi}\ln\frac{z}{R} + \Phi_0 \qquad (24.14)$$

The expression for the discharge function W is obtained upon differentiation,

$$W = -\frac{d\Omega}{dz} = Q_{x0}\left[1 + \frac{R^2}{z^2}\right] - \frac{Q_0}{2\pi}\frac{1}{z} \qquad (24.15)$$

The stagnation point occurs at a point z_s, where both components Q_x and Q_y vanish, i.e., where $W = 0$. Hence,

$$\left[\frac{1}{z_s}\right]^2\left[-\frac{Q_0}{2\pi}z_s + Q_{x0}z_s^2 + Q_{x0}R^2\right] = 0 \qquad (24.16)$$

The stagnation point cannot be at $z = 0$, so that (24.16) may be multiplied by z_s^2. The roots of the remaining equation are

$$\left[\frac{z_s}{R}\right]_{1,2} = \frac{Q_0}{4\pi Q_{x0}R} \pm \sqrt{\left[\frac{Q_0}{4\pi Q_{x0}R}\right]^2 - 1} \qquad (24.17)$$

This equation is identical in form to that obtained earlier, see (20.55), but provides more information. Whereas (20.55) is valid only for $\theta_s = 0$, and thereby applies only to cases where the stagnation points lie on the x axis, (24.17) also holds for the case that there are two stagnation points on the lake boundary. The latter case was referred to as case 2 in Sec. 20 and applies when the roots $(z_s)_{1,2}$ are complex, i.e., when

$$\left[\frac{Q_0}{4\pi R Q_{x0}}\right]^2 < 1 \qquad (24.18)$$

see (20.53). In that case the square root in (24.17) is purely imaginary so that

$$\left[\frac{z_s}{R}\right]_{1,2} = \frac{Q_0}{4\pi Q_{x0}R} \pm i\sqrt{1 - \left[\frac{Q_0}{4\pi Q_{x0}R}\right]^2} \qquad (24.19)$$

The moduli of $(z_s)_1$ and $(z_s)_2$ are R, as follows from (24.19), so that the stagnation points lie indeed on the lake boundary. The stagnation points are at $(z_s)_{1,2} = \pm Ri$ for the special case that $Q_0 = 0$.

Unconfined Flow Toward a Drain

As a final example, we will discuss the solution for a problem of unconfined flow over an impermeable base toward a horizontal drain, illustrated in Figure 3.40. The boundary conditions are that $1\mathcal{A}\text{-}2$ is an impermeable base, 2-3 is an equipotential with $\Phi = 0$, and $3\text{-}1\mathcal{B}$ is a phreatic surface. Point $(1\mathcal{A},1\mathcal{B})$ is at infinity and point 2 is at the origin. If we set $\Psi = 0$ along the impermeable base,

$$-\infty < x \leq 0 \qquad y = 0 \qquad \Psi = 0 \qquad (24.20)$$

and if the discharge per width B measured normal to the plane of flow is Q, then $\Psi = -Q$ along the phreatic surface. The potential for two-dimensional flow in the vertical plane is $kB\phi$ (see Sec. 17), $\phi = y$ along the phreatic surface (see Sec. 17), and the boundary condition along $3\text{-}1\mathcal{B}$ becomes:

$$\Psi = -Q \qquad \Phi = kBy \qquad (24.21)$$

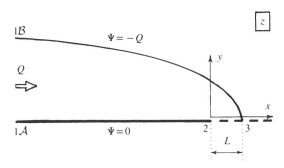

Figure 3.40 The Vreedenburgh parabola.

The condition along the equipotential 2-3, finally, is

$$0 \leq x \leq L \qquad y = 0 \qquad \Phi = 0 \qquad (24.22)$$

where L represents the x coordinate of point 3. The solution to this problem was obtained by Vreedenburgh (see Vreedenburgh [1929], De Vos [1929] and Kozeny [1931]) and has the following form:

$$\Omega = -i\sqrt{2kQBz} \qquad (24.23)$$

Inversion of (24.23) yields

$$z = -\frac{\Omega^2}{2a} \qquad a = kQB \qquad (24.24)$$

Separation of the real and imaginary parts yields

$$x = -\frac{1}{2a}(\Phi^2 - \Psi^2) \qquad y = -\frac{1}{a}\Phi\Psi \qquad (24.25)$$

We observe from the second equation that $y = 0$ both for $\Phi = 0$ and $\Psi = 0$, and from the first that $x < 0$ for $y = 0$, $\Psi = 0$ and $x > 0$ for $y = 0$, $\Phi = 0$ so that (24.20) and (24.22) are met. We see, in addition, that $\Phi = kBy$ along the streamline $\Psi = -Q$, so that (24.21) is also satisfied. Since the complex potential is analytic for all finite z except $z = 0$, and fulfills the boundary conditions, (24.23) is indeed the solution to the problem of Figure 3.40. It may be noted that point 2 is a singular point; the components of the discharge vector become infinite there, as a result of the boundary conditions. Indeed, the boundaries enforce at point 2 an unnatural condition: the streamline 1A-2 and the equipotential 2-3 are not mutually perpendicular as everywhere else, but collinear.

Elimination of Φ from the two equations in (24.25) yields the following expression for the streamlines:

$$y^2 = -\frac{2\Psi_j^2}{a}x + \frac{\Psi_j^4}{a^2} \qquad (24.26)$$

where Ψ_j represents a chosen constant value. The expressions for the equipotentials are obtained in a similar fashion by eliminating Ψ from (24.25) and setting Φ equal to the constant values Φ_j:

$$y^2 = \frac{2\Phi_j^2}{a}x + \frac{\Phi_j^4}{a^2} \qquad (24.27)$$

It is seen from (24.26) and (24.27) that the streamlines and equipotentials are a set of confocal parabolas.

We obtained the above solutions simply by combining real solutions determined earlier, except for the latter one, which we simply presented. Although the analytic function may, in principle, be considered as the general solution to groundwater flow problems governed by Laplace's equation, we have not yet addressed the problem of selecting the particular solution that meets any given set of boundary conditions. One of the tools available for solving such a problem is that of conformal mapping, introduced in Chapter 5.

Problem 24.1

Determine expressions for Q_x and Q_y for the problem illustrated in Figure 3.40, and determine the distance between points 2 and 3. Sketch a flow net for $\Delta\Psi = \Delta\Phi = Q/3$, using (24.26) and (24.27).

Problem 24.2

Answer question (b) in problem 21.3 by making use of the complex functions Ω and W.

25 COMPLEX LINE INTEGRALS

Complex integrals play an important role both in formulating exact solutions to groundwater flow problems and in approximate modeling by means of analytic functions. We will briefly discuss the most important integrals in this section, but defer their applications and generalization to later sections.

The Line-Sink

One of the most commonly used complex line integrals is the line-sink. Probably the first to apply the line-sink to a groundwater flow problem was Irmay [1960]. Applications of line-sinks to boundary value problems were given by Van der Veer [1976] in terms of a complex integral and by Ligget [1977] in terms of a real one. Strack and Haitjema [1981(a)] used line-sinks for modeling creeks and rivers in a regional flow model.

The line-sink is a distribution of wells along a line; as opposed to a well, the inflow takes place uniformly along a line and is not concentrated at a point. As for wells, the sign of the discharge governs whether inflow or outflow occurs. Thus, the line-source is a line-sink with negative discharge; it is not necessary to derive a separate complex potential for a line-source. A line-source may be used, for example, to model the infiltration through the bottom of a creek, as shown in Figure 3.41. The bottom of the creek lies above the groundwater table of a shallow unconfined aquifer, and the infiltration rate σ [m^2/s] per unit length of the creek is a function of the water level in the creek and the resistance of the creek bottom. We will neglect the width of the creek with respect to its length, and imagine the infiltration to be concentrated along the line.

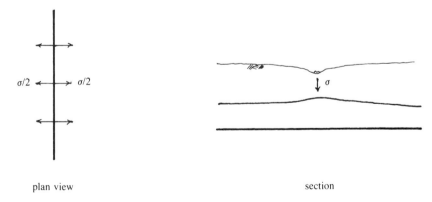

Figure 3.41 Infiltration through the bottom of a creek.

The Complex Potential

Consider a uniformly spaced row of n wells along a straight line of length L and inclination α to the x axis, as shown in Figure 3.42(b). The complex coordinates of the end points are $\overset{1}{z}$ and $\overset{2}{z}$. The wells all have a discharge $\sigma\Delta\xi$, where $\Delta\xi$ is the distance between the wells (see Figure 3.42[a]), and the mth well is located at $z = \delta_m$. The corresponding complex potential is

$$\Omega = \sum_{m=1}^{n} \frac{\sigma}{2\pi} \ln(z - \delta_m)\Delta\xi \qquad (25.1)$$

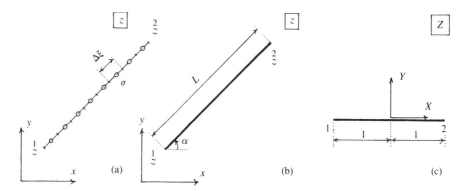

Figure 3.42 The line-sink.

In the limit, for $\Delta\xi \to 0$ and $n \to \infty$, the sum becomes the following integral:

$$\Omega = \int_{-\frac{1}{2}L}^{\frac{1}{2}L} \frac{\sigma}{2\pi} \ln(z - \delta)d\xi \qquad (25.2)$$

where ξ is taken to be $-\frac{1}{2}L$ at $\overset{1}{z}$ and $\frac{1}{2}L$ at $\overset{2}{z}$. The complex number δ denotes a point of the line and may be written in terms of ξ as follows:

$$\delta = \xi e^{i\alpha} + \frac{1}{2}\left(\overset{1}{z} + \overset{2}{z}\right) \qquad (25.3)$$

We introduce the following dimensionless variables:

$$Z = X + iY = \frac{z - \frac{1}{2}\left(\overset{1}{z} + \overset{2}{z}\right)}{\frac{1}{2}\left(\overset{2}{z} - \overset{1}{z}\right)} \qquad \Delta = \Delta_1 + i\Delta_2 = \frac{\delta - \frac{1}{2}\left(\overset{1}{z} + \overset{2}{z}\right)}{\frac{1}{2}\left(\overset{2}{z} - \overset{1}{z}\right)} \qquad (25.4)$$

where

$$\overset{2}{z} - \overset{1}{z} = Le^{i\alpha} \qquad (25.5)$$

The complex variables Z and Δ are real along the element and vary between -1 at $\overset{1}{z}$ to $+1$ and $\overset{2}{z}$ (see Figure 3.42[c]). Note that the complex number $\frac{1}{2}\left(\overset{1}{z} + \overset{2}{z}\right)$ corresponds to

Sec. 25 Complex Line Integrals

the center of the element, and that $\overset{2}{z} - \overset{1}{z}$ represents a vector pointing from $\overset{1}{z}$ to $\overset{2}{z}$, which has length L and inclination α. Introduction of these variables in (25.2) yields, writing $d\xi$ as $e^{-i\alpha}d\delta = \frac{1}{2}Ld\Delta$,

$$\Omega = \int_{-1}^{+1} \frac{\sigma}{2\pi} \ln\left[\tfrac{1}{2}\left(\overset{2}{z} - \overset{1}{z}\right)(Z - \Delta)\right] \tfrac{1}{2} Ld\Delta \tag{25.6}$$

The infiltration rate σ is constant, and (25.6) may be written as

$$\Omega = \frac{\sigma L}{4\pi}\left[\int_{-1}^{+1} \ln\left[\tfrac{1}{2}\left(\overset{2}{z} - \overset{1}{z}\right)\right] d\Delta - \int_{Z+1}^{Z-1} \ln(Z - \Delta) d(Z - \Delta)\right] \tag{25.7}$$

where $d\Delta$ is replaced by $-d(Z - \Delta)$ in the second integral, using that Z is constant during integration. Integration of (25.7) yields

$$\boxed{\Omega = \frac{\sigma L}{4\pi}\left\{(Z+1)\ln(Z+1) - (Z-1)\ln(Z-1) + 2\ln\left[\tfrac{1}{2}\left(\overset{2}{z} - \overset{1}{z}\right)\right] - 2\right\}} \tag{25.8}$$

which represents the complex potential for a line-sink of infiltration rate σ and length L, written in terms of the local complex variable Z. We refer to the infiltration rate σ as the strength of the line-sink.

The Far-Field. The behavior of (25.8) for large values of Z is obtained by first writing the complex potential as

$$\Omega = \frac{\sigma L}{4\pi}\left\{Z \ln\frac{1 + 1/Z}{1 - 1/Z} + \ln\left[Z^2\left(1 - \frac{1}{Z^2}\right)\right] + 2\ln\left[\tfrac{1}{2}\left(\overset{2}{z} - \overset{1}{z}\right)\right] - 2\right\} \tag{25.9}$$

and then expanding the function about $1/Z = 0$. This yields

$$\Omega = \frac{\sigma L}{4\pi}\left\{2 \ln Z + 2\ln\left[\tfrac{1}{2}\left(\overset{2}{z} - \overset{1}{z}\right)\right] + Z\left[\frac{2}{Z} + O\left(\frac{1}{Z^3}\right)\right] - 2 + O\left(\frac{1}{Z^2}\right)\right\}$$
$$(Z \to \infty) \tag{25.10}$$

where $O(1/Z^3)$ means "order of $1/Z^3$ or less." Disregarding all terms of order less than $1/Z^2$, we obtain

$$\Omega = \frac{\sigma L}{2\pi}\left\{\ln Z + \ln\left[\tfrac{1}{2}\left(\overset{2}{z} - \overset{1}{z}\right)\right]\right\} + O\left(\frac{1}{Z^2}\right)$$
$$= \frac{\sigma L}{2\pi}\ln\left[z - \tfrac{1}{2}\left(\overset{1}{z} + \overset{2}{z}\right)\right] + O\left(\frac{1}{Z^2}\right) \qquad (Z \to \infty) \tag{25.11}$$

where use is made of (25.4). We see that the complex potential approaches that of a well of discharge $Q = \sigma L$, located at the center of the line-sink. This is to be expected, since σL represents the total discharge of the line-sink.

The Discharge Function

We introduce the local coordinates ξ and η and determine the discharge vector components Q_ξ and Q_η, which are parallel and normal to the line-sink as shown in Figure 3.43. The complex variable $\zeta = \xi + i\eta$ may be expressed in terms of z as follows:

$$\zeta = \xi + i\eta = \left[z - \tfrac{1}{2}\left(\tfrac{1}{z} + \tfrac{2}{z}\right)\right]e^{-i\alpha} \quad (25.12)$$

so that $\eta = 0$ along the element, and the discharge function W^* is defined as

$$W^* = -\frac{d\Omega}{d\zeta} = Q_\xi - iQ_\eta \quad (25.13)$$

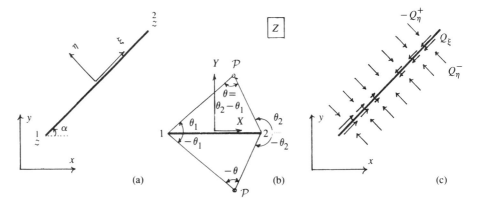

Figure 3.43 The normal and tangential discharge components along the element.

It follows from (25.12) and (25.4) that

$$\zeta = \tfrac{1}{2}\left(\tfrac{2}{z} - \tfrac{1}{z}\right)Ze^{-i\alpha} = \tfrac{1}{2}LZ \quad (25.14)$$

so that (25.13) becomes

$$Q_\xi - iQ_\eta = -\frac{d\Omega}{dZ}\frac{dZ}{d\zeta} = -\frac{2}{L}\frac{d\Omega}{dZ} \quad (25.15)$$

Using (25.8), we obtain

$$Q_\xi - iQ_\eta = \frac{\sigma}{2\pi}\ln\frac{Z-1}{Z+1} = \frac{\sigma}{2\pi}\left[\ln\left|\frac{Z-1}{Z+1}\right| + i\theta\right] \quad (25.16)$$

We choose the branch cut for the angle θ as shown in Figure 3.43(b), so that

$$\begin{aligned} Y \geq 0^+ & \quad 0 \leq \theta \leq \pi \\ Y \leq 0^- & \quad -\pi \leq \theta \leq 0 \end{aligned} \quad (25.17)$$

Sec. 25 Complex Line Integrals

where the superscripts $+$ and $-$ denote the side from which the point Z approaches the element. Denoting Q_η along the $+$ and $-$ sides of the element as Q_η^+ and Q_η^- (see Figure 3.43[c]), we obtain from (25.16) and (25.17):

$$Q_\eta^+ = -\frac{\sigma}{2} \qquad Q_\eta^- = \frac{\sigma}{2} \qquad Q_\eta^+ - Q_\eta^- = -\sigma \tag{25.18}$$

We observe that Q_η is discontinuous across the element; there is an inflow $-(Q_\eta^+ - Q_\eta^-)$ equal to σ. We further see that the component Q_ξ is continuous,

$$Q_\xi = \frac{\sigma}{2\pi} \ln\left|\frac{Z-1}{Z+1}\right| \tag{25.19}$$

but singular at the end points of the element: these end points are singular points. The tangential component Q_ξ is zero at the center of the element, and increases toward the tips, as illustrated in Figure 3.43(c).

The Stream Function

The stream function Ψ is obtained as the imaginary part of (25.8). We will examine the behavior of this function along the X axis, and obtain, for $Y = 0$,

$$\Psi = \frac{\sigma L}{4\pi}[(X+1)\theta_1 - (X-1)\theta_2 + 2\alpha]$$

$$= \frac{\sigma L}{4\pi}[-X(\theta_2 - \theta_1) + \theta_1 + \theta_2 + 2\alpha] \qquad (Y=0) \tag{25.20}$$

where θ_1 and θ_2 are the arguments of $(Z+1)$ and $(Z-1)$, respectively. With $\theta_2 - \theta_1 = \theta$, see Figure 3.43(b), we may write Ψ as

$$\Psi = \frac{\sigma L}{4\pi}(-X\theta + \theta_1 + \theta_2 + 2\alpha) \qquad (Y=0) \tag{25.21}$$

We choose the branch cuts for θ_1 and θ_2 such that

$$Y \geq 0^+ \quad \begin{cases} 0 \leq \theta_1 \leq \pi \\ 0 \leq \theta_2 \leq \pi \end{cases} \qquad Y \leq 0^- \quad \begin{cases} -\pi \leq \theta_1 \leq 0 \\ -\pi \leq \theta_2 \leq 0 \end{cases} \tag{25.22}$$

Denoting Ψ as Ψ^+ and Ψ^- along $Y = 0^+$ and $Y = 0^-$, we obtain with $\theta = 0$ for $|X| \geq 1$, $Y = 0$:

$$-\infty < X < -1 \qquad Y = 0^+ \qquad \Psi^+ = \frac{\sigma L}{4\pi}(+2\pi + 2\alpha) \tag{25.23}$$

$$-\infty < X < -1 \qquad Y = 0^- \qquad \Psi^- = \frac{\sigma L}{4\pi}(-2\pi + 2\alpha) \tag{25.24}$$

$$-1 \leq X \leq 1 \qquad Y = 0^+ \qquad \Psi^+ = \frac{\sigma L}{4\pi}(-X\pi + \pi + 2\alpha) \tag{25.25}$$

$$-1 \leq X \leq 1 \qquad Y = 0^- \qquad \Psi^- = \frac{\sigma L}{4\pi}(+X\pi - \pi + 2\alpha) \tag{25.26}$$

$$1 < X < \infty \quad \begin{Bmatrix} Y = 0^+ \\ Y = 0^- \end{Bmatrix} \quad \begin{Bmatrix} \Psi^+ \\ \Psi^- \end{Bmatrix} = \frac{\sigma L}{2\pi}\alpha \qquad (25.27)$$

It appears that the stream function jumps both across the negative X axis and across the element:

$$-\infty < X < -1 \quad Y = 0 \quad \Psi^+ - \Psi^- = \sigma L \qquad (25.28)$$
$$-1 \le X \le 1 \quad Y = 0 \quad \Psi^+ - \Psi^- = \tfrac{1}{2}\sigma L(-X+1) \qquad (25.29)$$

The jump σL along the negative X axis outside the element occurs because the element extracts an amount $Q = \sigma L$. The jump $\tfrac{1}{2}\sigma L(1-X)$ along the element varies linearly with X and corresponds to the constant rate of extraction. Indeed, this jump corresponds to the inflow:

$$Q_\eta^+ - Q_\eta^- = \frac{\partial}{\partial \xi}(\Psi^+ - \Psi^-) = \frac{2}{L}\frac{\partial}{\partial X}(\Psi^+ - \Psi^-) = -\sigma \qquad (25.30)$$

where use is made of the Cauchy-Riemann relation $\partial\Psi/\partial\xi = -\partial\Phi/\partial\eta = Q_\eta$.

It may be noted that the stream function is finite and double-valued at the tips of the element. For example, at $X = -1$, $Y = 0$, (25.20) becomes:

$$\Psi = \frac{\sigma L}{4\pi}(2\theta_2 + 2\alpha) = \frac{\sigma L}{2\pi}\theta_2 + \frac{\sigma L}{2\pi}\alpha \quad (X = -1, \ Y = 0) \qquad (25.31)$$

The angle θ_2 is π if $Y = 0^+$, and $-\pi$ if $Y = 0^-$.

The stream function jumps along the semi-infinite segment $-\infty < X \le 1$, $Y = 0$: the line-sink has a branch cut extending from the end point at $\overset{2}{z}$ via the end point at $\overset{1}{z}$ to infinity. The branch cut extends to infinity because the line-sink extracts a net amount of water from the aquifer.

The Potential

The potential is single-valued throughout the domain. This may be seen by first rearranging (25.8) as follows:

$$\Omega = \frac{\sigma L}{4\pi}\left\{-Z\ln\frac{Z-1}{Z+1} + \ln[(Z-1)(Z+1)] + 2\ln\left[\tfrac{1}{2}\left(\overset{2}{z}-\overset{1}{z}\right)\right] - 2\right\} \qquad (25.32)$$

and then taking the real part:

$$\Phi = \frac{\sigma L}{4\pi}\left[-X\ln\left|\frac{Z-1}{Z+1}\right| + Y\theta + \ln|Z^2-1| + 2\ln\frac{L}{2} - 2\right] \qquad (25.33)$$

Indeed, the only function that is discontinuous is θ; this function jumps only when Y passes through zero, and the factor Y in (25.33) ensures that this jump does not cause the potential to jump.

It may be noted that the complex potential is finite at the tips of the element; the terms $(Z-1)\ln(Z-1)$ and $(Z+1)\ln(Z+1)$ in (25.8) vanish at $Z = 1$ and $Z = -1$,

Sec. 25   Complex Line Integrals   289

respectively. The potential therefore is finite at the tips. The complex potential is not analytic at the end points, where none of the derivatives of Ω exist.

A plot of equipotentials and streamlines is reproduced in Figure 3.44, as an illustration.

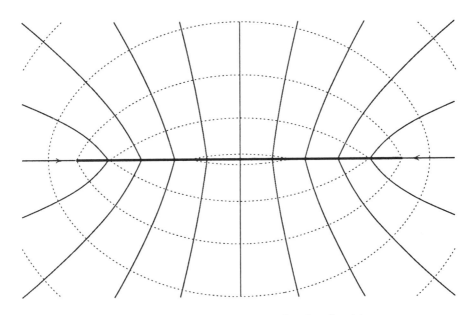

Figure 3.44   Equipotentials and streamlines for a line-sink.

Some Applications of Line-Sinks

Line-sinks may be used to model the infiltration through the bottoms of creeks that are not in direct contact with the groundwater, as is shown in Figure 3.41. In such cases the strengths of the line-sinks are known. For example, if there are n line-sinks between points $\overset{1}{z}_j$ and $\overset{2}{z}_j$ ($j = 1, 2, \ldots n$) of length L_j, and with strength σ_j, and if there is a uniform flow of discharge Q_0 in a direction α, then the complex potential becomes

$$\Omega = -Q_0 z e^{-i\alpha} + \sum_{j=1}^{n} \frac{\sigma_j L_j}{4\pi} \Big\{ (Z_j + 1)\ln(Z_j + 1) - (Z_j - 1)\ln(Z_j - 1)$$

$$+ 2\ln\left[\tfrac{1}{2}\left(\overset{2}{z}_j - \overset{1}{z}_j\right)\right] - 2 \Big\} + C \qquad (25.34)$$

where C is a real constant and Z_j is defined by (25.4), with $\overset{1}{z}_j$ and $\overset{2}{z}_j$ replacing $\overset{1}{z}$ and $\overset{2}{z}$

$$Z_j = \frac{z - \tfrac{1}{2}\left(\overset{2}{z}_j + \overset{1}{z}_j\right)}{\tfrac{1}{2}\left(\overset{2}{z}_j - \overset{1}{z}_j\right)} \qquad (25.35)$$

Often a creek or river is in direct contact with the groundwater, and may be taken to penetrate the aquifer fully. In such cases, line-sinks may be used to model the creek or river in an approximate fashion. We will briefly discuss how such an approximate model may be constructed; a more comprehensive treatment of such models will be given in Chapter 6. For reasons of brevity in writing, we introduce the real function $\Lambda_{ls}\left(z, \overset{1}{z}_j, \overset{2}{z}_j\right)$ as:

$$\Lambda_{ls}\left(z, \overset{1}{z}_j, \overset{2}{z}_j\right) = \frac{L_j}{4\pi} \Re\left\{(Z_j + 1)\ln(Z_j + 1) - (Z_j - 1)\ln(Z_j - 1) \right.$$
$$\left. + 2\ln\left[\tfrac{1}{2}\left(\overset{2}{z}_j - \overset{1}{z}_j\right)\right] - 2 \right\} \quad (25.36)$$

where Z_j is given by (25.35). The potential for a uniform far-field and n line-sinks then may be written as

$$\Phi = \sum_{j=1}^{n} \sigma_j \Lambda_{ls}\left(z, \overset{1}{z}_j, \overset{2}{z}_j\right) + C + \Phi_g(z) \quad (25.37)$$

where $\Phi_g(z)$ is defined as

$$\Phi_g(z) = \Re(-Q_0 z e^{-i\alpha}) \quad (25.38)$$

The strengths σ_j are unknown, as opposed to the strengths in (25.34), and are determined by requiring that the potential Φ be equal to the values Φ_m, $m = 1, 2, \ldots, n$ at the centers $\overset{c}{z}_m$ of the n line-sinks. Note that the values Φ_m are obtained from the given values of the head along the creeks; the result is n conditions:

$$\Phi_m = \sum_{j=1}^{n} \sigma_j \Lambda_{ls}\left(\overset{c}{z}_m, \overset{1}{z}_j, \overset{2}{z}_j\right) + C + \Phi_g\left(\overset{c}{z}_m\right) \quad (m = 1, 2, \ldots, n) \quad (25.39)$$

An additional constraint may be that the potential is equal to a fixed value Φ_0 at a reference point z_0,

$$\Phi_0 = \sum_{j=1}^{n} \sigma_j \Lambda_{ls}\left(z_0, \overset{1}{z}_j, \overset{2}{z}_j\right) + C + \Phi_g(z_0) \quad (25.40)$$

The equations (25.39) and (25.40) together constitute $n + 1$ equations for the $n + 1$ unknown parameters σ_j ($j = 1, 2, \ldots, n$) and C. These equations are usually solved numerically, for example, by Gaussian elimination. It may be noted that the function Φ_g is given; it does not contain any unknown parameters. In the present example, this function contains only a term due to uniform flow, but other given functions, such as wells or line-sinks with given discharges, could be added, if necessary. In some cases the equation (25.40) is replaced by a continuity equation. For example, if there is no

Sec. 25   Complex Line Integrals   291

net inflow from infinity, the sum of the strengths of the line-sinks is zero, and (25.40) would be replaced by

$$\sum_{j=1}^{n} \sigma_j L_j = 0 \qquad (25.41)$$

Although (25.40) and (25.41) are different in form, they both represent a far-field condition, and regulate the amount of flow from infinity.

At the end of this section, we will adapt the computer program SLW to a complex variable formulation, and include line-sinks in the program.

Problem 25.1
A well of given discharge Q is located at a point $z_w = d + id$ ($\Im d = 0$), and draws all of its water from a finite canal that extends from $z_s = -2d$ to $z_e = 2d$. There is no flow from infinity. The potential along the canal is Φ_0. Model the canal by two line-sinks of equal length and determine their strengths in terms of the aquifer parameters such that the potential at the center of each line-sink is Φ_0.

The Line-Dipole

The line-dipole is a continuous distribution of dipoles along a line. Accordingly, the complex potential for a line-dipole is obtained by integrating a dipole along a line, with the dipole oriented in the direction of the line as shown in Figure 3.45. The starting and end points of the line are denoted as $\overset{1}{z}$ and $\overset{2}{z}$, and the length and orientation of the element are given by L and α,

$$\overset{2}{z} - \overset{1}{z} = L e^{i\alpha} \qquad (25.42)$$

The integral is obtained using the complex potential (24.12) for a dipole, with α replacing β, δ replacing z_0, and $-\mu$ representing the strength rather than s,

$$\Omega = -\frac{1}{2\pi} \int_{-\frac{1}{2}L}^{\frac{1}{2}L} \frac{\mu e^{i\alpha}}{z - \delta} d\xi \qquad (25.43)$$

where ξ is defined in the same way as for the line-sink. It follows from (25.3) that $e^{i\alpha} d\xi = d\delta$, so that (25.43) may be written as:

$$\boxed{\Omega = -\frac{1}{2\pi} \int_{\overset{1}{z}}^{\overset{2}{z}} \frac{\mu}{z - \delta} d\delta} \qquad (25.44)$$

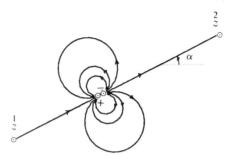

Figure 3.45 A line-dipole.

We use the dimensionless variables Z and Δ defined in (25.4), and obtain

$$\Omega = -\frac{1}{2\pi} \int_{-1}^{+1} \frac{\mu(\Delta)}{Z - \Delta} d\Delta \tag{25.45}$$

We introduce the function

$$\mu = \mu(Z) \tag{25.46}$$

which is analytic for all finite Z, with the possible exception of singular points outside the element, and is real along the element,

$$\Im \mu = 0 \quad -1 \leq X \leq 1 \quad Y = 0 \tag{25.47}$$

We restrict our analysis to density distributions μ that are equal to the above analytic function along the element, and write (25.45) in the following form:

$$\Omega = -\frac{1}{2\pi} \int_{-1}^{+1} \frac{\mu(Z)}{Z - \Delta} d\Delta + \frac{1}{2\pi} \int_{-1}^{+1} \frac{\mu(Z) - \mu(\Delta)}{Z - \Delta} d\Delta \tag{25.48}$$

Using that Z is constant during the integration, the first integration may be carried out, writing $-d\Delta$ as $d(Z - \Delta)$,

$$\Omega = \frac{\mu(Z)}{2\pi} \ln \frac{Z - 1}{Z + 1} + \frac{1}{2\pi} \int_{-1}^{+1} \frac{\mu(Z) - \mu(\Delta)}{Z - \Delta} d\Delta \tag{25.49}$$

The integrand in the remaining integral is analytic at $Z = \Delta$. This may be seen by expanding $\mu(Z)$ in a Taylor series about $Z = \Delta$:

$$\frac{1}{2\pi} \frac{\mu(Z) - \mu(\Delta)}{Z - \Delta} = \frac{\mu(\Delta) + \mu'(\Delta)(Z - \Delta) + \frac{\mu''(\Delta)}{2!}(Z - \Delta)^2 + \ldots - \mu(\Delta)}{2\pi(Z - \Delta)} \tag{25.50}$$

or

$$\frac{1}{2\pi}\frac{\mu(Z)-\mu(\Delta)}{Z-\Delta} = \frac{1}{2\pi}\left[\mu'(\Delta) + \frac{\mu''(\Delta)}{2!}(Z-\Delta) + ...\right]$$
$$= \frac{1}{2\pi}\sum_{m=1}^{\infty}\left[\frac{\mu^{(m)}(\Delta)}{m!}(Z-\Delta)^{m-1}\right] \quad (25.51)$$

which is a Taylor series, so that the integrand is indeed analytic at $Z = \Delta$, as asserted. Since the integral of an analytic function is analytic, the second term in (25.49) is analytic for all Z for which $\mu(Z)$ is analytic, and we represent this function symbolically as $q(Z)$ so that (25.49) becomes

$$\boxed{\Omega = \frac{\mu(Z)}{2\pi}\ln\frac{Z-1}{Z+1} + q(Z)} \quad (25.52)$$

We choose to represent the function $\mu(Z)$ as a polynomial of order n,

$$\mu(Z) = \sum_{j=0}^{n} a_j Z^j \quad \Im a_j = 0 \quad (j = 0, 1, \ldots, n) \quad (25.53)$$

where the coefficients a_j must be real in view of (25.47). An expression for $q(Z)$, then, may be obtained by substituting (25.51) for the integrand in (25.49) and integrating term by term. It is somewhat more convenient, however, to determine $q(Z)$ by an alternative approach as follows. It is seen from (25.45) that the line-dipole behaves as $1/Z$ near infinity: for large values of Z the integrand approaches $-\mu(\Delta)/(2\pi Z)$ so that the function approaches

$$\Omega \approx \frac{-1}{2\pi Z}\int_{-1}^{+1}\mu(\Delta)d\Delta \quad (Z \to \infty) \quad (25.54)$$

The first term in (25.52), however, behaves as a polynomial of order $n-1$ near infinity; the logarithm behaves as $1/Z$ for large values of Z and is multiplied by a polynomial of order n. In order that the complex potential behaves as $1/Z$ for large Z, the expansion of $q(Z)$ about infinity must contain a polynomial of order $n-1$. The coefficients in this polynomial are determined such that they cancel with the singular terms in the expansion about infinity of the first term in (25.52). The function $q(Z)$ is analytic along the element, and must be analytic for all other finite points Z since Ω is analytic for all finite Z except on the element, and the first term in (25.52) is analytic for all finite Z not on the element. A function that is analytic for all finite Z is a polynomial and therefore $q(Z)$ must be equal to:

$$q(Z) = \sum_{m=0}^{n-1} b_m Z^m \quad (25.55)$$

The expansion of the logarithm in (25.52) about infinity may be written as follows:

$$\ln \frac{Z-1}{Z+1} = \ln \frac{1-1/Z}{1+1/Z} = -2 \sum_{r=1}^{\infty} \frac{\beta_r}{Z^r} \qquad (25.56)$$

where

$$\beta_{2r-1} = \frac{1}{2r-1} \qquad \beta_{2r} = 0 \qquad (r=1,2,\ldots) \qquad (25.57)$$

Using (25.53), (25.56) and (25.52) we obtain the following expansion about infinity:

$$\Omega(Z) = -\frac{1}{\pi} \sum_{j=0}^{n} a_j Z^j \sum_{r=1}^{\infty} \frac{\beta_r}{Z^r} + \sum_{m=0}^{n-1} b_m Z^m$$

$$= -\frac{1}{\pi} \sum_{j=0}^{n} \sum_{r=1}^{\infty} a_j \beta_r Z^{j-r} + \sum_{m=0}^{n-1} b_m Z^m \qquad (25.58)$$

In order that all terms of order higher than Z^{-1} cancel, the following condition must be satisfied:

$$\sum_{m=0}^{n-1} b_m Z^m = \frac{1}{\pi} \sum_{j=1}^{n} \sum_{r=1}^{j} a_j \beta_r Z^{j-r} \qquad (25.59)$$

which may be solved for the coefficients b_m. It is noted that these coefficients are real, since both a_j and β_r are real:

$$\Im b_m = 0 \qquad m = 0, 1, 2, \ldots, n-1 \qquad (25.60)$$

We will defer the further details of determining the coefficients b_m for some special values of n to Chapter 6, and focus our attention presently on examining the behavior of $\Omega(Z)$ along the element. The strength μ, often referred to as the density distribution, is real along the element. Introducing the angle θ as (compare [25.17])

$$\theta = \arg \frac{Z-1}{Z+1} \qquad -\pi \leq \theta \leq \pi \qquad (25.61)$$

with a branch cut along the element as shown in Figure 3.43(b), we may represent Ω as follows along the $+$ and $-$ sides of the element:

$$-1 \leq X \leq 1 \quad Y = 0^+ \quad \Omega^+ = \frac{\mu(X)}{2\pi} \left[\ln \left| \frac{X-1}{X+1} \right| + i\pi \right] + q(X) \qquad (25.62)$$

$$-1 \leq X \leq 1 \quad Y = 0^- \quad \Omega^- = \frac{\mu(X)}{2\pi} \left[\ln \left| \frac{X-1}{X+1} \right| - i\pi \right] + q(X) \qquad (25.63)$$

Sec. 25 Complex Line Integrals

The potential Φ is continuous across the element, but the stream function jumps:

$$-1 \leq X \leq 1 \quad Y = 0 \quad \begin{Bmatrix} \Phi^+ - \Phi^- = 0 \\ \Psi^+ - \Psi^- = \mu(X) \end{Bmatrix} \tag{25.64}$$

The angle θ is zero along the X axis outside the element, and therefore the stream function does not jump there:

$$\begin{Bmatrix} -\infty \leq X < -1 \\ 1 < X < \infty \end{Bmatrix} \quad Y = 0 \quad \begin{Bmatrix} \Phi^+ - \Phi^- = 0 \\ \Psi^+ - \Psi^- = 0 \end{Bmatrix} \tag{25.65}$$

The behavior of the stream function along the element is illustrated in Figure 3.46, where the jump $\Psi^+ - \Psi^-$ is plotted along the x axis. There is no net inflow or outflow associated with a dipole, and the same is true for a line-dipole. This is the reason why the branch cut does not extend to infinity; the stream function only jumps along the element. The difference, therefore, between a line-dipole and a line-sink is that the line-dipole does not extract a net amount of fluid, whereas the line-sink does. It appears from (25.52) that the complex potential for a line-dipole has logarithmic singularities at the tips of the element. These singularities correspond to a well at $Z = 1$ of discharge $\mu(1)$ and a recharge well at $Z = -1$ of discharge $\mu(-1)$. If line-dipoles are strung together, and the density distribution μ is continuous at the nodal points where the elements meet, then the singularities cancel; the recharge well at the end point of the one line-dipole will cancel with the well at the adjoining starting point of the next one, as illustrated in Figure 3.47, where the jump $\Psi^+ - \Psi^-$ is plotted along two adjoining elements.

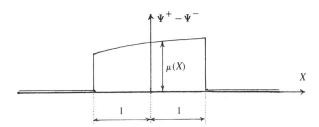

Figure 3.46 The jump in the stream function along the X axis.

We will next examine the relationship between the line-dipole and the line-sink. For reasons of convenience, we will use (25.44), rather than the dimensionless form (25.45). Integrating (25.44) by parts, we obtain

$$\Omega = -\frac{1}{2\pi} \int_{\frac{1}{z}}^{\frac{2}{z}} \frac{\mu}{z - \delta} d\delta = \frac{1}{2\pi} \int_{\frac{1}{z}}^{\frac{2}{z}} \mu \, d\ln(z - \delta)$$

$$= \frac{1}{2\pi} [\mu\!\left(\tfrac{2}{z}\right) \ln\!\left(z - \tfrac{2}{z}\right) - \mu\!\left(\tfrac{1}{z}\right) \ln\!\left(z - \tfrac{1}{z}\right)] - \frac{1}{2\pi} \int_{\frac{1}{z}}^{\frac{2}{z}} \frac{d\mu}{d\delta} \ln(z - \delta) d\delta \tag{25.66}$$

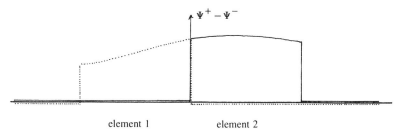

Figure 3.47 Two adjoining line-dipoles.

Using (25.3) we may write

$$\frac{d\mu}{d\delta} = \frac{d\mu}{d\xi} e^{-i\alpha} \qquad d\delta = e^{i\alpha} d\xi \qquad (25.67)$$

so that (25.66) becomes

$$\Omega = \frac{1}{2\pi}\left[\mu\binom{2}{z}\ln\left(z - \overset{2}{z}\right) - \mu\binom{1}{z}\ln\left(z - \overset{1}{z}\right)\right] - \frac{1}{2\pi}\int_{\overset{1}{z}}^{\overset{2}{z}} \frac{d\mu}{d\xi} \ln(z - \delta) d\xi \qquad (25.68)$$

Noticing that $\mu = \Psi^+ - \Psi^-$ along the element, and that $\partial(\Psi^+ - \Psi^-)/\partial\xi = -\sigma$ (see [25.30]), we obtain

$$\Omega = \frac{1}{2\pi}\left[\mu_2 \ln\left(z - \overset{2}{z}\right) - \mu_1 \ln\left(z - \overset{1}{z}\right)\right] + \frac{1}{2\pi}\int_{\overset{1}{z}}^{\overset{2}{z}} \sigma \ln(z - \delta) d\xi \qquad (25.69)$$

where

$$\sigma = -\frac{d\mu}{d\xi} \qquad \mu_1 = \mu\binom{1}{z} \qquad \mu_2 = \mu\binom{2}{z} \qquad (25.70)$$

The integral in this expression is the complex potential for a line-sink, (25.2). The first two terms represent a well of discharge μ_2 and a recharge well of discharge μ_1. Thus, the complex potential for a line-sink may be obtained from that for a line-dipole simply by adding the appropriate terms for wells at the end points:

$$\frac{1}{2\pi}\int_{\overset{1}{z}}^{\overset{2}{z}} \sigma \ln(z - \delta) d\xi = -\frac{1}{2\pi}\int_{\overset{1}{z}}^{\overset{2}{z}} \frac{\mu}{z - \delta} d\delta - \frac{1}{2\pi}\left[\mu_2 \ln\left(z - \overset{2}{z}\right) - \mu_1 \ln\left(z - \overset{1}{z}\right)\right] \qquad (25.71)$$

For example, the complex potential for a line-sink of constant strength may be obtained from (25.71) by letting μ be a linear function along the element; $\mu = -\sigma\xi + \mu_0$, where μ_0 is a constant.

The main difference between line-dipoles and line-sinks lies in the way the branch cut is handled. Consider, for example, a river that is to be approximated by a string of five straight elements with nodal points $z_1, z_2, \ldots, z_5$ as shown in Figure 3.48(a). The

Sec. 25 Complex Line Integrals   297

river may be modeled by using a line-sink for each element. The branch cut of each line-sink falls along the element and extends to infinity, as shown in Figure 3.48(b). Alternatively, the string may be modeled by a series of line-dipoles of linearly varying strength with $\mu_5 = 0$ plus a well at point 1 of discharge μ_1, where μ_1 equals the total amount of inflow into the river. In this case, however, the branch cut follows the river, and there exists only one branch cut in the field extending to infinity, namely the branch cut emanating from the well at point 1, as shown in Figure 3.48(c).

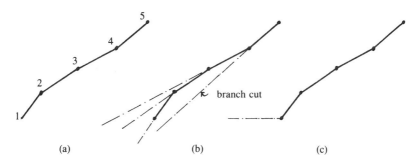

Figure 3.48 Modeling a river by a string of elements.

Some features in aquifers can only be modeled with line-dipoles. Thin zones of very high hydraulic conductivity, such as cracks, are examples of this. Along a crack, both inflow and outflow occur so that the stream-function jumps across the crack, but there is no net inflow or outflow.

Problem 25.2

Repeat problem 25.1, but now using line-dipoles of linearly varying strength and one well to model the canal.

The Line-Doublet

Like the line-dipole, the line-doublet is a continuous distribution of dipoles along a line, but now with the dipoles oriented normal to the line, as shown in Figure 3.49. It may be noted that the terms "line-dipole" and "line-doublet" are somewhat arbitrary. Both terms, dipole and doublet, are used in the literature for the isolated singularity. Since the behavior of the line-integral is quite different when the orientation of the dipoles is rotated over 90°, different names are appropriate, and we will use the above convention throughout this text.

We use the same notation as before, and integrate the dipole (24.12) along the line, taking β as $\alpha + \pi/2$, replacing s by λ, and z_0 by δ, which yields

$$\Omega = \frac{i}{2\pi} \int_{{}^1_z}^{{}^2_z} \frac{\lambda e^{i\alpha}}{z - \delta} d\xi = -\frac{1}{2\pi i} \int_{{}^1_z}^{{}^2_z} \frac{\lambda}{z - \delta} d\delta \qquad (25.72)$$

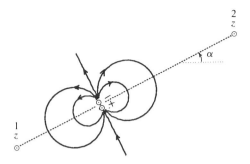

Figure 3.49 A line-doublet.

This equation is formally equal to the complex potential (25.43), except for a factor $1/i$. We may therefore use the results obtained for the line-dipole, and write Ω as follows (see [25.52]),

$$\boxed{\Omega = \frac{\lambda(Z)}{2\pi i} \ln \frac{Z-1}{Z+1} + ip(Z)} \qquad (25.73)$$

Because of the factor $1/i$, the potential Φ for a line-doublet will behave exactly as the stream function for a line-dipole; Φ jumps across the element and is continuous elsewhere, i.e., (25.65) applies and (25.64) becomes

$$-1 \leq X \leq 1 \qquad Y = 0 \qquad \left\{ \begin{array}{l} \Phi^+ - \Phi^- = \lambda(X) \\ \Psi^+ - \Psi^- = 0 \end{array} \right\} \qquad (25.74)$$

Line-doublets are used to model features in aquifers where the potential is discontinuous, but the stream function is continuous. Examples are thin impermeable walls, such as slurry walls, and thin leaky walls such as sheet pilings. Another important application, discussed in Chapter 6, is the modeling of inhomogeneities. The hydraulic conductivity jumps across the boundary of an inhomogeneity, but the head is continuous; the potential jumps, whereas the stream function is continuous, by continuity of flow.

Singular Cauchy Integrals

The three line-integrals discussed above are known as singular Cauchy integrals. Such integrals are discussed in detail by Muskhelishvili [1958], and are often referred to as integrals for single layers (line-sinks) and double layers (line-dipoles and line-doublets). We have already seen that these integrals are useful in modeling linear features in infinite aquifers; we will see below that they can also be used in solving boundary-value problems in domains with closed contours.

We begin by stating, without proof, that a definite complex integral of an analytic function is path-independent. Consider a finite domain $\mathcal{D}^+$ bounded by a closed contour $\mathcal{C}$ (see Figure 3.50). We integrate a function $\Omega(z)$, analytic in $\mathcal{D}^-$, along $\mathcal{C}$ in such a way

Sec. 25 Complex Line Integrals

that $\mathcal{D}^+$ is to the left with respect to the direction of integration. The Cauchy-Goursat integral theorem states that the contour integral of a function $\Omega(z)$ that is analytic in $\mathcal{D}^+$ vanishes:

$$\oint_C \Omega(\delta) d\delta = 0 \qquad (25.75)$$

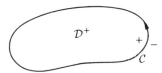

Figure 3.50 The domain $\mathcal{D}^+$.

This result may be understood from the path-independence of the complex integral. We next consider the integral

$$F(z) = -\frac{1}{2\pi i} \oint_C \frac{\Omega(\delta)}{z - \delta} d\delta \qquad (25.76)$$

where z denotes a point in D^+, and write it as follows:

$$F(z) = -\frac{1}{2\pi i} \Omega(z) \oint_C \frac{1}{z - \delta} d\delta + \frac{1}{2\pi i} \oint_C \frac{\Omega(z) - \Omega(\delta)}{z - \delta} d\delta \qquad (25.77)$$

The integrand in the second term is analytic in $\mathcal{D}^+$; the integral vanishes by the Cauchy-Goursat theorem. The first integration may be carried out, and we obtain

$$F(z) = \frac{1}{2\pi i} \Omega(z) \ln(z - \delta)\big|_C \qquad (25.78)$$

The real part of $\ln(z - \delta)$ has the same value at the beginning and end points of the contour, but the imaginary part, equal to $\arg(z - \delta)$, increases by 2π if C is traced by a complete revolution with $\mathcal{D}^+$ to the left, i.e., in counterclockwise direction. Hence, (25.78) becomes

$$F(z) = \Omega(z) \qquad (25.79)$$

The function $F(z)$ in (25.76) thus may be replaced by $\Omega(z)$ and we obtain, using a $+$ to refer to $\mathcal{D}^+$,

$$\Omega^+(z) = -\frac{1}{2\pi i} \oint_C \frac{\Omega^+(\delta)}{z - \delta} d\delta \qquad (z \text{ in } D^+) \qquad (25.80)$$

This equation is known as Cauchy's integral theorem and expresses the values of a function $\Omega^+(z)$ in a domain $\mathcal{D}^+$ in terms of its boundary values. Separating the real and imaginary parts, we obtain

$$\Omega^+(z) = -\frac{1}{2\pi i}\oint_\mathcal{C} \frac{\Phi^+(\delta)}{z-\delta}d\delta - \frac{1}{2\pi}\oint_\mathcal{C}\frac{\Psi^+(\delta)}{z-\delta}d\delta \qquad (25.81)$$

Comparison with (25.44) and (25.72) shows that the integrals represent continuous distributions of doublets and dipoles along $\mathcal{C}$.

The formulation (25.81) may be used to solve boundary value problems, and bears a close resemblance to the so-called direct method in boundary integral equation techniques. Suppose, for example, that $\Phi(\delta)$ is given along $\mathcal{C}$. Then we may write

$$\Phi^+(\delta^*) = \Re \lim_{z\to\delta^*}\left[-\frac{1}{2\pi i}\oint_\mathcal{C}\frac{\Phi^+(\delta)}{z-\delta}d\delta - \frac{1}{2\pi}\oint_\mathcal{C}\frac{\Psi^+(\delta)}{z-\delta}d\delta\right] \qquad (25.82)$$

where δ^* represents a point on the contour $\mathcal{C}$, and z approaches δ^* from the interior of $\mathcal{D}^+$. This represents an integral equation for the unknown function $\Psi^+(\delta)$. The integral equation is often solved in an approximate fashion by dividing the boundary $\mathcal{C}$ in a series of straight line segments, and imposing the condition (25.82) at a number of discrete points. The unknown values $\Psi^+(\delta)$ then are obtained by solving a system of linear equations. A similar procedure may be followed if $\Phi^+(\delta)$ is known along parts of the boundary and $\Psi^+(\delta)$ along others. Along each segment, then, either Φ^+ or Ψ^+ is known, and the unknowns are found in an approximate fashion as explained above.

Cauchy's integral theorem is not restricted to finite domains, but is also applicable to infinite domains, provided that the function is analytic at infinity; a function is defined as analytic at $z = \infty$, if it can be expanded in a Taylor series about $z^* = 1/z = 0$ (compare [23.18]). We label the domain outside $\mathcal{C}$ as $\mathcal{D}^-$ (see Figure 3.51), choose the point z in $\mathcal{D}^-$, and consider the function $\Omega^-(z)$. This function is analytic at $1/z = 0$, i.e., its expansion about infinity has the form

$$\Omega^-(z) = \sum_{j=1}^{\infty}\frac{a_j}{z^j} \qquad (25.83)$$

We assume for the sake of simplicity that $\Omega^-(z)$ is zero at infinity; the constant term is omitted from (25.83). Application of Cauchy's integral theorem to $\Omega^-(z)$ yields (see Muskhelishvili [1958])

$$\Omega^-(z) = +\frac{1}{2\pi i}\oint_\mathcal{C}\frac{\Omega^-(\delta)}{z-\delta}d\delta \qquad (25.84)$$

Note the difference in sign between (25.84) and (25.80), which results from the difference in orientation between $\mathcal{C}$ and $\mathcal{D}^-$; $\mathcal{D}^-$ is to the right with respect to the orientation of $\mathcal{C}$, whereas $\mathcal{D}^+$ is to the left.

The Cauchy integral (25.80) represents a function that jumps by an amount $\Omega^+(\delta)$ across the contour $\mathcal{C}$. Thus, if $\Omega^+(\delta)$ represents the boundary values of Ω^+ on the +

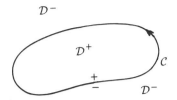

Figure 3.51 The domains $\mathcal{D}^+$ and $\mathcal{D}^-$.

side of $\mathcal{C}$, then the contour integral must vanish in $\mathcal{D}^-$. In fact, there exists a theorem, discussed in detail by Muskhelishvili [1958], that states that *the Cauchy integral of the form (25.80) vanishes for z in $\mathcal{D}^-$ if $\Omega^+(\delta)$ represents the boundary values of a function $\Omega^+(z)$ that is analytic in $\mathcal{D}^+$ and continuous on $\mathcal{C}$ from the inside.* Similarly, the integral in (25.84) vanishes for z in $\mathcal{D}^+$; *the Cauchy integral in (25.84) vanishes if $\Omega^-(\delta)$ represents the boundary values of a function that is analytic in $\mathcal{D}^-$, zero and analytic at infinity, and continuous along $\mathcal{C}$ from the outside.* It follows that

$$\Omega^+(z) = 0 \quad (z \text{ in } \mathcal{D}^-)$$
$$\Omega^-(z) = 0 \quad (z \text{ in } \mathcal{D}^+) \tag{25.85}$$

We may now define a function $\Omega(z)$ in the entire plane as

$$\Omega(z) = \Omega^+(z) + \Omega^-(z) \tag{25.86}$$

which equals $\Omega^+(z)$ in $\mathcal{D}^+$ and $\Omega^-(z)$ in $\mathcal{D}^-$. This function may be expressed as the sum of the two Cauchy integrals (25.80) and (25.84), which may be combined to a single integral as follows:

$$\boxed{\Omega(z) = \Omega^+(z) + \Omega^-(z) = -\frac{1}{2\pi i} \oint_\mathcal{C} \frac{\Omega^+(\delta) - \Omega^-(\delta)}{z - \delta} d\delta} \tag{25.87}$$

The function $\Omega^-(z)$ is zero in $\mathcal{D}^+$, regardless of the boundary values $\Omega^-(\delta)$, and therefore of the form of $\Omega^-(z)$ in $\mathcal{D}^-$: there exist many different integrals (25.87), all being equal in $\mathcal{D}^+$ and differing only in $\mathcal{D}^-$. The density distribution $\Omega^+(\delta) - \Omega^-(\delta)$ in the singular Cauchy integral that gives $\Omega = \Omega^+(z)$ in $\mathcal{D}^+$ is not unique; the component $\Omega^-(\delta)$ in this distribution may be changed without affecting the values of the integral in $\mathcal{D}^+$. For example, the Dirichlet problem in $\mathcal{D}^+$, where $\Phi^+(\delta)$ is given everywhere along $\mathcal{C}$, may be solved as follows. Choosing $\Phi^-(\delta) = \Phi^+(\delta)$, which is always possible, (the boundary values $\Phi^-(\delta)$ simply correspond to a Dirichlet problem in $\mathcal{D}^-$) we obtain for (25.87):

$$\Omega(z) = -\frac{1}{2\pi} \oint_\mathcal{C} \frac{\Psi^+(\delta) - \Psi^-(\delta)}{z - \delta} d\delta = -\frac{1}{2\pi} \oint_\mathcal{C} \frac{\mu(\delta)}{z - \delta} d\delta \tag{25.88}$$

Hence, a Dirichlet problem can be solved by a continuous distribution of dipoles. A formulation such as the one discussed above, where the density distribution does not equal the boundary values of $\Omega^+(z)$, is called an indirect formulation. Other indirect formulations may be obtained by choosing various admissible boundary values for $\Omega^-(z)$. It may be noted that the direct formulation (25.80) is a special case of (25.87), obtained when $\Omega^-(\delta) = 0$, so that $\Omega^-(z)$ vanishes for all z.

The applications of line-dipoles and line-doublets discussed earlier may be considered as special cases of the boundary element formulation, where the boundary C of the domain is the canal, river, or impermeable boundary. For such cases, however, either $\Phi^+ - \Phi^-$ is known to be zero (for rivers or creeks), or $\Psi^+ - \Psi^-$ is zero (for impermeable boundaries); in the former case line-dipoles must be used and in the latter case line-doublets must be used.

A special type of boundary-value problem exists when $\Omega(z)$ has a physical meaning in both $\mathcal{D}^+$ and $\mathcal{D}^-$. Examples are cases where a semi-permeable wall encloses part of the aquifer and cases where the hydraulic conductivities in $\mathcal{D}^+$ and $\mathcal{D}^-$ are different from each other. In such problems, the conditions along C are specified in terms of the jump $\Omega^+(\delta) - \Omega^-(\delta)$.

We will apply the Cauchy integrals discussed here to regional aquifers in Chapter 6, which we will treat as infinite domains, and in Chapter 8 to problems of flow domains with closed boundaries.

Problem 25.3

Use Cauchy's integral theorem to determine a function in the entire plane, with boundary values $\Omega^+(\delta) = \delta$ and $\Omega^-(\delta) = 1/\delta$, where C is the unit circle.

The Computer Program SLWL

We will convert the computer program SLW, developed in Chapter 2, to complex variables, and implement line-sinks with given strength and line-sinks with given head. We call the new program SLWL (Single Layer with Wells and Line-sinks). Adaptation of the program to complex variables affects all routines but is straightforward. We will discuss a few of these changes here; a complete listing of the program is given in Appendix C. FORTRAN supports complex variables. Throughout the program, we adopt the convention that all complex variables start with the letter C, and we use the declaration IMPLICIT COMPLEX (C) in each routine where complex variables are used. The operations for addition, subtraction, multiplication, and division of complex numbers are performed in the same way as for real variables. The real and imaginary parts of a complex variable $z = $ CZ can be obtained by the use of the built-in functions REAL(CZ) and AIMAG(CZ). Complex constants are entered by statements such as CZ = (1.,1.), which is equivalent to $z = 1 + i$. Two real variables can be combined into a complex one; CZ = CMPLX(RX,RY) is equivalent to $z = x + iy$. The complex conjugate of a complex variable is returned by the function CONJG(CZ). Most functions available for real variables also exist for complex variables. For example, the function CABS(CZ) returns the modulus of CZ, and the complex function CLOG(CZ) returns the complex logarithm of CZ.

Sec. 25 Complex Line Integrals

The complex potential Ω, the potential Φ, the stream function Ψ, the discharge function W, and the piezometric head ϕ all are functions of the complex variable $z = x + iy$. The listings of the corresponding function subprograms follow:

```
c     Complex potential.
      COMPLEX FUNCTION COMEGA(CZ)
      IMPLICIT COMPLEX (C)
      COMEGA=COMGVN(CZ)+RFRFC(CZ)+CFLSU(CZ)
      RETURN
      END
c     Complex discharge function.
      COMPLEX FUNCTION CW(CZ)
      IMPLICIT COMPLEX (C)
      CW=CWGVG(CZ)+CWWLSM(CZ)+CWLSSM(CZ)
      RETURN
      END
c     Potential function.
      REAL FUNCTION RFPOT(CZ)
      IMPLICIT COMPLEX (C)
      RFPOT=REAL(COMEGA(CZ))
      RETURN
      END
c     Stream function.
      REAL FUNCTION RFPSI(CZ)
      IMPLICIT COMPLEX (C)
      RFPSI=AIMAG(COMEGA(CZ))
      RETURN
      END
c     Piezometric head.
      REAL FUNCTION RFHEAD(CZ)
      IMPLICIT COMPLEX (C)
      RPOT=RFPOT(CZ)
      RFHEAD=RFHEDP(RPOT)
      RETURN
      END
c     Contributions to the complex potential of all functions with
c     given coefficients.
      COMPLEX FUNCTION COMGVN(CZ)
      IMPLICIT COMPLEX (C)
      COMGVN=CFGVG(CZ)+CFWLG(CZ)+CFLSG(CZ)
      RETURN
      END
```

We will focus our attention on the programming of the line-sink function, and on the implementation in the program of (25.39) and (25.40) for the determination of the strengths of the head-specified line-sinks.

The Module Line-Sink

The line-sink module contains a BLOCK DATA SUBPROGRAM, a function that returns the contribution of all line-sinks with given strengths to the complex potential (CFLSG), a function that returns the contribution of all line-sinks with given head (CFLSU), the contribution of a single line-sink of unit strength to the complex potential (COMLS), a function that returns the contribution of all line-sinks to the discharge function W (CWLSSM), the contribution of a single line-sink of unit strength to the discharge function, and a routine that produces the layout for all line-sinks. Also included is a logical function that may be used to determine whether a point is within a specified neighborhood of any of the line-sinks (LFATLS). The listings of these routines follow:

```
c      Variables contained in the common block BDLS.
c      NLS      = total number of line-sinks.
c      NLSS     = total number of line-sinks with given strength.
c      NLSF     = total number of line-sinks with given head.
c      CLSZS(I) = complex coordinate of the starting point of
c                 line-sink I.
c      CLSZC(I) = complex coordinate of the center of line-sink I.
c      CLSZE(I) = complex coordinate of the end point of line-sink I.
c      RLSDIS(I)= strength of line-sink I (discharge per unit length).
c      RLSFI(I) = head at the center of line-sink I.
c      ILSPTS(J)= pointer to the position of the discharge-specified
c                 line-sink number J in the arrays.
c      ILSPTF(J)= pointer to the position of the head-specified
c                 line-sink number J in the arrays.
c      Initialization of the common block with line-sink data:
c      Set the number of line-sinks equal to zero.
       BLOCK DATA BDLS
       IMPLICIT COMPLEX (C),LOGICAL (L)
       COMMON /CBLS/ NLS,NLSS,NLSF,
      .              CLSZS(200),CLSZE(200),CLSZC(200),RLSDIS(200),
      .              RLSFI(200),ILSPTS(200),ILSPTF(200)
       DATA NLS,NLSS,NLSF /3*0/
       END
c      Contribution of all strength-specified line-sinks to the
c      complex potential.
       COMPLEX FUNCTION CFLSG(CZ)
       IMPLICIT COMPLEX (C), LOGICAL (L)
       COMMON /CBLS/ NLS,NLSS,NLSF,
      .              CLSZS(200),CLSZE(200),CLSZC(200),RLSDIS(200),
      .              RLSFI(200),ILSPTS(200),ILSPTF(200)
       CFLSG=(.0,.0)
       DO 100 ILSS=1,NLSS
         IAD=ILSPTS(ILSS)
```

Sec. 25 Complex Line Integrals

```
              CFLSG=CFLSG+RLSDIS(IAD)*COMLS(CZ,CLSZS(IAD),CLSZE(IAD))
  100     CONTINUE
          RETURN
          END
C     Contribution of all head-specified line-sinks to the complex
C     potential.
          COMPLEX FUNCTION CFLSU(CZ)
          IMPLICIT COMPLEX (C), LOGICAL (L)
          COMMON /CBLS/ NLS,NLSS,NLSF,
         .              CLSZS(200),CLSZE(200),CLSZC(200),RLSDIS(200),
         .              RLSFI(200),ILSPTS(200),ILSPTF(200)
          CFLSU=(.0,.0)
          DO 100 ILSF=1,NLSF
            IAD=ILSPTF(ILSF)
            CFLSU=CFLSU+RLSDIS(IAD)*COMLS(CZ,CLSZS(IAD),CLSZE(IAD))
  100     CONTINUE
          RETURN
          END
C     Contribution of a single line-sink of unit strength to the
C     complex potential (see [25.8]).
          COMPLEX FUNCTION COMLS(CZ,CZS,CZE)
          IMPLICIT COMPLEX (C), LOGICAL (L)
          DATA RPI /3.1415926/
          CBZ=(2.0*CZ-(CZS+CZE))/(CZE-CZS)
          COM1=(.0,.0)
          COM2=(.0,.0)
          IF(CABS(CBZ+1.).GT..0001) COM1=(CBZ+1.)*CLOG(CBZ+1.)
          IF(CABS(CBZ-1.).GT..0001) COM2=(CBZ-1.)*CLOG(CBZ-1.)
          COMLS=COM1-COM2+2.*CLOG(.5*(CZE-CZS))-2.
          COMLS=.25/RPI*CABS(CZE-CZS)*COMLS
          RETURN
          END
C     Contribution of all line-sinks to the complex discharge (=Qx-iQy).
          COMPLEX FUNCTION CWLSSM(CZ)
          IMPLICIT COMPLEX (C), LOGICAL (L)
          COMMON /CBLS/ NLS,NLSS,NLSF,
         .              CLSZS(200),CLSZE(200),CLSZC(200),RLSDIS(200),
         .              RLSFI(200),ILSPTS(200),ILSPTF(200)
          CWLSSM=(.0,.0)
          DO 100 ILS=1,NLS
            CWLSSM=CWLSSM+RLSDIS(ILS)*CWLS(CZ,ILS)
  100     CONTINUE
          RETURN
          END
```

```
c      Complex discharge function due to a single line-sink of unit
c      strength (compare [25.16]).
       COMPLEX FUNCTION CWLS(CZ,ILS)
       IMPLICIT COMPLEX (C)
       COMMON /CBLS/ NLS,NLSS,NLSF,
      .              CLSZS(200),CLSZE(200),CLSZC(200),RLSDIS(200),
      .              RLSFI(200),ILSPTS(200),ILSPTF(200)
       DATA RPI /3.1415926/
       CZS=CLSZS(ILS)
       CZE=CLSZE(ILS)
       CBZD=1./(CZE-CZS)
       CBZ=(2.*CZ-(CZS+CZE))*CBZD
c      Avoid the end points of the line-sink because the discharge
c      function is singular there.
       IF(CABS(CBZ+1.).LT..0001) CBZ=-.9999
       IF(CABS(CBZ-1.).LT..0001) CBZ=.9999
       CWLS=.5/RPI*CABS(CZS-CZE)*CLOG((CBZ-1.)/(CBZ+1.))*CBZD
       RETURN
       END
c      Layout of all line-sinks.
       SUBROUTINE LSLAY
       IMPLICIT LOGICAL(L),COMPLEX(C)
       COMMON /CBLS/ NLS,NLSS,NLSF,
      .              CLSZS(200),CLSZE(200),CLSZC(200),RLSDIS(200),
      .              RLSFI(200),ILSPTS(200),ILSPTF(200)
       DO 10 I=1,NLS
         CALL DRAW(CLSZS(I),0)
         CALL DRAW(CLSZE(I),1)
   10  CONTINUE
       RETURN
       END
c      Logical function which is set true if CZ is less than RTOL away
c      from a line-sink of positive strength.
       LOGICAL FUNCTION LFATLS(CZ,RTOL)
       IMPLICIT COMPLEX (C), LOGICAL (L)
       COMMON /CBLS/ NLS,NLSS,NLSF,
      .              CLSZS(200),CLSZE(200),CLSZC(200),RLSDIS(200),
      .              RLSFI(200),ILSPTS(200),ILSPTF(200)
       LFATLS=.FALSE.
       DO 100 ILS=1,NLS
         IF(RLSDIS(ILS).LE.0) GOTO 100
         IF(CABS(CLSZS(ILS)-CZ).LT.RTOL
      .   .OR.CABS(CLSZE(ILS)-CZ).LT.RTOL) THEN
           LFATLS=.TRUE.
```

```
          ELSE
            CDZLS=CLSZE(ILS)-CLSZS(ILS)
            CCZ=(CZ+CZ-CLSZE(ILS)-CLSZS(ILS))/CDZLS
            IF(ABS(REAL(CCZ)).LT.1.
       .      .AND.ABS(AIMAG(CCZ))*CABS(CDZLS)*.5.LT.RTOL) LFATLS=.TRUE.
          ENDIF
  100   CONTINUE
        RETURN
        END
```

The Routine Solve

The strengths of the line-sinks with given head are determined from equations (25.39) and (25.40) in the subroutine SOLVE:

```
c     Solving.
c     RMAT   = coefficient matrix.
c     RKNVEC = vector with known coefficients.
c     RSOLUT = vector with computed coefficients, returned by SOLVE.
      SUBROUTINE SOLVE(IMXSZE,RMAT,RKNVEC,RSOLUT)
      IMPLICIT COMPLEX (C), LOGICAL (L)
      DIMENSION RMAT(IMXSZE,*),RKNVEC(*),RSOLUT(*)
      COMMON /CBRF/ CRFZ,RRFFI,RRFCON
      COMMON /CBLS/ NLS,NLSS,NLSF,
     .              CLSZS(200),CLSZE(200),CLSZC(200),RLSDIS(200),
     .              RLSFI(200),ILSPTS(200),ILSPTF(200)
c     Set number of equations to 1 + number of line-sinks (n in
c     [25.39]).
      NEQ=NLSF+1
c     Print error message if maximum number of equations is exceeded.
      IF(NEQ.GT.IMXSZE) THEN
        WRITE(*,9003) IMXSZE
        RETURN
      ENDIF
c     For columns 1 to NLSF = number of head specified line-sinks:
      IF(NLSF.GE.1) THEN
        DO 200 JCOL=1,NLSF
c         Make temporary storage for starting and end points.
c         These points are
          CZSLS=CLSZS(ILSPTF(JCOL))
          CZELS=CLSZE(ILSPTF(JCOL))
c         Refer to (25.39):
c         For each column, compute the coefficient of the strength of
c         the line-sink number JCOL (=j). The coefficient is computed
```

```
c           at the center (c / z_m) of line-sink number IROW (=m).
c           This coefficient is equal to the function
c           Λ_ls(c / z_m, 1 / z_j, 2 / z_j) and is stored as RMAT(IROW,JCOL).
            DO 100 IROW=1,NLSF
              RMAT(IROW,JCOL)=
     .           REAL(COMLS(CLSZC(ILSPTF(IROW)),CZSLS,CZELS))
 100        CONTINUE
c           Compute the coefficient of the strength of line-sink number
c           JCOL at the reference point (see (25.40)).
            RMAT(NEQ,JCOL)=REAL(COMLS(CRFZ,CZSLS,CZELS))
 200      CONTINUE
          ENDIF
c         Store the coefficients of the constants in equations (25.39)
c         and (25.40). These coefficients all are 1.
          DO 300 IROW=1,NEQ
            RMAT(IROW,NEQ)=1.
 300      CONTINUE
c         Store the known vector in RKNVEC. This known vector is equal to
c         Φ_m−Φ_g(c / z_m) in (25.39) and to Φ_0−Φ_g(z_0)
c         in (25.40). Note that Φ_g contains rainfall, uniform flow,
c         ponds, and line-sinks with given strengths.
          IF(NLSF.GE.1) THEN
            DO 1000 IROW=1,NLSF
              RKNVEC(IROW)=RFPOTH(RLSFI(ILSPTF(IROW)))-
     .           REAL(COMGVN(CLSZC(ILSPTF(IROW))))
 1000       CONTINUE
          ENDIF
          RKNVEC(NEQ)=RFPOTH(RRFFI)-REAL(COMGVN(CRFZ))
c         Solve the system of equations.
          CALL SSOLVE(IMXSZE,RMAT,RKNVEC,RSOLUT,NEQ)
c         Store all computed strengths in the array.
          IF(NLSF.GE.1) THEN
            DO 2000 IROW=1,NLSF
              RLSDIS(ILSPTF(IROW))=RSOLUT(IROW)
 2000       CONTINUE
          ENDIF
c         Retrieve and store the constant.
          RRFCON=RSOLUT(NEQ)
          RETURN
 9003   FORMAT('+MAXIMUM NUMBER OF EQUATIONS , ',I4,' EXCEEDED;'/
     .   ' SOLVING ABORTED'/)
          END
```

Sec. 25   Complex Line Integrals   **309**

The command CONTROL in the routine CHECK now includes the control points at the centers of the line-sinks, and causes both the specified and computed values of the heads to be displayed.

Problem 25.4

Modify SLWL so that the condition at the reference point is replaced by the condition that there is no flow from or toward infinity for cases without rainfall. Use this modified version of the program to test your answer to Problem 25.1.

Chapter 4

Fluid Particle Paths and Solute Transport

The subject matter covered in this chapter consists of (1) the average paths that fluid particles follow on their way through the soil and (2) the transport of solutes. The paths are average; deviations due to the discrete nature of the porous material are ignored. We will refer to these average particle paths as path lines.

For steady flow the path lines coincide with the streamlines. For transient flow the pattern of streamlines changes with time; the path lines are different from the streamlines. Closed form expressions for streamlines and path lines can be obtained only for a few cases; we will resort to numerical methods. We will first discuss the computation of points on streamlines and path lines, then cover an approximate technique for determining streamlines and path lines in three dimensions using the Dupuit-Forchheimer approximation. This chapter ends with a brief discussion of solute transport.

26 NUMERICAL DETERMINATION OF FLUID PARTICLE PATHS

The equations for the streamlines often are implicit in terms of the coordinates x and y which then must be obtained numerically. Streamlines may be determined by the use of the stream function if the potential is harmonic. The application of a contouring procedure to a set of values of Ψ computed at the mesh points of a grid is one way of computing points of a set of streamlines. A disadvantage of this approach is that jumps of the stream function along branch cuts appear in the plot as heavy black lines (see, for example, Sec. 19, Figure 3.16). This problem can be avoided only by using a special contouring routine that processes the jumps across the branch cuts correctly.

An alternative approach is to determine points of a given streamline in such a way that jumps in the stream function can be handled easily; we will discuss such a procedure in this section.

If the potential is not harmonic, as in shallow flow with infiltration and in shallow transient flow, then the stream function cannot be used. Procedures for computing points on streamlines and path lines applicable to such cases are covered separately.

A Procedure for Tracing Streamlines Using the Stream Function

We will determine points on a streamline by a process that we refer to as tracing streamlines: we start at any given point of a streamline $\Psi = \Psi_j$, and proceed by calculating complex coordinates z of points on the streamline one by one. Let $z = \underset{1}{z}$ be a point of the streamline $\Psi = \Psi_j$ (see Figure 4.1). We obtain a first estimate of the next point on the streamline, $\underset{2}{\overset{1}{z}}$, as follows:

$$\underset{2}{\overset{1}{z}} = \underset{1}{z} + e^{i\alpha}\Delta s \tag{26.1}$$

where Δs is some fixed quantity, equal to the distance between $\underset{1}{z}$ and $\underset{2}{\overset{1}{z}}$, and $e^{i\alpha}$ is tangent to the streamline at $\underset{1}{z}$. The angle α is equal to the argument of $\overline{W} = Q_x + iQ_y$, by definition, and $e^{i\alpha}$ may be represented as follows, provided that $W\left(\underset{1}{z}\right) \neq 0$,

$$e^{i\alpha} = \frac{\overline{W\left(\underset{1}{z}\right)}}{\left|W\left(\underset{1}{z}\right)\right|} \qquad \left(W\left(\underset{1}{z}\right) \neq 0\right) \tag{26.2}$$

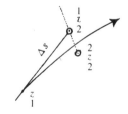

Figure 4.1 Tracing a streamline.

Sec. 26 Numerical Determination of Fluid Particle Paths

We obtain the second approximation $\overset{2}{\underset{2}{z}}$ of $\underset{2}{z}$ by application of a Newton-Raphson procedure along the equipotential through $\overset{1}{\underset{2}{z}}$. With $\Delta\Phi = 0$ along the equipotential, we may write the following approximation for W:

$$W\left(\overset{1}{\underset{2}{z}}\right) = -\left[\frac{d\Omega}{dz}\right]_{z=\overset{1}{\underset{2}{z}}} \approx -\frac{\Delta\Phi + i\Delta\Psi}{\overset{2}{\underset{2}{z}} - \overset{1}{\underset{2}{z}}} = -i\frac{\Psi_j - \Psi\left(\overset{1}{\underset{2}{z}}\right)}{\overset{2}{\underset{2}{z}} - \overset{1}{\underset{2}{z}}} \qquad (26.3)$$

Solving for $\overset{2}{\underset{2}{z}}$ we obtain

$$\overset{2}{\underset{2}{z}} = \overset{1}{\underset{2}{z}} + i\frac{\Psi\left(\overset{1}{\underset{2}{z}}\right) - \Psi_j}{W\left(\overset{1}{\underset{2}{z}}\right)} \qquad (26.4)$$

This relation may be used recursively until $\left|\overset{2}{\underset{2}{z}} - \overset{1}{\underset{2}{z}}\right|$ is less than a certain quantity ε. Tracing streamlines is done often in connection with graphical display; in such cases ε may be taken as the resolution of the graphics device.

Handling the Jumps in Ψ

Whenever the streamlines traced by the above procedure intersect a branch cut, the value of the stream function will jump. If such a jump occurs, the magnitude of $\Psi\left(\overset{1}{\underset{2}{z}}\right) - \Psi_j$ in (26.4) will be much larger than the magnitude due to the deviation from the streamline. This magnitude therefore should be checked in the program, and whenever it is larger than some preset value, a procedure may be called that determines which branch cut has been crossed. Assume, for example, that the branch cut is crossed of a well of discharge Q. The constant value Ψ_j then may be updated by subtracting or adding the amount Q. Whether to subtract or add may be decided on the basis of the sign of $\Psi\left(\overset{1}{\underset{2}{z}}\right) - \Psi_j$.

A Procedure for Tracing Streamlines and Path Lines Using the Velocity Vector

The stream function does not exist for flow problems where the potential is not harmonic, e.g., problems of shallow flow involving infiltration or leakage and problems of shallow transient flow. The procedure outlined above is not applicable to such problems, and points on the streamlines or path lines, then, may be determined as explained below. For transient flow the streamlines at any time give a graphical representation of the velocity field, whereas the path lines graphically represent the paths followed by the particles. The pattern of streamlines always occupies the entire flow domain and changes with time. A pattern of path lines may be generated by drawing the paths of individual particles that are at given positions at the same time, for example, particles injected through a well. We will discuss some simple procedures for determining streamlines and path lines in the following.

Steady Flow

Points on streamlines that correspond to the discharge vector (Q_x, Q_y) can be determined by a standard numerical technique as follows. The complex discharge $Q_x + iQ_y$ is given by

$$Q_x + iQ_y = \overline{W(z)} \tag{26.5}$$

The streamline is tangent to the discharge vector so that

$$\frac{dx}{ds} + i\frac{dy}{ds} = \frac{dz}{ds} = \frac{\overline{W(z)}}{|W(z)|} \tag{26.6}$$

where s represents the arc length measured along the streamline. This equation can be integrated numerically. We will discuss two elementary techniques: Euler's method is very simple but the errors associated with this technique are usually large and increase with the length of the streamline. If $\underset{1}{z}$ represents a point on the streamline then the next point $\overset{1}{\underset{2}{z}}$ is computed by Euler's method as follows:

$$\boxed{\overset{1}{\underset{2}{z}} = \underset{1}{z} + \frac{\overline{W\left(\underset{1}{z}\right)}}{\left|W\left(\underset{1}{z}\right)\right|}\Delta s} \tag{26.7}$$

This approximation amounts to the assumption that the fluid particle moves in the direction of the discharge vector at $\underset{1}{z}$. An improved technique, known as the modified Euler's method, is a second order Runge-Kutta scheme (e.g., Carnahan et al. [1969]). In this method, the fluid particle is assumed to move in the direction of the average discharge over the interval. This yields the approximation:

$$\boxed{\underset{2}{\overset{2}{z}} = \underset{1}{z} + \frac{\overline{W\left(\underset{1}{z}\right)} + \overline{W\left(\overset{1}{\underset{2}{z}}\right)}}{\left|W\left(\underset{1}{z}\right) + W\left(\overset{1}{\underset{2}{z}}\right)\right|}\Delta s} \tag{26.8}$$

Further improvement is possible by applying (26.8) again, replacing $\underset{2}{\overset{2}{z}}$ by $\underset{2}{\overset{3}{z}}$ and $\underset{2}{\overset{1}{z}}$ by $\underset{2}{\overset{2}{z}}$. This procedure may be repeated until the changes in subsequent approximations of $\underset{2}{z}$ are less than some specified amount.

The complex variable formulation (26.7) and (26.8) is not restricted to problems where the potential is harmonic, even though the complex potential for such cases does not exist. In the computer program SLWL, for example, the contributions of most elements to W are programmed as minus the derivative with respect to z of their complex potentials. The contributions to W of the rainfall function and the ponds (only for points underneath the pond) are obtained by the application of Darcy's law to the potential and included in the real and imaginary parts of $W = Q_x - iQ_y$. Note that the stream function is not single-valued for such cases so that (26.4) cannot be used.

The procedure for tracing streamlines according to (26.7) and (26.8) may be formulated in terms of real variables and generalized to three-dimensional flow as follows. We use a three-dimensional Cartesian coordinate system x_1, x_2, x_3 and represent the velocity vector as v_i ($i = 1, 2, 3$). It follows from the definition of a streamline that

$$\frac{dx_i}{ds} = \frac{v_i}{v} \quad (i = 1, 2, 3) \tag{26.9}$$

where

$$v = \sqrt{v_1^2 + v_2^2 + v_3^2} \tag{26.10}$$

adapting (26.7) to three dimensions we obtain:

$$\overset{1}{x}_i = \underset{1}{x_i} + \frac{v_i\left(\underset{1}{x_j}\right)}{v\left(\underset{1}{x_j}\right)} \Delta s \quad (i = 1, 2, 3) \tag{26.11}$$

where $v_i(\underset{1}{x_j})$ represents the velocity vector computed at $\underset{1}{x_j}$ ($j = 1, 2, 3$). Equation (26.8) is modified as follows:

$$\overset{2}{x}_i = \underset{1}{x_i} + \frac{\frac{1}{2}\left[v_i\left(\underset{1}{x_j}\right) + v_i\left(\overset{1}{x}_{2j}\right)\right]}{v^*} \Delta s \quad (i = 1, 2, 3) \tag{26.12}$$

where

$$v^* = \frac{1}{2}\sqrt{\sum_{i=1}^{3}\left[v_i\left(\underset{1}{x_j}\right) + v_i\left(\overset{1}{x}_{2j}\right)\right]^2} \tag{26.13}$$

The time increment Δt corresponding to the time in which the fluid moves from $\underset{1}{x_i}$ to $\overset{2}{x}_i$ may be estimated from

$$\Delta t = \frac{\Delta s}{v^*} \tag{26.14}$$

Because only the ratio of the velocities occurs in (26.9), formulations in terms of velocities and in terms of specific discharges are equivalent: we may use either q_i or $v_i = q_i/n$. The computation of travel times, see (26.14), must be done in terms of the velocity. It is important to remember that this velocity is an average (the seepage velocity).

Transient Flow

In the preceding formulation W and the velocity vector are considered to be independent of time. The equations can be applied to transient flow, but will yield a pattern of streamlines valid for a particular time if W or the velocity vector are kept constant with time. The same equations can be used to generate path lines, provided that the time dependency of W or v_i is taken into account. We will illustrate this for three-dimensional flow and write (26.11) as

$$\overset{1}{\underset{2}{x}}_i = \overset{1}{x}_i + \frac{v_i\left(\overset{1}{x}_j, \overset{1}{t}\right)}{v\left(\overset{1}{x}_j, \overset{1}{t}\right)} \Delta s \qquad (i = 1, 2, 3) \tag{26.15}$$

where $\overset{1}{t}$ corresponds to the time at which the fluid particle is at $\overset{1}{x}_i$. We estimate the time $\overset{1}{\underset{2}{t}}$ at which the particle is at $\overset{1}{\underset{2}{x}}_j$ as

$$\overset{1}{\underset{2}{t}} = \overset{1}{t} + \frac{\Delta s}{v\left(\overset{1}{x}_j, \overset{1}{t}\right)} \tag{26.16}$$

and use this estimate to compute v_i at $\overset{1}{\underset{2}{x}}_j$ so that (26.12) becomes

$$\overset{2}{\underset{2}{x}}_i = \overset{1}{x}_i + \frac{\frac{1}{2}\left[v_i\left(\overset{1}{x}_j, \overset{1}{t}\right) + v_i\left(\overset{1}{\underset{2}{x}}_j, \overset{1}{\underset{2}{t}}\right)\right]}{v^*} \Delta s \qquad (i = 1, 2, 3) \tag{26.17}$$

where

$$v^* = \tfrac{1}{2}\sqrt{\sum_{i=1}^{3}\left[v_i\left(\overset{1}{x}_j, \overset{1}{t}\right) + v_i\left(\overset{1}{\underset{2}{x}}_j, \overset{1}{\underset{2}{t}}\right)\right]^2} \tag{26.18}$$

The next estimate of $\overset{}{\underset{2}{t}}$ is obtained from (compare [26.14]):

$$\overset{2}{\underset{2}{t}} = \overset{1}{t} + \frac{\Delta s}{v^*} \tag{26.19}$$

An alternative procedure for determining path lines is to write the components of the velocity vector as

$$v_i = \frac{dx_i}{dt} \quad (i = 1, 2, 3) \tag{26.20}$$

where x_i represents the position of the fluid particle at time t. Numerical integration of (26.20) by Euler's method yields:

$$\boxed{\underset{2}{x}_i = \underset{1}{x}_i + v_i\left(\underset{1}{x}_j, \underset{1}{t}\right)\Delta t \quad (i = 1, 2, 3)} \tag{26.21}$$

A second estimate of $\underset{2}{x}_i$ is obtained as follows:

$$\boxed{\underset{2}{x}_i = \underset{1}{x}_i + \tfrac{1}{2}\left[v_i\left(\underset{1}{x}_j, \underset{1}{t}\right) + v_i\left(\underset{2}{x}_j, \underset{2}{t}\right)\right]\Delta t} \tag{26.22}$$

i.e., it is assumed that the particle moves from $\underset{1}{x}_i$ to $\underset{2}{x}_i$ with the estimated average velocity. The value of $\underset{2}{t}$ is obtained from

$$\underset{2}{t} = \underset{1}{t} + \Delta t \tag{26.23}$$

Equation (26.21) would yield the same answer as (26.15) if Δt in (26.21) were taken as

$$\Delta t = \frac{\Delta s}{v\left(\underset{1}{x}_j, \underset{1}{t}\right)} \tag{26.24}$$

Similarly, (26.22) would yield the same answer as (26.17) if Δt in (26.22) were updated as

$$\Delta t = \frac{\Delta s}{v^*} \tag{26.25}$$

with v^* given by (26.18). The two approaches differ in the step sizes taken along the path line: if a constant step in space is required then the former approach should be used, and if a constant step in time is desired the latter one should be applied.

It is important to note that path lines can be obtained by analytic integration of (26.9) only if the function $t = t(s)$ is known beforehand, i.e., if the time associated with each position of the particle on the path line is known, which is true only for trivial cases. In numerical integration, this function can be determined as the computation proceeds.

We deal in this text primarily with steady flow and will use the terms streamlines and path lines interchangeably for such flow. When dealing with transient flow, it will be stated explicitly whether streamlines or path lines are meant.

Problem 26.1

Consider the (purely imaginary) two-dimensional flow field $v_x = \overset{0}{v}_x e^{-\alpha t}$, $v_y = \overset{0}{v}_y e^{-\beta t}$ where α and β are real constants with the dimension of $[1/s]$.

Questions

(a) Integrate (26.20) to determine the equation for the path line with the initial condition that the particle is at $x = y = 0$ at $t = 0$.
(b) Determine t both as a function of x and as a function of y, and use these functions to generate the equation for the path line meant under (a) by integration of $dy/dx = v_y/v_x$. (The latter equation is obtained by dividing [26.9] for $i = 1$ into [26.9] for $i = 2$.)
(c) Determine the equation for the streamline through $x = y = 0$, valid for $t = 0$. Draw both the path line for $0 \leq t \leq \infty$ and the streamline and compare them.

Determination of Streamlines in SLWL

We will implement the procedure for tracing streamlines represented by (26.7) and (26.8) in SLWL. We add a command TRACE in SETUP, which will cause the layout to appear, and add the possibility to create a contour plot. We then activate the graphics cursor. Upon entering a "t" from the keyboard, the coordinates of the cursor are passed on to a routine TRACE, which computes points of the streamline and displays it. The user can terminate the entry of coordinates of starting points of streamlines by entering an "r" from the keyboard. The statements added to SETUP are elementary, and are contained in the listing in Appendix C.

The routine TRACE is an implementation of (26.7) and (26.8). For the sake of simplicity, we will refrain from the computation of travel times. The listing of TRACE is as follows:

```
c       Subroutine for plotting a streamline.
c       RSTEP  = step along the streamline (Δs).
c       CZ     = starting point of the streamline.
        SUBROUTINE TRACE(RSTEP,CZ)
        IMPLICIT COMPLEX(C),LOGICAL(L)
c       Tolerance used to determine whether a well or line-sink has
c       been reached.
        RTOL=1.1*RSTEP
        CZN=CZ
        CQN=CONJG(CW(CZN))
c       Move the pen to the first point of the streamline.
        CALL DRAW(CZN,0)
  100   CZO=CZN
        CQO=CQN
c       Proceed only if the point is not at a well, not at a line-sink,
c       and inside the window.
        IF(.NOT.LFATWL(CZO,RTOL).AND..NOT.LFATLS(CZO,RTOL)
       .         .AND.LFINWN(CZO)) THEN
c          Abort if a stagnation point is encountered.
```

```
            IF(CQO.EQ.(0.,0.)) THEN
               WRITE(*,*) 'DISCHARGE ZERO: STREAMLINE TERMINATED'
            ELSE
c              Determine the complex coordinate of the next point on the
c              discharge streamline, using (26.7).
               CZN=CZO+RSTEP*CQO/CABS(CQO)
c              Compute the complex discharge Qx+iQy at this point.
               CQN=CONJG(CW(CZN))
c              Estimate the average value of the complex discharge on the
c              interval.
               CQ=.5*(CQO+CQN)
c              Obtain the second approximation of the point on the
c              streamline using (26.8)
               CZN=CZO+RSTEP*CQ/CABS(CQ)
c              Plot the point.
               CALL DRAW(CZN,1)
c              Repeat the procedure for the next point.
               GOTO 100
            ENDIF
         ENDIF
         RETURN
         END
```

Problem 26.2

Use SLWL to determine streamlines for the case shown in Figure 3.8. First fill a grid with values of the stream function, then enter the command TRACE and obtain a contour plot. Second, start streamlines at points to the extreme left of each streamline produced by contouring. Determine the step size (as a fraction of the width of your window) for which you judge that the errors become unacceptable.

Problem 26.3

Modify the routine TRACE such that iterations are continued until either the error is within a predefined range, or a specified maximum number of iterations is surpassed. Repeat the numerical experiment of Problem 26.2.

27 THREE-DIMENSIONAL STREAMLINES AND PATH LINES IN DUPUIT-FORCHHEIMER MODELS

As stated in Sec. 7, the Dupuit-Forchheimer assumption consists of neglecting the resistance to flow in the vertical direction. We noted that the vertical component of flow is not neglected but rather that it is not associated with any head loss in the vertical direction. In this section, we will apply an analysis, which is a generalization of Strack [1984], to determine streamlines and path lines in three-dimensional space using Dupuit-Forchheimer models. This section consists of three parts. The first part is concerned with steady shallow flow in homogeneous aquifers, the second part with layered aquifers, and the third part with transient flow.

Steady Shallow Flow

Consider a case of shallow one-dimensional flow in a semi-infinite aquifer of constant thickness H, where infiltration occurs at a constant rate N along the horizontal upper boundary of the aquifer (see Figure 4.2). We employ a Cartesian x, x_3 coordinate system as shown; we use x_3 to denote the vertical coordinate rather than z, in order to avoid confusion with the complex variable $z = x + iy$. The potential $\Phi = kH\phi$ fulfills the differential equation (compare [9.4])

$$\frac{d^2\Phi}{dx^2} = -N \qquad (27.1)$$

with the solution

$$\Phi = -\tfrac{1}{2}Nx^2 + Ax + B \qquad (27.2)$$

The discharge vector component Q_x equals

$$Q_x = Hq_x = -\frac{\partial \Phi}{\partial x} = Nx - A \qquad (27.3)$$

The boundary $x = 0$ is impermeable, so that $Q_x = 0$ at $x = 0$; A must be zero. If the potential at $x = 0$ is denoted as Φ_0, then (27.2) becomes

$$\Phi = -\tfrac{1}{2}Nx^2 + \Phi_0 \qquad (27.4)$$

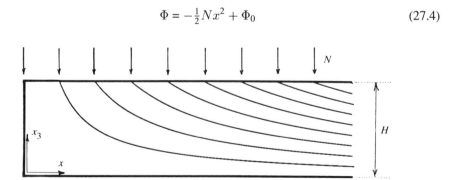

Figure 4.2 Shallow flow with constant infiltration.

The head ϕ is constant over the height of the aquifer because the resistance to flow in the vertical direction is neglected. Therefore q_x is constant over the height of

Sec. 27 3-D Streamlines and Path Lines in Dupuit-Forchheimer Models

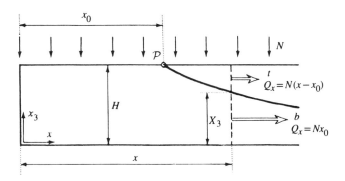

Figure 4.3 The path of a fluid particle.

the aquifer, an observation which makes it possible to determine the elevation of a fluid particle that entered the aquifer at a point $\mathcal{P}$ (x_0, H), see Figure 4.3.

We define the discharge $\overset{t}{Q}_x$ at any point x as the amount of water flowing between the streamline and the upper boundary. This amount equals the infiltration over the distance $x - x_0$,

$$\overset{t}{Q}_x = N(x - x_0) \tag{27.5}$$

The total discharge Q_x at x equals Nx and the discharge $\overset{b}{Q}_x$ flowing between the base and the streamline equals $Q_x - \overset{t}{Q}_x$,

$$Q_x = Nx \qquad \overset{b}{Q}_x = Nx_0 \tag{27.6}$$

Because q_x is constant over the height of the aquifer at x, the elevation X_3 of the streamline above the base is found from

$$\frac{X_3}{H} = \frac{q_x X_3}{q_x H} = \frac{\overset{b}{Q}_x}{Q_x} = \frac{x_0}{x} \tag{27.7}$$

This solution corresponds to the exact solution of the problem sketched in Figure 4.2, as will be shown below. We define the complex variable z as

$$z = x + ix_3 \tag{27.8}$$

and examine the following complex potential,

$$\Omega = -\frac{NB}{2H}z^2 + \Phi_0 \tag{27.9}$$

where $\Phi = \Re\Omega$ is the potential for two-dimensional flow in the vertical plane,

$$\Phi = kB\phi \tag{27.10}$$

and B is the width normal to the plane of flow. It follows from (27.9) that

$$W = Q_x - iQ_3 = Bq_x - iBq_3 = -\frac{d\Omega}{dz} = \frac{NB}{H}z \qquad (27.11)$$

where the index 3 refers to the vertical coordinate. It follows that

$$q_x = \frac{N}{H}x \qquad q_3 = -\frac{N}{H}x_3 \qquad (27.12)$$

and $q_x = 0$ along $x = 0$, $q_3 = 0$ along $x_3 = 0$, and $q_3 = -N$ along $x_3 = H$, so that the boundary conditions are met. The stream function Ψ equals

$$\Psi = -\frac{NB}{H}xx_3 \qquad (27.13)$$

and the value of Ψ along the streamline through $x = x_0$, $x_3 = H$ equals Ψ_0 with

$$\Psi_0 = -NBx_0 \qquad (27.14)$$

We obtain the equation for the streamline $\Psi = \Psi_0$ from (27.13) and (27.14) as follows, replacing x_3 by X_3 to indicate that we consider a particular streamline:

$$-NBx_0 = -\frac{NB}{H}xX_3 \qquad (27.15)$$

or

$$\frac{X_3}{H} = \frac{x_0}{x} \qquad (27.16)$$

which corresponds exactly to (27.7). It is important to note that the Dupuit-Forchheimer model gives the exact equation for the streamline only if both N and H are constants. The interested reader is referred to Strack [1984] for validations in cases where N varies with position.

The derivation given above for confined flow may be applied equally well if the thickness of the aquifer equals $h(x)$, and varies with position such as is the case for shallow unconfined flow. For such cases the thickness H in (27.7) must be replaced by h,

$$\frac{X_3}{h} = \frac{\overset{b}{Q_x}}{Q_x} = \frac{x_0}{x} \qquad (27.17)$$

The Equation for a Streamline

The above analysis is valid only for the case that the potential for shallow flow is a function of x alone; we will next generalize the approach to cases where this potential is a function of the two coordinates x and y defined in the horizontal plane. We do this for the general case that the aquifer thickness h is a function of position, and that there

Sec. 27 3-D Streamlines and Path Lines in Dupuit-Forchheimer Models

are infiltration rates $\overset{t}{N}$ and $\overset{b}{N}$ along the top and bottom boundaries (see Figure 4.4). We define the total infiltration rate N as

$$N = \overset{t}{N} + \overset{b}{N} \tag{27.18}$$

and assume that the solution to the flow problem is known so that the components Q_x and Q_y of the discharge vector are given in terms of x and y. The discharge vector fulfills the equation of continuity,

$$\frac{\partial Q_x}{\partial x} + \frac{\partial Q_y}{\partial y} = N = \overset{t}{N} + \overset{b}{N} \tag{27.19}$$

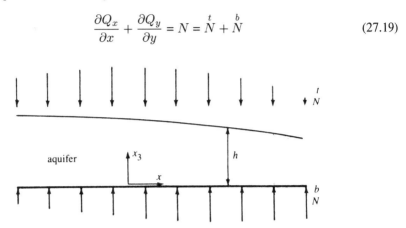

Figure 4.4 An unconfined aquifer with infiltration through the upper and lower boundaries.

The discharge vector $(\overset{b}{Q}_x, \overset{b}{Q}_y)$ corresponds to the flow below the streamline under consideration; the components of this vector fulfill the following equation of continuity:

$$\frac{\partial \overset{b}{Q}_x}{\partial x} + \frac{\partial \overset{b}{Q}_x}{\partial y} = \overset{b}{N} \tag{27.20}$$

The discharge vectors (Q_x, Q_y) and $(\overset{b}{Q}_x, \overset{b}{Q}_y)$ are collinear at any point x, y: we may introduce a scalar quantity μ such that

$$\overset{b}{Q}_x = \mu Q_x \qquad \overset{b}{Q}_y = \mu Q_y \tag{27.21}$$

Using (27.21) with (27.20) we obtain

$$\frac{\partial(\mu Q_x)}{\partial x} + \frac{\partial(\mu Q_y)}{\partial y} = Q_x \frac{\partial \mu}{\partial x} + Q_y \frac{\partial \mu}{\partial y} + \mu \left[\frac{\partial Q_x}{\partial x} + \frac{\partial Q_y}{\partial y} \right] = \overset{b}{N} \tag{27.22}$$

We introduced the term *discharge streamline* in Sec. 6 to denote the curve in the x, y plane that is tangent at each point to the discharge vector. We will use this term whenever confusion may arise with the streamlines in three-dimensional space. The arc length s

is measured along the discharge streamline and Q_s is the magnitude of the discharge vector, which points in the direction s. We observe from Figure 4.5 that

$$Q_x = Q_s \cos \alpha = Q_s \frac{dx}{ds} \qquad Q_y = Q_s \sin \alpha = Q_s \frac{dy}{ds} \qquad (27.23)$$

Combining this result with (27.22) and using (27.19), we obtain

$$Q_s \left[\frac{\partial \mu}{\partial x} \frac{dx}{ds} + \frac{\partial \mu}{\partial y} \frac{dy}{ds} \right] + \mu N = \overset{b}{N} \qquad (27.24)$$

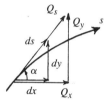

Figure 4.5 A discharge streamline.

The expression between the brackets represents the derivative of μ along s, and (27.24) may be simplified to

$$\boxed{Q_s \frac{d\mu}{ds} + \mu N = \overset{b}{N}} \qquad (27.25)$$

This is an ordinary differential equation in terms of μ, and has the following solution:

$$\mu(s) = e^{-f(s)} \left[\mu(s_0) + \int_{s_0}^{s} \frac{\overset{b}{N}}{Q_s} e^{f(s)} ds \right] \qquad (Q_s \neq 0)$$

$$\mu(s) = \frac{\overset{b}{N}}{N} \qquad (Q_s = 0) \qquad (27.26)$$

where s_0 is the value of s at the starting point of the streamline, $\mu(s_0)$ is the value of μ at s_0, and $f(s)$ is defined as

$$f(s) = \int_{s_0}^{s} \frac{N}{Q_s} ds \qquad (27.27)$$

The expression for X_3, the elevation of the streamline above the base of the aquifer, is obtained by noticing that uniformity of the specific discharge over the height of the aquifer implies that

$$\boxed{\frac{X_3}{h} = \frac{\overset{b}{Q_s}}{Q_s} = \mu}$$
(27.28)

where use is made of (27.21). The final result is obtained by combining (27.26) and (27.28):

$$\frac{X_3}{h} = e^{-f(s)}\left[\frac{X_3(s_0)}{h(s_0)} + \int_{s_0}^{s}\frac{\overset{b}{N}}{Q_s}e^{f(s)}ds\right] \qquad (Q_s \ne 0)$$
$$\frac{X_3}{h} = \frac{\overset{b}{N}}{N} \qquad (Q_s = 0)$$
(27.29)

The above expression for X_3/h is applicable also to cases where the base of the flow domain is curved, as in problems of interface flow. It is important to note, however, that for such problems X_3 is measured from the interface, as shown in Figure 4.6.

If the streamlines are determined numerically, then μ can be computed by numerical integration of (27.25). The integration of (27.29), then, is computationally less efficient.

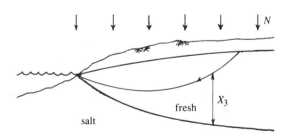

Figure 4.6 Definition of X_3 for interface flow.

Problem 27.1
Derive (27.17) by the use of (27.29).

The Component q_3

An expression for the vertical component q_3 of the specific discharge vector may be obtained as follows. For simplicity, we restrict the following analysis to cases where the base of the aquifer is horizontal. The derivative of X_3 with respect to s then yields the inclination of the streamline to the horizontal plane. Differentiation of (27.28) with respect of s yields, after multiplying both sides by h:

$$\frac{dX_3}{ds} = \frac{d}{ds}(\mu h) = \mu \frac{dh}{ds} + h\frac{d\mu}{ds} \qquad (27.30)$$

It follows from the definition of a streamline, and from (27.25), that

$$\frac{dX_3}{ds} = \frac{q_3}{q_s} = \frac{hq_3}{Q_s} = \mu \frac{dh}{ds} + h\frac{\overset{b}{N}}{Q_s} - h\mu\frac{N}{Q_s} \qquad (27.31)$$

We solve this equation for q_3 and use (27.28):

$$q_3 = \frac{X_3}{h}\left[\frac{Q_s}{h}\frac{dh}{ds} - N\right] + \overset{b}{N} \qquad (27.32)$$

We see that q_3 varies linearly with X_3: it equals $\overset{b}{N}$ along the base, and, if $h = H =$ constant, q_3 equals $\overset{b}{N} - N = -\overset{t}{N}$ for $X_3 = H$. The first term in (27.32) represents the effect on q_3 of the sloping upper boundary. It is important to remember that the above expressions for X_3 and q_3 are approximate because the resistance to flow in the vertical direction is neglected.

Leakage from a Circular Pond

We will discuss the problem of leakage of contaminated water through the bottom of a circular pond above an unconfined aquifer as an example. We assume that the contaminant does not affect the physical properties of the groundwater, and that the contamination has occurred for a long time. We refer to the area of the aquifer that is filled with contaminated water as the *plume*. The problem is illustrated in Figure 4.7, where both a plan view and a cross section through the center of the pond are given. Our objective is to determine, in three-dimensional space, the streamlines that start at points on the bottom of the pond. The origin of an x, y, x_3 coordinate system is at the center of the pond, and there is a uniform flow of discharge Q_{x0} in the x direction. The infiltration rate through the bottom of the pond is N and the radius of the pond is R. Using the function G_p introduced in Sec. 9 (see [9.72] and [9.73]), we may write the potential for this flow problem as follows:

$$\Phi = -Q_{x0}x + NG_p(x, y, 0, 0, R) + C \qquad (27.33)$$

where

$$G_p(x, y, 0, 0, R) = -\tfrac{1}{4}(x^2 + y^2 - R^2) \qquad (x^2 + y^2 \leq R^2) \qquad (27.34)$$

$$G_p(x, y, 0, 0, R) = -\frac{R^2}{4}\ln\frac{x^2 + y^2}{R^2} \qquad (x^2 + y^2 > R^2) \qquad (27.35)$$

Sec. 27 3-D Streamlines and Path Lines in Dupuit-Forchheimer Models

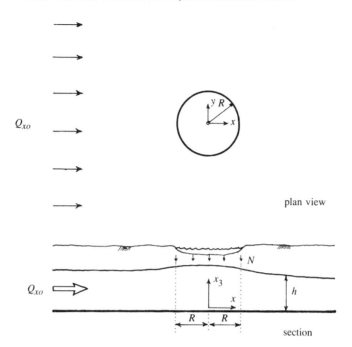

Figure 4.7 Leakage from a circular pond.

If Φ equals Φ_0 at $x = y = 0$, then the constant C is found to be $-NR^2/4 + \Phi_0$ and we obtain for $x^2 + y^2 \leq R$:

$$\Phi = -Q_{x0}x - \frac{N}{4}(x^2 + y^2) + \Phi_0 \qquad (x^2 + y^2 \leq R) \qquad (27.36)$$

Outside the pond the potential is harmonic; we may write the solution there in terms of a complex potential Ω of the complex variable $z = x + iy$. It follows from (27.33) and (27.35), with $C = -NR^2/4 + \Phi_0$, that

$$\Omega = -Q_{x0}z - \frac{NR^2}{2}\ln\frac{z}{R} - \frac{NR^2}{4} + \Phi_0 \qquad (x^2 + y^2 > R) \qquad (27.37)$$

Before examining the streamline pattern, we will determine whether stagnation points exist, and if so, where these are located.

The Stagnation Points. The stagnation points may be either below or outside the pond; we first examine whether a stagnation point exists outside the pond. The discharge function W is obtained from (27.37) upon differentiation and we obtain, denoting the location of the stagnation point as $\overset{o}{z}_s$, where the superscript o refers to outside,

$$Q_{x0} + \frac{NR^2}{2}\frac{1}{\overset{o}{z}_s} = 0 \qquad (|\overset{o}{z}_s| > R) \qquad (27.38)$$

or

$$\frac{\overset{o}{z}_s}{R} = -\frac{NR}{2Q_{x0}} \qquad (|\overset{o}{z}_s| > R) \qquad (27.39)$$

It follows that $\overset{0}{y}_s = 0$, and that the stagnation point exists outside the pond only if

$$\left|\frac{NR}{2Q_{x0}}\right| > 1 \qquad (27.40)$$

The expressions for Q_x and Q_y in the aquifer below the pond are obtained by differentiation of (27.36):

$$Q_x = Q_{x0} + \frac{N}{2}x \qquad (x^2 + y^2 < R^2) \qquad (27.41)$$

$$Q_y = \frac{N}{2}y \qquad (x^2 + y^2 < R^2) \qquad (27.42)$$

We denote the coordinates of the stagnation point in the aquifer below the pond as $\overset{i}{x}_s$ and $\overset{i}{y}_s$ and obtain the following by setting Q_x and Q_y equal to zero at $(\overset{i}{x}_s, \overset{i}{y}_s)$:

$$\frac{\overset{i}{x}_s}{R} = -\frac{2Q_{x0}}{NR} \qquad \overset{i}{y}_s = 0 \qquad (|\overset{i}{x}_s| < R) \qquad (27.43)$$

so that the stagnation point lies below the pond only if

$$\left|\frac{2Q_{x0}}{NR}\right| \leq 1 \qquad (27.44)$$

We observe from (27.40) and (27.44) that the following three cases may occur:

$$\text{case 1}: \quad \left|\frac{2Q_{x0}}{NR}\right| < 1 \quad \left\{\begin{array}{ll} \dfrac{\overset{o}{x}_s}{R} = -\dfrac{NR}{2Q_{x0}} & \overset{o}{y}_s = 0 \\[2mm] \dfrac{\overset{i}{x}_s}{R} = -\dfrac{2Q_{x0}}{NR} & \overset{i}{y}_s = 0 \end{array}\right\} \qquad (27.45)$$

$$\text{case 2}: \quad \left|\frac{2Q_{x0}}{NR}\right| = 1 \quad \left\{\begin{array}{ll} \dfrac{\overset{o}{x}_s}{R} = -1 & \overset{o}{y}_s = 0 \\[2mm] \dfrac{\overset{i}{x}_s}{R} = -1 & \overset{i}{y}_s = 0 \end{array}\right\} \qquad (27.46)$$

$$\text{case 3}: \quad \left|\frac{2Q_{x0}}{NR}\right| > 1 \qquad \text{no stagnation points exist} \qquad (27.47)$$

Vertical Stream Surfaces.
As a result of the Dupuit-Forchheimer assumption each streamline lies on a single vertical surface. The intersection of these vertical stream surfaces with a horizontal plane are the discharge streamlines, i.e., the streamlines for the discharge vector (Q_x, Q_y). The discharge streamlines outside the pond may be determined by the use of the stream function Ψ, which equals the imaginary part of (27.37),

$$\Psi = -Q_{x0}y - \frac{NR^2}{2}\theta \qquad (x^2 + y^2 > R^2) \tag{27.48}$$

where $\theta = \arg(z)$.

The stream function does not exist in the aquifer below the pond and the equations for the streamlines are obtained by integration of (19.5), which may be written as

$$Q_x dy - Q_y dx = 0 \tag{27.49}$$

Using (27.41) and (27.42), this becomes

$$\left(Q_{x0} + \frac{N}{2}x\right)dy - \frac{N}{2}y dx = 0 \tag{27.50}$$

or

$$\frac{dy}{y} = \frac{dx}{x + \frac{2Q_{x0}}{N}} \tag{27.51}$$

Integration gives

$$\ln y = \ln\left[C\left(x + \frac{2Q_{x0}}{N}\right)\right] \tag{27.52}$$

where C is a constant. The equations for the streamlines become, writing $-2Q_{x0}/N$ as x_s^*, and C as $\tan\beta$,

$$y = (x - x_s^*)\tan\beta \qquad (x^2 + y^2 \leq R^2) \tag{27.53}$$

where

$$x_s^* = -\frac{2Q_{x0}}{N} \tag{27.54}$$

and β is the angle enclosed between the discharge streamline and the x axis. If stagnation points exist, x_s^* is the x coordinate of the stagnation point below the pond; otherwise x_s^* has no physical meaning. We observe from (27.53) that all streamlines inside the pond lie on lines through the point $(x_s^*, 0)$.

Discharge streamlines are shown in Figure 4.8 for the cases that (a) stagnation occurs and (b) no stagnation points exist. The discharge streamlines that bound the plume, the dividing streamlines, may be identified in Figure 4.8 as follows. For case (a) they pass through S_2 and for case (b) they are the curves that touch the pond at the points T_1 and T_2. The data for the case of Figure 4.8(a) are $N/k = 0.001$, $Q_{x0}/(NR) = 0.4$, $\phi_0/R = 0.2$, and for the case of Figure 4.8(b) they are $N/k = 0.001$, $Q_{x0}/(NR) = 1$, $\phi_0/R = 0.2$.

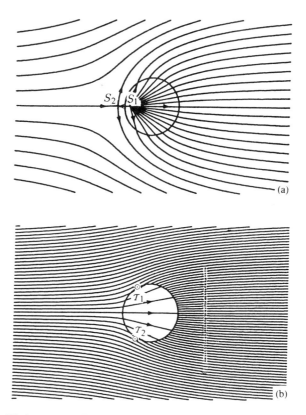

Figure 4.8 Discharge streamlines for case (a) with stagnation points, and case (b) without stagnation points. (Source: Strack [1984])

Streamlines. The equations for the streamlines in the vertical stream surfaces are obtained by the use of (27.29). We let s be the distance measured along each discharge streamline, with $s = 0$ at $x = x_s^*$, $y = 0$, and obtain from (27.53)

$$s = \frac{x - x_s^*}{\cos \beta} = \frac{y}{\sin \beta} \tag{27.55}$$

see Figure 4.9. We use (27.54) to express Q_x and Q_y, see (27.41) and (27.42), in terms of s and obtain

$$Q_x = \frac{N}{2}x + Q_{x0} = \frac{N}{2}(x - x_s^*) = \frac{N}{2}s\cos\beta \qquad (x^2 + y^2 \leq R^2) \tag{27.56}$$

$$Q_y = \frac{N}{2}y = \frac{N}{2}s\sin\beta \qquad (x^2 + y^2 \leq R^2) \tag{27.57}$$

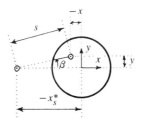

Figure 4.9 The definition of s.

We see that the component Q_s of the discharge vector along s equals

$$Q_s = \sqrt{Q_x^2 + Q_y^2} = \frac{N}{2}s \qquad (x^2 + y^2 \leq R^2) \tag{27.58}$$

For this case infiltration occurs only from above so that $\overset{b}{N} = 0$ and $N = \overset{t}{N}$, and (27.29) becomes

$$\frac{X_3(s)}{h(s)} = \frac{X_3(s_0)}{h(s_0)} e^{-f(s)} \qquad (x^2 + y^2 \leq R^2) \tag{27.59}$$

where

$$f(s) = \int_{s_0}^{s} \frac{N}{Q_s} ds = \int_{s_0}^{s} \frac{2ds}{s} = \ln\left[\frac{s}{s_0}\right]^2 \qquad (x^2 + y^2 \leq R^2) \tag{27.60}$$

Combining (27.59) and (27.60) we obtain

$$\frac{X_3(s)}{h(s)} = \frac{X_3(s_0)}{h(s_0)} \left[\frac{s_0}{s}\right]^2 \qquad (x^2 + y^2 \leq R^2) \tag{27.61}$$

Using this expression, we may calculate X_3 at any point of a streamline starting at any point s_0 with elevation $X_3(s_0)$ in the aquifer below the pond. Outside the pond, the infiltration rates $\overset{b}{N}$ and N are zero so that (27.29) becomes

$$\frac{X_3(s)}{h(s)} = \frac{X_3(\overset{p}{s_0})}{h(\overset{p}{s_0})} \qquad (x^2 + y^2 > R^2) \tag{27.62}$$

where $\overset{p}{s_0}$ is the value of s at the intersection of the discharge streamline with the boundary of the pond.

For the case of Figure 4.8(a), all discharge streamlines contained between the dividing discharge streamlines emanate from point $S_1(x_s^*, 0)$ inside the pond. The vertical surfaces through these discharge streamlines therefore each contain one streamline that starts at S_1. It follows from (27.57) that s is zero at S_1. Furthermore, all streamlines starting at $s = s_0 = 0$ lie on the aquifer base, since X_3/h in (27.61) vanishes for $s_0 = 0$.

Hence, for the case of Figure 4.8(a) the infiltrated water fills the entire space contained between the dividing discharge streamlines, the phreatic surface, and the aquifer base. The plot of streamlines in a vertical plane through the x axis reproduced in Figure 4.10(a) corresponds to the case of Figure 4.8(a). The plot extends from $x/R = -1.25$ to $x/R = 1.25$, and contains the two stagnation points, which lie at $x/R = -1.25$ and $x/R = -0.8$. The streamlines start at equal intervals between $x/R = -1$ and $x/R = 1$; there are 21 streamlines. A similar plot is shown in Figure 4.10(b), and corresponds to the case of Figure 4.8(b). For this case, stagnation does not occur, $x_s^*/R = -2$, and the bottom of the plume is the lowest curved streamline. The plot extends between $x/R = -1$ and $x/R = 1.5$. For both cases, the value of Φ_0 equals $0.02\ kR^2$. It may be noted that a discontinuity in slope of the streamlines occurs below the boundary of the pond. This discontinuity results from coupling the component q_3 directly to the infiltration rate, which is discontinuous across the boundary of the pond.

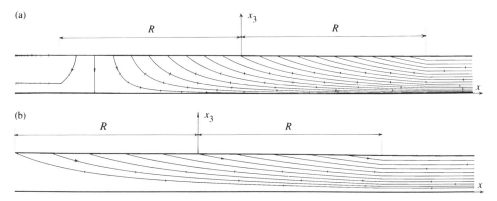

Figure 4.10 Streamlines in the x, x_3 plane. (Source: Strack [1984])

Only when a stagnation point exists outside the pond will the plume extend over the full height of the aquifer; otherwise the bottom of the plume is a curved surface above the base. A cross section through the plume parallel to the y, x_3 plane is shown in Figure 4.11. The location of this cross section is marked by the dotted line in Figure 4.8(b), and is at $x/R = x_1/R = 2$. Points of the curved bottom of the plume are determined as follows. First, the width of the plume at the given location is determined by intersecting the dividing discharge streamlines with the plane $x = x_1$. Second, each discharge streamline $\Psi = \Psi_j$ inside the plume is intersected with the downstream side of the circle, which yields a value for $\overset{p}{s}_0 = \overset{p}{s}_{0j}$ with the aid of (27.55). Third, the discharge streamline is traced below the pond until it intersects the circle on the upstream side, giving s_{0j}. Fourth, s_{0j} and $\overset{p}{s}_{0j}$ are substituted for s_0 and s, respectively, in (27.61) which yields the elevation X_{3j}/h_j of the bottom of the plume at $\overset{p}{s}_{0j}$. The elevation of the plume at the desired cross section is found, finally, as a function of the saturated thickness h by the use of (27.62). The above procedure is implemented in an elementary computer program, written merely for the purpose of illustration. This program is used to show, in Figure 4.12, how the cross section of the plume varies as a function of x_s^*/R.

When x_s^*/R equals -1, the plume is rectangular and fills the entire space between the dividing discharge streamlines. As the ratio $2Q_{x0}/(NR) = -x_s^*/R$ increases, the area of the plume decreases; the plume becomes both shallower and narrower. The data used for this case are the same as those for Figure 4.8(b), except that the saturated aquifer thickness was increased, $(\phi_0/R = 1)$ in order to make the curves more visible.

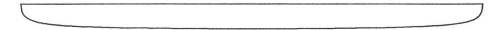

Figure 4.11 Cross section through the plume at $x/R = 2$. (Source: Strack [1984].)

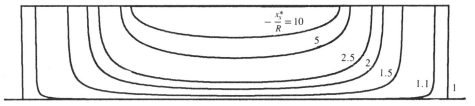

Figure 4.12 Cross sections through the plume for various values of $-x_s^*/R = 2Q_{x0}/NR$ at $X/R = 2$. (Source: Strack [1984].)

The plume attains the geometry described above after the pollution has occurred for a long time. It is possible to determine the front of the plume as it moves through the aquifer as a function of the time recorded since the pollution began. The marks shown on the streamlines in Figure 4.10 correspond to constant intervals of the dimensionless time $t^* = Nt/(nR)$, where n is the porosity. The front of the plume at any time, therefore, is obtained by connecting corresponding marks.

The elevation of streamlines may be determined numerically by including the integration of (27.25) as part of the procedure used for computing points on the discharge streamlines. Alternatively, equations of the form (26.11) and (26.12) may be applied in three dimensions, using the expression (27.32) for q_3. It should be noted that the technique is approximate and gives a simplified picture of the real streamline pattern. For example, the vertical faces of the plume shown in Figure 4.12 for $x_s^*/R = -1$ do not occur in reality, because both horizontal components of the specific discharge vector vary over the thickness of the aquifer near the bottom of the pond. Haitjema [1987] compares the results presented here with those obtained from a true three-dimensional analysis, and reports that the errors are small for ratios of R/H greater than 5. For cases where three-dimensional effects are too important to be neglected, a truly three-dimensional solution may be imbedded inside the Dupuit-Forchheimer model (see Haitjema [1985]).

For an application of the approach described in this section to a case of semiconfined flow, see Strack [1985].

Problem 27.2

For the case of Figure 4.8(b) determine:
1. the value of Ψ at point T_1
2. the elevation at $x = R$, $y = 0$ of the streamline that starts at the center of the pond.

Vertically Varying Hydraulic Conductivity

The technique discussed above may be applied to flow in systems of aquifers separated by leaky layers (see Sec. 15). Points on the streamlines may be computed in each aquifer, until the streamline intersects a leaky layer. The flow in the leaky layer is downward and the travel time over this vertical section of streamline is computed from the difference in the heads above and below the leaky layer and the hydraulic conductivity and thickness of the layer. The computations are then resumed in the next aquifer.

Another application is concerned with aquifers with vertically varying hydraulic conductivity as occurs, for example, if the aquifer is composed of layers of different hydraulic conductivities. If the contrasts in conductivity are not too large, such aquifers may be modeled as discussed in Sec. 11.

We will extend the procedure for determining the elevation of streamlines to include aquifers with a hydraulic conductivity that varies in the vertical direction. The differential equation (27.25) for μ and its solution (27.26) still apply, but μ no longer equals the ratio of the elevation of the streamline above the base to the aquifer thickness. We may express Q_s and $\overset{b}{Q}_s$ as follows (compare [11.4] and [11.7])

$$Q_s = -T\frac{d\phi}{ds} \tag{27.63}$$

$$\overset{b}{Q}_s = -\tau(X_3)\frac{d\phi}{ds} \tag{27.64}$$

where T is the transmissivity of the aquifer given by (11.6)

$$T = \int_{\overset{b}{X}_3}^{\overset{t}{X}_3} k(x_3)dx_3 \tag{27.65}$$

$\overset{b}{X}_3$ and $\overset{t}{X}_3$ stand for the values of x_3 at the lower and upper boundaries of the aquifer, respectively. The function τ in (27.64) is defined as

$$\boxed{\tau(X_3) = \int_{\overset{b}{X}_3}^{X_3} k(x_3)dx_3} \tag{27.66}$$

This function represents the transmissivity of a vertical section of an aquifer extending from the base to X_3. It follows from the definition of μ with (27.63) and (27.64) that

$$\mu = \frac{\overset{b}{Q}_s}{Q_s} = \frac{\tau(X_3)}{T} \tag{27.67}$$

Hence,

$$\tau(X_3) = \mu T \tag{27.68}$$

We know for physical reasons that the function τ given by (27.66) is single-valued. It can therefore be inverted:

$$X_3 = X_3(\tau) \tag{27.69}$$

The elevation X_3 of the streamline can be computed from (27.68) and (27.69) as

$$\boxed{X_3 = X_3(\mu T)} \tag{27.70}$$

As an example, consider the case of the layered aquifer discussed in Sec. 11. If X_3 corresponds to a point in layer m ($m = 1, 2, 3, \ldots, n$) then τ may be expressed as follows:

$$\tau = \overset{m-1}{T} + k_m(X_3 - b_m) \qquad (b_m \leq X_3 \leq b_{m+1}) \tag{27.71}$$

where the layers are numbered from the bottom up; k_m is the hydraulic conductivity of layer m, b_m is the elevation of the base of layer m, $\overset{m}{T}$ is defined for $1 \leq m \leq n$ by (11.27), and $\overset{0}{T}$ is zero:

$$\overset{m}{T} = \sum_{j=1}^{m} k_j(b_{j+1} - b_j) \qquad \overset{0}{T} = 0 \tag{27.72}$$

Inversion of (27.71) yields X_3 as a function of τ:

$$X_3 = b_m - \frac{\overset{m-1}{T}}{k_m} + \frac{\tau}{k_m} \qquad (b_m \leq X_3 \leq b_{m+1}) \tag{27.73}$$

Combination of (27.68) and (27.73) yields X_3 as a function of μ:

$$\boxed{X_3 = b_m - \frac{\overset{m-1}{T}}{k_m} + \mu \frac{T}{k_m} \qquad (b_m \leq X_3 \leq b_{m+1})} \tag{27.74}$$

where the transmissivity T of the aquifer is given by:

$$\begin{aligned} T &= \overset{m-1}{T} + k_m(\phi - b_m) & (b_m \leq \phi \leq b_{m+1}) & \quad (m = 1, 2, \ldots, n) \\ T &= \overset{n}{T} & (b_{n+1} \leq \phi) & \end{aligned} \tag{27.75}$$

where b_{n+1} is the elevation of the impermeable upper boundary of the aquifer.

Streamlines in a Layered Aquifer. Consider the aquifer consisting of two layers shown in Figure 4.13. The lower and upper layers are labeled as 1 and 2, respectively, and their hydraulic conductivities and thicknesses are k_1, H_1, k_2, and H_2. The boundary conditions are the same as for the case of Figure 4.2, and we choose the same x_3, x coordinate system. The expression for μ is the same as that obtained for the case of uniform hydraulic conductivity (compare [27.7])

$$\mu = \frac{Q_x^b}{Q_x} = \frac{x_0}{x} \qquad (27.76)$$

where x_0 is the x coordinate of a point on the upper boundary. It follows from (27.74) with $m = 2$ that

$$X_3 = b_2 - \frac{\frac{1}{T}}{k_2} + \frac{T}{k_2}\frac{x_0}{x} \qquad (b_2 \leq X_3 \leq b_3) \qquad (27.77)$$

and with $m = 1$ that

$$X_3 = b_1 + \frac{T}{k_1}\frac{x_0}{x} \qquad (b_1 \leq X_3 \leq b_2) \qquad (27.78)$$

where use is made of (27.72). Equations (27.77) and (27.78) define the streamlines as a function of position. With (compare [11.13] and [11.14])

$$\begin{array}{ccc} b_1 = 0 & b_2 = H_1 & b_3 = H_1 + H_2 \\ T_1 = k_1 H_1 & T_2 = k_2 H_2 & T = k_1 H_1 + k_2 H_2 \end{array} \qquad (27.79)$$

the equations for the streamlines may be written as

$$X_3 = H_1 - \frac{k_1 H_1}{k_2} + \frac{k_1 H_1 + k_2 H_2}{k_2}\frac{x_0}{x} \qquad (b_2 \leq X_3 \leq b_3) \qquad (27.80)$$

$$X_3 = \frac{k_1 H_1 + k_2 H_2}{k_1}\frac{x_0}{x} \qquad (b_1 \leq X_3 \leq b_2) \qquad (27.81)$$

The x coordinate of the point of intersection of the streamline with the interface between the layers, x_1, is obtained by substituting $b_2 = H_1$ for X_3 in either (27.80) or (27.81):

$$x_1 = \frac{k_1 H_1 + k_2 H_2}{k_1 H_1} x_0 \qquad (27.82)$$

It is of interest to note that equations (27.80) and (27.81) agree with the equations obtained from the exact solution to the problem (see Strack [1984]). Note that this is only true for uniform infiltration.

Problem 27.3

Determine expressions for the streamlines in a shallow unconfined aquifer with three layers, enclosed between two parallel rivers with heads $\phi_0 > H_1$. The distance between the rivers is L, the hydraulic conductivities of the three layers are k_1, k_2, and k_3, and the thicknesses of the lower two layers are H_1 and H_2. There is a uniform infiltration rate N.

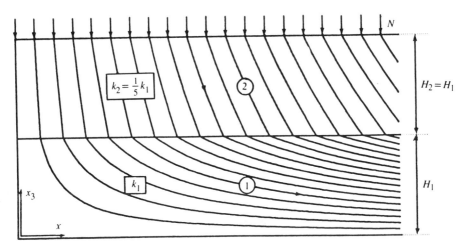

Figure 4.13 Streamlines in a layered aquifer. (Source: Strack [1984].)

Transient Shallow Flow

As the conclusion of this section, we will discuss a procedure for determining path lines in three dimensions. For the sake of brevity, we will make restriction to cases where the hydraulic conductivity is homogeneous. Path lines for problems of transient flow, then, may be determined either by computing μ by numerical integration of (27.25) and using the result in (27.28), or by using the procedure for determining path lines in three dimensions discussed in Sec. 26, with v_3 computed by dividing (27.32) by the porosity n. In either case, the arc length s refers to a discharge path line rather than to a discharge streamline and the values of all time-dependent variables ($Q_s, N, \overset{b}{N}$, and h) must be updated as the particle moves. It will be shown in the following that the infiltration rate N must be corrected for cases of transient shallow unconfined flow both in (27.25) and in (27.32) by including a term $(n - S_p)\partial \phi / \partial t$. This term thus needs to be included only if the phreatic storage coefficient S_p differs from the porosity n.

If s is measured along a path line then the term dh/ds in (27.32) represents the change of the saturated thickness of the aquifer with respect to the infinitesimal distance ds traveled by a particle during an infinitesimal time increment dt.

We write the first term inside the brackets in (27.32) as

$$\frac{Q_s}{h}\frac{dh}{ds} = nv\frac{dh}{ds} = n\frac{dh}{ds}\frac{ds}{dt} = n\frac{Dh}{Dt} \tag{27.83}$$

Dh/Dt is called the material-time derivative; it represents the rate of change of h with time, moving along the path line with the particle. The material-time derivative may be written as

$$\frac{Dh}{Dt} = \frac{\partial h}{\partial t} + \frac{\partial h}{\partial x}\frac{dx}{dt} + \frac{\partial h}{\partial y}\frac{dy}{dt} = \frac{\partial h}{\partial t} + \frac{\partial h}{\partial x}v_x + \frac{\partial h}{\partial y}v_y \tag{27.84}$$

where x and y are the coordinates of the position of the particle in the horizontal plane. Multiplication of both sides of (27.84) by n gives

$$n\frac{Dh}{Dt} = n\frac{\partial h}{\partial t} + q_x\frac{\partial h}{\partial x} + q_y\frac{\partial h}{\partial y} \qquad (27.85)$$

It follows from (27.32), (27.83), and (27.85) that

$$q_3 = \frac{x_3}{h}\left[n\frac{\partial h}{\partial t} + q_x\frac{\partial h}{\partial x} + q_y\frac{\partial h}{\partial y} - N\right] + \overset{b}{N} \qquad (27.86)$$

where X_3 is replaced by x_3 because this equation does not contain any variables defined in terms of a particular streamline, as opposed to (27.32) which contains s. The latter equation is valid for transient flow, provided that N is corrected. We will demonstrate this by deriving an expression for q_3 from the continuity equation in two dimensions in terms of the discharge vector, in combination with the continuity equation in three dimensions in terms of the specific discharge vector.

Consider the general case of transient unconfined flow in a compressible aquifer. The continuity equation in terms of the discharge vector may then be written as

$$\frac{\partial Q_x}{\partial x} + \frac{\partial Q_y}{\partial y} = -\nabla^2\Phi = E + N - S_p\frac{\partial \phi}{\partial t} \qquad (27.87)$$

This equation is a combination of (16.2), which applies to a compressible aquifer and (16.15), which takes phreatic storage and rainfall into account. The function $E = N^*$ represents the rate of change in elastic storage of the aquifer per unit volume, ϵ, integrated over the aquifer thickness (compare[16.1]). Since ϵ is assumed to be constant over the height of the aquifer, we may write E as

$$E = h\epsilon \qquad (27.88)$$

The right-hand side of the continuity equation in three dimensions, (4.3), is now equal to ϵ as a result of the elastic storage:

$$\frac{\partial q_x}{\partial x} + \frac{\partial q_y}{\partial y} + \frac{\partial q_3}{\partial x_3} = \epsilon \qquad (27.89)$$

With $(Q_x, Q_y) = h(q_x, q_y)$, (27.87) may be written as

$$h\left[\frac{\partial q_x}{\partial x} + \frac{\partial q_y}{\partial y}\right] + q_x\frac{\partial h}{\partial x} + q_y\frac{\partial h}{\partial y} = h\epsilon + N - S_p\frac{\partial \phi}{\partial t} \qquad (27.90)$$

We solve this equation for $\partial q_x/\partial x + \partial q_y/\partial y$ and substitute the result in (27.89):

$$\frac{\partial q_3}{\partial x_3} = -\epsilon + \frac{1}{h}\left[S_p\frac{\partial \phi}{\partial t} + q_x\frac{\partial h}{\partial x} + q_y\frac{\partial h}{\partial y} - N\right] + \epsilon \qquad (27.91)$$

We observe that the terms containing the rate of change in elastic storage cancel, and that the right-hand side of (27.91) is independent of x_3. Integration with respect to x_3 yields

$$q_3 = \frac{x_3}{h}\left[S_p\frac{\partial \phi}{\partial t} + q_x\frac{\partial h}{\partial x} + q_y\frac{\partial h}{\partial y} - N\right] + C \tag{27.92}$$

The constant of integration, C, is determined from the condition that $q_3 = \overset{b}{N}$ for $x_3 = 0$. Furthermore, $\phi = h$, and we obtain

$$\boxed{q_3 = \frac{x_3}{h}\left[S_p\frac{\partial h}{\partial t} + q_x\frac{\partial h}{\partial x} + q_y\frac{\partial h}{\partial y} - N\right] + \overset{b}{N}} \tag{27.93}$$

This equation is equal to (27.86) only if the phreatic storage coefficient S_p is equal to the porosity n. We write (27.93) as follows for the case that $S_p \neq n$:

$$q_3 = \left\{\frac{x_3}{h}\left[n\frac{\partial h}{\partial t} + q_x\frac{\partial h}{\partial x} + q_y\frac{\partial h}{\partial y} - N\right] + \overset{b}{N}\right\} - \frac{x_3}{h}N_p \tag{27.94}$$

where

$$\boxed{N_p = (n - S_p)\frac{\partial h}{\partial t}} \tag{27.95}$$

The term inside the braces in (27.94) is identical to the expression (27.86) for q_3. It follows that the infiltration rate N in (27.25), (27.32), and (27.86) must be corrected by adding N_p to it if these equations are to be used to determine the elevations of path lines for transient flow.

A physical interpretation of this somewhat surprising result follows. The phreatic storage coefficient S_p will be less than n if the soil above the water table is not entirely dry but contains water. This water is released into the aquifer when the water table rises and the soil becomes fully saturated. By the same reasoning, water is left behind when the water table lowers. This amount of water is represented by the term N_p given by (27.95). Consider the case that there is no infiltration into the aquifer, i.e., $N = N_p$ in (27.25) and (27.32). If $S_p = n$, the path line of any particle at the phreatic surface will follow the rising and lowering of this surface: the particle remains at the phreatic surface. Next, consider the case that $S_p < n$ and the phreatic surface rises from $h(x,y,t)$ to $h(x,y,t+\Delta t) = h(x,y,t) + \Delta h$ over a time interval Δt at a rate $\partial h/\partial t$. The volume $\Delta V = \Delta x \Delta y \Delta h$ is unsaturated at time t and contains an amount $(n - S_p)\Delta V$ of immobile water. At time $t + \Delta t$, this water is released as the volume ΔV becomes saturated. This is in accordance with expression (27.95): there is inflow at a rate N_p, where $N_p > 0$ because $\partial h/\partial t > 0$. Thus, particles at the phreatic surface will move downward with respect to the moving phreatic surface as a result of the correction term N_p. Similarly, particles at the phreatic surface will move out of the saturated zone if the phreatic surface lowers and $N_p < 0$ because $\partial h/\partial t < 0$. In either case the path lines intersect the locus of points on the phreatic surface on the vertical through the moving particle; this locus represents the transient position of the phreatic surface with respect to the particle.

Although the mechanism described here is a severe simplification of the complicated process occurring at the interface between the saturated and unsaturated zones, it highlights an effect that is important for problems of contaminant transport: if an aquifer is contaminated and the phreatic surface lowers for a period of time, then contaminated water is left behind in the unsaturated zone. This contaminated water will be released again when the water table rises.

If streamlines are considered rather than path lines, then the term dh/ds in (27.32) does not contain the change of h with respect to time because the streamline applies to a given fixed time. For such cases a term $-S_p \partial \phi/\partial t$ must be included as part of the rate of infiltration N in (27.25) and (27.32). Equation (27.93) is valid in either case.

28 SOLUTE TRANSPORT

Solute transport is governed by a number of different phenomena: advection, chemical reactions, diffusion, and dispersion. Transport of solutes may thus be subdivided into different components: an advective component, a chemical component, a diffusive component, and a dispersive component. In this text, we will concentrate on the first two: advective transport and the chemical interaction between water and solid particles.

Simplifying Assumptions

The following analysis is based on three simplifying assumptions:

1. The solutes are present in the flow only in small concentrations and do not influence the physical properties of the groundwater.
2. The porosity and hydraulic conductivity are not affected by the solutes.
3. The chemical interaction between solutes and solid particles is described fully by instantaneous linear sorption.

As a result of assumptions (1) and (2) the flow and the transport are independent of one another: the transport problem can be solved *after* determining the flow field.

Solute Movement

Advective transport is the main component of solute transport in most applications. Advection is the movement of solutes with the fluid, with the dissolved particles behaving as water particles.

If advection is the only form of transport, then water with a concentration c_0 moves as a plug into the aquifer if injected from some time t_0 onward, over a given area A (see

Sec. 28 Solute Transport

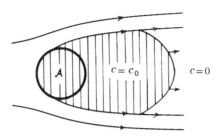

Figure 4.14 Plug flow.

Figure 4.14). This plug is bounded on the sides by path lines, and its front moves with the velocity of the groundwater.

In reality, the front moves slower than the groundwater. This is the result of chemical reactions that take place between solutes and solid particles; the reduction in velocity is called retardation.

The front is usually not sharp: there is a gradual transition in concentration from $c = c_0$ to $c = 0$. This may be caused by decay such as in radioactive materials or by another effect, called hydrodynamic dispersion.

Hydrodynamic dispersion is caused by inhomogeneities in the porous material, and results from differences in velocity of neighboring water particles. Dispersion caused by the inherent inhomogeneous nature of the porous medium on the particle scale is called microscopic dispersion; the effect of microscopic dispersion in most porous media is negligible, but is important in fractured rock. Dispersion caused by inhomogeneities on a larger scale, such as gravel or clay pockets, and, in particular, layering, is much more important and is called macroscopic dispersion. Dispersion has a smoothing effect on both the front and the sides of the plug that would exist if there were no dispersion. It is important to note that the concentration outside this plug will now be non-zero, since some particles will move faster than average. As stated above, the effect of hydrodynamic dispersion will not be included here.

Transport Equation

We will derive the transport equation in three dimensions in terms of a Cartesian coordinate system x, y, x_3, adhering to the notation used in the preceding section. Consider an elementary volume $\Delta \mathcal{V} = \Delta x \Delta y \Delta x_3$ (see Figure 4.15), with flow occurring only in the x direction. We introduce the symbols $c(x, y, x_3, t)$ [kg/m^3] and $c_s(x, y, x_3, t)$ [kg/m^3], representing the concentrations of the mobile phase and the sorbed phase of the solute, respectively. These concentrations are expressed in terms of mass per unit volume of fluid:

c = mass of mobile phase per unit volume of fluid

c_s = mass of sorbed phase per unit volume of fluid

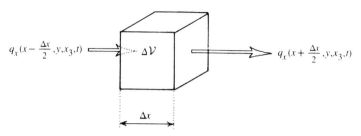

Figure 4.15 Mass balance.

If the rate of decay is proportional to the total mass of solute present in ΔV with a proportionality constant λ, and if the porosity is n, then the mass balance for flow in the x direction may be written as

$$\begin{aligned}
&-[c(x + \tfrac{\Delta x}{2}, y, x_3, t)\, q_x(x + \tfrac{\Delta x}{2}, y, x_3, t) \\
&\quad - c(x - \tfrac{\Delta x}{2}, y, x_3, t)\, q_x(x - \tfrac{\Delta x}{2}, y, x_3, t)]\Delta y \Delta x_3 \Delta t \\
&- [c(x, y, x_3, t) + c_s(x, y, x_3, t)]\lambda n \Delta x \Delta y \Delta x_3 \Delta t \\
&= [c(x, y, x_3, t + \tfrac{\Delta t}{2}) + c_s(x, y, x_3, t + \tfrac{\Delta t}{2}) \\
&\quad - c(x, y, x_3, t - \tfrac{\Delta t}{2}) - c_s(x, y, x_3, t - \tfrac{\Delta t}{2})](n \Delta x \Delta y \Delta x_3)
\end{aligned} \qquad (28.1)$$

The terms in the first two lines above the equal sign represent the net inflow of solute into ΔV due to advection, the remaining terms above the equal sign represent minus the decrease in mass due to decay, and the terms below the equal sign represent the increase in mass. By dividing both sides by $\Delta x \Delta y \Delta x_3 \Delta t$, and passing to the limit for $\Delta x \to 0$, $\Delta y \to 0$, $\Delta x_3 \to 0$, and $\Delta t \to 0$ we obtain:

$$-\frac{\partial}{\partial x}(cq_x) - n\lambda(c + c_s) = n\frac{\partial}{\partial t}(c + c_s) \qquad (28.2)$$

If there are components of flow in all three directions, this becomes

$$-\left[\frac{\partial}{\partial x}(cq_x) + \frac{\partial}{\partial y}(cq_y) + \frac{\partial}{\partial x_3}(cq_3)\right] - n\lambda(c + c_s) = n\frac{\partial}{\partial t}(c + c_s) \qquad (28.3)$$

The assumption of instantaneous linear sorption (see for example Javandel et al. [1984]) implies that

$$c = \frac{1}{R}(c + c_s) \qquad (28.4)$$

where R is a dimensionless constant called the retardation factor. If R is independent of time, then

$$\frac{\partial}{\partial t}(c + c_s) = R\frac{\partial c}{\partial t} \qquad (28.5)$$

Sec. 28 Solute Transport

so that (28.3) may be written as

$$nR\frac{\partial c}{\partial t} + \left[\frac{\partial}{\partial x}(cq_x) + \frac{\partial}{\partial y}(cq_y) + \frac{\partial}{\partial x_3}(cq_3)\right] = -n\lambda Rc \qquad (28.6)$$

Differentiating the terms within the brackets by parts and dividing both sides by a factor nR, this becomes

$$\frac{\partial c}{\partial t} + \frac{1}{nR}\left[q_x\frac{\partial c}{\partial x} + q_y\frac{\partial c}{\partial y} + q_3\frac{\partial c}{\partial x_3} + c\left(\frac{\partial q_x}{\partial x} + \frac{\partial q_y}{\partial y} + \frac{\partial q_3}{\partial x_3}\right)\right] = -\lambda c \qquad (28.7)$$

For the sake of simplicity we make restriction to incompressible flow. According to (4.3), the term in parentheses, then, vanishes and the velocity components v_x, v_y, and v_3 may be introduced:

$$\frac{\partial c}{\partial t} + \frac{1}{R}\left[v_x\frac{\partial c}{\partial x} + v_y\frac{\partial c}{\partial y} + v_3\frac{\partial c}{\partial x_3}\right] = -\lambda c \qquad (28.8)$$

This equation is the transport equation for solute transport with retardation and decay in an incompressible porous medium, with the effect of hydrodynamic dispersion neglected. The transport equation, then, is a first order partial differential equation, which can best be integrated along its characteristics.

Integration Along Characteristics

We introduce the apparent velocity with components (v_x^*, v_y^*, v_3^*) as

$$v_x^* = \frac{v_x}{R} \qquad v_y^* = \frac{v_y}{R} \qquad v_3^* = \frac{v_3}{R} \qquad (28.9)$$

If x, y, and x_3 represent the coordinates of a point moving with the apparent velocity, then

$$v_x^* = \frac{dx}{dt} \qquad v_y^* = \frac{dy}{dt} \qquad v_3^* = \frac{dx_3}{dt} \qquad (28.10)$$

Integration of these differential equations yields a curve that represents the path of a point moving with the apparent velocity; we will call this curve the *apparent path line*. The differential equation simplifies if (28.9) and (28.10) are combined with (28.8):

$$\frac{\partial c}{\partial t} + \frac{\partial c}{\partial x}\frac{dx}{dt} + \frac{\partial c}{\partial y}\frac{dy}{dt} + \frac{\partial c}{\partial x_3}\frac{dx_3}{dt} = -\lambda c \qquad (28.11)$$

The left-hand side of this equation is the total derivative of c with respect to time along the apparent path line:

$$\frac{Dc}{Dt} = -\lambda c \qquad (28.12)$$

The term Dc/Dt is the material-time derivative of the concentration c. The apparent path line is a special curve: we observe from (28.12) that the differential equation gives information only regarding the change of the concentration along this curve. Such a curve is called the *characteristic* of the differential equation.

To illustrate the meaning of the material-time derivative Dc/Dt, we introduce the distance s traveled along the apparent path line and represent the magnitude of the apparent velocity as v^*. The differential equation

$$v^* = \frac{ds}{dt} \tag{28.13}$$

defines a curve in the s, t plane that represents the characteristic of the differential equation. This characteristic passes through the point (s_0, t_0) corresponding to the position s_0 that the point moving along the apparent path line occupies at time t_0. The slope of the curve represents the apparent velocity v^*.

The material-time derivative Dc/Dt represents the change in concentration with respect to time, moving along the apparent path line with the apparent velocity v^*. Thus, Dc/Dt is the limit, for $\Delta t \to 0$, of $\Delta c / \Delta t$ where Δc is the difference in concentration between points $\mathcal{A}$ and $\mathcal{B}$ on the path divided by the projection Δt of the segment $\mathcal{AB}$ on the t axis (see Figure 4.16). Note that the values of s at points $\mathcal{A}$ and $\mathcal{B}$ are not the same; Δs represents the distance moved with velocity v^* along the path over the time interval Δt. The differential equation (28.12) can be integrated:

$$c = c_0 e^{-\lambda(t-t_0)} \tag{28.14}$$

where c_0 is the concentration at time $t = t_0$ and position $s = s_0$:

$$c_0 = c(t_0) \tag{28.15}$$

In the absence of decay, i.e., if $\lambda = 0$, (28.14) reduces to

$$c = c_0 \tag{28.16}$$

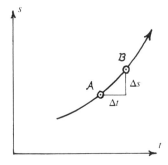

Figure 4.16 The characteristic.

which expresses that the concentration remains unchanged when moving along the apparent path line with the apparent velocity v^*: the front moves with this velocity and there is plug flow.

The retardation factor R is usually expressed in terms of yet another factor, called the distribution factor K_d.

The Distribution Factor

The distribution factor K_d is defined as

$$K_d = \frac{S}{c} \tag{28.17}$$

where S is defined as the ratio of the mass of sorbed material to the mass of solid material, both contained in some volume V of porous material. If ρ_b [kg/m^3] is the bulk density of solid material, then

$$S = \frac{c_s n V}{\rho_b V} = n \frac{c_s}{\rho_b} \tag{28.18}$$

The factor n occurs in the numerator of the fraction because c_s is defined as mass per unit volume of fluid. Equation (28.17) may be combined with (28.18):

$$K_d = \frac{n}{\rho_b} \frac{c_s}{c} \tag{28.19}$$

Using (28.4), it follows that R may be expressed in terms of K_d as

$$R = 1 + \frac{c_s}{c} = 1 + \frac{\rho_b K_d}{n} \tag{28.20}$$

The units of the distribution factor are [m^3/kg]. The distribution factor is the quantity associated with retardation that is usually measured in the laboratory.

Contaminant Transport in a Layered Aquifer

As an example, consider the case of steady flow in a layered aquifer solved in Sec. 27 and shown in Figure 4.13. We assume that contaminated water of concentration c_0 enters the aquifer over a section $a \leq x \leq a + b$, $y = H_1 + H_2$ of the upper boundary, from time $t = t_0$ onward (see Figure 4.17). The retardation factor is R. The question is to determine the time at which a point $\mathcal{P}$ on the front of the plume moving along the apparent path line defined by the initial condition

$$x = x_0 \qquad y = H_1 + H_2 \qquad t = t_0 \tag{28.21}$$

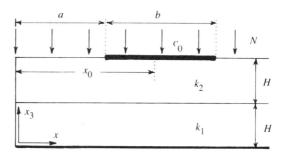

Figure 4.17 Contaminant transport in a layered aquifer.

reaches the plane $x = L$.

In solving this problem we note that the apparent path lines coincide with the streamlines, because the flow is steady. For simplicity we set the thicknesses of the layers equal:

$$H_1 = H_2 = H \tag{28.22}$$

The discharge Q_x is

$$Q_x = Nx \tag{28.23}$$

The discharges $\overset{1}{Q}_x$ and $\overset{2}{Q}_x$ in layers 1 and 2 are

$$\begin{aligned}\overset{1}{Q}_x &= -T_1 \frac{d\phi}{dx} = -k_1 H \frac{d\phi}{dx} \\ \overset{2}{Q}_x &= -T_2 \frac{d\phi}{dx} = -k_2 H \frac{d\phi}{dx}\end{aligned} \tag{28.24}$$

Since $\overset{1}{Q}_x + \overset{2}{Q}_x = Q_x$, by definition, we obtain from (28.24) that

$$\overset{1}{Q}_x = \frac{k_1}{k_1 + k_2} Nx \tag{28.25}$$

$$\overset{2}{Q}_x = \frac{k_2}{k_1 + k_2} Nx \tag{28.26}$$

The apparent velocities $\overset{1}{v}^*_x$ and $\overset{2}{v}^*_x$ in the two layers are obtained by dividing $\overset{1}{Q}_x$ and $\overset{2}{Q}_x$ by a factor nHR:

$$\overset{1}{v}^*_x = \frac{1}{nR} \frac{k_1}{k_1 + k_2} \frac{N}{H} x \tag{28.27}$$

$$\overset{2}{v}^*_x = \frac{1}{nR} \frac{k_2}{k_1 + k_2} \frac{N}{H} x \tag{28.28}$$

Sec. 28   Solute Transport

We have, by definition,

$$v^* = \frac{ds}{dt} \tag{28.29}$$

where s represents the arc length measured along the streamline. Because the streamline is tangent to the velocity vector, we may write:

$$v_x^* = \frac{dx}{ds}v^* = \frac{dx}{ds}\frac{ds}{dt} = \frac{dx}{dt} \tag{28.30}$$

where v_x^* is the projection of v^* on the x axis. The component v_x^* must be computed at each point of the streamline. In the present case, however, v_x^* is independent of x_3 in each layer, so that the travel time of point $\mathcal{P}$ may be computed by evaluating the following integrals:

$$\int_{t_0}^{t} dt = \int_{x_0}^{x} \frac{dx}{\overset{2}{v_x^*}} \qquad (H \leq X_3 \leq 2H) \tag{28.31}$$

$$\int_{t_1}^{t} dt = \int_{x_1}^{x} \frac{dx}{\overset{1}{v_x^*}} \qquad (0 \leq X_3 \leq H) \tag{28.32}$$

where X_3 is given by (27.80) and (27.81) and x_1 by (27.82). It follows from (27.82) and (28.22) that (28.31) applies when $x_0 \leq x \leq (k_1 + k_2)x_0/k_1$. Using (28.28) in (28.31) we obtain

$$t - t_0 = \frac{nRH}{N}\frac{k_1 + k_2}{k_2}\ln\frac{x}{x_0} \qquad x_0 \leq x \leq \frac{k_1 + k_2}{k_1}x_0 \tag{28.33}$$

Point $\mathcal{P}$ reaches the interface between the layers when $t = t_1$ and $x = x_1 = (k_1+k_2)x_0/k_1$:

$$t_1 - t_0 = \frac{nRH}{N}\frac{k_1 + k_2}{k_2}\ln\frac{k_1 + k_2}{k_1} \tag{28.34}$$

Assuming that this interface is reached before arriving at the boundary $x = L$, we next integrate (28.32) between x_1 and L using (28.27):

$$t_L - t_1 = \frac{nRH}{N}\frac{k_1 + k_2}{k_1}\ln\left[\frac{k_1}{k_1 + k_2}\frac{L}{x_0}\right] \tag{28.35}$$

The arrival time of point $\mathcal{P}$ of the plume, defined by the initial condition (28.21), is obtained by adding (28.34) and (28.35).

Problem 28.1

Consider the foregoing example. Introduce the dimensionless time $t^* = tN/(nRH)$ and compute the dimensionless time $t_L^* - t_0^*$ at which point $\mathcal{P}$ reaches the boundary at $x = L$ if

$$x = x_0 = a \qquad k_1 = k \qquad k_2 = 5k \qquad a = b = H \qquad L = 10H$$

Let Δt^* be $(t_L^* - t_0^*)/5$ and consider five points $\mathcal{P}$ of the front on apparent path lines defined by the initial conditions $t = t_0$, $y = 2H$, with $x = H$, $x = 5H/4$, $x = 3H/2$, $x = 7H/4$, $x = 2H$ for the five path lines, respectively.

Questions

(a) Plot the five apparent path lines and mark the points that correspond to $t_0^* + j\Delta t^*$ ($j = 1, 2, 3, 4, 5$). Sketch the position of the front of the plume for each of these values of t^*.

(b) Express the concentration c in terms of the parameters of the problem for all marked points if there is a decay rate λ.

Chapter 5

Applications of Conformal Mapping

Thus far, we have solved groundwater flow problems by superposition of elementary solutions. In this chapter we will discuss a technique that makes it possible to determine, at least in principle, exact solutions to a large class of groundwater flow problems. This method is based on a complex-variable technique called *conformal mapping*, which has been extensively used in the field of groundwater mechanics in the past. Before computers were in use, closed-form solutions, such as the ones obtained by conformal mapping, were the only way to obtain quantitative answers to groundwater flow problems, and many useful solutions have been published. The most important ones are found in the older textbooks, such as Muskat [1937], Polubarinova-Kochina [1952, 1962], Aravin and Numerov [1953, 1965], and Harr [1962]. Conformal mapping still is a useful tool in groundwater mechanics, and serves to obtain solutions to simplified versions of complex problems, which may be used to gain insight into the problem prior to using a computer program capable of generating a more detailed solution. The main reason, however, to discuss conformal mapping in detail is that conformal mapping techniques may be used to generate powerful elements that can be used in computer models. We refer to such elements as *analytic elements*, and will introduce them in Chapter 6.

We concentrate our efforts in this chapter on explaining how conformal mapping works, and on presenting the main formulae, rather than listing the many solutions that are already available in the literature.

29 THE PRINCIPLES OF CONFORMAL MAPPING

In this section we deal with the principles of conformal mapping. We begin by examining what a conformal mapping is, and how the geometrical properties are related to the functional behavior. We then proceed by explaining how conformal mapping can be used to solve boundary-value problems.

The Properties of a Conformal Map

We discussed the concept of a map already when introducing the complex functions in Sec. 23; the domain in the Ω plane of Figure 3.38 is a map of the domain in the z plane shown in the same figure. The map merely is a graphical way of presenting a boundary-value problem: the contours bounding the maps represent the boundary values in the physical and complex potential domains, and solving the boundary-value problem amounts to conforming these contours to prescribed shapes.

The maps produced by analytic functions are special: they are conformal. We will illustrate this concept of conformality by examining the properties of the mapping in the neighborhood of a point $\zeta = \zeta_0$ in the ζ plane of Figure 5.1(b). We examine how the analytic function

$$z = z(\zeta) \qquad (29.1)$$

maps a small circle C_ζ of radius ρ around ζ_0 onto some curve in the z plane. The function is analytic at ζ_0 and therefore can be expanded in a Taylor series as follows:

$$z = z(\zeta_0) + z'(\zeta_0)(\zeta - \zeta_0) + \frac{z''(\zeta_0)}{2!}(\zeta - \zeta_0)^2 + \ldots \qquad (29.2)$$

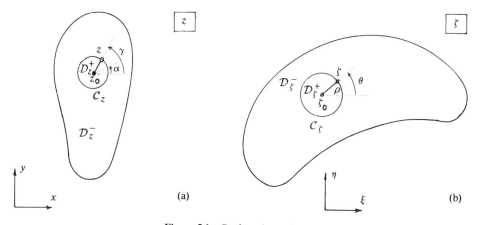

Figure 5.1 Conformal mapping.

If we assume that

$$z'(\zeta_0) \neq 0 \qquad (29.3)$$

and take $(\zeta - \zeta_0)$ sufficiently small, then (29.2) may be approximated by the first term:

$$z - z_0 \approx z'(\zeta_0)(\zeta - \zeta_0) \qquad (29.4)$$

Sec. 29   The Principles of Conformal Mapping   351

where $z_0 = z(\zeta_0)$. If ζ lies on $\mathcal{C}_\zeta$, we may write

$$\zeta - \zeta_0 = \rho e^{i\theta} \tag{29.5}$$

and representing $z'(\zeta_0)$ in polar form as

$$z'(\zeta_0) = |z'(\zeta_0)|e^{i\alpha} \tag{29.6}$$

(29.4) becomes

$$|z - z_0|e^{i\gamma} = |z'(\zeta_0)|\rho e^{i(\alpha+\theta)} \tag{29.7}$$

We observe that $|z - z_0|$ is constant; the circle in the ζ plane corresponds to a circle in the z plane, labeled as $\mathcal{C}_z$ in Figure 5.1(a). Furthermore, if $\theta = \arg(\zeta - \zeta_0)$ increases from 0 to 2π, then $\gamma = \arg(z - z_0)$ increases from α to $\alpha + 2\pi$. The conclusion is that a full circle in the ζ plane corresponds exactly to a full circle in the z plane, rotated over an angle $\alpha = \arg z'(\zeta_0)$: the mapping is conformal. Conformality of the mapping means, therefore, that the shape is preserved.

An important property of a conformal map can best be explained by the use of the concept of *tracing a contour around a domain $\mathcal{D}$ in positive sense*. Let $\mathcal{D}_\zeta^+$ and $\mathcal{D}_\zeta^-$ denote the domains inside and outside $\mathcal{C}_\zeta$. Then we say that ζ traces $\mathcal{C}_\zeta$ around $\mathcal{D}_\zeta^+$ in positive sense if $\mathcal{D}_\zeta^+$ lies to the left with respect to the direction in which ζ traces the circle, i.e., if ζ moves in counterclockwise direction. By the same token, $\mathcal{C}_\zeta$ may be considered to be the boundary of the infinite domain $\mathcal{D}_\zeta^-$; in that case ζ traces $\mathcal{C}_\zeta$ around $\mathcal{D}_\zeta^-$ in negative sense if $\mathcal{D}_\zeta^-$ lies to the right with respect to the direction in which ζ traces $\mathcal{C}_\zeta$, i.e., if ζ moves in counterclockwise direction. The reader may observe that these definitions tie in well with the orientation of contours discussed in the context of singular Cauchy integrals in Sec. 25; the contour $\mathcal{C}$ in Figure 3.51 is traced in positive sense with respect to the domain $\mathcal{D}^+$. We now state that if the map of domain $\mathcal{D}_\zeta^+$ onto the z plane is $\mathcal{D}_z^+$, then $z = z(\zeta)$ traces $\mathcal{C}_z$ in positive sense with respect to $\mathcal{D}_z^+$, if ζ traces $\mathcal{C}_\zeta$ in positive sense with respect to $\mathcal{D}_\zeta^+$. That this is true follows from the notion that $z_0 = z(\zeta_0)$ corresponds to ζ_0, and that ζ_0 lies inside $\mathcal{C}_\zeta$ and z_0 inside $\mathcal{C}_z$.

We did the above analysis under the assumption that $z'(\zeta_0) \neq 0$, see (29.3), and it is of interest to examine what happens if $z'(\zeta_0) = 0$. The coefficient of the term $(\zeta - \zeta_0)$ then vanishes, and the term $\frac{1}{2}z''(\zeta_0)(\zeta - \zeta_0)^2$ cannot be neglected; the mapping function in the neighborhood of ζ_0 now becomes

$$z - z_0 \approx \tfrac{1}{2}z''(\zeta_0)(\zeta - \zeta_0)^2 = \tfrac{1}{2}z''(\zeta_0)\rho^2 e^{2i\theta} \tag{29.8}$$

We observe that when ζ traces $\mathcal{C}_\zeta$ once, z traces $\mathcal{C}_z$ twice; the mapping is no longer conformal and an angle β between two radius vectors of $\mathcal{C}_\zeta$ is transformed into an angle 2β between the corresponding radius vectors of $\mathcal{C}_z$. In this case the point ζ_0 is called a branch point of order 1, and the point z_0 is called a winding point of order 1. In general, ζ_0 is a branch point of order m if the derivatives $z^{(j)}(\zeta_0)$ vanish for $j = 1, 2, \ldots, m$.

We will not consider branch points in detail in this section, and merely state that mappings by analytic functions are conformal everywhere, except at points where the derivative of the mapping function is zero.

Solving Boundary-Value Problems

It is possible to solve boundary-value problems in two different ways. The first consists of determining the boundaries of the domains in the physical and complex potential planes and then attempting to find the analytic function $\Omega = \Omega(z)$ that maps the domain in the z plane onto that in the Ω plane. The process of finding such a mapping function may consist of applying or modifying an existing mapping function as listed in the handbooks (e.g., Churchill [1960], Kober [1952], Nehari [1952], or Von Koppenfels and Stallmann [1959]). In other cases, it may be necessary to apply a method of conformal mapping to determine the function, as explained later in this chapter. We refer to this method as a direct application of conformal mapping.

An indirect way of using conformal mapping to solve a problem is based upon the notion that the analytic function of an analytic function is an analytic function. For example, if we succeed in mapping a flow domain in the z plane, bounded by some complex boundary $\mathcal{B}_z$ onto a domain in the ζ plane bounded by another boundary $\mathcal{B}_\zeta$, e.g., a circle or a straight line, then we know the function

$$z = z(\zeta) \qquad (29.9)$$

Depending upon the choice of the boundary $\mathcal{B}_\zeta$, we may now be able to solve the boundary value problem for $\Omega = \Omega(\zeta)$ in the ζ plane, where the boundary values of Φ and Ψ are transposed to the boundary segments of $\mathcal{B}_\zeta$. The complex potential

$$\Omega = \Omega(\zeta) \qquad (29.10)$$

together with the mapping function $z = z(\zeta)$ then constitutes the solution to our problem with ζ acting as a complex reference parameter.

The latter approach is more generally applicable than the first. The direct application method fails for any problem that is solved in terms of a reference parameter that cannot be eliminated.

30 AN APPLICATION OF THE BILINEAR TRANSFORMATION

The *bilinear transformation*, sometimes called the Möbius transformation, maps a circle onto a straight line. We will discuss this transformation, and then apply it to solve the problem of a circular lake in an aquifer with wells and a uniform flow field. Of course, we will obtain the same solution as that obtained earlier by the use of superposition; the application serves to illustrate the method of conformal mapping and to present a transformation that we will use in other applications later on.

Sec. 30 An Application of the Bilinear Transformation

The Bilinear Transformation

The bilinear transformation is a function of the form

$$\zeta = \xi + i\eta = \frac{az+b}{cz+d} \qquad (30.1)$$

where a, b, c, and d are complex constants. We consider the following special form of this transformation:

$$\zeta = i\frac{z+R}{z-R} \qquad (30.2)$$

and will prove that it maps the exterior of the circle of radius R around the origin in the z plane (see Figure 5.2[a]) onto the upper half plane $\Im\zeta \geq 0$. Four representative points on the circle in the z plane are marked as 1, 2, 3, and 4 as shown in the figure. In order to examine how the circle $|z| = R$ is transformed by (30.2), we choose an arbitrary point $z = Re^{i\theta}$ on the circle and determine the corresponding point in the ζ plane by the use of (30.2):

$$\zeta = i\frac{Re^{i\theta}+R}{Re^{i\theta}-R} = i\frac{e^{i\theta/2}+e^{-i\theta/2}}{e^{i\theta/2}-e^{-i\theta/2}} = i\frac{2\cos\frac{\theta}{2}}{2i\sin\frac{\theta}{2}} = \cot\frac{\theta}{2} \qquad (30.3)$$

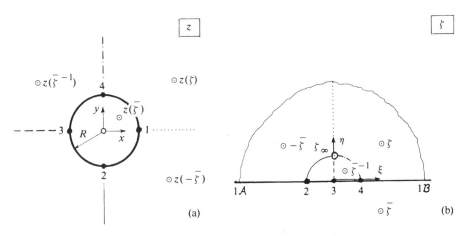

Figure 5.2 The bilinear transformation.

We observe that ζ is real for all values of θ; the circle corresponds to the real axis $\Im\zeta = 0$. Furthermore, point 1 ($\theta = 0$) corresponds to $\zeta = \infty$, point 2 ($\theta = -\pi/2$) corresponds to $\zeta = -1$, point 3 ($\theta = \pm\pi$) to $\zeta = 0$, and point 4 ($\theta = \pi/2$) to $\zeta = 1$. Infinity in the z plane corresponds to $\zeta = i$; the limit of (30.2) for $z \to \infty$ is i. The exterior of the circle in the z plane lies to the left with respect to the boundary 1-2-3-4-1, and because the sense is preserved, we conclude that this exterior domain corresponds to the upper half plane $\Im\zeta \geq 0$. By a similar reasoning, we see that the interior of the circle corresponds to the lower half plane.

It is possible to invert the transformation (30.2); solving this equation for z, we obtain

$$\frac{z}{R} = \frac{\zeta + i}{\zeta - i} \tag{30.4}$$

The η axis is an axis of symmetry in the ζ plane, and maps onto the section $|x| \geq R$ of the x axis in the z plane, which also is an axis of symmetry. We see this by substituting $i\eta$ for ζ in (30.4), which yields real values for z/R; the section of the η and x axes that correspond to each other are marked by dots and dashes in Figure 5.2. Points that lie symmetrically with respect to the η axis in the ζ plane correspond to points that lie symmetrically with respect to the x axis in the z plane, i.e.,

$$z(\zeta) = \overline{z(-\bar{\zeta})} \tag{30.5}$$

which may be verified by the use of (30.4). It is noted that $\zeta = \xi + i\eta$ and $-\bar{\zeta} = -\xi + i\eta$ lie symmetrically with respect to the η axis, and that z and $\bar{z}$ lie symmetrically with respect to the x axis.

Another interesting property of the transformation (30.4) is that

$$z(\zeta) = -\overline{z(1/\bar{\zeta})} \tag{30.6}$$

which equation is obtained as follows:

$$-\overline{z(1/\bar{\zeta})} = -R\overline{\left[\frac{1/\bar{\zeta} + i}{1/\bar{\zeta} - i}\right]} = -R\frac{1/\zeta - i}{1/\zeta + i} = -R\frac{\frac{1}{i} - \zeta}{\frac{1}{i} + \zeta} = R\frac{\zeta + i}{\zeta - i} = z(\zeta) \tag{30.7}$$

Equation (30.6) may be interpreted geometrically as follows: the points ζ and $1/\bar{\zeta}$ lie inversely with respect to the unit circle $|\zeta| = 1$, which means that they lie on a straight line through the center at distances that are each other's reciprocal, as indicated in Figure 5.2(b). The points $z = x + iy$ and $-\bar{z} = -x + iy$ lie symmetrically with respect to the y axis. In particular, if $\zeta = e^{i\alpha}$, then ζ equals $1/\bar{\zeta}$ and $z(\zeta)$ equals $z(1/\bar{\zeta})$; it follows from (30.6) that $z = -\bar{z}$, so that $x = 0$ and $z(e^{i\alpha})$ lies on the y axis: the unit circle $|\zeta| = 1$ corresponds to the y axis.

A third important property of the transformation (30.2) is the following:

$$\frac{z(\zeta)}{R} = \frac{R}{\overline{z(\bar{\zeta})}} \tag{30.8}$$

The proof of this equation is obtained from (30.4) as follows:

$$\frac{R}{\overline{z(\bar{\zeta})}} = \frac{1}{\frac{\overline{z(\bar{\zeta})}}{R}} = \overline{\left[\frac{\bar{\zeta} - i}{\bar{\zeta} + i}\right]} = \frac{\zeta + i}{\zeta - i} = \frac{z}{R} \tag{30.9}$$

If we denote $z(\overline{\zeta})$ as z_0 and $z(\zeta)$ as z, then it follows from (30.8) that

$$|z_0| = \frac{R^2}{|z|} \tag{30.10}$$

Since $z_0 = z(\overline{\zeta})$ and $z = z(\zeta)$ have, by (30.8), the same argument, these points lie on a straight line through $z = 0$; they are located inversely with respect to the circle of radius R about $z = 0$.

We observe from the above analysis that the transformation exhibits certain properties of symmetry. The axis of symmetry, the x axis, in the z plane that is transformed into a straight line in the ζ plane, remains an axis of symmetry after transformation. The axis of symmetry (the y axis) in the z plane that is transformed into the unit circle becomes a *curve of symmetry*; that is, points that are located symmetrically with respect to the y axis are transformed into points that are located "symmetrically" with respect to the circle $|\zeta| = 1$. The word symmetrically is placed between quotes because it is used in a broader sense; this circle symmetry means that points lie inversely with respect to the circle. The latter broad use of the term symmetry may be generalized to any curve, where the rule for finding points that lie symmetrically with respect to the curve are obtained from the transformation formula. The above concept, that points that lie symmetrically with respect to a line in the one plane are mapped onto points that lie symmetrically with respect to the transformed line, is a general property of conformal mapping known as the *Schwarz Reflection Principle*.

Problem 30.1

Determine a, b, c, and d in (30.1) if the center of the circle in the z plane is at z_0, and if point 1 corresponds to $\zeta = -1$, point 2 to $\zeta = 0$, point 3 to $\zeta = 1$, and point 4 to $\zeta = \infty$.

A Lake in an Aquifer with Wells and a Uniform Far-Field

We will next use the results obtained above to solve the problem of a fully penetrating lake of radius R in an infinite aquifer. The center of the lake is at the origin, the potential along the lake boundary is Φ_0, and there is a well of discharge Q at $z = z_w$ (see Figure 5.3[a]). We will initially limit the analysis to the case that the far-field is one of constant potential. We use the transformation derived above and solve the flow problem in $\Im\zeta \geq 0$. The boundary of the flow domain in the ζ plane is the real axis, with the boundary condition

$$-\infty < \xi < \infty \quad \eta = 0 \quad \Phi = \Phi_0 \tag{30.11}$$

The point in $\Im\zeta \geq 0$ corresponding to the well is δ, with (compare [30.2])

$$\delta = \zeta(z_w) = i\frac{z_w + R}{z_w - R} \tag{30.12}$$

The well of discharge Q at $z = z_w$ corresponds to a well of discharge Q at $\zeta = \delta$; that the discharge of the well remains the same after the transformation follows from the notion that the stream function must increase by Q if a point traces the well boundary, both before and after the transformation.

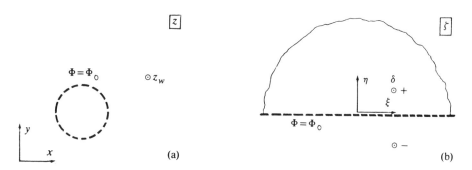

Figure 5.3 The z and ζ planes for a lake in an aquifer.

The flow problem in $\Im\zeta \geq 0$ is straightforward and may be solved by the method of images, which yields

$$\Omega = \frac{Q}{2\pi} \ln \frac{\zeta - \delta}{\zeta - \overline{\delta}} + \Phi_0 \tag{30.13}$$

Using (30.2) and (30.12) we obtain

$$\begin{aligned}\Omega &= \frac{Q}{2\pi} \ln \frac{i\frac{z+R}{z-R} - i\frac{z_w+R}{z_w-R}}{i\frac{z+R}{z-R} + i\frac{\overline{z}_w+R}{\overline{z}_w-R}} + \Phi_0 \\ &= \frac{Q}{2\pi} \ln \frac{zz_w - R(z-z_w) - R^2 - [zz_w + R(z-z_w) - R^2]}{z\overline{z}_w - R(z-\overline{z}_w) - R^2 + [z\overline{z}_w + R(z-\overline{z}_w) - R^2]} \\ &\quad - \frac{Q}{2\pi} \ln \frac{z_w - R}{\overline{z}_w - R} + \Phi_0 \end{aligned} \tag{30.14}$$

The term $Q/(2\pi) \ln[(z_w - R)/(\overline{z}_w - R)]$ is a purely imaginary constant since $|z_w - R| = |\overline{z}_w - R|$ and can be omitted since it affects only the reference level of the stream function, which is arbitrary. Simplifying (30.14) we obtain

$$\Omega = \frac{Q}{2\pi} \ln \frac{-2R(z - z_w)}{2z\overline{z}_w - 2R^2} + \Phi_0 = \frac{Q}{2\pi} \ln \left[\frac{(z - z_w)}{(z - R^2/\overline{z}_w)} \frac{-R}{\overline{z}_w} \right] + \Phi_0 \tag{30.15}$$

The real part of this equation corresponds exactly to the result obtained earlier for the potential Φ. In fact, we did not really need to go through the above analysis; we know by symmetry that wells that lie symmetrically with respect to the real axis $\eta = 0$ correspond to wells in the z plane that lie inversely with respect to the lake boundary, and therefore we could have concluded that the image well in the z plane must lie at $z = R^2/\overline{z}_w$.

Sec. 30   An Application of the Bilinear Transformation   357

If the lake draws an amount Q^* from infinity, then we must place a source of discharge Q^* at the point in $\Im\zeta \geq 0$ that corresponds to $z = \infty$. This point is $\zeta = \zeta_\infty = i$ (see Figure 5.2[b]), and we must place an image sink at $\zeta = -i$, so that (30.13) becomes

$$\Omega = \frac{Q}{2\pi} \ln \frac{\zeta - \delta}{\zeta - \bar{\delta}} - \frac{Q^*}{2\pi} \ln \frac{\zeta - i}{\zeta + i} + \Phi_0 \tag{30.16}$$

The first term may be written in terms of z by the use of (30.15), and the second term by the use of (30.4) so that (30.16) becomes

$$\Omega = \frac{Q}{2\pi} \ln \left[\frac{z - z_w}{z - R^2/\bar{z}_w} \frac{-R}{\bar{z}_w} \right] + \frac{Q^*}{2\pi} \ln \frac{z}{R} + \Phi_0 \tag{30.17}$$

We next consider the case that there are no wells and the far-field in the z plane is uniform, with a discharge Q_{x0} in the x direction. As opposed to the case with a source at infinity discussed above, we cannot say immediately what kind of singularity must be placed at $\zeta = \zeta_\infty = i$. In order to determine the behavior of $\Omega(\zeta)$ near $\zeta = i$, we examine the behavior of the mapping function (30.4) near $\zeta = i$ by writing it as follows:

$$z = R\frac{\zeta + i}{\zeta - i} = \frac{2iR}{\zeta - i} + R \tag{30.18}$$

A singularity of this kind is called a *pole of the first order*. (A function $f = f(\zeta)$ is said to have a pole of order m at a point ζ_0 if $(\zeta - \zeta_0)^m f(\zeta)$, but not $(\zeta - \zeta_0)^{m-1} f(\zeta)$, is analytic at ζ_0 (Korn and Korn [1968]). Since $(\zeta - i)z$ in (30.18) is analytic, the function $z(\zeta)$ has a pole of order 1 at $\zeta = i$. The expansion of a function about a pole is called a *Laurent expansion*.) We know that Ω behaves near infinity as $-Q_{x0}z$, so that $\Omega(\zeta)$ behaves near $\zeta = i$ as

$$\Omega(\zeta) = -2iQ_{x0}R\frac{1}{\zeta - i} - Q_{xo}R \tag{30.19}$$

The first term represents a dipole at $\zeta = i$, oriented toward the origin (point 3). In order to obtain a complex potential that renders an equipotential along the real axis, we place an image dipole, oriented the same way, at $\zeta = -i$ and obtain

$$\Omega(\zeta) = -2iQ_{x0}R\left[\frac{1}{\zeta - i} + \frac{1}{\zeta + i}\right] + C \tag{30.20}$$

where the real constant C must be chosen such that $\Phi = \Phi_0$ for $\eta = 0$. Along the real axis, (30.20) becomes

$$\Omega(\xi) = -2iQ_{x0}R\left[\frac{1}{\xi - i} + \frac{1}{\xi + i}\right] + C = -2iQ_{x0}R\frac{2\xi}{\xi^2 + 1} + C \tag{30.21}$$

The first term is purely imaginary, so that $C = \Phi_0$, and (30.20) becomes

$$\Omega(\zeta) = -2iQ_{x0}R\left[\frac{1}{\zeta - i} + \frac{1}{\zeta + i}\right] + \Phi_0 \tag{30.22}$$

In order to eliminate ζ, we use (30.18) to express $1/(\zeta - i)$ in terms of z,

$$\frac{1}{\zeta - i} = \frac{1}{2iR}(z - R) \qquad (30.23)$$

and it follows from (30.4) that

$$\frac{R}{z} = \frac{\zeta - i}{\zeta + i} = 1 - \frac{2i}{\zeta + i} \qquad (30.24)$$

or

$$\frac{1}{\zeta + i} = -\frac{1}{2iR}\left[\frac{R^2}{z} - R\right] \qquad (30.25)$$

Using these results in (30.22) we obtain

$$\Omega(z) = -Q_{x0}\left[z - \frac{R^2}{z}\right] + \Phi_0 \qquad (30.26)$$

which has a real part that is identical to the result obtained in Sec. 20 (compare [20.33]).

The results obtained above are not new; however, we will see in the next section that they may be generalized, by mapping the exterior of the circle onto the exterior of an ellipse, to problems with elliptic lakes.

31 AN ELLIPTIC LAKE IN AN INFINITE AQUIFER WITH WELLS AND A UNIFORM FAR-FIELD

We will in this section map the exterior of an elliptic lake in the z plane onto the exterior of a circle in the plane $\chi = \tau + i\omega$. After discussing the mapping function, we solve the problem of flow in the χ plane by the method of images. Elimination of the complex reference parameter χ then renders the complex potential as a function of z.

The Mapping Function

The axes of the elliptic lake have lengths $2a$ and $2b$, where $a \geq b$, and the longer axis is oriented at an angle α to the x axis as shown in Figure 5.4(a). The center of the ellipse is at z_0, and the distance from the center to the focus is f, where

$$f = \sqrt{a^2 - b^2} \qquad (31.1)$$

We map the flow domain onto the exterior of an ellipse in the Z plane,

$$Z = \frac{z - z_0}{a + b}e^{-i\alpha} \qquad (31.2)$$

Sec. 31 An Elliptic Lake in an Infinite Aquifer with Wells and a Uniform Far-Field

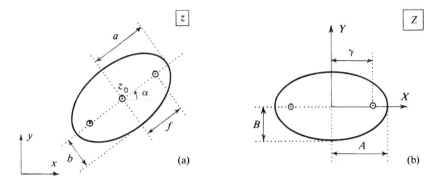

Figure 5.4 The ellipse in the z and Z planes.

so that the ellipse in the Z plane is centered at the origin, and is oriented with its major axis along the real axis. The lengths of the axes of the transformed ellipse are denoted as A and B, and we observe from (31.2) that

$$A = \frac{a}{a+b} \qquad B = \frac{b}{a+b} \tag{31.3}$$

We see from (31.3) that

$$A + B = 1 \tag{31.4}$$

and introduce γ as the distance from the origin to either focal point of the ellipse in the Z plane, where

$$\gamma^2 = A^2 - B^2 = (A-B)(A+B) = A - B \tag{31.5}$$

so that

$$A = \tfrac{1}{2}(1 + \gamma^2) \qquad B = \tfrac{1}{2}(1 - \gamma^2) \tag{31.6}$$

Now consider the transformation

$$\boxed{Z = \tfrac{1}{2}(A+B)\chi + \tfrac{1}{2}(A-B)\frac{1}{\chi} = \tfrac{1}{2}\left[\chi + \frac{\gamma^2}{\chi}\right]} \tag{31.7}$$

where χ is $\tau + i\omega$,

$$\chi = \tau + i\omega \tag{31.8}$$

This transformation maps the unit circle in the χ plane onto the ellipse in the Z plane, as will be shown below. If χ represents a point on the unit circle, then

$$\chi = e^{i\theta} \tag{31.9}$$

and (31.7) becomes

$$Z = X + iY = \tfrac{1}{2}\left(e^{i\theta} + \gamma^2 e^{-i\theta}\right) = \tfrac{1}{2}(1+\gamma^2)\cos\theta + \tfrac{i}{2}(1-\gamma^2)\sin\theta$$
$$= A\cos\theta + iB\sin\theta \tag{31.10}$$

Hence,

$$\left[\frac{X}{A}\right]^2 + \left[\frac{Y}{B}\right]^2 = 1 \tag{31.11}$$

which indeed represents the ellipse in the Z plane. The mapping is illustrated in Figure 5.5, where four representative points are labeled on both boundaries. It follows from (31.7) that infinity in the χ plane corresponds to infinity in the Z plane; for large values of χ, the transformation approaches the form $Z = \tfrac{1}{2}\chi$.

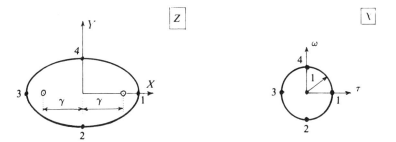

Figure 5.5 The transformation $Z = Z(\chi)$.

We may invert the transformation (31.7) by solving for χ, which yields

$$\chi = Z \pm \sqrt{(Z-\gamma)(Z+\gamma)} \tag{31.12}$$

Since infinity in the Z plane corresponds to infinity in the χ plane, the plus sign applies, so that

$$\boxed{\chi = Z + \sqrt{(Z-\gamma)(Z+\gamma)}} \tag{31.13}$$

It may be noted that this function is not analytic at $Z = \pm\gamma$; square root singularities exist at the two focal points of the ellipse. Because the square roots are many-valued, we must be careful in choosing the arguments of $Z-\gamma$ and $Z+\gamma$ such that χ is continuous everywhere outside the circle. We choose these arguments such that

$$0 \le \arg(Z+\gamma) \le \pi \qquad 0 \le arg(Z-\gamma) \le \pi \qquad Y \ge 0^+ \tag{31.14}$$
$$-\pi \le \arg(Z+\gamma) \le 0 \qquad -\pi \le arg(Z-\gamma) \le 0 \qquad Y \le 0^- \tag{31.15}$$

where the superscripts $+$ and $-$ indicate whether the x axis is approached from the positive or the negative side. With the above definitions, the arguments of $Z-\gamma$ and $Z+\gamma$

are each equal to either $+\pi$ or $-\pi$ along the negative real axis $Y = 0$, $-\infty < X < -\gamma$, so that the square root always is negative there:

$$-\infty < X < -\gamma \qquad Y = 0^+ \qquad \chi = X + \sqrt{|X^2 - \gamma^2|e^{2\pi i}} = X - \sqrt{X^2 - \gamma^2} \tag{31.16}$$

$$-\infty < X < -\gamma \qquad Y = 0^- \qquad \chi = X + \sqrt{|X^2 - \gamma^2|e^{-2\pi i}} = X - \sqrt{X^2 - \gamma^2} \tag{31.17}$$

and the mapping function is indeed continuous across $Y = 0$, $-\infty < X < -\gamma$. It is important to note that care must be taken in using a standard FORTRAN routine for evaluating the square root; this gives proper results only if the square roots of $Z - \gamma$ and $Z + \gamma$ are calculated first and then multiplied; calculating the square root of $Z^2 - \gamma^2$ directly will give erroneous results unless a correction is made depending upon the arguments of $Z - \gamma$ and $Z + \gamma$.

An Elliptic Lake in a Uniform Flow

We first consider the case of a uniform far-field of discharge Q_0, oriented at an angle β to the x axis in the z plane (see Figure 5.6). In order to determine how this far-field transforms in the Z plane, we examine the relation (31.2) between Z and z for large z:

$$Z \to \frac{z}{a+b} e^{-i\alpha} \qquad z \to \infty \tag{31.18}$$

Since for $z \to \infty$, Ω behaves as

$$\Omega \to -Q_0 z e^{-i\beta} \qquad z \to \infty \tag{31.19}$$

we obtain with (31.18) the far-field behavior of Ω in terms of Z,

$$\Omega \to -(a+b)Q_0 e^{-i(\beta-\alpha)} Z \qquad Z \to \infty \tag{31.20}$$

Thus, the flow is oriented the same way with respect to the major axis of the ellipse in both planes, as we expect for a conformal mapping. The next step is to determine the far-field behavior of Ω in terms of χ. Since Z approaches $\frac{1}{2}\chi$ for $\chi \to \infty$ (compare [31.7]), we obtain from (31.20):

$$\Omega \to -\tfrac{1}{2}(a+b)Q_0 e^{-i(\beta-\alpha)} \chi \qquad \chi \to \infty \tag{31.21}$$

The problem in the χ plane thus amounts to solving the problem of a circular lake in a field of a uniform flow at an angle $(\beta - \alpha)$ with the τ axis, and of discharge $\frac{1}{2}(a+b)Q_0$. The solution consists of the sum of a term for uniform flow, and a dipole

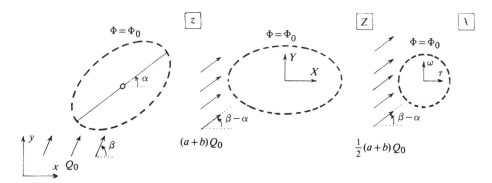

Figure 5.6 A uniform far-field.

at the origin oriented in the direction of flow. Using expression (24.12) for the complex potential for a dipole, and noticing that the flow is oriented at an angle $\beta - \alpha$, we obtain

$$\Omega = -\tfrac{1}{2}(a+b)Q_0 \left[\chi e^{-i(\beta-\alpha)} - \frac{e^{i(\beta-\alpha)}}{\chi} \right] + \Phi_0 \qquad (31.22)$$

where Φ_0 is the potential along the circle. The complex potential (31.22) represents the solution for the case that the lake does not draw any water from infinity; assuming that the lake draws an amount Q^* from infinity, (31.22) becomes

$$\Omega = -\tfrac{1}{2}(a+b)Q_0 \left[\chi e^{-i(\beta-\alpha)} - \frac{e^{i(\beta-\alpha)}}{\chi} \right] + \frac{Q^*}{2\pi} \ln \chi + \Phi_0 \qquad (31.23)$$

It is possible to eliminate the reference parameter χ, and write the complex potential in terms of the dimensionless variable Z by the use of (31.13). It follows from (31.7) that $\gamma^2/\chi = 2Z - \chi$ so that we obtain with (31.13)

$$\frac{\gamma^2}{\chi} = Z - \sqrt{(Z-\gamma)(Z+\gamma)} \qquad (31.24)$$

and (31.23) may be written as follows, provided that $\gamma \neq 0$,

$$\Omega = -\tfrac{1}{2}(a+b)Q_0 \left\{ Z \left[e^{-i(\beta-\alpha)} - \frac{1}{\gamma^2} e^{i(\beta-\alpha)} \right] \right.$$
$$\left. + \sqrt{(Z-\gamma)(Z+\gamma)} \left[e^{-i(\beta-\alpha)} + \frac{1}{\gamma^2} e^{i(\beta-\alpha)} \right] \right\}$$
$$+ \frac{Q^*}{2\pi} \ln \left[Z + \sqrt{(Z-\gamma)(Z+\gamma)} \right] + \Phi_0 \qquad (\gamma \neq 0) \qquad (31.25)$$

It may be noted that, for numerical evaluation, the form (31.23) may be retained; there is no particular advantage in writing the complex potential in terms of Z rather than of χ.

An Elliptic Lake with a Well

The presence of a well of discharge Q at $z = z_w$ may be accounted for as follows. We first determine the points Z_w and χ_w in the Z and χ planes that correspond to z_w, using the transformations (31.2) and (31.13):

$$Z_w = Z(z_w) = \frac{z_w - z_0}{a + b} e^{-i\alpha} \tag{31.26}$$

and

$$\chi_w = Z_w + \sqrt{(Z_w - \gamma)(Z_w + \gamma)} \tag{31.27}$$

Second, we solve the problem in the χ plane with a well of discharge Q at χ_w. The image recharge well lies inversely to χ_w with respect to the circle $|\chi| = 1$, so that

$$\Omega = \frac{Q}{2\pi} \ln\left[\frac{\chi - \chi_w}{\chi - 1/\overline{\chi}_w} \frac{1}{|\chi_w|}\right] + \Phi_0 \tag{31.28}$$

We obtain the complete solution for an elliptic lake in an infinite aquifer with a well by combining (31.23) with (31.28):

$$\Omega = -\tfrac{1}{2}(a+b)Q_0 \left[\chi e^{-i(\beta-\alpha)} - \frac{e^{i(\beta-\alpha)}}{\chi}\right] + \frac{Q^*}{2\pi} \ln \chi$$
$$+ \frac{Q}{2\pi} \ln\left[\frac{\chi - \chi_w}{\chi - 1/\overline{\chi}_w} \frac{1}{|\chi_w|}\right] + \Phi_0 \tag{31.29}$$

where χ is given in terms of Z by (31.13). The discharge Q^* represents the flow from infinity and may be determined from the given potential at some reference point.

Problem 31.1

Determine the stagnation point and points on the streamlines for an elliptic lake in a field of uniform flow with $a = 2b$ and $Q^*/\{(a+b)Q_0\} = 14$, and $\alpha = 30°$, $\beta = 45°$. You may determine points on the streamlines first in the χ plane, and then map these on the corresponding streamlines in the z plane.

A Finite Canal

If the width of the lake is small with respect to its length, i.e., if the lake is a finite canal, the transformation simplifies as follows. We replace a by $\tfrac{1}{2}L$, where L is the length of the canal, set b equal to zero, and apply (31.3) through (31.5) to determine A, B, and γ:

$$a = \tfrac{1}{2}L \quad b = 0 \quad A = 1 \quad B = 0 \quad \gamma = 1 \tag{31.30}$$

We denote the end points of the canal as z_1 and z_2, so that z_0 equals $\frac{1}{2}(z_1 + z_2)$, and $(a+b)e^{i\alpha}$ may be written as $\frac{1}{2}(z_2 - z_1)$. The transformations (31.2), (31.7) and (31.13) thus become

$$Z = \frac{z - \frac{1}{2}(z_1 + z_2)}{\frac{1}{2}(z_2 - z_1)} \tag{31.31}$$

$$Z = \frac{1}{2}(\chi + \frac{1}{\chi}) \tag{31.32}$$

$$\chi = Z + \sqrt{(Z-1)(Z+1)} \tag{31.33}$$

The complex potential (31.25) for the case without a well reduces as follows:

$$\Omega = -\frac{1}{2}LQ_0\left[\cos(\beta - \alpha)\sqrt{(Z-1)(Z+1)} - i\sin(\beta - \alpha)Z\right]$$
$$+ \frac{Q^*}{2\pi}\ln\left[Z + \sqrt{(Z-1)(Z+1)}\right] + \Phi_0 \tag{31.34}$$

It is of interest to note that the stream function Ψ jumps across the canal, and along the entire negative real axis.

Problem 31.2
Determine an expression for the jump $\Psi^+ - \Psi^-$ along the portion $Y = 0$, $-\infty < X < 1$ of the X axis for a finite canal. Determine an expression for the jump in the component of the discharge vector normal to the canal along the interval $Y = 0$, $-1 \leq X \leq 1$. Plot this jump versus X.

An Elliptic Impermeable Object

We may use the conformal transformation discussed above to determine the solution for flow around an impermeable object with an elliptic boundary. The functions $Z = Z(z)$ and $\chi = \chi(Z)$ remain unchanged, but the complex potential for flow in the χ plane now must fulfill the boundary condition that the circle is impermeable. The far-field is identical to that of the potential (31.22), but the orientation of the dipole at $\chi = 0$ is reversed, and we obtain

$$\Omega = -\frac{1}{2}(a+b)Q_0\left[\chi e^{-i(\beta - \alpha)} + \frac{e^{i(\beta - \alpha)}}{\chi}\right] + C \tag{31.35}$$

where C is a real constant. To verify that (31.35) indeed meets the condition that Ψ is constant along the circle, we set $\chi = e^{i\theta}$ in (31.35) and obtain

$$\Omega = -\frac{1}{2}(a+b)Q_0\left[e^{i(\theta - \beta + \alpha)} + e^{-i(\theta - \beta + \alpha)}\right] + C$$
$$= -(a+b)Q_0\cos(\theta - \beta + \alpha) + C \qquad (|\chi| = 1) \tag{31.36}$$

Sec. 31 An Elliptic Lake in an Infinite Aquifer with Wells and a Uniform Far-Field

which is purely real so that the boundary condition is met, as asserted. The complex potential may be written in terms of Z by the use of (31.13), and becomes

$$\Omega = -\tfrac{1}{2}(a+b)Q_0 \left\{ Z\left[e^{-i(\beta-\alpha)} + \frac{1}{\gamma^2} e^{i(\beta-\alpha)} \right] \right.$$
$$\left. + \sqrt{(Z-\gamma)(Z+\gamma)} \left[e^{-i(\beta-\alpha)} - \frac{1}{\gamma^2} e^{i(\beta-\alpha)} \right] \right\} + C \qquad (31.37)$$

For the special case of a thin impermeable barrier, i.e., $a = \tfrac{1}{2}L$, $b = 0$, $\gamma = 1$, this becomes

$$\Omega = \tfrac{1}{2} L Q_0 \left[i \sin(\beta-\alpha) \sqrt{(Z-1)(Z+1)} - \cos(\beta-\alpha) Z \right] + C \qquad (31.38)$$

Flow in Two Confined Aquifers Separated by an Impermeable Layer with an Elliptic Opening

We will consider the case of flow in a system of two confined aquifers separated by a thin impermeable bed with an elliptic opening as a final application of the transformation presented above. We will make use of the comprehensive potential introduced in Sec. 12, restricting the analysis to the case that the flow in both aquifers is confined in the region of interest. The aquifer configuration is shown in Figure 5.7(a) in cross section and in plan in Figure 5.7(b). The origin of an x, y, x_3 Cartesian coordinate system is chosen at the center of the elliptic opening with x_3 pointing vertically upward. The lengths of the axes of the ellipse are $2a$ and $2b$ and the major axis lies on the x axis. The thicknesses of the upper and lower aquifers are denoted as $\overset{u}{H}$ and $\overset{l}{H}$, respectively, and the potentials $\overset{u}{\Phi}$ and $\overset{l}{\Phi}$ are defined as

$$\overset{u}{\Phi} = k \overset{u}{H} \overset{u}{\phi} \qquad \overset{l}{\Phi} = k \overset{l}{H} \overset{l}{\phi} \qquad (31.39)$$

It may be noted that these definitions are valid only if the aquifers are everywhere confined:

$$\overset{u}{\phi} > \overset{u}{H} + \overset{l}{H} \qquad \overset{l}{\phi} > \overset{l}{H} \qquad (31.40)$$

We consider the case that the far-fields in the upper and lower aquifers are combinations of uniform and radial flows; the latter flow accounts for flow from infinity in the one aquifer via the opening toward infinity in the other aquifer. The components of the uniform flows in the upper and lower aquifers are denoted as $\overset{u}{Q}_{x0}$, $\overset{u}{Q}_{y0}$, $\overset{l}{Q}_{x0}$, and $\overset{l}{Q}_{y0}$, and the components Q_{x0} and Q_{y0} of the comprehensive discharge vector due to the uniform flows are

$$Q_{x0} = \overset{u}{Q}_{x0} + \overset{l}{Q}_{x0} \qquad Q_{y0} = \overset{u}{Q}_{y0} + \overset{l}{Q}_{y0} \qquad (31.41)$$

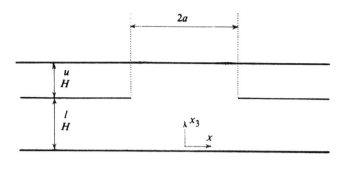

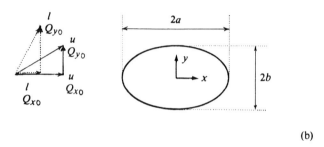

Figure 5.7 The aquifer system.

The comprehensive potential

$$\Phi = \overset{u}{\Phi} + \overset{l}{\Phi} \tag{31.42}$$

is the real part of the comprehensive complex potential. By continuity of flow the radial components of the far-fields in the upper and lower aquifers are opposite in direction and equal in magnitude, so that the comprehensive potential contains only uniform flow:

$$\Omega = -(Q_{x0} - iQ_{y0})z + C \tag{31.43}$$

If it is given that the heads in the upper and lower aquifers at a point $z = L$ are

$$z = L \quad \left\{ \begin{array}{l} \overset{u}{\phi} = \overset{u}{\phi_0} > \overset{u}{H} + \overset{l}{H} \\ \overset{l}{\phi} = \overset{l}{\phi_0} > \overset{l}{H} \end{array} \right\} \quad (L > a) \tag{31.44}$$

then the potentials at $z = L$ become

$$z = L \quad \left\{ \begin{array}{l} \overset{u}{\Phi} = \overset{u}{\Phi_0} = kH\overset{u}{\phi_0} \\ \overset{l}{\Phi} = \overset{l}{\Phi_0} = kH\overset{l}{\phi_0} \end{array} \right\} \quad \Phi = \Phi_0 = \overset{u}{\Phi_0} + \overset{l}{\Phi_0} \tag{31.45}$$

Application of this boundary condition to (31.43) yields $C = (Q_{x0} - iQ_{y0})L + \Phi_0$ so that the comprehensive potential becomes

$$\Omega = -(Q_{x0} - iQ_{y0})(z - L) + \Phi_0 \qquad (31.46)$$

We will use the dimensionless variable Z which for this case becomes

$$Z = \frac{z}{a+b} \qquad (31.47)$$

and we introduce the constant λ as

$$\lambda = \frac{L}{a+b} \qquad (31.48)$$

so that (31.46) may be written in terms of Z and λ as

$$\Omega = -\frac{(Q_{x0} - iQ_{y0})}{a+b}(Z - \lambda) + \Phi_0 \qquad (31.49)$$

The head along the boundary of the ellipse may be obtained from (31.49):

$$\overset{u}{\Phi} = k(\overset{u}{H} + \overset{l}{H})\overset{}{\phi} = -\frac{Q_{x0}(X_e - \lambda) + Q_{y0}Y_e}{a+b} + \Phi_0 \qquad (31.50)$$

where X_e and Y_e are the coordinates of a point on the ellipse in the Z plane.

The Complex Potential $\overset{l}{\Omega}$

The complex potential $\overset{l}{\Omega}$ is obtained by application of the boundary condition that $\overset{l}{\phi} = \overset{}{\phi}$ along the ellipse and that $\overset{l}{\Omega}$ produces the prescribed far-field. It follows from (31.39) and (31.42) that

$$\overset{l}{\Phi} = \frac{\overset{l}{H}}{\overset{u}{H} + \overset{l}{H}} \overset{}{\Phi} = \nu \overset{}{\Phi} \qquad (31.51)$$

where

$$\nu = \frac{\overset{l}{H}}{\overset{u}{H} + \overset{l}{H}} \qquad (31.52)$$

We write $\overset{l}{\Omega}$ as the sum of $\nu \Omega$ and a second complex potential $\overset{l}{\Omega}{}^*$ as follows:

$$\overset{l}{\Omega} = -\frac{\nu}{a+b}(Q_{x0} - iQ_{y0})(Z - \lambda) + \nu \Phi_0 + \overset{l}{\Omega}{}^* \qquad (31.53)$$

This function meets the boundary condition (31.51) (compare [31.49]), provided that the real part of $\overset{l}{\Omega}{}^*$ vanishes along the edge of the opening:

$$\overset{l}{\underset{e}{\Phi}}{}^* = \Re\overset{l}{\underset{e}{\Omega}}{}^* = 0 \tag{31.54}$$

Furthermore, $\overset{l}{\Omega}$ must have a far-field consisting of a uniform and a radial component:

$$Z \to \infty \qquad \overset{l}{\Omega} \to \frac{-\left(\overset{l}{Q}_{x0} - i\overset{l}{Q}_{y0}\right)}{a+b} Z + \frac{Q^*}{2\pi} \ln Z \tag{31.55}$$

It follows with (31.53) that $\overset{l}{\Omega}{}^*$ must be of the following form near infinity:

$$Z \to \infty \qquad \overset{l}{\Omega}{}^* \to -\frac{\left[\left(\overset{l}{Q}_{x0} - \nu Q_{x0}\right) - i\left(\overset{l}{Q}_{y0} - \nu Q_{y0}\right)\right]}{a+b} Z + \frac{Q^*}{2\pi} \ln Z \tag{31.56}$$

We observe from the above that $\overset{l}{\Omega}{}^*$ is formally equal to the complex potential for a lake in a uniform flow, with a potential of zero along the lake; the first term in (31.56) represents the uniform flow and the second one the far-field due to the net flow into the lake. Introducing Q_0^* and β such that

$$Q_0^* e^{-i\beta} = \overset{l}{Q}_{x0} - \nu Q_{x0} - i\left(\overset{l}{Q}_{y0} - \nu Q_{y0}\right) \tag{31.57}$$

we may apply (31.23) to $\overset{l}{\Omega}{}^*$, replacing Q_0 by Q_0^*, and α and Φ_0 by 0. This yields

$$\overset{l}{\Omega}{}^* = -\tfrac{1}{2}(a+b)Q_0^* \left[\chi e^{-i\beta} - \frac{e^{i\beta}}{\chi}\right] + \frac{Q^*}{2\pi} \ln \chi \tag{31.58}$$

Substitution of this expression for $\overset{l}{\Omega}{}^*$ in (31.53) yields the complex potential $\overset{l}{\Omega}$:

$$\overset{l}{\Omega} = -\frac{\nu}{a+b}(Q_{x0} - iQ_{y0})(Z - \lambda) + \nu\Phi_0$$

$$-\tfrac{1}{2}(a+b)Q_0^* \left[\chi e^{-i\beta} - \frac{e^{i\beta}}{\chi}\right] + \frac{Q^*}{2\pi} \ln \chi \tag{31.59}$$

The magnitude of the discharge Q^* through the opening is determined by applying the boundary condition (31.45) to $\overset{l}{\Phi}$, which gives with $Z = \lambda$ and $\chi = \chi_0$

$$\overset{l}{\Phi}_0 = \nu\Phi_0 - \tfrac{1}{2}(a+b)Q_0^* \cos\beta \left[\chi_0 - \frac{1}{\chi_0}\right] + \frac{Q^*}{2\pi} \ln \chi_0 \tag{31.60}$$

where χ_0 is expressed in terms of λ by the use of (31.13)

$$\chi_0 = \lambda + \sqrt{\lambda^2 - \gamma^2} \qquad (\lambda > \gamma) \tag{31.61}$$

This expression for χ_0 is real and positive because the point $Z = \lambda$ is outside the ellipse, so that $\lambda > \gamma$. Solving (31.60) for Q^* we obtain

$$Q^* = 2\pi \frac{\overset{l}{\Phi}_0 - \nu \Phi_0 + \frac{1}{2}(a+b)Q_0^* \cos\beta \left[\chi_0 - \frac{1}{\chi_0}\right]}{\ln \chi_0} \tag{31.62}$$

The Complex Potential $\overset{u}{\Omega}$

The complex potential $\overset{u}{\Omega}$ is found from the relation

$$\overset{u}{\Omega} = \Omega - \overset{l}{\Omega} \tag{31.63}$$

with Ω and $\overset{l}{\Omega}$ given by (31.49) and (31.59). It is left as an exercise for the reader to verify that (31.63) fulfills the boundary conditions along the ellipse and at infinity, and the condition that $\overset{u}{\Phi} = \overset{u}{\Phi}_0$ for $Z = \lambda$.

Problem 31.3

Prove that the complex potential $\overset{u}{\Omega}$ fulfills the conditions along the ellipse, at infinity, and at $Z = \lambda$.

Example 31.1

As an example, the flow pattern is shown in Figure 5.8 for the special case that the ellipse is a circle ($a = b = R$), that $\overset{u}{Q}_{y0} = \overset{l}{Q}_{x0} = \overset{l}{Q}_{y0} = 0$ and $Q^* = 0$, and that $\overset{u}{H} = \overset{l}{H} = H$, $\overset{u}{\phi}_0 = 2H$, $L = 10H$, $R = 2.5H$, and $\overset{l}{\phi}_0$ is such that the head in the lower aquifer at infinity is $4H$. The streamlines and piezometric contours are shown for the upper and lower aquifers in Figures 5.8(a) and (b), respectively. It is of interest to note that the flow in the lower aquifer is from the upstream side to the downstream side of the circle. The flow pattern in Figure 5.8(b) illustrates the effect of an opening connecting an aquifer with uniform flow to an unused aquifer.

32 APPLICATIONS OF THE SCHWARZ-CHRISTOFFEL TRANSFORMATION

The boundaries of flow domains are often composed of segments of straight lines. Such problems may be solved by the use of a general transformation, the Schwarz-Christoffel transformation, which maps the upper half plane $\Im\zeta \geq 0$ onto domains bounded by a

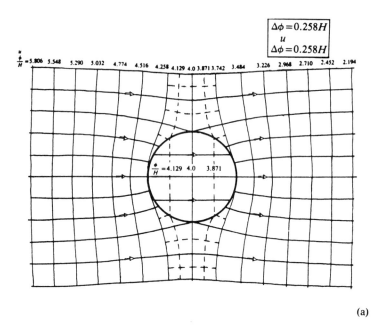

(a)

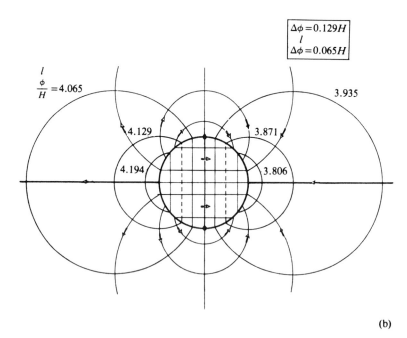

(b)

Figure 5.8 Flow nets for (a) the upper and (b) the lower aquifers. (Source: Strack [1981(a)].)

Sec. 32 Applications of the Schwarz-Christoffel Transformation

polygon of straight lines. We will discuss the Schwarz-Christoffel transformation in this section and use it to solve several problems of groundwater flow.

The Transformation

The Schwarz-Christoffel transformation maps the upper half plane $\Im \zeta \geq 0$ ($\zeta = \xi + i\eta$) of Figure 5.9(b) onto a domain bounded by a polygon in the z plane in such a way that the $n-1$ vertices $z_1, z_2, \ldots, z_{n-1}$ of the polygon correspond to the points $\xi_1, \xi_2, \ldots \xi_{n-1}$ of the real axis in the ζ plane. Point z_n corresponds to infinity in the ζ plane. We express the fact that the boundary in the z plane is bounded by a polygon by the equation

$$\arg(dz)_\nu - \arg(dz)_{\nu-1} = \pi k_\nu \qquad (\nu = 2, \ldots, n) \tag{32.1}$$

where $(dz)_\nu$ represents the complex increment of z along the straight line between z_ν and $z_{\nu+1}$. Thus, (32.1) expresses that the argument of dz is constant along each side of the polygon, but increases abruptly by an amount πk_ν if z passes the corner point z_ν while tracing the boundary in positive sense, i.e., with the domain to the left. By itself, the increment dz is undefined in magnitude, but its magnitude in relation to that of the corresponding increment $d\zeta$ is defined. When z traces the boundary of the polygon, ζ traces the boundary of the upper half plane, the real axis, in positive sense. The argument of $d\zeta$ therefore remains zero and the argument of $dz/d\zeta$ equals $\arg(dz) - \arg(d\zeta) = \arg(dz)$; we may write (32.1) as

$$\arg\left(\frac{dz}{d\zeta}\right)_\nu - \arg\left(\frac{dz}{d\zeta}\right)_{\nu-1} = \pi k_\nu \qquad (\nu = 2, \ldots, n) \tag{32.2}$$

We introduce the function $\chi = \tau + i\omega$ as

$$\chi = \tau + i\omega = \ln \frac{dz}{d\zeta} \tag{32.3}$$

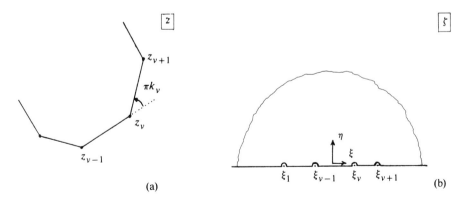

Figure 5.9 The Schwarz-Christoffel transformation.

so that we may write (32.2) as

$$\omega_\nu - \omega_{\nu-1} = \pi k_\nu \qquad (\nu = 2, \ldots, n) \qquad (32.4)$$

Since the mapping is conformal everywhere inside $\Im \zeta > 0$, $dz/d\zeta$ will be analytic and unequal to zero in $\Im \zeta > 0$; it follows from (32.3) that χ is analytic in $\Im \zeta > 0$. The function χ is fully determined, therefore, by the boundary conditions (32.4) complemented with the value of $dz/d\zeta$ along any one of the boundary segments, e.g., $(dz/d\zeta)_{n-1}$. These boundary conditions appear to be equal to the values that the stream function assumes when a set of wells of discharges $-2\pi k_\nu$ are placed at the points ξ_1, $\xi_2, \ldots, \xi_{n-1}$ of the real axis. We will make use of this analogy with a groundwater flow problem and write the function χ accordingly:

$$\chi = \sum_{\nu=1}^{n-1} -k_\nu \ln(\zeta - \xi_\nu) + C \qquad (32.5)$$

where C is a complex constant. We denote by φ the argument of $dz/d\zeta$ along the last interval of the real axis, i.e., the interval $\xi_{n-1} < \xi < \infty$, $\eta = 0$. Writing C as $\ln \left[|A| e^{i \arg A} \right]$, and noticing that the logarithms are all real along this interval, we obtain

$$\varphi = \arg A = \arg \left(\frac{dz}{d\zeta} \right) \qquad (\xi_{n-1} < \xi < \infty \qquad \eta = 0) \qquad (32.6)$$

The transformation (32.5) may now be written as

$$\chi = \ln \left[|A| e^{i\varphi} \prod_{\nu=1}^{n-1} (\zeta - \xi_\nu)^{-k_\nu} \right] \qquad (32.7)$$

and using (32.3) we obtain

$$\frac{dz}{d\zeta} = |A| e^{i\varphi} \prod_{\nu=1}^{n-1} (\zeta - \xi_\nu)^{-k_\nu} \qquad (32.8)$$

so that the Schwarz-Christoffel integral becomes

$$\boxed{z = |A| e^{i\varphi} \int \prod_{\nu=1}^{n-1} (\zeta - \xi_\nu)^{-k_\nu} d\zeta + B} \qquad (32.9)$$

where B is a complex constant of integration. The case where the nth corner point corresponds to a finite point, ξ_n, is covered simply by letting the product in (32.9) run over n terms, rather than $n - 1$, and by letting φ be defined as the argument of $dz/d\zeta$ on the interval $\eta = 0$, $\xi_n \leq \xi < \infty$:

$$\varphi = \arg \frac{dz}{d\zeta} \qquad (\xi_n < \xi < \infty \qquad \eta = 0) \qquad (32.10)$$

Sec. 32 Applications of the Schwarz-Christoffel Transformation

$$z = |A|e^{i\varphi} \int \prod_{\nu=1}^{n} (\zeta - \xi_\nu)^{-k_\nu} d\zeta + B \qquad (32.11)$$

Determination of the Parameters

The transformation (32.11) contains as parameters the n unknowns ξ_ν ($\nu = 1, 2, \ldots, n$), the modulus $|A|$, and the complex constant B. In order to establish how the choice of these parameters affects the mapping, we consider the case of a triangle T with corner points z_1, z_2, and z_3, which correspond to ξ_1, ξ_2, and ξ_3 (see Figure 5.10). For clarity, we write the transformation as a definite integral, with ξ_1 as the lower bound:

$$z = |A|e^{i\varphi} \int_{\xi_1}^{\zeta} \prod_{\nu=1}^{3} (\zeta^* - \xi_\nu)^{-k_\nu} d\zeta^* + B \qquad (32.12)$$

This function maps the real axis on a triangle T^* with interior angles $\pi(1 - k_\nu)$. The integral vanishes for $\zeta = \xi_1$, so that

$$z_1 = B_1 + iB_2 \qquad (32.13)$$

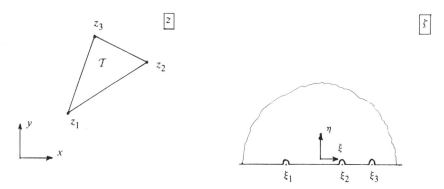

Figure 5.10 The parameters in the Schwarz-Christoffel transformation.

Thus, the triangle T^* has the same shape and orientation as the given triangle T; it can differ only in size and location from the given one. For any choice of ξ_1, ξ_2, and ξ_3, therefore, $|A|$ and B can be chosen such that T^* is mapped exactly onto T. We arrive at the conclusion that the three parameters ξ_1, ξ_2, and ξ_3 may be chosen anywhere on the ξ axis, including infinity, provided that $\xi_1 < \xi_2 < \xi_3$.

The shape of the polygon is no longer independent of the choice of ξ_ν ($\nu = 1, 2, \ldots, n$) when there are more than three sides; if ξ_1, ξ_2, and ξ_3 are chosen, the parameters $\xi_4, \xi_5, \ldots, \xi_n$ must be determined such that the polygon has the desired shape.

Problem 32.1
Determine the value of ω along the interval $\eta = 0$, $-\infty < \xi < \xi_1$ (using [32.7]) for the two cases covered by (32.9) and (32.11).

Mapping the Upper Half Plane onto the Exterior of a Polygon

A special case arises when $\Im\zeta \geq 0$ is mapped onto the exterior of a polygon with finite sides, as illustrated in Figure 5.11. In that case, the function χ defined in (32.3) is singular inside $\Im\zeta \geq 0$, and we must examine this singular behavior in order to determine the transformation. We already encountered a mapping where infinity in the z plane corresponds to a finite point in the ζ plane, when mapping the exterior of a circle onto $\Im\zeta \geq 0$ (see Figure 5.2). We observed (compare [30.18]) that the mapping function has a first-order pole at the point $\zeta = \delta$, if δ corresponds to $z = \infty$, and may therefore be written as

$$z(\zeta) = \frac{1}{\zeta - \delta} f(\zeta) \qquad (\zeta \to \delta) \tag{32.14}$$

where $f(\zeta)$ is analytic in $\Im\zeta > 0$. Differentiating (32.14), we obtain

$$z'(\zeta) = \frac{1}{(\zeta - \delta)^2}[f'(\zeta)(\zeta - \delta) - f(\zeta)] = \frac{h(\zeta)}{(\zeta - \delta)^2} \tag{32.15}$$

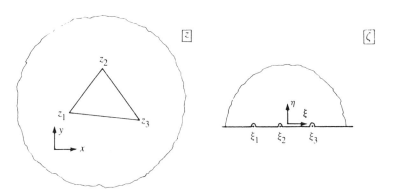

Figure 5.11 Mapping $\Im\zeta \geq 0$ onto the exterior of a polygon.

where $g(\zeta)$ is analytic in $\Im\zeta > 0$. We now return to the groundwater flow analogy: it follows that the function $\chi = \ln dz/d\zeta$ may be written as

$$\chi = -2\ln(\zeta - \delta) + \ln g(\zeta) \tag{32.16}$$

and the singularity of χ at $\zeta = \delta$ is equal to that of a recharge well of strength 4π. Thus, the function χ in (32.5) must be modified such that it exhibits a singularity of the form

(32.16) while the values of ω along the real axis $\Im\zeta = 0$ remain unchanged. This is achieved by adding an "image recharge well" at $\zeta = \bar{\delta}$, so that (32.5) becomes

$$\chi = \sum_{\nu=1}^{n-1}[-k_\nu \ln(\zeta - \xi_\nu)] - 2\ln\left[(\zeta - \delta)(\zeta - \bar{\delta})\right] + \ln|A|e^{i\varphi} \tag{32.17}$$

and, with $\chi = \ln dz/d\zeta$, we obtain

$$z = |A|e^{i\varphi} \int \frac{\prod_{\nu=1}^{n-1}(\zeta - \xi_\nu)^{-k_\nu}}{(\zeta - \delta)^2(\zeta - \bar{\delta})^2} d\zeta + B \tag{32.18}$$

It is important to note that, in general, the transformation (32.18) will have a logarithmic singularity at $\zeta = \delta$; the singularity at $\bar{\delta}$ is outside the domain and need not be considered. This may be seen as follows: let $h(\zeta)$ be defined as

$$h(\zeta) = |A|e^{i\varphi}\frac{\prod_{\nu=1}^{n-1}(\zeta - \xi_\nu)^{-k_\nu}}{(\zeta - \bar{\delta})^2} \tag{32.19}$$

so that the function $z'(\zeta)$ may be written as (compare [32.15]

$$z' = \frac{h(\zeta)}{(\zeta - \delta)^2} \tag{32.20}$$

We observe from (32.19) that $h(\zeta)$ is analytic at $\zeta = \delta$, and thus may be expanded in a Taylor series about that point:

$$z' = \frac{1}{(\zeta - \delta)^2}\left[h(\delta) + h'(\delta)(\zeta - \delta) + \frac{h''(\delta)}{2!}(\zeta - \delta)^2 + \ldots\right] \tag{32.21}$$

so that

$$z = \frac{-h(\delta)}{\zeta - \delta} + h'(\delta)\ln(\zeta - \delta) + \frac{h''(\delta)}{2!}(\zeta - \delta) + \ldots \tag{32.22}$$

This behavior will correspond to the required behavior (32.14) only if

$$h'(\delta) = 0 \tag{32.23}$$

This condition yields an additional complex equation, which may be used to determine δ; the condition (32.23) is known as the *residue condition* (see Von Koppenfels and Stallmann [1959]). In practice, the condition (32.23) is best applied by first carrying out the integration and then determining the parameters in the solution in such way that the logarithmic singularity is removable, as expressed by (32.23).

Flow Toward a Well Between Two Long Parallel Rivers

As a first application of the Schwarz-Christoffel transformation, we will solve the problem of flow toward a well between two long parallel rivers that are a distance d apart (see

Figure 5.12). The origin is chosen as shown in the figure, with the x axis coinciding with one of the rivers. We first consider the case that the potential along each river is Φ_0. The flow domain is the infinite strip $1\mathcal{A}$-2-$3\mathcal{A}$-$3\mathcal{B}$-4-$1\mathcal{B}$, and points 1 and 3 are chosen at infinity and the origin of the ζ plane, respectively. The values for πk_1 and πk_3 are both π; the argument of dz increases by an amount π if z moves through either point 1 (infinity) or point 3 (infinity) with the flow domain to the left. With $\varphi = \arg(dz/d\xi) = 0$ along $3\mathcal{B}$-4-$1\mathcal{B}$ (compare [32.6]), the Schwarz-Christoffel transformation becomes, leaving out the term $(\zeta - \xi_1)^{-k_1}$ (point 1 corresponds to infinity in the ζ plane),

$$z = |A| \int \frac{d\zeta}{\zeta} + B = |A| \ln \zeta + B \qquad (32.24)$$

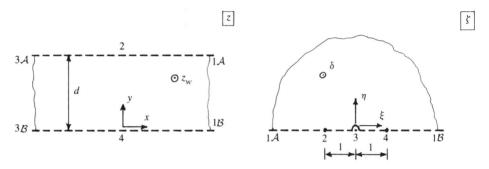

Figure 5.12 Flow toward a well between two long parallel rivers.

Equations for the constants $|A|$ and $B = B_1 + iB_2$ are obtained by application of the boundary conditions, which are

$$\eta = 0 \qquad 0 < \xi < \infty \qquad y = 0 \qquad -\infty < x < \infty \qquad (32.25)$$
$$\eta = 0 \qquad -\infty < \xi < 0 \qquad y = d \qquad -\infty < x < \infty \qquad (32.26)$$

Application of these conditions to (32.24) yields

$$x = |A| \ln |\zeta| + B_1 + iB_2 \qquad (0 < \xi < \infty \qquad \eta = 0) \qquad (32.27)$$

and, writing ζ as $|\zeta|e^{i\pi}$ along the negative real axis,

$$x + id = |A| \ln |\zeta| + |A|i\pi + B_1 + iB_2 \qquad (-\infty < \xi < 0 \qquad \eta = 0) \qquad (32.28)$$

It follows from the first condition that $B_2 = 0$, and from the second one that $|A| = d/\pi$, so that the transformation becomes, choosing point 4 ($z = 0$) at $\zeta = 1$,

$$z = \frac{d}{\pi} \ln \zeta \qquad (32.29)$$

This transformation may be readily inverted:

$$\zeta = e^{\pi z/d} \qquad (32.30)$$

Sec. 32 Applications of the Schwarz-Christoffel Transformation

The complex potential for flow toward a well at $z = z_w$ is determined by the method of images in $\Im \zeta \geq 0$. Denoting the location of the well in the ζ plane as δ, with

$$\delta = e^{\pi z_w/d} \tag{32.31}$$

the complex potential becomes

$$\Omega = \frac{Q}{2\pi} \ln \frac{\zeta - \delta}{\zeta - \bar{\delta}} + \Phi_0 \tag{32.32}$$

It is possible to write the complex potential in terms of z, which gives

$$\Omega = \frac{Q}{2\pi} \ln \frac{e^{\pi z/d} - e^{\pi z_w/d}}{e^{\pi z/d} - e^{\pi \bar{z}_w/d}} + \Phi_0 \tag{32.33}$$

The case that the values of the potential are different along the two rivers, e.g., the case where

$$y = d \quad -\infty < x < \infty \quad \Phi = \Phi_1 \tag{32.34}$$

$$y = 0 \quad -\infty < x < \infty \quad \Phi = \Phi_0 \tag{32.35}$$

may be covered by adding a potential for uniform flow to (32.33):

$$\Omega = -\frac{\Phi_1 - \Phi_0}{d} iz + \frac{Q}{2\pi} \ln \frac{e^{\pi z/d} - e^{\pi z_w/d}}{e^{\pi z/d} - e^{\pi \bar{z}_w/d}} + \Phi_0 \tag{32.36}$$

If there is infiltration of a rate N in the aquifer, then the complex potential cannot be used; the solution for this case is obtained by adding to the real part of (32.36) the elementary solution for infiltration, $\Phi = -\frac{1}{2} N y (y - d)$, which vanishes at $y = 0$ and $y = d$ and fulfills the Poisson equation $\nabla^2 \Phi = -N$:

$$\Phi = -\frac{1}{2} N y(y - d) + \frac{\Phi_1 - \Phi_0}{d} y + \frac{Q}{2\pi} \ln \left| \frac{e^{\pi z/d} - e^{\pi z_w/d}}{e^{\pi z/d} - e^{\pi \bar{z}_w/d}} \right| + \Phi_0 \tag{32.37}$$

This solution may, of course, be generalized for an arbitrary number of wells by superposition.

Flow Toward a Well Between Two Intersecting Rivers

As a second application of the Schwarz-Christoffel transformation, we consider the case of flow toward a well of discharge Q in an aquifer bounded by two long rivers which enclose an angle α; see Figure 5.13(a). The coordinate system is chosen as shown in the figure, and the well is located at $z = z_w$. The potential along each river is Φ_0. The boundary in the z plane is $1A$-2-$1B$, where point 1 is at infinity and point 2 is the point

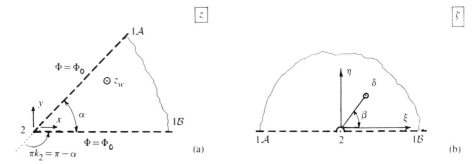

Figure 5.13 Flow toward a well between two intersecting rivers.

of intersection of the rivers. These points correspond in the ζ plane to infinity (point 1) and the origin (point 2). The Schwarz-Christoffel transformation takes the form

$$z = |A| \int \zeta^{-(1-\alpha/\pi)} d\zeta + B = \frac{\pi}{\alpha}|A|\zeta^{\alpha/\pi} + B \tag{32.38}$$

In order that point 2 corresponds to $\zeta = 0$, B must be zero,

$$z = \frac{\pi}{\alpha}|A|\zeta^{\alpha/\pi} \tag{32.39}$$

and we choose $|A|$ such that $\zeta = \delta = e^{i\beta}$ corresponds to the well, $z = z_w$:

$$z_w = \frac{\pi}{\alpha}|A|e^{i\alpha\beta/\pi} \tag{32.40}$$

so that

$$\frac{\pi}{\alpha}|A| = |z_w| \tag{32.41}$$

and the transformation becomes

$$\frac{z}{|z_w|} = \zeta^{\alpha/\pi} \tag{32.42}$$

The location of the well in the ζ plane is found from

$$\delta = e^{i\beta} = \left[\frac{z_w}{|z_w|}\right]^{\pi/\alpha} \tag{32.43}$$

The complex potential is obtained in terms of ζ by application of the method of images:

$$\Omega = \frac{Q}{2\pi} \ln \frac{\zeta - \delta}{\zeta - \bar{\delta}} + \Phi_0 \tag{32.44}$$

Sec. 32   Applications of the Schwarz-Christoffel Transformation

and may be written in terms of z by the use of (32.42),

$$\Omega = \frac{Q}{2\pi} \ln \frac{[z/|z_w|]^{\pi/\alpha} - [z_w/|z_w|]^{\pi/\alpha}}{[z/|z_w|]^{\pi/\alpha} - [\bar{z}_w/|z_w|]^{\pi/\alpha}} + \Phi_0 \tag{32.45}$$

or

$$\Omega = \frac{Q}{2\pi} \ln \frac{z^{\pi/\alpha} - z_w^{\pi/\alpha}}{z^{\pi/\alpha} - \bar{z}_w^{\pi/\alpha}} + \Phi_0 \tag{32.46}$$

Problem 32.2
Compute six points on each of five streamlines and equipotentials for the problem illustrated in Figure 5.13, and draw the flow net through these points.

Flow Underneath an Impermeable Dam

Consider the case of two-dimensional flow underneath a dam separating two lakes above a deep aquifer illustrated in Figure 5.14(a). The aquifer is the lower half plane $\Im z \leq 0$; point 1 is at infinity, and points 2 and 3 are at $z = L/2$ and $z = -L/2$, respectively. The heads along 1A-2 and 3-1B are ϕ_2 and ϕ_1, respectively, and the corresponding values of the potential are denoted as Φ_2 and Φ_1; the base of the dam is a streamline. The boundary conditions may be written as

$$y = 0 \quad -\infty < x < -\tfrac{1}{2}L \quad \Phi = \Phi_1 \tag{32.47}$$

$$y = 0 \quad -\tfrac{1}{2}L \leq x \leq \tfrac{1}{2}L \quad \Psi = 0 \tag{32.48}$$

$$y = 0 \quad \tfrac{1}{2}L < x < \infty \quad \Phi = \Phi_2 \tag{32.49}$$

We map the lower half plane $\Im z \leq 0$ onto the upper half plane $\Im \zeta \geq 0$ by the linear transformation

$$\zeta = -\frac{2z}{L} \tag{32.50}$$

so that points 2 and 3 are at $\zeta = -1$ and $\zeta = +1$ and point 1 is at $\zeta = \infty$.

We will solve this problem in a somewhat different way than we did for the two previous ones: we determine the boundary in the Ω plane that corresponds to that in the physical plane and map $\Im \zeta \geq 0$ onto the domain in the Ω plane. The two equipotentials 1A-2 and 3-1B correspond to two lines parallel to the Ψ axis in the Ω plane, and the base of the dam (2-3) is mapped onto the segment $\Phi_2 \leq \Phi \leq \Phi_1$ of the Φ axis. Point 1 is at the intersection of the two parallel lines 1A-2 and 1B-3, and therefore must be at infinity (see Figure 5.14[c]). Application of the Schwarz-Christoffel transformation yields

$$\Omega = |A|i \int \frac{d\zeta}{\sqrt{(\zeta - 1)(\zeta + 1)}} + B = |A|i \ln[\zeta + \sqrt{(\zeta - 1)(\zeta + 1)}] + B \tag{32.51}$$

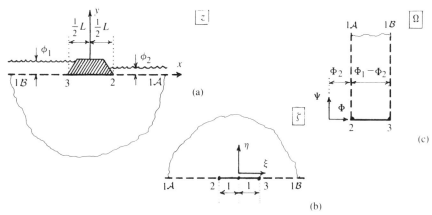

Figure 5.14 Flow underneath a dam.

ζ is defined in $\Im\zeta \geq 0$, and we choose the arguments of ζ, $\zeta - 1$, and $\zeta + 1$ accordingly,

$$0 \leq \arg \zeta \leq \pi \qquad 0 \leq \arg(\zeta - 1) \leq \pi \qquad 0 \leq \arg(\zeta + 1) \leq \pi \qquad (32.52)$$

Ω equals Φ_1 at point 3, $\zeta = 1$, so that

$$B = \Phi_1 \qquad (32.53)$$

and Ω equals Φ_2 at point 2, $\zeta = -1 = e^{i\pi}$, so that by (32.51) and (32.52),

$$\Phi_2 = |A|i^2\pi + B = -|A|\pi + \Phi_1 \qquad (32.54)$$

or

$$|A| = \frac{\Phi_1 - \Phi_2}{\pi} \qquad (32.55)$$

The transformation (32.51) thus becomes

$$\Omega = \frac{\Phi_1 - \Phi_2}{\pi} i \ln[\zeta + \sqrt{(\zeta - 1)(\zeta + 1)}] + \Phi_1 \qquad (32.56)$$

and may be written in an alternative form as follows:

$$\Omega = \frac{\Phi_1 - \Phi_2}{\pi} i \operatorname{arccosh} \zeta + \Phi_1 \qquad (32.57)$$

(This form is obtained by substituting $\cosh \lambda$ for ζ in (32.56); the natural logarithm then reduces to $\lambda = \operatorname{arccosh} \zeta$). Inversion of (32.57) yields, using (32.50),

$$\zeta = -\frac{2x}{L} - i\frac{2y}{L} = \cosh\left[\frac{\pi(\Omega - \Phi_1)}{i(\Phi_1 - \Phi_2)}\right] \qquad (32.58)$$

Sec. 32   Applications of the Schwarz-Christoffel Transformation   381

The reader may verify (problem 32.3) that the streamlines and equipotentials are ellipses and hyperbolae.

Problem 32.3

Determine the equations for the streamlines and equipotentials by the use of (32.58).

Flow Toward a Narrow Partially Penetrating Ditch in a Confined Aquifer

The Dupuit-Forchheimer assumption, which states that all equipotentials may be approximated by vertical surfaces, often is in direct contradiction with the boundary conditions. Common examples are partially penetrating wells, rivers, or creeks (compare Sec. 6, Figure 2.9). We will use an exact solution to a simplified problem of partial penetration to estimate the magnitude of the errors made by assuming full penetration.

The Mapping Function

The problem is illustrated in Figure 5.15, where flow occurs in a confined aquifer toward a narrow partially penetrating ditch. The flow takes place only in the vertical plane, and occurs from point 5 ($x = -\infty$) toward the ditch 2-3-4 and point 1 ($x = +\infty$). The width of the ditch is neglected, the ditch is an equipotential, the thickness of the aquifer is H, and the depth of the ditch is d. The origin is chosen at point 6, which lies on the base of the aquifer vertically below the ditch; the x axis points along the base toward point 1.

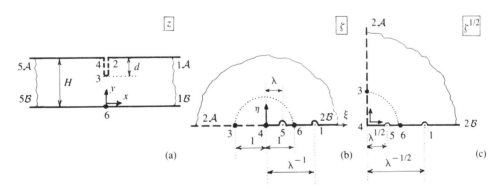

Figure 5.15 Flow toward a narrow partially penetrating ditch.

The corresponding points in $\Im \zeta \geq 0$ are chosen as shown in Figure 5.15(b): point 2 is at infinity, point 3 at $\zeta = -1$, point 4 at the origin, and points 5, 6, and 1 at $\zeta = \xi_5$, $\zeta = \xi_6$, and $\zeta = \xi_1$. The line 3-6 in the z plane is an axis of symmetry, which corresponds in the ζ plane to the unit circle, so that $\xi_6 = 1$. Points 1 and 5 lie symmetrically with respect to the axis of symmetry in the z plane and therefore are mapped in the ζ plane onto points that lie inversely with respect to the unit circle. Thus, if $\xi_5 = \lambda$, then $\xi_1 = \lambda^{-1}$, where $\lambda < 1$. The values of πk_ν ($\nu = 1, 2, \ldots, 6$) are found from the geometry of Figure 5.15(a): $\pi k_1 = \pi$, $\pi k_2 = \pi/2$, $\pi k_3 = -\pi$, $\pi k_4 = \pi/2$, $\pi k_5 = \pi$, $\pi k_6 = 0$.

Arg $dz/d\zeta$ is π along 1-2 so that $\varphi = \pi$, and the Schwarz-Christoffel transformation becomes:

$$\begin{aligned} z &= -|A| \int \frac{\zeta + 1}{(\zeta - \lambda)(\zeta - 1/\lambda)\zeta^{1/2}} d\zeta + B \\ &= -2|A| \int \frac{\zeta + 1}{(\zeta - \lambda)(\zeta - 1/\lambda)} d\zeta^{1/2} + B \end{aligned} \quad (32.59)$$

The integration may be performed after expanding the integrand in partial fractions with the following result:

$$z = \frac{|A|}{\lambda^{-1/2} - \lambda^{1/2}} \left[\ln \frac{\zeta^{1/2} - \lambda^{1/2}}{\zeta^{1/2} + \lambda^{1/2}} - \ln \frac{\zeta^{1/2} - \lambda^{-1/2}}{\zeta^{1/2} + \lambda^{-1/2}} \right] + B \quad (32.60)$$

The domain in the $\zeta^{1/2}$ plane is shown in Figure 5.15(c) and consists of the upper right quadrant. The angles θ_1 and θ_2 are introduced as

$$\theta_1 = \arg \frac{\zeta^{1/2} - \lambda^{1/2}}{\zeta^{1/2} + \lambda^{1/2}} \qquad \theta_2 = \arg \frac{\zeta^{1/2} - \lambda^{-1/2}}{\zeta^{1/2} + \lambda^{-1/2}} \quad (32.61)$$

and the range of values of these arguments for $\zeta^{1/2}$ in the upper right quadrant are chosen as

$$0 \leq \theta_1 \leq \pi \qquad 0 \leq \theta_2 \leq \pi \quad (32.62)$$

The four real constants $|A|$, $B_1 = \Re(B)$, $B_2 = \Im(B)$, and λ are determined from the boundary conditions as follows. We observe from Figure 5.15 that $y = H$ along $1A$-2, where both θ_1 and θ_2 are zero, so that

$$H = \Im z = B_2 \quad (32.63)$$

and $y = 0$ along $5B$-6-$1B$, where $\theta_1 = 0$ and $\theta_2 = \pi$, so that

$$0 = -\frac{|A|}{\lambda^{-1/2} - \lambda^{1/2}} \pi + H \quad (32.64)$$

or

$$\frac{|A|}{\lambda^{-1/2} - \lambda^{1/2}} = \frac{H}{\pi} \quad (32.65)$$

Application of the condition that $z = 0$ (point 6) corresponds to $\zeta = 1$, yields, using (32.63) and (32.64),

$$z(1) = 0 = \frac{H}{\pi} \left[\ln \left| \frac{1 - \lambda^{1/2}}{1 + \lambda^{1/2}} \frac{1 + \lambda^{-1/2}}{1 - \lambda^{-1/2}} \right| - i\pi \right] + B_1 + iH \quad (32.66)$$

The term between the modulus signs equals 1, so that (32.66) yields $B_1 = 0$ and the mapping function (32.60) becomes

$$z = \frac{H}{\pi}\left[\ln\frac{\zeta^{1/2} - \lambda^{1/2}}{\zeta^{1/2} + \lambda^{1/2}} - \ln\frac{\zeta^{1/2} - \lambda^{-1/2}}{\zeta^{1/2} + \lambda^{-1/2}}\right] + iH \qquad (32.67)$$

$\zeta^{1/2}$ is purely imaginary along 2-3-4 so that the real part of (32.67) vanishes, and

$$z = \frac{H}{\pi}i(\theta_1 - \theta_2) + iH \qquad (\Re\zeta^{1/2} = 0 \quad 0 \leq \Im\zeta^{1/2} < \infty) \qquad (32.68)$$

We expect, by symmetry, that point 3 ($\zeta_3^{1/2} = i$) corresponds to the lowest point of the ditch; that this is indeed true may be verified by calculating $dz/d\zeta$ at point 3: the derivative of the mapping function is zero there (compare [32.59], $dz/d\zeta = 0$ at $\zeta = -1$). Differentiation of (32.67) with respect to ζ indeed gives a function that vanishes at $\zeta = -1$. Requiring that z be $i(H - d)$ at point 3, $\zeta^{1/2} = i$, gives an expression that contains λ as the only unknown:

$$iH - id = \frac{H}{\pi}i\left(2\arctan\lambda^{1/2} - 2\arctan\lambda^{-1/2}\right) + iH \qquad (32.69)$$

where use is made of Figure 5.16 to express θ_1 and θ_2 at point 3 in terms of λ. Equation (32.69) may be written as follows:

$$\frac{\pi d}{H} = 2\arctan\frac{\lambda^{-1/2} - \lambda^{1/2}}{2} \qquad (32.70)$$

or

$$\lambda^{-1/2} - \lambda^{1/2} = 2\mu = 2\tan\frac{\pi d}{2H} \qquad (32.71)$$

where

$$\mu = \tan\frac{\pi d}{2H} \qquad (32.72)$$

Solving this equation for $\lambda^{1/2}$, we obtain with $0 < \lambda^{1/2} < 1$,

$$\lambda^{1/2} = -\mu + \sqrt{\mu^2 + 1} = \frac{1 - \sin\frac{\pi d}{2H}}{\cos\frac{\pi d}{2H}} = \tan\left[\frac{\pi}{4} - \frac{\pi d}{4H}\right] \qquad (32.73)$$

The only boundary condition that remains to be applied is the one along 4-5A, where $y = H$ and $\theta_1 = \theta_2 = \pi$:

$$H = \frac{H}{\pi}[\pi - \pi] + H \qquad (32.74)$$

which is indeed met.

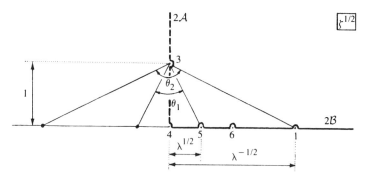

Figure 5.16 The angles θ_1 and θ_2 at point 3 ($\zeta^{1/2} = i$).

Inversion of the Mapping Function

The mapping function (32.67) may be written in a form suitable for inversion as follows:

$$\pi \frac{z - iH}{H} = Z = \ln \frac{(\zeta^{1/2} - \lambda^{1/2})(\zeta^{1/2} + \lambda^{-1/2})}{(\zeta^{1/2} + \lambda^{1/2})(\zeta^{1/2} - \lambda^{-1/2})}$$

$$= \ln \frac{\zeta + (\lambda^{-1/2} - \lambda^{1/2})\zeta^{1/2} - 1}{\zeta - (\lambda^{-1/2} - \lambda^{1/2})\zeta^{1/2} - 1} \quad (32.75)$$

or, after taking the exponential of both sides and using (32.71),

$$\left[\zeta - 2\zeta^{1/2} \tan \frac{\pi d}{2H} - 1 \right] e^Z = \zeta + 2\zeta^{1/2} \tan \frac{\pi d}{2H} - 1 \quad (32.76)$$

Solving this equation for $\zeta^{1/2}$ we obtain

$$\zeta^{1/2}_{1,2} = \tan \frac{\pi d}{2H} \coth \frac{Z}{2} \pm \sqrt{\left(\tan \frac{\pi d}{2H} \coth \frac{Z}{2} - i \right) \left(\tan \frac{\pi d}{2H} \coth \frac{Z}{2} + i \right)} \quad (32.77)$$

In order that point 2, $z = iH$, $Z = 0$, corresponds to $\zeta^{1/2} = \infty$ the $+$ sign must apply, and (32.77) becomes

$$\zeta^{1/2} = \chi + \sqrt{(\chi - i)(\chi + i)} \quad (32.78)$$

where

$$\chi = \mu \coth \frac{Z}{2} = \tan \frac{\pi d}{2H} \coth \frac{Z}{2} = \tan \frac{\pi d}{2H} \coth \frac{\pi(z - iH)}{H} \quad (32.79)$$

It is left as an exercise for the reader to verify that the boundary of the domain in the χ plane is as shown in Figure 5.17(b). With this region of definition of χ, we choose the arguments of $\chi - i$ and $\chi + i$ such that the argument of the square root in (32.78) does not jump inside the domain, but only across the boundary 2-3-4. If χ represents a point moving inside the boundaries from 2-3 (where $\arg(\chi - i)$ equals $\pi/2$) toward

3-4, then $\arg(\chi - i)$ decreases by an amount 2π (see Figure 5.17[b]). We therefore must assign the value $-3\pi/2$ to $\arg(\chi - i)$ along 3-4. The ranges of values that $\arg(\chi + i)$ and $\arg(\chi - i)$ may assume are

$$0 < \arg(\chi + i) < \pi \qquad -\frac{3\pi}{2} \le \arg(\chi - i) \le \frac{\pi}{2} \tag{32.80}$$

That point 3 corresponds to $\chi = i$ is seen as follows: separation of χ into real and imaginary parts gives

$$\chi = \tan\frac{\pi d}{2H} \coth\frac{Z}{2} = \tan\left[\frac{\pi d}{2H}\right] \frac{\sinh X - i\sin Y}{\cosh X - \cos Y} \tag{32.81}$$

with $Z_3 = \pi(z_3 - iH)/H$, (32.81) becomes, using that $z_3 = i(H - d)$,

$$\chi_3 = i\tan\left[\frac{\pi d}{2H}\right] \frac{\sin\frac{\pi d}{H}}{1 - \cos\frac{\pi d}{H}} = i\tan\frac{\pi d}{2H} \cot\frac{\pi d}{2H} = i \tag{32.82}$$

as asserted.

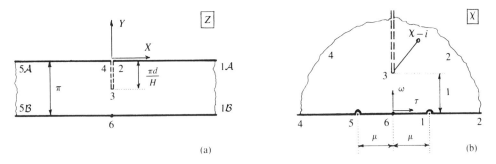

Figure 5.17 The domains in the Z and χ planes.

The Complex Potential $\Omega = \Omega(\zeta)$

The complex potential as a function of ζ may best be determined in the $\zeta^{1/2}$ plane of Figure 5.15(c). If the flow from point 5 is denoted as Q_5 and the flow toward point 1 as Q_1, then points 5 and 1 in the $\zeta^{1/2}$ plane represent a source and a sink of strengths $2Q_5$ and $2Q_1$, respectively. (Note that Q_5 and Q_1 represent the amounts flowing into or out of the upper half plane; the strengths are twice these amounts.) We use the method of images to meet the condition that 2-3-4, the imaginary axis in the $\zeta^{1/2}$ plane, is an equipotential. Denoting the potential along the ditch as Φ_0, we obtain

$$\Omega = -\frac{Q_5}{\pi} \ln \frac{\zeta^{1/2} - \lambda^{1/2}}{\zeta^{1/2} + \lambda^{1/2}} + \frac{Q_1}{\pi} \ln \frac{\zeta^{1/2} - \lambda^{-1/2}}{\zeta^{1/2} + \lambda^{-1/2}} + \Phi_0 \tag{32.83}$$

We will apply the conditions that Φ be Φ_5 and Φ_1 at $z = -L + iH$ and $z = L + iH$, respectively, where $L \gg H$. These points will correspond to points close to points 5

($\zeta^{1/2} = \lambda^{1/2}$) and 1 ($\zeta^{1/2} = \lambda^{-1/2}$) in the $\zeta^{1/2}$ plane, and to points close to $\chi = -\mu$ and $\chi = \mu$ in the χ plane. With $Z = -\pi L/H$ for $z = -L + iH$ (see [32.75]), we obtain for χ, with (32.79):

$$\chi = -\mu \coth \frac{\pi L}{2H} = -\mu \frac{1 + e^{-\pi L/H}}{1 - e^{-\pi L/H}} \approx -\mu\left(1 + 2e^{-\pi L/H}\right)$$
$$(L/H \gg 1 \quad Z = -\pi L/H) \tag{32.84}$$

so that

$$\chi + \mu \approx -2\mu e^{-\pi L/H} \quad (L/H \gg 1 \quad Z = -\pi L/H) \tag{32.85}$$

Next, we expand the function $\zeta^{1/2} = \zeta^{1/2}(\chi)$ (see [32.78]) about $\chi = -\mu$:

$$\zeta^{1/2} - \lambda^{1/2} = (\chi + \mu)\left[\frac{d\zeta^{1/2}}{d\chi}\right]_{\chi=-\mu} + O[(\chi + \mu)^2] \quad \left(Z = -\frac{\pi L}{H}\right) \tag{32.86}$$

It follows from (32.78) that

$$\frac{d\zeta^{1/2}}{d\chi} = \frac{\chi + \sqrt{\chi^2 + 1}}{\sqrt{\chi^2 + 1}} \tag{32.87}$$

so that

$$\left[\frac{d\zeta^{1/2}}{d\chi}\right]_{\chi=-\mu} = \frac{-\mu + \sqrt{\mu^2 + 1}}{\sqrt{\mu^2 + 1}} = \frac{\lambda^{1/2}}{\frac{1}{2}(\lambda^{1/2} + \lambda^{-1/2})} = \frac{2\lambda}{1 + \lambda} \tag{32.88}$$

where use is made of (32.73). Using (32.85), (32.86), and (32.88) we obtain

$$\zeta^{1/2} - \lambda^{1/2} \approx \frac{2\lambda}{1 + \lambda}(\chi + \mu) \approx -\frac{4\lambda}{1 + \lambda}\mu e^{-\pi L/H} \quad (Z = -\pi L/H) \tag{32.89}$$

Using this expression for $\zeta^{1/2} - \lambda^{1/2}$ and approximating $\zeta^{1/2}$ as $\lambda^{1/2}$ for all other terms in (32.83), we obtain the following for Φ_5:

$$\Phi_5 \approx -\frac{Q_5}{\pi} \ln \frac{\frac{4\lambda}{1+\lambda}\mu e^{-\pi L/H}}{2\lambda^{1/2}} + \frac{Q_1}{\pi} \ln \left|\frac{\lambda^{1/2} - \lambda^{-1/2}}{\lambda^{1/2} + \lambda^{-1/2}}\right| + \Phi_0 \tag{32.90}$$

or, using expression (32.71) for μ to write $2\lambda^{1/2}\mu/(1 + \lambda)$ as $(1 - \lambda)/(1 + \lambda)$,

$$\Phi_5 \approx \frac{Q_5 L}{H} - \frac{Q_5}{\pi} \ln \frac{1 - \lambda}{1 + \lambda} + \frac{Q_1}{\pi} \ln \frac{1 - \lambda}{1 + \lambda} + \Phi_0 \tag{32.91}$$

A similar procedure may be followed to obtain an expression for Φ_1, the potential at $z = L + iH$, which yields

$$\Phi_1 \approx -\frac{Q_1 L}{H} + \frac{Q_1}{\pi} \ln \frac{1 - \lambda}{1 + \lambda} - \frac{Q_5}{\pi} \ln \frac{1 - \lambda}{1 + \lambda} + \Phi_0 \tag{32.92}$$

Sec. 32 Applications of the Schwarz-Christoffel Transformation

Subtraction of (32.92) from (32.91) yields an expression for $Q_1 + Q_5$:

$$Q_1 + Q_5 = \frac{\Phi_5 - \Phi_1}{L} H \qquad (32.93)$$

It appears that the average discharge through the aquifer, $\frac{1}{2}(Q_1 + Q_5)$, is exactly the same as one obtains assuming full penetration. The discharge into the ditch, Q_0, equals $Q_5 - Q_1$, and we obtain, adding (32.91) to (32.92):

$$Q_0 = Q_5 - Q_1 = \frac{\Phi_5 + \Phi_1 - 2\Phi_0}{\frac{L}{H} - \frac{2}{\pi} \ln \frac{1-\lambda}{1+\lambda}} \qquad (32.94)$$

Using (32.71) and (32.73), $(1-\lambda)/(1+\lambda)$ may be written as $\sin \frac{\pi d}{2H}$, so that

$$Q_0 = Q_5 - Q_1 = \frac{\Phi_5 + \Phi_1 - 2\Phi_0}{\frac{L}{H} - \frac{2}{\pi} \ln \sin \frac{\pi d}{2H}} \qquad (32.95)$$

We observe that there is no flow into the ditch if Φ_0 equals the average of Φ_1 and Φ_5. Expressions for the individual discharges are readily obtained from (32.93) and (32.95):

$$Q_1 = \frac{1}{2} \left[\frac{\Phi_5 - \Phi_1}{L} H - \frac{\Phi_5 + \Phi_1 - 2\Phi_0}{\frac{L}{H} - \frac{2}{\pi} \ln \sin \frac{\pi d}{2H}} \right] \qquad (32.96)$$

and

$$Q_5 = \frac{1}{2} \left[\frac{\Phi_5 - \Phi_1}{L} H + \frac{\Phi_5 + \Phi_1 - 2\Phi_0}{\frac{L}{H} - \frac{2}{\pi} \ln \sin \frac{\pi d}{2H}} \right] \qquad (32.97)$$

Comparison with the Dupuit-Forchheimer Solution

We will first consider the case that $\Phi_5 = \Phi_1$; for this case the flow occurs from both sides toward the ditch, and the portion of the y axis below the ditch becomes a streamline by symmetry; see Figure 5.18(a). We denote the inflow into the ditch from the right by $\overset{r}{Q}_0$, and obtain from (32.95) with $\overset{r}{Q}_0 = \frac{1}{2} Q_0$, and $\Phi_5 = \Phi_1$:

$$\overset{r}{Q}_0 = \frac{(\Phi_1 - \Phi_0) H}{L - \frac{2H}{\pi} \ln \sin \frac{\pi d}{2H}} \qquad (32.98)$$

The discharge obtained by assuming full penetration is $\overset{r}{Q}_0^*$, with

$$\overset{r}{Q}_0^* = \frac{(\Phi_1 - \Phi_0) H}{L} \qquad (32.99)$$

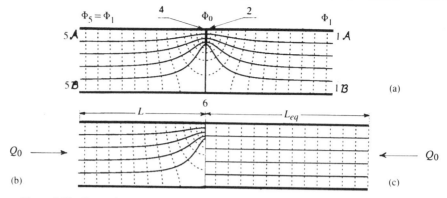

Figure 5.18 Comparison with the Dupuit-Forchheimer solution for the case that $\Phi_1 = \Phi_5$.

We observe that $\overset{r}{Q_0^*}$ is always larger than $\overset{r}{Q_0}$, since $\ln \sin \frac{\pi d}{2H}$ is less than 0. The aquifer of length L with the partially penetrating ditch in Figure 5.18(b) yields the same discharge as the aquifer of length L_{eq} with the fully penetrating ditch in Figure 5.18(c), where

$$L_{eq} = L - \frac{2H}{\pi} \ln \sin \frac{\pi d}{2H} = L + \frac{2H}{\pi} \ln \frac{1}{\sin \frac{\pi d}{2H}} \qquad (32.100)$$

The length L_{eq} is called the *equivalent length* (see Verruijt [1970]). When $d = H$, the ditch penetrates the aquifer fully and $L_{eq} = L$, whereas when $d = 0$ the ditch does not penetrate at all and $L = \infty$.

The relative error, ε, caused by assuming full penetration, may be expressed as follows:

$$\varepsilon = \frac{\overset{r}{Q_0^*} - \overset{r}{Q_0}}{\overset{r}{Q_0}} = -\frac{2H}{\pi L} \ln \sin \frac{\pi d}{2H} \qquad (32.101)$$

We see that the error is proportional to the ratio H/L; the error is small if the length of the aquifer is large with respect to its thickness. For such cases the resistance due to partial penetration is small with respect to the resistance over the distance L. We obtain for a ditch that penetrates only $0.1H$ into the aquifer, for example,

$$\varepsilon = 1.18 \frac{H}{L} \qquad (d/H = 0.1) \qquad (32.102)$$

This error is still only about 12% if $L = 10H$. For the general case that $\Phi_5 \neq \Phi_1$, the assumption of full penetration yields, denoting the total inflow into the ditch as Q_0^*,

$$Q_0^* = \frac{(\Phi_1 + \Phi_5 - 2\Phi_0)H}{L} \qquad (32.103)$$

whereas the exact solution predicts (see [32.95] and [32.100]),

$$Q_0 = \frac{(\Phi_1 + \Phi_5 - 2\Phi_0)H}{L_{eq}} \qquad (32.104)$$

Of special interest is the case that the flow occurs from the left to the right toward the ditch, which captures all the flow ($Q_0 = Q_5$, $Q_1 = 0$). The approximate solution predicts that this is the case when

$$\Phi_1 = \Phi_0 \qquad (32.105)$$

but in reality Q_1 will not be zero when $\Phi_1 = \Phi_0$; there is a discharge Q_1 not entering the ditch, which is found from (32.96) with $\Phi_1 = \Phi_0$:

$$Q_1 = -\frac{H}{L}(\Phi_5 - \Phi_0)\frac{\frac{1}{\pi}\ln\sin\frac{\pi d}{2H}}{\frac{L}{H} - \frac{2}{\pi}\ln\sin\frac{\pi d}{2H}} \qquad (32.106)$$

and we obtain from (32.97) for Q_5,

$$Q_5 = \frac{H}{L}(\Phi_5 - \Phi_0)\frac{\frac{L}{H} - \frac{1}{\pi}\ln\sin\frac{\pi d}{2H}}{\frac{L}{H} - \frac{2}{\pi}\ln\sin\frac{\pi d}{2H}} \qquad (32.107)$$

With (32.107), Q_1 may be written in terms of Q_5:

$$Q_1 = Q_5 \frac{-\frac{1}{\pi}\ln\sin\frac{\pi d}{2H}}{\frac{L}{H} - \frac{1}{\pi}\ln\sin\frac{\pi d}{2H}} \qquad (32.108)$$

For example, when the ditch penetrates $0.1H$ into the aquifer, (32.108) becomes

$$Q_1 = Q_5 \frac{0.59}{\frac{L}{H} + 0.59} \qquad (32.109)$$

If $L/H = 10$, this yields

$$Q_1 = 0.057\, Q_5 \qquad (32.110)$$

so that about 6% of the flow is not captured by the ditch.

33 TRANSFORMATIONS WITH PIECEWISE CONSTANT ARGUMENT

In this section we will consider transformations that map the real axis $\Im\zeta = 0$ onto polygons whose sides lie on lines through the same point, which is taken as the origin. Such transformations have the property that their argument is equal to a constant along each interval of $\Im\zeta = 0$. These constant values of the argument of the mapping function are not the same: the argument is piecewise constant and we call the mapping function a *transformation with piecewise constant argument*. Transformations with piecewise constant argument have a form similar to that of the derivative of the Schwarz-Christoffel transformation.

We distinguish three different categories of polygons, each requiring its own form of transformation:

1. Polygons that contain both the origin and infinity.
2. Polygons that contain infinity and not the origin.
3. Polygons that contain the origin and not infinity.

We will discuss three examples, one for each category, skipping some of the derivations for which the interested reader is referred to handbooks on conformal mapping such as Von Koppenfels and Stallmann [1959] and Lawretjew and Schabat [1967].

Polygons That Contain Both the Origin and Infinity

As we observed in deriving the Schwarz-Christoffel transformation, a function whose argument is piecewise constant along the real axis $\Im\tau = 0$ and is analytic elsewhere, can be written as a product of factors of the form $(\tau - \tau_\nu^*)^{\mu_\nu}$, where the real numbers τ_ν^* denote the points of $\Im\tau = 0$ where the argument of the mapping function jumps. In order to determine the exponents μ_ν, we examine the behavior of the transformation for the case of a corner point at the origin, such as point 2 in Figure 5.19(a), and for the case of a corner point at infinity, such as point 1 in this figure.

We applied in Sec. 32 the Schwarz-Christoffel transformation to determine the function that maps the upper half plane onto the domain $1A$-2-$1B$ in the z plane of Figure 5.13(a) and obtained a function of the form

$$z = |C|\tau^{\alpha/\pi} \tag{33.1}$$

where C is a constant and τ is defined in the upper half plane $\Im\tau \geq 0$ shown in Figure 5.19(b). The argument of τ changes as τ passes $\tau = 0$ and the exponent of the term $(\tau - 0)$ is α/π, i.e., equal to $1/\pi$ times the angle enclosed by the boundary at the corner point in the z plane. If we denote the corresponding point in the τ plane as τ_2^* and replace α by α_2, then (33.1) may be written as

$$z = |C|(\tau - \tau_2^*)^{\alpha_2/\pi} \tag{33.2}$$

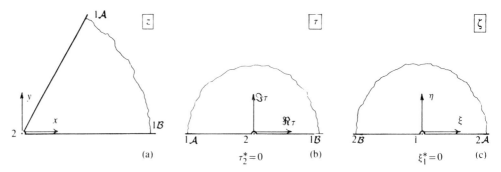

Figure 5.19 Mapping the upper half plane onto a polygon that contains both the origin and infinity.

Sec. 33 Transformations with Piecewise Constant Argument

Next, consider the domain in the ζ plane of Figure 5.19(c), where point 1 is mapped onto $\zeta = \xi_1^* = 0$ and point 2 onto infinity. The transformation $\zeta = \zeta(\tau)$ is

$$\zeta - \xi_1^* = e^{i\pi}\frac{1}{\tau - \tau_2^*} \tag{33.3}$$

as may be readily verified. The transformation that maps $\Im\zeta \geq 0$ onto the region in the z plane is obtained from (33.2) and (33.3) as

$$z = |C|e^{i\alpha_2}(\zeta - \xi_1^*)^{-\alpha_2/\pi} \tag{33.4}$$

We observe that the exponent of the term $\zeta - \xi_1^* = \zeta$ equals minus $1/\pi$ times the angle between the sides of the vertex at infinity in the z plane. We may conclude from the above, by analogy with the Schwarz-Christoffel transformation, that the general form of the transformation with piecewise constant argument for polygons of category 1 has the following form:

$$\boxed{z = |A|e^{i\varphi^*}\prod_{\nu=1}^{n}(\zeta - \xi_\nu^*)^{\mu_\nu}} \tag{33.5}$$

where, again in analogy with the Schwarz-Christoffel transformation, φ^* is the argument of $z(\zeta)$ along the rightmost section on the ξ axis, and where the μ_ν ($\nu = 1, 2, \ldots, n$) are defined as

$$\begin{aligned} z_\nu &= 0 & \mu_\nu &= \alpha_\nu/\pi \\ z_\nu &= \infty & \mu_\nu &= -\alpha_\nu/\pi \end{aligned} \tag{33.6}$$

The angle α_ν equals the angle enclosed by the νth and $\nu + 1^{th}$ boundary segments of the polygon in the z plane; μ_ν equals the increment of $\arg z(\xi)$ as ξ passes point ξ_ν^*, moving in the positive direction along the ξ axis. It is important to note that the point mapped onto $\xi_\nu^* = \infty$ is left out of the transformation.

The argument of the transformation (33.5) is constant between the points ξ_ν^*, where it jumps by an amount $\mu_\nu\pi$. Thus, the points ξ_ν^* correspond only to those points in the z plane where the argument of $z(\xi)$ jumps.

As an example, we will apply (33.5) to map the ζ plane in Figure 5.20(b) onto the domain bounded by the semi-infinite straight lines $1A$-2 and 2-$1B$ which enclose angle α. Point 1 is at infinity and point 2 at the origin of the z plane of Figure 5.20(a). We choose $\xi_1^* = -1$ and $\xi_2^* = 1$, and observe from the figure and (33.6) that $\mu_1 = -\alpha_1/\pi$, $\mu_2 = \alpha_1/\pi$, and φ^* equals zero, so that

$$z = |A|(\zeta + 1)^{-\alpha/\pi}(\zeta - 1)^{\alpha/\pi} = |A|\left[\frac{\zeta - 1}{\zeta + 1}\right]^{\alpha/\pi} \tag{33.7}$$

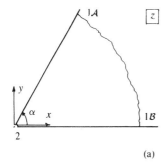 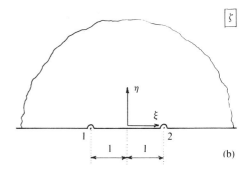

Figure 5.20 Application of the transformation with piecewise constant argument.

In addition, we will map the ζ plane in Figure 5.20(b) onto the τ plane in Figure 5.19(b) to determine the function $\tau = \tau(\zeta)$. This function has a discontinuous argument at $\tau = 0$ (point 2) and $\tau = \infty$ (point 1); $\alpha_1 = \alpha_2 = \pi$. Hence, $\mu_1 = -1$ and $\mu_2 = 1$. The angle φ^* equals the argument of $\tau(\zeta)$ along section 2-1$\mathcal{B}$ and equals 0. Hence, (33.5) becomes

$$\tau = |A_0| \frac{\zeta - 1}{\zeta + 1} \tag{33.8}$$

Taking the constant $|A_0|$ in (33.8) as 1 and combining the latter equation with (33.7) we obtain

$$z = |A| \tau^{\alpha/\pi} \tag{33.9}$$

which corresponds to (33.1).

Polygons That Contain Infinity and Not the Origin

If the polygon does not contain the origin there will be a point in $\Im \zeta > 0$ for which the function $z = z(\zeta)$ is zero. The mapping will be conformal at this point, denoted as $\zeta = \sigma$, and therefore the behavior of $z = z(\zeta)$ near $\zeta = \sigma$ will be as follows:

$$z(\zeta) = z'(\sigma)(\zeta - \sigma) + \frac{z''(\sigma)}{2!}(\zeta - \sigma)^2 + \ldots \tag{33.10}$$

The transformation (33.5) fulfills the conditions along the boundary, but lacks the behavior expressed by (33.10); we must provide this behavior without affecting the argument of $z(\zeta)$ along the real axis. This may be done by multiplying the transformation by a factor $(\zeta - \sigma)(\zeta - \overline{\sigma})$ which is real and positive for $\zeta = \xi$, provides the desired behavior at $\zeta = \sigma$ and adds a zero to the transformation in $\Im \zeta < 0$, i.e., outside the domain. The function $z = z(\zeta)$ thus becomes

$$\boxed{z = |A| e^{i\varphi^*} (\zeta - \sigma)(\zeta - \overline{\sigma}) \prod_{\nu=1}^{n} (\zeta - \xi_\nu^*)^{\mu_\nu}} \tag{33.11}$$

Sec. 33   Transformations with Piecewise Constant Argument

Flow in a System of Two Aquifers Separated by an Impermeable Layer with a Gap

We will solve the problem of flow in the vertical plane illustrated in Figure 5.21(a) as an example. Two thick aquifers are separated by two semi-infinite impermeable layers. The far-field in the upper aquifer is a constant potential, $\Phi = \Phi_0$, and the far-field in the lower aquifer is a uniform flow of discharge Q_{x0} in the x direction as shown in the figure:

$$\begin{aligned} y > 0 && z \to \infty && \Omega \to \Phi_0 \\ y < 0 && z \to \infty && \Omega \to -Q_{x0} z \end{aligned} \qquad (33.12)$$

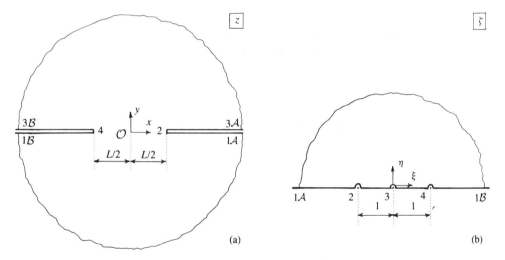

Figure 5.21 Flow in a system of two aquifers separated by an impermeable layer with a gap.

We will solve the flow problem by first mapping the upper half plane $\Im \zeta \geq 0$ onto the flow domain and then determine Ω as a function of ζ. The points where the argument of $z(\zeta)$ jumps are 1 and 3 and we choose $\xi_1^* = \infty$ and $\xi_3^* = 0$. The argument of $z(\zeta)$ does not change at points 2 and 4, which therefore do not enter explicitly into the transformation. The origin in the z plane does not lie on the boundary and therefore we apply (33.11). Since $\alpha_3 = \pi$ and point 3 lies at infinity in the z plane, μ_3 equals -1. The argument of $z(\zeta)$ along section 4-1$\mathcal{B}$, the rightmost section of the ξ axis, equals π and the transformation becomes

$$z = -|A| \frac{(\zeta - \sigma)(\zeta - \overline{\sigma})}{\zeta} \qquad (33.13)$$

By symmetry, the y axis corresponds to the η axis, so that the origin $\mathcal{O}$ in the z plane corresponds to a point of the η axis, i.e., $\Re \sigma = 0$. We have chosen two of the three degrees of freedom in the transformation: ξ_1^* and ξ_3^*. We use the remaining degree of freedom by choosing $\Im\sigma = 1$ so that $\sigma = i$ and (33.13) reduces to

$$z = -|A|\frac{\zeta^2 + 1}{\zeta} = -|A|\left(\zeta + \frac{1}{\zeta}\right) \tag{33.14}$$

We noted that points 2 and 4 do not enter explicitly into the transformation. The locations of these points on the ξ axis are found as follows. The mapping is not conformal at points 2 and 4: angles of π in the ζ plane are transformed into angles of 2π in the z plane. We saw in Sec. 29 that at such points the first order derivative of the mapping function vanishes (compare [29.8]):

$$z'(\xi_2) = z'(\xi_4) = 0 \tag{33.15}$$

where the prime denotes differentiation with respect to ζ. We use these equations to determine ξ_2 and ξ_4. We observe from (33.14) that

$$z' = -|A|\left(1 - \frac{1}{\zeta^2}\right) \tag{33.16}$$

This function is zero at the points $\zeta = 1$ and $\zeta = -1$ so that

$$\xi_2 = -1 \qquad \xi_4 = 1 \tag{33.17}$$

Application of the conditions $z(\xi_2) = \frac{1}{2}L$ and $z(\xi_4) = -\frac{1}{2}L$ (see Figure 5.21) gives,

$$|A| = \frac{L}{4} \tag{33.18}$$

so that the transformation becomes

$$z = -\frac{L}{4}\left(\zeta + \frac{1}{\zeta}\right) \tag{33.19}$$

The Complex Potential. The complex potential is obtained as a function of ζ by application of the conditions that the real axis is a streamline, that the far-field near point 1 is uniform and that the potential is finite at point 3. It follows from (33.19) that $z \to -L\zeta/4$ for $\zeta \to \infty$ so that the complex potential near $\zeta = \infty$ is obtained from (33.12) as

$$\zeta \to \infty \qquad \Omega \to \tfrac{1}{4} Q_{x0} L \zeta \tag{33.20}$$

The complex potential with this far-field that meets the boundary condition is

$$\Omega = \tfrac{1}{4} Q_{x0} L \zeta + \Phi_0 \tag{33.21}$$

where Φ_0 represents the potential along the η axis, and therefore along the y axis. The reference parameter ζ may be eliminated by solving (33.19) for ζ and substituting the resulting expression in (33.21). Equation (33.19) may be written as

$$\zeta^2 + \frac{4z}{L}\zeta + 1 = 0 \tag{33.22}$$

with the solutions

$$\zeta_{1,2} = -\frac{2}{L}\left[z \pm \sqrt{(z - \tfrac{1}{2}L)(z + \tfrac{1}{2}L)}\right] \tag{33.23}$$

The arguments of $z - \tfrac{1}{2}L$ and $z + \tfrac{1}{2}L$ are chosen such that the branch cuts coincide with the impermeable layer:

$$0 \leq \arg(z - \tfrac{1}{2}L) \leq 2\pi \qquad -\pi \leq \arg(z + \tfrac{1}{2}L) \leq \pi \tag{33.24}$$

We determine whether the $+$ or the $-$ sign applies in (33.23) by examining the behavior of $\zeta(z)$ along the boundary. For example, along 1A-2 $\arg(z - \tfrac{1}{2}L) = 2\pi$ and $\arg(z + \tfrac{1}{2}L) = 0$, so that $(z - \tfrac{1}{2}L)(z + \tfrac{1}{2}L)^{1/2} = -(x^2 - \tfrac{1}{4}L^2)^{1/2}$. Since $\zeta = \xi$ must approach $-\infty$ as x approaches ∞, the $-$ sign must apply and we obtain

$$\zeta = -\frac{2}{L}\left[z - \sqrt{(z - \tfrac{1}{2}L)(z + \tfrac{1}{2}L)}\right] \tag{33.25}$$

and the complex potential (33.21) becomes

$$\Omega = -\tfrac{1}{2}Q_{x0}\left[z - \sqrt{(z - \tfrac{1}{2}L)(z + \tfrac{1}{2}L)}\right] + \Phi_0 \tag{33.26}$$

The pattern of streamlines shown in Figure 5.22 is obtained by contouring the imaginary part of the function (33.26).

Problem 33.1

Determine the complex potential for the aquifer system if there are uniform far-fields in the x direction with discharges $\overset{u}{Q}_{x0}$ and $\overset{l}{Q}_{x0}$ in the upper and lower aquifers, and if the head at point $x = -R$, $y = 0^+$ is $\overset{u}{\phi}_0$.

Polygons That Contain the Origin and Not Infinity

If the polygon does not contain infinity there will be a point $\zeta = \delta$ inside $\Im\zeta > 0$ for which $z = z(\zeta)$ becomes infinite. We considered the behavior near such a point $\zeta = \delta$ of the mapping function in Sec. 32, and obtained (32.14):

$$z(\zeta) = \frac{1}{\zeta - \delta}f(\zeta) \qquad (\zeta \to \delta) \tag{33.27}$$

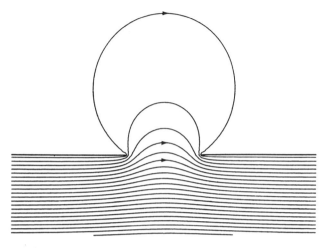

Figure 5.22 Flow pattern for uniform flow below an impermeable layer with a gap.

where $f(\zeta)$ is analytic inside $\Im\zeta > 0$. By analogy to (33.11), the following form of the transformation is obtained:

$$z = |A|e^{i\varphi^*} \frac{\prod_{\nu=1}^{n}(\zeta - \xi_\nu^*)^{\mu_\nu}}{(\zeta - \delta)(\zeta - \bar{\delta})} \tag{33.28}$$

As an application, we will consider the case of a fully penetrating radial collector well with n collectors in a field of uniform flow.

A Radial Collector Well in a Field of Uniform Flow

A radial collector well with three collectors is shown in Figure 5.23. The collectors are equally spaced and all have a length L. We first apply the transformation

$$Z = \frac{z - z_0}{L} e^{i\alpha} \tag{33.29}$$

with α defined in Figure 5.23(a) and where z_0 is the center of the well, so that one collector is mapped onto a section of the negative Y axis, and all collectors in the Z plane are of unit length. We label the collectors as $\mathcal{A}, \mathcal{B}, \mathcal{C} \ldots$ going clockwise around the well, and label the points at the intersection of the collectors as $1, 2, \ldots, n$ as shown in the figure.

We map the boundary of the collector well onto the upper half plane $\Im\zeta \geq 0$ in such a way that point $\mathcal{A}$ corresponds to $\zeta = \infty$ and that $Z = \infty$ corresponds to a point $\zeta = \delta$ on the η axis, which corresponds to the Y axis by symmetry. We further choose $|\delta|$ to be 1, so that

$$\delta = i \tag{33.30}$$

Sec. 33 Transformations with Piecewise Constant Argument

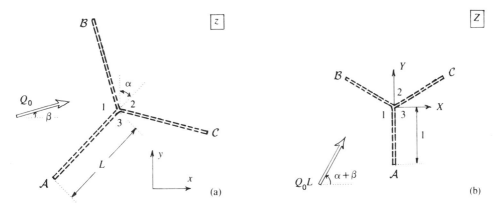

Figure 5.23 A fully penetrating radial collector well.

The angles between the collectors are denoted as γ, where

$$\gamma = \frac{2\pi}{n} \qquad (33.31)$$

The argument of $Z(\zeta)$ jumps at the points $1, 2, \ldots, n$, and we denote the corresponding points on the ξ axis as ξ_ν^*. The points $1, 2, \ldots, n$ lie at $Z = 0$ so that $\mu_1 = \mu_2 = \ldots = \mu_n = \gamma/\pi = 2/n$, and the argument φ^* of $Z(\zeta)$ along the rightmost section of the ξ axis equals $-\pi/2$. Application of the transformation (33.28) to the function $Z(\zeta)$ yields

$$Z = |A|e^{-i\pi/2}\frac{\prod_{\nu=1}^n (\zeta - \xi_\nu^*)^{2/n}}{\zeta^2 + 1} \qquad (33.32)$$

Point $\mathcal{A}$, $\zeta = \infty$, corresponds to $Z = -i$; we take the limit for $\zeta \to \infty$, so that

$$|A| = 1 \qquad (33.33)$$

and (33.32) reduces to

$$Z = -i\frac{\prod_{\nu=1}^n (\zeta - \xi_\nu^*)^{2/n}}{\zeta^2 + 1} \qquad (33.34)$$

This transformation contains the n unknown parameters ξ_ν^*. For this special case they may be determined by the use of symmetry. The symmetry properties of the mapping are difficult to establish in the ζ plane and therefore we use an auxiliary transformation to map the upper half plane onto the exterior of the unit circle in the χ plane of Figure 5.24(c). The latter plane is chosen such that the radii $\mathcal{OA}$, $\mathcal{OB}$, and $\mathcal{OC}$ are parallel to the collectors in the Z plane. By symmetry, points 1, 2, and 3 correspond to the centers of the arcs $\mathcal{AB}$, $\mathcal{BC}$, and $\mathcal{CA}$. Denoting the values of χ at these points as χ_ν^* ($\nu = 1, 2, \ldots, n$), we obtain from Figure 5.24(c) that

$$\chi_\nu^* = e^{i\gamma\nu} \qquad (\nu = 1, 2, \ldots, n) \qquad (33.35)$$

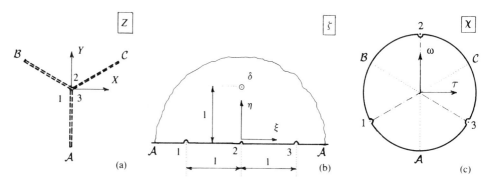

Figure 5.24 The transformations $Z(\zeta)$ and $\chi(\zeta)$.

where

$$\gamma_1 = -\frac{\pi}{2} - \frac{\gamma}{2} \qquad \gamma_\nu = \gamma_1 - (\nu - 1)\gamma \qquad (\nu = 1, 2, \ldots, n) \tag{33.36}$$

Expressions for the points ξ_ν^* may now be obtained by mapping $\Im\zeta \geq 0$ onto the exterior of the circle $|\chi| = 1$. We discussed a similar transformation in Sec. 30 (see Figure 5.2). The χ plane in Figure 5.24(b) is obtained from the z plane in Figure 5.2 by multiplication by a factor $1/R$ and rotation over $\pi/2$ in clockwise direction. Thus, we replace z in (30.2) and (30.4) by $iR\chi$ and obtain

$$\zeta = i\frac{\chi - i}{\chi + i} = i + \frac{2}{\chi + i} \qquad \chi = -i\frac{\zeta + i}{\zeta - i} = -i + \frac{2}{\zeta - i} \tag{33.37}$$

Expressions for ξ_ν^* are obtained by the use of (33.35) and (33.37) as

$$\xi_\nu^* = i\frac{e^{i\gamma_\nu} - e^{i\pi/2}}{e^{i\gamma_\nu} + e^{i\pi/2}} = i\frac{e^{i(\gamma_\nu + \pi/2)/2}}{e^{i(\gamma_\nu + \pi/2)/2}}\frac{\left[e^{i(\gamma_\nu - \pi/2)/2} - e^{-i(\gamma_\nu - \pi/2)/2}\right]}{\left[e^{i(\gamma_\nu - \pi/2)/2} + e^{-i(\gamma_\nu - \pi/2)/2}\right]} \tag{33.38}$$

or

$$\xi_\nu^* = -\tan\left[\tfrac{1}{2}(\gamma_\nu - \pi/2)\right] \tag{33.39}$$

It may be noted that there was no need to determine the locations of points $B, C \ldots$ in the ζ plane by examining at which points $dZ/d\zeta$ vanishes; the locations of these points are obtained by symmetry, and the known location of point A was sufficient to determine the constant $|A|$ in the transformation. The proof that the derivative of the mapping function indeed vanishes may be given by writing the transformation in terms of χ. This expression contains the product $(\chi - \chi_1^*)(\chi - \chi_2^*)\ldots(\chi - \chi_n^*)$, which may be written as $\chi^n - e^{-i\pi(1+n/2)}$. The proof is rather involved and not essential and is therefore omitted. The mapping function is now fully determined, and we will proceed with the determination of the complex potential in terms of χ.

Sec. 33 Transformations with Piecewise Constant Argument

The Complex Potential as a Function of χ. We will determine the complex potential in terms of χ rather than of ζ so that we may use results obtained in Sec. 31. Let the uniform far-field in the z plane be oriented at an angle β with respect to the x axis. It follows from (33.29) that for $Z \to \infty$ the term due to uniform flow in the far-field is

$$z \to \infty \qquad Z \to \infty \qquad \Omega \to -Q_0 z e^{-i\beta} \qquad \Omega \to -Q_0 L Z e^{-i(\alpha+\beta)} \qquad (33.40)$$

$Z = \infty$ corresponds to $\zeta = \delta = i$ and the first term in the expansion of $Z = Z(\zeta)$ about $\zeta = i$ is

$$\zeta \to i \qquad Z \to -\frac{\prod_{\nu=1}^{n}(i-\xi_\nu^*)^{2/n}}{2(\zeta-i)} \qquad (33.41)$$

Using (33.37) this may be written as

$$\chi \to \infty \qquad Z \to -\tfrac{1}{4}\chi \prod_{\nu=1}^{n}(i-\xi_\nu^*)^{2/n} \qquad (33.42)$$

Hence the uniform component of the far-field of $\Omega = \Omega(\chi)$ becomes

$$\chi \to \infty \qquad \Omega \to -\tfrac{1}{4}Q_0 L e^{-i(\alpha+\beta)}\chi\left[-\prod_{\nu=1}^{n}(i-\xi_\nu^*)^{2/n}\right] \qquad (33.43)$$

The constant

$$B = -\prod_{\nu=1}^{n}(i-\xi_\nu^*)^{2/n} \qquad (33.44)$$

is real and positive. This may be seen by observing that the points ξ_ν^* either lie in pairs symmetrically with respect to the η axis or lie at the origin. For example, for the case of Figure 5.24(b) $\arg(i-\xi_1)(i-\xi_3) = \pi$. For the case that n is odd

$$\arg \prod_{\nu=1}^{n}(i-\xi_\nu^*)^{2/n} = \tfrac{2}{n}\left[\tfrac{1}{2}(n-1)\pi + \tfrac{\pi}{2}\right] = \pi \qquad (33.45)$$

and for the case that n is even

$$\arg \prod_{\nu=1}^{n}(i-\xi_\nu^*)^{2/n} = \tfrac{2}{n}\left[\tfrac{1}{2}n\pi\right] = \pi \qquad (33.46)$$

so that B is real and positive, as asserted,

$$B = |B| \qquad (33.47)$$

and (33.43) becomes

$$\chi \to \infty \qquad \Omega \to -\tfrac{1}{4}|B|Q_0 L e^{-i(\alpha+\beta)}\chi \qquad (33.48)$$

The flow problem in the χ plane is similar to that of a lake in a field of uniform flow; the circle is an equipotential and $\Omega(\chi)$ produces a field of uniform flow near infinity. The complex potential thus becomes (compare [31.21] through [31.23]):

$$\Omega = -\tfrac{1}{4}|B|Q_0 L \left[\chi e^{-i(\alpha+\beta)} - \frac{e^{i(\alpha+\beta)}}{\chi}\right] + \frac{Q}{2\pi}\ln\chi + \Phi_0 \qquad (33.49)$$

where Φ_0 is the potential along the boundary of the collector well and Q is the discharge of the collector well.

Streamlines are shown in Figure 5.25 for the case that $n = 6$ and $Q/(Q_0 L) = 15$. Points of these streamlines are found by determining them in the χ plane, obtaining ζ, Z, and z for each point from (33.37), (33.34), and (33.29).

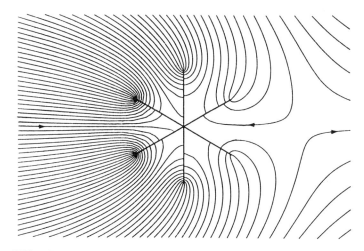

Figure 5.25 Flow pattern for a fully penetrating radial collector well in a field of uniform flow.

Problem 33.2

Solve the problem for flow in a deep aquifer underneath a hydraulic structure with a cutoff wall. The hydraulic structure separates two lakes with heads ϕ_1 and ϕ_2, and is of width L. The cutoff wall penetrates a distance d vertically into the aquifer and is located at the downstream edge of the base of the structure which is flush with the lake bottoms (see Figure 5.26).

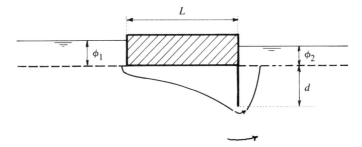

Figure 5.26 Flow underneath a hydraulic structure with a cutoff wall (Problem 33.2).

Chapter 6

Regional Aquifer Modeling Using the Analytic Element Method

The geology in regional aquifer systems is usually insufficiently known and far too complex to be duplicated in a model. We use a simplified model aquifer system that ideally contains the main features of the real system. Even the simplified aquifer model is generally quite complex, and approximate numerical methods are used for the mathematical description of the flow. Various numerical methods exist, both for transient and for steady-state modeling. Although conditions in reality are rarely steady-state, it is often advantageous to use a steady-state model for the initial phase of modeling where an attempt is being made to reproduce observed average conditions. The disadvantages of transient models are twofold: in the first place a transient model needs as input, besides the hydraulic conductivity, another material constant. In the second place a transient model requires an initial condition: heads must be provided throughout the aquifer system in order to start the model. As a result, initially the model predictions reproduce exactly the observations, and perhaps gradually begin to deviate from them. Steady-state models appear to be more sensitive: the model is required to reproduce observed data with only boundary conditions and aquifer parameters as input. It is important to note, in this context, that steady-state models require few input data; only the basic geology and boundary conditions are needed for running the program so that the numerical model may be used from early on for guidance in data collection.

The numerical methods discussed in this text are the analytic element method, discussed in this chapter, the boundary element method (Chapter 8), and the finite element and finite difference methods (Chapter 9). The analytic element method is based on the superposition of suitable closed-form analytic functions and is applied usually to infinite aquifers. By "suitable" is meant that the function is constructed to model a particular feature in an efficient way. The analytic elements presented in Secs. 37 through 40 are examples of such functions and it is of interest to note that there is considerable freedom in the choice of the element: the more of the necessary special features (such as the singularities at the tips of the double-root elements of Sec. 39) the element contains, the more powerful and efficient it will be in the numerical model. The boundary element method is somewhat similar to the analytic element method, but leaves less freedom in determining the elements, and is primarily applied to problems with closed boundaries. As opposed to analytic elements, the boundary elements are always given in the form of integrals, which are often evaluated numerically. The finite element and finite difference methods rely on a complete discretization of the domain and require that the aquifer be fully bounded.

This chapter is devoted to the modeling of regional aquifer systems by superposition of functions referred to here as analytic elements. Each of these functions represents a particular feature of the aquifer system to be modeled, and contains a certain number of degrees of freedom. These degrees of freedom are embodied in constant factors, called coefficients, to be determined in such a way that the complete solution meets the boundary conditions at chosen points, called constraint points. Each of these conditions gives rise to an equation: a constraint equation. By taking the number of constraint points equal to the number of coefficients (degrees of freedom) a system of equations is obtained from which all coefficients can be determined. An example of this approach was discussed in Sec. 25 for the modeling of an aquifer with creeks. The creeks were modeled by means of line-sinks. Each line-sink function had one degree of freedom (the strength), determined such that the head assumed prescribed values at the centers of the line-sinks: the control points.

Line-sinks are only one example of functions that may be used in regional aquifer modeling. The analytic elements discussed in Secs. 37 through 40 are other examples, and we will see how the line-dipoles and line-doublets discussed in Sec. 25 may be used to model such things as thin but highly permeable fissures, inhomogeneities in the hydraulic conductivity, and leaky sheet pilings or slurry walls. A very useful element is the area-sink, which is used, for example, to model the leakage through the bottom of rivers or lakes, and the leakage between stacked aquifers. A method is presented in Sec. 42 for deriving analytic elements that exhibit jumps in either the potential or the stream function across a closed boundary of any shape. This method is applied to develop an analytic element for inhomogeneities with elliptic boundaries.

34 MODELING A SINGLE HOMOGENEOUS AQUIFER

We consider in this section the case of a single homogeneous shallow aquifer, assuming at first that the entire aquifer is unconfined. The aquifer is infinite; boundary conditions only are applied in the interior. We consider the occurrence of uniform infiltration due to rainfall, a far-field that contains uniform flow, ponds with given infiltration through their bottoms (see Sec. 9), narrow creeks above the groundwater table, wells, narrow creeks in direct contact with the groundwater, and lakes that may be considered fully penetrating.

Sec. 34 Modeling a Single Homogeneous Aquifer

We first construct a model for such an aquifer by superposition of analytic elements presented in the preceding chapters. Line-sinks of constant strength, called first-order line-sinks in this text, are used for modeling the creeks and lake boundaries. The choice of these first-order elements is made for the sake of simplicity; higher order elements are discussed in Sec. 38.

An example of a regional aquifer is given in Figure 6.1. There is infiltration from rainfall at a rate N. The creeks are approximated as strings of segments of straight lines, each representing a line-sink. Although the aquifer is modeled as an infinite one, meaningful results are obtained only in the area where elements are included. It was found by experimenting with models of this kind that the results often improve when elements are included outside the area of interest. The influence of the elements upon the heads in the area of interest decreases as their distance from the latter area increases; the extent to which modeling needs to be done outside the area of interest is found by trial and error.

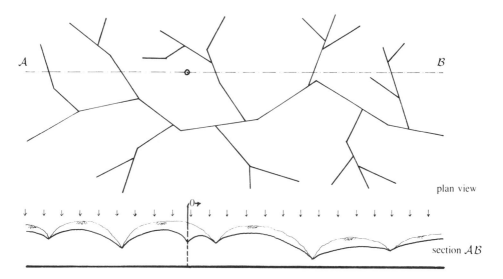

Figure 6.1 A regional unconfined aquifer.

The location of wells, ponds, creeks, and lake boundaries may be obtained from a map. The wells are assumed to have either given discharges or given heads. The rates of infiltration through the bottoms of the ponds are known; the same is true for those creeks that are elevated above the groundwater. The heads along creeks that are in direct contact with the groundwater are taken to be equal to the elevation of the creek bottom, and the heads along the boundaries of fully penetrating lakes are set equal to the lake levels.

Some of the analytic functions have coefficients that are given, whereas others contain coefficients that are not known a priori. For some wells, for example, the discharge is given, whereas for others the head at the well is prescribed. The discharge of the well serves as a coefficient in the solution; in the former case this coefficient is given, whereas in the latter case it must be solved for. For the purpose of setting up the system of equations for the unknown coefficients in the solution, it is useful to write the potential as a sum of a function V and a function U. The coefficients of all analytic elements collected in V are known a priori, whereas U contains all functions with coefficients that must be solved for

$$\Phi = V + U \tag{34.1}$$

Both of these functions will be considered in detail, prior to explaining how to determine the unknown coefficients. All individual potentials are represented as the real parts of a complex potential, with the exception of the potentials for infiltration from rainfall and from circular ponds.

Notation

It is necessary to adopt a unique notation in order to distinguish between the parameters for the many different functions. Throughout this section, we use the * to label parameters of functions with given coefficients. The index of a parameter refers to the element number of any given type. Superscripts are used for certain elements to indicate whether a coordinate represents a starting point or an end point (e.g., $\overset{1}{z}{}_j^*$ represents the starting point of line-sink j of given strength). Subscripts are used for various purposes in conjunction with parameters other than coordinates, as indicated in the text. The symbol Λ is used to represent real functions due to a single element of unit strength (coefficient functions); the indices refer to the type of function: wl stands for well in Λ_{wl}.

Functions with Given Coefficients

The function V in (34.1) is the sum of potentials for infiltration, cross-flow, wells, ponds, and line-sinks of given strength:

$$V = V_{rn} + V_{cf} + V_{wl} + V_{pd} + V_{ls} \tag{34.2}$$

where the indices refer to the feature that each function represents.

Rainfall

The function V_{rn} is an elementary solution to the Poisson equation,

$$\nabla^2 V_{rn} = -N \tag{34.3}$$

The following function V_{rn} exhibits some freedom in the shape of equipotentials it generates and yet is elementary (compare [9.29]):

$$V_{rn} = -\frac{\frac{1}{2}N}{a^2 + b^2} \left[b^2(x - x_0)^2 + a^2(y - y_0)^2 \right] \tag{34.4}$$

The equipotentials produced by V_{rn} are ellipses, centered at (x_0, y_0), and with axes in a ratio a/b. The function has four independent degrees of freedom: N, a/b, x_0, and y_0. It may be noted that (34.4) may be written entirely in terms of the ratio a/b; a and b together represent a single degree of freedom.

The values of the four parameters of V_{rn} affect the far-field behavior of the potential. The function V_{rn} further has an effect on the internal boundaries, such as streams. The analytic elements must compensate for this effect. The four independent parameters in (34.4) are usually chosen such that V_{rn} disturbs the boundary conditions as little as possible.

The Cross-Flow

The complex potential for cross-flow is given by (24.7),

$$\Omega_{cf} = -Q_0 z e^{-i\alpha} \qquad (34.5)$$

where Q_0 is the discharge per unit width of aquifer, and α gives the orientation of the cross-flow with respect to the x axis. The function V_{cf} is the real part of (34.5):

$$V_{cf}(z) = \Re \Omega_{cf}(z) \qquad (34.6)$$

The Wells

The complex potential for a well of discharge Q_j^* and centered at w_j^* is obtained from (24.8). We choose the constant C as $-[Q_j^*/(2\pi)] \ln R$, where R is an arbitrary parameter with the dimension of length, and introduce a real function Λ_{wl}, defined as follows:

$$\Lambda_{wl}(z, w_j^*) = \frac{1}{2\pi} \Re \ln \frac{z - w_j^*}{R} \qquad (34.7)$$

If there are n_w^* wells of given discharge Q_j^* located at w_j^* then the function V_{wl} may be written as

$$V_{wl}(z) = \sum_{j=1}^{n_w^*} Q_j^* \Lambda_{wl}(z, w_j^*) \qquad (34.8)$$

The Ponds

If there are n_p^* circular ponds of infiltration rates N_{pj}^* ($j = 1, 2, \ldots, n_p^*$), centered at (x_{pj}^*, y_{pj}^*) and of radii R_{pj}^*, then the function V_{pd} takes the form:

$$V_{pd}(z) = \sum_{j=1}^{n_p^*} N_{pj}^* G_p(x, y, x_{pj}^*, y_{pj}^*, R_{pj}^*) \qquad (34.9)$$

where the function G_p is defined by (9.70) and (9.71).

The Line-Sinks

We introduce a function Λ_{ls} as the real part of a line-sink with strength $\sigma = 1$ (see [25.8]),

$$\Lambda_{ls}(z, \overset{1}{z}, \overset{2}{z}) = \frac{|\overset{2}{z}-\overset{1}{z}|}{4\pi}\Re\{(Z+1)\ln(Z+1)-(Z-1)\ln(Z-1)+2\ln[\tfrac{1}{2}(\overset{2}{z}-\overset{1}{z})]-2\} \quad (34.10)$$

where Z is given by (25.4),

$$Z = Z(z, \overset{1}{z}, \overset{2}{z}) = \frac{z - \tfrac{1}{2}(\overset{1}{z}+\overset{2}{z})}{\tfrac{1}{2}(\overset{2}{z}-\overset{1}{z})} \quad (34.11)$$

If there are n_{ls}^* line-sinks of given strengths σ_j^* with end points $\overset{1}{z}_j^*$ and $\overset{2}{z}_j^*$ ($j = 1, 2, \ldots, n_{ls}^*$) then the function V_{ls} may be written as

$$V_{ls}(z) = \sum_{j=1}^{n_{ls}^*} \sigma_j^* \Lambda_{ls}(z, \overset{1}{z}_j^*, \overset{2}{z}_j^*) \quad (34.12)$$

Functions With Unknown Coefficients

The function U in (34.1) contains all elements with unknown coefficients and may be decomposed in contributions of wells, line-sinks and an unknown real constant C_1,

$$U = U_{wl} + U_{ls} + C_1 \quad (34.13)$$

The function U_{wl} is added for the sake of completeness; in practice most wells have given discharges, but there are instances where the head is given and the discharge is unknown.

The Wells

If there are n_w wells of unknown discharge Q_j, radius r_j and located at $z = w_j$, then the function U_{wl} becomes

$$U_{wl}(z) = \sum_{j=1}^{n_w} Q_j \Lambda_{wl}(z, w_j) \quad (34.14)$$

where the function Λ_{wl} is defined by (34.7).

The Line-Sinks

If the number of line-sinks of unknown strength σ_j is n_{ls} ($j = 1, 2, \ldots, n_{ls}$) and their end points are $\overset{1}{z}_j$ and $\overset{2}{z}_j$, then the function U_{ls} becomes

$$U_{ls}(z) = \sum_{j=1}^{n_{ls}} \sigma_j \Lambda_{ls}(z, \overset{1}{z}_j, \overset{2}{z}_j) \quad (34.15)$$

The System of Equations

The potential U contains a total of $n_w + n_{ls} + 1$ unknowns: the discharges of the wells, the strengths of the line-sinks and the constant C_1. There are $n_w + n_{ls} + 1$ conditions needed to provide the equations from which the unknowns can be obtained. The first n_w of these conditions are that the potential at each well equals a given value $\underset{w}{\Phi_\nu}$, $(\nu = 1, 2, \ldots, n_w)$, where the subscript w refers to well. It is noted that the functions used here model the wells only approximately: the well boundary is not exactly an equipotential. We therefore apply the condition only at one point of each well boundary: $z = w_\nu + r_\nu$ and obtain the following n_w equations using (34.1), (34.13), and (34.14):

$$\underset{w}{\Phi_\nu} = \sum_{j=1}^{n_w} Q_j \Lambda_{wl}(w_\nu + r_\nu, w_j) + U_{ls}(w_\nu + r_\nu) + C_1 + V(w_\nu + r_\nu)$$

$$(\nu = 1, 2, \ldots, n_w) \qquad (34.16)$$

A set of n_{ls} additional real equations is obtained by requiring that the potential be equal to some prescribed value $\underset{c}{\Phi_\nu}$, $(\nu = 1, 2, \ldots, n_{ls})$ (the subscript c stands for creek) at the center $\overset{c}{z}_\nu = \frac{1}{2}(\overset{1}{z}_\nu + \overset{2}{z}_\nu)$ of each line-sink:

$$\underset{c}{\Phi_\nu} = U_{wl}\left[\tfrac{1}{2}(\overset{1}{z}_\nu + \overset{2}{z}_\nu)\right] + \sum_{j=1}^{n_{ls}} \sigma_j \Lambda_{ls}\left[\tfrac{1}{2}(\overset{1}{z}_\nu + \overset{2}{z}_\nu), \overset{1}{z}_j, \overset{2}{z}_j\right]$$

$$+ C_1 + V\left[\tfrac{1}{2}(\overset{1}{z}_\nu + \overset{2}{z}_\nu)\right] \qquad (\nu = 1, 2, \ldots, n_{ls}) \qquad (34.17)$$

The final equation, which brings the total to $n_w + n_{ls} + 1$, may either be that the potential equals a given value $\underset{0}{\Phi}$ at some point z_0, called the reference point, or may be a continuity equation. Which one of these two conditions is chosen for any given problem depends upon the problem, but each one of them affects the amount of flow from infinity. If the condition at the reference point is chosen, one usually selects some point where the head is known outside the area of interest, but within the area containing elements. This condition is

$$\underset{0}{\Phi} = U(z_0) + C_1 + V(z_0) \qquad (34.18)$$

The continuity equation expresses that the infiltration entering the system over some area A both as rainfall and through the bottoms of infiltrating creeks or ponds equals the total amount of water taken out of the aquifer by wells and creeks:

$$\sum_{j=1}^{n_w^*} Q_j^* + \sum_{j=1}^{n_w} Q_j - \sum_{j=1}^{n_p^*} N_j^* \pi \overset{*}{R}_{p_j}^2 + \sum_{j=1}^{n_{ls}^*} \sigma_j^* \left|\overset{2}{z}_j^* - \overset{1}{z}_j^*\right| + \sum_{j=1}^{n_{ls}} \sigma_j \left|\overset{2}{z}_j - \overset{1}{z}_j\right| = NA \qquad (34.19)$$

The complete system of equations thus consists of the n_w equations (34.16), the n_{ls} equations (34.17) and either (34.18) or (34.19), and may be represented symbolically as follows:

$$b_i = \sum_{j=1}^{n_w+n_{ls}+1} A_{ij} a_j \qquad (i = 1, 2, \ldots, n_w + n_{ls} + 1) \qquad (34.20)$$

The matrix A_{ij} is called the coefficient matrix, the b_i represent the known terms (b_i is called the known vector), and the a_j are the unknown constants:

$$\begin{aligned} a_j &= Q_j & j &= 1, 2, \ldots, n_w \\ a_j &= \sigma_j & j &= n_w + 1, n_w + 2, \ldots, n_w + n_{ls} \\ a_j &= C_1 & j &= n_w + n_{ls} + 1 \end{aligned} \qquad (34.21)$$

The $n_w + n_{ls} + 1$ constants Q_j, σ_j, and C_1 may be determined by solving this system of equations by Gaussian elimination.

Type of Flow

The model described above is formulated entirely in terms of potentials, and is applicable to unconfined aquifers, confined aquifers, coastal aquifers, or aquifers consisting of two or more layers of different hydraulic conductivities. The type of flow affects only the relation between potentials and heads, which may be written symbolically as

$$\Phi = \tilde{\Phi}(\phi) \qquad \phi = \tilde{\phi}(\Phi) \qquad (34.22)$$

where the wavy superscript serves to distinguish these functions from the functions of position. If the aquifer is confined or unconfined, the forms of equations (34.22) are given in Sec. 8. If the flow is interface flow they are found in Sec. 10, and in the case of layers of different hydraulic conductivities the potentials are given in Sec. 11.

With the exception of wells of given head, the above functions are implemented in the computer program SLWL described in Sec. 25.

Problem 34.1

Use SLWL to model flow in an infinite aquifer with a well of discharge Q at $z = z_w$ and a finite narrow canal extending from $z = -\frac{1}{2}L$ to $z = +\frac{1}{2}L$. The head along the canal is ϕ_0. The head at a point x_R, y_R is ϕ_R. Model the canal by using 8 line-sinks. Produce a plot of piezometric contours for the case that $Q/(kH^2) = 8$, $L/H = 5$, $z_w/H = 5i$, $\phi_0/H = 15$, $z_R/H = 10^6 i$, $\phi_R/H = 15$, with the same levels as shown in Figure 6.2. Determine the exact solution to this problem using (31.29) along with (31.33), and implement it in the module GIVEN of SLWL. Produce piezometric contours corresponding to the exact solution and compare them with the approximate ones. Your results should look similar to Figure 6.2, where the solid and dotted curves correspond to the exact and approximate solutions, respectively; the minimum level is $10H$ and the contour interval is $0.2H$.

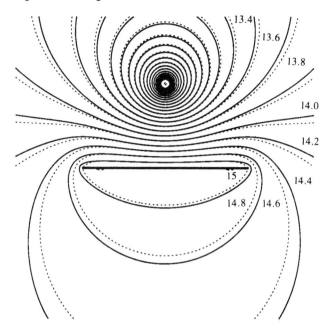

Figure 6.2 Flow toward a finite canal: comparison between approximate and exact solutions.

35 MODELING AREAL INHOMOGENEITIES

Real aquifers are rarely homogeneous, and it often is necessary to take discontinuities of the hydraulic conductivity, the aquifer thickness, or the elevation of the aquifer base into account. We will see in this section that many inhomogeneities may be modeled efficiently by the use of strings of line-doublets.

Boundary Conditions for an Inhomogeneity in the Hydraulic Conductivity

There are two conditions along the boundary $\mathcal{B}$ of an inhomogeneity in the hydraulic conductivity (see Figure 6.3): (1) continuity of pressure and (2) continuity of flow. The first condition implies continuity of head, and we may write,

$$\boxed{\phi^+ = \phi^- \qquad Q_n^+ = Q_n^-} \tag{35.1}$$

where the + and − signs refer to the + and − sides of $\mathcal{B}$ as shown in Figure 6.3, and Q_n is the normal component of the discharge vector.

All potentials discussed thus far, with the exception of those presented in Sec. 11, are linear functions of the hydraulic conductivity that may be written in the form:

$$\Phi = k f(\phi) \tag{35.2}$$

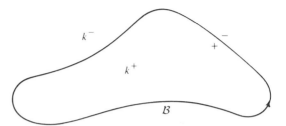

Figure 6.3 An inhomogeneity in the hydraulic conductivity.

where $f(\phi)$ does not depend upon k. The head is continuous across $\mathcal{B}$ and it follows from (35.2) that the following expression holds for the potentials Φ^+ and Φ^- at neighboring points on either side of $\mathcal{B}$:

$$\frac{\Phi^+}{\Phi^-} = \frac{k^+}{k^-} \qquad (35.3)$$

The jump in the potential thus may be expressed in terms of Φ^- as

$$\boxed{\lambda = \Phi^+ - \Phi^- = \frac{k^+ - k^-}{k^-} k^- f(\phi) = \frac{k^+ - k^-}{k^-}\Phi^-} \qquad (35.4)$$

The second condition in (35.1) may be written in terms of the stream function as follows: Since $Q_n = -\partial \Phi/\partial n = +\partial \Psi/\partial s$, where s is tangent to $\mathcal{B}$, the condition $Q_n^+ = Q_n^-$ is fulfilled if Ψ is continuous:

$$\Psi^+ - \Psi^- = 0 \qquad (35.5)$$

We observe that the potential for a line-doublet is suitable for modeling an inhomogeneity in the hydraulic conductivity: across a line-doublet the potential jumps, but the stream function is continuous. The inhomogeneity may be modeled by discretizing its boundary into linear elements, each representing a line-doublet. The condition (35.4) is applied at control points, and unknown constants in the line-doublet function are determined using these conditions. This way of modeling inhomogeneities in the hydraulic conductivity was used in a regional model by Strack and Haitjema [1981(b)].

A better way of modeling an inhomogeneity is by constructing a special analytic element that satisfies the jump conditions (35.4) and (35.5) along a continuous boundary. A method for deriving such analytic elements by the use of conformal mapping is described in Sec. 42.

Boundary Conditions for an Inhomogeneity in the Hydraulic Conductivity, the Aquifer Thickness, and the Base Elevation

Discontinuities in the base elevation or aquifer thickness can be handled in a similar way as discontinuities in the hydraulic conductivity. We will next consider a general

Sec. 35 Modeling Areal Inhomogeneities

inhomogeneity in the hydraulic conductivity k, the aquifer thickness H, and the base elevation b, as illustrated in Figure 6.4. The potential Φ is defined in $\mathcal{D}^+$ as

$$\begin{cases} \Phi = k^+ H^+ (\phi - b^+) - \frac{1}{2} k^+ (H^+)^2 & (\phi > b^+ + H^+) \\ \Phi = \frac{1}{2} k^+ (\phi - b^+)^2 & (\phi \leq b^+ + H^+) \end{cases} \quad (z \text{ in } \mathcal{D}^+) \quad (35.6)$$

where k^+, H^+, and b^+ represent the hydraulic conductivity, the aquifer thickness, and the base elevation in $\mathcal{D}^+$, respectively. The potential Φ is defined in $\mathcal{D}^-$ as

$$\begin{cases} \Phi = k^- H^- (\phi - b^-) - \frac{1}{2} k^- (H^-)^2 & (\phi > b^- + H^-) \\ \Phi = \frac{1}{2} k^- (\phi - b^-)^2 & (\phi \leq b^- + H^-) \end{cases} \quad (z \text{ in } \mathcal{D}^-) \quad (35.7)$$

where k^-, H^-, and b^- represent the aquifer properties in $\mathcal{D}^-$.

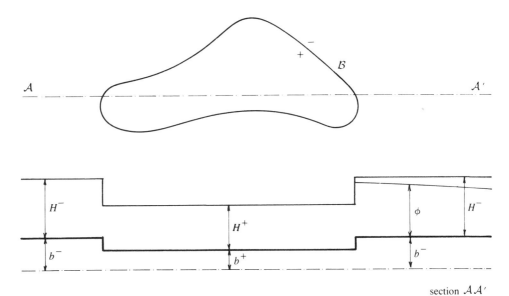

section $\mathcal{A}\mathcal{A}'$

Figure 6.4 An inhomogeneity in the hydraulic conductivity, the aquifer thickness, and the base elevation.

The boundary condition (35.1) remains valid so that the boundary $\mathcal{B}$ can be modeled by the use of line-doublets. We observe from (35.6) and (35.7) that the expression for λ along $\mathcal{B}$ is a nonlinear equation in terms of Φ^- if the flow on either side or on both sides of $\mathcal{B}$ is unconfined. We will deal with this problem here in a way suggested by Fitts [1985], which has the advantage of being easy to implement in a computer program but suffers from the drawback that an iterative procedure is always necessary if either H or b is discontinuous. The expression for λ is written in the form

$$\lambda = \Phi^+ - \Phi^- = A\Phi^- \quad (35.8)$$

where the constant A is defined as

$$A = \frac{\Phi^+ - \Phi^-}{\Phi^-}, \qquad (35.9)$$

This constant is independent of ϕ only if $H^+ = H^-$ and $b^+ = b^-$; in that case (35.8) reduces to (35.4). Otherwise, estimates of Φ^+ and Φ^- are made and used in (35.9). With A estimated, the jump condition (35.8) is linear so that new estimates for Φ^+ and Φ^- can be obtained. This iterative procedure is continued until the jump condition is met with a preset accuracy. The first estimate may be made, for example, by leaving the inhomogeneity out. This procedure appears to work well in practice but, as stated above, can be improved with regard to numerical efficiency.

Fitts [1985] compared solutions obtained in this manner with exact solutions for flow in the vertical plane with discontinuous base elevations and aquifer thicknesses, both for confined and unconfined flow. The errors caused by adopting the Dupuit-Forchheimer approximation appear to be acceptable for most practical problems.

A similar iterative procedure may be adopted if the hydraulic conductivity varies in the vertical direction.

A Cylindrical Inhomogeneity

The case of a cylindrical inhomogeneity in the hydraulic conductivity in an aquifer with a uniform far-field will serve to illustrate the behavior of the complex potential along the boundary of an inhomogeneity. We denote by $\mathcal{D}_1$ and $\mathcal{D}_2$ the domains inside and outside the boundary of the inhomogeneity, which is a cylinder with radius R, and choose the origin of the z plane at the center of the cylinder. The hydraulic conductivities in $\mathcal{D}_1$ and $\mathcal{D}_2$ are k_1 and k, and the far-field is a uniform flow in the x direction of discharge Q_{x0}. The complex potentials $\overset{1}{\Omega}$ and $\overset{2}{\Omega}$, valid in $\mathcal{D}_1$ and $\mathcal{D}_2$ are, respectively:

$$\overset{1}{\Omega} = -Q_{x0}\frac{2k_1}{k+k_1}z + \overset{1}{\Phi}_0 \qquad (35.10)$$

and

$$\overset{2}{\Omega} = -Q_{x0}\left[z + \frac{k-k_1}{k+k_1}\frac{R^2}{z}\right] + \frac{k}{k_1}\overset{1}{\Phi}_0 \qquad (35.11)$$

We will demonstrate that these functions fulfill the two conditions (35.3) and (35.5) along $|z| = R$. Denoting $\overset{1}{\Omega}$ as Ω^+ and $\overset{2}{\Omega}$ as Ω^- for $z = Re^{i\theta}$, we obtain

$$\Omega^+ = -Q_{x0}\frac{2k_1}{k+k_1}(R\cos\theta + iR\sin\theta) + \overset{1}{\Phi}_0 \qquad (35.12)$$

Sec. 35   Modeling Areal Inhomogeneities   415

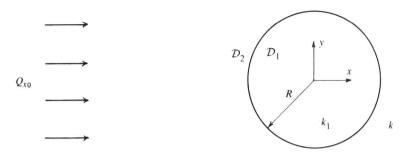

Figure 6.5 A cylindrical inhomogeneity.

and

$$\Omega^- = -Q_{x0}\left[Re^{i\theta} + \frac{k-k_1}{k+k_1}Re^{-i\theta}\right] + \frac{k}{k_1}\overset{1}{\Phi}_0$$

$$= -Q_{x0}\left[\frac{2k}{k+k_1}R\cos\theta + i\frac{2k_1}{k+k_1}R\sin\theta\right] + \frac{k}{k_1}\overset{1}{\Phi}_0 \qquad (35.13)$$

And it follows that $\Phi^+/\Phi^- = k_1/k$ and $\Psi^+ - \Psi^- = 0$ so that both the conditions of continuity of head and continuity of flow are satisfied.

The First-Order Line-Doublet

Consider a line-doublet with strength λ extending from $z = \delta_m$ to $z = \delta_{m+1}$. We define the dimensionless local variable Z_m as

$$Z_m = Z(z, \delta_m, \delta_{m+1}) = \frac{z - \frac{1}{2}(\delta_m + \delta_{m+1})}{\frac{1}{2}(\delta_{m+1} - \delta_m)} \qquad (35.14)$$

The complex potential for a line-doublet is given in (25.73) with Z_m replacing Z:

$$\Omega = \frac{\lambda(Z_m)}{2\pi i}\ln\frac{Z_m - 1}{Z_m + 1} + ip(Z_m) \qquad (35.15)$$

We limit ourselves for the time being to a line-doublet with linear strength distribution and refer to such doublets as being of the first order. If λ_m and λ_{m+1} represent the strengths at the nodes δ_m and δ_{m+1}, then $\lambda(Z_m)$ may be written as

$$\lambda(Z_m) = \lambda_m f(Z_m) + \lambda_{m+1} g(Z_m) \qquad (35.16)$$

where $f(Z_m)$ and $g(Z_m)$ are influence functions:

$$f(Z_m) = -\tfrac{1}{2}(Z_m - 1) \qquad g(Z_m) = \tfrac{1}{2}(Z_m + 1) \qquad (35.17)$$

We define the functions $F(Z_m)$ and $G(Z_m)$ as follows:

$$F(Z_m) = f(Z_m)\ln\frac{Z_m - 1}{Z_m + 1} + a(Z_m)$$
$$G(Z_m) = g(Z_m)\ln\frac{Z_m - 1}{Z_m + 1} + b(Z_m)$$
(35.18)

and represent Ω as

$$\boxed{\Omega = \frac{1}{2\pi i}\left[\lambda_m F(Z_m) + \lambda_{m+1} G(Z_m)\right]}$$
(35.19)

We saw in Sec. 25 that the complex potential (35.15) must be of order $1/Z_m$ near infinity, and that the coefficients of the polynomial $p(Z_m)$ are determined from this condition. We further saw that the order of $p(Z_m)$ is one less than the order of λ, i.e., zero for this case. $p(Z_m)$ is split into two functions in the present formulation, a and b, which reduce to constants. As $\ln[(Z_m - 1)/(Z_m + 1)]$ approaches $-2Z_m^{-1} + O(Z_m^{-3})$ for $Z_m \to \infty$, it follows from (35.17) and (35.18) that

$$a(Z_m) = -1 \qquad b(Z_m) = 1$$
(35.20)

With these values for a and b, both F and G vanish at infinity and become

$$F(Z_m) = -\tfrac{1}{2}(Z_m - 1)\ln\frac{Z_m - 1}{Z_m + 1} - 1$$
$$G(Z_m) = \tfrac{1}{2}(Z_m + 1)\ln\frac{Z_m - 1}{Z_m + 1} + 1$$
(35.21)

Far-Field Expansions

A far-field expansion is used for the functions (35.21) for large values of $|Z_m|$ (e.g., $|Z_m| > 3$); in this way the accumulation of numerical inaccuracies in the evaluation of the functions is avoided. The expansion about $1/Z_m = 0$ of the logarithmic terms in (35.21) is

$$\ln\frac{Z_m - 1}{Z_m + 1} = \ln\frac{1 - 1/Z_m}{1 + 1/Z_m} = -2\sum_{j=1}^{\infty}\frac{Z_m^{-2j+1}}{2j - 1}$$
(35.22)

so that the expansions of F and G become:

$$F = \sum_{j=1}^{\infty}\frac{Z_m^{-2j}}{2j + 1} - \sum_{j=1}^{\infty}\frac{Z_m^{-2j+1}}{2j - 1}$$
(35.23)

and

$$G = -\sum_{j=1}^{\infty}\frac{Z_m^{-2j}}{2j + 1} - \sum_{j=1}^{\infty}\frac{Z_m^{-2j+1}}{2j - 1}$$
(35.24)

Behavior at a Node

The boundary of the inhomogeneity is modeled by a string of line-doublets, and it will be necessary to evaluate the potential at the nodes. We examine the contributions at $z = \delta_m$ of two elements that have the node δ_m in common. If we denote the potential due to these two elements as Ω^*, then

$$\Omega^*(z) = \frac{1}{2\pi i}[\lambda_{m-1}F(Z_{m-1}) + \lambda_m G(Z_{m-1}) + \lambda_m F(Z_m) + \lambda_{m+1}G(Z_m)] \quad (35.25)$$

For convenience, we express Z_{m-1} and Z_m in terms of $z - \delta_m$ as follows:

$$Z_{m-1} = \frac{z - \frac{1}{2}(\delta_m + \delta_{m-1})}{\frac{1}{2}(\delta_m - \delta_{m-1})} = 2\frac{z - \delta_m}{\delta_m - \delta_{m-1}} + 1$$

$$Z_m = \frac{z - \frac{1}{2}(\delta_{m+1} + \delta_m)}{\frac{1}{2}(\delta_{m+1} - \delta_m)} = 2\frac{z - \delta_m}{\delta_{m+1} - \delta_m} - 1 \quad (35.26)$$

We observe from (35.21) and (35.26) that $F(Z_{m-1})$ and $G(Z_m)$ approach -1 and $+1$, respectively, for $z \to \delta_m$. Using (35.26), Ω^* may be written as follows, in the limit for $z \to \delta_m$:

$$\Omega^*(\delta_m) = \frac{1}{2\pi i}(\lambda_{m+1} - \lambda_{m-1}) + \frac{1}{2\pi i}\lim_{z \to \delta_m}\lambda_m \left\{ \left[\frac{z - \delta_m}{\delta_m - \delta_{m-1}} + 1 \right] \ln \frac{z - \delta_m}{z - \delta_{m-1}} \right.$$
$$\left. + \left[-\frac{z - \delta_m}{\delta_{m+1} - \delta_m} + 1 \right] \ln \frac{z - \delta_{m+1}}{z - \delta_m} \right\} \quad (35.27)$$

Since the limit for $z \to 0$ of $z \ln z$ is zero, (35.27) reduces to

$$\Omega^*(\delta_m) = \frac{1}{2\pi i}(\lambda_{m+1} - \lambda_{m-1}) + \frac{1}{2\pi i}\lambda_m \ln \frac{\delta_m - \delta_{m+1}}{\delta_m - \delta_{m-1}} \quad (35.28)$$

We must remember that the branch cuts of the line-doublets fall along the boundary of the inhomogeneity; that is where the jump in the potential occurs. The branch cut of the logarithmic term in (35.28) is taken accordingly (see Figure 6.6), and we define the angle $\theta_{m-1,m+1}$ as

$$\theta_{m-1,m+1} = \arg \frac{\delta_m - \delta_{m+1}}{\delta_m - \delta_{m-1}} \quad (35.29)$$

The values of $\theta^+_{m-1,m+1}$ and $\theta^-_{m-1,m+1}$ on either side of the cut are obtained with the aid of Figure 6.6 as

$$\theta^+_{m-1,m+1} = \beta + \pi$$
$$\theta^-_{m-1,m+1} = \beta - \pi \quad (35.30)$$

where the angle β is defined as

$$\beta = \arg(\delta_{m+1} - \delta_m) - \arg(\delta_m - \delta_{m-1}) \qquad -\pi < \beta \leq \pi \quad (35.31)$$

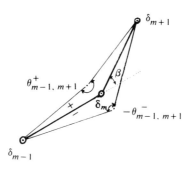

Figure 6.6 Branch cut for $\ln[(z - \delta_{m+1})/(z - \delta_{m-1})]$.

The latter inequality implies that the arguments of $\delta_{m+1} - \delta_m$ and $\delta_m - \delta_{m-1}$ must be measured in the same sense with respect to the x axis. The angle β is the difference in orientation of elements $m-1$ and m and is positive for the case shown in Figure 6.6.

The System of Equations

The line-doublets are included as part of the function U, which is now represented as

$$U = U_{wl} + U_{ls} + U_{db} + C_1 \tag{35.32}$$

where U_{db} represents the contribution of all line-doublets:

$$U_{db} = \sum_{r=1}^{N_i} \sum_{j=1}^{m_r} \lambda_j \Lambda_{db}(z, \delta_{j-1}, \delta_j, \delta_{j+1}) \tag{35.33}$$

The number N_i represents the number of inhomogeneities. The hydraulic conductivity in the rth inhomogeneity is k_r. The boundary of the rth inhomogeneity consists of m_r line-doublets, and the function $\Lambda_{db}(z, \underset{r}{\delta_{j-1}}, \underset{r}{\delta_j}, \underset{r}{\delta_{j+1}})$ represents the coefficient function of the jump $\underset{r}{\lambda_j}$ at node j of inhomogeneity r. It is noted that $\underset{r}{\delta_{m_r+1}} = \underset{r}{\delta_1}$ since each string of line-doublets is closed (see Figure 6.7), and that $\underset{r}{\delta_0} = \underset{r}{\delta_{m_r}}$:

$$\underset{r}{\delta_{m_r+1}} = \underset{r}{\delta_1} \qquad \underset{r}{\delta_0} = \underset{r}{\delta_{m_r}} \qquad (r = 1, 2 \ldots, N_i) \tag{35.34}$$

The function $\Lambda_{db}(z, \underset{r}{\delta_{j-1}}, \underset{r}{\delta_j}, \underset{r}{\delta_{j+1}})$ contains the contribution of elements $j-1$ and j to the coefficient function of $\underset{r}{\lambda_j}$ so that, by (35.19),

$$\Lambda_{db}(z, \underset{r}{\delta_{j-1}}, \underset{r}{\delta_j}, \underset{r}{\delta_{j+1}}) = \Re \left\{ \frac{1}{2\pi i} \left[G(\underset{r}{Z_{j-1}}) + F(\underset{r}{Z_j}) \right] \right\} \tag{35.35}$$

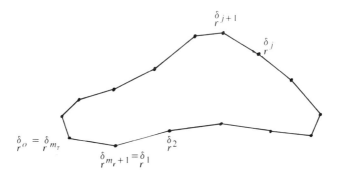

Figure 6.7 The boundary of an inhomogeneity.

where

$$Z_{\underset{r}{j}} = \frac{z - \frac{1}{2}(\underset{r}{\delta}_{j+1} + \underset{r}{\delta}_j)}{\frac{1}{2}(\underset{r}{\delta}_{j+1} - \underset{r}{\delta}_j)} \tag{35.36}$$

The conditions (35.8) may now be written for each inhomogeneity. The left-hand side of the equation represents the jump $\Phi^+ - \Phi^-$ at a node, and the right-hand side equals a factor times Φ^-, the potential just outside the node, i.e., at $\underset{r}{\delta}_m^-$:

$$\underset{r}{\lambda}_m A^* = U_{wl}(\underset{r}{\delta}_m^-) + U_{ls}(\underset{r}{\delta}_m^-) + \sum_{s=1}^{N_i}\sum_{j=1}^{m_s} \underset{s}{\lambda}_j \Lambda_{db}(\underset{r}{\delta}_m^-, \underset{s}{\delta}_{j-1}, \underset{s}{\delta}_j, \underset{s}{\delta}_{j+1}) + C_1 + V(\underset{r}{\delta}_m^-)$$

$$(r = 1, 2, \ldots, N_i \qquad m = 1, 2, \ldots, m_r) \tag{35.37}$$

where $\underset{r}{\lambda}_m$ equals the jump $\Phi^+ - \Phi^-$ at $\underset{r}{\delta}_m^-$ and A^* is the inverse of the factor in (35.9):

$$A^* = \frac{1}{A} \tag{35.38}$$

The value of Λ_{db} at $z = \underset{rj}{\delta}^-$ equals the contribution of the coefficient function of $\underset{r}{\lambda}_j$ at node $\underset{rj}{\delta}^-$, and is obtained by the use of (35.28), (35.29), and (35.30) as

$$\Lambda_{db}(\underset{rj}{\delta}^-, \underset{r}{\delta}_{j-1}, \underset{r}{\delta}_j, \underset{r}{\delta}_{j+1}) = \frac{1}{2\pi}\theta^-_{m-1,m+1} \tag{35.39}$$

Equations (35.37) represent M equations, where

$$M = \sum_{r=1}^{N_i} m_r \tag{35.40}$$

that are added to the $n_w + n_{ls} + 1$ equations determined earlier. The system of equations may be solved for the $n_w + n_{ls} + 1 + M$ unknowns, which includes the M constants λ_j^r.

Problem 35.1

Adapt SLWL to handle a single inhomogeneity in the hydraulic conductivity. Model a cylindrical inhomogeneity centered at $z = 0$ and of radius R in a field of uniform flow of discharge Q_{x0} using 18 line-doublets. Use $k^+ = 10k^-$, $Q_{x0} = k^- H$, $R = 2H$, and $\phi_0 = 10H$ at $z = 0$, and compare your results to the exact solution given by (35.10) and (35.11). Your plot should look similar to the one displayed in Figure 6.8; the contour interval is $0.2H$, the solid curves correspond to the exact solution, and the dotted ones to the approximate solution.

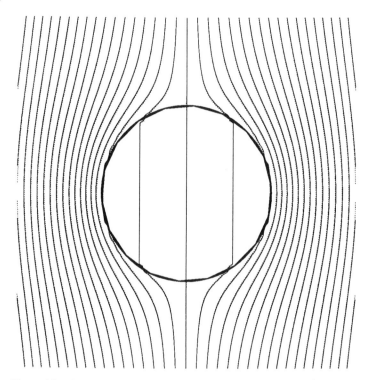

Figure 6.8 Piezometric contours for a circular inhomogeneity with $k^+ = 10k^-$.

36 LEAKY OR DRAINING OBJECTS

Leaky or draining objects in aquifers may be either natural or man-made. An example of a leaky object is a thin slurry wall, constructed in order to divert flow. Draining objects may be highly permeable thin zones in an aquifer, such as cracks filled with highly porous material. If the objects are wide, they may be modeled as inhomogeneities in the hydraulic conductivity as explained in Sec. 35; if the objects are narrow, they

Sec. 36 Leaky or Draining Objects

may be more efficiently modeled by the use of strings of line-doublets or line-dipoles as explained below, or by using the special analytic elements discussed in Sec. 39.

The Boundary Condition Along a Leaky Object

An example of a thin slurry wall is given in Figure 6.9(a), where a U-shaped wall is shown in plan view. Because the slurry wall resists the flow through it, there will be a difference in head across the wall. We first consider the case that the slurry wall has a finite width b^* and hydraulic conductivity k^*, where $k^* \ll k$. If the aquifer is unconfined then the flow Q_n through the slurry wall is obtained from the Dupuit-formula (see Figure 6.9[b]):

$$Q_n = \frac{\frac{1}{2}k^*[(\phi^+)^2 - (\phi^-)^2]}{b^*} \tag{36.1}$$

where Q_n is positive if pointing from the $+$ side of the wall to the $-$ side. The above equation may be expressed in terms of the jump $\Phi^+ - \Phi^-$ of the potential across the wall as follows:

$$\boxed{Q_n = \frac{k^*}{kb^*}\frac{1}{2}k[(\phi^+)^2 - (\phi^-)^2] = \frac{k^*}{kb^*}(\Phi^+ - \Phi^-)} \tag{36.2}$$

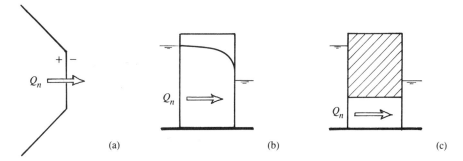

Figure 6.9 A slurry wall.

For convenience, we neglect the width of the slurry wall insofar as the flow in the aquifer is concerned; (36.2) is applied across a line representing the slurry wall. If the aquifer is confined, then the expression for Q_n becomes (see Figure 6.9[c]):

$$Q_n = \frac{k^*H(\phi^+ - \phi^-)}{b^*} \tag{36.3}$$

which may be expressed in terms of the jump in potential:

$$\boxed{Q_n = \frac{k^*}{kb^*}kH(\phi^+ - \phi^-) = \frac{k^*}{kb^*}(\Phi^+ - \Phi^-)} \tag{36.4}$$

which is identical to the equation for the case that the aquifer is unconfined, compare (36.2). It may be noted that the rightmost expression in (36.4) also applies to the case that the flow is confined at one side of the wall and unconfined at the other side.

The Boundary Condition Along a Draining Object

Consider a narrow crack with transmissivity T, as shown in Figure 6.10, and with end points A and B. We first imagine that the crack has a certain width b^*, and may be viewed as an inhomogeneity with high hydraulic conductivity. The difference in the stream function on either side of the crack, then, equals the amount flowing in the crack as follows:

$$\Psi^+ - \Psi^- = -\overset{c}{Q}_s \tag{36.5}$$

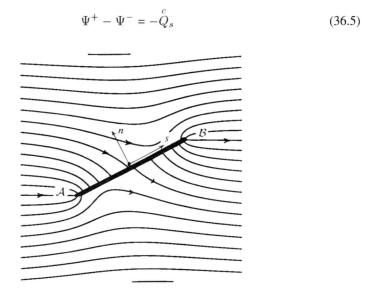

Figure 6.10 A narrow crack.

This equation is obtained from (19.12) where reference is made to Figure 3.3 in which points B and D would correspond to points on the boundary of the crack. Equation (36.5) still holds true if we neglect the width of the crack, keeping its transmissivity the same. If we assume that there is no flow through the end points A and B of the crack, then

$$(\Psi^+ - \Psi^-)_A = (\Psi^+ - \Psi^-)_B = 0 \tag{36.6}$$

If the hydraulic conductivity and width of the crack are k^* and b^*, then $\overset{c}{Q}_s$ may be expressed as

$$\overset{c}{Q}_s = -\frac{k^*}{k} b^* \frac{d\Phi}{ds} \tag{36.7}$$

where $d\Phi/ds$ is the tangential component of the gradient of the potential just outside the crack. If Q_s represents the tangential discharge just outside the crack, i.e., $Q_s = -d\Phi/ds$, (36.5) and (36.7) may be combined as follows:

$$Q_s = -\frac{k}{k^*b^*}(\Psi^+ - \Psi^-)$$ (36.8)

It may be noted that the width of the crack, b^*, is used to compute its transmissivity k^*b^*, but is neglected geometrically; the approximation implies that the effect of the crack width on the flow outside the crack is considered unimportant.

Modeling Thin Leaky Objects by Using Line-Doublets

Line-doublets cause the potential to jump but the normal component of discharge to be continuous across the element, and therefore may be used to model the leaky objects. The condition (36.2) is applied at a number of control points equal to the number of unknown coefficients λ. For the sake of brevity we will describe the system of equations in words rather than using symbols, and assume that first-order line-doublets are used. The left-hand side of the equation at a control point then represents the unknown coefficients λ multiplied by k^*/kb^* and the right-hand side, Q_n, is the normal component of the gradient of the potential. Thus, in order to apply the condition (36.2), the coefficient functions Λ_{wl}, Λ_{ls}, and Λ_{do} must be differentiated and their contribution to Q_n determined. As before, the right-hand side contains all unknown coefficients, and each control point contributes one equation to the system.

The contributions of two adjoining first-order line-doublets to Q_n are infinite at their common node. If the control points are to be chosen at the nodes, therefore, the contribution of the two adjoining line-doublets to Q_n must be approximated in some way. Haitjema [1977] applied line-doublets to model the moving interface between two fluids using a method proposed by De Josselin de Jong [1960]. Haitjema computed the contribution of the adjoining elements $\mathcal{AB}$ and $\mathcal{BC}$ to Q_n at node $\mathcal{B}$ in Figure 6.11 by replacing the two elements by the three elements $\mathcal{AB}'$, $\mathcal{B}'\mathcal{B}''$, and $\mathcal{B}''\mathcal{C}$ as shown. The approximation appeared to be satisfactory.

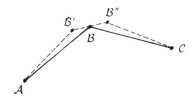

Figure 6.11 Computation of Q_n at a nodal point.

An alternative approach that eliminates the need for estimating Q_n is to use higher order elements as discussed in Sec. 38, joined by the small circular elements presented in Sec. 41. He [1987] implemented this approach in a computer program and obtained excellent agreement with exact solutions. If the walls are straight, then the double-root elements discussed in Sec. 39 can be used. These have no nodal points so that the control points can be chosen anywhere on the elements.

Modeling Thin Impermeable Walls Using Line-Doublets

A thin impermeable object is a special case of a leaky object and may be modeled by the use of line-doublets. The hydraulic conductivity of an impermeable wall is zero, and (36.2) becomes

$$\boxed{Q_n = 0} \qquad (36.9)$$

Or, since $Q_n = -d\Psi/ds$, this may be written as

$$\Psi = \text{constant} \qquad (36.10)$$

This boundary condition is usually applied by requiring that the difference in value of Ψ at any point of the object minus the value of Ψ at one of the end points, $\Psi_\mathcal{A}$, is zero:

$$\Psi - \Psi_\mathcal{A} = 0 \qquad (36.11)$$

The equations thus are obtained by subtracting two equations for the imaginary parts of the complex potential evaluated at two different points. Equation (36.11) applies only if there is no infiltration and if no branch cuts intersect the element. Equation (36.9) is always applicable and therefore is usually preferable over (36.11).

Modeling Narrow Draining Objects Using Line-Dipoles

The stream function jumps but the potential is continuous across a line-dipole, which therefore may be used to model draining objects with boundary conditions (36.8). Since the complex potential for a line-dipole differs from that for a line-doublet only by a factor of i, the complex potential for a line-dipole is readily obtained from that for a line-doublet. If ρ_j and ρ_{j+1} are the end points of a first-order line-dipole, and if μ_j and μ_{j+1} represent its strengths at ρ_j and ρ_{j+1}, then the potential for a line-dipole is obtained from (35.19) as follows:

$$\boxed{\Omega = \frac{1}{2\pi}\left[\mu_j F(Z_j) + \mu_{j+1} G(Z_j)\right]} \qquad (36.12)$$

where

$$Z_j = \frac{z - \frac{1}{2}(\rho_j + \rho_{j+1})}{\frac{1}{2}(\rho_{j+1} - \rho_j)} \qquad (36.13)$$

Sec. 37   Variable Infiltration

and the functions F and G are given by (35.18).

The internal boundary conditions (36.8) yield equations for each control point with the jump μ at the left-hand side of the equation and the tangential component of the derivative of the potential at the right-hand side. The remarks made for the leaky objects regarding the singularities of the discharge vector at the nodal points apply to draining objects also.

Modeling Narrow Ideal Drains

If the resistance of the crack or drain is zero, i.e., if the factor $1/k^*$ in (36.8) becomes zero, the internal boundary condition simplifies to

$$\boxed{Q_s = 0} \tag{36.14}$$

which may be replaced by the condition

$$\boxed{\Phi - \Phi_{\mathcal{A}} = 0} \tag{36.15}$$

where $\Phi_{\mathcal{A}}$ represents the value of the potential at an end point $\mathcal{A}$ of the drain.

37   VARIABLE INFILTRATION

There are many sources of infiltration into groundwater, rainfall being the most common one. The actual rate of infiltration reaching the groundwater varies: it not only depends upon the average rainfall but also on such factors as evaporation, surface runoff, vegetation, and the type of soil. The rate of infiltration therefore varies from place to place, and a desirable refinement of the model discussed above would be the capability to incorporate a rate of infiltration that varies with position. The area-sink is useful for modeling such a variable rate of infiltration, and is a distribution of sinks over an area, just as a line-sink is a distribution of sinks along a line. We will discuss in this section how the potential for an area-sink is determined and consider several applications.

Infiltration Through River- and Lake-Bottoms

Rivers and lakes usually have deposits along their bottoms, which often are less permeable than the aquifer underneath. If the river bottom is above the groundwater table, the leakage through the bottom is downward and independent of the flow in the aquifer (see Figure 6.12[a]), assuming that the pressure below the river bottom is atmospheric. We measure elevations with respect to the aquifer base and denote the elevation of points in the aquifer just below the river bottom as H_b, the head in the river as ϕ^*, and the resistance of the river bottom as c. The rate of infiltration N^* then may be expressed as follows:

$$N^* = \frac{\phi^* - H_b}{c} \tag{37.1}$$

which is indeed independent of the head ϕ in the aquifer. If ϕ is greater than H_b, the leakage depends upon the head in the aquifer, and

$$N^* = \frac{\phi^* - \phi}{c} \tag{37.2}$$

see Figure 6.12(b).

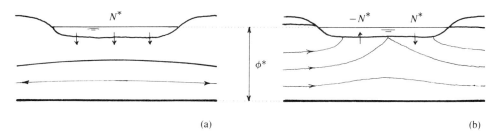

Figure 6.12 Interaction between a river and an aquifer.

The Potential for an Area-Sink

The potential for an area-sink may be obtained by integrating the potential for a sink over an area. This integration is cumbersome, and may be carried out analytically for only a few cases; we elect to determine the desired function by an alternative procedure as follows.

Statement of the Problem

Consider the contour $\mathcal{C}$ enclosing a domain $\mathcal{D}^+$; s and n are defined along and normal to $\mathcal{C}$, with n pointing inward (see Figure 6.13). If we use the symbol γ to denote the generally variable distribution strength, then the problem may be posed as follows: a potential Φ is to be determined such that it fulfills the differential equation

$$\nabla^2 \Phi = \gamma \qquad z \text{ in } \mathcal{D}^+ \tag{37.3}$$

inside the domain $\mathcal{D}^+$, and the differential equation

$$\nabla^2 \Phi = 0 \qquad z \text{ in } \mathcal{D}^- \tag{37.4}$$

in the domain $\mathcal{D}^-$, which is the exterior of the contour $\mathcal{C}$.

Sec. 37 Variable Infiltration

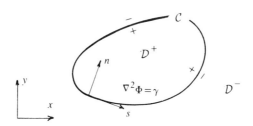

Figure 6.13 A general area-sink.

Decomposition of the Potential. We will solve the problem by writing the potential Φ as a sum of two functions:

$$\boxed{\Phi = \overset{i}{\Phi} + \overset{e}{\Phi}} \qquad (37.5)$$

where $\overset{i}{\Phi}$ is a discontinuous function, which is zero in $\mathcal{D}^-$:

$$\nabla^2 \overset{i}{\Phi} = \gamma \qquad z \text{ in } \mathcal{D}^+ \qquad (37.6)$$

$$\overset{i}{\Phi} = 0 \qquad z \text{ in } \mathcal{D}^- \qquad (37.7)$$

The function $\overset{e}{\Phi}$ is harmonic for all z:

$$\nabla^2 \overset{e}{\Phi} = 0 \qquad (37.8)$$

except, possibly, along $\mathcal{C}$.

The function $\overset{i}{\Phi}$ may be any suitable solution to (37.6); an example is the case that both γ and $\overset{i}{\Phi}$ are polynomials. The function $\overset{i}{\Phi}$ by itself violates the conditions of continuity of both pressure and flow across $\mathcal{C}$; the function $\overset{e}{\Phi}$ is used to eliminate these discontinuities. In order that the pressure, and therefore the head and the potential, are continuous across $\mathcal{C}$, the function $\overset{e}{\Phi}$ must jump as follows:

$$\overset{e}{\Phi}{}^+ - \overset{e}{\Phi}{}^- = -(\overset{i}{\Phi}{}^+ - \overset{i}{\Phi}{}^-) = -\overset{i}{\Phi} \qquad (37.9)$$

representing this jump as $\overset{e}{\lambda}$, we have

$$\overset{e}{\lambda} = -\overset{i}{\Phi} \qquad (37.10)$$

If $\overset{i}{Q}_n$ represents the normal component of the discharge vector due to $\overset{i}{\Phi}$, then continuity of flow implies that

$$\overset{e}{Q}{}_n^+ - \overset{e}{Q}{}_n^- = -\left[\overset{i}{Q}{}_n^+ - \overset{i}{Q}{}_n^-\right] = -\overset{i}{Q}{}_n^+ \tag{37.11}$$

This expression represents $-\overset{e}{\sigma}$, the production that must be generated by $\overset{e}{\Phi}$ along the contour $\mathcal{C}$ in order to maintain continuity of flow, and (37.11) may be written as

$$\overset{e}{\sigma} = \overset{i}{Q}{}_n^+ \tag{37.12}$$

We represent the harmonic function $\overset{e}{\Phi}$ as the real part of a complex potential $\overset{e}{\Omega}$:

$$\overset{e}{\Phi} = \Re \overset{e}{\Omega} \tag{37.13}$$

and determine $\overset{e}{\Omega}$ as a boundary integral.

The Complex Potential $\overset{e}{\Omega}$ as a Boundary Integral. It follows from (37.10) and (37.12) that $\overset{e}{\Omega}$ may be obtained as a combination of distributed doublets (density $\overset{e}{\lambda}$) and distributed sinks (density $\overset{e}{\sigma}$). This complex potential may therefore be written as the following complex contour-integral (compare [25.72] and [25.2]):

$$\overset{e}{\Omega} = -\frac{1}{2\pi i} \oint_{\mathcal{C}} \frac{\overset{e}{\lambda}(\delta)}{z - \delta} d\delta + \frac{1}{2\pi} \oint_{\mathcal{C}} \overset{e}{\sigma}(\delta) \ln(z - \delta) ds \tag{37.14}$$

Using (25.70) to express $\overset{e}{\sigma}(\delta) ds$ in the second integral as $-d\overset{e}{\mu}$ and applying partial integration to the second term, we obtain

$$\overset{e}{\Omega} = -\frac{1}{2\pi i} \oint_{\mathcal{C}} \frac{\overset{e}{\lambda}(\delta)}{z - \delta} d\delta - \frac{1}{2\pi} \overset{e}{\mu}(\delta) \ln(z - \delta)\Big|_{\mathcal{C}} - \frac{1}{2\pi} \oint_{\mathcal{C}} \frac{\overset{e}{\mu}(\delta)}{z - \delta} d\delta \tag{37.15}$$

It is seen from (37.12), continuity of flow, and the property that $\overset{i}{Q}{}_n^+$ is positive if pointing into $\mathcal{D}^+$, that the total inflow into the contour $\mathcal{C}$ is equal to the total discharge of the area sink:

$$\oint_{\mathcal{C}} \overset{e}{\sigma} ds = -\oint_{\mathcal{C}} \frac{d\mu}{ds} ds = \int_{\mathcal{D}^+} \gamma dx\, dy = \overset{e}{Q}_0 \tag{37.16}$$

The discharge, $\overset{e}{Q}_0$, of the area-sink is generally not zero, and therefore the jump in Ψ, μ, will decrease along the contour; $\overset{e}{\mu} = \Psi^+ - \Psi^-$ is double-valued at the starting point of integration, z_1. If we choose $\Psi^+ - \Psi^-$ as zero at z_1, then μ will decrease to $-\overset{e}{Q}_0$

Sec. 37 Variable Infiltration

when the integration is completed, as is seen from the second integral in (37.16). The second term in (37.15) now may be evaluated, and we obtain:

$$\overset{e}{\Omega} = -\frac{1}{2\pi i}\oint_{\mathcal{C}}\frac{\overset{e}{\lambda}(\delta)}{z-\delta}d\delta - \frac{1}{2\pi}\oint_{\mathcal{C}}\frac{\overset{e}{\mu}(\delta)}{z-\delta}d\delta + \frac{\overset{e}{Q_0}}{2\pi}\ln(z-z_1) \qquad (37.17)$$

We will limit ourselves to the case that the boundary $\mathcal{C}$ is a polygon of n segments with corner points $\overset{e}{z}_j$ ($j = 1, 2, \ldots, n$). The complex potential $\overset{e}{\Omega}$ then is the sum of potentials for line-doublets with strength $\overset{e}{\lambda}$, line-dipoles of strength $\overset{e}{\mu}$, and a well of discharge $\overset{e}{Q}_0$ at z_1.

An Area-Sink of Constant Strength Bounded by a Polygon

The vertices at $\overset{e}{z}_j$, $j = 1, 2, \ldots, n$, of the polygon are numbered in such a way that the side between vertices $\overset{e}{z}_1$ and $\overset{e}{z}_2$ is the longest one and $\mathcal{D}^+$ lies to the left with respect to the vector pointing from $\overset{e}{z}_j$ to $\overset{e}{z}_{j+1}$. We denote the length and orientation of the first side as L_1 and α_1 (see Figure 6.14)

$$\overset{e}{z}_2 - \overset{e}{z}_1 = L_1 e^{i\alpha_1} \qquad (37.18)$$

For convenience, we introduce a complex variable ζ such that

$$\zeta = \xi + i\eta = (z - \overset{e}{z}_0)e^{-i\alpha_1} \qquad (37.19)$$

where $\overset{e}{z}_0$ is defined as

$$\overset{e}{z}_0 = \frac{1}{n}\sum_{j=1}^{n}\overset{e}{z}_j \qquad (37.20)$$

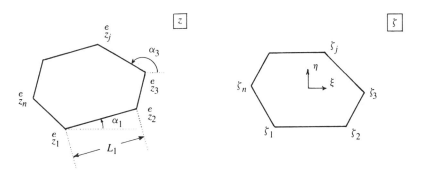

Figure 6.14 The boundary polygon.

The values of ζ at the corner points of the polygon are denoted as ζ_j,

$$\zeta_j = (\overset{e}{z}_j - \overset{e}{z}_0)e^{-i\alpha_1} \qquad (j = 1, 2, \ldots, n) \tag{37.21}$$

and the real and imaginary parts of ζ are obtained from (37.19) as follows:

$$\begin{aligned}\xi + i\eta &= [(x - \overset{e}{x}_0) + i(y - \overset{e}{y}_0)][\cos\alpha_1 - i\sin\alpha_1] \\ &= (x - \overset{e}{x}_0)\cos\alpha_1 + (y - \overset{e}{y}_0)\sin\alpha_1 - i[(x - \overset{e}{x}_0)\sin\alpha_1 - (y - \overset{e}{y}_0)\cos\alpha_1]\end{aligned} \tag{37.22}$$

We restrict ourselves for the time being to the case that γ is constant, and choose the potential $\overset{i}{\Phi}$ as

$$\overset{i}{\Phi} = \tfrac{1}{2}\gamma\xi^2 = \tfrac{1}{2}\gamma[(x - \overset{e}{x}_0)\cos\alpha_1 + (y - \overset{e}{y}_0)\sin\alpha_1]^2 \tag{37.23}$$

so that

$$\nabla^2 \overset{i}{\Phi} = \gamma \tag{37.24}$$

The components $\overset{i}{Q}_\xi$ and $\overset{i}{Q}_\eta$ of the discharge vector are obtained from (37.23) as

$$\overset{i}{Q}_\xi = -\gamma\xi \qquad \overset{i}{Q}_\eta = 0 \tag{37.25}$$

We introduce the length L_j and orientation α_j of side j between the vertices $\overset{e}{z}_j$ and $\overset{e}{z}_{j+1}$,

$$\overset{e}{z}_{j+1} - \overset{e}{z}_j = L_j e^{i\alpha_j} \qquad (j = 1, 2, \ldots, n) \tag{37.26}$$

where $\overset{e}{z}_{n+1}$ is equal to $\overset{e}{z}_1$,

$$\overset{e}{z}_{n+1} = \overset{e}{z}_1 \tag{37.27}$$

It follows from (37.21) that the orientation of the corresponding side in the ζ plane with respect to the ξ axis is $\alpha_j - \alpha_1$:

$$\zeta_{j+1} - \zeta_j = L_j e^{i(\alpha_j - \alpha_1)} \qquad (j = 1, 2, \ldots, n) \tag{37.28}$$

where

$$\zeta_{n+1} = \zeta_1 \tag{37.29}$$

Sec. 37 Variable Infiltration

The component $\overset{i}{Q}_n{}_j$ normal to side j then may be written as follows (see Figure 6.15),

$$\overset{i}{Q}_n{}_j = -\overset{i}{Q}_\xi \cos(\alpha_j - \alpha_1 - \pi/2) - \overset{i}{Q}_\eta \sin(\alpha_j - \alpha_1 - \pi/2)$$

$$= -\overset{i}{Q}_\xi \sin(\alpha_j - \alpha_1) + \overset{i}{Q}_\eta \cos(\alpha_j - \alpha_1) = \gamma \xi \sin(\alpha_j - \alpha_1) \qquad (37.30)$$

According to (37.10) and (37.12) the line-doublet distribution $\overset{e}{\lambda}_j$ and the line-sink distribution $\overset{e}{\sigma}_j$ for side j become

$$\overset{e}{\lambda}_j = -\overset{i}{\Phi}_j = -\tfrac{1}{2}\gamma \xi^2 \qquad (37.31)$$

and

$$\overset{e}{\sigma}_j = \overset{i}{Q}{}^+_{n\,j} = \gamma \xi \sin(\alpha_j - \alpha_1) \qquad (37.32)$$

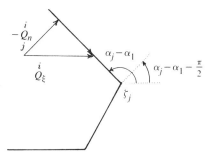

Figure 6.15 The component $\overset{i}{Q}_n{}_j$ for side j.

We will write the complex potentials for the jth line-doublet and line-dipole in terms of local complex variables Z_j, defined as

$$Z_j = X_j + iY_j = \frac{z - \tfrac{1}{2}(\overset{e}{z}_{j+1} + \overset{e}{z}_j)}{\tfrac{1}{2}(\overset{e}{z}_{j+1} - \overset{e}{z}_j)} = \frac{\zeta - \tfrac{1}{2}(\zeta_{j+1} + \zeta_j)}{\tfrac{1}{2}(\zeta_{j+1} - \zeta_j)} \qquad (37.33)$$

In order to express the distributions $\overset{e}{\lambda}_j$ and $\overset{e}{\sigma}_j$ in terms of Z_j, we solve (37.33) for ζ:

$$\zeta = \tfrac{1}{2}(\zeta_{j+1} - \zeta_j)Z_j + \tfrac{1}{2}(\zeta_{j+1} + \zeta_j) \qquad (37.34)$$

Y_j equals 0 along the element, and we obtain

$$\xi = \tfrac{1}{2}(\xi_{j+1} - \xi_j)Z_j + \tfrac{1}{2}(\xi_{j+1} + \xi_j) \qquad (Y_j = 0) \qquad (37.35)$$

and
$$\eta = \tfrac{1}{2}(\eta_{j+1} - \eta_j)Z_j + \tfrac{1}{2}(\eta_{j+1} + \eta_j) \qquad (Y_j = 0) \qquad (37.36)$$

We may now continue the strengths $\overset{e}{\lambda}_j$ and $\overset{e}{\sigma}_j$ analytically such that these functions reduce to (37.31) and (37.32) along the element ($Y_j = 0$). This gives for $\overset{e}{\lambda}_j$

$$\overset{e}{\lambda}_j = -\tfrac{1}{2}\gamma \left[\tfrac{1}{2}(\xi_{j+1} - \xi_j)Z_j + \tfrac{1}{2}(\xi_{j+1} + \xi_j)\right]^2 \qquad (37.37)$$

or

$$\overset{e}{\lambda}_j = -\tfrac{1}{8}\gamma(\xi_{j+1} - \xi_j)[(\xi_{j+1} - \xi_j)(Z_j^2 - 1) + 2(\xi_{j+1} + \xi_j)(Z_j + 1)] - \tfrac{1}{2}\gamma\xi_j^2 \qquad (37.38)$$

and we obtain for $\overset{e}{\sigma}_j$

$$\overset{e}{\sigma}_j = \gamma \left[\tfrac{1}{2}(\xi_{j+1} - \xi_j)Z_j + \tfrac{1}{2}(\xi_{j+1} + \xi_j)\right] \sin(\alpha_j - \alpha_1) \qquad (37.39)$$

According to (25.70), $\sigma = -d\mu/ds$ along the element. Z_j is real along the element and it follows from (37.33) that $dZ_j = (2/L_j)dze^{-i\alpha_j} = 2ds/L_j$. Defining the dipole distribution $\overset{e}{\mu}_j$ as an analytic function that is real along the element and satisfies (25.70), we may write

$$\overset{e}{\sigma}_j = -\frac{2}{L_j} \frac{d\overset{e}{\mu}_j}{dZ_j} \qquad (37.40)$$

We will choose the function $\overset{e}{\mu}$ such that it is zero at $\zeta = \zeta_1$,

$$\left[\overset{e}{\mu}\right]_{\zeta=\zeta_1} = \overset{e}{\mu}_1 = 0 \qquad (37.41)$$

so that integration of (37.40) gives, with (37.39),

$$\overset{e}{\mu}_j = -\tfrac{1}{2}L_j \int_{-1}^{Z_j} \tfrac{1}{2}\gamma \sin(\alpha_j - \alpha_1) \left[(\xi_{j+1} - \xi_j)Z_j + (\xi_{j+1} + \xi_j)\right] dZ_j + \overset{e}{\mu}_j \qquad (37.42)$$

where $\overset{e}{\mu}_j$ represents the value of $\overset{e}{\mu}$ at node j:

$$\overset{e}{\mu}_j = \left[\overset{e}{\mu}\right]_{\zeta=\zeta_j} \qquad (37.43)$$

Evaluation of (37.42) yields

$$\overset{e}{\mu}_j = -\frac{\gamma L_j \sin(\alpha_j - \alpha_1)}{8(\xi_{j+1} - \xi_j)} \left\{\left[(\xi_{j+1} - \xi_j)Z_j + (\xi_{j+1} + \xi_j)\right]^2 - 4\xi_j^2\right\} + \overset{e}{\mu}_j \qquad (37.44)$$

Sec. 37 Variable Infiltration

With

$$L_j \sin(\alpha_j - \alpha_1) = \eta_{j+1} - \eta_j \tag{37.45}$$

this may be written as

$$\overset{e}{\mu}_j = -\tfrac{1}{8}\gamma(\eta_{j+1} - \eta_j)\left[(\xi_{j+1} - \xi_j)(Z_j^2 - 1) + 2(\xi_{j+1} + \xi_j)(Z_j + 1)\right] + \overset{e}{\mu}_j \tag{37.46}$$

The complex potential $\overset{e}{\Omega}_j$ due to side j may now be written as follows:

$$\overset{e}{\Omega}_j = \frac{1}{2\pi i}(\overset{e}{\lambda}_j + i\overset{e}{\mu}_j)\ln\frac{Z_j - 1}{Z_j + 1} + \underset{j}{P}(Z_j) \tag{37.47}$$

where the function $\underset{j}{P}(Z_j)$ is such that $\overset{e}{\Omega}_j$ is of order Z_j^{-1} near infinity; $\underset{j}{P}(Z_j)$ is the far-field correction for element j. The complex function $\overset{e}{\lambda}_j + i\overset{e}{\mu}_j$ is obtained from (37.37) and (37.46), which yields, writing $\xi_{j+1} - \xi_j + i(\eta_{j+1} - \eta_j)$ as $\zeta_{j+1} - \zeta_j$,

$$\overset{e}{\lambda}_j + i\overset{e}{\mu}_j = -\tfrac{1}{8}\gamma(\zeta_{j+1} - \zeta_j)[(\xi_{j+1} - \xi_j)(Z_j^2 - 1) + 2(\xi_{j+1} + \xi_j)(Z_j + 1)] - \tfrac{1}{2}\gamma\xi_j^2 + i\overset{e}{\mu}_j \tag{37.48}$$

A recursive relation for the constants $\overset{e}{\mu}_j$ is obtained by using that $\overset{e}{\mu}_{j+1}$ equals the value of $\overset{e}{\mu}_j$ at node ζ_{j+1}, where $Z_j = 1$, which gives with (37.46)

$$\overset{e}{\mu}_{j+1} = \overset{e}{\mu}_j(1) = -\tfrac{1}{2}\gamma(\xi_{j+1} + \xi_j)(\eta_{j+1} - \eta_j) + \overset{e}{\mu}_j \qquad (j \geq 1) \tag{37.49}$$

With (37.41),

$$\overset{e}{\mu}_1 = 0 \tag{37.50}$$

equation (37.49) may be used recursively to determine all values $\overset{e}{\mu}_j$.

An expression for the complex potential $\overset{e}{\Omega}$ may now be obtained from (37.17) by replacing the two integrals by a sum over all functions $\overset{e}{\Omega}_j$,

$$\overset{e}{\Omega} = \sum_{j=1}^{n} \overset{e}{\Omega}_j + \frac{\overset{e}{Q}_0}{2\pi}\ln(z - \overset{e}{z}_1) \tag{37.51}$$

where $\overset{e}{z}_1$ replaces z_1 and $\overset{e}{Q}_0$ is the total discharge of the area-sink, equal to the product of the strength γ and the area A:

$$\overset{e}{Q}_0 = \gamma A \tag{37.52}$$

The function (37.51) represents the complex potential valid outside the area sink; the potential valid inside the area sink is obtained by adding the function $\overset{i}{\Phi}$, (37.23), to the real part of (37.51).

Equipotentials for a square area-sink are shown in Figure 6.16 as an illustration. It may be noted that the far-field of an area-sink approaches that of a well of discharge γA. This is illustrated in Figure 6.17 where equipotentials are shown for a square area-source with a well at its center, capturing all of the water produced by the area-source: the system has little effect outside the square.

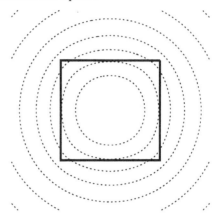

Figure 6.16 Equipotentials for an area-sink.

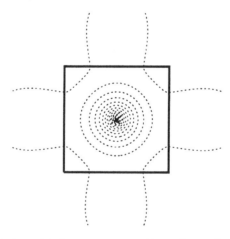

Figure 6.17 Equipotentials for an area-source with a well capturing all of its water.

Efficient application of area-sinks in a computer program requires some further examination of the functions presented above. The reader not interested in following the corresponding mathematical elaborations is encouraged to proceed to the subsection entitled *applications of area-sinks*.

Sec. 37   Variable Infiltration   **435**

The Complex Potential $\overset{e}{\delta}$ as a Function of ζ.   For reasons of computational efficiency, we express the n functions (37.47) in terms of a single variable ζ. We observe from (37.33) that

$$Z_j + 1 = \frac{\zeta - \zeta_j}{\frac{1}{2}(\zeta_{j+1} - \zeta_j)} \qquad Z_j - 1 = \frac{\zeta - \zeta_{j+1}}{\frac{1}{2}(\zeta_{j+1} - \zeta_j)} \qquad Z_j^2 - 1 = \frac{(\zeta - \zeta_j)(\zeta - \zeta_{j+1})}{\frac{1}{4}(\zeta_{j+1} - \zeta_j)^2}$$
(37.53)

We introduce n complex functions $\overset{e}{\chi}_j$ as

$$\overset{e}{\chi}_j = \frac{1}{\gamma}(\overset{e}{\lambda}_j + i\overset{e}{\mu}_j)$$
(37.54)

and define n constants $\overset{e}{\nu}_j$ as

$$\overset{e}{\nu}_j = \frac{1}{\gamma}\overset{e}{\mu}_j$$
(37.55)

so that (compare [37.49] and [37.50])

$$\begin{aligned}\overset{e}{\nu}_1 &= 0 \\ \overset{e}{\nu}_j &= -\tfrac{1}{2}(\xi_j + \xi_{j-1})(\eta_j - \eta_{j-1}) + \overset{e}{\nu}_{j-1} \qquad (j > 1)\end{aligned}$$
(37.56)

The complex potential $\overset{e}{\Omega}_j$, (37.47), may now be written in terms of ζ as

$$\overset{e}{\Omega}_j = \frac{\gamma}{2\pi i}\left[\overset{e}{\chi}_j(\zeta) \ln \frac{\zeta - \zeta_{j+1}}{\zeta - \zeta_j} + p_j(\zeta)\right]$$
(37.57)

where $p_j(\zeta) = (2\pi i/\gamma)P[Z_j(\zeta)]$. An expression for the function $\overset{e}{\chi}_j(\zeta)$ is obtained from (37.48), (37.53), (37.54), and (37.55) after some elaboration:

$$\overset{e}{\chi}_j = -\tfrac{1}{2}\left[\frac{\xi_{j+1} - \xi_j}{\zeta_{j+1} - \zeta_j}(\zeta - \zeta_j)(\zeta - \zeta_{j+1}) + (\xi_{j+1} + \xi_j)(\zeta - \zeta_j) + \xi_j^2\right] + i\overset{e}{\nu}_j$$
(37.58)

This function is a polynomial of order 2:

$$\overset{e}{\chi}_j = \underset{j}{c_2}\zeta^2 + \underset{j}{c_1}\zeta + \underset{j}{c_0}$$
(37.59)

where the coefficients $\underset{j}{c_m}$, $(m = 0, 1, 2)$ are obtained from (37.58) as

$$\underset{j}{c_2} = -\tfrac{1}{2}\frac{\xi_{j+1} - \xi_j}{\zeta_{j+1} - \zeta_j} \qquad \underset{j}{c_1} = \frac{\xi_{j+1} - \xi_j}{\zeta_{j+1} - \zeta_j}\zeta_j - \xi_j$$

$$\underset{j}{c_0} = -\tfrac{1}{2}\frac{\xi_{j+1} - \xi_j}{\zeta_{j+1} - \zeta_j}\zeta_j\zeta_{j+1} + \tfrac{1}{2}(\xi_{j+1} + \xi_j)\zeta_j - \tfrac{1}{2}\xi_j^2 + i\overset{e}{\nu}_j$$
(37.60)

The coefficients of the polynomials $p(\zeta)\atop j$ are obtained from the condition that $\Omega\atop j$^{e} is of order ζ^{-1} near infinity. Writing

$$\ln\frac{\zeta-\zeta_{j+1}}{\zeta-\zeta_j} = \ln\frac{1-\zeta_{j+1}/\zeta}{1-\zeta_j/\zeta} = \ln(1-\frac{\zeta_{j+1}}{\zeta}) - \ln(1-\frac{\zeta_j}{\zeta}) \tag{37.61}$$

and expanding the two logarithms about $\zeta^{-1} = 0$ we obtain

$$\ln\frac{\zeta-\zeta_{j+1}}{\zeta-\zeta_j} = -\sum_{m=1}^{\infty}\frac{1}{m}\left[\frac{\zeta_{j+1}}{\zeta}\right]^m + \sum_{m=1}^{\infty}\frac{1}{m}\left[\frac{\zeta_j}{\zeta}\right]^m$$

$$= -\sum_{m=1}^{\infty}\frac{\zeta_{j+1}^m - \zeta_j^m}{m\zeta^m} \qquad (\zeta^{-1} \to 0) \tag{37.62}$$

The far-field expansion of the function (37.57) now may be determined:

$$\Omega_j^e = -\frac{\gamma}{2\pi i}\left[c_2\zeta^2 + c_1\zeta + c_0\right]\sum_{m=1}^{\infty}\frac{\zeta_{j+1}^m - \zeta_j^m}{m\zeta^m} + \frac{\gamma}{2\pi i}p_j^e(\zeta) \qquad (\zeta^{-1} \to 0) \tag{37.63}$$

We write the first term as the sum of a polynomial of order one and a series that vanishes at $\zeta^{-1} = 0$; this gives

$$\Omega_j^e = -\frac{\gamma}{2\pi i}\left[c_2(\zeta_{j+1}-\zeta_j)\zeta + \tfrac{1}{2}c_2(\zeta_{j+1}^2-\zeta_j^2) + c_1(\zeta_{j+1}-\zeta_j)\right]$$

$$+ \frac{\gamma}{2\pi i}p_j^e(\zeta) + \frac{\gamma}{2\pi i}\sum_{m=1}^{\infty}b_m\zeta^{-m} \qquad (\zeta^{-1} \to 0) \tag{37.64}$$

where

$$b_m\atop j = -\tfrac{1}{m+2}c_2(\zeta_{j+1}^{m+2}-\zeta_j^{m+2}) - \tfrac{1}{m+1}c_1(\zeta_{j+1}^{m+1}-\zeta_j^{m+1}) - \tfrac{1}{m}c_0(\zeta_{j+1}^m-\zeta_j^m) \tag{37.65}$$

In order that the function (37.64) vanishes at infinity the polynomial $p(\zeta)\atop j$ must be equal to the polynomial inside the brackets in (37.64):

$$p(\zeta)\atop j = a_1\atop j\zeta + a_0\atop j \tag{37.66}$$

where

$$a_1 = c_2(\zeta_{j+1}-\zeta_j) = -\tfrac{1}{2}(\xi_{j+1}-\xi_j) \tag{37.67}$$

and

$$a_0 = \tfrac{1}{2}c_2(\zeta_{j+1}^2-\zeta_j^2) + c_1(\zeta_{j+1}-\zeta_j)$$

$$= -\tfrac{1}{4}(\xi_{j+1}^2-\xi_j^2) - \tfrac{i}{4}(\xi_{j+1}\eta_{j+1} - \xi_j\eta_j + 3\xi_j\eta_{j+1} - 3\xi_{j+1}\eta_j) \tag{37.68}$$

where use is made of (37.60).

Sec. 37 Variable Infiltration

The complex potential (37.51) may now be written as follows:

$$\overset{e}{\Omega} = \frac{\gamma}{2\pi i} \sum_{j=1}^{n} \overset{e}{\chi}(\zeta) \ln \frac{\zeta - \zeta_{j+1}}{\zeta - \zeta_j} + \frac{\gamma A}{2\pi} \ln(z - \overset{e}{z}_1) + \frac{\gamma}{2\pi i} \sum_{j=1}^{n} \overset{e}{p}(\zeta) \qquad (37.69)$$

We observe from (37.67) and (37.68) that

$$\sum_{j=1}^{n} \overset{e}{a_1} = \tfrac{1}{2} \sum_{j=1}^{n} (\xi_{j+1} - \xi_j) = 0 \qquad (37.70)$$

and

$$\sum_{j=1}^{n} \overset{e}{a_0} = -\tfrac{1}{4} \sum_{j=1}^{n} \{(\xi_{j+1}^2 - \xi_j^2) + i[(\xi_{j+1}\eta_{j+1} - \xi_j\eta_j) + 3(\xi_j\eta_{j+1} - \xi_{j+1}\eta_j)]\} \qquad (37.71)$$

The first two terms between parentheses vanish after summation and the last term between parentheses represents the cross-product of the vectors (ξ_j, η_j) and (ξ_{j+1}, η_{j+1}). Hence, $\tfrac{1}{2}(\xi_j \eta_{j+1} - \xi_{j+1}\eta_j)$ equals the shaded area in Figure 6.18 and (37.71) becomes

$$\sum_{j=1}^{n} \overset{e}{a_0} = -\tfrac{3}{2} i A \qquad (37.72)$$

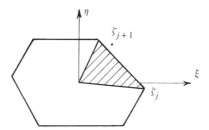

Figure 6.18 The area of the polygon.

With (37.59), (37.66), (37.70), and (37.72), the complex potential (37.69) becomes

$$\overset{e}{\Omega} = \frac{\gamma}{2\pi i} \sum_{j=1}^{n} (\overset{e}{c_2}\zeta^2 + \overset{e}{c_1}\zeta + \overset{e}{c_0}) \ln \frac{\zeta - \zeta_{j+1}}{\zeta - \zeta_j} + \frac{\gamma A}{2\pi} \ln(z - \overset{e}{z}_1) - \frac{3\gamma}{4\pi} A \qquad (37.73)$$

where the constants $\overset{e}{c_m}$ are given by (37.60). The far-field expansion is obtained by the use of (37.64); we note that the terms of order higher than ζ^{-1} cancel as a result of the choice of $\overset{e}{p}(\zeta)$:

$$\overset{e}{\Omega} = \frac{\gamma}{2\pi i} \sum_{j=1}^{n} \sum_{m=1}^{\infty} \overset{e}{b_m} \zeta^{-m} + \frac{\gamma A}{2\pi} \ln(z - \overset{e}{z}_1) \qquad (\zeta^{-1} \to 0) \qquad (37.74)$$

The potential Φ due to the area sink is obtained as the real part of the complex potential (37.73) for all points outside the polygon. For any point inside the polygon, the function $\overset{i}{\Phi} = \frac{1}{2}\gamma\xi^2$ must be added to the real part of (37.73). The imaginary part of the latter complex potential has physical meaning only outside the polygon, where it represents the stream function. It is possible to represent the potential for the area-sink, including the function $\overset{i}{\Phi}$, as the real part of a complex function defined for all z. Consider the function

$$\overset{i}{\Phi} = \frac{\gamma}{4\pi i}\xi^2 \sum_{j=1}^{n} \ln \frac{\zeta - \zeta_{j+1}}{\zeta - \zeta_j} \tag{37.75}$$

Because the polygon is closed, the real part of the sum of logarithms vanishes and the imaginary part equals $2\pi i$ for ζ inside the polygon and 0 for ζ outside the polygon, as is illustrated in Figure 6.19. For any point inside the polygon shown, the imaginary part of each logarithm equals a positive angle and the sum of these angles is 2π. For the point shown outside the polygon, however, $n-2$ of the angles are positive but two of them are negative and equal in magnitude to the sum of all the others, so that the algebraic sum of the angles is zero. The logarithm equals $2\pi i$ inside the polygon and vanishes outside; (37.75) represents the potential $\overset{i}{\Phi}$, which may now be added to the complex potential (37.73):

$$\boxed{\Omega = \frac{\gamma}{2\pi i} \sum_{j=1}^{n} \left(c_{2j}\zeta^2 + c_{1j}\zeta + c_{0j} + \tfrac{1}{2}\xi^2 \right) \ln \frac{\zeta - \zeta_{j+1}}{\zeta - \zeta_j} + \frac{\gamma A}{2\pi} \ln(z - \overset{e}{z}_1) - \frac{3\gamma}{4\pi} A} \tag{37.76}$$

It is important to note that this potential is analytic only outside the polygon.

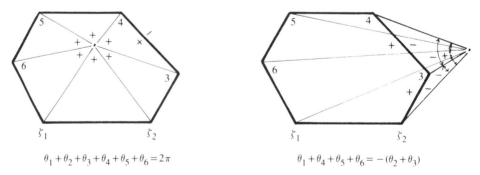

Figure 6.19 Behavior of the imaginary part of the sum of logarithms inside and outside the polygon.

Sec. 37   Variable Infiltration

The Discharge Function. The contribution of the function $\overset{i}{\Phi}$ to the components of the discharge vector (Q_x, Q_y) is obtained upon differentiation of (37.23) which gives

$$\overset{i}{Q}_x = -\gamma \xi \cos \alpha_1 \qquad \overset{i}{Q}_y = -\gamma \xi \sin \alpha_1 \tag{37.77}$$

The contribution of $\overset{e}{\Omega}$ to the components Q_x and Q_y may be obtained by complex differentiation:

$$\overset{e}{W} = \overset{e}{Q}_x - i\overset{e}{Q}_y = -\frac{d\overset{e}{\Omega}}{dz} = -\frac{d\overset{e}{\Omega}}{d\zeta}\frac{d\zeta}{dz} \tag{37.78}$$

This gives with (37.73) and (37.19)

$$\overset{e}{W} = -\frac{\gamma e^{-i\alpha_1}}{2\pi i} \sum_{j=1}^{n} \left[\left(2\underset{j}{c_2}\zeta + \underset{j}{c_1}\right) \ln \frac{\zeta - \zeta_{j+1}}{\zeta - \zeta_j} \right.$$
$$\left. + \left(\underset{j}{c_2}\zeta^2 + \underset{j}{c_1}\zeta + \underset{j}{c_0}\right)\left(\frac{1}{\zeta - \zeta_{j+1}} - \frac{1}{\zeta - \zeta_j}\right) \right] - \frac{\gamma A}{2\pi} \frac{1}{z - \overset{e}{z}_1} \tag{37.79}$$

The far-field expansion for $\overset{e}{W}$ is obtained by differentiating (37.74),

$$\overset{e}{W} = \frac{\gamma e^{-i\alpha_1}}{2\pi i} \sum_{j=1}^{n} \sum_{m=1}^{\infty} m\underset{j}{b_m}\zeta^{-m-1} - \frac{\gamma A}{2\pi} \frac{1}{z - \overset{e}{z}_1} \qquad (\zeta^{-1} \to 0, \; z^{-1} \to 0) \tag{37.80}$$

The components Q_x and Q_y, finally, are found from

$$Q_x = \overset{i}{Q}_x + \overset{e}{Q}_x = -\gamma \xi \cos \alpha_1 + \Re \overset{e}{W}$$
$$Q_y = \overset{i}{Q}_y + \overset{e}{Q}_y = -\gamma \xi \sin \alpha_1 - \Im \overset{e}{W} \tag{37.81}$$

The Behavior of $\overset{e}{\Omega}$ and $\overset{e}{W}$ Near the Corner Points

We observe from (37.73) and (37.79) that neither $\overset{e}{\Omega}$ nor $\overset{e}{W}$ is analytic at the corner points of the polygon. However, the singularities are removable, as will be shown below. Let n functions $\underset{j}{\overset{e}{F}}(\zeta)$ $(j = 1, 2, \ldots, n)$ be defined as follows:

$$\underset{j}{\overset{e}{F}}(\zeta) = \frac{\gamma}{2\pi i} \underset{j}{\overset{e}{\chi}}(\zeta) \ln \frac{\zeta - \zeta_{j+1}}{\zeta - \zeta_j} \tag{37.82}$$

and let ζ_j be a corner point other than ζ_1. Then the terms that are singular at ζ_j are $\overset{e}{F}_{j-1}$, and $\overset{e}{F}_j$. The functions $\overset{e}{\chi}_{j-1}$, and $\overset{e}{\chi}_j$ are polynomials of the second order and may be written as

$$\overset{e}{\chi}_{j-1}(\zeta) = \overset{e}{\chi}_{j-1}(\zeta_j) + \overset{e}{\chi}'_{j-1}(\zeta_j)(\zeta - \zeta_j) + \tfrac{1}{2}\overset{e}{\chi}''_{j-1}(\zeta_j)(\zeta - \zeta_j)^2 \qquad (37.83)$$

and

$$\overset{e}{\chi}_j(\zeta) = \overset{e}{\chi}_j(\zeta_j) + \overset{e}{\chi}'_j(\zeta_j)(\zeta - \zeta_j) + \tfrac{1}{2}\overset{e}{\chi}''_j(\zeta_j)(\zeta - \zeta_j)^2 \qquad (37.84)$$

where the primes stand for differentiation with respect to ζ. It follows from (37.58) that

$$\overset{e}{\chi}_{j-1}(\zeta_j) = \overset{e}{\chi}_j(\zeta_j) = -\tfrac{1}{2}\xi_j^2 + i\nu_j = \overset{e}{\chi}_j \qquad (37.85)$$

$$\overset{e}{\chi}'_{j-1}(\zeta_j) = \overset{e}{\chi}'_j(\zeta_j) = -\xi_j = \overset{e}{\chi}'_j \qquad (37.86)$$

$$\overset{e}{\chi}''_{j-1}(\zeta_j) = -\frac{\xi_j - \xi_{j-1}}{\zeta_j - \zeta_{j-1}} \qquad \overset{e}{\chi}''_j(\zeta_j) = -\frac{\xi_{j+1} - \xi_j}{\zeta_{j+1} - \zeta_j} \qquad (37.87)$$

The sum of $\overset{e}{F}_{j-1}$ and $\overset{e}{F}_j$ thus may be written as

$$\overset{e}{F}_{j-1}(\zeta) + \overset{e}{F}_j(\zeta) = \frac{\gamma}{2\pi i}[\overset{e}{\chi}_j + \overset{e}{\chi}'_j(\zeta - \zeta_j)]\left[\ln\frac{\zeta - \zeta_j}{\zeta - \zeta_{j-1}} + \ln\frac{\zeta - \zeta_{j+1}}{\zeta - \zeta_j}\right]$$
$$- \frac{\gamma}{4\pi i}\frac{\xi_j - \xi_{j-1}}{\zeta_j - \zeta_{j-1}}(\zeta - \zeta_j)^2 \ln\frac{\zeta - \zeta_j}{\zeta - \zeta_{j-1}} \qquad (37.88)$$
$$- \frac{\gamma}{4\pi i}\frac{\xi_{j+1} - \xi_j}{\zeta_{j+1} - \zeta_j}(\zeta - \zeta_j)^2 \ln\frac{\zeta - \zeta_{j+1}}{\zeta - \zeta_j}$$

The last two terms vanish in the limit for $\zeta \to \zeta_j$ and the singular logarithms between the brackets cancel so that

$$\overset{e}{F}_{j-1}(\zeta_j) + \overset{e}{F}_j(\zeta_j) = \frac{\gamma}{2\pi i}\overset{e}{\chi}_j \ln\frac{\zeta_j - \zeta_{j+1}}{\zeta_j - \zeta_{j-1}} \qquad (37.89)$$

Sec. 37   Variable Infiltration   441

It is to be noted in evaluating the logarithm that the branch cuts of all logarithmic terms fall along the polygon; refer to Sec. 35, (35.28) through (35.31). The value of $\overset{e}{\Omega}$ at node ζ_j now may be written as

$$\overset{e}{\Omega}(\zeta_j) = \frac{\gamma}{2\pi i} \sum_{\substack{m=1 \\ m \neq j-1 \\ m \neq j}}^{n} \left[\left(\underset{m}{c_2} \zeta_j^2 + \underset{m}{c_1} \zeta_j + \underset{m}{c_0} \right) \ln \frac{\zeta_j - \zeta_{m+1}}{\zeta_j - \zeta_m} \right]$$

$$+ \frac{\gamma}{2\pi i} \left(-\tfrac{1}{2} \xi_j^2 + i\overset{e}{\nu}_j \right) \ln \frac{\zeta_j - \zeta_{j+1}}{\zeta_j - \zeta_{j-1}}$$

$$+ \frac{\gamma A}{2\pi} \ln(\overset{e}{z}_j - \overset{e}{z}_1) + \frac{3\gamma}{4\pi} A \qquad (37.90)$$

A study of the singularity at ζ_1 requires some special attention. The functions that contain terms that are singular at ζ_1 are $\overset{e}{F}_1(\zeta)$ and $\overset{e}{F}_n(\zeta)$ and the jump in the stream function, $\overset{e}{\mu}$, is discontinuous at ζ_1. It follows from (37.56) and (37.58) that

$$\underset{1}{\overset{e}{\chi}}(\zeta_1) = -\tfrac{1}{2} \xi_1^2 \qquad (37.91)$$

and from (37.58) that

$$\underset{n}{\overset{e}{\chi}}(\zeta_{n+1}) = -\tfrac{1}{2} \xi_n^2 - \tfrac{1}{2}(\xi_{n+1} + \xi_n)(\zeta_{n+1} - \zeta_n) + i\overset{e}{\nu}_n \qquad (37.92)$$

or, using (37.56) to eliminate $\overset{e}{\nu}_n$,

$$\underset{n}{\overset{e}{\chi}}(\zeta_{n+1}) = -\tfrac{1}{2} \xi_n^2 - \tfrac{1}{2}(\xi_{n+1}^2 - \xi_n^2) - \tfrac{i}{2} \sum_{j=1}^{n} (\xi_{j+1} + \xi_j)(\eta_{j+1} - \eta_j) \qquad (37.93)$$

This may be written as

$$\underset{n}{\overset{e}{\chi}}(\zeta_{n+1}) = -\tfrac{1}{2} \xi_{n+1}^2 - \tfrac{i}{2} \sum_{j=1}^{n} (\xi_{j+1} \eta_{j+1} - \xi_j \eta_j) - \tfrac{i}{2} \sum_{j=1}^{n} (\xi_j \eta_{j+1} - \xi_{j+1} \eta_j) \qquad (37.94)$$

The second term vanishes and the third one equals $-iA$ (see Figure 6.18) so that, with $\zeta_{n+1} = \zeta_1$,

$$\underset{n}{\overset{e}{\chi}}(\zeta_{n+1}) = -\tfrac{1}{2} \xi_1^2 - iA \qquad (37.95)$$

The derivatives of $\overset{e}{\underset{1}{\chi}}(\zeta)$ and $\overset{e}{\underset{n}{\chi}}(\zeta)$ at ζ_1 are equal and we obtain

$$\overset{e}{\underset{1}{F}}(\zeta) + \overset{e}{\underset{n}{F}}(\zeta) = \frac{\gamma}{2\pi i}[-\tfrac{1}{2}\xi_1^2 + \overset{e'}{\chi_1}(\zeta - \zeta_1)]\left[\ln\frac{\zeta - \zeta_2}{\zeta - \zeta_1} + \ln\frac{\zeta - \zeta_1}{\zeta - \zeta_n}\right]$$

$$-\frac{\gamma}{4\pi i}\frac{\xi_2 - \xi_1}{\zeta_2 - \zeta_1}(\zeta - \zeta_1)^2 \ln\frac{\zeta - \zeta_2}{\zeta - \zeta_1}$$

$$-\frac{\gamma}{4\pi i}\frac{\xi_1 - \xi_n}{\zeta_1 - \zeta_n}(\zeta - \zeta_1)^2 \ln\frac{\zeta - \zeta_1}{\zeta - \zeta_n}$$

$$+\frac{\gamma}{2\pi i}(-iA)\ln\frac{\zeta - \zeta_1}{\zeta - \zeta_n} \tag{37.96}$$

The singular logarithms in the first term cancel, and the last term has a singularity at $\zeta = \zeta_1$ that cancels with the term $\gamma A/(2\pi)\ln(z - \overset{e}{z}_1)$ in expression (37.73) for $\overset{e}{\Omega}$. It follows from (37.19) that the latter term may be written as

$$\frac{\gamma A}{2\pi}\ln(z - \overset{e}{z}_1) = \frac{\gamma A}{2\pi}\ln(\zeta - \zeta_1) + \frac{\gamma A}{2\pi}i\alpha_1 \tag{37.97}$$

and the function $\overset{e}{\Omega}$ at ζ_1 becomes

$$\overset{e}{\Omega}(\zeta_1) = \frac{\gamma}{2\pi i}\sum_{j=2}^{n-1}(\underset{j}{c_2}\zeta_1^2 + \underset{j}{c_1}\zeta_1 + \underset{j}{c_0})\ln\frac{\zeta_1 - \zeta_{j+1}}{\zeta_1 - \zeta_j} - \frac{\gamma}{4\pi i}\xi_1^2 \ln\frac{\zeta_1 - \zeta_2}{\zeta_1 - \zeta_n}$$

$$+\frac{\gamma}{2\pi}A\ln(\zeta_1 - \zeta_n) + \frac{\gamma A}{2\pi}i\alpha_1 + \frac{3\gamma}{4\pi}A \tag{37.98}$$

The singularities of the discharge function at ζ_j also are removable; this is seen by differentiating (37.88) and (37.96). It is left as an exercise to determine the values of $\overset{e}{W}$ at ζ_j, $j = 1, 2, \ldots, n$.

Problem 37.1

Determine expressions for $\overset{e}{W}$ at the corner points of the polygon, and demonstrate that the singularities are removable.

Applications of Area-Sinks

The potential for an area-sink presented above may be used to model given infiltration from rainfall, or through river and lake bottoms above the groundwater table. For such cases the distribution strength γ is given, and the potentials for the area-sinks are simply added to the given potential V. If the area-sinks are used to model infiltration through river or lake bottoms where the head in the aquifer is higher than the lake bottom, equations of the form (37.2) are applied at the centers of the area-sinks, providing one

Sec. 37   Variable Infiltration   443

equation for each unknown strength γ. For cases where the aquifer is confined, equation (37.2) may be written in terms of the potential $\Phi = kH\phi - \frac{1}{2}kH^2$:

$$N^* = -\gamma = \frac{kH\phi^* - \frac{1}{2}kH^2 - \Phi}{kHc} \qquad (37.99)$$

If the aquifer is unconfined and becomes confined below the river bottom, the transition in aquifer thickness causes the potential to jump along the river banks. This jump may be taken into account by adding strings of line-doublets along the river bank. If the decrease in aquifer thickness is small relative to the thickness, an approximate solution may be obtained by approximating the aquifer thickness below the river as the head ϕ. With this approximation, the use of line-doublets along the river bank is not necessary, but the conditions at the control points become nonlinear. We write the head ϕ as the sum of an average value and a deviation ε:

$$\phi = \overline{\phi} + \varepsilon \qquad (37.100)$$

The potential becomes

$$\Phi = \tfrac{1}{2}k\phi^2 = \tfrac{1}{2}k(\overline{\phi}^2 + 2\overline{\phi}\varepsilon + \varepsilon^2) \qquad (37.101)$$

If ε is small, we may neglect the term ε^2 and obtain

$$\varepsilon = \frac{\Phi}{k\overline{\phi}} - \tfrac{1}{2}\overline{\phi} \qquad (37.102)$$

Using (37.100) and (37.102), the condition (37.2) may be approximated as

$$N^* = -\gamma = \frac{\phi^* - \overline{\phi} - \varepsilon}{c} = \frac{\phi^* - \tfrac{1}{2}\overline{\phi}}{c} - \frac{\Phi}{kc\overline{\phi}} \qquad (37.103)$$

The problem is solved iteratively: $\overline{\phi}$ is first estimated, and then updated after each iteration until the desired accuracy is reached.

Example 37.1: Modeling Flow in the Aquifers Underneath Grey Cloud Island.
As an example of the modeling with area-sinks, we will consider the flow in the aquifer system underneath Grey Cloud Island, an island in the Mississippi river in Minnesota, about 10 miles downstream from St. Paul. The study was sponsored by the J. L. Shiely Company in Minnesota, with the objective to determine the impact on the aquifer system of lowering the bottom of a quarry on the island. A map of the area is given in Figure 6.20, which shows the main rivers in the neighborhood of the island.

We will consider this problem mainly to illustrate the use of area-sinks, and give only a brief summary of the geologic features. The system consists of two aquifers that act as a single hydraulic unit, and which have approximately the same hydraulic conductivity. The upper aquifer is unconfined and consists mainly of highly fractured dolomite with sandy layers, and the lower one is a sandstone aquifer with a thickness of 100 feet. The hydraulic conductivity of the upper aquifer is taken as 46 feet per day, and

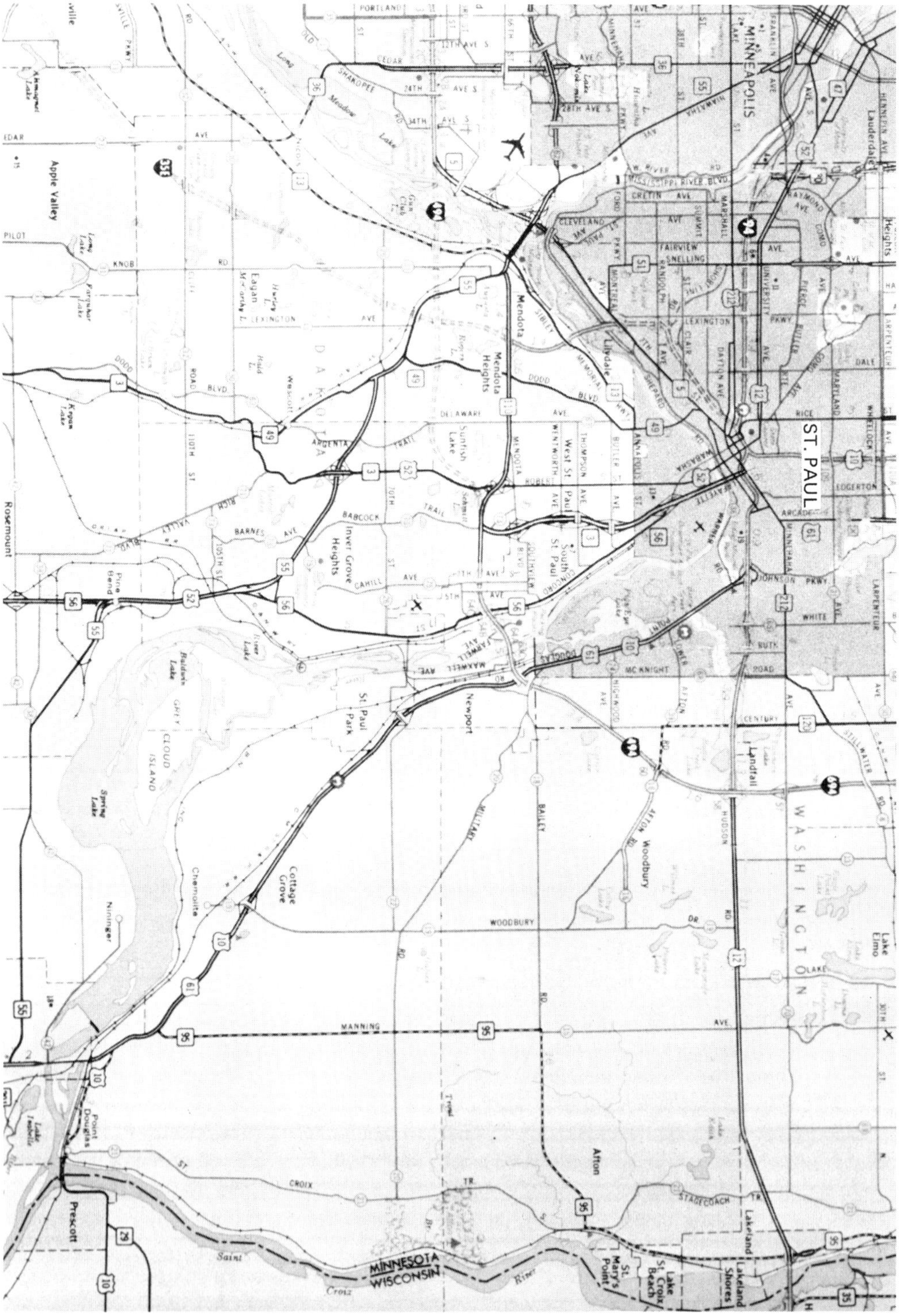

Figure 6.20 The area near Grey Cloud Island.

that of the lower one as 59 feet per day. The lower boundary of the aquifer is bedrock at an elevation of 495 feet above mean sea level. There is an average infiltration into the aquifer system of $0.34 * 10^{-3}$ feet per day. The rivers are not in direct contact with the aquifers; there is some resistance to flow caused by deposits on the river bottoms.

The pumping rate of the quarry is much more than the total amount of infiltration from rainfall on the island; the quarry draws water either from the rivers, or from the recharge area on the other side of the rivers bounding the island. The high quality of the water pumped from the quarry, however, indicates that most of the pumped water flows underneath the rivers and is supplied from the recharge area. Although the main interest is to make accurate predictions for the aquifers underneath the island, it will be necessary to include the recharge areas in the model, at least in an approximate fashion.

The model aquifer system is infinite in extent, and the functions included in the model are that for uniform flow (see [34.5]), rainfall (see [34.4]), wells (see [34.7]) line-sinks (see [34.15]), and area-sinks. The conditions applied along the bottoms of the rivers and lakes are of the form (37.103). The model aquifer is infinite in extent, but features such as rivers and lakes are included only so far as they influence the piezometric contours in the area of interest. The computer program used (called SLAEM for Single Layer Analytic Element Model) is highly interactive; data are entered by the use of a digitizer, marking the corner points of the area-sinks used to model the river and lake bottoms by pressing a pen onto a map positioned on the digitizer, and the results are obtained in the form of contour maps and plots of streamlines. An estimate is made of the size of the area to be modeled, and elements are entered inside this area. The size of the area then is increased in steps until the effect of the additional elements on the area of interest becomes negligible. The elements entered into the program are shown in Figure 6.21. It may be noted that the elements are smaller in the neighborhood of the island, i.e., in the area where detail is important. The infiltration from rainfall varies somewhat due to differences in top soils, and a few large area-sinks with given strength are added to account for this; the latter elements are omitted from the layout in Figure 6.21 to avoid confusion.

Data available for tuning the model are few, and not all of them are reliable. The points where heads are known are shown on the map in Figure 6.22. The predicted contours are shown in Figure 6.23; it is seen from these contours that mounds are formed between the rivers, controlling to a large extent the amount of water entering the aquifer system below the island.

The model being analytic, contour maps may be generated for any part of the aquifer system. A map showing both piezometric contours and the data points is given in Figure 6.24; points where the deviations are over 20 feet are marked with a star. Predicted contours for the aquifers underneath the island are shown in Figure 6.25 (the solid curves), along with the contours that apply to the case that the quarry bottom is lowered (the dotted contours). It is of interest to note that there is some resistance between the quarry bottom and the main aquifer; the magnitude of this resistance is chosen such that the discharge from the quarry as predicted from the model matches the discharge actually pumped from the quarry. It appears that the discharge to be pumped from the quarry with a lowered bottom is about twice that pumped at present.

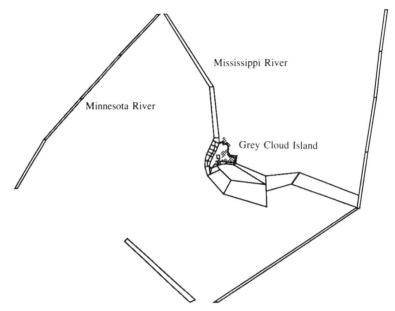

Figure 6.21 Layout of the wells, rivers, and lakes; the small circles are wells and the quadrilaterals along the rivers are area-sinks (area-sources used for infiltration from rainfall are not shown).

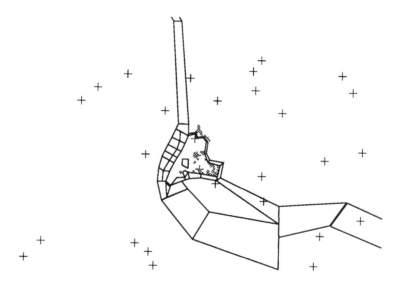

Figure 6.22 Layout showing points where groundwater levels are known.

Sec. 37 Variable Infiltration 447

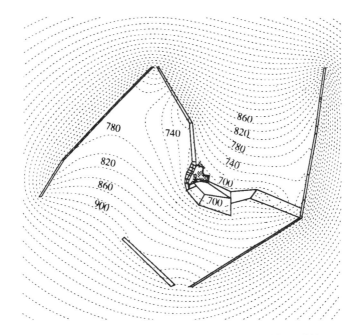

Figure 6.23 Predicted contours for the area shown in Figure 6.21.

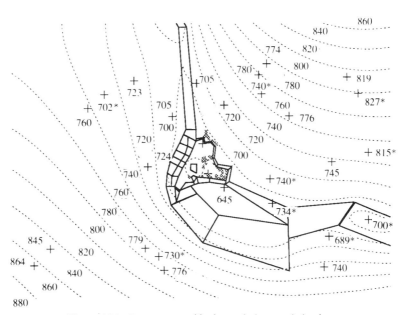

Figure 6.24 Contour map with observed piezometric levels.

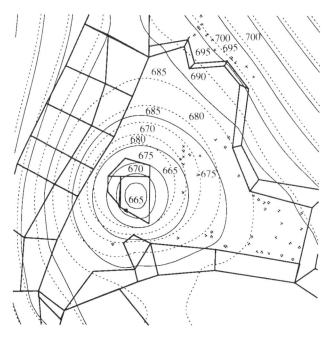

Figure 6.25 Piezometric contours in the main aquifer below Grey Cloud Island for both the present conditions (solid curves) and the conditions with the lowered quarry bottom (dotted curves).

The quarry draws water both from the recharge area and from the Mississippi river. This is illustrated in Figures 6.26(a) and (b), which apply to the flow prior to lowering the quarry bottom. A discharge streamline crossing the river is shown in Figure 6.26(a) and a cross-sectional view along this discharge streamline is shown in Figure 6.26(b). The elevations of the streamlines in the latter figure were computed by the use of the procedure outlined in Sec. 27, which is implemented in SLAEM. Note that the vertical scale in Figure 6.26(b) is exaggerated by a factor of 5.

The example given above may serve to illustrate the use of a simple analytic element model. Additional features, such as inhomogeneities in the hydraulic conductivity, often need to be included, as we will see in an example of a case of a two-layer aquifer problem discussed in Sec. 43. The model used in the above example contains elements of constant strength only. For some cases it may be desirable to use elements of higher order, where the density distribution varies within the element. Such elements are discussed in Sec. 38. Aquifers such as the one discussed in the present example often exhibit, besides a continuous system of minor fractures, some large isolated fractures or solution channels. We will discuss in Sec. 39 how such thin isolated features may be modeled by the use of special analytic elements.

Sec. 38 Elements of Order Higher Than One

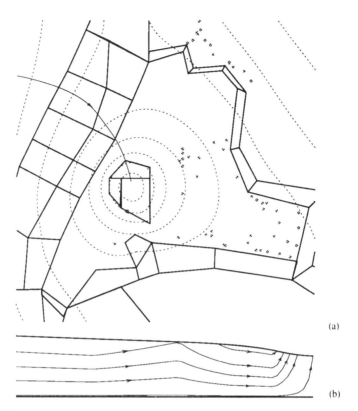

Figure 6.26 Cross-sectional view (b) of streamlines in the vertical plane through the discharge streamlines shown in (a).

38 ELEMENTS OF ORDER HIGHER THAN ONE

All linear elements discussed so far are characterized by a jump in either the potential or the stream function across the element, where the jump is a polynomial in terms of the local dimensionless variable Z. We identify the order of the element with the order of this polynomial. Thus, the line-doublet discussed in Sec. 35 is of the first order because the jump in the potential varies linearly along the element, see (35.16). By the same token, we refer to the line-sink presented in Sec. 25 as a line-sink of order one because the jump in the stream function varies linearly along the element (see [25.25] and [25.26]), even though the density distribution σ is a constant. This anomaly in nomenclature is the result of the difference in order between the rate of infiltration σ and the jump in the stream function; the former is the derivative of the latter. By analogy with the line-sink, we refer to an area-sink of constant density distribution γ as an area-sink of order 1.

We derived the complex potentials for the various elements in sufficiently general terms that density distributions of various forms and orders may be readily obtained.

Rather than presenting in this section scores of functions for higher order density distributions, we will select a few examples and emphasize the procedure rather than the results. The reader may then obtain the complex potential for an element of desired density distribution by following a similar procedure.

The Second-Order Line-Sink

The second-order line-sink has a density distribution that varies linearly between the value $\overset{1}{\sigma}$ at the starting point $\overset{1}{z}$ and the value $\overset{2}{\sigma}$ at the end point $\overset{2}{z}$ of the element. The complex potential (25.6) for a line-sink of length L and density distribution σ,

$$\Omega = \int_{-1}^{+1} \frac{\sigma}{2\pi} \ln[\tfrac{1}{2}(\overset{2}{z}-\overset{1}{z})(Z-\Delta)]\tfrac{1}{2}L d\Delta \tag{38.1}$$

is written in terms of the dimensionless variables Z and Δ defined by (25.4),

$$Z = X + iY = \frac{z - \tfrac{1}{2}(\overset{1}{z}+\overset{2}{z})}{\tfrac{1}{2}(\overset{2}{z}-\overset{1}{z})} \qquad \Delta = \Delta_1 + i\Delta_2 = \frac{\delta - \tfrac{1}{2}(\overset{1}{z}+\overset{2}{z})}{\tfrac{1}{2}(\overset{2}{z}-\overset{1}{z})} \tag{38.2}$$

The density distribution σ may be written in terms of Δ as

$$\sigma = \tfrac{1}{2}(\overset{2}{\sigma}-\overset{1}{\sigma})\Delta + \tfrac{1}{2}(\overset{1}{\sigma}+\overset{2}{\sigma}) \tag{38.3}$$

This density distribution is the sum of one that varies linearly and a constant one. The complex potential may thus be written as the sum of two complex potentials:

$$\Omega = \Omega_1 + \Omega_2 \tag{38.4}$$

where the first one is due to the variable term in (38.3)

$$\Omega_1 = \frac{L}{8\pi}\int_{-1}^{+1}(\overset{2}{\sigma}-\overset{1}{\sigma})\Delta \ln(Z-\Delta)d\Delta + \frac{L}{8\pi}\int_{-1}^{+1}(\overset{2}{\sigma}-\overset{1}{\sigma})\Delta \ln\left[\tfrac{1}{2}(\overset{2}{z}-\overset{1}{z})\right]d\Delta \tag{38.5}$$

and the second complex potential is obtained from (25.8) by replacing σ by $\tfrac{1}{2}(\overset{1}{\sigma}+\overset{2}{\sigma})$,

$$\Omega_2 = \frac{(\overset{1}{\sigma}+\overset{2}{\sigma})L}{8\pi}\left\{(Z+1)\ln(Z+1) - (Z-1)\ln(Z-1) + 2\ln\left[\tfrac{1}{2}(\overset{2}{z}-\overset{1}{z})\right] - 2\right\} \tag{38.6}$$

The second integral in (38.5) vanishes. Writing $(\overset{2}{\sigma}-\overset{1}{\sigma})\Delta$ as $-(\overset{2}{\sigma}-\overset{1}{\sigma})(Z-\Delta)+(\overset{2}{\sigma}-\overset{1}{\sigma})Z$ and $d\Delta$ as $-d(Z-\Delta)$ in (38.5), we obtain

$$\Omega_1 = \frac{L}{8\pi}(\overset{2}{\sigma}-\overset{1}{\sigma})\int_{Z+1}^{Z-1}(Z-\Delta)\ln(Z-\Delta)d(Z-\Delta)$$

$$-\frac{L}{8\pi}(\overset{2}{\sigma}-\overset{1}{\sigma})Z\int_{Z+1}^{Z-1}\ln(Z-\Delta)d(Z-\Delta) \tag{38.7}$$

Sec. 38   Elements of Order Higher Than One

Writing $(Z-\Delta)\ln(Z-\Delta)d(Z-\Delta)$ as $\frac{1}{4}\ln(Z-\Delta)^2 d(Z-\Delta)^2$ and integrating, we get

$$\Omega_1 = \frac{L}{8\pi}(\overset{2}{\sigma}-\overset{1}{\sigma})\{\tfrac{1}{2}(Z-1)^2\ln(Z-1) - \tfrac{1}{2}(Z+1)^2\ln(Z+1) + Z \\ - Z[(Z-1)\ln(Z-1) - (Z+1)\ln(Z+1)] + 2\} \quad (38.8)$$

Combining terms, this simplifies to

$$\Omega_1 = -\frac{L}{16\pi}(\overset{2}{\sigma}-\overset{1}{\sigma})\left[(Z^2-1)\ln\frac{Z-1}{Z+1} + 2Z\right] \quad (38.9)$$

The complex potential is obtained by adding the term (38.6) to (38.9),

$$\boxed{\begin{aligned}\Omega = \frac{L}{16\pi}\Big\{&-(\overset{2}{\sigma}-\overset{1}{\sigma})\left[(Z^2-1)\ln\frac{Z-1}{Z+1}+2Z\right] \\ &+2(\overset{1}{\sigma}+\overset{2}{\sigma})\{(Z+1)\ln(Z+1)-(Z-1)\ln(Z-1) \\ &+2\ln\left[\tfrac{1}{2}(\overset{2}{z}-\overset{1}{z})\right]-2\}\Big\}\end{aligned}} \quad (38.10)$$

An examination of the behavior of the function Ω_1 is left as an exercise for the reader.

Problem 38.1
Prove that the function Ω_1 is of order $1/Z$ near infinity and determine a far-field expansion for this function. Examine the behavior of both Ω and W near the points $\overset{1}{z}$ and $\overset{2}{z}$.

Second-order line-sinks may be used either by themselves, or may be combined in strings in the same way as was done for line-doublets in Sec. 35; the strengths σ_j, then, are associated with the nodal points z_j of the string. For example, if a line-sink is the jth element in the string, then $\overset{1}{z}=z_j$, $\overset{1}{\sigma}=\sigma_j$, $\overset{2}{z}=z_{j+1}$, and $\overset{2}{\sigma}=\sigma_{j+1}$. Especially when used in strings, second-order line-sinks are economical; a string of n second-order line-sinks has $n+1$ degrees of freedom as compared to the n degrees of freedom of a string of n first-order line-sinks. The string of second-order line-sinks has the advantage that the strength distribution is continuous at the nodes.

The Second-Order Line-Doublet

The line-doublet of the first order used for modeling inhomogeneities (see Sec. 35) does not always yield accurate results. For example, if the hydraulic conductivity inside the inhomogeneity is much lower than outside (e.g., if $k^+/k^- = 100$), a higher accuracy for a given number of degrees of freedom is obtained by using second-order line-doublets. A detailed discussion of the reasons for this gain in accuracy is given by Strack and Haitjema [1981(b)]; here it will suffice to present plots of piezometric contours for modeling a circular inhomogeneity in a confined aquifer with 28 line-doublets of the first order (Figure 6.27) and of the second order (Figure 6.28). The solid piezometric contours

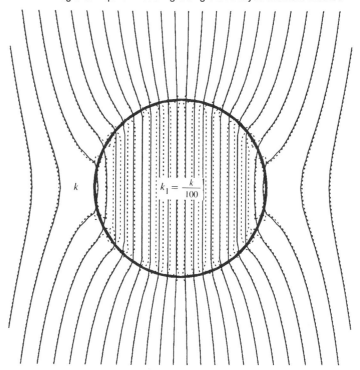

Figure 6.27 Piezometric contours obtained from the exact solution (solid curves) and by the use of first-order line-doublets (dotted curves).

were obtained from an exact solution; the dotted piezometric contours were obtained by the use of the computer program SLAEM.

The general complex potential for a line-doublet is given by (25.73)

$$\Omega = \frac{\lambda(Z)}{2\pi i} \ln \frac{Z-1}{Z+1} + ip(Z) \qquad (38.11)$$

The complex potential for the second-order line-doublet is obtained by adding a term Ω^* to the complex potential (35.19) that contributes to the strengths λ with a second-order polynomial that vanishes at the end points of the element:

$$\Omega^* = -\frac{1}{2\pi i} \lambda_m^* \left[(Z_m^2 - 1) \ln \frac{Z_m - 1}{Z_m + 1} + p^*(Z_m) \right] \qquad (38.12)$$

where λ_m^* represents the jump in the real part of Ω^* at the center of the mth element. The function $p^*(Z_m)$ is a polynomial chosen such that Ω^* is of order $1/Z_m$ near infinity. Since the logarithm behaves near infinity as $-2/Z_m + O(1/Z_m^3)$, it follows that $p^*(Z_m)$ is of order one, and (38.12) may be written as

$$\Omega^* = -\frac{1}{2\pi i} \lambda_m^* \left[(Z_m^2 - 1) \ln \frac{Z_m - 1}{Z_m + 1} + 2Z_m \right] \qquad (38.13)$$

Sec. 38   Elements of Order Higher Than One   453

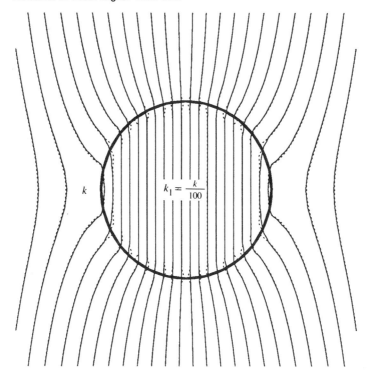

Figure 6.28   Piezometric contours obtained from the exact solution (solid curves) and by the use of second-order line-doublets (dotted curves).

The far-field expansion of (38.13) is obtained by expanding the logarithmic term according to (35.22):

$$\Omega^* = -\frac{1}{2\pi i}\lambda_m^* \left[-2\sum_{j=1}^{\infty}\frac{Z_m^{-2j+1+2}}{2j-1} - 2\sum_{j=1}^{\infty}\frac{Z_m^{-2j+1}}{2j-1} + 2Z_m\right] \qquad (38.14)$$

or

$$\Omega^* = \frac{1}{\pi i}\lambda_m^* \left\{\sum_{j=1}^{\infty}\left[\frac{1}{2j+1} - \frac{1}{2j-1}\right]Z_m^{-2j+1}\right\} \qquad (38.15)$$

Functions of the form (38.13) are added to the line-doublets of linear density distribution presented in Sec. 35. The value $\overset{c}{\lambda}_m$ of λ at the center of the element, $Z_m = 0$, is equal to the sum of λ_m^* and $\frac{1}{2}(\lambda_m + \lambda_{m+1})$:

$$\overset{c}{\lambda}_m = \lambda_m^* + \tfrac{1}{2}(\lambda_m + \lambda_{m+1}) \qquad (38.16)$$

so that (38.13) may be written as:

$$\Omega^* = -\frac{1}{2\pi i}\left[\overset{c}{\lambda}_m F^*(Z_m) - \tfrac{1}{2}\lambda_m F^*(Z_m) - \tfrac{1}{2}\lambda_{m+1}F^*(Z_m)\right] \qquad (38.17)$$

where

$$F^*(Z_m) = (Z_m^2 - 1)\ln\frac{Z_m - 1}{Z_m + 1} + 2Z_m \tag{38.18}$$

The complete complex potential for a second-order line-doublet is obtained by adding (38.17) to the complex potential (35.19) for a first-order line-doublet:

$$\Omega = \frac{1}{2\pi i}\left\{\lambda_m[F(Z_m) + \tfrac{1}{2}F^*(Z_m)] - \overset{c}{\lambda}_m F^*(Z_m) + \lambda_{m+1}[G(Z_m) + \tfrac{1}{2}F^*(Z_m)]\right\}$$

(38.19)

The condition (35.8) is applied at the nodes, which yields a set of equations similar to (35.37), with Λ_{db} now defined as (compare [35.35]):

$$\Lambda_{db}(z, \underset{r}{\delta}_{j-1}, \underset{r}{\delta}_j, \underset{r}{\delta}_{j+1}) = \Re\left\{\frac{1}{2\pi i}\left[G(\underset{r}{Z}_{j-1}) + \tfrac{1}{2}F^*(\underset{r}{Z}_{j-1}) + F(\underset{r}{Z}_j) + \tfrac{1}{2}F^*(\underset{r}{Z}_j)\right]\right\}$$

(38.20)

It may be noted that the value of $\Lambda_{db}(\underset{r}{\delta}_j, \underset{r}{\delta}_{j-1}, \underset{r}{\delta}_j, \underset{r}{\delta}_{j+1})$ may still be computed by the use of (35.28) since $F^*(-1) = F^*(+1) = 0$. An additional set of equations is obtained by applying the boundary condition (35.8) at the center of each element, where the jump in potential equals $\overset{c}{\lambda}_m$.

The improvement resulting from this addition is evident by comparing the dotted piezometric contours in Figure 6.28 with those in Figure 6.27. It may be noted that the improvement obtained in this way is much more than is achieved by doubling the number of first-order line-doublets.

It is possible to write the density distributions of a string of elements in such a way that the derivatives of the distributions up to any given order are continuous at the nodes. For example, in Strack [1982(a)] density distributions are presented of order 5 for which the distribution along with its first- and second-order derivatives are continuous. These functions have the disadvantage, however, of being complex, particularly in view of implementation in a computer program, and will not be discussed here.

The Second-Order Triangular Area-Sink

We will discuss a second-order area-sink as a final example of a second-order element. The area-sink is bounded by a triangle with vertices at $\overset{e}{z}_1$, $\overset{e}{z}_2$, and $\overset{e}{z}_3$, where the side between $\overset{e}{z}_1$ and $\overset{e}{z}_2$ is the longest one. The density distribution varies linearly between the values $\overset{e}{\gamma}_1$, $\overset{e}{\gamma}_2$, and $\overset{e}{\gamma}_3$ at the nodes with corresponding indices.

The triangle is transformed into a triangle in the ζ plane by the transformation

$$\zeta = \left[z - \tfrac{1}{3}(\overset{e}{z}_1 + \overset{e}{z}_2 + \overset{e}{z}_3)\right]e^{-i\alpha_1} \tag{38.21}$$

Sec. 38 Elements of Order Higher Than One

where α_1 is the angle between the first side and the x axis (see Figure 6.29). The reader may verify that the density distribution may be written as follows:

$$\gamma = a_1 \xi + a_2 \eta + a_3 \tag{38.22}$$

where

$$a_1 = \frac{\overset{e}{\gamma}_1(\eta_2 - \eta_3) + \overset{e}{\gamma}_2(\eta_3 - \eta_1) + \overset{e}{\gamma}_3(\eta_1 - \eta_2)}{2A} \tag{38.23}$$

$$a_2 = -\frac{\overset{e}{\gamma}_1(\xi_2 - \xi_3) + \overset{e}{\gamma}_2(\xi_3 - \xi_1) + \overset{e}{\gamma}_3(\xi_1 - \xi_2)}{2A} \tag{38.24}$$

$$a_3 = \frac{\overset{e}{\gamma}_1(\xi_2\eta_3 - \xi_3\eta_2) + \overset{e}{\gamma}_2(\xi_3\eta_1 - \xi_1\eta_3) + \overset{e}{\gamma}_3(\xi_1\eta_2 - \xi_2\eta_1)}{2A} \tag{38.25}$$

and A is the area of the triangle,

$$A = \tfrac{1}{2}[(\xi_1\eta_2 - \xi_2\eta_1) + (\xi_2\eta_3 - \xi_3\eta_2) + (\xi_3\eta_1 - \xi_1\eta_3)] \tag{38.26}$$

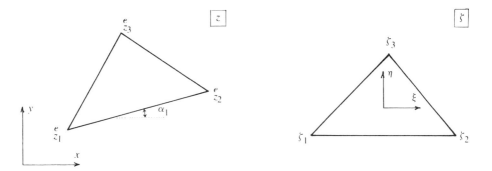

Figure 6.29 Mapping the area-sink onto the ζ plane.

The function $\overset{i}{\Phi}$ introduced in Sec. 37 (compare [37.5]) is chosen as follows:

$$\overset{i}{\Phi} = \tfrac{1}{6}a_1\xi^3 + \tfrac{1}{6}a_2\eta^3 + \tfrac{1}{2}a_3\xi^2 \tag{38.27}$$

This function satisfies the Poisson equation $\nabla^2 \overset{i}{\Phi} = \gamma$, with γ given by (38.22). The third term corresponds to an area-sink of the first order, and we may use the solution determined in Sec. 37 to incorporate that term. We write $\overset{i}{\Phi}$ and $\overset{e}{\Omega}$ as the sum of two parts each:

$$\overset{i}{\Phi} = \overset{i}{\Phi}{}^* + \overset{i}{\Phi}_0 \qquad \overset{e}{\Omega} = \overset{e}{\Omega}{}^* + \overset{e}{\Omega}_0 \tag{38.28}$$

where

$$\overset{i}{\Phi}{}^* = \tfrac{1}{6}a_1\xi^3 + \tfrac{1}{6}a_2\eta^3 \qquad \overset{i}{\Phi}{}_0 = \tfrac{1}{2}a_3\xi^2 \qquad (38.29)$$

The complex potential $\overset{e}{\Omega}{}^*$ is determined in much the same way as was done in Sec. 37; we will discuss the derivation below, leaving out some of the details of the algebraic manipulations.

The Complex Potential $\overset{e}{\Omega}{}^*$

We will follow the steps outlined in Sec. 37, and refer back to equations used for the derivation for the case of constant density distribution. It is to be understood that the symbols and functions in this section are somewhat different from those in Sec. 37; it is left to the reader to substitute the proper symbols and functions without this being mentioned in the text. We first determine the components of the discharge vector $\overset{i}{Q}{}^*_\xi$ and $\overset{i}{Q}{}^*_\eta$ due to $\overset{i}{\Phi}{}^*$:

$$\overset{i}{Q}{}^*_\xi = -\tfrac{1}{2}a_1\xi^2 \qquad \overset{i}{Q}{}^*_\eta = -\tfrac{1}{2}a_2\eta^2 \qquad (38.30)$$

and the component normal to side j is found by the use of (37.30),

$$\overset{i}{Q}{}^*_{n\atop j} = \tfrac{1}{2}a_1\xi^2 \sin(\alpha_j - \alpha_1) - \tfrac{1}{2}a_2\eta^2 \cos(\alpha_j - \alpha_1) \qquad (38.31)$$

An expression for the doublet strength $\overset{e}{\lambda}{}^*_j$ is obtained by the use of the first equality in (37.31), replacing $\overset{i}{\Phi}$ by $\overset{i}{\Phi}{}^*$ (see [38.29]),

$$\overset{e}{\lambda}{}^*_j = -\overset{i}{\Phi}{}^* = -\tfrac{1}{6}a_1\xi^3 - \tfrac{1}{6}a_2\eta^3 \qquad (38.32)$$

and the strength $\overset{e}{\sigma}{}^*_j$ is found from the first equality in (37.32), using (38.31),

$$\overset{e}{\sigma}{}^*_j = \tfrac{1}{2}a_1\xi^2 \sin(\alpha_j - \alpha_1) - \tfrac{1}{2}a_2\eta^2 \cos(\alpha_j - \alpha_1) \qquad (38.33)$$

These functions may be expressed in terms of the local complex variable Z_j by the use of (37.35) and (37.36),

$$\overset{e}{\lambda}{}^*_j = -\tfrac{1}{6}a_1\left[\tfrac{1}{2}(\xi_{j+1} - \xi_j)(Z_j + 1) + \xi_j\right]^3$$
$$\quad - \tfrac{1}{6}a_2\left[\tfrac{1}{2}(\eta_{j+1} - \eta_j)(Z_j + 1) + \eta_j\right]^3 \qquad (38.34)$$

Sec. 38   Elements of Order Higher Than One

and

$$\overset{e}{\sigma}{}^*_j = \tfrac{1}{2}a_1\left[\tfrac{1}{2}(\xi_{j+1} - \xi_j)(Z_j + 1) + \xi_j\right]^2 \sin(\alpha_j - \alpha_1)$$
$$- \tfrac{1}{2}a_2\left[\tfrac{1}{2}(\eta_{j+1} - \eta_j)(Z_j + 1) + \eta_j\right]^2 \cos(\alpha_j - \alpha_1) \qquad (38.35)$$

Integration of (37.40) gives with (38.35), after some elaboration,

$$\overset{e}{\mu}{}^*_j = -\tfrac{1}{6}a_1(\eta_{j+1} - \eta_j)\left[\tfrac{1}{8}(\xi_{j+1} - \xi_j)^2(Z_j + 1)^3\right.$$
$$\left. + \tfrac{3}{4}(\xi_{j+1} - \xi_j)\xi_j(Z_j + 1)^2 + \tfrac{3}{2}\xi_j^2(Z_j + 1)\right]$$
$$+ \tfrac{1}{6}a_2(\xi_{j+1} - \xi_j)\left[\tfrac{1}{8}(\eta_{j+1} - \eta_j)^2(Z_j + 1)^3\right.$$
$$\left. + \tfrac{3}{4}(\eta_{j+1} - \eta_j)\eta_j(Z_j + 1)^2 + \tfrac{3}{2}\eta_j^2(Z_j + 1)\right]$$
$$+ \overset{e}{\mu}{}^*_j \qquad (38.36)$$

where $\overset{e}{\mu}{}^*_j$ represents the value of $\overset{e}{\mu}{}^*_j$ at vertex j, where $Z_j = -1$. A recursive relation for the value of $\overset{e}{\mu}{}^*_j$ at vertex $j + 1$ is found from (38.36) by substituting $+1$ for Z_j,

$$\overset{e}{\mu}{}^*_{j+1} = -\tfrac{1}{6}a_1(\eta_{j+1} - \eta_j)(\xi_{j+1}^2 + \xi_j\xi_{j+1} + \xi_j^2)$$
$$+ \tfrac{1}{6}a_2(\xi_{j+1} - \xi_j)(\eta_{j+1}^2 + \eta_j\eta_{j+1} + \eta_j^2) + \overset{e}{\mu}{}^*_j \qquad (j \geq 1) \qquad (38.37)$$

where

$$\overset{e}{\mu}{}^*_1 = 0 \qquad (38.38)$$

An expression for the strength $\overset{e}{\lambda}{}^*_j + i\overset{e}{\mu}{}^*_j$ is obtained from (38.34) and (38.36) which yields, after elaboration,

$$\overset{e}{\lambda}{}^*_j + i\overset{e}{\mu}{}^*_j = \tfrac{1}{6}(\zeta_{j+1} - \zeta_j)\{-\tfrac{1}{8}[a_1(\xi_{j+1} - \xi_j)^2 - ia_2(\eta_{j+1} - \eta_j)^2](Z_j + 1)^3$$
$$- \tfrac{3}{4}[a_1(\xi_{j+1} - \xi_j)\xi_j - ia_2(\eta_{j+1} - \eta_j)\eta_j](Z_j + 1)^2$$
$$- \tfrac{3}{2}[a_1\xi_j^2 - ia_2\eta_j^2](Z_j + 1)\} - \tfrac{1}{6}a_1\xi_j^3 - \tfrac{1}{6}a_2\eta_j^3 + i\overset{e}{\mu}{}^*_j$$
$$(38.39)$$

This complex density distribution may be written in terms of ζ by the use of (37.53) as follows:

$$\overset{e}{\lambda}{}^*_j + i\overset{e}{\mu}{}^*_j = \tfrac{1}{6}\frac{1}{(\zeta_{j+1} - \zeta_j)^2}\{-[a_1(\xi_{j+1} - \xi_j)^2 - ia_2(\eta_{j+1} - \eta_j)^2](\zeta - \zeta_j)^3$$
$$- 3(\zeta_{j+1} - \zeta_j)[a_1(\xi_{j+1} - \xi_j)\xi_j - ia_2(\eta_{j+1} - \eta_j)\eta_j](\zeta - \zeta_j)^2$$
$$- 3(\zeta_{j+1} - \zeta_j)^2[a_1\xi_j^2 - ia_2\eta_j^2](\zeta - \zeta_j)\} - \tfrac{1}{6}a_1\xi_j^3 - \tfrac{1}{6}a_2\eta_j^3 + i\overset{e}{\mu}{}^*_j$$
$$(38.40)$$

In view of (38.37) we may write $\overset{e}{\mu}{}^*_j$ as

$$\overset{e}{\mu}{}^*_j = a_1 \overset{1}{\tau}_j + a_2 \overset{2}{\tau}_j \qquad (38.41)$$

where expressions for $\overset{1}{\tau}_j$ and $\overset{2}{\tau}_j$ are obtained from (38.37) and (38.38) with $\eta_1 = \eta_2$:

$$\overset{1}{\tau}_1 = 0 \qquad \overset{2}{\tau}_1 = 0 \qquad (38.42)$$

$$\overset{1}{\tau}_2 = 0 \qquad \overset{2}{\tau}_2 = \tfrac{1}{2}(\xi_2 - \xi_1)\eta_1^2 \qquad (38.43)$$

$$\overset{1}{\tau}_3 = -\tfrac{1}{6}(\eta_3 - \eta_2)(\xi_3^2 + \xi_2\xi_3 + \xi_2^2)$$

$$\overset{2}{\tau}_3 = \tfrac{1}{6}(\xi_3 - \xi_2)(\eta_3^2 + \eta_2\eta_3 + \eta_2^2) + \tfrac{1}{2}(\xi_2 - \xi_1)\eta_1^2 \qquad (38.44)$$

We may now add the density distribution due to the constant strength a_3 to the function (38.40) and obtain the following for the total density distribution (compare [37.60]):

$$\overset{e}{\lambda}_j + i\overset{e}{\mu}_j = \sum_{m=1}^{3} a_m \left[\overset{m}{c}_{3_j}\zeta^3 + \overset{m}{c}_{2_j}\zeta^2 + \overset{m}{c}_{1_j}\zeta + \overset{m}{c}_{0_j} \right] \qquad (38.45)$$

where

$$\overset{1}{c}_{3_j} = -\tfrac{1}{6}\left(\frac{\xi_{j+1} - \xi_j}{\zeta_{j+1} - \zeta_j}\right)^2$$

$$\overset{2}{c}_{3_j} = \tfrac{i}{6}\left(\frac{\eta_{j+1} - \eta_j}{\zeta_{j+1} - \zeta_j}\right)^2$$

$$\overset{3}{c}_{3_j} = 0 \qquad (38.46)$$

$$\overset{1}{c}_{2_j} = \tfrac{1}{2}\frac{\xi_{j+1} - \xi_j}{(\zeta_{j+1} - \zeta_j)^2}[\zeta_j(\xi_{j+1} - \xi_j) - \xi_j(\zeta_{j+1} - \zeta_j)]$$

$$\overset{2}{c}_{2_j} = -\tfrac{i}{2}\frac{\eta_{j+1} - \eta_j}{(\zeta_{j+1} - \zeta_j)^2}[\zeta_j(\eta_{j+1} - \eta_j) - [\eta_j(\zeta_{j+1} - \zeta_j)]$$

$$\overset{3}{c}_{2_j} = -\tfrac{1}{2}\frac{\xi_{j+1} - \xi_j}{\zeta_{j+1} - \zeta_j} \qquad (38.47)$$

$$\overset{1}{c}_{1_j} = -\tfrac{1}{2}\frac{\xi_{j+1} - \xi_j}{(\zeta_{j+1} - \zeta_j)^2} \zeta_j \left[\zeta_j(\xi_{j+1} - \xi_j) - 2\xi_j(\zeta_{j+1} - \zeta_j)\right] - \tfrac{1}{2}\xi_j^2$$

$$\overset{2}{c}_{1_j} = \tfrac{i}{2}\frac{\eta_{j+1} - \eta_j}{(\zeta_{j+1} - \zeta_j)^2} \zeta_j \left[\zeta_j(\eta_{j+1} - \eta_j) - 2\eta_j(\zeta_{j+1} - \zeta_j)\right] + \tfrac{i}{2}\eta_j^2$$

$$\overset{3}{c}_{1_j} = \frac{\xi_{j+1} - \xi_j}{\zeta_{j+1} - \zeta_j} \zeta_j - \xi_j \qquad (38.48)$$

Sec. 38   Elements of Order Higher Than One

$$\overset{1}{\underset{j}{c_0}} = \tfrac{1}{6}\frac{\xi_{j+1}-\xi_j}{(\zeta_{j+1}-\zeta_j)^2}\zeta_j^2\left[\zeta_j(\xi_{j+1}-\xi_j)-3\xi_j(\zeta_{j+1}-\zeta_j)\right]$$
$$+\tfrac{1}{2}\xi_j^2\zeta_j - \tfrac{1}{6}\xi_j^3 + i\overset{1}{\tau}_j$$

$$\overset{2}{\underset{j}{c_0}} = -\tfrac{i}{6}\frac{\eta_{j+1}-\eta_j}{(\zeta_{j+1}-\zeta_j)^2}\zeta_j^2\left[\zeta_j(\eta_{j+1}-\eta_j)-3\eta_j(\zeta_{j+1}-\zeta_j)\right]$$
$$-\tfrac{i}{2}\eta_j^2\zeta_j - \tfrac{1}{6}\eta_j^3 + i\overset{2}{\tau}_j$$

$$\overset{3}{\underset{j}{c_0}} = -\tfrac{1}{2}\frac{\xi_{j+1}-\xi_j}{\zeta_{j+1}-\zeta_j}\zeta_j\zeta_{j+1} + \tfrac{1}{2}(\xi_{j+1}+\xi_j)\zeta_j - \tfrac{1}{2}\xi_j^2 + i\overset{3}{\tau}_j \qquad (38.49)$$

where $\overset{3}{\tau}_j$ is identical to the variable $\overset{e}{\nu}_j$ introduced in Sec. 37,

$$\overset{3}{\tau}_j = \overset{e}{\nu}_j \qquad (38.50)$$

The density distribution may now be expressed in terms of the three strengths $\overset{e}{\gamma}_j$. Writing (38.23) through (38.26) as

$$a_m = \sum_{r=1}^{3}\overset{m}{b}_r\overset{e}{\gamma}_r \qquad (38.51)$$

the function (38.45) becomes

$$\overset{e}{\lambda}_j + i\overset{e}{\mu}_j = \sum_{r=1}^{3}\overset{e}{\gamma}_r\sum_{m=1}^{3}\overset{m}{b}_r\left[\overset{m}{\underset{j}{c_3}}\zeta^3 + \overset{m}{\underset{j}{c_2}}\zeta^2 + \overset{m}{\underset{j}{c_1}}\zeta + \overset{m}{\underset{j}{c_0}}\right] \qquad (38.52)$$

We may now introduce for each side three functions $\overset{e}{\underset{j}{\chi}}_r(\zeta)$ ($r=1,2,3$) as follows:

$$\overset{e}{\underset{j}{\chi}}_r = \sum_{m=1}^{3}\overset{m}{b}_r\left[\overset{m}{\underset{j}{c_3}}\zeta^3 + \overset{m}{\underset{j}{c_2}}\zeta^2 + \overset{m}{\underset{j}{c_1}}\zeta + \overset{m}{\underset{j}{c_0}}\right] = \sum_{\nu=0}^{3}\underset{j}{d_{r\nu}}\zeta^\nu \qquad (38.53)$$

where the constants $\underset{j}{d_{r\nu}}$ are

$$\underset{j}{d_{r\nu}} = \sum_{m=1}^{3}\overset{m}{b}_r\overset{m}{\underset{j}{c_\nu}} \qquad (38.54)$$

with $\overset{m}{\underset{j}{c_\nu}}$ given by (38.46) through (38.49) and $\overset{m}{b}_r$ obtained from (38.23) through (38.26) and (38.51). The functions $\overset{e}{\underset{j}{\chi}}_r$ do not depend upon the values of $\overset{e}{\gamma}_r$ and the density distribution (38.52) may be expressed in terms of these functions as

$$\overset{e}{\lambda}_j + i\overset{e}{\mu}_j = \sum_{r=1}^{3}\overset{e}{\gamma}_r\overset{e}{\underset{j}{\chi}}_r(\zeta) \qquad (38.55)$$

The complex potential $\overset{e}{\underset{j}{\Omega}}$ may be written in a form similar to (37.57):

$$\overset{e}{\underset{j}{\Omega}} = \frac{1}{2\pi i}\sum_{r=1}^{3}\overset{e}{\gamma}_r\left[\overset{e}{\underset{j}{\chi}}_r(\zeta)\ln\frac{\zeta-\zeta_{j+1}}{\zeta-\zeta_j}+\overset{e}{\underset{j}{p}}_r(\zeta)\right] \qquad (38.56)$$

Expressions for the functions $\overset{e}{\underset{j}{p}}_r(\zeta)$ are obtained from the far-field expansion of (38.56) (compare [37.62]),

$$\overset{e}{\underset{j}{p}}_r = \underset{j}{d}_{r3}(\zeta_{j+1}-\zeta_j)\zeta^2 + (\zeta_{j+1}-\zeta_j)\left[\frac{1}{2}\underset{j}{d}_{r3}(\zeta_{j+1}+\zeta_j)+\underset{j}{d}_{r2}\right]\zeta$$
$$+\frac{1}{3}\underset{j}{d}_{r3}(\zeta_{j+1}^3-\zeta_j^3) + \frac{1}{2}\underset{j}{d}_{r2}(\zeta_{j+1}^2-\zeta_j^2) + \underset{j}{d}_{r1}(\zeta_{j+1}-\zeta_j) \qquad (38.57)$$

It is left as an exercise for the reader to determine the far-field expansion for $\overset{e}{\underset{j}{\Omega}}$, following the procedure outlined in Sec. 37 (compare [37.63]).

The complete complex potential $\overset{e}{\Omega}$ is obtained by the use of (37.51) and (38.56)

$$\overset{e}{\Omega} = \sum_{j=1}^{3}\overset{e}{\underset{j}{\Omega}} + \frac{\overset{e}{Q}_0}{2\pi}\ln(z-\overset{e}{z}_1)$$

$$= \frac{1}{2\pi i}\sum_{r=1}^{3}\overset{e}{\gamma}_r\sum_{j=1}^{3}\left[\overset{e}{\underset{j}{\chi}}_r(\zeta)\ln\frac{\zeta-\zeta_{j+1}}{\zeta-\zeta_j}+\overset{e}{\underset{j}{p}}_r(\zeta)\right] + \frac{\overset{e}{Q}_0}{2\pi}\ln(z-\overset{e}{z}_1) \qquad (38.58)$$

As for the area-sink of order 1, the potential is obtained for all points outside the triangle as the real part of $\overset{e}{\Omega}$. For points inside the triangle, the potential is found from

$$\Phi = \overset{e}{\Phi}+\overset{i}{\Phi} = \Re\overset{e}{\Omega} + \frac{1}{6}a_1\xi^3 + \frac{1}{6}a_2\eta^3 + \frac{1}{2}a_3\xi^2 \qquad (38.59)$$

or, using (38.51),

$$\Phi = \Re\overset{e}{\Omega} + \frac{1}{6}\sum_{r=1}^{3}\left[\overset{1}{b}_r\overset{e}{\gamma}_r\xi^3 + \overset{2}{b}_r\overset{e}{\gamma}_r\eta^3 + \overset{3}{b}_r\overset{e}{\gamma}_r\xi^2\right] \qquad (38.60)$$

Formally, the potential $\overset{i}{\Phi}$ may be obtained from (compare [37.75]):

$$\overset{i}{\Phi} = \frac{1}{12\pi i}\sum_{r=1}^{3}[\overset{1}{b}_r\overset{e}{\gamma}_r\xi^3 + \overset{2}{b}_r\overset{e}{\gamma}_r\eta^3 + \overset{3}{b}_r\overset{e}{\gamma}_r\xi^2]\left[\ln\frac{\zeta-\zeta_2}{\zeta-\zeta_1}+\ln\frac{\zeta-\zeta_3}{\zeta-\zeta_2}+\ln\frac{\zeta-\zeta_1}{\zeta-\zeta_3}\right] \qquad (38.61)$$

Sec. 39 Analytic Elements for Thin Isolated Features 461

and a complex potential Ω may be introduced (compare [37.76]) as

$$\Omega = \frac{1}{2\pi i} \sum_{r=1}^{3} \overset{e}{\gamma}_r \sum_{j=1}^{3} \left\{ \left[\overset{e}{\chi}_{r}(\zeta) + \tfrac{1}{6}(\overset{1}{b}_r \xi^3 + \overset{2}{b}_r \eta^3 + \overset{3}{b}_r \xi^2) \right] \ln \frac{\zeta - \zeta_{j+1}}{\zeta - \zeta_j} + \overset{e}{p}_{r} \right\} \\ + \frac{\overset{e}{Q}_0}{2\pi} \ln(z - \overset{e}{z}_1) \qquad (38.62)$$

with a real part equal to the potential Φ for all points in the plane.

Problem 38.2

Determine the far-field expansion of the complex potential $\overset{e}{\Omega}$ in (38.58).

Problem 38.3

Determine an expression for the discharge function $\overset{e}{W}$.

Numerical Implementation

All coefficients involved in the functions $\overset{e}{\chi}_{r}(\zeta)$ and $\overset{e}{p}_{r}(\zeta)$ are independent of the values of $\overset{e}{\gamma}_r$ and therefore may be computed at the beginning of the execution of the program and stored for further use. If the values of $\overset{e}{\gamma}_r$ are unknown, the complex potential (38.62) may be used to generate the coefficients of the unknowns at the control points. Once the values of $\overset{e}{\gamma}_r$ have been determined, an increase in computational efficiency is obtained by interchanging the order of the summations in (38.62); the time-consuming computations of the logarithmic terms then need be performed three times per element rather than nine times, taking advantage of the independence of the latter terms of the index r.

39 ANALYTIC ELEMENTS FOR THIN ISOLATED FEATURES

There are a number of thin straight isolated features that may occur in an aquifer. Examples of such features are narrow canals or drainage ditches, drains, cracks or narrow highly permeable gravel pockets, slurry walls, and sheet pilings. These features have in common that high flow rates occur near the end points; it is important to include this singular behavior in the function used to model the feature.

We discussed in Sec. 31 the complex potential for flow toward a finite canal (see [31.34]), and for flow around an impermeable barrier (see [31.38]). It is possible to generalize these complex potentials to a set of two functions that contain the proper singular behavior near the tips of the element and possess degrees of freedom that can be used to meet boundary conditions at a number of control points on the element.

We will limit ourselves to the case that the element is straight, and refer to the analytic elements as a double-root doublet and a double-root dipole. These two double-root elements will be presented below, followed by a discussion of their application to various groundwater flow problems.

The Double-Root Dipole

Let the end points of the element in the physical plane $z = x + iy$ be denoted as $\overset{1}{z}$ and $\overset{2}{z}$ and let Z be defined, as before, as

$$Z = \frac{z - \frac{1}{2}(\overset{1}{z} + \overset{2}{z})}{\frac{1}{2}(\overset{2}{z} - \overset{1}{z})} \tag{39.1}$$

see Figure 6.30. The function $\chi = \chi(Z)$ that maps the exterior of the element in the Z plane onto the exterior of the unit circle in the χ plane was determined in Sec. 31 (see [31.33]), and has the form

$$\chi = Z + \sqrt{(Z-1)(Z+1)} \tag{39.2}$$

with its inverse

$$Z = \frac{1}{2}(\chi + \frac{1}{\chi}) \tag{39.3}$$

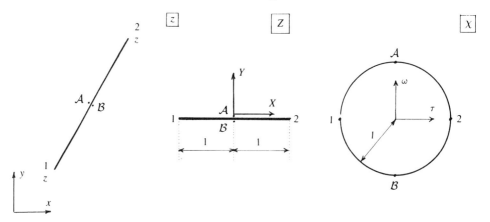

Figure 6.30 Transformation of the element into a circle.

The complex potential for a line-dipole is defined such that its real part is continuous and its imaginary part discontinuous across the element. If we represent this complex potential as $\Omega = \Phi + i\Psi$, then the conditions are

$$-1 \leq X \leq 1 \qquad Y = 0 \qquad \left\{ \begin{array}{l} \Phi^+ - \Phi^- = 0 \\ \Psi^+ - \Psi^- = \mu(X) \end{array} \right\} \tag{39.4}$$

Sec. 39 Analytic Elements for Thin Isolated Features

where μ is the density distribution of the line-dipole. Furthermore, the complex potential must be analytic everywhere outside the element including infinity; it must behave near infinity as

$$Z \to \infty \qquad \Omega \to \frac{A}{Z} + O(Z^{-2}) \qquad (39.5)$$

where A is a constant and O stands for "order of." In addition, we desire to create a complex potential with a singular behavior near the tips of the element, similar to that of the complex potentials derived in Sec. 31 for a thin canal and a thin impermeable object. These singularities have the form of square roots, and enter into the complex potential through the mapping function $\chi = \chi(Z)$. Hence the desired complex potential may be obtained as a function that is analytic for all χ with $|\chi| \geq 1$ and meets the conditions (39.4) and (39.5). For this purpose we first write the latter two conditions in terms of χ. Points on opposite sides of the slot are mapped onto points of the unit circle in the χ plane that are each other's complex conjugate. This is seen from (39.2) as follows:

$$-1 \leq X \leq 1 \qquad Y = 0^+ \qquad \begin{cases} \arg(Z+1) = 0 \\ \arg(Z-1) = \pi \end{cases} \qquad \chi^+ = X + i\sqrt{1-X^2} \qquad (39.6)$$

and

$$-1 \leq X \leq 1 \qquad Y = 0^- \qquad \begin{cases} \arg(Z+1) = 0 \\ \arg(Z-1) = -\pi \end{cases} \qquad \chi^- = X - i\sqrt{1-X^2} \qquad (39.7)$$

The condition (39.4) may now be written in terms of χ as

$$-1 \leq X \leq 1 \qquad Y = 0 \qquad \begin{cases} \Re(\Omega^+ - \Omega^-) = \Re[\Omega(\chi^+) - \Omega(\overline{\chi^+})] = 0 \\ \Im(\Omega^+ - \Omega^-) = \Im[\Omega(\chi^+) - \Omega(\overline{\chi^+})] = \mu^*(\chi^+) \end{cases} \qquad (39.8)$$

where

$$\mu^*(\chi^+) = \mu(X(\chi^+)) \qquad (39.9)$$

These conditions will be met if $\Omega(\chi)$ is chosen such that

$$\Omega(\chi) = \overline{\Omega(\overline{\chi})} \qquad (39.10)$$

The real parts of $\Omega(\chi)$ and $\Omega(\overline{\chi})$ will then be equal but their imaginary parts will be opposite in sign. We see from (39.3) that the condition (39.5) implies that

$$\chi \to \infty \qquad \Omega = \frac{2A}{\chi} + O(1/\chi^2) \qquad (39.11)$$

Consider the following function:

$$f_j = \frac{1}{\chi^j} \qquad (39.12)$$

This function fulfills the conditions (39.10) and (39.11) and is analytic for all χ except $\chi = 0$ which lies inside the unit circle, i.e., outside the region of interest. This function therefore meets all requirements, and a complex potential for a double-root dipole with n degrees of freedom may be written as

$$\Omega = \sum_{j=1}^{n} a_j f_j(\chi) = \sum_{j=1}^{n} \frac{a_j}{\chi^j} \qquad (\Im a_j = 0) \qquad (39.13)$$

where a_j are real constants. This function yields the following expression for $\mu^*(\chi^+)$:

$$\mu^*(\chi) = \Im \sum_{j=1}^{n} a_j \left[\frac{1}{(\chi^+)^j} - \frac{1}{(\chi^-)^j} \right] \qquad (39.14)$$

where χ^+ may be written as $e^{i\theta}$ and χ^- as $e^{-i\theta}$ (χ^+ and χ^- both are points on the unit circle) so that (39.14) becomes

$$\mu^*(\chi) = \Im \sum_{j=1}^{n} a_j \left(e^{-ij\theta} - e^{ij\theta} \right) = \Im \left[-2i \sum_{j=1}^{n} a_j \sin(j\theta) \right] = -2 \sum_{j=1}^{n} a_j \sin(j\theta) \quad (39.15)$$

Further degrees of freedom may be obtained by adding Laurent expansions about a selected number of poles at points δ_r, $(r = 1, 2, \ldots, m)$, located inside the unit circle and therefore outside the flow domain. Consider the functions $F_j(\chi, \delta_r)$ and $G_j(\chi, \delta_r)$ defined as follows:

$$F_j(\chi, \delta_r) = \frac{\delta_r}{(\chi - \delta_r)^j} + \frac{\overline{\delta}_r}{(\chi - \overline{\delta}_r)^j} \qquad (|\delta_r| < 1) \qquad (39.16)$$

and

$$G_j(\chi, \delta_r) = i \left[\frac{\delta_r}{(\chi - \delta_r)^j} - \frac{\overline{\delta}_r}{(\chi - \overline{\delta}_r)^j} \right] \qquad (|\delta_r| < 1) \qquad (39.17)$$

Both functions fulfill conditions of the form (39.10) and (39.11) and are analytic for all χ outside the unit circle and thus for all z in the flow domain. The complete complex potential thus may be written as

$$\Omega = \sum_{j=1}^{n} a_j f_j(\chi) + \sum_{r=1}^{m} \sum_{j=1}^{m_r} \left[b_j^r F_j(\chi, \delta_r) + c_j^r G_j(\chi, \delta_r) \right] \qquad (39.18)$$

where m_r is the number of terms in the Laurent expansions about δ_r and $\overline{\delta}_r$, and where the coefficients a_j, b_j^r, and c_j^r are real:

$$\Im a_j = 0 \quad (j = 1, 2, \ldots, n) \qquad \Im b_\nu^r = \Im c_\nu^r = 0 \qquad (\nu = 1, 2, \ldots, m_r) \qquad (39.19)$$

Sec. 39 Analytic Elements for Thin Isolated Features

Using (39.13), (39.16), and (39.17) the complex potential becomes

$$\Omega = \sum_{j=1}^{n} \frac{a_j}{\chi^j} + \sum_{r=1}^{m}\sum_{j=1}^{m_r} \left[\frac{\underset{r}{\beta_j}\underset{r}{\delta_r}}{(\chi - \delta_r)^j} + \frac{\overline{\underset{r}{\beta_j}\underset{r}{\delta_r}}}{(\chi - \overline{\delta}_r)^j} \right] \tag{39.20}$$

where

$$\underset{r}{\beta_j} = \underset{r}{b_j} + i\underset{r}{c_j} \tag{39.21}$$

If the analytic element is used to model a feature from which a total amount Q_0 of water is being pumped, then a term $Q_0/(2\pi)\ln\chi$ must be added to the complex potential (compare [31.23]) and we obtain

$$\Omega = \sum_{j=1}^{n} \frac{a_j}{\chi^j} + \sum_{r=1}^{m}\sum_{j=1}^{m_r} \left[\frac{\underset{r}{\beta_j}\underset{r}{\delta_r}}{(\chi - \delta_r)^j} + \frac{\overline{\underset{r}{\beta_j}\underset{r}{\delta_r}}}{(\chi - \overline{\delta}_r)^j} \right] + \frac{Q_0}{2\pi}\ln\chi \tag{39.22}$$

It is left as an exercise for the reader to write an expression for $\mu^*(\chi^+)$.

Problem 39.1
Determine an expression for $\mu^*(\chi^+) = \Psi^+ - \Psi^-$ using the complex potential (39.22).

The Discharge Function

Many applications require an expression for the discharge function, which may be obtained by the use of the chain rule:

$$W = -\frac{d\Omega}{dz} = -\frac{d\Omega}{d\chi}\frac{d\chi}{dZ}\frac{dZ}{dz} \tag{39.23}$$

It follows from (39.3) and (39.1) that

$$\frac{d\chi}{dZ} = \frac{1}{\frac{dZ}{d\chi}} = \frac{1}{\frac{1}{2}(1 - \frac{1}{\chi^2})} = \frac{2\chi^2}{\chi^2 - 1} \qquad \frac{dZ}{dz} = \frac{2}{\frac{2}{z} - \frac{1}{z}} \tag{39.24}$$

and we obtain the following for (39.23), after differentiation of (39.22) with respect to χ,

$$W = \left\{ \sum_{j=1}^{n} \frac{ja_j}{\chi^{j+1}} + \sum_{r=1}^{m}\sum_{j=1}^{m_r} \left[\frac{j\underset{r}{\beta_j}\underset{r}{\delta_r}}{(\chi - \delta_r)^{j+1}} + \frac{j\overline{\underset{r}{\beta_j}\underset{r}{\delta_r}}}{(\chi - \overline{\delta}_r)^{j+1}} \right] - \frac{Q_0}{2\pi\chi} \right\} \frac{4\chi^2}{(\frac{2}{z} - \frac{1}{z})(\chi^2 - 1)} \tag{39.25}$$

It is possible to eliminate the function χ from the complex potential and the discharge function, but it is more efficient for numerical applications to retain χ as a reference parameter.

The Double-Root Doublet

The only difference between a double-root doublet and a double-root dipole is that the real part of the complex potential jumps across the element rather than the imaginary part. The complex potential thus is obtained by dividing the right-hand side in (39.20) by i:

$$\Omega = \frac{1}{i}\sum_{j=1}^{n}\frac{a_j}{\chi^j} + \frac{1}{i}\sum_{r=1}^{m}\sum_{j=1}^{m_r}\left[\frac{\frac{\beta_j \delta_r}{r}}{(\chi - \delta_r)^j} + \frac{\frac{\overline{\beta_j \delta_r}}{r}}{(\chi - \overline{\delta}_r)^j}\right] \qquad (39.26)$$

The discharge function for the double-root doublet is obtained by dividing (39.25) by i, with $Q_0 = 0$:

$$W = \frac{1}{i}\left\{\sum_{j=1}^{n}\frac{ja_j}{\chi^{j+1}} + \sum_{r=1}^{m}\sum_{j=1}^{m_r}\left[\frac{\frac{j\beta_j \delta_r}{r}}{(\chi - \delta_r)^{j+1}} + \frac{\frac{j\overline{\beta_j \delta_r}}{r}}{(\chi - \overline{\delta}_r)^{j+1}}\right]\right\}\frac{4\chi^2}{(\bar{z} - \frac{1}{z})(\chi^2 - 1)}$$

(39.27)

The two complex potentials (39.22) and (39.26) may be used to model the various aquifer features mentioned earlier in this chapter. We will discuss such applications below.

Applications of Double-Root Elements

The unknown constants a_j, b_j, and c_j in the complex potentials for double-root elements may be determined from conditions applied at control points on the elements. We will refer to the first Laurent expansions in (39.22) and (39.26) as the central expansions and to the remaining ones as the expansions about poles. The central expansion behaves quite differently from the expansions about poles, depending upon the distance between the poles and the boundary (the unit circle). The former expansions are equally effective along the entire boundary, whereas the latter ones affect the behavior of the potential primarily along the section of the boundary close to the poles. This effect is the more pronounced the closer the poles are to the boundary. The combination of all expansions therefore provides us with a complex potential that is useful for dealing with both local influences as caused by a neighboring element that mainly affects a small section of the element, and global effects due to uniform flow or elements that are far away.

Implementation of the double-root elements amounts to adding a number of complex potentials of the form (39.22) and (39.26) to those for wells, line-sinks, area-sinks, line-doublets, uniform flow, and global infiltration. For each element the number of control points is set equal to the chosen number of degrees of freedom, and boundary

Sec. 39 Analytic Elements for Thin Isolated Features   467

conditions are applied at each control point according to the type of element. If the element is a straight leaky wall the boundary conditions are of the form (36.4), and the complex potential used to model the wall is that for a double-root doublet, see (39.26). It is important to note that the application of this boundary condition requires that the discharge function W be programmed for each function in the model. For the special case that the wall is impermeable the boundary condition (36.4) is replaced by either (36.9) or (36.11).

Figure 6.31 Flow in the vertical plane between thin leaky layers. The factor $k^*/(kb^*)$ equals 40.

Although the elements are intended primarily for use in a regional aquifer model, they may be used equally well to model two-dimensional flow in the vertical plane. As an example, consider the patterns of streamlines shown in Figures 6.31 and 6.32, where flow in the vertical plane is depicted between closely-spaced thin horizontal layers. For both cases, the flow domain is bounded by two impermeable parallel vertical walls which may be considered as axes of symmetry in a pattern of layers that repeats itself ad infinitum. For the case of Figure 6.31 the layers are leaky, i.e., the boundary conditions are of the form (36.4); for the case of Figure 6.32 the layers are impermeable and the boundary conditions are of the form (36.11). The vertical walls have 14 degrees of freedom; the central expansions are of order 6 and account for 6 degrees of freedom. There are 5 function pairs F_j and G_j ($m = 5$ in [39.18]), corresponding to 8 poles of order 1 ($m_r = 1$), two of which lie on the real axis. Since G_j does not contribute to expansions about points on the real axis ($G_j = 0$ if $\Im \delta_r = 0$), these 5 function pairs correspond to $2 + 2 * 3 = 8$ degrees of freedom. The other elements all have 10 degrees of freedom with 3 function pairs F_j and G_j accounting for 4 degrees of freedom and a central polynomial of order 6. All poles are spaced uniformly in the χ plane and are a distance 0.5 from the origin. The distribution of poles both in the z and χ planes is shown in Figure 6.33. Note that the poles in the physical plane are not part of the flow domain but lie on a different sheet of a Riemann surface; for a discussion on Riemann surfaces, see Von Koppenfels and Stallmann [1959] or Strack [1972, 1973].

It may be noted that the boundary conditions along narrow cracks are rather similar to those along thin layers, except that Q_n is replaced by Q_s and $\Phi^+ - \Phi^-$ by $\Psi^+ - \Psi^-$. The plot in Figure 6.31 therefore may also be interpreted as flow with a system of narrow cracks with the curves representing equipotentials rather than streamlines, and the two vertical walls equipotential boundaries.

Figure 6.32 Flow in the vertical plane between thin impermeable layers.

Problem 39.2

Figure 6.31 corresponds to flow with leaky walls characterized by $kb^*/(k^*) = 40$ (compare [36.4]). Interpreting this figure as flow with a system of narrow cracks, determine the transmissivity of the cracks, k^*b^*, in terms of the hydraulic conductivity of the aquifer.

It may be noted that the formulation presented here does not allow for the possibility that the cracks intersect each other. Intersecting cracks may be modeled by adding a

Sec. 40 Tip Elements   469

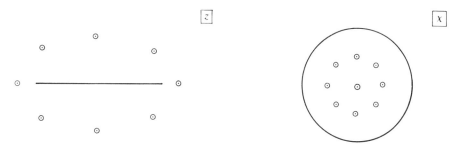

Figure 6.33 The distribution of poles in the z and χ planes used to generate flow patterns in Figure 6.31 and Figure 6.32.

special term to the analytic elements for each of the intersecting cracks. A discussion of intersecting cracks is beyond the scope of this text; the interested reader is referred to Strack [1982(a)] .

Thin ideal drains may be modeled by the use of the double-root dipoles (39.20); the condition applied at the control points is of the form (36.15). If water is being pumped from the drains, a logarithmic term is added to the potential (see [39.22]) to account for the pumpage. The additional condition corresponding to the unknown discharge Q_0 is that the potential Φ_A in (36.15) be equal to the value of the potential maintained along the drain.

An example of such a case is illustrated in Figure 6.34 where piezometric contours are shown for infiltration in an unconfined aquifer through a number of straight ditches. The infiltrated water is captured by a set of wells that are spaced in such a way that the travel times from the ditches to the wells vary, so that an average water quality is obtained. Such systems are used, for example, in the Netherlands to recover water with a chloride content that is the average of that of the river water infiltrated in the ditches, which has a much higher chloride content in the summer than in the winter (see Huisman and Olsthoorn [1983]). The number of degrees of freedom used for modeling each canal is 10; the central polynomial is of order 6, and there are 3 function pairs F_j and G_j accounting for 4 degrees of freedom.

40 TIP ELEMENTS

The double-root elements discussed in Sec. 39 have square-root singularities at their end points. The singular behavior of the complex potential near the tips of the elements is built into the function in order to obtain a better approximation of the high flow rates occurring near the end point of the feature modeled by the element. The double-root elements are useful for modeling straight isolated features with two end points, but are not suitable for stringing together. We will construct in this section an element that has a square-root singularity at one end point and at the other one behaves as the line-doublets and line-dipoles (see Secs. 35 and 36). The tip elements may be used as the first and last elements of strings of line-dipoles or line-doublets. These elements were used by

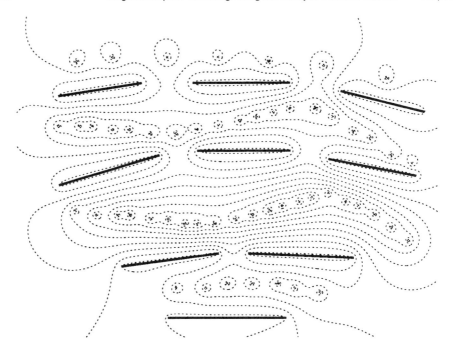

Figure 6.34 Piezometric contours for flow from ditches toward drains in an unconfined aquifer.

Strack [1982(a)] in combination with fifth order line-doublets to model flow in cracks. Characteristic for the tip elements is the single square-root singularity; we therefore refer to these elements as single-root elements.

The Single-Root Line-Dipole

The complex potential for a single-root line-dipole suitable for modeling the tip of a crack or the end point of a drain is determined in a transformed domain as follows (see Figure 6.35). The element extends from the tip $\overset{1}{z}$ to $\overset{2}{z}$ and has an orientation α and a length L,

$$\overset{2}{z} - \overset{1}{z} = Le^{i\alpha} \tag{40.1}$$

We apply the transformation

$$\chi_1 = \tau_1 + i\omega_1 = \frac{z - \overset{1}{z}}{\overset{2}{z} - \overset{1}{z}} \tag{40.2}$$

to map the element onto the segment

$$0 \leq \tau_1 \leq 1 \qquad \omega_1 = 0 \tag{40.3}$$

Sec. 40 Tip Elements

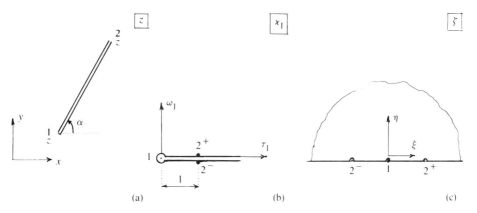

Figure 6.35 The tip element in the z, χ_1, and ζ planes.

in the χ_1 plane of Figure 6.35(b). We establish the range of definition of $\arg \chi_1$ such that it is continuous throughout the slotted domain, i.e., with the branch cut along the positive real axis:

$$0 \leq \arg \chi_1 \leq 2\pi \tag{40.4}$$

The transformation

$$\zeta = \xi + i\eta = \sqrt{\chi_1} \tag{40.5}$$

maps the slotted χ_1 plane onto the upper half plane $\Im \zeta \geq 0$ shown in Figure 6.35(c). The + side of the element is mapped onto the interval $\eta = 0$, $0 \leq \xi \leq 1$ and the − side onto the interval $\eta = 0$, $-1 \leq \xi \leq 0$. The range of definition of $\arg \zeta$ is obtained from (40.4) and (40.5) as

$$0 \leq \arg \zeta \leq \pi \tag{40.6}$$

We examine the complex potential Ω_1 for a line-dipole along the interval

$$-1 \leq \xi \leq 1 \qquad \eta = 0 \tag{40.7}$$

with an analytic density distribution $\mu(\zeta)$ (compare [25.52]),

$$\Omega_1 = \frac{\mu(\zeta)}{2\pi} \ln \frac{\zeta - 1}{\zeta + 1} + q(\zeta) \tag{40.8}$$

where $\mu(\zeta)$ and $q(\zeta)$ are polynomials with real coefficients, so that

$$\eta = 0 \qquad \Im \mu(\zeta) = 0 \qquad \Im q(\zeta) = 0 \tag{40.9}$$

The coefficients of $q(\zeta)$ may be expressed in terms of those of $\mu(\zeta)$ by requiring that Ω_1 vanish at infinity. The density distribution $\mu(\zeta)$ must be such that the real part Φ_1 of

Ω_1 is continuous and the imaginary part Ψ_1 is discontinuous across the element in the z plane:

$$\Phi_1^+ - \Phi_1^- = 0 \qquad \Psi_1^+ - \Psi_1^- \neq 0 \qquad (40.10)$$

We separate the real and imaginary part of the logarithm in (40.8) and obtain

$$\Phi_1(\zeta) + i\Psi_1(\zeta) = \frac{\mu(\zeta)}{2\pi}\left[\ln\left|\frac{\zeta-1}{\zeta+1}\right| + i\theta\right] + q(\zeta) \qquad (40.11)$$

where

$$\theta = \arg(\zeta - 1) - \arg(\zeta + 1) \qquad (40.12)$$

The variable ζ is restricted to $\Im\zeta \geq 0$ and the argument θ is taken accordingly:

$$0 \leq \theta \leq \pi \qquad (40.13)$$

Along the element, θ equals π and we obtain for the real part of Ω_1

$$\eta = 0 \quad -1 \leq \xi \leq 1 \quad \Phi_1(\xi) = \frac{\mu(\xi)}{2\pi}\ln\left|\frac{\xi-1}{\xi+1}\right| + q(\xi) \qquad (40.14)$$

and for the imaginary part

$$\eta = 0 \quad -1 \leq \xi \leq 1 \quad \Psi_1(\xi) = \frac{\mu(\xi)}{2} \qquad (40.15)$$

Opposite points at the $+$ and $-$ sides of the element in the physical plane are mapped onto points ζ^+ and ζ^- of the ξ axis that lie symmetrically with respect to the origin (see Figure 6.35):

$$\xi^+ = -\xi^- \qquad \eta^+ = \eta^- = 0 \qquad (40.16)$$

The condition that Φ is continuous across the element thus implies that

$$\Phi^+ - \Phi^- = 0 = \frac{\mu(\xi^+)}{2\pi}\ln\left|\frac{\xi^+ - 1}{\xi^+ + 1}\right| - \frac{\mu(-\xi^+)}{2\pi}\ln\left|\frac{-\xi^+ - 1}{-\xi^+ + 1}\right| + q(\xi^+) - q(-\xi^+) \quad (40.17)$$

It follows that the functions $\mu(\xi)$ and $q(\xi)$ must fulfill the following conditions:

$$\mu(\xi) = -\mu(-\xi) \qquad q(\xi) = q(-\xi) \qquad (40.18)$$

Both conditions are met if $\mu(\zeta)$ and $q(\zeta)$ are polynomials of the form

$$\mu(\zeta) = \sum_{m=0}^{n} B_m \zeta^{2m+1} \qquad q(\zeta) = \sum_{m=0}^{n} d_m \zeta^{2m} \qquad (40.19)$$

where B_m and d_m are real:

$$\Im B_m = 0 \qquad \Im d_m = 0 \qquad (m = 0, 1, 2, \ldots, n) \tag{40.20}$$

With (40.5) this may be written as

$$\mu = \sqrt{\chi_1} \sum_{m=0}^{n} B_m \chi_1^m \qquad q = \sum_{m=0}^{n} d_m \chi_1^m \tag{40.21}$$

and the complex potential (40.8) becomes

$$\Omega_1 = \frac{\mu^*(\chi_1)}{2\pi} \sqrt{\chi_1} \ln \frac{\sqrt{\chi_1} - 1}{\sqrt{\chi_1} + 1} + q^*(\chi_1) \tag{40.22}$$

where

$$\mu^*(\chi_1) = \sum_{m=0}^{n} B_m \chi_1^m \qquad q^*(\chi_1) = \sum_{m=0}^{n} d_m \chi_1^m \tag{40.23}$$

An expression for the jump in Ψ_1 across the element is obtained by the use of (40.15), (40.16), and (40.18):

$$\Psi_1^+ - \Psi_1^- = \tfrac{1}{2}\left[\mu(\xi^+) - \mu(-\xi^+)\right] = \mu(\xi^+) = \sqrt{\tau_1}\mu^*(\tau_1) \tag{40.24}$$

which indeed possesses a square-root singularity at the tip of the element. Both Φ_1 and Ψ_1 are continuous along the τ_1 axis outside the element as may be verified by the use of (40.22).

Expressions for the coefficients d_m in terms of B_m are obtained by expanding the complex potential (40.8) about infinity, requiring that all terms in the expansion of order higher than ζ^{-1} vanish. The first term in (40.8) yields an expansion of the form (compare [35.22])

$$\frac{\mu(\zeta)}{2\pi} \ln \frac{\zeta - 1}{\zeta + 1} = -2 \sum_{j=1}^{\infty} \sum_{m=0}^{n} \frac{B_m}{2j - 1} \zeta^{2(m-j)+2} \qquad (\zeta \to \infty) \tag{40.25}$$

which contains even powers of ζ only, with the highest power being $2n$. The polynomial $q(\zeta)$ also is of order $2n$ and has even powers, so that the d_m can indeed be chosen such that all terms in (40.25) of order higher than ζ^{-1} vanish.

Problem 40.1
Prove that Φ_1 and Ψ_1 are continuous outside the element.

Application

The tip elements discussed above were used by Strack [1982(a)] to model flow through fissures in a permeable medium. The cracks were modeled by using line-dipoles with polynomial density distributions of order 5. Tip elements of order 3 ($n = 3$ in [40.21]) were joined to the line-doublets at either end of the cracks. The approximate solution to the problem of a single crack in a field of uniform flow, obtained by using two tip elements and a third line-dipole in between, was nearly indistinguishable from the exact solution.

The tip elements are particularly useful if the feature is not straight but modeled by a string of elements that have different orientations.

The Single-Root Line-Doublet

The complex potential for a single-root line-doublet is obtained from that for the single-root line-dipole simply by division by a factor of i and by replacing μ^* by λ^* and q^* by p^*,

$$\Omega_1 = \frac{\lambda^*(\chi_1)}{2\pi i} \sqrt{\chi_1} \ln \frac{\sqrt{\chi_1} - 1}{\sqrt{\chi_1} + 1} - ip^*(\chi_1) \tag{40.26}$$

where λ^* and p^* are formally equal to μ^* and q^*.

Problem 40.2
Determine the coefficients in the polynomial $q^*(\chi_1)$ for the case that $n = 1$.

41 CURVILINEAR ELEMENTS

The complex potentials for line-dipoles and line-doublets were derived in Sec. 25 for straight elements. The restriction to straight elements is not necessary and complex potentials for curvilinear elements may be determined without much difficulty. The need for curvilinear elements is limited in regional aquifer modeling, but there are applications where the use of such elements is desirable.

In this section we will briefly discuss how curvilinear analytic elements may be constructed and present the complex potential for a circular line-dipole as an example. Circular analytic elements are useful for joining straight elements in modeling leaky walls and cracks. In this way a complex potential may be obtained whose derivative is finite at the nodal points, which is a desirable property because the derivative occurs in the boundary condition. He [1987] modeled leaky walls by combining single-root line-doublets, curvilinear line-doublets and straight line-doublets, and obtained excellent agreement with an exact solution for flow in an aquifer with a leaky wall consisting of two noncollinear finite straight segments.

The Complex Potential for a General Line-Dipole

The complex potential (25.52) for a straight line-dipole is written in terms of a dimensionless variable Z. The derivation given between equations (25.43) and (25.52) may

Sec. 41 Curvilinear Elements

be carried out without reference to such a dimensionless variable and without specifying the shape of the element. Such a derivation yields

$$\Omega = \frac{\mu(z)}{2\pi} \ln \frac{z - \overset{2}{z}}{z - \overset{1}{z}} + q(z) \tag{41.1}$$

where $\overset{1}{z}$ and $\overset{2}{z}$ are the end points of the curvilinear element (see Figure 6.36[a]). We separate the logarithm into its real and imaginary parts and obtain

$$\Omega = \frac{\mu(z)}{2\pi} \left[\ln \left| \frac{z - \overset{2}{z}}{z - \overset{1}{z}} \right| + i\theta \right] + q(z) \tag{41.2}$$

where

$$\theta = \arg(z - \overset{2}{z}) - \arg(z - \overset{1}{z}) \tag{41.3}$$

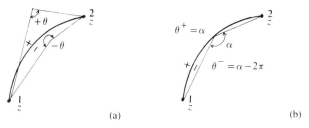

(a)   (b)

Figure 6.36 A curvilinear element.

As opposed to the case of a straight element, the angle θ will vary along the element, see Figure 6.36(b). The jump in θ, $\theta^+ - \theta^-$, however, will be constant and equal to 2π along the element, as is seen from Figure 6.36(b),

$$\theta^+ - \theta^- = 2\pi \tag{41.4}$$

The jump in complex potential across the element thus may be written as follows:

$$\Omega^+ - \Omega^- = i\mu(z_e) \tag{41.5}$$

where z_e represents an arbitrary point on the element. For a dipole distribution this jump must be purely imaginary, so that the function $\mu(z)$ must be real for $z = z_e$,

$$\Im \mu(z_e) = 0 \tag{41.6}$$

The construction of a curvilinear analytic element, therefore, amounts to determining a density distribution $\mu(z)$ that is real along the element. The function $q(z)$, then, is used to remove any singularities that may have been introduced outside the element through the function $\mu(z)$. Curvilinear analytic elements may, for example, be constructed along segments of streamlines of any flow net. Along such streamlines there exists a function (namely the complex potential Ω^* used to generate the flow net) that has a constant imaginary part. The function μ may thus be obtained as a polynomial in terms of a variable $Z = \Omega^* - i\Psi_0^*$ where Ψ_0^* is the value of Ψ^* along the streamline selected from the flow net. Since Z will be both real and analytic along the element, representing μ as a polynomial in terms of Z with real coefficients meets the requirement (41.6). Depending on the flow net selected, the variable Z may have singularities outside the element. The analytic element may be constructed in this way only if the singularities may be removed by the use of the function $q(z)$; the complex potential finally obtained must be analytic everywhere outside the element, including infinity.

Certain analytic elements may be constructed by the use of conformal mapping: the curve is transformed into a straight line along the real axis and the function $\mu(z)$ written in terms of the new variable, which is real along the element. We will follow the latter procedure for the case that the curve is a circular arc.

Mapping a Circular Element on a Straight Line

The analytic element lies on the circular arc between points $\overset{1}{z}$ and $\overset{2}{z}$ shown in Figure 6.37(a). The radius is R, z_c is the complex coordinate of the center of the circular arc, and $z = \delta$ denotes the point located on the circle diametrically opposite to z_c. The angle at point $z = \delta$ spanning the arc from $\overset{1}{z}$ to $\overset{2}{z}$ is β and the line through $\overset{1}{z}$ and $\overset{2}{z}$ is oriented at an angle γ with respect to the x axis. It may be noted that the angle β need not be positive; we define β such that it is positive if the center of the circle lies at the positive side of the element (as in Figure 6.37[a]), and negative if the center of the circle lies at the negative side of the element.

The transformation (see Sec. 30)

$$Z = \frac{z - \delta}{2R} e^{-i\gamma} \qquad (41.7)$$

maps the circle in the z plane onto a circle of radius $\frac{1}{2}$ in the Z plane. It follows from (41.7) that $z = \delta$ corresponds to $Z = 0$, and it may be seen by inspection of Figure 6.37(a) that

$$z_c - \delta = 2R e^{i(\gamma - \pi/2)} \qquad (41.8)$$

so that $z = z_c$ corresponds to $Z = -i$, as follows from (41.7). Consider the transformation

$$\zeta = \cot \tfrac{\beta}{2} \left[\frac{1}{Z} - i \right] \qquad (41.9)$$

In order to investigate how the element is mapped by this transformation we represent Z corresponding to a point ζ of the circle by the use of Figure 6.37(b) as

$$Z = \sin\theta e^{-i\theta} \qquad (41.10)$$

Sec. 41   Curvilinear Elements   477

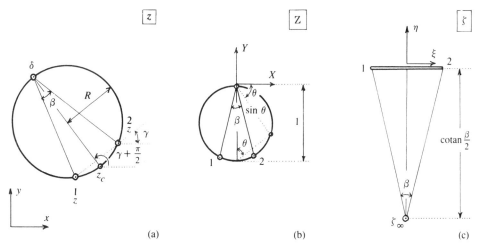

Figure 6.37 The element in the z, Z, and ζ planes.

Substitution of this expression for Z in (41.9) yields

$$\zeta = \cot \tfrac{\beta}{2} \left[\frac{e^{i\theta}}{\sin \theta} - i \right] = \cot \tfrac{\beta}{2} \cot \theta \qquad (41.11)$$

so that the element is mapped onto the real axis. At end points 1 and 2 the angle θ equals the respective values θ_1 and θ_2 with

$$\theta_1 = \tfrac{\pi}{2} + \tfrac{\beta}{2} \qquad \theta_2 = \tfrac{\pi}{2} - \tfrac{\beta}{2} \qquad (41.12)$$

and the corresponding values for ζ_1 and ζ_2 are obtained from (41.11) as

$$\zeta_1 = \cot \tfrac{\beta}{2} \cot(\tfrac{\pi}{2} + \tfrac{\beta}{2}) = -1 \qquad \zeta_2 = \cot \tfrac{\beta}{2} \cot(\tfrac{\pi}{2} - \tfrac{\beta}{2}) = 1 \qquad (41.13)$$

Infinity, $z = \infty$ and $Z = \infty$, corresponds in the ζ plane to ζ_∞ with

$$\zeta_\infty = -i \cot \tfrac{\beta}{2} \qquad (41.14)$$

as follows from (41.9) by taking the limit for $Z \to \infty$.

The Complex Potential

The variable ζ is real along the element; a density distribution that is real along the element may be obtained by writing it as a polynomial in terms of ζ with real coefficients:

$$\mu(\zeta) = \sum_{m=0}^{n} a_m \zeta^m \qquad \Im a_m = 0 \qquad (41.15)$$

Substitution of this expression for $\mu(\zeta)$ in (41.1) yields,

$$\Omega = \frac{1}{2\pi} \sum_{m=0}^{n} a_m \zeta^m \ln \frac{z - \overset{2}{z}}{z - \overset{1}{z}} + q(z) \qquad (41.16)$$

where ζ may be expressed in terms of z by the use of (41.9) and (41.7),

$$\zeta = \cot \frac{\beta}{2} \left[\frac{2Re^{i\gamma}}{z - \delta} - i \right] \qquad (41.17)$$

It follows that the polynomial (41.15) may be viewed as the sum of the first n terms of a Laurent expansion about $z = \delta$. Since the logarithm is analytic at $z = \delta$, the singular behavior of the first term in (41.16) in turn may be expressed as a Laurent expansion about $z = \delta$. The function $q(z)$ is chosen such that it cancels the n singular terms in the Laurent expansion.

For the present case the analysis is best carried out entirely in terms of ζ. The reader may verify by the use of (41.17) that

$$\frac{z - \overset{2}{z}}{z - \overset{1}{z}} = e^{i\beta} \frac{\zeta - 1}{\zeta + 1} \qquad (41.18)$$

so that (41.16) may be written entirely in terms of ζ as

$$\Omega = \frac{1}{2\pi} \left\{ \sum_{m=0}^{n} a_m \zeta^m \left[\ln \frac{\zeta - 1}{\zeta + 1} + i\beta \right] + q^*(\zeta) \right\} \qquad (41.19)$$

where

$$q^*(\zeta) = 2\pi q[z(\zeta)] \qquad (41.20)$$

Infinity in the z plane corresponds to $\zeta = -i \cot \frac{\beta}{2}$. In order that the complex potential vanishes at infinity in the z plane, it must be zero at $\zeta = -i \cot \frac{\beta}{2}$:

$$\zeta = -i \cot \frac{\beta}{2} \qquad \Omega = 0 \qquad (41.21)$$

The complex potential must be analytic at $z = \delta$. This point corresponds to $\zeta = \infty$ and the latter condition will be satisfied if Ω has an expansion about infinity of the following form:

$$\Omega = c_0 + \frac{c_1}{\zeta} + \frac{c_2}{\zeta^2} + \ldots \qquad \zeta \to \infty \qquad (41.22)$$

where $c_0, c_1, \ldots$, are complex constants. The condition (41.22) implies with (41.19) that $q^*(\zeta)$ may be written as follows:

$$q^*(\zeta) = -i\beta \sum_{m=0}^{n} a_m \zeta^m + c_0 + \sum_{j=1}^{n-1} b_j \zeta^j \qquad (\Im b_j = 0) \qquad (41.23)$$

where the real coefficients b_j are to be determined from the condition (41.22). The constant c_0 is chosen such that (41.21) is fulfilled; substitution of the expression for c_0 obtained from (41.19), (41.21), and (41.22) in (41.19) yields, with $\ln((\zeta_\infty - 1)/(\zeta_\infty + 1)) = -i\beta$ (see Figure 6.37[c]), and (41.23),

$$\Omega = \frac{1}{2\pi} \left\{ \sum_{m=0}^{n} a_m \zeta^m \ln \frac{\zeta - 1}{\zeta + 1} + i\beta \sum_{m=0}^{n} a_m (-i \cot \tfrac{\beta}{2})^m \right. \\ \left. + \sum_{j=1}^{n-1} b_j \left[\zeta^j - (-i \cot \tfrac{\beta}{2})^j \right] \right\} \tag{41.24}$$

Expressions for the real coefficients b_j in terms of a_j may be obtained from the condition (41.22) in the usual manner.

An expression for the jump $\Omega^+ - \Omega^-$ is obtained from (41.24), where the logarithm jumps by an amount $2\pi i$ across the element:

$$\Omega^+ - \Omega^- = i(\Psi^+ - \Psi^-) = i \sum_{m=0}^{n} a_m \xi^m \tag{41.25}$$

It may be noted that the behavior of this jump in the z plane is affected by the transformation $\zeta = \zeta(z)$. Using (41.11), $\Psi^+ - \Psi^-$ may be expressed in terms of the angle θ as follows:

$$\Psi^+ - \Psi^- = \sum_{m=0}^{n} a_m \left[\cot \tfrac{\beta}{2} \cot \theta \right]^m \tag{41.26}$$

Choosing a linearly varying density distribution in the ζ plane, $n = 1$, the density distribution in the z plane is nonlinear:

$$\Psi^+ - \Psi^- = a_0 + a_1 \cot \tfrac{\beta}{2} \cot \theta \tag{41.27}$$

The element presented above may be used to join two straight line-dipoles that enclose an angle β. The density distribution may then be chosen in such a way that the discharge function is finite at the nodal points, thus obtaining an approximation that is better than that obtained by omitting the circular element; the singularity in the discharge function at the nodal points of strings of straight elements can be removed only if the elements have the same inclinations. Detournay [1985] applied the above elements in an approximate method for solving problems involving a free surface.

Problem 41.1

Determine the coefficients b_j in (41.24) for the case that $n = 1$. Express the constants a_0 and a_1 in terms of μ at the center of the element and $d\mu/ds$ at points $\overset{1}{z}$ and $\overset{2}{z}$, where s is measured along the arc, in terms of the real constants a_0 and a_1.

42 ANALYTIC ELEMENTS FOR CLOSED BOUNDARIES

We saw in Sec. 40 how analytic elements can be constructed for straight isolated thin aquifer features. In this section we will generalize the approach to analytic elements for isolated features with boundaries $\mathcal{B}$ of any shape. We refer to such elements as analytic elements for closed boundaries. We call the elements one-sided if the flow in the interior of $\mathcal{B}$ need not be modeled. One-sided elements are useful, for example, for modeling impermeable objects or fully penetrating lakes. We call the elements two-sided if the flow both outside and inside $\mathcal{B}$ must be considered. Two-sided analytic elements are useful, for example, for modeling inhomogeneities and closed leaky walls or narrow drainage ditches with finite hydraulic conductivity.

One-sided analytic elements can be constructed by first mapping the exterior of $\mathcal{B}$ onto the exterior of the unit circle $\mathcal{C}$ in the χ plane. An analytic element with the desired number of degrees of freedom is obtained as the sum of Laurent expansions about points δ_m ($m = 1, 2, \ldots, n$) inside the unit circle. These elements are analytic throughout the flow domain because the points δ_m lie outside the map of the flow domain onto the χ plane. The derivation of the analytic elements in Sec. 39 relies on this procedure.

Two-sided analytic elements are more difficult to determine because they must meet a specific jump condition; unless special measures are taken, the analytic elements will exhibit a singular behavior at points inside the flow domain.

We will present in this section a general procedure for deriving two-sided analytic elements, and apply this procedure to the case that $\mathcal{B}$ is the elliptical boundary of an inhomogeneity.

Two-Sided Analytic Elements for Closed Boundaries

We will derive an analytic element for a boundary $\mathcal{B}$ that divides the z plane into domains $\mathcal{D}^+$ and $\mathcal{D}^-$, see Figure 6.38, for the case that the jump conditions are

$$Q_n^+ - Q_n^- = 0 \tag{42.1}$$

and

$$\Phi^+ - \Phi^- = \lambda \neq 0 \tag{42.2}$$

The linear transformation

$$Z = Az + B \tag{42.3}$$

where A and B are complex constants, maps the infinite z plane of Figure 6.39(a) onto the infinite dimensionless Z plane of Figure 6.39(b); $|A|$ has the dimension of $[L^{-1}]$ and $|B|$ is dimensionless. $\mathcal{B}_z$ is the map of $\mathcal{B}$ onto the Z plane.

The function

$$\chi = \chi(Z) \tag{42.4}$$

maps the exterior of the boundary $\mathcal{B}_Z$ onto the exterior of the unit circle $\mathcal{C}$ in the χ plane of Figure 6.39(c). The function $\chi = \chi(Z)$ can be obtained either by the use

Sec. 42   Analytic Elements for Closed Boundaries   481

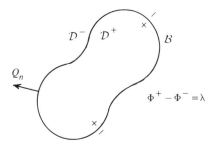

Figure 6.38 A two-sided analytic element.

of the Schwarz-Christoffel transformation if $\mathcal{B}_Z$ is a polygon, or by using an existing transformation, or by using an approximate method of conformal mapping (e.g., Gaier [1964], Bisshopp [1983], Detournay [1985]). We will assume that the function (42.4) is known. The objective is to determine a complex potential that fulfills condition (42.1) exactly, and possesses degrees of freedom that make it possible to match condition (42.2) at a given number of control points. Consider a function $\overset{1}{\Omega} = \overset{1}{\Phi} + i\overset{1}{\Psi} = \overset{1}{\Omega}(z)$ chosen such that

$$z \text{ in } \mathcal{D}^- \qquad \overset{1}{\Omega} = \overset{1}{\Omega}(z) \qquad (42.5)$$

$$z \text{ on } \mathcal{B} \qquad \overset{1}{\Omega} = i\Psi_0 \qquad (42.6)$$

$$z \text{ in } \mathcal{D}^+ \qquad \overset{1}{\Omega} = i\Psi_0 \qquad (42.7)$$

where Ψ_0 is a constant. This complex potential fulfills (42.1) in view of (42.6) (it makes the boundary $\mathcal{B}$ into a streamline) and creates a jump $\Phi^+ - \Phi^-$:

$$z \text{ on } \mathcal{B} \qquad \Phi^+ - \Phi^- = -\overset{1}{\Phi}(z) \qquad (42.8)$$

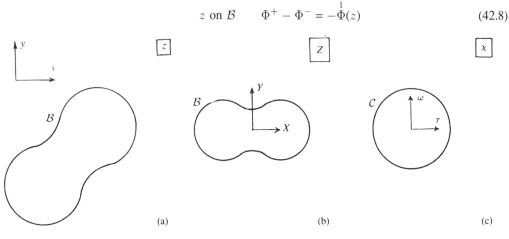

Figure 6.39 Mapping the boundary $\mathcal{B}$ of the element onto the Z and χ planes.

We will construct the potential $\overset{1}{\Omega}$ in such a way that it has singularities in $\mathcal{D}^-$, so that we must add a potential $\overset{2}{\Omega}$ that cancels the singularities of $\overset{1}{\Omega}$. Hence, the complete complex potential for the analytic element is of the form

$$z \text{ in } \mathcal{D}^- \qquad \Omega = \overset{1}{\Omega}(z) + \overset{2}{\Omega}(z) \qquad (42.9)$$

$$z \text{ in } \mathcal{D}^+ \qquad \Omega = i\Psi_0 + \overset{2}{\Omega}(z) \qquad (42.10)$$

The complex potential $\overset{2}{\Omega}(z)$ is continuous across $\mathcal{B}$ so that the condition (42.1) is not disturbed and (42.8) remains valid. Finally, we impose the condition that $\Omega(z)$ is of order $1/z$ at infinity:

$$z \to \infty \qquad \Omega(z) \to A/z + O(1/z^2) \qquad (42.11)$$

where A is a constant.

The complex potential $\overset{1}{\Omega}(z)$ may be determined by the use of conformal mapping; the method is applicable to any boundary $\mathcal{B}$ that can be mapped onto the unit circle.

The Function $\overset{1}{\Omega}(z)$

We choose a function $\overset{1}{\Omega}$ such that it fulfills the condition that $\overset{1}{\Psi} = \Psi_0$ along the unit circle $\mathcal{C}$ in the χ plane,

$$|\chi| = 1 \qquad \overset{1}{\Psi} = \Psi_0 \qquad (42.12)$$

and introduce a function $\overset{1}{h}_m$ as

$$\overset{1}{h}_m = \underset{m}{a}\chi^m + \underset{m}{\overline{a}}\chi^{-m} \qquad (42.13)$$

where $\underset{m}{a}$ is a complex constant,

$$\underset{m}{a} = \underset{m}{a_1} + i\underset{m}{a_2} \qquad (42.14)$$

The imaginary part of the function $\overset{1}{h}_m$ is zero along $\mathcal{C}$ as is seen by substituting $e^{i\theta}$ for χ,

$$\overset{1}{h}_m(e^{i\theta}) = \underset{m}{a}e^{im\theta} + \underset{m}{\overline{a}}e^{-im\theta} = 2\Re(\underset{m}{a}e^{im\theta}) \qquad (42.15)$$

We introduce functions $\overset{1}{f}_m$ and $\overset{1}{g}_m$ such that

$$\overset{1}{h}_m = \underset{m}{a_1}\overset{1}{f}_m + \underset{m}{a_2}\overset{1}{g}_m \qquad (42.16)$$

Sec. 42 Analytic Elements for Closed Boundaries

thus

$$\overset{1}{f}_m = \chi^m + \frac{1}{\chi^m} \qquad \overset{1}{g}_m = i\left[\chi^m - \frac{1}{\chi^m}\right] \tag{42.17}$$

Another function that yields a constant imaginary part along $\mathcal{C}$ is the following one, which has poles of the first order at points δ_j and $\overline{\delta_j}^{-1}$:

$$\overset{1}{H}_m = e^{i\alpha_m}\left[\frac{\underset{m}{b}}{\chi - \delta_m} - \frac{\overline{\underset{m}{b}}}{\rho_m^2}\frac{1}{\chi - \overline{\delta_m}^{-1}}\right] \tag{42.18}$$

where δ_m and $\underset{m}{b}$ are complex constants,

$$\delta_m = \rho_m e^{i\alpha_m} \qquad \rho_m > 1 \tag{42.19}$$

$$\underset{m}{b} = \underset{m}{b}_1 + i\underset{m}{b}_2 \tag{42.20}$$

The value of $\Im \overset{1}{H}_m$ along the unit circle is obtained by substituting $e^{i\theta}$ for χ in (42.18), which yields, combining terms and using (42.19),

$$\overset{1}{H}_m(e^{i\theta}) = e^{i\alpha_m}\frac{(\underset{m}{b} - \overline{\underset{m}{b}}/\rho_m^2)e^{i\theta} + \overline{\underset{m}{b}}/\rho_m e^{i\alpha_m} - \underset{m}{b}/\rho_m e^{i\alpha_m}}{e^{2i\theta} - (\rho_m + 1/\rho_m)e^{i(\theta+\alpha_m)} + e^{2i\alpha_m}} \tag{42.21}$$

Division of both numerator and denominator by $e^{i(\theta+\alpha_m)}$ yields

$$\overset{1}{H}_m(e^{i\theta}) = \frac{\underset{m}{b} - \overline{\underset{m}{b}}/\rho_m^2 + 1/\rho_m(\overline{\underset{m}{b}} - \underset{m}{b})e^{-i(\theta-\alpha_m)}}{2\cos(\theta - \alpha_m) - (\rho_m + 1/\rho_m)} \tag{42.22}$$

and we obtain for the imaginary part, writing $\underset{m}{b}$ as $\underset{m}{b}_1 + i\underset{m}{b}_2$,

$$\Im \overset{1}{H}_m(e^{i\theta}) = -\frac{\underset{m}{b}_2}{\rho_m} \tag{42.23}$$

so that the imaginary part of $\overset{1}{H}_m$ is constant along $\mathcal{C}$, as asserted. We introduce functions $\overset{1}{F}_m + \overset{1}{G}_m$ such that

$$\overset{1}{H}_m = \underset{m}{b}_1 \overset{1}{F}_m + \underset{m}{b}_2 \overset{1}{G}_m \tag{42.24}$$

where

$$\overset{1}{F}_m = e^{i\alpha_m}\left[\frac{1}{\chi-\delta_m} - \frac{1}{\rho_m^2}\frac{1}{\chi-\overline{\delta_m^{-1}}}\right]$$

$$\overset{1}{G}_m = ie^{i\alpha_m}\left[\frac{1}{\chi-\delta_m} + \frac{1}{\rho_m^2}\frac{1}{\chi-\overline{\delta_m^{-1}}}\right]$$

(42.25)

It may be noted that these functions represent dipoles at $\chi = \delta_m$ and image dipoles at $\chi = \overline{\delta_m^{-1}}$, oriented normal ($\overset{1}{F}_m$) and tangential ($\overset{1}{G}_m$) to the circle.

We define $\overset{1}{\Omega}$ as a sum of n functions $\overset{1}{f}_m$, $\overset{1}{g}_m$, and m_r functions $\overset{1}{F}_m$ and $\overset{1}{G}_m$ as follows:

$$\boxed{\overset{1}{\Omega} = \sum_{m=1}^{n}\left[\underset{m}{a}_1 \overset{1}{f}_m(\chi) + \underset{m}{a}_2 \overset{1}{g}_m(\chi)\right] + \sum_{m=1}^{m_r}\left[\underset{m}{b}_1 \overset{1}{F}_m(\chi) + \underset{m}{b}_2 \overset{1}{G}_m(\chi)\right] + \Phi_0^*}$$

(42.26)

The constant Φ_0^* is added to provide the capability of modeling a constant jump in the potential. The functions $\overset{1}{f}_m$ and $\overset{1}{g}_m$ are given by (42.17) and $\overset{1}{F}_m$ and $\overset{1}{G}_m$ by (42.25). Each of these functions has singular points inside the flow domain: the functions $\overset{1}{f}_m$ and $\overset{1}{g}_m$ are of order Z^m near infinity and the functions $\overset{1}{F}_m$ and $\overset{1}{G}_m$ have poles of the first order at $\chi = \delta_m$ (these points lie outside the unit circle (see [42.19]) and therefore correspond to points inside the flow domain). We will determine the function $\overset{2}{\Omega}$ such that the above singularities are removable.

The Complex Potential $\overset{2}{\Omega}$

We write $\overset{2}{\Omega}$ as a sum of functions analogous to (42.26):

$$\boxed{\overset{2}{\Omega} = \sum_{m=1}^{n}\left[\underset{m}{a}_1 \overset{2}{f}_m(Z) + \underset{m}{a}_2 \overset{2}{g}_m(Z)\right] + \sum_{m=1}^{m_r}\left[\underset{m}{b}_1 \overset{2}{F}_m(Z) + \underset{m}{b}_2 \overset{2}{G}_m(Z)\right]}$$

(42.27)

where the functions $\overset{2}{f}_m$, $\overset{2}{g}_m$, $\overset{2}{F}_m$, and $\overset{2}{G}_m$ are chosen such that the singularities of $\overset{1}{f}_m$, $\overset{1}{g}_m$, $\overset{1}{F}_m$, and $\overset{1}{G}_m$ are removable. The function $\overset{1}{\Omega}$ is defined as a constant inside the inhomogeneity (see [42.7]) and therefore the singularities of (42.26) inside the unit circle are of no concern. The functions $\overset{1}{f}_m$ and $\overset{1}{g}_m$ are analytic for all remaining χ, except near infinity where

$$\chi \to \infty \qquad \overset{1}{f}_m \to \chi^m \qquad \overset{1}{g}_m \to i\chi^m$$

(42.28)

Sec. 42   Analytic Elements for Closed Boundaries

The singularities of the functions $\overset{1}{f}_m$ and $\overset{1}{g}_m$ differ by a factor of i so that

$$\overset{2}{g}_m = i\overset{2}{f}_m \tag{42.29}$$

Infinity in the Z plane corresponds to infinity in the χ plane in such a way that a large circle in the Z plane corresponds to a large circle in the χ plane. Therefore the function χ may be represented as follows for $Z \to \infty$:

$$\chi = Z \sum_{j=0}^{\infty} \beta_j^* Z^{-j} \tag{42.30}$$

where the coefficients β_j^* are obtained by expanding the mapping function $\chi = \chi(Z)$ about $1/Z = 0$. Raising both sides of this equation to the power m, we obtain an expansion of the form

$$\chi^m = Z^m \sum_{j=0}^{\infty} \overset{m}{\beta}_j Z^{-j} \tag{42.31}$$

The coefficients $\overset{m}{\beta}_j$ may be expressed in terms of the coefficients β_j^* by the use of an algorithm employing a technique outlined in Appendix A. The function $\overset{2}{f}_m$ is chosen such that it cancels all terms in (42.31) of an order higher than Z^{-1}. This yields

$$\overset{2}{f}_m = -\sum_{j=0}^{m} \overset{m}{\beta}_j Z^{m-j} \tag{42.32}$$

The functions $\overset{2}{F}_m$ and $\overset{2}{G}_m$ must remove the singularities of $\overset{1}{F}_m$ and $\overset{1}{G}_m$. The points δ_m lie outside the unit circle (see [42.19]) and correspond to points in the flow domain, whereas the points $\overline{\delta_m^{-1}}$ lie inside the unit circle and correspond to points outside the flow domain; the singularities of $\overset{1}{F}_m$ and $\overset{1}{G}_m$ that need be removed are those at δ_m. The coefficients of the singularities of $\overset{1}{G}_m$ differ from those of $\overset{1}{F}_m$ by a factor of i so that

$$\overset{2}{G}_m = i\overset{2}{F}_m \tag{42.33}$$

We denote as d_m the point in the Z plane that corresponds to $\chi = \delta_m$,

$$\delta_m = \chi(d_m) \tag{42.34}$$

The function $\chi = \chi(Z)$ is analytic at d_m; it may be expanded in a Taylor series about that point:

$$\chi - \delta_m = (Z - d_m)\left[\chi'(d_m) + \frac{\chi''(d_m)}{2!}(Z - d_m) + \ldots\right] \tag{42.35}$$

Because the mapping is conformal at $Z = d_m$, the derivative of the mapping function is unequal to zero there:

$$\chi'(d_m) \neq 0 \tag{42.36}$$

The function $\overset{1}{F}_m$ in (42.25) may thus be expanded about $Z = d_m$ as

$$\overset{1}{F}_m = e^{i\alpha_m} \left\{ \frac{1}{Z - d_m} [B_0 + B_1(Z - d_m) + \ldots] - \frac{1}{\rho_m^2} \frac{1}{\chi - \overline{\delta_m^{-1}}} \right\} \tag{42.37}$$

where

$$B_0 + B_1(Z - d_m) + \ldots = \left[\chi'(d_m) + \frac{\chi''(d_m)}{2!}(Z - d_m) + \ldots \right]^{-1} \tag{42.38}$$

For the present case we only need to compute the coefficients B_0 and B_1 because the poles at d_m are of order 1. It follows that

$$B_0 = \frac{1}{\chi'(d_m)} \qquad B_1 = -\frac{1}{2} \frac{\chi''(d_m)}{[\chi'(d_m)]^2} \tag{42.39}$$

The functions $\overset{2}{F}_m$ must cancel both the singular term and the constant in the expansion of $\overset{1}{F}_m$ about d_m so that

$$\overset{2}{F}_m = -e^{i\alpha_m} \left[\frac{B_0}{Z - d_m} + B_1 \right] \tag{42.40}$$

The functions $\overset{2}{g}_m$ and $\overset{2}{G}_m$ are obtained in terms of $\overset{2}{f}_m$ and $\overset{2}{F}_m$ by the use of (42.29) and (42.33), so that all functions are known. It may be noted that functions exhibiting higher order poles at δ_m and $\overline{\delta_m^{-1}}$ may be added if necessary.

For numerical applications it is advisable to determine a far-field expansion and expansions about the poles, to be used for large values of Z, and in the neighborhood of the poles. The singular terms in these expansions are to be eliminated analytically, i.e., prior to numerical evaluation.

The complex potential presented so far exhibits a jump in potential across $\mathcal{B}$ while having a continuous imaginary part. The analytic element can be used to model any feature that requires such a jump condition, such as an inhomogeneity or a closed leaky wall. A complex potential that exhibits a continuous potential and an imaginary part that jumps across $\mathcal{B}$ is obtained simply by dividing the complex potential by a factor of i.

The analytic element presented in this section has the advantage over strings of line-doublets and line-dipoles that all derivatives exist on either side of $\mathcal{B}$, and that the number of degrees of freedom is independent of the shape of $\mathcal{B}$.

As an illustration, we will construct an analytic element for an elliptic inhomogeneity.

An Analytic Element for an Elliptic Inhomogeneity

If the boundary B is an ellipse, as shown in Figure 6.40(a), then the functions $Z = Z(z)$, $Z = Z(\chi)$, and $\chi = \chi(Z)$ are given by (31.2), (31.7), and (31.13) (compare Figures 5.4 and 5.5).

$$Z = \frac{z - z_0}{a + b} e^{-i\alpha} \qquad Z = \frac{1}{2}\left[\chi + \frac{\gamma^2}{\chi}\right] \qquad \chi = Z + \sqrt{(Z - \gamma)(Z + \gamma)} \qquad (42.41)$$

The constants A, B, and γ are given by (31.3) and (31.5):

$$A = \frac{a}{a + b} \qquad B = \frac{b}{a + b} \qquad \gamma^2 = A^2 - B^2 = A - B \qquad (42.42)$$

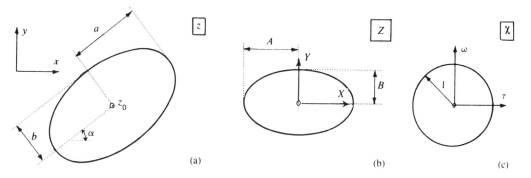

Figure 6.40 Mapping the exterior of the inhomogeneity onto the exterior of the unit circle.

Equation (42.30) becomes with (42.41):

$$\chi = \left[Z + \sqrt{Z^2 - \gamma^2}\right] = Z\left[1 + \sqrt{1 - \left[\frac{\gamma}{Z}\right]^2}\right] \qquad (42.43)$$

The term between the brackets may be expanded in a Taylor series about $Z^{-2} = 0$:

$$1 + \sqrt{1 - \left[\frac{\gamma}{Z}\right]^2} = \sum_{j=0}^{\infty} \beta_j^* Z^{-2j} \qquad (42.44)$$

The coefficients β_j^* are obtained from expansions given in most mathematical handbooks (e.g., Spiegel [1968]) and may be obtained from the following recursive relation:

$$\beta_0^* = 2 \qquad \beta_1^* = -\tfrac{1}{2}\gamma^2 \qquad \beta_j^* = \beta_{j-1}^* \frac{j - \tfrac{3}{2}}{j}\gamma^2 \qquad (j = 2, 3, \ldots) \qquad (42.45)$$

The expansion (42.31) about $Z^{-1} = 0$ is obtained by raising the expansion (42.44) to the power m and multiplying the result by Z^m:

$$\chi^m = Z^m \sum_{j=0}^{\infty} \overset{m}{\beta}_j Z^{-2j} = \sum_{j=0}^{\infty} \overset{m}{\beta}_j Z^{m-2j} \qquad (Z \to \infty) \qquad (42.46)$$

As mentioned, the coefficients $\overset{m}{\beta}_j$ may be determined by the use of an elementary computer routine employing a technique outlined in Appendix A.

The function $\overset{2}{f}_m$ must cancel all terms in (42.46) that are of an order higher than Z^{-1}. Considering even and odd values of m separately, we obtain

$$\overset{2}{f}_m = \overset{2}{f}_{2r} = -\sum_{j=0}^{r} \overset{m}{\beta}_j Z^{2r-2j} = -\sum_{\nu=0}^{r} \overset{m}{\beta}_{r-\nu} Z^{2\nu} \qquad (r = 0, 1, 2, \ldots)$$
(42.47)

$$\overset{2}{f}_m = \overset{2}{f}_{2r+1} = -\sum_{j=0}^{r} \overset{m}{\beta}_j Z^{2r-2j+1} = -\sum_{\nu=0}^{r} \overset{m}{\beta}_{r-\nu} Z^{2\nu+1} \qquad (r = 0, 1, 2, \ldots)$$
(42.48)

The point in the Z plane that corresponds to $\chi = \delta_m$ is d_m with

$$d_m = \tfrac{1}{2}(\delta_m + \frac{\gamma^2}{\delta_m}) \qquad \delta_m = d_m + \sqrt{d_m^2 - \gamma^2} \qquad (42.49)$$

The function $\overset{2}{F}_m$ is given by (42.40) with (42.39), which involves the first- and second-order derivatives of χ with respect to Z. The latter derivatives are obtained from (42.41) as

$$\chi' = \frac{Z + \sqrt{Z^2 - \gamma^2}}{\sqrt{Z^2 - \gamma^2}} \qquad \chi'' = \frac{-\gamma^2}{(Z^2 - \gamma^2)^{3/2}} \qquad (42.50)$$

so that

$$\chi'(d_m) = \frac{d_m + \sqrt{d_m^2 - \gamma^2}}{\sqrt{d_m^2 - \gamma^2}} = \frac{\delta_m}{\delta_m - d_m} \qquad (42.51)$$

and

$$\chi''(d_m) = \frac{-\gamma^2}{(d_m^2 - \gamma^2)^{3/2}} = \frac{-\gamma^2}{(\delta_m - d_m)^3} \qquad (42.52)$$

where use is made of (42.49). Combining these expressions with equations (42.39) for the coefficients B_0 and B_1 and using the result in (42.40), we obtain

$$\overset{2}{F}_m = -\frac{e^{i\alpha_m}}{\delta_m}\left[\frac{\delta_m - d_m}{Z - d_m} + \tfrac{1}{2}\frac{\gamma^2/\delta_m}{\delta_m - d_m}\right] \qquad (42.53)$$

Sec. 42   Analytic Elements for Closed Boundaries   489

The functions $\overset{2}{g}_m$ and $\overset{2}{G}_m$ are obtained from (42.29) and (42.32), (42.33) and (42.53), so that all functions in the expression (42.27) for $\overset{2}{\Omega}$ are known. The far-field expansion of Ω is obtained by combining each pair of functions $\overset{1}{f}_m$ and $\overset{2}{f}_m$, writing the term χ^m in $\overset{1}{f}_m$ in terms of Z by the use of (42.46). The constant and all singular terms in the expansion of $\overset{1}{f}_m$ will be canceled by $\overset{2}{f}_m$. Expansions of $\overset{1}{F}_m$ about the poles at $Z = d_m$ should be used for evaluating Ω in the neighborhood of these poles; note that the singular terms in this expansion will be cancelled by $\overset{2}{F}_m$.

Applications

The complex potential for the analytic element is given by (42.9) and (42.10). The only term in these expressions that needs yet be determined is the constant Ψ_0, which equals the constant value of $\overset{1}{\Psi}$ generated by $\overset{1}{\Omega}$ along the ellipse. The contribution to Ψ_0 of the first sum in (42.26) is zero and the contributions of the second sum are found from (42.23) and (42.24) so that

$$\Psi_0 = -\sum_{m=1}^{m_r} \frac{\underset{m}{b_2}}{\rho_m} \qquad (42.54)$$

The complex potential has $2n + 2m_r + 1$ degrees of freedom: the real constants $\underset{m}{a_1}, \underset{m}{a_2}, \underset{m}{b_1}, \underset{m}{b_2}$, and Φ_0^*. Boundary conditions therefore are applied at $2n + 2m_r + 1$ control points chosen on the ellipse. The analytic elements may be added together to model flow with a number of elliptic inhomogeneities, and may be combined with the functions presented earlier in this chapter.

In Figure 6.41, a plot is shown of piezometric contours and streamlines for flow in a confined aquifer with a uniform far field and seven elliptic inhomogeneities, each with a hydraulic conductivity ten times higher than that of the remaining portion of the aquifer. For each analytic element, the order n of the central polynomial is four, and each element has four poles, so that $m_r = 4$. Thus, each element has $2n + 2m_r + 1 = 17$ degrees of freedom. A detail of the plot is shown in Figure 6.42.

As an illustration, we will apply the analytic element to the case of an elliptic inhomogeneity in a field of uniform flow. We will see that an exact solution is obtained for that case.

We solve the problem by adding the complex potential Ω_g for uniform flow to the analytic element. The complex potential Ω_g for uniform flow with discharge Q_0 inclined at an angle β to the x axis is

$$\Omega_g = -Q_0 z e^{-i\beta} + C_z \qquad (42.55)$$

where C_z is an arbitrary complex constant, chosen such that the complex potential Ω_g may be written in terms of Z as (compare [42.41])

$$\Omega_g = -(a+b)Q_0 e^{-i(\beta-\alpha)} Z + \Phi_0 \qquad (42.56)$$

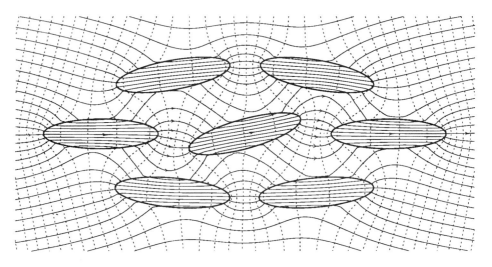

Figure 6.41 Flow pattern for seven elliptic inhomogeneities in a field of uniform flow.

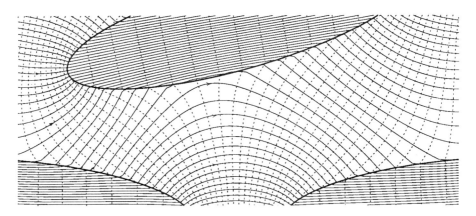

Figure 6.42 Detail of the flow pattern shown in Figure 6.41.

where Φ_0 is a real constant. Addition of the complex potential for the analytic element to (42.56) yields

$$z \text{ in } \mathcal{D}^- \qquad \Omega = \overset{1}{\Omega} + \overset{2}{\Omega} - (a+b)Q_0 e^{-i(\beta-\alpha)} Z + \Phi_0 \qquad (42.57)$$

$$z \text{ in } \mathcal{D}^+ \qquad \Omega = i\Psi_0 + \overset{2}{\Omega} - (a+b)Q_0 e^{-i(\beta-\alpha)} Z + \Phi_0 \qquad (42.58)$$

We will choose a function $\overset{1}{\Omega}$ with only three degrees of freedom:

$$\overset{1}{\Omega} = \underset{1}{a_1} f_1(\chi) + \underset{1}{a_2} g_1(\chi) + \Phi_0^* = \underset{1}{a_1} \left[\chi + \frac{1}{\chi} \right] + \underset{1}{a_2} i \left[\chi - \frac{1}{\chi} \right] + \Phi_0^* \qquad (42.59)$$

Sec. 42   Analytic Elements for Closed Boundaries   491

so that

$$\Psi_0 = 0 \qquad (42.60)$$

and $\overset{2}{\Omega}$ becomes, with (42.48) and (42.29),

$$\overset{2}{\Omega} = a_1 \overset{2}{\underset{1}{f_1}} + a_2 \overset{2}{\underset{1}{g_1}} = -\underset{1}{a_1}\beta_0 Z - i\underset{1}{a_2}\beta_0 Z \qquad (42.61)$$

where (see [42.45])

$$\beta_0 = \beta_0^* = 2 \qquad (42.62)$$

so that

$$\overset{2}{\Omega} = -2\underset{1}{a_1} Z - 2i\underset{1}{a_2} Z \qquad (42.63)$$

There are three degrees of freedom and we choose the corresponding control points at $\chi = 1$, $\chi = -1$, and $\chi = i$. The values for Z at these points are obtained from (42.41) as $Z = \frac{1}{2}(1+\gamma^2) = A$, $Z = -\frac{1}{2}(1+\gamma^2) = -A$ and $Z = \frac{i}{2}(1-\gamma^2) = iB$. Application of the condition (35.4) at these three points yields with (42.57) through (42.61), and with $\Phi^+ - \Phi^- = -\overset{1}{\Phi}$ and $(k^+ - k^-)/k^- = \nu$

$$-\Phi_0^* - 2\underset{1}{a_1} = \nu\left[\, 2\underset{1}{a_1}(1-A) - (a+b)AQ_0\cos(\beta-\alpha) + \Phi_0^* + \Phi_0 \right] \qquad (42.64)$$

$$-\Phi_0^* + 2\underset{1}{a_1} = \nu\left[-2\underset{1}{a_1}(1-A) + (a+b)AQ_0\cos(\beta-\alpha) + \Phi_0^* + \Phi_0 \right] \qquad (42.65)$$

$$-\Phi_0^* + 2\underset{1}{a_2} = \nu\left[-2\underset{1}{a_2}(1-B) - (a+b)BQ_0\sin(\beta-\alpha) + \Phi_0^* + \Phi_0 \right] \qquad (42.66)$$

We use that $1 - A = B$, $1 - B = A$, $(a+b)A = a$, $(a+b)B = b$ (see [42.42]) and solve this system of three equations in three unknowns, which yields

$$\Phi_0^* = -\frac{\nu}{1+\nu}\Phi_0 \qquad \underset{1}{a_1} = \frac{a\nu Q_0 \cos(\beta-\alpha)}{2(1+B\nu)} \qquad \underset{1}{a_2} = -\frac{b\nu Q_0 \sin(\beta-\alpha)}{2(1+A\nu)} \qquad (42.67)$$

where

$$\nu = \frac{k^+ - k^-}{k^-} \qquad (42.68)$$

It is of interest to note that this solution is exact. The reader may verify this by setting $\chi = e^{i\theta}$ in the expressions for Ω, using (42.41) to express Z for points on the ellipse as $\frac{1}{2}(e^{i\theta} + \gamma^2 e^{-i\theta}) = A\cos\theta + iB\sin\theta$.

Problem 42.1

Prove that the complex potential given by (42.57) and (42.58) with the constants given by (42.67) and (42.68) is an exact solution for an elliptic inhomogeneity in a field of uniform flow.

Problem 42.2

Write a computer program to model a set of two elliptic inhomogeneities in a field of uniform flow, with $m_r = 0$.

Problem 42.3

Apply the technique outlined in equations (42.5) through (42.11) to solve the problem of a well of discharge Q and located at $x = 0$, $y = d$ in an infinite aquifer. The hydraulic conductivity is k^+ in the upper half plane $y \geq 0$ and k^- in the lower half plane $y < 0$. Demonstrate that the solution exactly meets the boundary conditions. (Solve the problem by setting $\overset{1}{\Omega}$ equal to $A\ln(z^2 + d^2) + \Phi_0^*$ in $y \leq 0$ and to zero in $y > 0$.)

43 MODELING FLOW IN TWO AQUIFERS SEPARATED BY A THIN LAYER

The analytic element method is not restricted to single aquifers; it may be applied equally well to systems of one or more stacked aquifers separated by layers that are either impermeable or leaky. In this section we will discuss the modeling of a system of two aquifers separated by a horizontal thin discontinuous bed that extends throughout the aquifer system. The separating bed is impermeable, but may locally be leaky. The upper boundary of the aquifer system is a phreatic surface; the upper aquifer is always unconfined and the lower one may either be confined or unconfined as shown in Figure 6.43. The analytic elements used in the model are the same as those presented in this chapter, but the formulation of the problem and the application of boundary conditions are different from the single-layer model.

We formulate the problem in terms of three potentials: a comprehensive potential Φ, a potential $\overset{u}{\Phi}$ defined in the upper aquifer, and a potential $\overset{l}{\Phi}$ defined in the lower aquifer. The formulation of problems of flow in systems of two aquifers separated by a thin impermeable layer by the use of the latter three potentials is discussed in Sec. 12; the reader should be familiar with the material covered in that section before proceeding to read the material presented below, which closely follows Strack and Haitjema [1981(a)] and [1981(b)].

We will consider the case that the aquifer system consists of two layers with different hydraulic conductivities, with the interface between these layers at the same elevation as the separating bed. The latter elevation is represented as H and the hydraulic conductivities in the upper and lower aquifers as $\overset{u}{k}$ and $\overset{l}{k}$:

$$0 \leq x_3 \leq H \qquad k = \overset{l}{k} \qquad (43.1)$$

$$H < x_3 \qquad k = \overset{u}{k} \qquad (43.2)$$

Adhering to the notation used in Sec. 12, we represent the comprehensive potential in the zone without separating bed as $\overset{s}{\Phi}$ and in the zone with separating bed as $\overset{d}{\Phi}$, where

Sec. 43 Modeling Flow in Two Aquifers Separated by a Thin Layer

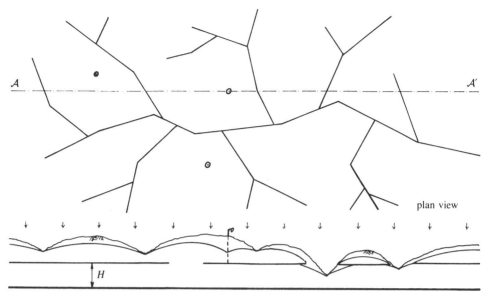

Figure 6.43 The aquifer system in plan view.

$$\overset{s}{\Phi} = \tfrac{1}{2}\overset{u}{k}(\phi - H)^2 + \overset{l}{k}H\phi - \tfrac{1}{2}\overset{l}{k}H^2 \qquad \phi > H \tag{43.3}$$

$$\overset{s}{\Phi} = \tfrac{1}{2}\overset{l}{k}\phi^2 \qquad \phi \leq H \tag{43.4}$$

$$\overset{d}{\Phi} = \overset{u}{\Phi} + \overset{l}{\Phi} \tag{43.5}$$

where the potentials in the upper and lower aquifers are given as

$$\overset{u}{\Phi} = \tfrac{1}{2}\overset{u}{k}(\overset{u}{\phi} - H)^2 \tag{43.6}$$

and

$$\overset{l}{\Phi} = \overset{l}{k}H\overset{l}{\phi} - \tfrac{1}{2}\overset{l}{k}H^2 \qquad \overset{l}{\phi} > H \tag{43.7}$$

$$\overset{l}{\Phi} = \tfrac{1}{2}\overset{l}{k}\overset{l}{\phi}^2 \qquad \overset{l}{\phi} \leq H \tag{43.8}$$

The comprehensive potential is defined in the zone where the separating bed is absent as

$$\Phi = \overset{s}{\Phi} \tag{43.9}$$

and in the zone where the separating bed is present as

$$\Phi = \overset{d}{\Phi} \tag{43.10}$$

with $\overset{s}{\Phi}$ and $\overset{d}{\Phi}$ given by (43.3) through (43.8).

Formulation of the Problem

The problem is solved in four steps. First, the comprehensive potential is determined as a function of position by solving a boundary value problem formulated in terms of this potential. Second, the comprehensive potential is used to determine the heads along the boundaries of the openings in the separating bed. Third, the potential $\overset{u}{\Phi}$ is determined by solving the boundary-value problem in the upper aquifer, using the heads along the boundaries of the openings as internal boundary conditions along with the other ones specified for the upper aquifer. Fourth, the potential in the lower aquifer is obtained as the difference between the potentials in the upper and lower aquifers.

Restriction is made at first to the case that the separating layer is impermeable; zones where leakage occurs through this layer are treated toward the end of this section.

Homogeneous Hydraulic Conductivity

We will present the systems of equations for the unknown parameters in the functions Φ and $\overset{u}{\Phi}$ for the case that the hydraulic conductivity is homogeneous throughout the aquifer system.

The Comprehensive Potential

The comprehensive potential is separated into two functions U and V, as was done in Sec. 34 for the case of a single aquifer. U contains the functions with unknown coefficients and V those with known ones,

$$\Phi = U + V \tag{43.11}$$

These functions consist of a summation of analytic elements. For simplicity we consider the case here that U contains only line-sinks; the addition of other elements and their corresponding boundary conditions is straightforward and will not be discussed. We will assume initially that n_{ls} line-sinks are used to model creeks with given heads, and that these creeks penetrate both aquifers. The heads in both the upper and lower aquifers then are known at the centers of these line-sinks, and the corresponding values of the comprehensive potential, Φ_ν, $\nu = 1, 2, \ldots, n_{ls}$, may be computed by the use of (43.9) and (43.10). This leads to the following n_{ls} equations (compare [34.17]):

$$\underset{c}{\Phi_\nu} = \sum_{j=1}^{n_{ls}} \sigma_j \Lambda_{ls}\left[\tfrac{1}{2}(\overset{1}{z}_\nu + \overset{2}{z}_\nu), \overset{1}{z}_j, \overset{2}{z}_j\right] + C_1 + V\left[\tfrac{1}{2}(\overset{1}{z}_\nu + \overset{2}{z}_\nu)\right] \qquad (\nu = 1, 2, \ldots, n_{ls}) \tag{43.12}$$

where C_1 is a real constant. If the heads in both aquifers at some reference point z_0 are known and the corresponding value of Φ is $\underset{0}{\Phi}$, then the condition at the reference point is

$$\underset{0}{\Phi} = U(z_0) + C_1 + V(z_0) \tag{43.13}$$

The above system of $n_{ls} + 1$ equations may be solved for the $n_{ls} + 1$ unknowns σ_j ($j = 1, 2, \ldots, n_{ls}$) and C_1. Afterward the comprehensive potential may be computed for any point in the aquifer system. The head ϕ, however, can be determined from the comprehensive potential only inside the openings in the separating bed; this is done by the use of (43.3) and (43.4).

The Potential in the Upper Aquifer

The potential $\overset{u}{\Phi}$ is written as the sum of two functions $\overset{u}{U}$ and $\overset{u}{V}$,

$$\overset{u}{\Phi} = \overset{u}{U} + \overset{u}{V} \tag{43.14}$$

The comprehensive potential is used to determine the heads along the openings in the separating layer. The boundary of each of these openings is discretized into a number of straight line segments, treated as boundary segments with given heads and modeled with line-sinks. We denote the number of line-sinks along the boundaries of the openings as n_o.

The heads in the upper aquifer at the centers of the n_{ls} line-sinks used to model the creeks are known, and we represent the corresponding values of the potential in the upper aquifer as $\overset{u}{\underset{c}{\Phi}}_\nu$ ($\nu = 1, 2, \ldots, n_{ls}$). The conditions at the centers of these line-sinks yield a system of n_{ls} equations:

$$\overset{u}{\underset{c}{\Phi}}_\nu = \sum_{j=1}^{n_{ls}+n_o} \overset{u}{\sigma}_j \Lambda_{ls}\left[\tfrac{1}{2}(\overset{1}{z}_\nu + \overset{2}{z}_\nu), \overset{1}{z}_j, \overset{2}{z}_j\right] + \overset{u}{C}_1 + \overset{u}{V}\left[\tfrac{1}{2}(\overset{1}{z}_\nu + \overset{2}{z}_\nu)\right] \qquad (\nu = 1, 2, \ldots, n_{ls})$$

$$\tag{43.15}$$

where the $\overset{u}{\sigma}_j$ represent the strengths of the line-sinks in the upper aquifer and $\overset{u}{C}_1$ is a real constant.

The values of $\overset{u}{\Phi}$ at the centers of the line-sinks along the boundaries of the openings are computed from the heads by the use of (43.6) and are represented as $\overset{u}{\underset{o}{\Phi}}_\nu$ ($\nu = n_{ls} + 1, n_{ls} + 2, \ldots, n_{ls} + n_o$). The corresponding n_o equations are

$$\overset{u}{\underset{o}{\Phi}}_\nu = \sum_{j=1}^{n_{ls}+n_o} \overset{u}{\sigma}_j \Lambda_{ls}\left[\tfrac{1}{2}(\overset{1}{z}_\nu + \overset{2}{z}_\nu), \overset{1}{z}_j, \overset{2}{z}_j\right] + \overset{u}{C}_1 + \overset{u}{V}\left[\tfrac{1}{2}(\overset{1}{z}_\nu + \overset{2}{z}_\nu)\right]$$

$$(\nu = n_{ls} + 1, n_{ls} + 2, \ldots, n_{ls} + n_o) \tag{43.16}$$

If the value of $\overset{u}{\Phi}$ at $z = z_0$ is $\overset{u}{\underset{0}{\Phi}}$, then the condition at this reference point becomes

$$\overset{u}{\underset{0}{\Phi}} = \overset{u}{U}(z_0) + \overset{u}{C}_1 + \overset{u}{V}(z_0) \qquad (43.17)$$

The system of $n_{ls} + n_o + 1$ equations given by (43.15) through (43.17) is solved for the $n_{ls} + n_o$ unknown strengths $\overset{u}{\sigma}_j$ and the constant $\overset{u}{C}_1$, so that $\overset{u}{\Phi}$ can be determined at any point of the upper aquifer. The potential in the lower aquifer, $\overset{l}{\Phi}$, is obtained from

$$\overset{l}{\Phi} = \Phi - \overset{u}{\Phi} \qquad (43.18)$$

Heads are computed from the potentials by the use of equations (43.3) through (43.10).

A complication arises if some of the creeks do not intersect the separating layer; in that case they occur in the upper aquifer only, and the heads in the lower aquifer underneath these line-sinks are unknown so that the values Φ_{ν_c} in (43.12) cannot be determined a priori. Such cases are treated by the use of an iterative procedure as follows. Let there be $\overset{u}{n}_{ls}$ line-sinks in the upper aquifer only. For the first iteration, the heads at points in the lower aquifer underneath the centers of these line-sinks are set equal to the heads in the upper aquifer (i.e., the creeks are assumed to be connected to both aquifers). These line-sinks then are included in the equations (43.12), which together with (43.13) are solved for the unknown coefficients in the expression for the comprehensive potential. Second, the heads along the edges of the openings are determined, and the number of equations in (43.15) is increased by $\overset{u}{n}_{ls}$. Solving these equations together with (43.16) and (43.17) yields the unknown parameters in the upper potential. The solution will be inaccurate: the heads at the control points on the boundaries of the openings computed from the comprehensive potential and from the potential for the upper aquifer will be different. For the second and further iterations, the line-sinks in the upper aquifer are entered in the function V of the comprehensive potential, treating the $\overset{u}{n}_{ls}$ strengths $\overset{u}{\sigma}_j$ as knowns. New values are obtained for the heads at the control points along the boundaries of the openings, and the problem in the upper aquifer is solved again. The iteration cycle is continued, computing the potentials Φ and $\overset{u}{\Phi}$ in turn, until the heads along the openings obtained from these two potentials are sufficiently close.

Inhomogeneous Hydraulic Conductivity

Inhomogeneities in the hydraulic conductivity are modeled in much the same way as for the single-aquifer model. We will highlight the differences between the latter model and the present one, rather than presenting a complete discussion of the approach, covering two cases: (1) inhomogeneities lie only in the upper aquifer, and (2) they lie only in the lower one.

Sec. 43   Modeling Flow in Two Aquifers Separated by a Thin Layer

Upper Aquifer

The conditions along the boundaries of the inhomogeneities in terms of the potential $\overset{u}{\Phi}$ are obtained from (35.4),

$$\overset{u}{\lambda} = \overset{u}{\Phi}^+ - \overset{u}{\Phi}^- = \frac{\overset{u}{k}^+ - \overset{u}{k}^-}{\overset{u}{k}^-}\overset{u}{\Phi}^- \qquad (43.19)$$

where $\overset{u}{\lambda}$ represents the jump in potential across the boundary of the inhomogeneity in the upper aquifer, and the $-$ refers to the exterior of the inhomogeneity. The boundaries of the inhomogeneities are discretized as explained in Sec. 35, and line-doublets are added for each boundary section both to the comprehensive potential and to the potential for the upper aquifer. An iterative procedure is applied, similar to the one used for line-sinks in the upper aquifer. The jump λ in Φ is equal to the jump $\overset{u}{\lambda}$ in $\overset{u}{\Phi}$ since $\overset{l}{\Phi}$ is continuous,

$$\lambda = \overset{u}{\lambda} \qquad (43.20)$$

and for the first iteration the heads in the upper and lower aquifers are assumed to be the same at the control points of the inhomogeneities. The line-doublets are added to the system of equations for Φ, and the problem for $\overset{u}{\Phi}$ is solved using Φ to compute the heads at the control points on the boundaries of the openings, with the line-doublets included in the system of equations for $\overset{u}{\Phi}$. For the second and further iterations, the heads in the lower aquifer below the control points on the boundaries of the inhomogeneities are taken from the results of the previous iteration. The iteration cycle is continued until the heads along the boundaries of the openings meet some prescribed accuracy.

Lower Aquifer

Inhomogeneities in the lower aquifer are treated in a similar fashion. However, the inhomogeneities now enter only in the comprehensive potential; they do not enter in the expression for $\overset{u}{\Phi}$. The problem is solved iteratively, initially taking the heads in the upper aquifer above the control points on the boundaries of the inhomogeneities equal to those in the lower aquifer. For further iterations, the heads in the upper aquifer above the latter control points are taken from the previous iteration.

Leakage Through the Separating Bed

Leakage through the thin layer that separates the two aquifers may be incorporated by the use of area-sinks. Since, by continuity of flow, the amount of water leaking from the upper aquifer must enter the lower one, and vice versa, the strengths $\overset{u}{\gamma}$ and $\overset{l}{\gamma}$ of the area-sinks must be equal and opposite in sign:

$$\overset{u}{\gamma} = -\overset{l}{\gamma} \qquad (43.21)$$

The leakage therefore does not affect the comprehensive potential and only enters in the problem for flow in the upper aquifer. The leakage through the dividing layer is equal to (compare [14.5]):

$$\overset{u}{\gamma} = \frac{\overset{u}{\phi} - \overset{l}{\phi}}{c} \qquad (\overset{l}{\phi} \geq H) \tag{43.22}$$

where c is the resistance of the clay layer and $\overset{u}{\gamma}$ is positive if the leakage is from the upper aquifer into the lower one. The potential in the upper aquifer is a quadratic function of the head $\overset{u}{\phi}$. In order to obtain a system of linear equations, we linearize the potential $\overset{u}{\Phi}$ and apply an iterative procedure as explained below. We define F as

$$F = \overset{u}{\phi} - H \tag{43.23}$$

and denote as F^* an estimate of F so that

$$F = F^* + \varepsilon \tag{43.24}$$

where we assume that ε is small with respect to F. The potential $\overset{u}{\Phi}$ may now be written as

$$\overset{u}{\Phi} = \tfrac{1}{2}\overset{u}{k}\left[(F^*)^2 + 2F^*\varepsilon + \varepsilon^2\right] \tag{43.25}$$

We neglect ε^2 with respect to the other terms between the brackets; the equation becomes

$$\overset{u}{\Phi} \approx \tfrac{1}{2}\overset{u}{k}\left[(F^*)^2 + 2F^*\varepsilon\right] \tag{43.26}$$

Solving this for $\varepsilon = F - F^*$ we obtain

$$F - F^* = \frac{\overset{u}{\Phi}}{\overset{u}{k}F^*} - \tfrac{1}{2}F^* \tag{43.27}$$

Using (43.23) to write $\overset{u}{\phi}$ as $F + H$, with F given by (43.27), and introducing a constant $\overset{u}{\Phi}^*$ such that

$$F^* = \sqrt{\frac{2\overset{u}{\Phi}^*}{\overset{u}{k}}} \tag{43.28}$$

we obtain the following linearized expression for the head $\overset{u}{\phi}$ in terms of $\overset{u}{\Phi}$:

$$\overset{u}{\phi} = \overset{u}{a}\overset{u}{\Phi} + \overset{u}{b} \tag{43.29}$$

Sec. 43 Modeling Flow in Two Aquifers Separated by a Thin Layer

where

$$\overset{u}{a} = \frac{1}{\sqrt{2\overset{u}{k}\overset{u}{\Phi}^*}} \qquad \overset{u}{b} = H + \sqrt{\frac{\overset{u}{\Phi}^*}{2\overset{u}{k}}} \tag{43.30}$$

Assuming at first that the lower aquifer is confined, the head $\overset{l}{\phi}$ may be expressed in terms of $\overset{l}{\Phi}$ by the use of (43.7) as follows:

$$\overset{l}{\phi} = \overset{l}{a}\overset{l}{\Phi} + \overset{l}{b} \tag{43.31}$$

where

$$\overset{l}{a} = \frac{1}{\overset{l}{k}H} \qquad \overset{l}{b} = \tfrac{1}{2}H \tag{43.32}$$

The potential in the lower aquifer may be expressed as the difference between the comprehensive potential and the potential in the upper aquifer as follows:

$$\overset{l}{\Phi} = \Phi - \overset{u}{\Phi} \tag{43.33}$$

The leakage $\overset{u}{\gamma}$ may now be expressed in terms of Φ and $\overset{u}{\Phi}$ by the use of (43.22), (43.29), (43.31), and (43.33) as follows:

$$\overset{u}{\gamma} = A\overset{u}{\Phi} + B\Phi + C \tag{43.34}$$

where

$$A = \frac{\overset{u}{a} + \overset{l}{a}}{c} \qquad B = -\frac{\overset{l}{a}}{c} \qquad C = \frac{\overset{u}{b} - \overset{l}{b}}{c} \qquad (\overset{l}{\phi} \geq H) \tag{43.35}$$

If the lower aquifer is unconfined, the head below the separating layer equals the elevation head, H, and the expression for $\overset{u}{\gamma}$ becomes

$$\overset{u}{\gamma} = \frac{\overset{u}{\phi} - H}{c} \qquad (\overset{l}{\phi} < H) \tag{43.36}$$

Using (43.29) the expressions for A, B, and C in (43.34) become for this case

$$A = \frac{\overset{u}{a}}{c} \qquad B = 0 \qquad C = \frac{\overset{u}{b} - H}{c} \qquad (\overset{l}{\phi} < H) \tag{43.37}$$

Leaky zones may be modeled by the use of area-sinks, which are added to the expression for the potential $\overset{u}{\Phi}$. First, the comprehensive potential is determined as a function of position as explained above, noticing that the expression for the comprehensive potential is not affected by the leakage. Second, an initial value for $F^* = \overset{u}{\phi}{}^* - H$ is chosen, and conditions of the form (43.34) are applied at the control points of the area-sinks, using computed values for Φ at these points. A new value for F^* is obtained using the potential $\overset{u}{\Phi}$, which now may be evaluated at all control points, and a value for $\overset{l}{\phi}$ is computed to determine whether the flow in the lower aquifer below the control points is confined or unconfined. The constants A, B, and C in (43.34) may now be updated by the use of either (43.35) or (43.37). This procedure is set forth until a desired accuracy is reached.

The Regional Effect of the Tennessee-Tombigbee Waterway

The analysis covered in this section is implemented for the case of an impermeable separating bed in the computer program SYLENS written for the Nashville District of the U.S. Army Corps of Engineers. Results obtained with the model are reported in Strack, Haitjema and Melnyk [1980]. The following is extracted from the latter publication.

The model has been applied to simulate the impact of the divide-cut section of the Tennessee-Tombigbee Waterway on the neighboring aquifer system. This waterway is located in northeastern Mississippi and western-central Alabama, and connects the Tennessee River with the Tombigbee River to facilitate southbound barge traffic to Mobile, Alabama. The divide-cut section runs from Pickwick Lake, Tennessee, toward the Bay Springs Reservoir, 40 km south of Pickwick Lake. The waterway alignment follows the major creekbeds between Pickwick Lake and Bay Springs Reservoir.

Groundwater Regime

The groundwater flow in the area of the divide-cut section takes place primarily in two formations: the Eutaw and the Gordo formations. The Eutaw formation contains two members, both shown on the map of the areal geology reproduced in Figure 6.44. The Gordo formation lies below the Eutaw except in the area to the east where the Eutaw is absent. The creekbeds lie in alluvium deposits. The projected waterway is marked by a black solid line running north-south through two of the major creekbeds.

A cross section through the aquifer system is sketched in Figure 6.45. The discontinuous layer between the Eutaw and the Gordo aquifers is the McShan. Mississippian rock forms the base of the Gordo.

The average flow in the Eutaw is controlled to a large extent by known conditions. The aquifer body is fairly uniform and infiltration from rainfall is the main source of recharge. Many creeks are connected to the Eutaw and provide a natural control of the groundwater regime.

The flow in the Gordo aquifer is determined only to a limited extent by known conditions. The aquifer body is nonuniform, with considerable variations in confined thickness and hydraulic conductivity. Only a fraction of the creeks cut through the separating McShan and into the Gordo, thus providing less natural control than in the

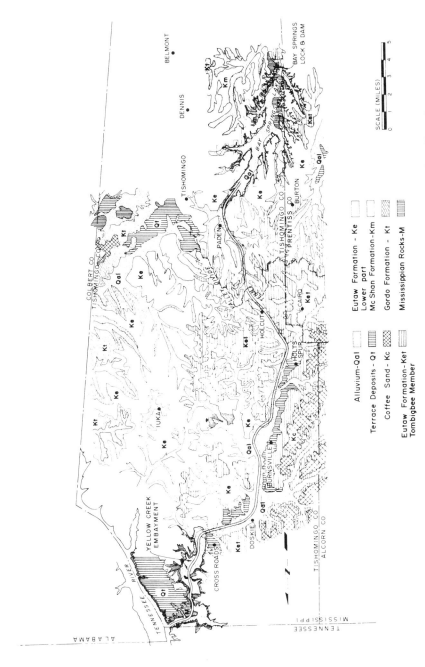

Figure 6.44 Areal geology, Tennessee-Tombigbee Waterway. Divide-cut section. (Reproduction of a map in color, prepared by the U.S. Army Engineer District, Nashville, Tennessee, 1972.)

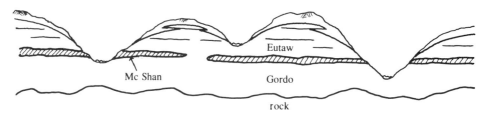

Figure 6.45 The aquifer system in sectional view.

Eutaw. The recharge is less direct than in the Eutaw, and results from infiltration at the outcrops and from leakage from the Eutaw in areas where the McShan is absent. Thus, the groundwater regime in the Gordo is greatly affected by factors that are not known in advance. The main two of these factors are (1) the locations of openings in the McShan, and (2) the variations in hydraulic conductivity.

The draw-downs due to the waterway cut are expected to occur over larger distances in the Gordo than in the Eutaw as a result of the lack of natural control in the former aquifer. Therefore, the main thrust of the study has been toward obtaining insight in the flow regime in the Gordo. However, the two aquifers interact hydraulically and the entire system needs to be modeled as a unit.

The Model Aquifer System

The model aquifer system consists of two aquifers separated by a thin and discontinuous aquiclude. The effect of the thin clay strata on the flow in the Eutaw is neglected; this aquifer is modeled as a single unconfined aquifer. The McShan is approximated as a horizontal aquiclude with openings and of infinitesimal thickness. The actual aquifer bases dip at 1:200, but the flow is governed primarily by heads at creekbeds and outcrops, rather than by the sloping beds. Thus, the beds may be taken to be horizontal without introducing gross errors.

Inhomogeneities in the hydraulic conductivity are included for the Gordo aquifer. Creeks are modeled by using line-sinks. For the initial runs, the creeks are taken to be effluent streams, i.e., the groundwater is assumed to flow toward the creeks. If the model predicts some creeks to be influent rather than effluent streams, the conditions along those creeks may be changed to account for the possibility that the groundwater table is below the creekbeds. The creeks may be specified either to be connected to the Eutaw or to both the Eutaw and the Gordo. Wells are incorporated in the model and may be screened either in the Eutaw, or in the Gordo, or in both.

The layout of the creeks and openings as approximated in the model is given in Figure 6.46(a). The solid lines represent creeks that are connected only to the Eutaw, whereas creeks connected to both aquifers are reproduced as dashed lines. The five

Sec. 43 Modeling Flow in Two Aquifers Separated by a Thin Layer

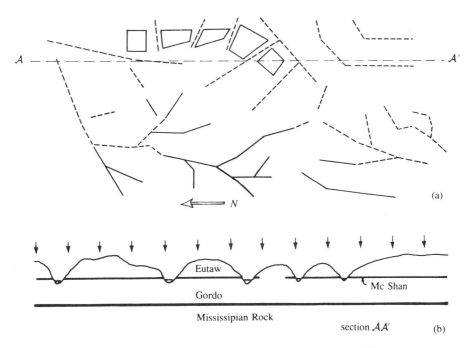

Figure 6.46 Layout of creeks (a) and cross section (b) of the model aquifer system.

polygons along the eastern boundary correspond to openings in the McShan. A cross section through the model aquifer system is sketched in Figure 6.46(b).

Review of the Modeling

The model predictions have evolved from a process of gradual refinement of input data on the basis of experience gained during the course of the modeling. We will discuss the main steps in this process to illustrate how the current view of the aquifer configuration developed. The main effort of the modeling has been toward reproducing the conditions that prevailed before excavation started; we will discuss only the modeling of the latter conditions.

Modeling with Homogeneous Hydraulic Conductivity. The main parameters adjusted during the initial stages of the modeling are the average hydraulic conductivity, the average infiltration from rainfall, the openings in the dividing layer, and the creeks that cut into the Gordo aquifer. Estimates of the average hydraulic conductivity were obtained from several pumping tests and the average infiltration from rainfall was estimated from field observations. The openings in the dividing layer shown in Figure 6.46 were known to cut into the Gordo aquifer. Others were assumed to be connected to the Gordo during a process of interactive modeling.

The piezometric levels produced for the Eutaw aquifer are proportional to the ratio of rainfall (N) to hydraulic conductivity (k). It was attempted to select N/k such that the piezometric levels in the Eutaw are below the soil surface at distances that match observations. It appeared during the course of the modeling, however, that it was not possible to obtain contour levels that fit well throughout the area modeled; the levels are reproduced in Figure 6.47, which is to the same scale as the layout of Figure 6.46(a), and corresponds to an infiltration rate of 11 inches per year and a hydraulic conductivity of 7.78 feet per day, or a ratio of N/k of $3.23 * 10^{-4}$. It may be noted that the above ratio for N/k was obtained without the benefit of the measured values for N and k, and appeared afterward to be close to that estimated from field data.

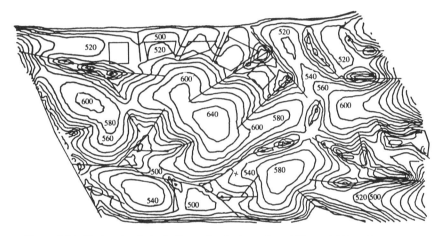

Figure 6.47 Contour levels in the Eutaw for the input data of Figure 6.46(a). (Source: Strack, Haitjema, and Melnyk [1980].)

The deviation between predicted and expected contour levels in the central portion of the Eutaw indicated that the conditions in that area are different from those assumed in the model.

Measured piezometric levels in the Gordo aquifer have led to estimated contour levels represented as dashed-dotted curves in Figure 6.48. The shape of these contour lines deviates significantly from those produced by the model and plotted as solid curves in Figure 6.48. The bending of the dashed-dotted contours around the central area may be explained by assuming that the two aquifers are connected in the central area. This fits well with the high values predicted for piezometric levels in that area of the Eutaw.

On the basis of the above considerations, an opening in the separating layer was assumed to exist in the central area and the size and position of this opening were adjusted until the contour lines in the Gordo approximated those obtained from field observations. The introduction of this opening into the model caused the piezometric levels in the Eutaw to drop by about 20 feet and the contour lines in the Gordo approached those observed (see Figure 6.49). The dashed-dotted lines in Figures 6.48 and 6.49 were obtained by interpolation of data measured at the locations marked by stars in the figures.

Modeling with a Single Inhomogeneity in the Gordo. It appears from Figure 6.49 that an opening in the separating layer causes the predicted contours to bend in much the same way as those obtained from field data. Thus, the bending of contours

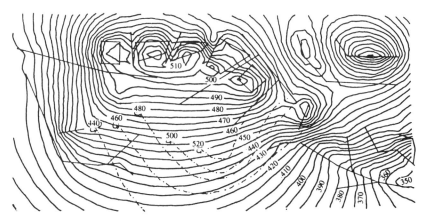

Figure 6.48 Contour levels in the Gordo for the input data of Figure 6.46(a). (Source: Strack, Haitjema, and Melnyk [1980].)

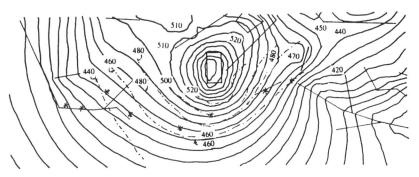

Figure 6.49 Contour levels in the central portion of the Gordo with an opening in the dividing layer. (Source: Strack, Haitjema, and Melnyk [1980].)

may be caused by recharge. There exists, however, an alternative explanation for the curved shape of the contours. This explanation is that the transmissivity in the Gordo would be high in an area that connects the eastern outcrop of the Gordo, a recharge area, to the location where the sharp bend occurs in the observed 520 foot contour (see Figure 6.49). The inhomogeneity with high transmissivity would act as a shortcut to the recharge area, thus raising the levels and bending the contours around the tip of the inhomogeneity.

A set of runs was made in order to investigate whether the inhomogeneity, without recharge through an opening in the dividing layer, could lead to the observed contour levels. The inhomogeneity was assumed to connect the recharge area to the area where the contours bend, and to be highly permeable: 50 times as permeable as the remainder of the Gordo. The contours obtained in this fashion indeed show the observed bending, but the levels are too low. The results therefore indicate that there must be at least some leakage from the Eutaw into the Gordo to cause the observed piezometric levels.

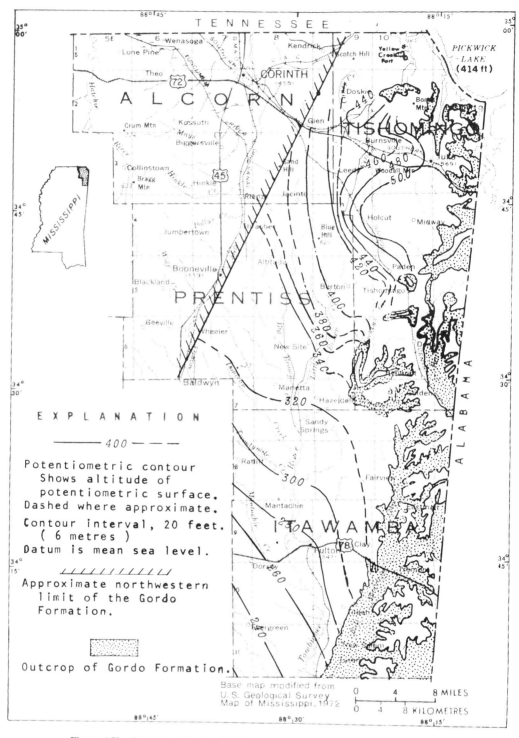

Figure 6.50 Water levels in the Gordo Formation, 1973. (Source: Water for industrial development in Alcorn, Itawamba, Prentiss, and Tishomingo Counties, Mississippi. A cooperative study sponsored by the USGS and the Mississippi Research and Development Center, Jackson, Mississippi, 1975.)

Sec. 43 Modeling Flow in Two Aquifers Separated by a Thin Layer

Predictions Obtained for the Conditions Without the Waterway Present.
The measured contours discussed thus far could be matched by modeling either with an opening in the dividing layer alone, or with a combination of an opening with an inhomogeneity in the Gordo. There is evidence, however, in support of the latter supposition. The Gordo aquifer is known to be rather inhomogeneous in transmissivity, a fact that is substantiated by the irregularities in a plot of the piezometric contours in the Gordo determined by the United States Geological Survey.

A map prepared by the survey is reproduced in Figure 6.50. The map represents an area that is considerably larger than that considered thus far, and the scale of the modeling is increased accordingly. The bending contours discussed above appear at the left of the map where they bend around the village of Holcut. The contours do not close, but bend back toward the west in the areas to the north and south of Holcut. It has appeared from modeling experiments that this shape of the contours can be obtained by assuming the presence of a large inhomogeneity of high hydraulic conductivity that roughly covers the area between Holcut, Midway and Iuka, and a smaller one near Burnsville. It is further seen from Figure 6.50 that the space between the contours increases in several areas. Such widening of contours may be caused either by leakage or by a high transmissivity in the areas of widening, or by a combination of both. It is inferred from the geology, however, that the presence of inhomogeneities is the more likely cause for the anomalies in the contours. Openings in the dividing layer therefore are included only where necessary to obtain observed piezometric levels. The boundaries of the inhomogeneities and the hydraulic conductivities were chosen such that the predicted contours roughly resemble the observed ones. A diamond-shaped opening in the separating layer is introduced in the central area near Holcut, in accordance with the supposition that a combination of leakage and high transmissivity is responsible for the high piezometric levels in the Gordo near Holcut.

Figure 6.51 Contour levels predicted for the Eutaw. (Source: Strack, Haitjema, and Melnyk [1980].)

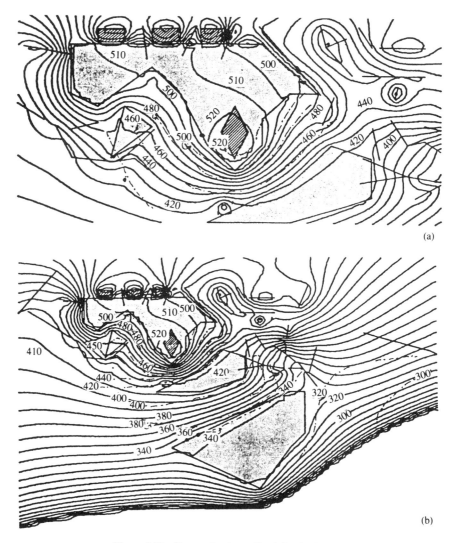

Figure 6.52 Contour levels predicted for the Gordo.

The contours obtained for the above conditions are reproduced in Figure 6.51 and Figures 6.52(a) and 6.52(b). The areas covered in Figures 6.51 and 6.52(a) correspond to that represented in the map of the areal geology of Figure 6.44, whereas the scale of Figure 6.52(b) matches that of Figure 6.50. The contour levels obtained for the Eutaw are reproduced in Figure 6.51 where the straight lines represent creeks. The shaded areas correspond to openings in the dividing layer. It is noted that the layout has been changed somewhat from that used initially. The contour levels predicted for the Gordo are reproduced in Figures 6.52(a) and (b), where the straight lines indicate creeks, the shaded areas the openings in the dividing layer, and the areas surrounded by solid boundaries the inhomogeneities. The dashed-dotted lines correspond to the contours shown on the map of Figure 6.50.

Uniform flows in both aquifers and some line-sinks were added for the final predictions, in an attempt to simulate the effect of the aquifers surrounding the area modeled. These were chosen to some extent on the basis of the geology and the topology and to some extent by trial and error, and consist of long line elements that represent rivers, and of uniform flows that correspond to gradients that are assumed to be present.

The model predictions reproduced in Figure 6.52 were judged to be sufficiently realistic to conclude that the model could be used to predict the draw-downs caused by the waterway. These predictions themselves, however, are of no particular interest for the purposes of this text and will not be discussed.

It is of interest to note that an excavation was made near the small triangular inhomogeneity shown in Figure 6.52. Large quantities of water had to be pumped in order to leave the excavation dry, which provides evidence that the zone of high hydraulic conductivity is indeed present.

The process of computer interactive modeling of the large aquifer system appeared to be highly effective. The graphics make communication possible between experts of entirely different backgrounds. This communication leads toward an insight into the interaction between the various components of a complex system that may otherwise be difficult to achieve.

The process requires very little initial field experimentation. Ideally, the field experiments would be designed during the course of the modeling, thus limiting these to a few critical ones. For example, a limited number of experiments would be sufficient to prove or disprove the occurrence of inhomogeneities as assumed for the final predictions.

44 TRANSIENT EFFECTS, THREE-DIMENSIONAL EFFECTS, AND COMPUTER PROGRAMS

This section contains some comments on the incorporation of transient effects and three-dimensional effects in the analytic element method. We conclude this chapter with some notes about the implementation of the analytic element method in computer models.

Transient Effects

As was shown in Sec. 16, transient and steady-state solutions may be superimposed without violation of the differential equation. Thus, potentials for well functions of the form (16.24) may be added to the analytic elements presented in this chapter, provided that the density distributions of the analytic elements remain constant in time. This implies that transient wells can be included as long as the transient effects on the analytic elements can be neglected. Full generalization of the method to transient flow may be realized by obtaining transient solutions for line-sinks, line-doublets, and area-sinks by the use of the Theis solution, and superimposing these solutions to their steady-state counterparts. Transient effects often are limited in extent, so that the transient solutions might be added or removed dynamically, as the area of influence of the transient phenomena increases or decreases in time. Alternatively, transient flow may be modeled by the use of area-sinks as discussed by Haitjema and Strack [1985].

Three-dimensional Effects

Three-dimensional effects in regional aquifer systems are often of importance only in limited areas. Haitjema [1982, 1985] suggested a technique for including three-dimensional effects in models of regional confined flow. The idea behind this technique is that three-dimensional solutions valid for a confined aquifer of infinite extent can be superimposed to a two-dimensional shallow-flow model. The three-dimensional solution correctly incorporates the vertical components of flow where necessary, but reduces with good approximation to a two-dimensional solution at some distance away from the source of the three-dimensional effect. This distance may be computed in the program for each three-dimensional solution. For example, the potential for three-dimensional flow toward a partially penetrating well in a confined aquifer (see Muskat [1937]) may be included for confined flow to account for partially penetrating wells.

The superposition of potentials for two- and three-dimensional flow leads to an efficient model that is truly three-dimensional only where necessary, but two-dimensional elsewhere. It is beyond the scope of this text to discuss the various three-dimensional solutions that are useful for incorporation in a shallow-flow model; reference is made to Haitjema [1982, 1985] for further reading on the subject.

Implementation in Computer Models

The analytic element method has been implemented in a variety of computer programs, all written in FORTRAN. Most of these programs were written for a specific purpose, e.g., to test three-dimensional solutions, curvilinear elements, or elliptic inhomogeneities. The usefulness of these programs is too limited to warrant description here. The computer program SYLENS, discussed in the preceding section, was developed for the Nashville District of the U.S. Army Corps of Engineers, specifically to simulate the flow in the aquifer system near the Tennessee-Tombigbee Waterway. This program runs on a Perkin-Elmer 3220 minicomputer with 1 Mb of memory and is written in Perkin-Elmer FORTRAN VI. Because this program was designed for a specific purpose and is machine-dependent, its usefulness for other applications is limited. It has been archived and is no longer in use.

The computer program SLAEM is written in ANSI FORTRAN 77, is highly interactive, and is as machine-independent as possible with present technology. It currently includes all analytic elements discussed in this chapter with the exception of the single-root elements (Sec. 40), the curvilinear elements (Sec. 41), and the analytic elements for closed boundaries (Sec. 42); the latter three elements were tested either in other programs or in special versions of SLAEM. The program does include transient wells, with the stipulation that the effect of these wells on the other elements must be negligible. The program was written originally in Perkin-Elmer FORTRAN VI on a Perkin-Elmer 3220, but has been rewritten in ANSI FORTRAN 77 and runs on the Perkin-Elmer, the IBM PC and compatibles with 640 Kb of memory and with a mathematics co-processor, and on the 32000 and 68020 co-processor boards developed by Difinicon Systems. These boards reside in the IBM PC. The program uses about 500 Kb of memory, of which about 400 Kb is taken up by the source code, not including space reserved for data storage in common blocks and argument lists. Because of the large memory space available (up to 16 Megabytes), and the speed (up to 25 Mhz), the versions running on the Difinicon systems' co-processor boards are currently the most powerful.

The program SLAEM supports extensive graphics capabilities; plots of piezometric contours and of three-dimensional path lines displayed both in plan and in section can be generated on all graphics devices supported by the GSS/GKS graphics development toolkit. Data entry is possible through the use of a cross-hair; such data entry is useful, for example, for entering starting points of path lines. Nearly all of the plots of streamlines and piezometric contours in this book were generated on an Apple Laserwriter by the use of SLAEM.

A difficulty in implementing the analytic element method in a computer program is the variety of functions and aquifer features supported. In order to include new analytic elements in the program without the need for major modification, a special modular structure was designed. This structure is based on the concept of combining all routines associated with a particular type of analytic element in a single module. Each module contains all functions associated with the analytic element in question, a subroutine for the entry of data, a subroutine for storing data in binary form on disk, subroutines for data retrieval, a subroutine for plotting layouts, a subroutine for communicating data to the module for path line determination, and subroutines for generating contributions to the system of equations. The modular structure makes it possible to implement a new module in the program by adding or altering a limited number of statements; installation of a completed module typically requires about a day.

The system of equations is solved in SLAEM by the use of a method of transpose elimination suggested by Wassyng [1982]. For solving a system of n equations in n unknowns, this method requires $\frac{1}{4}n^2 + 2n + 4$ real elements to be stored, as compared to $n^2 + n$ real elements in Gaussian elimination; Wassyng's method is about four times as efficient in terms of memory requirements as Gaussian elimination.

The computer program SLWL, discussed in detail in this text, is a scaled-down version of SLAEM, and is contained on the diskette enclosed. It does not have the modular structure of SLAEM to keep it limited both in size and complexity.

Chapter 7

Solving Problems with Free Boundaries by Conformal Mapping

This chapter deals exclusively with two-dimensional flow in the vertical plane $z = x+iy$, where y points vertically upward. We have seen in Chapter 5 how conformal mapping may be applied to solve boundary value problems for Laplace's equation. The method consists of mapping the upper half plane $\Im\zeta \geq 0$ conformally onto both the regions $\mathcal{R}_z$ in the physical plane (z plane) and $\mathcal{R}_\Omega$ in the complex potential plane (Ω plane). As a result, both z and Ω are obtained as functions of the complex reference parameter ζ.

The procedure breaks down if one or more segments of the boundaries of $\mathcal{R}_z$ and $\mathcal{R}_\Omega$ are not known a priori. Examples of such segments are a phreatic surface, whose position in the z plane is unknown, and a seepage face. In the latter case the boundary condition is that the pressure is zero along the boundary segment. This condition yields a relation between Φ and y, but does not define a curve in the Ω plane that is known a priori. We will refer to such problems as problems with free boundaries, thereby indicating that the boundaries in either the z or the Ω plane contain segments that are free to adjust themselves geometrically to the boundary conditions.

Problems with free boundaries form an important class of groundwater flow problems, and various procedures for solving such problems for cases of two-dimensional flow in the vertical plane have been suggested. Zhukovski [1920] was the first to propose a method of solution that is applicable to a limited class of problems with free boundaries. The hodograph method, developed in the field of hydraulics by Helmholz [1868] and Kirchhoff [1869], was adapted to groundwater flow by Hamel [1934] and Davison [1936]. The hodograph method is applicable to a large class of groundwater flow problems, and consists of using the discharge function $W = -d\Omega/dz$ as an auxiliary function. The benefit of the hodograph, which is the complex conjugate of W, is that the boundary in the hodograph plane is a combination of segments of straight lines and circles for an important class of problems with free boundaries.

A second auxiliary function is needed, however, to solve problems by conformal mapping. Classically, functions such as the complex potential and the function introduced by Zhukovski have been used as second auxiliary functions, but these functions are not generally applicable. Given a general boundary value problem with a known boundary of the region $\mathcal{R}_W$ in the W plane, an appropriate second auxiliary function needs to be determined. The reference function $R = R(\zeta)$, introduced by Strack and Asgian [1978], has the property that the region $\mathcal{R}_R$ in its plane is bounded by straight line segments for any problem for which the boundary of $\mathcal{R}_W$ is known. The functions R and W form a pair that we will call the pair of *basic auxiliary functions*. Given any problem with free boundaries with a known boundary in the hodograph plane, a pair of other auxiliary functions may be derived from R and W that will lead to an efficient method of solution.

Many solutions to problems with free boundaries have been determined since Zhukovski solved the first one. It will not be attempted in this text to present a catalogue of such solutions: these are adequately listed or described in such classical handbooks on groundwater flow as Polubarinova-Kochina [1962], Muskat [1937], Aravin and Numerov [1965], and Harr [1962]. Instead, we will present a general approach to solving a class of problems with free boundaries, based on determining the appropriate set of auxiliary functions beforehand, minimizing the necessary mathematical operations. The problems solved in this chapter were selected with the purpose in mind of illustrating a procedure, rather than for their significance in view of applications in practice.

This chapter is organized as follows. In Sec. 45 the boundary conditions are listed that define the class of problems considered. The hodograph is discussed in Sec. 46, along with the representation in the hodograph plane of the boundary conditions listed in Sec. 45. The reference function is covered in a similar way in Sec. 47, along with a procedure for selecting appropriate auxiliary functions. The remaining sections of this chapter are devoted to solving boundary value problems, using various pairs of auxiliary functions.

45 BOUNDARY TYPES

We will express seven types of boundary conditions in terms of the real and imaginary parts of the pair of functions

$$z = z(\zeta) \qquad \Omega = \Omega(\zeta) \tag{45.1}$$

of the complex variable $\zeta = \xi + i\eta$, which is defined in the upper half plane $\Im \zeta \geq 0$. We will refer to this variable as the reference parameter, and indicate derivatives with respect to ζ by primes:

$$z' = \frac{dz}{d\zeta} \qquad \Omega' = \frac{d\Omega}{d\zeta} \tag{45.2}$$

Sec. 45 Boundary Types

The first type of boundary segment is a straight equipotential. Because this boundary segment is straight, $\arg z'$ is equal to a constant (α_1), obtained from its inclination with respect to the x axis. The potential is constant ($\Phi = \Phi_0$) along the segment, which may be characterized as follows:

$$\boxed{\Phi(\xi) = \Phi_0 \qquad \arg z'(\xi) = \alpha_1 \qquad \text{(type 1)}} \tag{45.3}$$

The second type of boundary segment is a straight streamline. If the value of $\arg z'(\xi)$ along the segment is α_2 and the constant value of the stream function along the streamline is Ψ_0, then

$$\boxed{\Psi(\xi) = \Psi_0 \qquad \arg z'(\xi) = \alpha_2 \qquad \text{(type 2)}} \tag{45.4}$$

The third type of boundary segment is a straight seepage face, with the boundary condition that the pressure is zero, or $\phi = Y$, where Y is the elevation head with respect to the reference level of ϕ (compare [17.28]). The reference level for the head is parallel to the x axis; if the y coordinate of the reference level is y_r then

$$y = Y + y_r \tag{45.5}$$

If the value of $\arg z'(\xi)$ along the seepage face is α_3 then this boundary segment may be characterized in terms of the potential (compare [17.4]) as

$$\boxed{\Phi(\xi) = kBy(\xi) - kBy_r \qquad \arg z'(\xi) = \alpha_3 \qquad \text{(type 3)}} \tag{45.6}$$

The fourth type of boundary segment is a straight outflow face of a coastal aquifer, as illustrated in Figure 7.1. The condition along a straight outflow face is that the pressure in the salt water just outside the outflow face equals that in the fresh groundwater just inside the outflow face,

$$p_f = \rho_f g(\phi - Y) = p_s = \rho_s g(\phi_s - Y) \tag{45.7}$$

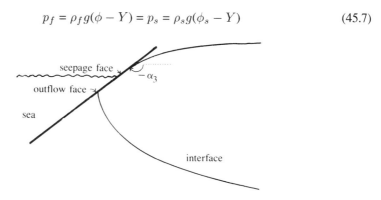

Figure 7.1 A straight outflow face.

where ϕ_s is the saltwater head. Solving (45.7) for ϕ we obtain

$$\phi = \frac{\rho_s}{\rho_f}\phi_s - Y\frac{\rho_s - \rho_f}{\rho_f} \qquad (45.8)$$

Multiplication by kB yields, using (45.5),

$$\Phi = -k^*By + (k^*y_r + k\frac{\rho_s}{\rho_f}\phi_s)B \qquad (45.9)$$

where the factor k^* has the dimension of a hydraulic conductivity and is defined by

$$k^* = k\frac{\rho_s - \rho_f}{\rho_f} \qquad (45.10)$$

Denoting the value of arg $z'(\xi)$ along the outflow face as α_4, we characterize this boundary segment as

$$\boxed{\Phi(\xi) = -k^*By(\xi) + (k^*y_r + k\frac{\rho_s}{\rho_f}\phi_s)B \qquad \arg z'(\xi) = \alpha_4 \qquad \text{(type 4)}} \qquad (45.11)$$

The fifth type of boundary segment is a phreatic surface without infiltration or evaporation. The boundary conditions are that the pressure is zero so that $\Phi = kB(y-y_r)$ (compare [45.6]) and that the phreatic surface is a streamline so that $\Psi = \Psi_0 =$ constant. Hence, the conditions along a phreatic surface are

$$\boxed{\Phi(\xi) = kBy(\xi) - kBy_r \qquad \Psi(\xi) = \Psi_0 \qquad \text{(type 5)}} \qquad (45.12)$$

The sixth type of boundary segment is a phreatic surface with an infiltration rate N. In this case the stream function changes along the phreatic surface due to the infiltration (see Figure 7.2). Streamlines start at the phreatic surface, and we consider a flow channel bounded by the streamlines starting at points $\mathcal{A}$ and $\mathcal{C}$ that are horizontally a distance Δx apart. The flow ΔQ through the channel equals the infiltration entering the aquifer over a width Δx,

$$\Delta Q = NB\Delta x \qquad (45.13)$$

It may be seen by comparison of Figure 7.2 with Figure 3.3 and from (19.12) that

$$\Psi_\mathcal{C} - \Psi_\mathcal{B} = -\Delta Q = -NB\Delta x \qquad (45.14)$$

Points $\mathcal{A}$ and $\mathcal{B}$ lie on the same streamline so that $\Psi_\mathcal{B} = \Psi_\mathcal{A}$ and

$$\Psi_\mathcal{C} - \Psi_\mathcal{A} = \Delta\Psi = -NB\Delta x \qquad (45.15)$$

Sec. 45   Boundary Types   517

Points $\mathcal{A}$ and $\mathcal{C}$ correspond to points on the ξ axis in the ζ plane as shown in Figure 7.2(b), and we may modify (45.15) as follows

$$\frac{\Delta\Psi}{\Delta\xi} = -NB\frac{\Delta x}{\Delta\xi} \qquad (45.16)$$

Passing to the limit for $\Delta\xi \to 0$ we obtain

$$\Psi'(\xi) = -NBx'(\xi) \qquad (45.17)$$

The second boundary condition is that the pressure is zero which leads to the first equation in (45.12). The type 6 boundary segment thus may be characterized as

$$\boxed{\Phi(\xi) = kBy(\xi) - kBy_r \qquad \Psi'(\xi) = -NBx'(\xi) \qquad \text{(type 6)}} \qquad (45.18)$$

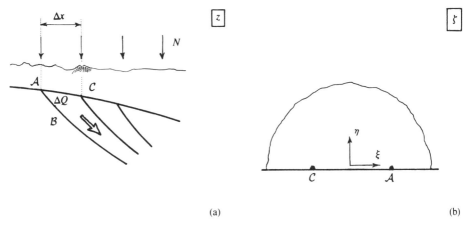

Figure 7.2 A phreatic surface with an infiltration rate N.

The seventh and final type of boundary segment is an interface between flowing fresh and stationary salt groundwater. The condition that the pressures in the fresh and salt water are equal leads to (45.9) and the condition that the interface is a streamline is expressed as $\Psi = \Psi_0$. Hence, the conditions along a type 7 boundary segment are

$$\boxed{\Phi(\xi) = -k^*By(\xi) + (k^*y_r + k\frac{\rho_s}{\rho_f}\phi_s)B \qquad \Psi(\xi) = \Psi_0 \qquad \text{(type 7)}} \qquad (45.19)$$

The above seven types of boundary segment together define the class of problems considered in this chapter: we consider only those problems where any boundary segment is one of the above types. We will see in the next two sections that each of the seven types corresponds to either a straight line or a circular arc in the W plane and to a straight line through the origin in the plane of the reference function R.

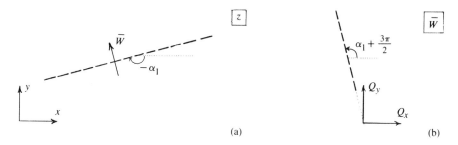

Figure 7.3 The map of a straight equipotential onto the hodograph plane.

46 THE HODOGRAPH

The hodograph is the complex representation, $Q_x + iQ_y$, of the discharge vector. The hodograph plane is a complex plane with Q_x and Q_y as axes. We will determine in this section the maps of the seven types of boundary segment onto the hodograph plane. These maps are obtained by plotting in the hodograph plane the discharge vector occurring at points of the boundary segment.

Before determining the seven maps of the boundary types we first recall the relationships presented in Sec. 24 between the hodograph, the discharge function, and the complex potential.

Basic Equations

The hodograph $Q_x + iQ_y$ is the complex conjugate of the discharge function W,

$$\overline{W} = Q_x + iQ_y \tag{46.1}$$

where (compare [24.2])

$$W = -\frac{d\Omega}{dz} \tag{46.2}$$

The objective of the hodograph method is to establish z and Ω as functions of the reference parameter ζ. Using the chain rule, (46.2) may be written as

$$W = -\frac{\Omega'(\zeta)}{z'(\zeta)} \tag{46.3}$$

Thus, the basic auxiliary function W provides one relationship between $\Omega(\zeta)$ and $z(\zeta)$: once $W(\zeta)$ is determined by conformal mapping, the ratio of $\Omega'(\zeta)$ and $z'(\zeta)$ is known.

A Straight Equipotential

A straight equipotential is sketched in Figure 7.3(a). At any point of the equipotential, $\overline{W}$ is perpendicular to the boundary segment, which therefore corresponds in the hodograph plane to a segment of the straight line through the origin and inclined at an angle $\alpha_1 + 3\pi/2$ to the Q_x axis, as shown in Figure 7.3(b).

A Straight Streamline

A straight streamline is sketched in Figure 7.4(a). The specific discharge vector is tangent to the streamline and therefore corresponds in the hodograph plane to a segment of a straight line through the origin and inclined at an angle $\alpha_2 + \pi$, as shown in Figure 7.4(b).

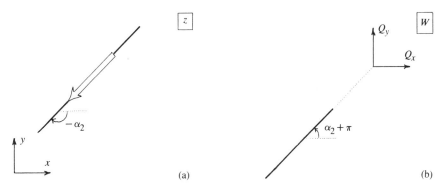

Figure 7.4 The map of a straight streamline onto the hodograph plane.

A Straight Seepage Face

A straight seepage face is shown in Figure 7.5(a). We introduce a local coordinate s pointing along the seepage face as shown in the figure, and observe that

$$dy = -\sin\alpha_3 \, ds \qquad (46.4)$$

It follows from the boundary condition (45.6) for type 3 and (46.4) that

$$Q_s = -\frac{d\Phi}{ds} = -kB\frac{dy}{ds} = kB\sin\alpha_3 \qquad (46.5)$$

The component Q_s thus points downward along the seepage face and is constant. As shown in Figure 7.5(b), the seepage face corresponds to a segment of the straight line through $\overline{W} = -ikB$ and normal to the seepage face.

A Straight Outflow Face

The boundary condition along an outflow face is rather similar to that along a seepage face (compare [45.6] to [45.11]). The component Q_s for this case is

$$Q_s = -\frac{d\Phi}{ds} = k^*B\frac{dy}{ds} = -k^*B\sin\alpha_4 \qquad (46.6)$$

which is a constant. The outflow face corresponds in the hodograph plane to a segment of the straight line through $\overline{W} = ik^*B$ and normal to the outflow face, as shown in Figure 7.6.

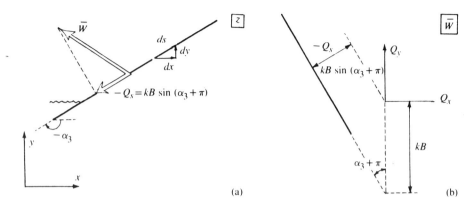

Figure 7.5 The map of a straight seepage face onto the hodograph plane.

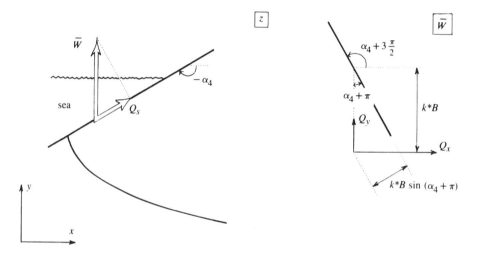

Figure 7.6 The map of a straight outflow face onto the hodograph plane.

A Phreatic Surface Without Infiltration or Evaporation

A phreatic surface without infiltration or evaporation is a streamline. The discharge vector at any point of the phreatic surface therefore is tangent to the surface. We take s to represent a local coordinate tangent to the phreatic surface and denote the angle between the increment ds and the x axis as $-\alpha$, as shown in Figure 7.7(a). It follows that, along the phreatic surface,

$$dy = \sin \alpha \, ds \tag{46.7}$$

We obtain an expression for the component Q_s of the discharge vector tangent to the phreatic surface from (45.12) and (46.7) as

Sec. 46 The Hodograph

$$Q_s = -kB\frac{dy}{ds} = -kB\sin\alpha = kB\sin(-\alpha) \tag{46.8}$$

The discharge vector is inclined at an angle $-\alpha$ with respect to the x axis and its end point is plotted as point Q in Figure 7.7(b). The lengths of $\mathcal{O}Q$ and $\mathcal{O}S$ are $kB\sin(-\alpha)$ and kB so that the angles $\mathcal{O}QS$ and $QS\mathcal{O}$ are $\pi/2$ and $-\alpha$, respectively. Point Q therefore lies on a circle through $\mathcal{O}$ and centered at the midpoint of $\mathcal{O}S$; this circle is the locus of end points of the discharge vectors that correspond to points of a phreatic surface. This may be seen also by resolving Q_s into components Q_x and Q_y by the use of Figure 7.7(a),

$$Q_x = Q_s\cos\alpha \qquad Q_y = -Q_s\sin(-\alpha) = Q_s\sin\alpha \tag{46.9}$$

It follows from (46.8) and (46.9) that

$$Q_x = -kB\sin\alpha\cos\alpha = -\tfrac{1}{2}kB\sin(2\alpha) \tag{46.10}$$

$$Q_y = -kB\sin^2\alpha = -\tfrac{1}{2}kB + \tfrac{1}{2}kB\cos(2\alpha) \tag{46.11}$$

Hence,

$$Q_x^2 + (Q_y + \tfrac{1}{2}kB)^2 = (\tfrac{1}{2}kB)^2 \tag{46.12}$$

which indeed is the equation for the circle shown in Figure 7.7(b).

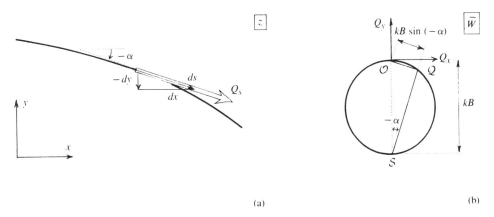

Figure 7.7 The map of a phreatic surface without infiltration or evaporation onto the hodograph plane.

A Phreatic Surface with an Infiltration Rate N

A phreatic surface with an infiltration rate N is shown in Figure 7.8(a). We introduce at a point of the phreatic surface the local coordinates s and n, tangent and normal to the phreatic surface as shown in Figure 7.8(a). The component $-Q_n$ equals the projection of the infiltration rate N onto the n axis, multiplied by the width B normal to the plane of flow,

$$Q_n = -NB\cos\alpha \qquad (46.13)$$

The component Q_s is the same as for the case without infiltration (compare [46.8]):

$$Q_s = kB\sin(-\alpha) \qquad (46.14)$$

The components Q_x and Q_y may be expressed in terms of Q_n and Q_s as

$$Q_x = Q_s\cos\alpha - Q_n\sin\alpha \qquad Q_y = Q_s\sin\alpha + Q_n\cos\alpha \qquad (46.15)$$

Using (46.13) and (46.14), this becomes

$$Q_x = -kB\sin\alpha\cos\alpha + NB\sin\alpha\cos\alpha = -\tfrac{1}{2}(k-N)B\sin(2\alpha) \qquad (46.16)$$

and

$$Q_y = -kB\sin^2\alpha - NB\cos^2\alpha = -\tfrac{1}{2}(k+N)B + \tfrac{1}{2}(k-N)B\cos(2\alpha) \qquad (46.17)$$

Using the relation $\sin^2(2\alpha) + \cos^2(2\alpha) = 1$ we obtain

$$Q_x^2 + \left[Q_y + \tfrac{1}{2}(k+N)B\right]^2 = \left[\tfrac{1}{2}(k-N)B\right]^2 \qquad (46.18)$$

which represents the equation for a circle centered at $Q_x = 0$, $Q_y = -\tfrac{1}{2}(k+N)B$ and of radius $\tfrac{1}{2}(k-N)B$, as shown in Figure 7.8(b).

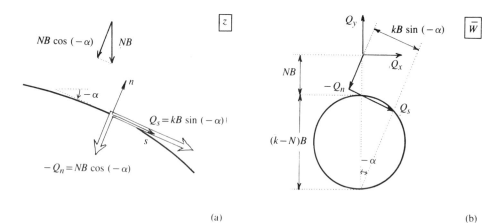

Figure 7.8 The map of a phreatic surface with an infiltration rate N onto the hodograph plane.

An Interface Between Fresh Groundwater and Salt Groundwater at Rest

An interface between fresh groundwater and salt groundwater at rest is a streamline, as shown in Figure 7.9(a). We denote the inclination of the interface as α and use (46.7) and (45.19) to write the following expression for Q_s:

$$Q_s = k^* B \frac{dy}{ds} = k^* B \sin \alpha \qquad (46.19)$$

Using (46.9) to express Q_x and Q_y in terms of Q_s, we obtain

$$\begin{aligned} Q_x &= \tfrac{1}{2} k^* B \sin(2\alpha) \\ Q_y &= \tfrac{1}{2} k^* B - \tfrac{1}{2} k^* B \cos(2\alpha) \end{aligned} \qquad (46.20)$$

It follows that the locus of the end points Q of the discharge vector is a circle through the origin, centered at $Q_x = 0$, $Q_y = \tfrac{1}{2} k^* B$, and of radius $\tfrac{1}{2} k^* B$:

$$Q_x^2 + (Q_y - \tfrac{1}{2} k^* B)^2 = (\tfrac{1}{2} k^* B)^2 \qquad (46.21)$$

This circle is shown in Figure 7.9(b).

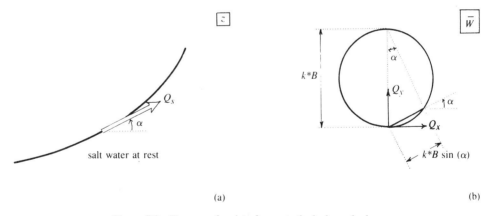

Figure 7.9 The map of an interface onto the hodograph plane.

Example 46.1: Unconfined Flow Toward a Drain

As an example, we will draw the boundary in the hodograph plane for the case of unconfined flow over an impermeable base toward a horizontal drain. We considered this problem in Sec. 24, where the solution was given without derivation (see Figure 3.40). The boundary of the flow domain is sketched in Figure 7.10(a). The boundary segments are 1$\mathcal{A}$-2, a horizontal streamline (type 2), 2-3, a horizontal equipotential (type 1) and 3-1$\mathcal{B}$, a phreatic surface (type 5). These segments correspond in the hodograph plane of Figure 7.10(b) to segments of the Q_x axis, the Q_y axis, and a circle. Point 1 is at the intersection of the Q_x axis and the circle and therefore corresponds to the origin. At point 3 the phreatic surface is vertical so that this point corresponds to $\overline{W} = -ikB$. At point 2 both components of the discharge vector are infinite so that this point is mapped onto $\overline{W} = \infty$. The hodograph is the complex conjugate of an analytic function of z. The domain in the hodograph plane therefore lies on the opposite side with respect to the boundary as the domain in the z plane does; the domain lies to the right with respect to a point tracing the boundary in Figure 7.10(b) in the sequence $1A$-2-3-$1B$. The domain in the W plane is obtained from that in the $\overline{W}$ plane by reflection through the real axis. Note that the orientation is preserved in the mapping from the z plane to the W plane; W is an analytic function of z.

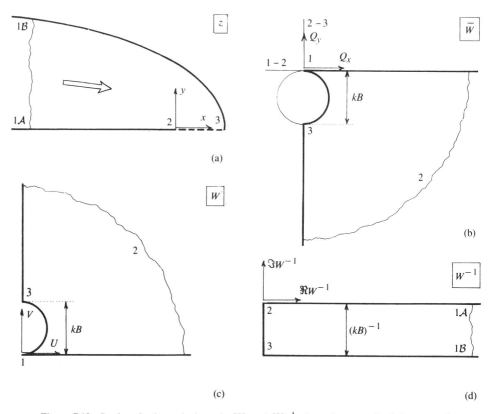

Figure 7.10 Regions in the z, hodograph, W, and W^{-1} planes for unconfined flow toward a drain.

The conformal mapping of domains bounded by a combination of straight line segments and circular arcs is much more complicated than the conformal mapping of domains bounded exclusively by segments of straight lines. We have seen in Sec. 30, however, that circular arcs passing through the origin are transformed into straight lines by inversion. In addition, an inversion maps straight lines through the origin onto straight

Sec. 47   The Reference Function

lines through the origin. Thus, if there exists a point τ_j in the W plane that is common to all straight lines or circular arcs on which the boundary segments lie, then inversion of the function $W - \tau_j$ transforms the boundary in the W plane into a polygon of segments of straight lines in the $(W - \tau_j)^{-1}$ plane. We denote this function as $\underset{j}{W}$, i.e.,

$$\underset{j}{W} = \frac{1}{W - \tau_j} \qquad (46.22)$$

For the W plane of Figure 7.10(c) this common point is the origin, $\tau_j = 0$, and the function $\underset{j}{W}$ is

$$\underset{j}{W} = \frac{1}{W} \qquad (46.23)$$

The domain in the W^{-1} plane is sketched in Figure 7.10(d). The straight lines 1-2 and 2-3 remain straight lines, point 1 ($W = 0$) corresponds to $W^{-1} = \infty$, and point 2 ($W = \infty$) to $W^{-1} = 0$. Along the circular arc corresponding to the phreatic surface we may write (compare [46.3])

$$W^{-1} = \frac{1}{W(\zeta)} = \frac{1}{-\Omega'/z'} = -\frac{z'}{\Omega'} = -\frac{x' + iy'}{\Phi' + i\Psi'} \qquad (46.24)$$

It follows from the boundary condition (45.12) along a phreatic surface that

$$W^{-1} = -\frac{x' + iy'}{kBy'} = -\frac{x'}{kBy'} - i\frac{1}{kB} \qquad (46.25)$$

so that the free surface is mapped onto a line parallel to the $\Re W^{-1}$ axis and a distance $1/(kB)$ below it.

The function $W^{-1} = W^{-1}(\zeta)$ may be obtained by mapping $\Im \zeta \geq 0$ onto the domain in the W^{-1} plane of Figure 7.10(d). This function by itself is not sufficient to obtain a solution, however; we must determine a second auxiliary function in terms of ζ. We choose this function with the aid of the reference function R, discussed in the next section. The point $W = \tau_j$ plays an important role in this process; we will refer to such a point as a point of inversion.

47 THE REFERENCE FUNCTION

The reference function R is defined as a function of the reference parameter ζ (whence the name) as follows:

$$\boxed{R = [z'(\zeta)]^2 W'(\zeta)} \qquad (47.1)$$

It has the property that its argument is a constant for any of the seven types of boundary segments listed in Sec. 45. Formally, therefore, any problem with a boundary consisting of any combination of these seven types may be solved by mapping the upper half plane $\Im \zeta \geq 0$ onto the regions in the W and R planes. In practice, however, the procedure for solving general problems in this way is prohibitively complex, unless an approximate procedure of conformal mapping is used (Detournay [1985]). There exist many problems that may be solved with comparative ease using a pair of auxiliary functions determined by the use of points of inversion $W = \tau_j$ and the reference function, as explained below. First, however, we will prove that $\arg R(\zeta)$ is indeed constant for any of the seven boundary types.

Straight Boundary Segments

Along the four straight boundary segments the argument of $z'(\xi)$ is constant because the segment is straight. All four straight segments correspond in the hodograph plane to straight lines. Therefore, $\arg W'(\xi)$ is constant along these segments, and

$$\arg R(\xi) = 2 \arg z'(\xi) + \arg W'(\xi) = \text{constant} \qquad \text{(types 1, 2, 3, 4)} \qquad (47.2)$$

A Phreatic Surface

We will demonstrate that $\arg R(\xi)$ is constant along a phreatic surface with infiltration rate N. A section $\mathcal{AB}$ of a phreatic surface along with its maps in the W and ζ planes is shown in Figure 7.11. The vector $\mathcal{OQ}$ inclined at an angle $-\alpha$ ($\alpha < 0$) to the real axis in the W plane represents the discharge vector at point $\mathcal{Q}$ (note that the W plane is the complex conjugate of the hodograph so that $\mathcal{OQ}$ is not parallel to the tangent to the phreatic surface at point $\mathcal{Q}$, but to the reflection of this tangent through the x axis). Consider an increment dz that is tangent to the phreatic surface at point $\mathcal{Q}$. The phreatic surface is mapped onto a circle in the W plane and therefore the corresponding increment dW is tangent to the circle. Since both the angles $-\alpha$ and β shown in Figure 7.11(b) span the arc $\mathcal{OQ}$, β equals $-\alpha$, and

$$\arg dW = -2\alpha \qquad (47.3)$$

Furthermore, dz is tangent to the phreatic surface at $\mathcal{Q}$ so that

$$\arg dz = \alpha \qquad (47.4)$$

The orientation is preserved in conformal mapping, so that (see Figure 7.11[c])

$$\arg d\zeta = \arg d\xi = \pi \qquad (47.5)$$

It follows from the latter three equations that

$$\arg z'(\xi) = \arg dz - \arg d\zeta = \alpha - \pi \qquad (47.6)$$

Sec. 47 The Reference Function

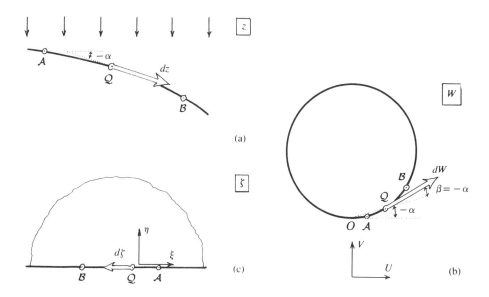

Figure 7.11 Determination of arg $R(\xi)$ along a phreatic surface with infiltration rate N.

and

$$\arg W'(\xi) = \arg dW - \arg d\zeta = -2\alpha - \pi \qquad (47.7)$$

Hence,

$$\arg R(\xi) = 2\arg z'(\xi) + \arg W'(\xi) = 2\alpha - 2\pi - 2\alpha - \pi = -3\pi \qquad (47.8)$$

which is constant, as asserted. The above derivation remains valid if $N = 0$, so that

$$\arg R(\xi) = \text{constant} \qquad \text{(types 5, 6)} \qquad (47.9)$$

An Interface

The proof that $\arg R$ is constant along an interface is similar to the one given above. A section of an interface along with its maps in the W and ζ planes is sketched in Figure 7.12. If the inclination of the interface at a point Q is α, then $\arg dz$ and $\arg dW$ are equal to α and -2α, respectively. In this case $\arg d\zeta$ is positive, given the increment dz as shown in the figure, and

$$\arg R = 2\arg z'(\xi) + \arg W'(\xi) = 2\alpha - 2\alpha = 0 \qquad (47.10)$$

so that $\arg R(\xi)$ is indeed constant along an interface:

$$\arg R(\xi) = \text{constant} \qquad \text{(type 7)} \qquad (47.11)$$

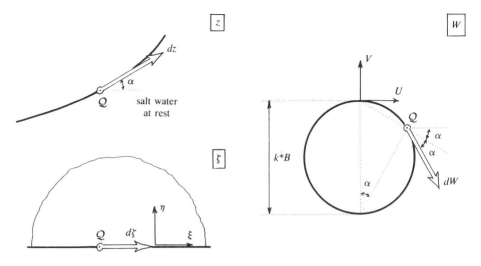

Figure 7.12 Determination of arg $R(\xi)$ along an interface.

Other Auxiliary Functions

If one or more points of inversion exist in the W plane, auxiliary functions other than W and R may be used to solve the problem. We will consider separately the case that there exists one point of inversion and the case that there exist two points of inversion.

One Point of Inversion

Let $W = \tau_1$ denote the only point in the W plane that is common to all circles and straight lines on which the boundary segments lie. The function

$$W_1 = \frac{1}{W - \tau_1} \tag{47.12}$$

then transforms all of the boundary segments into segments of straight lines. The argument of the derivative W_1' therefore will be piecewise constant along the ξ axis:

$$\arg W_1'(\xi) = \text{constant} \tag{47.13}$$

It follows from (47.12) that

$$W' = -\frac{1}{W_1^2} W_1' \tag{47.14}$$

Sec. 47 The Reference Function

Substitution of this expression for W' in the definition (47.1) of the reference function gives

$$R = -\left[\frac{z'}{W_1}\right]^2 W_1' \qquad (47.15)$$

Both the arguments of W_1' and R are piecewise constant along the real axis $\Im\zeta = 0$ and it follows from (47.15) that there exists an auxiliary function T_1 with a derivative T_1' defined as

$$T_1' = \frac{z'}{-W_1} = z'(-W + \tau_1) \qquad (47.16)$$

whose argument is piecewise constant along $\Im\zeta = 0$. Using (46.3) the following expression is obtained for T_1':

$$T_1' = z'\left[\frac{\Omega'}{z'} + \tau_1\right] = \Omega' + \tau_1 z' \qquad (47.17)$$

Since the argument of $T_1' = dT_1/d\zeta$ is piecewise constant along the boundary, the boundary in the plane of the function T_1 is a polygon of segments of straight lines. It follows from (47.17) that

$$\boxed{T_1 = \Omega + \tau_1 z} \qquad (47.18)$$

where the constant of integration is taken as zero.

Problems where one point of inversion exists may thus be solved as follows. First, the value of τ_1 is determined and the polygons drawn that form the boundaries in the W_1 and $\Omega + \tau_1 z$ planes. Next, the upper half plane $\Im\zeta \geq 0$ is mapped onto the regions in these planes, which gives the functions

$$W_1 = W_1(\zeta) \qquad T_1 = T_1(\zeta) \qquad (47.19)$$

The function $z(\zeta)$ is obtained by integration of (47.16),

$$\boxed{z = -\int W_1(\zeta) T_1'(\zeta) d\zeta} \qquad (47.20)$$

and $\Omega = \Omega(\zeta)$ is found from (47.18),

$$\boxed{\Omega(\zeta) = T_1(\zeta) - \tau_1 z(\zeta)} \qquad (47.21)$$

For the problem illustrated in Figure 7.10, one point of inversion exists, the origin $W = 0$. The auxiliary functions to be used for this problem therefore are the functions

$$W_1 = W^{-1} \qquad (47.22)$$

and

$$T_1 = \Omega \qquad (47.23)$$

It is left as an exercise for the reader to draw the boundary in the Ω plane and solve the problem by conformal mapping. In this special case the boundaries in the W^{-1} and Ω planes are identical, except for position and scale, so that the reference parameter need not be used.

Van Deemter [1949] solved problems with one point of inversion, using a general function $T_1 = \Omega + iCz$; he stated that C must be chosen such that the boundary in the T_1 plane is composed of segments of straight lines. Van Deemter did not have access to the reference function; the constant C had to be found by trial and error.

Problem 47.1

Draw the boundary in the Ω plane for the flow problem illustrated in Figure 7.10(a). Determine Ω as a function of z by conformal mapping and compare your result with the solution given in Sec. 24.

Problem 47.2

Solve the problem of confined interface flow in a coastal aquifer underneath a horizontal impermeable layer toward a horizontal sea bottom at the same level as the impermeable layer. The discharge flowing toward the coast is Q m³/s, and sea level is ϕ_s m above the sea bottom. The problem is illustrated in Figure 7.13.

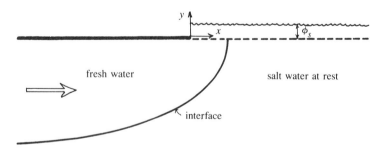

Figure 7.13 Confined interface flow toward the sea (Problem 47.2).

Problem 47.3

Solve the problem of unconfined interface flow in a coastal aquifer, approximating the sea bottom by a horizontal slot with $\phi = (\rho_s/\rho_f)\phi_s$, as shown in Figure 7.14.

Sec. 47   The Reference Function   531

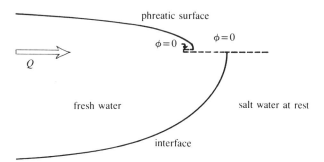

Figure 7.14 Unconfined interface flow toward the sea (Problem 47.3).

Problem 47.4

Solve problems 47.2 and 47.3 using the Dupuit-Forchheimer solution and compare your results for the case that $\rho_s = 1.02\rho_f$.

Two Points of Inversion

In some cases the circles and straight lines on which the boundary segments lie in the W plane have two points in common: $W = \tau_1$ and $W = \tau_2$. For such cases there exist two functions $\underset{1}{W}$ and $\underset{2}{W}$,

$$\underset{1}{W} = \frac{1}{W - \tau_1} \qquad \underset{2}{W} = \frac{1}{W - \tau_2} \qquad (47.24)$$

with the property that their boundaries are composed of segments of straight lines. There exist, then, two auxiliary functions $\underset{1}{T}$ and $\underset{2}{T}$, defined as (compare [47.18])

$$\underset{1}{T} = \Omega + \tau_1 z \qquad \underset{2}{T} = \Omega + \tau_2 z \qquad (47.25)$$

The functions

$$\underset{1}{T} = \underset{1}{T}(\zeta) \qquad \underset{2}{T} = \underset{2}{T}(\zeta) \qquad (47.26)$$

may be determined by the use of the Schwarz-Christoffel transformation. The function $z = z(\zeta)$ is obtained by first subtracting $\underset{2}{T}$ from $\underset{1}{T}$ to eliminate Ω and then dividing the result by $\tau_1 - \tau_2$:

$$z = \frac{\underset{1}{T}(\zeta) - \underset{2}{T}(\zeta)}{\tau_1 - \tau_2} \qquad (47.27)$$

With $z = z(\zeta)$ known, the function $\Omega = \Omega(\zeta)$ is obtained from (47.25):

$$\boxed{\Omega = \underset{1}{T}(\zeta) - \tau_1 z(\zeta)} \qquad (47.28)$$

When applicable, this procedure has the advantage over using the function $\underset{1}{W}(\zeta)$ that the integration of (47.20) is not necessary. In some cases the boundary in the z plane is known. If a point of inversion exists, such problems may be solved by the use of $z = z(\zeta)$ and a single auxiliary function $\underset{1}{T}(\zeta)$.

Special cases of the function $\underset{1}{T}$ and $\underset{2}{T}$ were introduced by Davison and Rosenhead [1940], namely for $\tau_1 = -ik$ and $\tau_2 = -iN$. List [1964] applied these functions to obtain approximate solutions for unconfined flow toward horizontal drains.

No Points of Inversion

If there are no points of inversion, the boundary in the W plane cannot be transformed into a polygon of straight line segments. For such cases, the function $W = W(\zeta)$ may be determined by integration of the differential equation of Schwarz, discussed in some of the handbooks on conformal mapping (e.g., Von Koppenfels and Stallmann [1959] and Nehari [1975]). The reference function may be determined in a straightforward manner (see Strack and Asgian [1978]); the function $z = z(\zeta)$ may be obtained from the relation

$$z' = \sqrt{\frac{R(\zeta)}{W(\zeta)}} \qquad (47.29)$$

and the complex potential by integrating (46.3). The complexity of the above procedure limits its practical applicability, and the author is aware of only three solutions to problems without points of inversion. The first solution is that to the classical Dupuit problem of flow through a dam with vertical faces, and was determined independently by Hamel [1934] and Polubarinova-Kochina [1940]. The second solution applies to overland flow over an inclined bed and was determined by Wooding [1966]. The solutions to the latter two problems were obtained without using the reference function. The third solution is that of flow near the outflow face of a dam and was presented by Strack and Asgian [1978], who introduced the reference function as the second basic auxiliary function. Because of its limited practical significance, we will not present the procedure for solving problems without points of inversion in this text, and refer the reader to the afore mentioned references for details of the approach.

48 OUTFLOW FROM A LAYERED SLOPE

The first example of a problem with a free boundary is associated with a problem of piping in an embankment consisting of horizontal sand layers separated by thin impermeable

Sec. 48   Outflow from a Layered Slope   533

clay laminae (see Figure 7.15). If the flow rates in the sand layers exceed a certain critical value, piping may occur just below the dividing layers. The high flow rates responsible for the piping may be the result of interconnections with a deep artesian aquifer with high pressures, as is illustrated in the figure. This problem was solved by Strack and Curtis [1980]; the following analysis is taken largely from the latter publication.

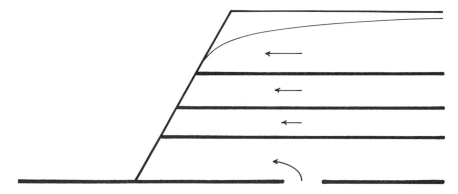

Figure 7.15   Outflow from a layered slope.

Boundary Conditions

We consider a semi-infinite confined aquifer of thickness H that intersects the slope with a seepage face of inclination α (see Figure 7.16). The discharge flowing per width B is Q m³/s. We choose the x axis to coincide with the aquifer base and the origin at the intersection of this base with the seepage face.

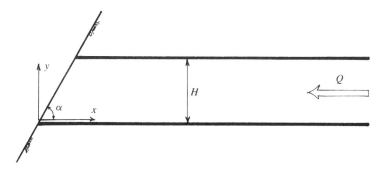

Figure 7.16   The model aquifer. (Source: Strack and Curtis [1980].)

Two different cases of flow may occur: an unconfined case and a confined case. For the unconfined case a phreatic surface occurs near the seepage face and the velocities are finite throughout the flow domain. For the confined case the flow is everywhere confined and infinite velocities occur at the intersection of the seepage face and the upper confining boundary. We will solve the problem only for the confined case but draw the boundary in the hodograph plane for the unconfined case as an exercise.

The Hodograph for the Unconfined Case

The domains in the z and hodograph planes for the unconfined case are shown in Figures 7.17(a) and (b). The streamlines $1A$-2 and 3-$1B$ correspond to two sections of the negative Q_x axis with point 1, $z = \infty$, located at $Q_x = -Q/H$, $Q_y = 0$. The phreatic surface is mapped onto the circular arc 2-2* and the seepage face onto the straight line 2*-3 (compare Figure 7.5). The domain in the hodograph plane lies inside the curvilinear triangle, and we observe from this that only finite flow rates occur in the flow domain. The highest flow rate is at point 3. It is of interest to note that this flow rate is independent of the value of Q; the components of the discharge vector at point 3 are

$$\overset{3}{Q_x} = -kB \tan \alpha \qquad \overset{3}{Q_y} = 0 \qquad (48.1)$$

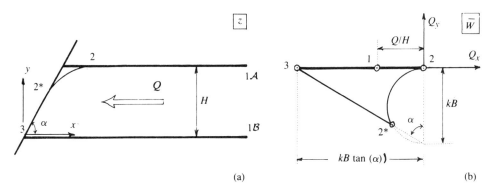

Figure 7.17 The domains in the z and hodograph planes for the unconfined case. (Source: Strack and Curtis [1980].)

The unconfined case may occur only if point 1 lies between points 2 and 3 in the hodograph plane, i.e., if

$$0 < Q < kBH \tan \alpha \qquad (48.2)$$

A special case occurs when

$$Q = kBH \tan \alpha \qquad (48.3)$$

The entire aquifer for this case is saturated and the flow is uniform, as will be shown next. Consider the complex potential

$$\Omega = kBz \tan \alpha \qquad (48.4)$$

which has the real and imaginary parts

$$\Phi = kBx\tan\alpha \qquad \Psi = kBy\tan\alpha \qquad (48.5)$$

Expressions for Q_x and Q_y are obtained upon differentiation:

$$Q_x = -kB\tan\alpha \qquad Q_y = 0 \qquad (48.6)$$

which are constants; $Q_x = -Q/H$ as follows from (48.3). The boundary conditions that $y = 0$ and $y = H$ are streamlines are fulfilled, as follows from (48.5). y equals $x\tan\alpha$ along the seepage face and it follows from the first equation in (48.5) that the boundary condition that Φ be kBy (compare [45.6]) is met along the seepage face, provided that the reference level for the head coincides with the x axis. The complex potential (48.4) thus represents the solution for the case that $Q = kBH\tan\alpha$. It may be noted that for this case $W = -kB\tan\alpha$, so that the entire flow domain corresponds to a single point in the hodograph plane, point 3 in Figure 7.17(b). The flow net for this special case is shown in Figure 7.18.

It appears that confined conditions occur if

$$Q \geq kBH\tan\alpha \qquad (48.7)$$

We will solve the flow problem for the case that the $>$ sign in (48.7) applies.

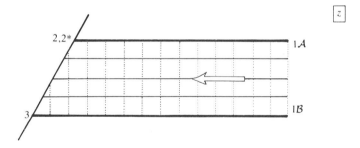

Figure 7.18 Flow net for the case that $Q = kBH\tan\alpha$. (Source: Strack and Curtis [1980].)

The Confined Case

The confined case involves a free boundary in the Ω plane: the seepage face. We will solve the problem in steps as follows: (1) the boundary in the hodograph plane is drawn, (2) the auxiliary function T_1 is chosen, to be used together with $z(\zeta)$, (3) the boundary in the T_1 plane is drawn, and (4) the functions $T_1 = T_1(\zeta)$ and $z = z(\zeta)$ are determined by mapping the upper half plane $\Im\zeta \geq 0$ onto the regions in these planes.

The Hodograph

The domains in the z and hodograph planes are shown in Figures 7.19(a) and (b). The streamlines 3-1B and 1A-2 correspond to segments of the negative Q_x axis, with point 1 ($z = \infty$) mapped onto $(-Q/H, 0)$. The seepage face is mapped onto the straight line 2-3 and point 3 lies at the intersection of the two straight lines that form the boundary in the hodograph plane. It is of interest to note that point 3 has the same position as for the unconfined case. Point 2 also lies on both straight lines in the hodograph plane, and therefore must be at infinity: the outflow velocity is infinite at point 2.

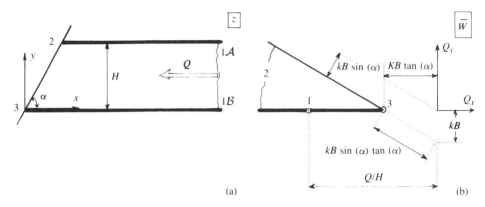

Figure 7.19 The domains in the z and hodograph planes for the confined case. (Source: Strack and Curtis [1980].)

Point 3 is a point of inversion. Since W at point 3, W_3, equals

$$W_3 = -kB \tan \alpha \tag{48.8}$$

it follows that

$$\tau_1 = -kB \tan \alpha \tag{48.9}$$

The Auxiliary Function T_1

An expression for the auxiliary function T_1 is obtained from (47.18) and (48.9) as

$$T_1 = \Omega + \tau_1 z = \Omega - kBz \tan \alpha \tag{48.10}$$

Note that the use of the point of inversion for determining an auxiliary function does not require that the boundary in the W plane contain circular arcs; this procedure is based solely on the requirement that the boundary in the W_1 plane consist of segments of straight lines.

We examine the values of T_1 along the various boundary segments. Separation of T_1 into real and imaginary parts yields

$$T_1 = T_1 + iT_2 = \Phi - kBx \tan \alpha + i(\Psi - kBy \tan \alpha) \tag{48.11}$$

Sec. 48 Outflow from a Layered Slope

Along the seepage face condition (45.6) applies. Using that $y = x\tan\alpha$ and $y_r = 0$, it follows that

$$y = x\tan\alpha \qquad 0 \leq y \leq H \qquad T_{\underset{1}{1}} = 0 \qquad (48.12)$$

We choose Ψ to be zero along the aquifer base, $1\mathcal{B}$-3, and obtain with $y = 0$

$$y = 0 \qquad 0 < x < \infty \qquad \Psi = 0 \qquad T_{\underset{1}{2}} = 0 \qquad (48.13)$$

Along the upper boundary $\Psi = Q$ and $y = H$, so that

$$y = H \qquad H\cot\alpha < x < \infty \qquad \Psi = Q \qquad T_{\underset{1}{2}} = Q - kBH\tan\alpha \qquad (48.14)$$

Near infinity, Φ approaches the potential Qx/H for uniform flow. Hence, the real part $T_{\underset{1}{1}}$ behaves as

$$x \to \infty \qquad T_{\underset{1}{1}} \to (Q - kBH\tan\alpha)x/H \qquad (48.15)$$

In view of (48.7) the coefficient of x is positive, except for the special case shown in Figure 7.18 so that $T_{\underset{1}{1}}$ approaches infinity at point 1. The domain in the $T_{\underset{1}{}}$ plane can now be drawn, and is shown in Figure 7.20(b); it is the interior of the semi-finite strip $1\mathcal{A}$-2-3-$1\mathcal{B}$.

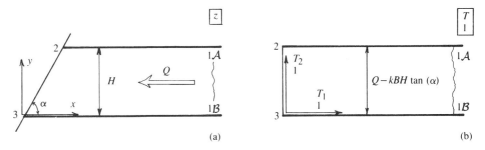

Figure 7.20 The domains in the z and $T_{\underset{1}{}}$ planes.

Mapping $\Im\zeta \geq 0$ onto the Domains in the z and $T_{\underset{1}{}}$ Planes

We choose the upper half plane $\Im\zeta \geq 0$ as shown in Figure 7.21, with point 1 at $\zeta = \infty$, point 2 at $\zeta_2 = -1$, and point 3 at $\zeta_3 = 1$. We first apply the Schwarz-Christoffel transformation (32.9) to map the upper half plane $\Im\zeta \geq 0$ onto the domain in the z plane. Noticing that φ equals the argument of $z'(\xi)$ along segment 3-1, and that πk_2 and πk_3 are the increments of $\arg z'(\xi)$ as a point $\zeta = \xi$ passes points 2 and 3 so that

$$\varphi = 0 \qquad \pi k_2 = \alpha \qquad \pi k_3 = \pi - \alpha \qquad (48.16)$$

we obtain, omitting point 1 from the transformation,

$$z = \frac{H}{\pi} \int_1^\zeta \left[\frac{\zeta^* - 1}{\zeta^* + 1}\right]^{\alpha/\pi} \frac{d\zeta^*}{\zeta^* - 1} \tag{48.17}$$

where the lower bound of the integral is chosen at point 3 so that $z(\zeta_3) = 0$. We must choose the arguments of $\zeta - 1$ and $\zeta + 1$ such that they are continuous inside $\Im \zeta \geq 0$:

$$0 \leq \arg(\zeta - 1) \leq \pi \qquad 0 \leq \arg(\zeta + 1) \leq \pi \tag{48.18}$$

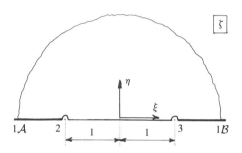

Figure 7.21 The domain in the ζ plane.

A closed-form expression for the integral in (48.17) does not exist; it may be evaluated either numerically or by expanding the integrand in series about the singular points and about some regular points in such a way that the entire domain of interest is covered by the combination of series. The series may then be integrated term by term.

The mapping of $\Im \zeta \geq 0$ onto the T_1 plane is similar to that performed in Sec. 32; see Figures 5.14(b) and (c) and (32.58). The transformation is

$$\zeta = \cosh \frac{\pi T_1}{Q - kBH \tan \alpha} \tag{48.19}$$

with the inverse

$$T_1 = \frac{Q - kBH \tan \alpha}{\pi} \operatorname{arccosh} \zeta = \frac{Q - kBH \tan \alpha}{\pi} \ln\left[\zeta + \sqrt{\zeta^2 - 1}\right] \tag{48.20}$$

Using (48.10) we obtain the following expression for the complex potential:

$$\Omega = \frac{Q - kBH \tan \alpha}{\pi} \ln\left[\zeta + \sqrt{\zeta^2 - 1}\right] + kBz \tan \alpha \tag{48.21}$$

Two flow nets are shown in Figure 7.22 as an illustration, both corresponding to $\alpha = 30°$; Q equals $4kBH \tan \alpha$ for the case of Figure 7.22(a) and Q equals $15kBH \tan \alpha$ for that of Figure 7.22(b). Points on the streamlines were determined in $\Im \zeta \geq 0$ by the use of the procedure discussed in Sec. 26; points on the equipotentials were determined

Sec. 48 Outflow from a Layered Slope

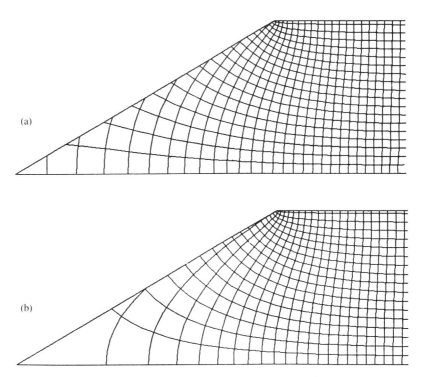

Figure 7.22 Flow nets for $\alpha = 30^0$, $Q = 4kBH \tan \alpha$ (a) and $\alpha = 30^0$, $Q = 15kBH \tan \alpha$ (b). (Source: Strack and Curtis [1980].)

in a similar fashion. The corresponding points in the z plane were computed by the use of (48.17), which was evaluated by integrating the expansions of the integrand about the singular points term by term.

We will next derive an expression for the normal component of the specific discharge vector along the seepage face. An expression for the discharge function $W = -\Omega'/z'$ is obtained from (48.17) and (48.21) as

$$W = -\left[\frac{Q}{H} - kB \tan \alpha\right] \left[\frac{\zeta - 1}{\zeta + 1}\right]^{\frac{1}{2} - \frac{\alpha}{\pi}} - kB \tan \alpha \qquad (48.22)$$

The specific discharge function is obtained by dividing this expression by B,

$$w = q_x - iq_y = \frac{W}{B} \qquad (48.23)$$

In order to compute the normal component q_n of the specific discharge vector along the seepage face we make the following substitution. We observe from Figure 7.21 that the section $-1 \leq \xi \leq 1$ of the real axis corresponds to the seepage face; we set

$$\eta = 0 \qquad -1 \leq \xi \leq 1 \qquad \zeta = \cos(2\theta) \qquad 0 \leq \theta \leq \pi/2 \qquad (48.24)$$

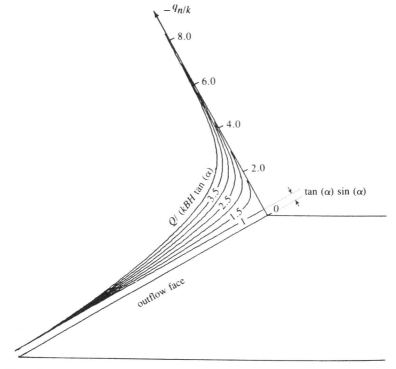

Figure 7.23 Plot of $-q_n/k$ along the outflow face for various values of $Q/(kBH\tan\alpha)$. (Source: Strack and Curtis [1980].)

With this substitution, we obtain the following for $|z|$ and $|W + kB\tan\alpha|$ (see [48.17] and [48.22]):

$$|z| = \frac{2H}{\pi}\int_0^\theta (\tan\lambda)^{2\alpha/\pi - 1}d\lambda \qquad (48.25)$$

and

$$|W + kB\tan\alpha| = \left[\frac{Q}{H} - kB\tan\alpha\right](\tan\theta)^{1-2\alpha/\pi} \qquad (48.26)$$

we observe from Figure 7.19(b) that the component Q_n of the discharge vector equals

$$Q_n = |W + kB\tan\alpha| + kB\sin\alpha\tan\alpha \qquad (48.27)$$

so that

$$q_n = k\sin\alpha\tan\alpha + \left[\frac{Q}{BH} - k\tan\alpha\right](\tan\theta)^{1-2\alpha/\pi} \qquad (48.28)$$

The plot of q_n/k along the seepage face shown in Figure 7.23 is obtained by letting θ vary between 0 and $\pi/2$ in (48.25) and (48.28).

49 FLOW TOWARD A VERTICAL EMBANKMENT

As a second application of the hodograph method, we consider the case of unconfined flow near a vertical embankment. The problem is illustrated in Figure 7.24; the aquifer is infinitely thick and the upper boundary consists of the horizontal seepage face $1\mathcal{A}$-2, the vertical seepage face 2-3, and the phreatic surface 3-$1\mathcal{B}$. It is assumed that the groundwater leaving the seepage faces runs off, so that the pressure remains zero along $1\mathcal{A}$-2-3. As opposed to the previous problem, part of the boundary of the flow domain is not known a priori, so that two auxiliary functions need be used. We begin our analysis by drawing the domain in the hodograph plane and by determining the point of inversion.

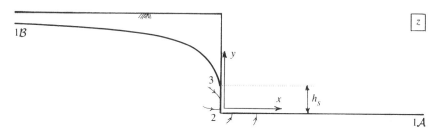

Figure 7.24 Flow toward a vertical embankment.

The Hodograph and the Function $\underset{1}{W} = \underset{1}{W}(\zeta)$

The domain in the hodograph plane is shown in Figure 7.25(b) and is obtained from the boundary of the flow domain in Figure 7.25(a) as follows. The seepage face $1\mathcal{A}$-2 is an equipotential, because the pressure is zero and the face is horizontal. This seepage face therefore is mapped onto the Q_y axis with point 1 corresponding to the origin. The vertical seepage face 2-3 is mapped in the hodograph plane onto the straight line $Q_y = -kB$. The latter two segments ($1\mathcal{A}$-2 and 2-3) have point 2 in common, which therefore is mapped onto infinity in the hodograph plane; point 2 is a singularity. The boundary of the domain in the hodograph plane is completed by the circle 3-1, which corresponds to the phreatic surface.

There is one point of inversion, $\overline{W} = -ikB$, so that

$$\tau_1 = ikB \tag{49.1}$$

and the function $\underset{1}{W}$ (see [47.12]) becomes

$$\underset{1}{W} = \frac{1}{W - ikB} \tag{49.2}$$

The boundary of the domain in the $\underset{1}{W}$ plane is shown in Figure 7.25(c). The phreatic surface corresponds to the straight line $\Im \underset{1}{W} = 1/(kB)$, the equipotential 1-2 to a section

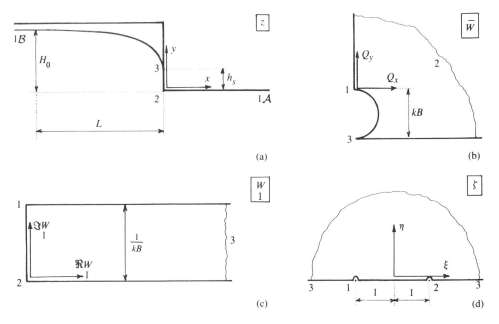

Figure 7.25 Domains in the z, hodograph, W_1, and ζ planes.

of length $1/(kB)$ of the imaginary axis, and the seepage face 2-3 to the positive real axis.

There is one point of inversion and we use the functions W_1 and T_1 as the auxiliary functions. We choose the ζ plane as shown in Figure 7.25(d), with point 3 at $\zeta = \infty$ and points 1 and 2 at $\zeta = -1$ and $\zeta = 1$, respectively. The mapping of $\Im \zeta \geq 0$ onto the domain in the W_1 plane is achieved by elementary modification of the function used to map the upper half plane onto the T_1 plane of Figure 7.20(b),

$$\zeta = \cosh(\pi k B W_1) \tag{49.3}$$

An expression for the auxiliary function T_1 is obtained from (47.18) and (49.1) as

$$T_1 = Z = \Omega + ikBz \tag{49.4}$$

This function is the one originally proposed by Zhukovski [1920], and is known as the Zhukovski function Z.

The Zhukovski Function

Separation of the Zhukovski function into its real and imaginary parts gives

$$Z = X + iY = (\Phi - kBy) + i(\Psi + kBx) \tag{49.5}$$

Sec. 49 Flow Toward a Vertical Embankment

We choose the origin of the x, y coordinate system at the bottom of the embankment with the x axis coinciding with the seepage face 1-2 (see Figure 7.25[a]). Application of the boundary condition (45.12) along the phreatic surface gives, with $y_r = 0$ and the ζ plane of Figure 7.25(d),

$$\eta = 0 \quad -\infty < \xi \leq -1 \quad X = 0 \tag{49.6}$$

The same condition applies along the two seepage faces 1A-2 and 2-3 (compare [45.6]) so that

$$\eta = 0 \quad -1 < \xi < \infty \quad X = 0 \tag{49.7}$$

Hence, the real part of the Zhukovski function is zero along the entire boundary. If we choose the value of Ψ to be zero along the phreatic surface, then Y equals zero at point 3,

$$Y_3 = 0 \tag{49.8}$$

so that point 3 corresponds to the origin of the Z plane. Near point 1B, x approaches $-\infty$ and $\Psi = 0$ so that Y approaches $-\infty$ there. Near point 1A, x approaches $+\infty$ and Ψ is positive so that Y approaches $+\infty$. The domain in the Z plane thus is the half plane shown in Figure 7.26(a). The upper half plane $\Im\zeta \geq 0$ is mapped onto the domain in the Z plane by the use of the transformation of piecewise constant argument for polygons that contain both the origin and infinity, see (33.5). Z approaches ∞ for $\zeta = -1$ and the angle enclosed at infinity in the Z plane is π so that $\xi_1^* = -1$ and $\mu_1 = -1$. The origin, $Z = 0$, corresponds to $\zeta = \infty$; this point is not included in the transformation. The argument of Z along the rightmost section 2-3 of the ξ axis is $\pi/2$, so that $\varphi^* = \pi/2$ and we obtain

$$Z = \frac{i|A|}{\zeta + 1} \tag{49.9}$$

The solution may be conveniently expressed in terms of a complex variable ω defined as

$$\omega = \tfrac{1}{2}\pi k B W_1 \tag{49.10}$$

The domain in the ω plane is the semi-infinite strip of width $\pi/2$ shown in Figure 7.26(b), and is obtained from that in the W_1 plane in Figure 7.25(c) by multiplying all coordinates by $\pi kB/2$. Using (49.3), (49.9), and (49.10) we obtain

$$Z = \frac{i|A|}{1 + \cosh(2\omega)} = \frac{i|A|}{2\cosh^2\omega} \tag{49.11}$$

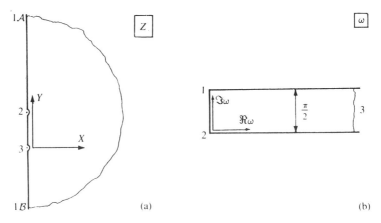

Figure 7.26 The domains in the Z and ω planes.

We determine the constant $|A|$ from the condition that the phreatic surface is at a given height H_0 above the y axis at $x = -L$:

$$x = -L \qquad y = H_0 \tag{49.12}$$

Using (49.5) and (49.11) and denoting the value of ω corresponding to this point as ω_0 we obtain, taking into account that $\Psi = 0$ along the phreatic surface,

$$-ikBL = \frac{i|A|}{2\cosh^2 \omega_0} \tag{49.13}$$

Noticing from Figure 7.26(b) that

$$\omega_0 = \lambda + i\pi/2 \qquad \lambda > 0 \tag{49.14}$$

where λ is a positive number, we obtain with $\cosh(\lambda + i\pi/2) = i\sinh\lambda$,

$$|A| = 2kBL\sinh^2 \lambda \tag{49.15}$$

The Function $z = z(\omega)$

We obtain an expression for z by integrating (47.20), with Z replacing T,

$$z = -\int_1 W(\zeta)Z'(\zeta)d\zeta = -\int_1 W\,dZ = -\frac{2}{\pi kB}\int \omega\,dZ \tag{49.16}$$

where use is made of (49.10). With (49.11) and (49.15) this becomes

$$z = -\frac{2iL}{\pi}\sinh^2 \lambda \int \omega\,d\left[\frac{1}{\cosh^2 \omega}\right] \tag{49.17}$$

Sec. 49   Flow Toward a Vertical Embankment

Integrating by parts we obtain

$$z = -\frac{2iL}{\pi} \sinh^2 \lambda \frac{\omega}{\cosh^2 \omega} + \frac{2iL}{\pi} \sinh^2 \lambda \tanh \omega + C \qquad (49.18)$$

where C is a constant of integration. Both z and ω are zero at point 2 so that C is zero. Writing $(\cosh^2 \omega)^{-1}$ as $1 - \tanh^2(\omega)$ we obtain

$$z = -\frac{2iL}{\pi} \sinh^2 \lambda [\omega(1 - \tanh^2 \omega) - \tanh \omega] \qquad (49.19)$$

Application of the condition (49.12) yields, using (49.14), and writing $\tanh(\lambda + i\pi/2)$ as $\coth \lambda$,

$$-L + iH_0 = -\frac{2iL}{\pi} \sinh^2 \lambda \left[(\lambda + i\tfrac{\pi}{2})(1 - \coth^2 \lambda) - \coth \lambda \right] \qquad (49.20)$$

The real parts of this equation match and we obtain the following equation for λ from the imaginary part:

$$H_0 = \frac{2L}{\pi} \sinh^2 \lambda \left[\frac{\lambda}{\sinh^2 \lambda} + \coth \lambda \right] \qquad (49.21)$$

or

$$\frac{H_0}{L} = \frac{2\lambda}{\pi} + \frac{1}{\pi} \sinh(2\lambda) \qquad (49.22)$$

Given H_0/L, the value for the parameter λ is obtained by solving this equation numerically, for example, by the use of a Newton-Raphson procedure.

The Height of the Seepage Face

An expression for the elevation of the uppermost point of the seepage face, h_s, is obtained from (49.19) by taking the limit for $\omega \to \infty$ (see Figures 7.24 and 7.26[b]). The function $\tanh \omega$ approaches 1 for $\omega \to \infty$ and the term between parentheses in (49.19) approaches zero as $\omega/\cosh^2 \omega$ for $\omega \to \infty$. Hence, we obtain

$$h_s = \frac{2L}{\pi} \sinh^2 \lambda \qquad (49.23)$$

Points on the Phreatic Surface

The complex coordinates of points on the phreatic surface are obtained from (49.19) by noticing that ω may be written as $\varrho + i\pi/2$ ($\varrho \geq 0$) there (see Figures 7.24 and 7.26[b]). The resulting expression is (compare [49.20])

$$z = \frac{2iL}{\pi} \sinh^2 \lambda \left[\frac{\varrho + i\tfrac{\pi}{2}}{\sinh^2 \varrho} + \coth \varrho \right] \qquad (49.24)$$

Problem 49.1

Normalize z with respect to h_s and compute points of z/h_s on the phreatic surface for $-10 \leq x/h_s \leq 0$ with intervals of 5. Plot the phreatic surface.

Problem 49.2

Determine an expression for the total amount of flow through the vertical seepage face.

Problem 49.3

Compute the dimensionless specific discharge q_x/k as a function of y/h_s along the vertical seepage face, and plot q_x/k versus y/h_s.

50 UNCONFINED FLOW WITH A SEMI-INFINITE IMPERMEABLE LAMINA

As a third application of the hodograph method, we consider the case of flow in an unconfined aquifer with a semi-infinite impermeable lamina. This problem was solved by Strack [1981(b)] in order to validate the comprehensive potential approach discussed in Sec. 12. The flow domain is sketched in Figure 7.27. The base of the aquifer system is horizontal and impervious. The lamina is elevated at a distance H above the base of the aquifer system and is sufficiently thin that it can be approximated as the semi-infinite slot 1-2-3 of infinitesimal thickness. The phreatic surface 3-4-5 extends from $x = +\infty$ (point 3) to $x = -\infty$ (point 5). The x axis points along the base toward point 1 and the y axis points vertically upward. The origin lies vertically below the edge of the lamina, point 2. Sufficiently far from the edge of the lamina the equipotentials may with good approximation be taken as vertical. We define vertical planes $\mathcal{A}$ and $\mathcal{B}$ at $x = -L_a \ll -H$ and at $x = L_b \gg H$, and denote the heads in the three aquifers at these planes as ϕ_a, $\overset{u}{\phi_b}$, and $\overset{l}{\phi_b}$:

$$x = -L_a \qquad \phi = \phi_a \qquad x = L_b \qquad \phi = \overset{u}{\phi_b} \qquad \phi = \overset{l}{\phi_b} \qquad (50.1)$$

where the superscripts u and l refer to the upper and lower aquifers. We further denote the discharges flowing through these cross sections as Q_a, $\overset{u}{Q_b}$ and $\overset{l}{Q_b}$, taken positive as shown in Figure 7.27. It follows from continuity of flow that

$$Q_a = \overset{u}{Q_b} + \overset{l}{Q_b} \qquad (50.2)$$

The objective of the analysis is to determine an expression for the phreatic surface, to produce flow nets, and to establish a relationship between the heads ϕ_a, $\overset{u}{\phi_b}$, $\overset{l}{\phi_b}$, and the discharges.

The Hodograph and the Function $\underset{1}{W} = \underset{1}{W}(\zeta)$

The boundary of the region in the hodograph plane is shown in Figure 7.28(b) and was constructed by tracing the boundary in the z plane of Figure 7.28(a) in the sequence

Sec. 50   Unconfined Flow with a Semi-Infinite Impermeable Lamina   547

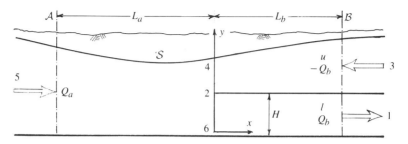

Figure 7.27 Unconfined flow with a semi-finite impermeable lamina. (Source: Strack [1981(b)].)

1-2-3-4-$\mathcal{S}$-5-6-1 while plotting the end points of the discharge vector in the hodograph plane. At point 1, infinity, Q_x equals $\overset{l}{Q_b}/H$ and Q_y is zero. Tracing the lamina from 1 to 2, Q_y remains zero and Q_x positive, with Q_x increasing to infinity upon reaching point 2. Along 2-3, Q_y is zero and Q_x negative. The phreatic surface is mapped onto the arc 3-4-$\mathcal{S}$-5: point 3 is at the intersection of the Q_x axis and the circle and therefore lies at the origin. Tracing the phreatic surface from point 3 toward $\mathcal{S}$, the discharge vector first increases in magnitude, pointing downward and to the left, until a maximum is reached and then decreases to zero at the stagnation point $\mathcal{S}$. In moving from $\mathcal{S}$ toward point 5, a similar behavior of the discharge vector is observed, but the vector now points downward and to the right. Tracing the aquifer base from 5 via 6 to 1, the discharge vector is horizontal and Q_x increases from zero at 5 to $\overset{l}{Q_b}/H$ at 1, thus closing the boundary in the hodograph plane.

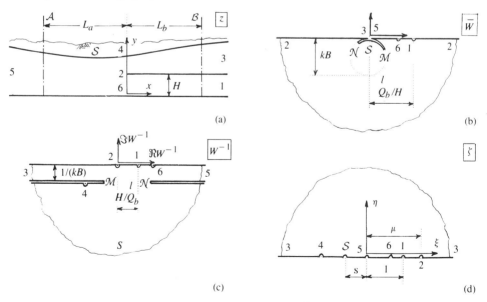

Figure 7.28 Flow domain and regions in the hodograph, W^{-1}, and ζ planes (Source: Strack [1981(b)].)

We observe from the hodograph that there is one point of inversion: the origin. Hence,

$$\tau_1 = 0 \qquad (50.3)$$

and

$$W_1 = W^{-1} \qquad (50.4)$$

The region in the W^{-1} plane is shown in Figure 7.28(c). As a result of the inversion, the circular arcs 3-4-S and S-5 are transformed into the two semi-infinite slots 3-4-S and S-5, lying a distance $1/(kB)$ below the real axis and with point S, $Q_x = Q_y = 0$, corresponding to $W^{-1} = \infty$. The streamlines 1-2-3 and 5-6-1 are mapped together onto the real axis with points 3 and 5, which were at the origin in the hodograph plane, mapped onto $W^{-1} = \infty$ and point 2, which was at infinity in the hodograph plane, mapped onto the origin, $W^{-1} = 0$.

The ζ plane is chosen as shown in Figure 7.28(d) with point 3 at infinity, point S at $\zeta_s = -s$, point 5 at the origin and points 1 and 2 at $\zeta = 1$ and $\zeta = \mu$, respectively. Only three points can be chosen freely in the transformation (points 3, 5, and 1), and therefore s and μ represent parameters that are unknown at this stage of the analysis.

The function $W^{-1} = W^{-1}(\zeta)$ is determined by application of the Schwarz-Christoffel transformation. It is seen from Figure 7.28(c) that

$$\pi k_1 = \pi k_2 = \pi k_4 = \pi k_6 = 0 \qquad \pi k_3 = \pi k_5 = \pi$$
$$\pi k_M = \pi k_N = -\pi \qquad \pi k_s = 2\pi \qquad (50.5)$$

where πk_M and πk_N correspond to the end points M and N of the slots 3-4-S and S-5. The argument of $dW^{-1}/d\zeta$ equals $-\pi$ along the rightmost segment 2-3 of the ξ axis so that $\varphi = -\pi$. Denoting the points of the ξ axis corresponding to points M and N as ξ_M and ξ_N, the Schwarz-Christoffel transformation (32.9) becomes

$$W^{-1} = |A|e^{-i\pi} \int \frac{(\zeta - \xi_M)(\zeta - \xi_N)}{\zeta(\zeta + s)^2} d\zeta + C \qquad (50.6)$$

where C is a complex constant. Expanding the integrand in terms of fractions we obtain

$$W^{-1} = \int \left[\frac{a_1}{\zeta} + \frac{a_2}{\zeta + s} + \frac{a_3}{(\zeta + s)^2} \right] d\zeta + C \qquad (\Im a_1 = \Im a_2 = \Im a_3 = 0) \qquad (50.7)$$

where a_1, a_2 and a_3 are newly introduced real constants. Integration yields

$$W^{-1} = a_1 \ln \zeta + a_2 \ln(\zeta + s) - \frac{a_3}{\zeta + s} + C \qquad (50.8)$$

The term $a_2 \ln(\zeta + s)$ contributes an amount $a_2 \pi$ to the imaginary part of W^{-1} for $\zeta = \xi$, $\xi < -s$ and an amount zero for $\zeta = \xi$, $\xi > -s$. Since the imaginary parts of W^{-1} on the two sides of s are equal, the term $a_2 \ln(\zeta + s)$ cannot be present, so that

$$a_2 = 0 \qquad (50.9)$$

Sec. 50 Unconfined Flow with a Semi-Infinite Impermeable Lamina

The imaginary part of W^{-1} must increase by an amount $1/(Bk)$ if ζ passes point 5, the origin, while moving in the positive ξ direction. The only term in (50.8) that creates such a jump is the first one, and it follows that

$$a_1 = -\frac{1}{\pi k B} \tag{50.10}$$

The function W^{-1} equals $H/\overset{l}{Q_b}$ at point 1, $\zeta = 1$. Applying this condition to (50.8), we obtain, using (50.9) and (50.10),

$$\frac{H}{\overset{l}{Q_b}} = -\frac{a_3}{1+s} + C \tag{50.11}$$

so that

$$C = \frac{H}{\overset{l}{Q_b}} + \frac{a_3}{1+s} \tag{50.12}$$

Using (50.9), (50.10), and (50.12), (50.8) may be written in the following form:

$$W^{-1} = -\frac{1}{\pi k B}\ln\zeta + a\frac{\zeta-1}{\zeta+s} + \frac{H}{\overset{l}{Q_b}} \tag{50.13}$$

where

$$a = \frac{a_3}{1+s} \tag{50.14}$$

The auxiliary function $\underset{1}{T}$ equals Ω, as follows from (50.3) and (47.18). We determine this function by solving the boundary value problem in $\Im\zeta \geq 0$.

The Function $\Omega = \Omega(\zeta)$

The boundary conditions in the ζ plane are that the entire real axis is a streamline, and that there is a source emitting a discharge Q_a into $\Im\zeta \geq 0$ at point 5, $\zeta = 0$, and a sink removing a discharge $\overset{l}{Q_b}$ from $\Im\zeta \geq 0$ at point 1, $\zeta = 1$. The complex potential that meets these conditions is

$$\Omega = -\frac{Q_a}{\pi}\ln\zeta + \frac{\overset{l}{Q_b}}{\pi}\ln(\zeta-1) + \Phi_0 \tag{50.15}$$

where Φ_0 is a real constant.

The Function $z = z(\zeta)$

The function $z' = z'(\zeta)$ is obtained by the use of (46.3),

$$z' = -\Omega'(\zeta)W^{-1}(\zeta) \tag{50.16}$$

Differentiation of (50.15) yields the following expression for $\Omega'(\zeta)$:

$$\Omega' = -\frac{Q_a}{\pi\zeta} + \frac{\overset{l}{Q_b}}{\pi}\frac{1}{\zeta-1} = \frac{\overset{l}{Q_b} - Q_a}{\pi}\frac{\zeta + Q_a/(\overset{l}{Q_b} - Q_a)}{\zeta(\zeta-1)} \quad (50.17)$$

Using this expression and the expression (50.13) for W^{-1} in (50.16), we obtain

$$z' = -\frac{Q_a}{\pi^2 kB}\frac{\ln\zeta}{\zeta} + \frac{\overset{l}{Q_b}}{\pi^2 kB}\frac{\ln\zeta}{\zeta-1} - a\frac{\overset{l}{Q_b} - Q_a}{\pi}\frac{\zeta + Q_a/\overset{l}{Q_b} - Q_a)}{\zeta(\zeta+s)} - \frac{H}{\overset{l}{Q_b}}\Omega'(\zeta) \quad (50.18)$$

The stagnation point is a finite point in the z plane, and therefore the singularity at $\zeta = -s$ of the third term in (50.18) must be a removable one. All other terms are analytic at $\zeta = -s$ and the condition that z is finite at $\zeta = -s$ will be fulfilled only if

$$\frac{Q_a}{\overset{l}{Q_b} - Q_a} = s \quad (50.19)$$

It is of interest to note that point $\zeta = -s$ is a stagnation point for the flow in $\Im\zeta \geq 0$; (50.19) and (50.17) imply that the derivative Ω', which equals minus the discharge vector for flow in $\Im\zeta \geq 0$, vanishes at $\zeta = -s$. Combining (50.19) with (50.18) and integrating, we obtain

$$z = -\frac{Q_a}{2\pi^2 kB}\ln^2\zeta + A_0\ln\zeta - \frac{H}{\pi}\ln(\zeta-1) - \frac{\overset{l}{Q_b}}{\pi^2 kB}\text{Li}_2(1-\zeta) + C_1 + iC_2 \quad (50.20)$$

where

$$A_0 = -\frac{\overset{l}{Q_b} - Q_a}{\pi}a + \frac{H}{\pi}\frac{Q_a}{\overset{l}{Q_b}} \quad (50.21)$$

The function $\text{Li}_2(1-\zeta)$ is the dilogarithm (see Lewin [1958]), defined as

$$\text{Li}_2(1-\zeta) = -\int_0^{1-\zeta}\frac{\ln(1-t)}{t}dt \quad (50.22)$$

This function is discussed in some detail in Appendix B.

Determination of the Parameters in the Solution

In order to evaluate the various logarithms and the dilogarithm in (50.20), we must define the ranges of values of the arguments of ζ, $\zeta - 1$, and $1 - \zeta$. Using that ζ is defined in $\Im\zeta \geq 0$ and that the arguments must be continuous for ζ inside the upper half plane, including the real axis, we obtain

$$0 \leq \arg\zeta \leq \pi \quad 0 \leq \arg(\zeta - 1) \leq \pi \quad -\pi \leq \arg(1-\zeta) \leq 0 \quad (50.23)$$

Sec. 50 Unconfined Flow with a Semi-Infinite Impermeable Lamina

The solution to the problem is given by the functions $\Omega = \Omega(\zeta)$ and $z = z(\zeta)$ in (50.15) and (50.20), which together contain the five constants A_0, C_1, C_2, Φ_0, and μ as unknowns. An expression for C_2 is obtained by requiring that $y = 0$ along boundary segment 5-6-1, which corresponds to $\eta = 0$, $0 < \xi < 1$. The argument of $1 - \zeta$ is zero along this section so that the dilogarithm (50.22) is real there. All terms in (50.20) are real along 5-6-1, except the third term, which has an imaginary part equal to $-H$, so that

$$C_2 = H \tag{50.24}$$

Choosing the y axis as the reference level for the heads, the condition along the phreatic surface becomes $\Phi = kBy$. It is shown in Appendix B that this condition is met if

$$A_0 = \frac{\Phi_0}{\pi k B} \tag{50.25}$$

Point 2, $\zeta = \mu$, corresponds to $z = iH$, so that

$$iH = z(\mu) \tag{50.26}$$

The mapping is not conformal at point 2; the derivative of the mapping function vanishes there,

$$z'(\mu) = 0 \tag{50.27}$$

Equations (50.24) through (50.27) represent four equations for the five unknown constants. The remaining condition is obtained by requiring that the head be $\overset{l}{\phi_b}$ at $z = L_b$:

$$z = L_b \qquad \Phi = kB\overset{l}{\phi_b} \tag{50.28}$$

We apply this condition in an approximate fashion, taking into account that $L_b \gg H$. The image in $\Im\zeta \geq 0$ of the point $z = L_b$, then, will be on the segment 5-6-1 and close to $\zeta = 1$. Representing ζ there as ζ_b with

$$\zeta_b = 1 + \varepsilon e^{i\pi} \qquad (0 < \varepsilon \ll 1) \tag{50.29}$$

we may approximate (50.20) as follows, noticing that all terms are nearly zero except the third one and the constants,

$$z = L_b \approx -\frac{H}{\pi}\ln\varepsilon - iH + C_1 + iC_2 \tag{50.30}$$

or, using (50.24),

$$L_b \approx -\frac{H}{\pi}\ln\varepsilon + C_1 \tag{50.31}$$

The potential at $\zeta = \zeta_b$ may be approximated similarly (see [50.15]),

$$\overset{l}{\Phi}_b \approx \frac{\overset{l}{Q}_b}{\pi} \ln \varepsilon + \Phi_0 \tag{50.32}$$

We eliminate $\ln \varepsilon$ from the latter two equations and solve for C_1:

$$C_1 \approx L_b + \frac{H}{\overset{l}{Q}_b}(\overset{l}{\Phi}_b - \Phi_0) \tag{50.33}$$

Using (50.24), (50.25), and (50.33), all constants may be eliminated from the solution, with the exception of μ and Φ_0. The latter two constants are determined by solving (50.26) and (50.27) by the use of a Newton-Raphson procedure.

Once all constants are determined, the phreatic surface may be plotted by computing values of z from (50.20), letting ζ vary along the negative real axis. A flow net may be generated by first determining points of the net in $\Im \zeta \geq 0$, and then computing the corresponding values of z. Such a flow net is shown in Figure 7.29 as an illustration.

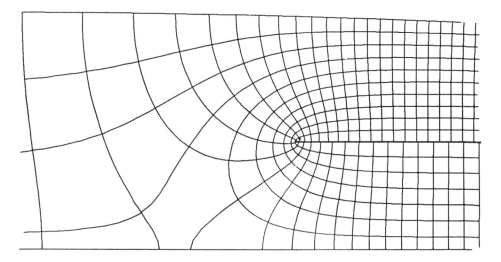

Figure 7.29 Flow net for unconfined flow with a lamina. (Source: Strack [1981(b)].)

The solution presented above may be readily adapted to the case of interface flow shown in Figure 7.30, which is obtained by rotating the domain in the z plane 180 degrees about the origin. Replacing z in the above transformations by $-z^* = -(x^* + iy^*)$ and k by k^*, the function $z^*(\zeta)$ meets the condition that $\Phi = -k^* y^*$. With an appropriate choice of the reference level, this condition corresponds to that along the interface.

Problem 50.1

Determine the function $z'(\zeta)$, and write a computer program to determine the parameters μ and Φ_0, using the series expansion given in Appendix B to compute the dilogarithm.

Sec. 51 Interface Flow from an Equipotential Toward Drains   553

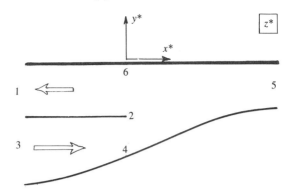

Figure 7.30 Interface flow with a lamina.

51 INTERFACE FLOW FROM AN EQUIPOTENTIAL TOWARD DRAINS

We will solve some problems of interface flow toward a drain in this section by an analysis that follows Strack [1972]. An example of a coastal aquifer is shown in Figure 7.31(a), where the lower boundary is an interface between flowing fresh water and salt water at rest, and the upper boundary is a phreatic surface. We approximate the phreatic surface by a single horizontal equipotential at the same elevation as the sea bottom (see Figure 7.31[b]), with a head equal to the average estimated elevation of the phreatic surface. This approximation is crude, but may give an insight into the shape and position of the interface in aquifers where the average elevation of the phreatic surface is controlled by irrigation or pumping. We will examine, in particular, the effect of pumpage from a horizontal drain with its axis normal to the plane of flow. Two cases will be treated separately. For the first case the section of the aquifer where the pumping takes place is sufficiently far from the coast that its influence may be neglected. For the second case the presence of the coast is taken into account.

Flow from a Single Horizontal Equipotential Toward a Drain

Consider the case of flow from a single infinite equipotential with $\phi = \phi_0$ toward a drain located above an interface (see Figure 7.32[a]). The x axis coincides with the equipotential, the head in the stationary salt water is ϕ_s, and all heads are measured with respect to the x axis. The boundary in the hodograph plane is shown in Figure 7.32(b). At the two points $1A$ and $2A$ (infinity) of the interface the discharge is zero: these points are mapped onto the origin in the hodograph plane. Moving along the interface from $1A$ toward $2A$, the discharge vector first increases in magnitude, pointing upward and to the right, then decreases to zero at point S vertically below the drain, increases again in magnitude but now pointing upward and to the left, and finally decreases to zero at point $2A$. Moving from $2B$ along the equipotential via point 3 toward point $1B$, the discharge vector first increases in magnitude, pointing downward, until a maximum is reached at point 3 above the well, then decreases in magnitude to zero at point $1B$. The boundary in the hodograph plane thus obtained is the slot 1-S-2-3-1; the domain is the exterior of

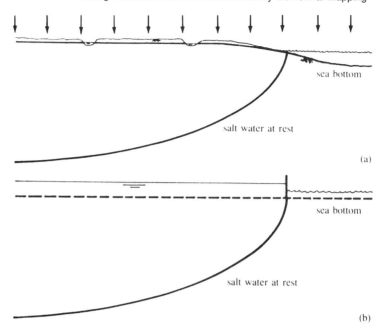

Figure 7.31 Approximation of the upper boundary of a coastal aquifer.

the slot and contains infinity. Indeed, the drain lies inside the flow domain and infinite values for the discharge vector occur at the center of the drain: the drain is mapped onto infinity in the hodograph plane.

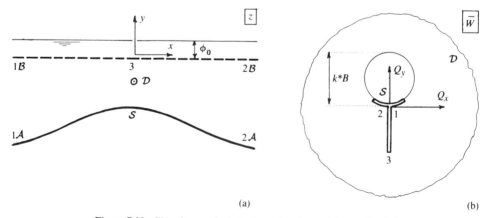

Figure 7.32 Flow from a single horizontal equipotential toward a drain.

We observe from the boundary in the hodograph plane that there are two points of inversion:

$$\overline{W} = \overline{\tau}_1 = 0 \qquad \overline{W} = \overline{\tau}_2 = ik^*B \tag{51.1}$$

Sec. 51 Interface Flow from an Equipotential Toward Drains

We may therefore solve this flow problem without resorting to the use of the hodograph, and use the two auxiliary functions (compare [47.25])

$$\underset{1}{T} = \Omega + \tau_1 z = \Omega \qquad \underset{2}{T} = \Omega + \tau_2 z = \Omega - ik^* B z \tag{51.2}$$

We will solve the problem by determining the functions $\underset{1}{T}$ and $\underset{2}{T}$ in terms of ζ, using the method of images.

The Complex Potential $\Omega = \Omega(\zeta)$

We choose the reference half plane shown in Figure 7.33(a). Point 1 corresponds to infinity, point S to $\zeta = -s$, point 2 to $\zeta = 0$, and point 3 to $\zeta = \xi_3$. The center of the drain, $z = z_d$, corresponds to $\zeta = \delta$. The boundary conditions for the complex potential are that the negative real axis is a streamline:

$$\eta = 0 \qquad -\infty < \xi \leq 0 \qquad \Psi = \Psi_0 \tag{51.3}$$

where Ψ_0 is a constant, and that the positive real axis is the equipotential $\Phi = \Phi_0$,

$$\eta = 0 \qquad 0 \leq \xi < \infty \qquad \Phi = kB\phi_0 = \Phi_0 \tag{51.4}$$

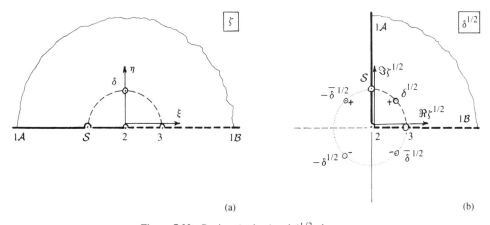

Figure 7.33 Regions in the ζ and $\zeta^{1/2}$ planes.

As for the flow problem solved in Sec. 32 (see Figure 5.15), we determine the complex potential by applying the method of images in the $\zeta^{1/2}$ plane shown in Figure 7.33(b), taking advantage of the normality of the streamline and equipotential boundaries in the latter plane. The complex potential for flow from the equipotential 2-1B toward the drain at $\zeta^{1/2} = \delta^{1/2}$ becomes

$$\Omega = \frac{Q}{2\pi} \ln \frac{(\zeta^{1/2} - \delta^{1/2})(\zeta^{1/2} + \overline{\delta}^{1/2})}{(\zeta^{1/2} - \overline{\delta}^{1/2})(\zeta^{1/2} + \delta^{1/2})} + \Phi_0 + i\underset{1}{C_2} \tag{51.5}$$

where Q is the discharge of the drain and C_2^1 is a real constant.

The Function $T_2 = T_2(\zeta)$

Separation of the real and imaginary parts of the function T_2 given by (51.2) yields

$$T_2^1 + iT_2^2 = \Phi + k^*By + i(\Psi - k^*Bx) \tag{51.6}$$

The boundary condition along an interface is given by (45.19):

$$\eta = 0 \quad -\infty < \xi \leq 0 \quad \Phi = -k^*By + (k^*y_r + k\frac{\rho_s}{\rho_f}\phi_s)B \quad \Psi = \Psi_0 \tag{51.7}$$

The reference level for Φ coincides with the x axis so that $y_r = 0$. We denote the thickness of the freshwater zone at infinity as H_0; y in (51.7) approaches $-H_0$ near infinity. Since there is no flow at infinity (the effect of the drain, which draws all of its water from the equipotential, is negligible there), the potential in the freshwater zone equals Φ_0 at infinity; (51.7) becomes, with $y_r = 0$ and $y = -H_0$,

$$\eta = 0 \quad \xi \to -\infty \quad \Phi_0 = k^*BH_0 + kB\frac{\rho_s}{\rho_f}\phi_s \tag{51.8}$$

so that

$$kB\frac{\rho_s}{\rho_f}\phi_s = \Phi_0 - k^*BH_0 \tag{51.9}$$

and (51.7) may be written as

$$\eta = 0 \quad -\infty < \xi \leq 0 \quad \Phi = -k^*By - k^*BH_0 + \Phi_0 \quad \Psi = \Psi_0 \tag{51.10}$$

This boundary condition may be expressed in terms of the real part of T_2 by the use of (51.6) as

$$\eta = 0 \quad -\infty < \xi \leq 0 \quad T_2^1 = \Phi_0 - k^*BH_0 \tag{51.11}$$

Along the equipotential 2-3-1$\mathcal{B}$ the potential equals Φ_0 and $y = 0$ so that

$$\eta = 0 \quad 0 \leq \xi \leq \infty \quad T_2^1 = \Phi_0 \tag{51.12}$$

The function $z = z(\zeta)$ is analytic at the drain, and therefore the singular behavior of T_2 near $\zeta = \delta$ is governed by that of the complex potential. The real part of T_2 is constant along the entire real axis, with a jump occurring at $\zeta = 0$. Thus, the function T_2 may be viewed as a complex potential for flow in $\Im\zeta \geq 0$ with a vortex at $\zeta = 0$ and

Sec. 51 Interface Flow from an Equipotential Toward Drains

a drain of discharge Q at $\zeta = \delta$. Placing an image drain at $\zeta = \bar{\delta}$ to ensure that $\Re \underset{2}{T}$ is constant along the real axis, we obtain

$$\underset{2}{T} = \frac{k^* B H_0}{\pi} i \ln \zeta + \frac{Q}{2\pi} \ln \frac{\zeta - \delta}{\zeta - \bar{\delta}} + \Phi_0 + i \underset{2}{C_2} \tag{51.13}$$

where $\underset{2}{C_2}$ is a real constant. It is left as an exercise for the reader to verify that (51.13) indeed fulfills the boundary conditions (51.11) and (51.12).

The Function $z = z(\zeta)$

An expression for z as a function of ζ is obtained from (51.2), (51.5), and (51.13):

$$ik^* B z = \Omega - \underset{2}{T} = \frac{Q}{2\pi} \ln \frac{(\zeta^{1/2} - \delta^{1/2})(\zeta^{1/2} + \bar{\delta}^{1/2})}{(\zeta^{1/2} - \bar{\delta}^{1/2})(\zeta^{1/2} + \delta^{1/2})} + \Phi_0 + i\underset{1}{C_2}$$

$$- \frac{k^* B H_0}{\pi} i \ln \zeta - \frac{Q}{2\pi} \ln \frac{(\zeta^{1/2} - \delta^{1/2})(\zeta^{1/2} + \delta^{1/2})}{(\zeta^{1/2} - \bar{\delta}^{1/2})(\zeta^{1/2} + \bar{\delta}^{1/2})} - \Phi_0 - i\underset{2}{C_2} \tag{51.14}$$

where $\zeta - \delta$ is written as $(\zeta^{1/2} - \delta^{1/2})(\zeta^{1/2} + \delta^{1/2})$ and $\zeta - \bar{\delta}$ similarly. Eliminating the logarithmic terms that cancel and introducing a new real constant C_1 we obtain, after division by $k^* B i$,

$$z = -\frac{H_0}{\pi} \ln \zeta + \frac{Q}{\pi k^* B} i \ln \frac{\zeta^{1/2} + \delta^{1/2}}{\zeta^{1/2} + \bar{\delta}^{1/2}} + C_1 \tag{51.15}$$

where C_1 equals $(\underset{1}{C_2} - \underset{2}{C_2})/(k^* B)$.

Thus far we have chosen only two parameters in the ζ plane, whereas the transformation contains three degrees of freedom. We use the third one by choosing

$$|\delta| = 1 \tag{51.16}$$

so that δ may be written as

$$\delta = e^{i\alpha} \qquad (0 < \alpha < \pi) \tag{51.17}$$

By symmetry, the unit circle in the ζ plane will correspond to the vertical through the drain (compare Figures 7.32[a] and 7.33[a]). Point 3 therefore corresponds to $\zeta = 1$,

$$\xi_3 = 1 \tag{51.18}$$

and the condition that $z = 0$ at point 3 gives with (51.15) and (51.17)

$$0 = \frac{Q}{\pi k^* B} i \ln \frac{1 + e^{i\alpha/2}}{1 + e^{-i\alpha/2}} + C_1 = \frac{Q}{\pi k^* B} i \ln \frac{e^{i\alpha/2}(1 + e^{-i\alpha/2})}{1 + e^{-i\alpha/2}} + C_1 \tag{51.19}$$

or
$$C_1 = -\frac{Q}{\pi k^* B} i^2 \frac{\alpha}{2} = \frac{Q\alpha}{2\pi k^* B} \tag{51.20}$$

so that (51.15) becomes

$$z = -\frac{H_0}{\pi} \ln \zeta + \frac{Q}{\pi k^* B} i \ln \frac{\zeta^{1/2} + e^{i\alpha/2}}{\zeta^{1/2} + e^{-i\alpha/2}} + \frac{Q\alpha}{2\pi k^* B} \tag{51.21}$$

The parameter α is determined by requiring that z be z_d for $\zeta = \delta = e^{i\alpha}$:

$$z_d = -\frac{H_0}{\pi} i\alpha + \frac{Q}{\pi k^* B} i \ln \frac{2e^{i\alpha/2}}{e^{i\alpha/2} + e^{-i\alpha/2}} + \frac{Q\alpha}{2\pi k^* B} \tag{51.22}$$

Denoting the distance between the drain and the x axis as h_1, so that $z_d = -ih_1$, this becomes

$$\frac{h_1}{H_0} = \frac{\alpha}{\pi} + \frac{q^*}{\pi} \ln \cos \frac{\alpha}{2} \tag{51.23}$$

where the dimensionless constant q^* is defined as

$$q^* = \frac{Q}{k^* B H_0} \tag{51.24}$$

Once an expression for α is obtained by solving (51.23) numerically, points of the interface may be determined by setting ζ equal to $|\zeta|e^{i\pi}$ in (51.21). Of particular interest is the highest point of the interface, point S (see Figure 7.32[a]), where ζ equals $e^{i\pi}$ (see Figure 7.33[a]). Denoting the distance between the x axis and point S as h_s, we obtain

$$z_s = -ih_s = -iH_0 + \frac{Q}{\pi k^* B} i \ln \frac{e^{i\pi/2} + e^{i\alpha/2}}{e^{i\pi/2} + e^{-i\alpha/2}} + \frac{Q\alpha}{2\pi k^* B} \tag{51.25}$$

or

$$\frac{h_s}{H_0} = 1 - \frac{q^*}{\pi} \ln \left[\frac{e^{i\pi/4-i\alpha/4} + e^{-i\pi/4+i\alpha/4}}{-e^{-i\pi/4+i\alpha/4} + e^{i\pi/4-i\alpha/4}} \frac{e^{i\pi/4}e^{i\alpha/4}}{e^{-i\pi/4}e^{-i\alpha/4}} \right] + i\frac{q^*\alpha}{2\pi} \tag{51.26}$$

which may be simplified to

$$\frac{h_s}{H_0} = 1 - \frac{q^*}{\pi} \ln \cot \frac{\pi - \alpha}{4} \tag{51.27}$$

Instability of the Interface

If the discharge of the well exceeds some critical value, the salt water will start flowing and the solution given above no longer applies. We will determine this critical value, for which the interface is unstable and the salt water is about to start flowing toward the well. The interface is unstable if it possesses a vertical tangent at some

Sec. 51 Interface Flow from an Equipotential Toward Drains

point; any increase in slope would cause salt water to be above fresh water, a situation that cannot be maintained. Instability will occur first at point S with a cusp developing as shown in Figure 7.34(a). The mapping is then no longer conformal at point S; the boundary in the z plane encloses an angle of 2π whereas it encloses an angle of π at point S in the ζ plane: $z'(\zeta)$ is zero at $\zeta = e^{i\pi}$. An expression for $z'(\zeta)$ is obtained from (51.21) upon differentiation:

$$z' = -\frac{H_0}{\pi \zeta} + \frac{Q}{\pi k^* B} i \left[\frac{1}{\zeta^{1/2} + e^{i\alpha/2}} - \frac{1}{\zeta^{1/2} + e^{-i\alpha/2}} \right] \frac{\frac{1}{2}}{\zeta^{1/2}} \qquad (51.28)$$

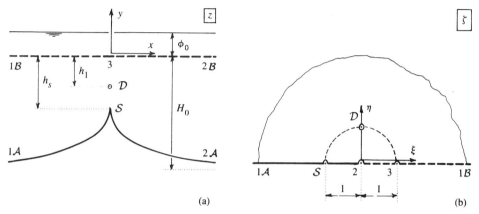

Figure 7.34 Unstable interface: cusp at point S.

Setting $z' = 0$ at $\zeta = e^{i\pi}$ we obtain, combining the two terms between the brackets:

$$0 = \frac{H_0}{\pi} + \frac{Q}{\pi k^* B} i \left[\frac{e^{-i\alpha/2} - e^{i\alpha/2}}{i(e^{i\alpha/2} + e^{-i\alpha/2})} \right] \frac{1}{2i} = \frac{H_0}{\pi} - \frac{Q}{2\pi k^* B} \tan \frac{\alpha}{2} \qquad (51.29)$$

Hence, (see [51.24])

$$\alpha = 2 \arctan \frac{2}{q^*} \qquad (51.30)$$

The value for h_s, $(h_s)_{\text{crit}}$, corresponding to this unstable situation is obtained by combining (51.27) with (51.30). Writing $\tan \frac{\alpha}{2}$ as

$$\tan \frac{\alpha}{2} = \cot \frac{\pi - \alpha}{2} = \frac{1}{2} \left[\cot \frac{\pi - \alpha}{4} - \tan \frac{\pi - \alpha}{4} \right] \qquad (51.31)$$

in (51.29) and solving for $\cot \frac{\pi - \alpha}{4}$ we obtain

$$\cot \frac{\pi - \alpha}{4} = \frac{2}{q^*} + \sqrt{\left[\frac{2}{q^*} \right]^2 + 1} \qquad (51.32)$$

where the positive root is chosen because $\cot \frac{\pi-\alpha}{4}$ is positive. Using this in (51.27) the following results:

$$\left(\frac{h_s}{H_0}\right)_{\text{crit}} = 1 - \frac{q^*}{\pi} \ln\left[\frac{2}{q^*} + \sqrt{\left[\frac{2}{q^*}\right]^2 + 1}\right] = 1 - \frac{q^*}{\pi} \operatorname{arcsinh} \frac{2}{q^*} \quad (51.33)$$

Equations (51.23) and (51.27) together give a relationship between q^*, h_s/H_0, and h_1/H_0. Curves of $(H_0 - h_s)/H_0$ versus q^* for various values of h_1/H_0 are reproduced in Figure 7.35. The dotted line which bounds the graph on the upper right represents unstable conditions and is obtained from (51.33). It is seen from the graph that instability occurs well before point S has reached the drain. This explains why sometimes wells in coastal aquifers draw brackish water unexpectedly, particularly in view of the sensitivity of the rise of the interface to the discharge: it is seen from the graph that the curves for constant values of h_1/H_0 become steep near the dotted line.

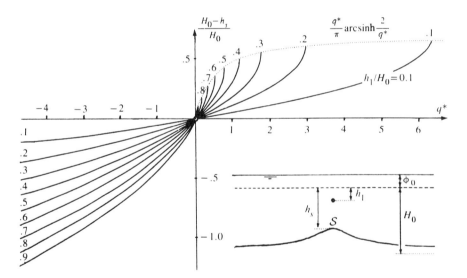

Figure 7.35 The relation between the rise $H_0 - h_s$ of the interface at S, the position of the drain, h_1, and the dimensionless discharge q^*. (Source: Strack [1972].)

Flow Toward the Sea

The problem of interface flow toward the coast is illustrated in Figure 7.36(a): the flow occurs from the horizontal equipotential 3-1B ($\Phi = \Phi_0$) toward the horizontal sea bottom 2-3. The domain in the ζ plane is shown in Figure 7.36(b) and is the same as for the previous problem, with point 2 corresponding now to the point of intersection of the interface with the sea bottom, rather than infinity. The head is approximated to be discontinuous at the coast, point 3, which is taken to be at the origin in the z plane and

Sec. 51 Interface Flow from an Equipotential Toward Drains

at $\zeta = 1$ in the ζ plane. The head $\phi_{2,3}$ along the sea bottom is expressed in terms of the sea level ϕ_s by requiring that the pressure p_f along 2-3 equals $p_s = \rho_s \phi_s$:

$$\phi_{2,3} = \frac{\rho_s}{\rho_f} \phi_s \tag{51.34}$$

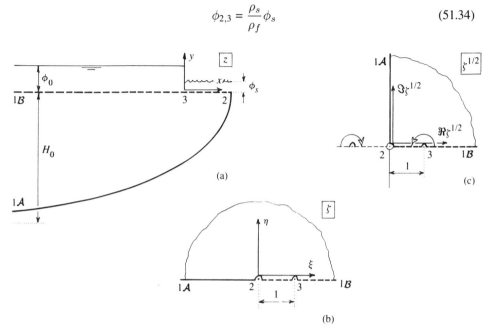

Figure 7.36 Flow toward the sea.

The Complex Potential

The boundary conditions for $\Omega = \Omega(\zeta)$ are (compare [51.3] and [51.4])

$$\eta = 0 \quad -\infty < \xi \leq 0 \quad \Psi = \Psi_0 \tag{51.35}$$

$$\eta = 0 \quad 0 \leq \xi \leq 1 \quad \Phi = \Phi_{2,3} = kB\frac{\rho_s}{\rho_f}\phi_s = \Phi_0 - k^*BH_0 \tag{51.36}$$

$$\eta = 0 \quad 1 \leq \xi < \infty \quad \Phi = \Phi_0 \tag{51.37}$$

where use is made of (51.9). The flow problem is solved in the $\zeta^{1/2}$ plane of Figure 7.36(c). The jump in Φ is accounted for by placing a vortex at $\zeta^{1/2} = 1$ and an image vortex at $\zeta^{1/2} = -1$ to meet the boundary condition that the $\Im\zeta^{1/2}$ axis is a streamline. The corresponding complex potential is

$$\Omega = \frac{k^*BH_0}{\pi} i \ln \frac{\zeta^{1/2} - 1}{\zeta^{1/2} + 1} + \Phi_0 + iC_2 \tag{51.38}$$

where C_2 is a real constant.

The Function $T_2 = T_2(\zeta)$

The boundary conditions for the function T_2, defined by (51.6), are as follows. Along the interface, (51.11) applies:

$$\eta = 0 \qquad -\infty < \xi \leq 0 \qquad T_{2_1} = \Phi_0 - k^* B H_0 \tag{51.39}$$

Along the boundary equipotentials, y is equal to zero so that T_{2_1} equals Φ; the conditions for T_{2_1} along these segments are obtained from (51.36) and (51.37) as

$$\eta = 0 \qquad 0 \leq \xi \leq 1 \qquad T_{2_1} = \Phi_0 - k^* B H_0 \tag{51.40}$$

$$\eta = 0 \qquad 1 \leq \xi < \infty \qquad T_{2_1} = \Phi_0 \tag{51.41}$$

The function T_2 for this case is formally equal to the complex potential with a vortex at $\zeta = 1$ that meets the above boundary conditions:

$$T_2 = \frac{k^* B H_0}{\pi} i \ln(\zeta - 1) + \Phi_0 + i C_{2_2} \tag{51.42}$$

where C_{2_2} is a real constant.

The Function $z = z(\zeta)$

The function $z = z(\zeta)$ is obtained by the use of the definition (51.2) of T_2 and (51.38) and (51.42) as

$$ik^* B z = \Omega - T_2 = -2 \frac{k^* B H_0}{\pi} i \ln(\zeta^{1/2} + 1) + i(C_{2_1} - C_{2_2}) \tag{51.43}$$

or

$$z = -2 \frac{H_0}{\pi} \ln(\zeta^{1/2} + 1) + C_{1_z} \tag{51.44}$$

where C_{1_z} is a new real constant, determined from the condition that z be zero at $\zeta = 1$:

$$C_{1_z} = 2 \frac{H_0}{\pi} \ln 2 \tag{51.45}$$

Substitution of this expression for C_{1_z} in (51.44) yields

$$z = -\frac{2 H_0}{\pi} \ln \frac{\zeta^{1/2} + 1}{2} \tag{51.46}$$

The solution given by (51.38) and (51.46) is found in Polubarinova-Kochina [1962] along with some solutions to similar interface flow problems. The latter author derived the solution by the use of the hodograph, rather than the auxiliary function T_2.

Problem 51.1

Derive an equation for the interface from (51.46). Using the dimensionless variables z/H_0 and $\Omega/(k^*BH_0)$, draw the interface, three streamlines, and three equipotentials by using the preceeding solution.

52 SUPERPOSITION; PROBLEMS WITH DRAINS

We determined two solutions in the previous section: one deals with flow from a single horizontal equipotential toward a drain, the other with flow from a horizontal equipotential toward the coast. It is the purpose of this section to demonstrate that such solutions may be superimposed, provided that certain conditions are fulfilled. By superposition, the two solutions may be combined to cover the case of flow from an equipotential toward both the coast and a drain. The method of superposition is applicable irrespective of how the solution was determined, but requires that each individual solution be given in the form of the basic pair of functions $\Omega = \Omega(\zeta)$ and $z = z(\zeta)$ of the same reference parameter ζ. An important application of superposition is the derivation of solutions to problems with an arbitrary number of drains from the one valid for one drain. At the end of this section a brief outline is given of a procedure for determining solutions with a single drain using the hodograph method.

Superposition in the Upper Half Plane $\Im\zeta \geq 0$

We will demonstrate that superposition is applicable to pairs of derivatives of the basic functions $\Omega = \Omega(\zeta)$ and $z = z(\zeta)$ with respect to ζ; it then follows that superposition is also applicable to the basic functions themselves. Consider the following two pairs of functions, each written in terms of the same reference parameter ζ:

$$\boxed{\Omega'_1 = \Omega'_1(\zeta) \qquad z'_1 = z'_1(\zeta)} \tag{52.1}$$

and

$$\boxed{\Omega'_2 = \Omega'_2(\zeta) \qquad z'_2 = z'_2(\zeta)} \tag{52.2}$$

The real axis $\Im\zeta = 0$ is subdivided into segments, each corresponding to a boundary segment in the physical plane of one of the seven types listed in Sec. 45. We will see that if each of the above pairs fulfills the boundary conditions, then these conditions are also satisfied by the pair

$$\boxed{\Omega' = a_1\Omega'_1(\zeta) + a_2\Omega'_2(\zeta) \qquad z' = a_1z'_1(\zeta) + a_2z'_2(\zeta)} \tag{52.3}$$

where a_1 and a_2 are real multipliers,

$$\Im a_1 = \Im a_2 = 0 \tag{52.4}$$

It will be shown below that these functions fulfill the boundary conditions along two typical types of boundary conditions; it is left to the reader to verify in a similar manner that the conditions along the remaining five types of boundary segments are satisfied as well.

Let the interval $\eta = 0$, $\alpha_r \leq \xi \leq \beta_r$ correspond to a phreatic surface with accretion. The pairs (52.1) and (52.2) each fulfill the boundary conditions (45.18) along this segment:

$$\eta = 0 \quad \alpha_r \leq \xi \leq \beta_r \quad \left\{ \begin{array}{l} \Phi'_j(\xi) = kBy'_j(\xi) \\ \Psi'_j(\xi) = -NBx'_j(\xi) \end{array} \right\} \quad (j = 1, 2) \tag{52.5}$$

Hence, since a_1 and a_2 are real,

$$\eta = 0 \quad \alpha_r \leq \xi \leq \beta_r \quad \left\{ \begin{array}{l} a_1\Phi'_1(\xi) + a_2\Phi'_2(\xi) = kB[a_1y'_1(\xi) + a_2y'_2(\xi)] \\ a_1\Psi'_1(\xi) + a_2\Psi'_2(\xi) = -NB[a_1x'_1(\xi) + a_2x'_2(\xi)] \end{array} \right\} \tag{52.6}$$

With (52.3), it follows that

$$\eta = 0 \quad \alpha_r \leq \xi \leq \beta_r \quad \left\{ \begin{array}{l} \Phi'(\xi) = kBy'(\xi) \\ \Psi'(\xi) = -NBx'(\xi) \end{array} \right\} \tag{52.7}$$

so that the pair of functions obtained by superposition meets the boundary conditions, as asserted. It is important to note that the above proof is valid only if the values of α_r and β_r are the same for the two independent solutions (52.1) and (52.2). The points α_r and β_r occur as parameters in the conformal transformations used to obtain the solution, and some of such parameters cannot be determined a priori, but depend upon the solution. It is wrong to determine all such parameters independently in the solutions (52.1) and (52.2); they must be kept as unknowns until the superposition is carried out.

As a second boundary type we consider a straight equipotential. If the corresponding boundary segment in $\Im \zeta \geq 0$ is $\eta = 0$, $\alpha_e \leq \xi \leq \beta_e$, then

$$\eta = 0 \quad \alpha_e \leq \xi \leq \beta_e \quad \Phi'_j = 0 \quad \arg z'_j(\xi) = \alpha_1 \quad (j = 1, 2) \tag{52.8}$$

so that

$$\eta = 0 \quad \alpha_e \leq \xi \leq \beta_e \quad \left\{ \begin{array}{l} a_1\Phi'_1(\xi) + a_2\Phi'_2(\xi) = 0 \\ \arg[a_1z'_1(\xi) + a_2z'_2(\xi)] = \alpha_1 + \lambda\pi \end{array} \right\} \quad (\lambda = 0 \text{ or } \pm 1) \tag{52.9}$$

The factor $\lambda\pi$ occurs because there may be values of a_1 and a_2 such that the sum of $a_1|z'_1(\xi)|$ and $a_2|z'_2(\xi)|$ is negative. Such values for the multipliers are usually not allowed, and we require that

$$\eta = 0 \quad \alpha_e \leq \xi \leq \beta_e \quad a_1|z'_1(\xi)| + a_2|z'_2(\xi)| > 0 \tag{52.10}$$

Sec. 52 Superposition; Problems with Drains

This condition is clearly always fulfilled if both a_1 and a_2 are positive.

The proofs that superposition is valid for the remaining boundary types are analogous; for all straight boundaries, conditions of the form (52.10) appear.

Superposition may be applied to any number of pairs of functions that fulfill the boundary conditions, and we may write in general,

$$\Omega'(\zeta) = \sum_{j=1}^{n} a_j \Omega'_j(\zeta) \qquad z'(\zeta) = \sum_{j=1}^{n} a_j z'_j(\zeta) \qquad (52.11)$$

The functions $\Omega(\zeta)$ and $z(\zeta)$ are obtained upon integration:

$$\Omega(\zeta) = \sum_{j=1}^{n} a_j \Omega_j(\zeta) + C_\Omega \qquad z(\zeta) = \sum_{j=1}^{n} a_j z_j(\zeta) + C_z \qquad (52.12)$$

where C_Ω and C_z are complex constants of integration. The solution given by (52.12) is such that the derivatives $\Omega'(\zeta)$ and $z'(\zeta)$ fulfill the boundary conditions. This does not necessarily imply, however, that $\Omega(\zeta)$ and $z(\zeta)$ fulfill their boundary conditions; they satisfy these except for a constant of integration.

The functions $\Omega(\zeta)$ and $z(\zeta)$ contain a number of parameters, such as the real numbers a_j, the constants of integration, the positions of drains in $\Im \zeta \geq 0$, and possibly points on the real axis $\eta = 0$. The values of these constants must be determined after superposition in such a way that all boundary conditions are fulfilled. It is worth emphasizing once more that the values of the parameters in the solution should not be determined prior to the superposition; the position of a drain in $\Im \zeta \geq 0$, for example, will change as a result of the addition of functions, and keeping such a position constant in $\Im \zeta \geq 0$ will result in an undesired shift of the position of the drain in the physical plane.

Application

As an illustration, we will modify the solution for interface flow from a horizontal equipotential toward a drain, determined in Sec. 51, to include an arbitrary number of drains. This solution has the form (compare [51.5] and [51.15]):

$$\Omega = \frac{Q}{2\pi} \ln \frac{(\zeta^{1/2} - \delta^{1/2})(\zeta^{1/2} + \overline{\delta}^{1/2})}{(\zeta^{1/2} - \overline{\delta}^{1/2})(\zeta^{1/2} + \delta^{1/2})} + \Phi_0 + iC_2 \qquad (52.13)$$

$$z = -\frac{H_0}{\pi} \ln \zeta + \frac{Q}{\pi k^* B} i \ln \frac{\zeta^{1/2} + \delta^{1/2}}{\zeta^{1/2} + \overline{\delta}^{1/2}} + C_1 \qquad (52.14)$$

The value of H_0 represents the thickness of the freshwater zone at infinity, Q the discharge of the drain, and δ its position in $\Im \zeta \geq 0$. We wish to superimpose solutions of this form to account for n wells, located at $\zeta = \delta_j$ in $\Im \zeta \geq 0$ and of discharge Q_j $(j = 1, 2, \ldots n)$, while keeping the thickness of the freshwater zone at infinity at H_0. This may be achieved by superimposing n solutions of the form (52.13) and (52.14), replacing Q by Q_j and δ by δ_j, and taking H_0 as zero for all solutions, except the first one. This yields

$$\Omega = \sum_{j=1}^{n} \frac{Q_j}{2\pi} \ln \frac{(\zeta^{1/2} - \delta_j^{1/2})(\zeta^{1/2} + \overline{\delta}_j^{1/2})}{(\zeta^{1/2} - \overline{\delta}_j^{1/2})(\zeta^{1/2} + \delta_j^{1/2})} + \Phi_0 + iC_{2_1} \tag{52.15}$$

$$z = -\frac{H_0}{\pi} \ln \zeta + \sum_{j=1}^{n} \frac{Q_j}{\pi k^* B} i \ln \frac{\zeta^{1/2} + \delta_j^{1/2}}{\zeta^{1/2} + \overline{\delta}_j^{1/2}} + C_z \tag{52.16}$$

As for the case of a single drain, there is one degree of freedom that is not used and we may choose

$$\delta_1 = e^{i\alpha_1} \tag{52.17}$$

The constant C_z may be determined by requiring that z is zero for $\zeta = 1$,

$$0 = z(1) \tag{52.18}$$

It is left as an exercise for the reader to demonstrate that the solution given by (52.15) and (52.16) indeed meets the boundary conditions, and to set up a system of $2n - 1$ nonlinear real equations for the $n - 1$ complex constants δ_j and the real constant α_1. This system of equations may be solved generally with good results by the use of a Newton-Raphson procedure for a system of nonlinear equations (see e.g., Carnahan et al. [1969]).

Problem 52.1

Demonstrate that the solution given by (52.15) and (52.16) meets the boundary conditions.

Problem 52.2

Determine an expression for the constant C_z in (52.16), using (52.18), and write a system of $2n - 1$ real equations for α_1 and the $2n$ real and imaginary parts of the n complex numbers δ_j.

The problem of flow from a horizontal equipotential toward the coast illustrated in Figure 7.36(a) may be generalized to include drains in a similar fashion. Adding m solutions of the form (52.13) and (52.14) with $H_0 = 0$ to the functions (51.38) and (51.46) yields

$$\Omega = \frac{k^* B H_0}{\pi} i \ln \frac{\zeta^{1/2} - 1}{\zeta^{1/2} + 1} + \sum_{j=1}^{n} \frac{Q_j}{2\pi} \ln \frac{(\zeta^{1/2} - \delta^{1/2})(\zeta^{1/2} + \overline{\delta}^{1/2})}{(\zeta^{1/2} - \overline{\delta}^{1/2})(\zeta^{1/2} + \delta^{1/2})} + \Phi_0 + iC_{2_1} \tag{52.19}$$

$$z = -2\frac{H_0}{\pi} \ln \frac{\zeta^{1/2} + 1}{2} + \sum_{j=1}^{n} \frac{Q_j}{\pi k^* B} i \ln \frac{\zeta^{1/2} + \delta_j^{1/2}}{\zeta^{1/2} + \overline{\delta}_j^{1/2}} + C_z \tag{52.20}$$

where the constant C_z may again be determined by the use of (52.18).

Sec. 52 Superposition; Problems with Drains

The first terms in (52.19) and in (52.20) are responsible for the jump in potential at the coastline. They represent a vortex at $\zeta = 1$ and together meet the boundary conditions along the interface and that Φ is piecewise constant along the x axis. This remains true if the vortex were located at a point other than $\zeta = 1$ (point 3). We therefore may add solutions of the form

$$\Omega = \frac{\Delta \Phi_j}{\pi} i \ln \frac{\zeta^{1/2} - \nu_j^{1/2}}{\zeta^{1/2} + \nu_j^{1/2}} \tag{52.21}$$

$$z = -2 \frac{\Delta \Phi_j}{\pi k^* B} \ln(\zeta^{1/2} + \nu_j^{1/2}) \tag{52.22}$$

to generate a solution that is valid for the case that the upper boundary consists of a number of equipotentials with jumps in the potential occurring at the points $\zeta = \nu_j$. We must realize in doing this, however, that H_0 in the coefficient of the first terms in (52.20) no longer represents the saturated thickness of the aquifer at infinity, but rather the jump in potential between the sea bottom and the adjacent equipotential. We should therefore replace $k^* B H_0$ in (52.19) by $\Delta \Phi_1$,

$$\Delta \Phi_1 = k^* B H_0 \tag{52.23}$$

and H_0 in (52.20) accordingly,

$$H_0 = \frac{\Delta \Phi_1}{k^* B} \tag{52.24}$$

As an illustration, streamlines are reproduced for two flow cases in Figure 7.37.

Problem 52.3

Write expressions for Ω and z applicable to the flow case reproduced in the lower half of Figure 7.37, and express the coefficients of the vortex terms in the solution in terms of the heads along the various sections of the upper boundary.

Problems with One Drain

The application of superposition as discussed above eliminates the need for applying the hodograph method to problems involving more than a single drain; it suffices to solve problems with a single drain. There are two types of such problems. The first type of problem occurs if there is flow only if the drain operates. Such problems may be solved by direct application of the methods outlined in this chapter. An example of the hodograph for such a case is given in Figure 7.32(b). The domain in the hodograph plane always includes infinity, as a result of the infinite velocities occurring at the center of the drain.

The second type of problem occurs if there is flow without the drain pumping. As a result of the addition of the drain to the flow, a separate flow domain is produced inside the existing flow. In this domain all possible magnitudes of the discharge vector

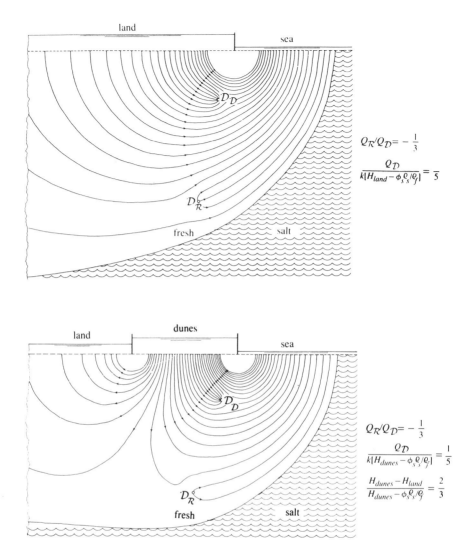

Figure 7.37 Streamlines for two flow cases. For both cases, $k = 4k^*$ and the difference in value of Ψ on neighboring streamlines is $0.025k\Delta\phi_1$. There is a drain at D_D ($Q_D > 0$) and a recharge drain at D_R ($Q_R < 0$). (Source: Strack [1973].)

will occur, from infinity at the drain to zero at the stagnation point. The hodograph will therefore become many-valued and the boundary in the hodograph plane lies on a Riemann surface, which includes an infinite sheet with the drain mapped onto infinity on that sheet and the stagnation point at the origin. It is beyond the scope of this text to discuss the details of such complex hodographs. The interested reader is referred to Strack [1972, 1973] for a detailed discussion on the subject.

It is possible, however, to present the generalized Schwarz-Christoffel transformation that is applicable to these problems without discussing the details of the many-valued hodograph. The transformations are first written for the case without drain, and the necessary modifications may be made without drawing the boundary in the many-valued hodograph that results from the addition of the drain.

We first consider the case that the problem may be solved by the use of the functions W^{-1} and Ω as the auxiliary pair. Let the functions $W^{-1}(\zeta)$ and $\Omega(\zeta)$ for the case without drain be given by the following two Schwarz-Christoffel transformations:

$$W^{-1} = |A_1|e^{i\overset{1}{\varphi}} \int \prod_{\nu=1}^{\overset{1}{n}} (\zeta - \overset{1}{\xi_\nu})^{-\overset{1}{k_\nu}} d\zeta + \overset{1}{C} \tag{52.25}$$

$$\Omega = |A_2|e^{i\overset{2}{\varphi}} \int \prod_{\nu=1}^{\overset{2}{n}} (\zeta - \overset{2}{\xi_\nu})^{-\overset{2}{k_\nu}} d\zeta + \overset{2}{C} \tag{52.26}$$

It was shown by Strack [1972], that the addition of a drain to an existing flow will cause a stagnation point to occur. If $\zeta = \sigma$ represents the location of this stagnation point in $\Im \zeta \geq 0$, then the function W will have a zero of the first order at $\zeta = \sigma$, and W^{-1} a pole of the first order at $\zeta = \sigma$. The derivative of W^{-1} will thus have a second order pole at $\zeta = \sigma$ and we modify (52.25) to account for this pole as follows:

$$W^{-1} = |A_1|e^{i\overset{1}{\varphi}} \int \frac{\prod_{j=1}^{2}(\zeta - \gamma_j)(\zeta - \overline{\gamma}_j)}{[(\zeta - \sigma)(\zeta - \overline{\sigma})]^2} \prod_{\nu=1}^{\overset{1}{n}} (\zeta - \overset{1}{\xi_\nu})^{-\overset{1}{k_\nu}} d\zeta + \overset{1}{C} \tag{52.27}$$

The factor added under the integral sign provides the desired singularity without affecting the argument of $dW^{-1}/d\zeta$ along the real axis, and without adding other singularities in the flow domain; $\zeta = \overline{\sigma}$ lies in $\Im \zeta < 0$. The numerator in the modifying factor ensures that the behavior of the function at $\zeta = \infty$ is not affected. The complex numbers γ_i represent winding points on the Riemann surface. For the purpose of integration, (52.27) is written as

$$W^{-1} = |A_1|e^{i\overset{1}{\varphi}} \int \left[\frac{a}{(\zeta - \sigma)^2} + \frac{\overline{a}}{(\zeta - \overline{\sigma})^2} + \frac{b}{(\zeta - \sigma)} + \frac{\overline{b}}{\zeta - \overline{\sigma}} + c \right]$$

$$\prod_{\nu=1}^{\overset{1}{n}} (\zeta - \overset{1}{\xi_\nu})^{-\overset{1}{k_\nu}} d\zeta + \overset{1}{C} \tag{52.28}$$

Upon integration of this expression, it will appear that W^{-1} has a logarithmic singularity at $\zeta = \sigma$. Since W is analytic at this point (it only has a zero there) no such singularity may occur; it must be removable. This condition is equivalent to the requirement that the residue of the transformation at $\zeta = 0$ vanish. (Compare the discussion of a similar singularity of the transformation [32.18] that maps $\Im \zeta \geq 0$ onto the exterior of a polygon.)

The complex potential (52.26) is modified in a similar manner. This function exhibits a logarithmic singularity at the drain, $\zeta = \delta$; its derivative has a first order pole at $\zeta = \delta$. In addition, the derivative $\Omega'(\zeta)$ will be zero at the map of the stagnation point in $\Im\zeta \geq 0$. We modify the transformation (52.26) to account for this behavior as follows:

$$\Omega = |A_2|e^{i\varphi} \int \frac{(\zeta-\sigma)(\zeta-\bar{\sigma})}{(\zeta-\delta)(\zeta-\bar{\delta})} \prod_{\nu=1}^{\frac{n}{2}}(\zeta-\xi_\nu)^{-\frac{2}{k_\nu}} d\zeta + \overset{2}{C} \qquad (52.29)$$

Also in this transformation, the modifying factor affects neither the behavior of arg $\Omega'(\xi)$ nor the behavior of $\Omega(\zeta)$ at infinity.

After carrying out the integrations, the function $z' = z'(\zeta)$ is obtained from (46.3):

$$z' = -W^{-1}(\zeta)\Omega'(\zeta) \qquad (52.30)$$

It is important to note that the singularity of W^{-1} at $\zeta = \sigma$ cancels with the zero of $\Omega'(\zeta)$ at the same point. The function $W = W(\zeta)$ is infinite at the drain and therefore $W^{-1} = W^{-1}(\zeta)$ is zero there; $z = z(\zeta)$ will be analytic at the drain because the pole of Ω' at $\zeta = \delta$ cancels with the zero that $W^{-1}(\zeta)$ exhibits there. That this is indeed the case is often not immediately evident from the functions, and considerable simplification is achieved by imposing this requirement directly on $z'(\zeta)$, rather than on the function $W^{-1}(\zeta)$:

$$\lim_{\zeta \to \delta} z'(\zeta) = M \qquad (52.31)$$

where M is some finite number.

The procedure is analogous for the case that the auxiliary functions are W_1 and T_1. We merely need replace W^{-1} by W_1, Ω by T_1 and σ by another parameter σ^*; the latter point no longer represents the stagnation point but rather the point where W equals τ_1. Replacing σ by σ^*, this gives

$$W(\sigma^*) = \tau_1 \qquad (52.32)$$

The function T_1, defined as

$$T_1 = \Omega + \tau_1 z \qquad (52.33)$$

exhibits a singularity at $\zeta = \delta$ equal to that of Ω, since z is analytic at that point. It follows from (47.16) that

$$W_1^{-1} = W - \tau_1 = -\frac{T_1'(\zeta)}{z'(\zeta)} \qquad (52.34)$$

Since the mapping is conformal at $\zeta = \sigma^*$, $z'(\zeta)$ is analytic and unequal to zero there so that, by (52.32) and (52.34),

$$T_1'(\sigma^*) = 0 \qquad (52.35)$$

Sec. 53   Flow from Recharge Drains Toward Drains in a Phreatic Aquifer   571

It appears that the functions $\underset{1}{W}$ and $\underset{1}{T}$ behave in the same way at $\zeta = \delta$ and $\zeta = \sigma^*$ as W^{-1} and Ω do at $\zeta = \delta$ and $\zeta = \sigma$. The transformations applicable to this case become (compare [52.27] and [52.29]):

$$\underset{1}{W} = |A_1|e^{i\overset{1}{\varphi}} \int \frac{\prod_{j=1}^{2}(\zeta - \gamma_j)(\zeta - \overline{\gamma}_j)}{[(\zeta - \sigma^*)(\zeta - \overline{\sigma}^*)]^2} \prod_{\nu=1}^{\overset{1}{n}}(\zeta - \overset{1}{\xi}_\nu)^{-\overset{1}{k}_\nu} d\zeta + \overset{1}{C} \qquad (52.36)$$

and

$$\underset{1}{T} = |A_2|e^{i\overset{2}{\varphi}} \int \frac{(\zeta - \sigma^*)(\zeta - \overline{\sigma}^*)}{(\zeta - \delta)(\zeta - \overline{\delta})} \prod_{\nu=1}^{\overset{2}{n}}(\zeta - \overset{2}{\xi}_\nu)^{-\overset{2}{k}_\nu} d\zeta + \overset{2}{C} \qquad (52.37)$$

The function $z' = z'(\zeta)$ is obtained by integration of (52.34); the singularity of $\underset{1}{W}$ at $\zeta = \sigma^*$ cancels with the zero of $\underset{1}{T}$; the singularity of Ω' at $\zeta = \delta$ cancels with the zero that $\underset{1}{W}$ exhibits there so that (52.31) may be applied.

The approach outlined above is often rather involved, and therefore no applications will be discussed. The reader who wishes to gain some experience with this method, however, may consider solving the problem of flow from a horizontal equipotential toward a drain and toward the sea bottom by the approach outlined above, using the functions W^{-1} and Ω. The hodograph for the case without drain is quite simple, and the operations necessary to obtain the solution are instructive; the result should be the function pair (52.19) and (52.20) with $n = 1$.

Pioneering work in solving problems with many-valued hodographs was done by De Josselin de Jong [1965] who solved a problem of interface flow toward the sea, with the upper boundary of the flow domain being composed of a horizontal streamline and the sea bottom. A drain was placed on the boundary streamline, resulting in a many-valued hodograph. De Josselin de Jong did not have to resort to the application of a modified Schwarz-Christoffel transformation, however, because the drain was located on the boundary, so that the necessary term appeared in the transformations (52.25) and (52.26).

53 FLOW FROM RECHARGE DRAINS TOWARD DRAINS IN A PHREATIC AQUIFER

As a final application of the hodograph method we will consider the problem of flow from recharge drains toward drains in a phreatic aquifer, with the recharge drains as the sole source of water. The work reported in this section was sponsored in part by the United States Bureau of Mines. The boundary of the flow domain is a phreatic surface that extends downward to infinity as shown in Figure 7.38. The problem is of interest in view of applications to solution mining, particularly if the hydraulic conductivity is anisotropic. In this section we will make restriction to cases of isotropic hydraulic conductivity, and cover anisotropy in Sec. 54. In either case, the objective of the analysis is to determine

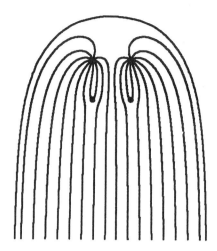

Figure 7.38 The flow domain with two recharge drains and two drains.

a solution from which spacings and flow rates of the drains may be determined such that a stable phreatic surface exists.

Flow From a Single Recharge Drain

We first solve the problem of flow with a single recharge drain. The solution for flow with an arbitrary number of recharge drains and drains is obtained by superposition.

The Hodograph

A sketch of the flow domain along with the corresponding domain in the hodograph plane is given in Figure 7.39. The origin in the z plane is chosen at the axis of the recharge drain, with the x and y axis horizontal and vertical as shown in Figure 7.39(a). The boundary of the domain in the hodograph plane of Figure 7.39(b) is the circle $1\mathcal{A}\text{-}\mathcal{S}\text{-}1\mathcal{B}$, with the domain being the exterior of the circle and the drain mapped onto infinity. Since the only boundary segment is a circle, any point on the circle may be used as a point of inversion. We choose points 1 and $\mathcal{S}$ as points of inversion so that

$$\tau_1 = 0 \qquad \tau_2 = ikB \tag{53.1}$$

with the corresponding auxiliary functions equal to Ω and the Zhukovski function Z (compare [49.47]),

$$\underset{1}{T} = \Omega \qquad \underset{2}{T} = Z = \Omega + ikBz \tag{53.2}$$

We will determine the complex potential Ω and the Zhukovski function Z in terms of the reference parameter ζ, chosen such that infinity corresponds to $\zeta = \infty$ and the recharge drain to $\zeta = \delta$ (see Figure 7.40).

Sec. 53 Flow from Recharge Drains Toward Drains in a Phreatic Aquifer

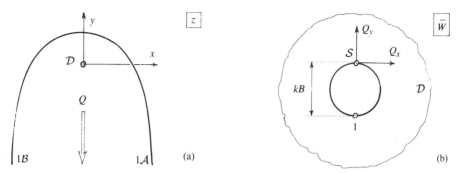

Figure 7.39 The flow domain and the hodograph.

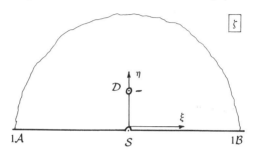

Figure 7.40 The reference half plane.

The Complex Potential

The boundary conditions for the complex potential are that the real axis $\Im\zeta = 0$ is a streamline and that there is a drain of recharge Q at $\zeta = \delta$. The function $\Omega = \Omega(\zeta)$ is obtained by the use of the method of images,

$$\Omega = -\frac{Q}{2\pi}\ln(\zeta - \delta)(\zeta - \bar{\delta}) + \underset{1}{C_1} \tag{53.3}$$

where $\underset{1}{C_1}$ is a real constant, and the value for Ψ along the ξ axis is chosen as zero.

The Zhukovski Function

We choose the reference level for the head to coincide with the x axis so that the boundary conditions (45.12) along the phreatic surface give with (53.2):

$$\eta = 0 \qquad -\infty < \xi < \infty \qquad X = \Re(Z) = \Phi - kBy = 0 \tag{53.4}$$

The Zhukovski function behaves as the complex potential near the recharge drain. Placing an image drain at $\zeta = \bar{\delta}$ in order to satisfy the boundary condition (53.4), we obtain

$$Z = -\frac{Q}{2\pi}\ln\frac{\zeta - \delta}{\zeta - \bar{\delta}} + i\underset{2}{C_2} \tag{53.5}$$

where $\underset{2}{C_2}$ is a real constant.

The Function $z = z(\zeta)$

An expression for the function $z = z(\zeta)$ is obtained from (53.2), (53.3), and (53.5) as follows:

$$ikBz = Z - \Omega = \frac{Q}{\pi}\ln(\zeta - \bar{\delta}) + ikBC \tag{53.6}$$

where

$$C = \frac{1}{kB}(C_2 + iC_1) \tag{53.7}$$

Division of (53.6) by ikB yields

$$z = -\frac{Q}{\pi kB}i\ln(\zeta - \bar{\delta}) + C \tag{53.8}$$

Flow with an Arbitrary Number of Drains and Recharge Drains

The functions $\Omega = \Omega(\zeta)$ and $z = z(\zeta)$ for the case of flow with n_i recharge drains and n_e drains is obtained by superposition of $n_i + n_e$ solutions of the form (53.3) and (53.6). Denoting the recharges and locations in $\Im\zeta \geq 0$ of the recharge drains as $\overset{i}{Q}_j$ and δ_j ($j = 1, 2, \ldots, n_i$) and the discharges and locations of the drains as $\overset{e}{Q}_j$ and ϱ_j ($j = 1, 2, \ldots, n_e$), we obtain, introducing new constants C_Ω and C_z,

$$\Omega = -\frac{1}{2\pi}\sum_{j=1}^{n_i}\overset{i}{Q}_j\ln[(\zeta - \delta_j)(\zeta - \bar{\delta}_j)] + \frac{1}{2\pi}\sum_{j=1}^{n_e}\overset{e}{Q}_j\ln[(\zeta - \varrho_j)(\zeta - \bar{\varrho}_j)] + \Re C_\Omega \tag{53.9}$$

and

$$z = -\frac{i}{\pi kB}\sum_{j=1}^{n_i}\overset{i}{Q}_j\ln(\zeta - \bar{\delta}_j) + \frac{i}{\pi kB}\sum_{j=1}^{n_e}\overset{e}{Q}_j\ln(\zeta - \bar{\varrho}_j) + C_z \tag{53.10}$$

We choose the origin of the z plane to be at the center of the first recharge drain:

$$0 = z(\delta_1) \tag{53.11}$$

Applying this condition to (53.10), solving for C_z and substituting the result in (53.10) we obtain:

$$z = -\frac{i}{\pi kB}\sum_{j=1}^{n_i}\overset{i}{Q}_j\ln\frac{\zeta - \bar{\delta}_j}{\delta_1 - \bar{\delta}_j} + \frac{i}{\pi kB}\sum_{j=1}^{n_e}\overset{e}{Q}_j\ln\frac{\zeta - \bar{\varrho}_j}{\delta_1 - \bar{\varrho}_j} \tag{53.12}$$

We set $\zeta = \xi$ in (53.9) and (53.12) and apply the condition that $\Phi = kBy$, which yields an expression for $\Re C_\Omega$. Substituting this expression in (53.9) we obtain:

$$\Omega = -\frac{1}{2\pi}\sum_{j=1}^{n_i} \overset{i}{Q}_j \ln\frac{(\zeta-\delta_j)(\zeta-\overline{\delta}_j)}{|\delta_1-\overline{\delta}_j|^2} + \frac{1}{2\pi}\sum_{j=1}^{n_e} \overset{e}{Q}_j \ln\frac{(\zeta-\varrho_j)(\zeta-\overline{\varrho}_j)}{|\delta_1-\overline{\varrho}_j|^2} \qquad (53.13)$$

Denoting the locations of the recharge drains and the drains in the physical plane as $\overset{i}{z}_m$ ($m = 2,\ldots,n_i$) and $\overset{e}{z}_m$ ($m = 1, 2,\ldots,n_e$), respectively, we obtain the following system of equations:

$$\overset{i}{z}_m = -\frac{i}{\pi kB}\sum_{j=1}^{n_i} \overset{i}{Q}_j \ln\frac{\delta_m-\overline{\delta}_j}{\delta_1-\overline{\delta}_j} + \frac{i}{\pi kB}\sum_{j=1}^{n_e} \overset{e}{Q}_j \ln\frac{\delta_m-\overline{\varrho}_j}{\delta_1-\overline{\varrho}_j} \qquad (m=2,3,\ldots,n_i)$$

(53.14)

and

$$\overset{e}{z}_m = -\frac{i}{\pi kB}\sum_{j=1}^{n_e} \overset{e}{Q}_j \ln\frac{\varrho_m-\overline{\delta}_j}{\delta_1-\overline{\delta}_j} + \frac{i}{\pi kB}\sum_{j=1}^{n_e} \overset{e}{Q}_j \ln\frac{\varrho_m-\overline{\varrho}_j}{\delta_1-\overline{\varrho}_j} \qquad (m=1,2,\ldots,n_i)$$

(53.15)

(53.14) and (15.15) provide $n_i + n_e - 1$ complex nonlinear equations for the $n_i + n_e - 1$ complex parameters δ_j ($j = 2, 3,\ldots, n_i$) and ϱ_j ($j = 1, 2,\ldots, n_e$).

The mapping contains three degrees of freedom, one of which has been used by choosing point 1 to be at $\zeta = \infty$. The remaining two degrees of freedom are used by choosing

$$\delta_1 = i \qquad (53.16)$$

The flow pattern shown in Figure 7.38 was produced by a computer program that uses a Newton-Raphson procedure for solving the above system of nonlinear equations. The phreatic surface is plotted by letting ζ assume real values while computing the corresponding points in the z plane by the use of (53.12). Points on the streamlines are determined in terms of ζ by tracing the curves Ψ = constant from the recharge drain in $\Im\zeta \geq 0$; the corresponding values of z are obtained from (53.12).

Of special interest are cases where the phreatic surface is on the verge of instability and unsaturated flow is about to occur. We will establish the criterion for stability of the phreatic surface, and examine the solution for the case of flow with one recharge drain and one drain in some detail.

Stability of the Phreatic Surface

Stability of the phreatic surface requires that a small amount of water that has left the saturated body of water will return to it. It is assumed in the present analysis that the pressure is constant (zero) outside the phreatic surface, so that the gradient of the potential there is vertical. If the phreatic surface is steeper than vertical so that air is below groundwater, then any water that has left the body of groundwater will not return. The condition for stability therefore may be formulated for the present problem as follows:

$$\eta = 0 \quad -\infty < \xi < \infty \quad \frac{\partial x}{\partial \xi} = x'(\xi) \leq 0 \qquad (53.17)$$

which may be interpreted as the requirement that a positive increment $d\xi$ must always correspond to a negative increment of x at any point of the real axis, which is the image of the phreatic surface (see Figure 7.41).

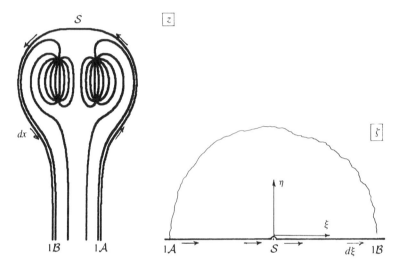

Figure 7.41 The stability criterion.

Solution for a System of One Recharge Drain and One Drain

We obtain the solution for a single pair of a drain vertically below a recharge drain (see Figure 7.42) from (53.12) and (53.13) by setting $n_i = n_e = 1$. By symmetry, the recharge drain is mapped onto a point of the imaginary axis $\xi = 0$. Writing ϱ_1 as $i\lambda$ ($\lambda > 1$), and using (53.16), the expression for $z(\zeta)$ becomes

$$z = -\frac{i}{\pi k B}\overset{i}{Q}\ln\frac{\zeta + i}{2i} + \frac{i}{\pi k B}\overset{e}{Q}\ln\frac{\zeta + i\lambda}{i + i\lambda} \qquad (53.18)$$

where the index 1 is dropped in Q and $\overset{e}{Q}$. If the drain lies in the z plane a distance h below the recharge drain, then $\overset{e}{z}$ equals $-ih$ for $\zeta = i\lambda$,

$$\overset{e}{z} = -ih = -\frac{i}{\pi k B}\overset{i}{Q}\ln\frac{i\lambda + i}{2i} + \frac{i}{\pi k B}\overset{e}{Q}\ln\frac{i\lambda + i\lambda}{i + i\lambda} \qquad (53.19)$$

Sec. 53   Flow from Recharge Drains Toward Drains in a Phreatic Aquifer   577

or, simplifying this,

$$h = \frac{1}{\pi kB}\overset{i}{Q}\ln\frac{\lambda+1}{2} - \frac{1}{\pi kB}\overset{e}{Q}\ln\frac{2\lambda}{\lambda+1} \qquad (53.20)$$

Application of the stability condition (53.17) requires that the derivative of $z(\zeta)$ with respect to ζ be determined:

$$\frac{dz}{d\zeta} = -\frac{i}{\pi kB}\overset{i}{Q}\frac{1}{\zeta+i} + \frac{i}{\pi kB}\overset{e}{Q}\frac{1}{\zeta+i\lambda} \qquad (53.21)$$

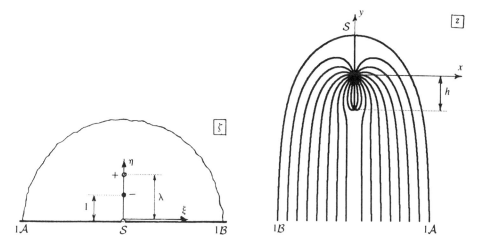

Figure 7.42   The case of flow with a single recharge drain and a single drain.

Along the real axis, $\zeta = \xi$, this may be written as

$$z'(\xi) = -\frac{i}{\pi kB}\overset{i}{Q}\frac{\xi-i}{\xi^2+1} + \frac{i}{\pi kB}\overset{e}{Q}\frac{\xi-i\lambda}{\xi^2+\lambda^2} \qquad (53.22)$$

Applying (53.17), we obtain

$$\eta = 0 \qquad -\infty < \xi < \infty \qquad x' = -\frac{i}{\pi kB}\overset{i}{Q}\frac{-i}{\xi^2+1} + \frac{i}{\pi kB}\overset{e}{Q}\frac{-i\lambda}{\xi^2+\lambda^2}$$

$$= -\frac{\overset{i}{Q}}{\pi kB}\frac{1}{\xi^2+1} + \frac{\overset{e}{Q}}{\pi kB}\frac{\lambda}{\xi^2+\lambda^2} \leq 0 \qquad (53.23)$$

or, with $\xi^2+1 > 0$ and $\xi^2+\lambda^2 > 0$,

$$-\overset{i}{Q}(\xi^2+\lambda^2) + \overset{e}{Q}\lambda(\xi^2+1) \leq 0 \qquad (53.24)$$

This becomes, after rearrangement,

$$-\xi^2(\overset{i}{Q} - \overset{e}{Q}\lambda) - \overset{i}{Q}\lambda^2 + \overset{e}{Q}\lambda \leq 0 \qquad (53.25)$$

This inequality must be fulfilled for all values of ξ; for $\xi \to \infty$ the inequality can be fulfilled only if the ratio, γ, of $\overset{e}{Q}$ to $\overset{i}{Q}$ is less than or equal to $1/\lambda$:

$$\gamma = \frac{\overset{e}{Q}}{\overset{i}{Q}} \leq \frac{1}{\lambda} \qquad (53.26)$$

With this condition fulfilled, (53.25) may be written as

$$\xi^2 + \frac{\overset{i}{Q}\lambda^2 - \overset{e}{Q}\lambda}{\overset{i}{Q} - \overset{e}{Q}\lambda} = \xi^2 + \frac{\lambda(\lambda - \gamma)}{1 - \gamma\lambda} \geq 0 \qquad (53.27)$$

In view of (53.26), $1 - \gamma\lambda$ is positive; λ is greater than one because the drain lies below the recharge drain in the physical plane (compare Figure 7.42) and γ is less than or equal to one because the drain cannot extract more water than the recharge drain produces:

$$\lambda > 1 \qquad \gamma = \frac{\overset{e}{Q}}{\overset{i}{Q}} \leq 1 \qquad (53.28)$$

It follows from (53.28) that (53.27) is met; equation (53.26) is the stability criterion.

Multiplication of both sides of (53.20) by $\pi k B / \overset{i}{Q}$ yields, with $\gamma = \overset{e}{Q}/\overset{i}{Q}$,

$$\frac{\pi k B h}{\overset{i}{Q}} = \ln \frac{\lambda + 1}{2} - \gamma \ln \frac{2\lambda}{\lambda + 1} \qquad (53.29)$$

The maximum rate of extraction, and therefore the maximum value of γ, $\gamma_{\max}$, corresponds to the case that the equal sign applies in the stability criterion (53.26), i.e., if $\gamma = 1/\lambda$:

$$\frac{\pi k B h}{\overset{i}{Q}} = \ln \frac{1 + \gamma_{\max}}{2\gamma_{\max}} - \gamma_{\max} \ln \frac{2}{1 + \gamma_{\max}} \qquad (53.30)$$

This equation yields the maximum rate of extraction, for which the phreatic surface is stable as a function of the ratio $kBh/\overset{i}{Q}$. This equation is represented graphically in Figure 7.43.

Problem 53.1

Solve the problem for flow from a single recharge drain for the case that there is an infiltration rate N ($N < k$) along the phreatic surface, assuming that the phreatic surface extends infinitely far downward as in Figure 7.38 and that the pressure along the phreatic surface is zero.

Sec. 54 Anisotropic Hydraulic Conductivity

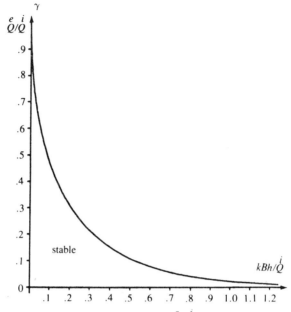

Figure 7.43 The maximum value for $\gamma = \overset{e}{Q}/\overset{i}{Q}$ as a function of $kBh/\overset{i}{Q}$.

54 ANISOTROPIC HYDRAULIC CONDUCTIVITY

Problems with free boundaries in aquifers with an anisotropic, but uniform, hydraulic conductivity may be solved in a transformed domain, as discussed in Sec. 17. In the latter section we did not consider the boundary conditions along free boundaries. We will see in this section that the conditions along free boundaries in the transformed domain are similar to those in an isotropic domain. We will limit the discussion to a phreatic surface; the boundary conditions along the other types of free boundaries may be established analogously.

Transformation of the Domain

We denote the Cartesian coordinates in the physical plane as $\overset{p}{x}$, $\overset{p}{y}$, where the $\overset{p}{x}$ and $\overset{p}{y}$ axes are horizontal and vertical, respectively, and where the superscript p stands for physical. The major principal direction of the hydraulic conductivity tensor makes an angle α with the $\overset{p}{x}$ axis, and we choose the Cartesian coordinates $\overset{p}{x}{}^*$, $\overset{p}{y}{}^*$ such that the $\overset{p}{x}{}^*$ axis is inclined at an angle α to the $\overset{p}{x}$ axis, as shown in Figure 7.44(a). The Cartesian coordinates $\overset{t}{x}{}^*$ and $\overset{t}{y}{}^*$ in the transformed domain of Figure 7.44(b), labeled by the superscript t, are chosen such that they correspond to the coordinates $\overset{p}{x}$, and $\overset{p}{y}$ in the physical plane. The Cartesian coordinate system $\overset{t}{x}$, $\overset{t}{y}$, finally, is introduced in the transformed domain such that the $\overset{t}{x}$ axis corresponds to the $\overset{p}{x}$ axis.

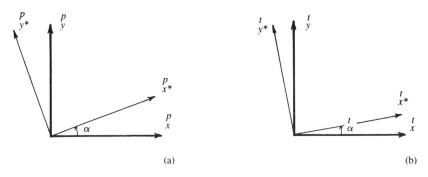

Figure 7.44 Transformation onto an isotropic domain.

The coordinates $\overset{p}{x}{}^*$ and $\overset{p}{y}{}^*$ are expressed in terms of $\overset{p}{x}$ and $\overset{p}{y}$ by (compare [3.17]):

$$\overset{p}{x} = \overset{p}{x}{}^* \cos\alpha - \overset{p}{y}{}^* \sin\alpha$$
$$\overset{p}{y} = \overset{p}{x}{}^* \sin\alpha + \overset{p}{y}{}^* \cos\alpha \qquad (54.1)$$

The inverse of this transformation is given by

$$\overset{p}{x}{}^* = \overset{p}{x}\cos\alpha + \overset{p}{y}\sin\alpha$$
$$\overset{p}{y}{}^* = -\overset{p}{x}\sin\alpha + \overset{p}{y}\cos\alpha \qquad (54.2)$$

The transformation into the isotropic domain is given by (17.35), where X and Y are to be replaced by $\overset{t}{x}{}^*$ and $\overset{t}{y}{}^*$, and x^* and y^* by $\overset{p}{x}{}^*$ and $\overset{p}{y}{}^*$,

$$\overset{t}{x}{}^* = \overset{p}{x}{}^*$$
$$\overset{t}{y}{}^* = \beta \overset{p}{y}{}^* \qquad (54.3)$$

where β is given by (17.38),

$$\beta = \sqrt{\frac{k_1}{k_2}} \qquad (54.4)$$

If the angle between the $\overset{t}{x}{}^*$ and $\overset{t}{x}$ axes is $\overset{t}{\alpha}$, then

$$\overset{t}{x} = \overset{t}{x}{}^* \cos\overset{t}{\alpha} - \overset{t}{y}{}^* \sin\overset{t}{\alpha}$$
$$\overset{t}{y} = \overset{t}{x}{}^* \sin\overset{t}{\alpha} + \overset{t}{y}{}^* \cos\overset{t}{\alpha} \qquad (54.5)$$

The $\overset{t}{x}$ axis corresponds to the $\overset{p}{x}$ axis in the physical plane. Since $\overset{t}{y} = 0$ along the $\overset{t}{x}$ axis we have from (54.5)

$$\frac{\overset{t}{y}{}^*}{\overset{t}{x}{}^*} = -\tan\overset{t}{\alpha} \qquad (\overset{t}{y} = 0) \qquad (54.6)$$

Sec. 54 Anisotropic Hydraulic Conductivity

and $\overset{p}{y} = 0$ along the $\overset{p}{x}$ axis so that, with (54.1)

$$\frac{\overset{p}{y}{}^*}{\overset{p}{x}{}^*} = -\tan\alpha \qquad (\overset{p}{y} = 0) \tag{54.7}$$

We obtain an expression for $\overset{t}{\alpha}$ in terms of α from (54.3), (54.6) and (54.7),

$$\tan\overset{t}{\alpha} = \beta\tan\alpha \tag{54.8}$$

We now express $\overset{t}{x}$ and $\overset{t}{y}$ in terms of $\overset{p}{x}$ and $\overset{p}{y}$ by the use of (54.5), (54.3), and (54.2),

$$\begin{aligned}\overset{t}{x} &= (\overset{p}{x}\cos\alpha + \overset{p}{y}\sin\alpha)\cos\overset{t}{\alpha} - \beta(-\overset{p}{x}\sin\alpha + \overset{p}{y}\cos\alpha)\sin\overset{t}{\alpha}\\ \overset{t}{y} &= (\overset{p}{x}\cos\alpha + \overset{p}{y}\sin\alpha)\sin\overset{t}{\alpha} + \beta(-\overset{p}{x}\sin\alpha + \overset{p}{y}\cos\alpha)\cos\overset{t}{\alpha}\end{aligned} \tag{54.9}$$

This may be written as

$$\begin{aligned}\overset{t}{x} &= \cos\alpha\cos\overset{t}{\alpha}\left[\overset{p}{x}(1 + \beta\tan\alpha\tan\overset{t}{\alpha}) + \overset{p}{y}(\tan\alpha - \beta\tan\overset{t}{\alpha})\right]\\ \overset{t}{y} &= \cos\alpha\cos\overset{t}{\alpha}\left[\overset{p}{x}(\tan\overset{t}{\alpha} - \beta\tan\alpha) + \overset{p}{y}(\tan\alpha\tan\overset{t}{\alpha} + \beta)\right]\end{aligned} \tag{54.10}$$

We can always choose the coordinate transformations in such a way that $-\pi/2 \le \alpha \le \pi/2$. By (54.8), $\overset{t}{\alpha}$ will then fall in the same range; we can limit our analysis without loss of generality to the case that

$$-\frac{\pi}{2} \le \alpha \le \frac{\pi}{2} \qquad -\frac{\pi}{2} \le \overset{t}{\alpha} \le \frac{\pi}{2} \tag{54.11}$$

With (54.11), $\cos\overset{t}{\alpha}$ is always ≥ 0 so that $\cos\overset{t}{\alpha} = (\tan^2\overset{t}{\alpha} + 1)^{-1/2}$. Using the latter expression with (54.8) in (54.10), we obtain after some elaboration

$$\boxed{\overset{t}{x} = \overset{p}{x}\sqrt{\cos^2\alpha + \beta^2\sin^2\alpha} + \overset{p}{y}\frac{(1-\beta^2)\sin\alpha\cos\alpha}{\sqrt{\cos^2\alpha + \beta^2\sin^2\alpha}}} \tag{54.12}$$

$$\boxed{\overset{t}{y} = \overset{p}{y}\frac{\beta}{\sqrt{\cos^2\alpha + \beta^2\sin^2\alpha}}} \tag{54.13}$$

Boundary Conditions Along a Phreatic Surface Without Infiltration

We may now write the boundary conditions along the map of a phreatic surface onto the transformed domain. Since the pressure is zero there, we have

$$\phi = \overset{p}{y} - \overset{p}{y}_r \tag{54.14}$$

where $\overset{p}{y}_r$ is the ordinate of the reference level for the head. Using the transformation (54.13), we obtain

$$\phi = \overset{t}{y}\frac{\sqrt{\cos^2\alpha + \beta^2\sin^2\alpha}}{\beta} - \overset{p}{y}_r \qquad (54.15)$$

The potential Φ is defined in the transformed domain by (17.44),

$$\Phi = B\phi\sqrt{k_1 k_2} = k_2\beta B\phi \qquad (54.16)$$

and the boundary condition may be expressed in terms of the potential as

$$\boxed{\Phi = k_2 B\overset{t}{y}\sqrt{\cos^2\alpha + \beta^2\sin^2\alpha} - B\overset{p}{y}_r\sqrt{k_1 k_2}} \qquad (54.17)$$

Usually $\overset{p}{y}_r$ is taken as zero; in that case the boundary condition (54.16) is obtained from that for flow in an isotropic aquifer by replacing k by a factor K, defined as

$$\boxed{K = k_2 B\sqrt{\cos^2\alpha + \beta^2\sin^2\alpha}} \qquad (54.18)$$

The second boundary condition along the phreatic surface is that it is a streamline; this boundary condition remains unchanged:

$$\Psi = \Psi_0 \qquad (54.19)$$

Boundary Conditions Along a Phreatic Surface with an Infiltration Rate N

We next consider the case that there is an infiltration rate N. We assume that the flow in the unsaturated zone is driven by gravity alone, and that the pressure is zero there. The head ϕ in the unsaturated zone then equals the elevation head:

$$\phi = \overset{p}{y} - \overset{p}{y}_r \qquad (54.20)$$

with a gradient with components

$$\frac{\partial \phi}{\partial x} = 0 \qquad \frac{\partial \phi}{\partial y} = 1 \qquad (54.21)$$

Expressions for the components Q_x and Q_y that occur under the influence of this gradient are obtained from Darcy's law (3.22),

$$\begin{aligned} Q_x &= -k_{xy}B \\ Q_y &= -k_{yy}B \end{aligned} \qquad (54.22)$$

Sec. 54 Anisotropic Hydraulic Conductivity

Using (3.23), this may be written as

$$Q_x = -(k_1 - k_2)B \sin \alpha \cos \alpha$$
$$Q_y = -B(k_1 \sin^2 \alpha + k_2 \cos^2 \alpha) \tag{54.23}$$

We introduce the angle ν as shown in Figure 7.45 and obtain

$$\tan \nu = \frac{-Q_y}{-Q_x} = \frac{k_1 \sin^2 \alpha + k_2 \cos^2 \alpha}{(k_1 - k_2) \sin \alpha \cos \alpha} \tag{54.24}$$

The magnitude of the discharge vector with components given by (54.23) represents the maximum rate of infiltration that may occur,

$$N_{\max} = \sqrt{Q_x^2 + Q_y^2}$$
$$= B\sqrt{k_1^2 \sin^4 \alpha + 2k_1 k_2 \sin^2 \alpha \cos^2 \alpha + k_2^2 \cos^4 \alpha + (k_1 - k_2)^2 \sin^2 \alpha \cos^2 \alpha} \tag{54.25}$$

Using (54.4) and combining terms, we obtain

$$N_{\max} = k_2 B \sqrt{\cos^2 \alpha + \beta^4 \sin^2 \alpha} \tag{54.26}$$

The components of the discharge vector resulting from an infiltration rate $N < N_{\max}$ are obtained by multiplying the right-hand sides of (54.23) by a factor A equal to

$$A = \frac{N}{N_{\max}} \tag{54.27}$$

The potential Φ in the isotropic domain in the unsaturated zone is identical to the one given by (54.17), and the corresponding components of the discharge vector in the transformed domain, $\overset{t}{Q}_x$ and $\overset{t}{Q}_y$, are obtained upon differentiation:

$$\overset{t}{Q}_x = 0 \qquad \overset{t}{Q}_y = -k_2 B \sqrt{\cos^2 \alpha + \beta^2 \sin^2 \alpha} \tag{54.28}$$

We observe that the infiltration, which occurs at an angle ν in the physical plane, corresponds to infiltration in the $\overset{t}{y}$ direction in the isotropic domain (the reader may verify this also by transforming a line at an angle ν in the physical plane to the isotropic domain). Equation (54.28) corresponds to the case that N equals $N_{\max}$. We represent the components of the discharge vector in the transformed domain due to an infiltration rate $N < N_{\max}$ as $\overset{t}{Q}_{x\,r}$, $\overset{t}{Q}_{y\,r}$ and obtain expressions for these components by multiplying the right-hand sides in (54.28) by a factor $N/N_{\max}$, which gives

$$\overset{t}{Q}_{x\,r} = 0 \qquad \overset{t}{Q}_{y\,r} = -NB \sqrt{\frac{\cos^2 \alpha + \beta^2 \sin^2 \alpha}{\cos^2 \alpha + \beta^4 \sin^2 \alpha}} \tag{54.29}$$

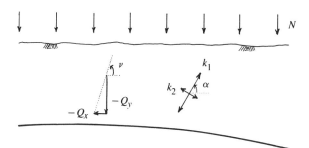

Figure 7.45 Inflow under the influence of gravity.

It follows that the infiltration along a phreatic surface may be handled in the transformed domain in a similar fashion as is done if the aquifer is isotropic, but the infiltration rate in the transformed domain equals N multiplied by the square root in (54.29). The complete set of boundary conditions along a phreatic surface mapped onto a section of the ξ axis in the upper half plane becomes (compare [45.18], [54.17], and [54.29]):

$$\Phi(\xi) = k_2 B \overset{t}{y}(\xi) \sqrt{\cos^2 \alpha + \beta^2 \sin^2 \alpha} + B \overset{p}{y}_r \sqrt{k_1 k_2}$$

$$\Psi'(\xi) = -NB \sqrt{\frac{\cos^2 \alpha + \beta^2 \sin^2 \alpha}{\cos^2 \alpha + \beta^4 \sin^2 \alpha}} x'(\xi)$$

(54.30)

Flow from One Recharge Drain Toward a Drain in an Anisotropic Aquifer

The results obtained in the previous section may be readily applied to the case that the hydraulic conductivity is anisotropic. The solutions presented in Sec. 53 meet the boundary condition that $\Phi = kBy$ along the phreatic surface. We adapt this solution by replacing z by

$$\overset{t}{z} = \overset{t}{x} + i\overset{t}{y}$$

(54.31)

and the factor k occurring in the expression for z in terms of ζ by K (see [54.18]),

$$K = k_2 B \sqrt{\cos^2 \alpha + \beta^2 \sin^2 \alpha}$$

(54.32)

The potential Φ is now defined in terms of ϕ by (54.16),

$$\Phi = B\phi\sqrt{k_1 k_2}$$

(54.33)

We will discuss the case of flow with a single recharge drain and a single drain in some detail. The stability criterion must be adapted; water leaving the saturated zone will

Sec. 54 Anisotropic Hydraulic Conductivity

now tend to flow downward along a line inclined at the angle ν given by (54.24). Since this line corresponds to one parallel to the $\overset{t}{y}$ axis in the transformed domain, the stability criterion in the latter domain is identical to that presented for an isotropic aquifer. It thus follows that the optimum location of the drain relative to the recharge drain is obtained by placing it on the $\overset{t}{y}$ axis, so that the line connecting the drains makes an angle ν with the $\overset{p}{x}$ axis in the physical plane.

We denote the coordinates of the drains in the physical and $\overset{t}{z}$ planes as

$$\overset{p}{x}_d = -\overset{p}{d} \qquad \overset{p}{y}_d = -\overset{p}{h} \qquad \overset{t}{z}_d = -i\overset{t}{h} \tag{54.34}$$

Using the transformation formulae (54.12) and (54.13), we obtain the following expressions for $\overset{p}{d}$ and $\overset{p}{h}$,

$$\overset{p}{d} = \overset{p}{h}(\beta^2 - 1)\frac{\sin\alpha\cos\alpha}{\cos^2\alpha + \beta^2\sin^2\alpha} \qquad \overset{p}{h} = \frac{\overset{t}{h}}{\beta}\sqrt{\cos^2\alpha + \beta^2\sin^2\alpha} \tag{54.35}$$

The stability criterion (53.30) may be applied to the present case, provided that h be replaced by $\overset{t}{h}$ and k by the factor K given in (54.18),

$$\frac{\pi k_2 B \overset{t}{h}\sqrt{\cos^2\alpha + \beta^2\sin^2\alpha}}{\overset{i}{Q}} = \ln\frac{1+\gamma_{\max}}{2\gamma_{\max}} - \gamma_{\max}\ln\frac{2}{1+\gamma_{\max}} \tag{54.36}$$

Using the second equation in (54.35) and the expression (54.4) for β, this becomes

$$\frac{\pi B \overset{p}{h}\sqrt{k_1 k_2}}{\overset{i}{Q}} = \ln\frac{1+\gamma_{\max}}{2\gamma_{\max}} - \gamma_{\max}\ln\frac{2}{1+\gamma_{\max}} \tag{54.37}$$

It follows that the graph shown in Figure 7.43 is valid for the anisotropic case, provided that the variable plotted along the horizontal axis is interpreted as $B\overset{p}{h}\sqrt{k_1 k_2}/\overset{i}{Q}$. It is of interest to note that, with this positioning of the recovery well, the recovery rate $\gamma_{\max}$ is independent of the orientation of the principal axes and the ratio of the principal values, even though the latter two variables greatly affect the shape of the phreatic surface.

Plots of the Phreatic Surface

The phreatic surface may be determined in the $\overset{t}{z}$ plane by the use of (53.18). Replacing z by $\overset{t}{z}$, k by K and dividing both sides of the equation by $\overset{t}{h}$, we obtain

$$\frac{\overset{t}{z}}{\overset{t}{h}} = -i\frac{\overset{i}{q}}{\pi}\ln\frac{\zeta+i}{2i} + i\frac{\overset{e}{q}}{\pi}\ln\frac{\zeta+i\lambda}{i+i\lambda} \tag{54.38}$$

where $\overset{i}{q}$ and $\overset{e}{q}$ are dimensionless quantities, defined as

$$\overset{i}{q} = \frac{\overset{i}{Q}}{KB\overset{t}{h}} = \frac{\overset{i}{Q}}{B\overset{p}{h}\sqrt{k_1 k_2}} \qquad \overset{e}{q} = \frac{\overset{e}{Q}}{B\overset{p}{h}\sqrt{k_1 k_2}} \tag{54.39}$$

where use is made of (54.4), (54.32), and (54.35). An expression from which λ can be computed is obtained from (53.20), where h is to be replaced by $\overset{t}{h}$ and k by K. Dividing this equation by $\overset{t}{h}$ we obtain

$$1 = \frac{\overset{i}{q}}{\pi} \ln \frac{\lambda+1}{2} - \frac{\overset{e}{q}}{\pi} \ln \frac{2\lambda}{\lambda+1} \tag{54.40}$$

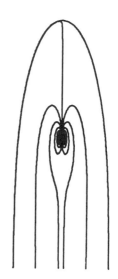

Figure 7.46 Phreatic surface and streamlines for $\alpha = 90°$.

The phreatic surface is plotted by letting ζ assume values on the real axis while computing $\overset{t}{z}/\overset{t}{h}$ from (54.38). The coordinates in the physical plane are then obtained by the use of (54.12) and (54.13). As an illustration, plots of the phreatic surface and streamlines are shown for three cases. For all three cases B equals 1, β^2 equals 10, $\overset{i}{q}$ equals 273, and $\gamma = \gamma_{max} = 0.88$. For the case of Figure 7.46 α equals 90°, for the case of Figure 7.47 α equals 60°, and for the case of Figure 7.48 α equals 0°.

Sec. 54 Anisotropic Hydraulic Conductivity

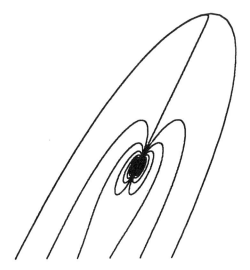

Figure 7.47 Phreatic surface and streamlines for $\alpha = 60°$.

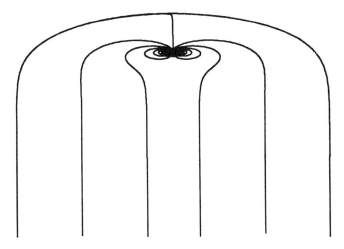

Figure 7.48 Phreatic surface and streamlines for $\alpha = 0°$.

Chapter 8

The Boundary Integral Equation Method

The boundary integral equation method consists of formulating boundary value problems in terms of an integral equation. In numerical applications, the integral equation is solved in an approximate fashion: the boundary is approximated by a series of straight lines or elementary curves and simplifying assumptions are made regarding the behavior of the solution along the boundary segments. The solution thus obtained satisfies the governing differential equation exactly, but meets the boundary conditions approximately. The boundary integral equation method as commonly applied to groundwater flow problems is formulated in terms of real variables and is referred to in the literature as the BIEM (Liggett and Liu [1983]). For two-dimensional problems the BIEM is closely related to the Cauchy integral, discussed briefly in Sec. 25. The BIEM has the advantage of being applicable to three-dimensional flow. For two-dimensional problems, however, the formulation in terms of complex variables (Hunt and Isaacs [1981], Strack [1982], Sidiropoulos et al. [1983]) is more elegant and compact and has the advantage that the solution is obtained in terms of a complex potential Ω, so that both the potential and the stream function are available for computational purposes.

Application of the boundary integral equation method to groundwater flow is discussed in detail by Liggett and Liu [1983]. This method and its relation to the Cauchy integral formulation will be discussed briefly in this chapter. The interested reader is referred to the latter reference for additional details of the BIEM.

55 GREEN'S SECOND IDENTITY

The boundary integral formulation is based on an integral formula known as Green's second identity, which in turn is derived from the divergence theorem (see Korn and

Korn [1968]). Application of this theorem to the divergence of a vector with components Q_i in a domain $\mathcal{D}$ bounded by a boundary $\mathcal{B}$ yields:

$$\int_\mathcal{D} \frac{\partial Q_i}{\partial x_i} dV = \int_\mathcal{B} Q_i n_i dA \tag{55.1}$$

where the Einstein summation convention is adopted (i.e., summation is to be taken over indices that occur twice in a term); we will use this convention throughout this chapter. x_i ($i = 1, 2, 3$) represents Cartesian coordinates, and n_i the components of the outward unit normal of the boundary $\mathcal{B}$. In terms of groundwater flow, (55.1) may be interpreted as an equation of continuity of flow applied to a domain $\mathcal{D}$: the volume integral over $\mathcal{D}$ represents the total amount of flow generated in $\mathcal{D}$ and the surface integral the total outflow through the boundary. The theorem is valid for both three ($i = 1, 2, 3$), and two ($i = 1, 2$) dimensions. As an example, consider the domain $\mathcal{D}$ to be a portion of an unconfined aquifer, with Q_i representing the discharge vector. The divergence $\partial Q_i/\partial x_i$ equals N in that case:

$$\frac{\partial Q_i}{\partial x_i} = \frac{\partial Q_1}{\partial x_1} + \frac{\partial Q_2}{\partial x_2} = N \tag{55.2}$$

If the unit normal n_i makes an angle β with the x_1 axis, then

$$Q_i n_i = Q_1 \cos\beta + Q_2 \sin\beta = Q_n \tag{55.3}$$

where Q_n is the component of Q_i normal to $\mathcal{B}$, taken positive if pointing out of $\mathcal{D}$. Hence, (55.1) becomes for that case

$$\iint_\mathcal{D} N(x_1, x_2) dx_1 dx_2 = \int_\mathcal{B} Q_n ds \tag{55.4}$$

which indeed expresses continuity of flow: the total amount of infiltrated water equals the outflow through the boundary.

The divergence theorem is valid for any vector or scalar function. If we choose

$$Q_i = \Phi \frac{\partial G}{\partial x_i} - G \frac{\partial \Phi}{\partial x_i} \tag{55.5}$$

then

$$\frac{\partial Q_i}{\partial x_i} = \frac{\partial \Phi}{\partial x_i}\frac{\partial G}{\partial x_i} + \Phi\frac{\partial^2 G}{\partial x_i \partial x_i} - \frac{\partial G}{\partial x_i}\frac{\partial \Phi}{\partial x_i} - G\frac{\partial^2 \Phi}{\partial x_i \partial x_i} \tag{55.6}$$

Writing $\partial^2/(\partial x_i \partial x_i)$ as ∇^2, and substituting (55.6) for $\partial Q_i/\partial x_i$ in (55.1), we obtain

$$\int_\mathcal{D} (\Phi \nabla^2 G - G \nabla^2 \Phi) dV = \int_\mathcal{B} \left[\Phi \frac{\partial G}{\partial x_i} - G \frac{\partial \Phi}{\partial x_i}\right] n_i dA \tag{55.7}$$

Sec. 56   Boundary Integral Formulation in Two Dimensions

The right-hand side is often written in terms of normal derivatives, replacing $n_i \partial/\partial x_i$ by $\partial/\partial n$,

$$\int_{\mathcal{D}} (\Phi \nabla^2 G - G \nabla^2 \Phi) dV = \oint_{\mathcal{B}} \left[\Phi \frac{\partial G}{\partial n} - G \frac{\partial \Phi}{\partial n} \right] dA \qquad (55.8)$$

This integral theorem is known as Green's second identity. If both Φ and G fulfill Laplace's equation, then (55.8) becomes

$$\oint_{\mathcal{B}} \left[\Phi \frac{\partial G}{\partial n} - G \frac{\partial \Phi}{\partial n} \right] dA = 0 \qquad (55.9)$$

This integral is the basis for the boundary integral equation method. The function Φ is taken to be the potential for a case of groundwater flow in a domain $\mathcal{D}$ bounded by a boundary $\mathcal{B}$, and for G a suitable choice is made, as discussed below.

56 BOUNDARY INTEGRAL FORMULATION IN TWO DIMENSIONS

For problems of two-dimensional flow the function G is chosen as

$$G = \ln r = \ln \sqrt{(x-\xi)^2 + (y-\eta)^2} \qquad (56.1)$$

where x and y are the coordinates of a point $\mathcal{P}$ in the flow region and (ξ, η) represents a point on $\mathcal{B}$. The boundary $\mathcal{B}$ is chosen such that point $\mathcal{P}$ is excluded from $\mathcal{D}$ as shown in Figure 8.1, so that application of (55.8) gives

$$\oint_{\mathcal{B}} \left[\Phi \frac{\partial}{\partial n} \ln \sqrt{(x-\xi)^2 + (y-\eta)^2} - \ln \sqrt{(x-\xi)^2 + (y-\eta)^2} \frac{\partial \Phi}{\partial n} \right] ds = 0 \qquad (56.2)$$

The boundary $\mathcal{B}$ is composed of a contour $\mathcal{C}$, a small circle c around $\mathcal{P}$ and two straight segments on a line connecting $\mathcal{C}$ to c. The contributions along the line segments cancel because the directions of n_i are opposite along these paths of integration. What remains of (56.2) is

$$\oint_{\mathcal{C}} \left[\Phi \frac{\partial}{\partial n} \ln r - \frac{\partial \Phi}{\partial n} \ln r \right] ds + \oint_{c} \left[\Phi \frac{\partial}{\partial n} \ln r - \frac{\partial \Phi}{\partial n} \ln r \right] ds = 0 \qquad (56.3)$$

where

$$r = \sqrt{(x-\xi)^2 + (y-\eta)^2} \qquad (56.4)$$

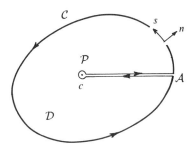

Figure 8.1 The boundary $\mathcal{B}$ composed of the contour $\mathcal{C}$ and the circle C.

The unit normal points out of the domain and thus toward $\mathcal{P}$ along C, and is opposite to the positive r direction, so that $\partial/\partial n = -\partial/\partial r$ along C. If the radius of the circle C is r_0, and θ decreases from a value θ_0 to $\theta_0 - 2\pi$ on C, then $ds = -r_0 d\theta$ (the minus sign occurs because ds is positive by definition) and (56.3) becomes

$$\oint_{\mathcal{C}} \left[\Phi \frac{\partial}{\partial n} \ln r - \frac{\partial \Phi}{\partial n} \ln r \right] ds + \int_{\theta_0}^{\theta_0 - 2\pi} \left\{ \frac{\Phi}{r_0} - \ln(r_0) \left[\frac{\partial \Phi}{\partial r}\right]_{r=r_0} \right\} r_0 \, d\theta = 0 \quad (56.5)$$

In the limit for $r_0 \to 0$ the second term in the second integral vanishes since $\partial \Phi/\partial r$ is finite because Φ is harmonic throughout $\mathcal{D}$, and we obtain

$$\boxed{\Phi(x,y) = \frac{1}{2\pi} \oint_{\mathcal{C}} \left[\Phi \frac{\partial}{\partial n} \ln r - \frac{\partial \Phi}{\partial n} \ln r\right] ds} \quad (56.6)$$

It is important to note that the contour $\mathcal{C}$ is fully closed; the end point of the contour, $\mathcal{A}$ in Figure 8.1, is the same as the starting point.

The second term in the integral represents a distribution of sinks along the boundary; the first one represents a distribution of doublets, as may be verified by working out the derivative $\partial \ln r/\partial n$. The boundary integral thus represents a combination of distributed doublets and sources along the boundary, with Φ and $\partial \Phi/\partial n$ serving as the density distributions. If both Φ and $\partial \Phi/\partial n$ are known along $\mathcal{B}$, then (56.6) may be applied to compute Φ at any point (x,y) of $\mathcal{D}$, including points on $\mathcal{B}$. In a boundary-value problem, however, either Φ, or $\partial \Phi/\partial n$, or a relation between them is known along the boundary, but not both. The boundary integral equation (56.6) may be used to generate a system of equations for the coefficients of $\partial \ln r/\partial n$ and $\ln r$ in (56.6) that are not known a priori. To determine the equations at the control points, (56.6) must be evaluated at points on the boundary.

Equation (56.6) offers an interesting interpretation of (55.8). The function G may be viewed as the potential for a well of discharge 2π centered at (x, y), computed at an arbitrary point (ξ, η). The Laplacian $\nabla^2 \Phi$ vanishes for all (ξ, η) and $\nabla^2 G$ is zero everywhere except at $(\xi, \eta) = (x, y)$. Thus, the only point where the integrand of the integral over $\mathcal{D}$ in (55.8) is not zero is (x, y); evidently, the contribution of that point to the integral yields $2\pi\Phi(x, y)$, which equals the discharge of a well at (x, y) represented by the potential $\Phi(x, y)\ln r$.

Evaluation of the Potential at Boundary Points

The integral (56.6) was obtained assuming that point $\mathcal{P}$ lies inside the boundary $\mathcal{B}$. The integral remains valid in the limit that point $\mathcal{P}$ lies on the boundary, as is illustrated in Figure 8.2(a). It must be remembered in evaluating the integral along $\mathcal{C}$, however, that point $\mathcal{P}$ approached the boundary from the inside; this affects the values of arctangents that appear after performing the integration. We will examine in detail how (56.6) is to be evaluated for cases that point $\mathcal{P}$ lies on the boundary after writing the integral in terms of complex variables in the next section.

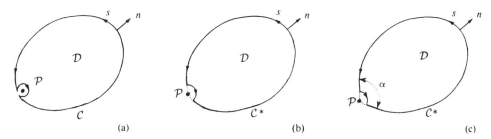

Figure 8.2 The case that point $\mathcal{P}$ lies on the boundary.

The boundary integral may be written in a special form, applicable only to the case that $\mathcal{P}$ is a boundary point. This formulation is more convenient when the integrations are carried out numerically, as is usually done in the BIEM. This special form of the boundary integral is obtained by excluding the boundary point $\mathcal{P}$ from the domain $\mathcal{D}$ as is illustrated in Figure 8.2(b); the boundary $\mathcal{B}$ is decomposed into a boundary segment $\mathcal{C}^*$ and a small circle c around $\mathcal{P}$. This may be done both for the case that the boundary at point $\mathcal{P}$ is smooth and for the case that $\mathcal{P}$ lies at a corner point where the tangents to the boundary at $\mathcal{P}$ enclose an angle α; see Figure 8.2(c). We must recognize the difference between the contour $\mathcal{C}$ and the contour $\mathcal{C}^*$. The contour $\mathcal{C}$ is fully closed; the integration ends at the same point as where it started, whereas the contour $\mathcal{C}^*$ is not closed; the integration starts at an infinitesimal distance on one side of $\mathcal{P}$, and it ends at an infinitesimal distance on the other side. The procedure for establishing the expression for Φ at point $\mathcal{P}$ is similar to that followed above, except that the upper bound of the second integral in (56.5) now becomes $\theta_0 - \alpha$ if the general case of Figure 8.2(c) is considered. The result therefore may be written as

$$\Phi(x, y) = \frac{1}{\alpha} \int_{\mathcal{C}^*} \left[\Phi \frac{\partial}{\partial n} \ln r - \frac{\partial \Phi}{\partial n} \ln r \right] ds \qquad (56.7)$$

where x and y are the coordinates of point $\mathcal{P}$. It is not obvious that (56.6) and (56.7) give the same result; that these two formulations are indeed equivalent will be demonstrated in Sec. 57.

Liggett and Liu [1983] explain in detail how to obtain from (56.7) a system of equations for the unknowns, and we will outline the procedure briefly below. First, the case is considered that Φ is given along $\mathcal{C}$ (the Dirichlet problem). Both the coefficient of $\partial \ln r/\partial n$ and the left-hand side of (56.7) then are known along $\mathcal{C}$: the latter equation is an integral equation for the unknown function $\partial \Phi/\partial n$. The integral equation is solved in an approximate fashion by discretizing $\mathcal{C}^*$ to a polygon of segments of straight lines, and the functions Φ and $\partial \Phi/\partial n$ along the boundary are approximated as polynomials, containing a given number, say m, of unknown parameters. The integration is performed either analytically or numerically, usually the latter, and (56.7) is applied at m control points on the boundary. The resulting system of equations is solved by the use of a standard technique, such as Gaussian elimination. Once all parameters in the solution have been determined, (56.6) may be used to compute the potential inside $\mathcal{D}$, as required. The procedure is quite similar in case $\partial \Phi/\partial n$ is given along some of the boundary segments. Then (56.7) is differentiated with respect to x and y so that $\partial \Phi/\partial n$ may be computed. In differentiating (56.7), it is to be noted that only r depends upon x and y; Φ and $\partial \Phi/\partial n$ are functions of ξ and η only. At control points where $\partial \Phi/\partial n$ is known, the differentiated integral equation is used, with $\partial \Phi/\partial n$ appearing at the left-hand side of the equation and Φ being the unknown.

57 BOUNDARY INTEGRAL EQUATION METHOD IN TERMS OF COMPLEX VARIABLES

In this section we write the boundary integrals (56.6) and (56.7) in terms of complex variables, and establish the link with singular Cauchy integrals. We further examine the difference between the formulations (56.6) and (56.7) as applied to boundary points, and apply each one in complex form to an elementary boundary value problem as an illustration of the method.

Complex Variable Formulation of the Boundary Integral

We write the function $\ln r$ in (56.6) as $\ln|z - \delta|$ where δ represents a point on the boundary and $z = x + iy$ is the independent complex variable:

$$\Phi(z) = \frac{1}{2\pi} \oint_{\mathcal{C}} \left[\Phi(\delta) \frac{\partial}{\partial n} \ln|z - \delta| - \frac{\partial \Phi}{\partial n} \ln|z - \delta| \right] ds \qquad (57.1)$$

We consider a point of the boundary $\mathcal{C}$ and denote the angle of the increment ds with the x axis as α (see Figure 8.3). The normal n then makes an angle $\beta = \alpha - \pi/2$ with the x axis and the normal derivative $\partial \Phi^*/\partial n$ of an arbitrary function Φ^* may be written as

$$\frac{\partial \Phi^*}{\partial n} = \frac{\partial \Phi^*}{\partial \delta_1} \cos(\alpha - \pi/2) + \frac{\partial \Phi^*}{\partial \delta_2} \sin(\alpha - \pi/2) = \frac{\partial \Phi^*}{\partial \delta_1} \sin \alpha - \frac{\partial \Phi^*}{\partial \delta_2} \cos \alpha \qquad (57.2)$$

where δ_1 and δ_2 are the real and imaginary parts of δ,

$$\delta = \delta_1 + i\delta_2 \tag{57.3}$$

If Φ^* is the real part of an analytic function Ω^*, then (57.2) may be expressed as

$$\frac{\partial \Phi^*}{\partial n} = -\Re\left[ie^{i\alpha}\frac{d\Omega^*}{d\delta}\right] \tag{57.4}$$

as may be verified by working out the term between the brackets. Applying this to the function $\Omega^* = \ln(z - \delta)$ we obtain

$$\frac{\partial}{\partial n}\left[\ln|z - \delta|\right] = \Re\left[\frac{ie^{i\alpha}}{z - \delta}\right] \tag{57.5}$$

so that (57.1) becomes

$$\Phi(z) = \frac{1}{2\pi}\oint_C \left[\Phi(\delta)\Re\frac{ie^{i\alpha}}{z - \delta} - \frac{\partial \Phi}{\partial n}\Re\ln(z - \delta)\right]ds \tag{57.6}$$

We represent Φ as the real part of a complex potential $\Omega(z)$, write $e^{i\alpha}ds$ as $d\delta$ and obtain

$$\boxed{\Omega(z) = -\frac{1}{2\pi i}\oint_C \frac{\Phi(\delta)}{z - \delta}d\delta - \frac{1}{2\pi}\oint_C \frac{\partial \Phi}{\partial n}\ln(z - \delta)ds} \tag{57.7}$$

It appears that the boundary integral is composed of a doublet distribution of strength $\Phi(\delta)$ and a sink distribution of strength $-\partial \Phi/\partial n$.

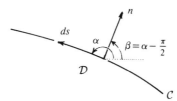

Figure 8.3 Expressing the normal component of the derivative in terms of complex variables.

Relation with Singular Cauchy Integrals

The boundary integral (57.7) may be expressed as the singular Cauchy integral discussed in Sec. 25 as follows. We use the Cauchy-Riemann equations to express $\partial \Phi/\partial n$ as $\partial \Psi/\partial s$,

$$\frac{\partial \Phi}{\partial n} = \frac{\partial \Psi}{\partial s} \tag{57.8}$$

so that the second term in (57.7) may be integrated by parts:

$$\Omega(z) = -\frac{1}{2\pi i} \oint_C \frac{\Phi(\delta)}{z-\delta} d\delta - \frac{1}{2\pi} \Psi(\delta) \ln(z-\delta) \bigg|_C - \frac{1}{2\pi} \oint_C \frac{\Psi(\delta)}{z-\delta} d\delta \qquad (57.9)$$

The potential Φ is harmonic and therefore the function $\Psi(\delta)$ is single-valued. The argument of $(z-\delta)$ increases by 2π upon completion of the contour C. If we denote the starting point of the path of integration as δ_0 then

$$-\frac{1}{2\pi} \Psi(\delta) \ln(z-\delta) \bigg|_C = -\frac{1}{2\pi} \Psi(\delta_0) 2\pi i = -i\Psi(\delta_0) = -i\Psi_0 \qquad (57.10)$$

and (57.9) becomes

$$\Omega(z) = -\frac{1}{2\pi i} \oint_C \frac{\Phi(\delta) + i\Psi(\delta)}{z-\delta} d\delta - i\Psi_0 \qquad (57.11)$$

If we choose the constant Ψ_0 as zero,

$$\Psi_0 = 0 \qquad (57.12)$$

and write $\Phi(\delta) + i\Psi(\delta)$ as $\Omega(\delta)$ then (57.11) becomes

$$\boxed{\Omega(z) = -\frac{1}{2\pi i} \oint_C \frac{\Omega(\delta)}{z-\delta} d\delta} \qquad (57.13)$$

which is equivalent to the singular Cauchy integral (25.80). It thus follows that the boundary integral applied in the BIEM is equivalent to the real part of the Cauchy integral (57.13). Formulation in terms of singular Cauchy integrals has the advantage of a compact formulation in terms of a single complex variable, so that the integrals can be evaluated analytically for a variety of element shapes and density distributions. A second advantage is that the solution yields both the potential and the stream function.

Equation (57.13) is used in the direct boundary integral formulation (see Hunt and Isaacs [1981]). As shown in Sec. 25, indirect formulations, where only one of the two distributions are used, are mathematically equivalent to the direct one. An indirect formulation using line-sinks is discussed by Van der Veer [1978] for problems of two-dimensional groundwater flow in the vertical plane.

The density distribution $\Omega(\delta)$ actually represents the jump in potential across the boundary:

$$\Omega(z) = -\frac{1}{2\pi i} \oint_C \frac{\Omega^+(\delta) - \Omega^-(\delta)}{z-\delta} d\delta \qquad (57.14)$$

where Ω^+ is the potential inside the domain $\mathcal{D}$ and Ω^- the one outside $\mathcal{D}$. In the direct formulation the potential $\Omega^-(\delta)$ is zero so that (57.14) reduces to (57.13) (see Sec. 25).

Evaluation of the Complex Potential at Boundary Points

In order to examine the behavior of the Cauchy integral (57.13) near points of the boundary we write the function $\Omega(\delta)$ as $\Omega(z) - [\Omega(z) - \Omega(\delta)]$ and obtain

$$\Omega(z) = -\frac{1}{2\pi i}\Omega(z) \oint_C \frac{d\delta}{z - \delta} + \frac{1}{2\pi i}\oint_C \frac{\Omega(z) - \Omega(\delta)}{z - \delta} d\delta \tag{57.15}$$

For simplicity, we make restriction in the following analysis to cases that $\Omega(z)$ is analytic inside $\mathcal{D}$ including the boundary $\mathcal{B}$. The integrand in the second integral in (57.15) then is analytic at $\delta = z$; this term vanishes by the Cauchy-Goursat theorem (25.75), and

$$\Omega(z) = \frac{1}{2\pi i}\Omega(z) \oint_C d\ln(z - \delta) \tag{57.16}$$

If point $\mathcal{P}$ lies inside $\mathcal{C}$, the argument of $\ln(z - \delta)$ increases by 2π upon completion of the contour, so that an identity is obtained in (57.16). This holds true if point $\mathcal{P}$ has reached the boundary from the inside, provided that it is taken into account that $\arg(z - \delta)$ still increases by an amount 2π upon completing the contour $\mathcal{C}$, as shown in Figure 8.2(a). Thus, the boundary integral behaves in a continuous fashion as $\mathcal{P}$ approaches the boundary and eventually reaches it, provided that the argument of the logarithmic term is computed taking into account that $\mathcal{P}$ approached the boundary from the inside. This remains true if $\mathcal{P}$ approaches a corner point, as shown in Figure 8.2(c).

We next examine the alternative formulation (56.7) and apply it to the case shown in Figure 8.2(c), maintaining the restriction that $\Omega(z)$ is analytic at $\mathcal{P}$. Written in the form of a Cauchy integral, this formulation becomes

$$\Omega(z_p) = -\frac{1}{\alpha i}\int_{\mathcal{C}^*} \frac{\Omega(\delta)}{z_p - \delta} d\delta \tag{57.17}$$

where z_p is the complex coordinate of point $\mathcal{P}$ and $\mathcal{C}^*$ is the open contour shown in Figure 8.2(c). We obtain by a procedure similar to the one used above,

$$\Omega(z_p) = \frac{1}{\alpha i}\Omega(z)\int_{\mathcal{C}^*} d\ln(z_p - \delta) \tag{57.18}$$

As is seen from Figure 8.2(c), the argument of $\ln(z - \delta)$ only increases by an amount α upon completion of the contour $\mathcal{C}^*$ so that again an identity is obtained.

We will apply the Cauchy integral formulation to an elementary boundary value problem, as an illustration, using the above two formulations.

Application: Flow in a Strip

We apply the Cauchy integral (57.13) to solve the problem of flow in a domain inside a rectangular boundary (see Figure 8.4). The length of the strip is 2 m, the height is 1 m; the boundaries 1-2 and 3-4 are streamlines, and the boundaries 4-1 and 2-3 equipotentials with $\Phi = \Phi_1$ and $\Phi = \Phi_2$, respectively. We approximate the density distributions as linear functions and write the Cauchy integral as the sum of four integrals:

$$\Omega(z) = -\frac{1}{2\pi i}\left[\int_{z_1}^{z_2}\overset{1}{\frac{\Omega(\delta)}{z-\delta}}d\delta + \int_{z_2}^{z_3}\overset{2}{\frac{\Omega(\delta)}{z-\delta}}d\delta + \int_{z_3}^{z_4}\overset{3}{\frac{\Omega(\delta)}{z-\delta}}d\delta + \int_{z_4}^{z_1}\overset{4}{\frac{\Omega(\delta)}{z-\delta}}d\delta\right] \quad (57.19)$$

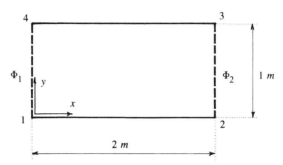

Figure 8.4 Flow in a strip.

We represent these four integrals as $\overset{n}{F}$ ($n = 1, 2, 3, 4$), so that

$$\Omega(z) = \sum_{n=1}^{4}\overset{n}{F}(z) \quad (57.20)$$

and use (25.73) and (25.52) to express each function $\overset{n}{F}$ in terms of a local variable Z_n as

$$\overset{n}{F} = -\frac{1}{2\pi i}\int_{z_n}^{z_{n+1}}\frac{\overset{n}{\Omega}(\delta)}{z-\delta}d\delta = \frac{\overset{n}{\Omega}(Z_n)}{2\pi i}\ln\frac{Z_n-1}{Z_n+1} + i\overset{n}{p}(Z_n) + \overset{n}{q}(Z_n) \quad (57.21)$$

where

$$Z_n = \frac{2z - z_n - z_{n+1}}{z_{n+1} - z_n} \quad (57.22)$$

We approximate the complex potential $\overset{n}{\Omega}$ as a linear function, varying between the values Ω_n at z_n and Ω_{n+1} at z_{n+1}:

$$\overset{n}{\Omega} = -\tfrac{1}{2}\Omega_n(Z_n - 1) + \tfrac{1}{2}\Omega_{n+1}(Z_n + 1) \quad (57.23)$$

The far-field corrections $\overset{n}{p}$ and $\overset{n}{q}$ in (57.16) are obtained by requiring that $\overset{n}{\Omega}$ vanish at infinity. This gives, using (57.23) and expanding $\ln\{(Z_n - 1)/(Z_n + 1)\}$ as $-2/Z_n + O(Z_n^{-3})$:

$$i\overset{n}{p} + \overset{n}{q} = \frac{1}{2\pi i}(\Omega_{n+1} - \Omega_n) \quad (57.24)$$

Sec. 57 Boundary Integral Equation Method in Terms of Complex Variables 599

The sum of the four functions $i\overset{n}{p}+\overset{n}{q}$ is zero and substitution of (57.21) for $\overset{n}{F}$ in (57.20) yields

$$\Omega(z) = \frac{1}{4\pi i} \sum_{n=1}^{4} [(\Omega_{n+1} - \Omega_n)Z_n + \Omega_{n+1} + \Omega_n] \ln \frac{Z_n - 1}{Z_n + 1} \qquad (57.25)$$

where $\Omega_5 = \Omega_1$. We take the stream function to be zero along $1-2$, and denote the unknown value of Ψ along $3-4$ as Ψ_3. We obtain the following values for Ω_n, $n = 1, 2, \ldots, 5$:

$$\Omega_1 = \Phi_1 \qquad \Omega_2 = \Phi_2 \qquad \Omega_3 = \Phi_2 + i\Psi_3$$
$$\Omega_4 = \Phi_1 + i\Psi_3 \qquad \Omega_5 = \Omega_1 = \Phi_1 \qquad (57.26)$$

The local complex variables Z_n may be expressed in terms of z by the use of (57.22) and Figure 8.4 as follows:

$$Z_1 = z - 1 \qquad Z_2 = -2iz + 4i - 1 \qquad Z_3 = -z + 1 + i \qquad Z_4 = 2iz + 1 \qquad (57.27)$$

There is only a single unknown in the problem, namely Ψ_3. We may solve for this unknown by applying (57.25) at any one of the four corner points. Two of the logarithms in (57.25) are singular at these points and we must examine the behavior of the potential there. Using (57.26) and (57.27) with (57.25) we obtain the following expression for Ω:

$$\Omega = \frac{(\Phi_2 - \Phi_1)z + 2\Phi_1}{4\pi i} \ln \frac{z-2}{z} + \frac{\Psi_3(2z-4) + 2\Phi_2}{4\pi i} \ln \frac{z-2-i}{z-2}$$
$$+ \frac{(\Phi_2 - \Phi_1)(z-i) + 2(\Phi_1 + i\Psi_3)}{4\pi i} \ln \frac{z-i}{z-2-i} + \frac{2\Psi_3 z + 2\Phi_1}{4\pi i} \ln \frac{z}{z-i}$$
$$(57.28)$$

We examine the behavior of this function near point 1, $z = 0$. The arguments of the four logarithms in (57.28) at point 1 are denoted as θ_1, θ_2, θ_3, and θ_4. It is seen from Figure 8.5(a) that

$$\theta_2 = \beta \qquad \theta_3 = \frac{\pi}{2} - \beta \qquad (57.29)$$

The arguments θ_1 and θ_4 are such that in the limit for $z \to 0$ (see Figure 8.5[b])

$$\theta_1 + \theta_4 = \frac{3\pi}{2} \qquad (57.30)$$

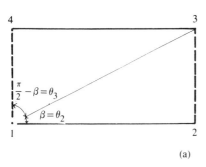 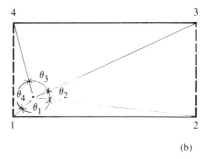

Figure 8.5 The arguments θ_1, θ_2, θ_3, and θ_4 for point $\mathcal{P}$ approaching point 1 from the inside.

The first term and the last one in (57.28) are singular at $z = 0$. In the limit for $z \to 0$, (57.28) may be written as follows, with $z \ln z \to 0$ for $z \to 0$,

$$\Phi_1 = \frac{2\Phi_1}{4\pi i} \lim_{z \to 0} \ln\left[\frac{z-2}{z}\frac{z}{z-i}\right] + \frac{-4\Psi_3 + 2\Phi_2}{4\pi i}\left[\ln\frac{\sqrt{5}}{2} + i\beta\right]$$
$$+ \frac{2\Phi_1 - i(\Phi_2 - \Phi_1) + 2i\Psi_3}{4\pi i}\left[-\ln\sqrt{5} + i\frac{\pi}{2} - i\beta\right] \qquad (57.31)$$

The singularity at $z = 0$ cancels in the first term, and the imaginary part of the logarithm approaches $3\pi/2$ in the limit for $z \to 0$, so that

$$\lim_{z \to 0} \ln\left[\frac{z-2}{z}\frac{z}{z-i}\right] = \ln 2 + i\frac{3\pi}{2} \qquad (57.32)$$

Collecting the coefficients of Φ_1, Φ_2, and Ψ_3 in (57.31) we obtain, using (57.32),

$$\frac{\Phi_1}{2\pi i}\left\{\ln 2 + i\frac{3\pi}{2} + \tfrac{1}{2}(2+i)\left[-\ln\sqrt{5} + i\left(\tfrac{\pi}{2} - \beta\right)\right] - 2\pi i\right\}$$
$$+ \frac{\Phi_2}{2\pi i}\left[\ln\frac{\sqrt{5}}{2} + i\beta + \frac{i}{2}\ln\sqrt{5} + \tfrac{1}{2}\left(\tfrac{\pi}{2} - \beta\right)\right]$$
$$+ \frac{\Psi_3}{2\pi i}\left[-2\ln\frac{\sqrt{5}}{2} - 2i\beta - i\ln\sqrt{5} - \left(\tfrac{\pi}{2} - \beta\right)\right] = 0 \qquad (57.33)$$

or, after simplification,

$$\frac{\Phi_2 - \Phi_1}{2\pi i}\left[\ln\frac{\sqrt{5}}{2} + \tfrac{1}{2}\left(\tfrac{\pi}{2} - \beta\right) + i\left(\tfrac{1}{2}\ln\sqrt{5} + \beta\right)\right]$$
$$= \frac{2\Psi_3}{2\pi i}\left[\ln\frac{\sqrt{5}}{2} + \tfrac{1}{2}\left(\tfrac{\pi}{2} - \beta\right) + i\left(\tfrac{1}{2}\ln\sqrt{5} + \beta\right)\right] \qquad (57.34)$$

Sec. 57 Boundary Integral Equation Method in Terms of Complex Variables

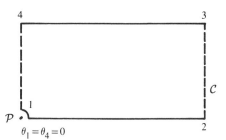

Figure 8.6 Formulation with point $\mathcal{P}$ considered to be a boundary point.

so that

$$\Psi_3 = -\frac{\Phi_1 - \Phi_2}{2} = -Q_{x0} \tag{57.35}$$

which is equal to the discharge Q_{x0} flowing in the strip as obtained directly from Darcy's law. Substituting this expression for Ψ_3 in (57.28) we obtain

$$\Omega = \frac{-\frac{1}{2}z(\Phi_1 - \Phi_2) + \Phi_1}{2\pi i} \left[\ln \frac{z - z_2}{z - z_1} + \ln \frac{z - z_3}{z - z_2} + \ln \frac{z - z_4}{z - z_3} + \ln \frac{z - z_1}{z - z_4} \right] \tag{57.36}$$

The real part of the term between the brackets is zero and the imaginary part equals the sum of the four arguments:

$$\Omega = \frac{\theta_1 + \theta_2 + \theta_3 + \theta_4}{2\pi} \left[-\frac{1}{2}z(\Phi_1 - \Phi_2) + \Phi_1 \right] \tag{57.37}$$

It may be seen from Figure 8.5(b) that $\theta_1 + \theta_2 + \theta_3 + \theta_4$ equals 2π for z inside $\mathcal{D}$ and zero for z outside $\mathcal{D}$ so that

$$\begin{aligned} \Omega &= -\tfrac{1}{2}z(\Phi_1 - \Phi_2) + \Phi_1 && (z \text{ in } \mathcal{D}) \\ \Omega &= 0 && (z \text{ not in } \mathcal{D}) \end{aligned} \tag{57.38}$$

which corresponds to the exact solution to the problem.

In numerical applications the logarithms must be programmed such that the singularities are handled correctly and the singularities at the nodes are removed. The same function subroutines may then be used both for computing the coefficients of the unknowns in the system of equations, and for evaluating Ω for points inside $\mathcal{D}$.

The same result is obtained if the alternative formulation (57.18) is used at the boundary points. The left-hand side in (57.31) must be replaced by $\alpha/(2\pi)\Phi_1 = \tfrac{1}{4}\Phi_1$. The boundary point $\mathcal{P}$ is now excluded from the contour $\mathcal{C}$ so that the arguments θ_1 and θ_4 are zero there (see Figure 8.6) and (57.32) is replaced by

$$\lim_{z \to 0} \ln \left[\frac{z-2}{z} \frac{z}{z-i} \right] = \ln 2 \tag{57.39}$$

The coefficient of Φ_1 is thus reduced on both sides of the equation by an amount $3/4 = (3\pi/2)/(2\pi)$, as compared to the previous case, so that the final result is the same, as asserted.

58 SOLVING BOUNDARY-VALUE PROBLEMS

Boundary-value problems may be solved using the singular Cauchy integral approach in much the same way as discussed for the analytic element method in Chapter 6. Both the Cauchy integral method and the BIEM are particularly suitable for problems with finite boundaries, such as in problems of two-dimensional flow in the vertical plane.

Division into Subdomains

Liggett and Liu [1983] as well as Hunt and Isaacs [1981] subdivide the flow domain into zones when solving problems involving thin features and inhomogeneities in the hydraulic conductivity. The former authors solve, for example, a problem with a cutoff wall by subdividing the flow domain into two zones as shown in Figure 8.7. They state that the system of equations becomes singular if two separate boundary points have the same coordinates, and circumvent this difficulty by subdividing the flow domain into zones 1 and 2 in Figure 8.7. This approach has the disadvantage that an artificial boundary is inserted between point B of the cutoff wall and the contour C, and that two layers of elements are used along $\mathcal{ABA}'$.

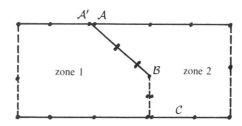

Figure 8.7 Discretization of the problem with a cutoff wall. (Source: J. A. Liggett/P. L.-F. Liu, *The Boundary Integral Equation Method for Porous Media Flow*, ©1983, p. 52. Reprinted by permission of the authors.)

An alternative solution is obtained by realizing that the potential jumps but the stream function is continuous across the cutoff wall $\mathcal{AB}$. We define the contour C as including a contour C^* and line integrals back and forth along the cutoff wall as shown in Figure 8.8. The corresponding Cauchy integral becomes

$$\Omega(z) = -\frac{1}{2\pi i}\int_{C^*}\frac{\Omega(\delta)}{z-\delta}d\delta - \frac{1}{2\pi i}\int_A^B\frac{\Omega^+(\delta)}{z-\delta}d\delta - \frac{1}{2\pi i}\int_B^A\frac{\Omega^-(\delta)}{z-\delta}d\delta \qquad (58.1)$$

where Ω^+ and Ω^- represent the complex potential along the $+$ and $-$ sides of the cutoff wall as shown in Figure 8.8. Reversing the limits of the third integral we obtain

$$\boxed{\Omega(z) = -\frac{1}{2\pi i}\int_{C^*}\frac{\Omega(\delta)}{z-\delta}d\delta - \frac{1}{2\pi i}\int_A^B\frac{\Omega^+(\delta)-\Omega^-(\delta)}{z-\delta}d\delta} \qquad (58.2)$$

Sec. 58 Solving Boundary-Value Problems

Across the cutoff wall the potential jumps and the stream function is continuous so that (58.2) becomes

$$\Omega(z) = -\frac{1}{2\pi i}\int_{C^*}\frac{\Omega(\delta)}{z-\delta}d\delta - \frac{1}{2\pi i}\int_A^B \frac{\lambda(\delta)}{z-\delta}d\delta \tag{58.3}$$

where

$$\lambda(\delta) = \Phi^+(\delta) - \Phi^-(\delta) \tag{58.4}$$

The integral along the cutoff wall contains only $\lambda(\delta)$ as the unknown function, and single nodes may be used. Application of the condition that Ψ be constant along the wall yields the desired equations. Sidiropoulos et al. [1983] suggest the use of distributed doublets for cutoff walls in an indirect complex boundary element formulation. It appears that the use of distributed doublets along thin walls follows naturally from a direct boundary integral formulation as well.

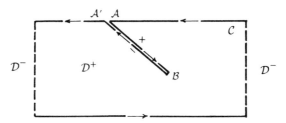

Figure 8.8 The contour C for the problem with a cutoff wall.

The latter approach has the advantages that subdivision of the domain is not necessary, that no conditions needs to be applied except along the actual boundary, that the condition of continuity of flow is exactly satisfied, and, finally, that fewer boundary elements and unknowns are needed.

The flow is singular near the tip of the cutoff wall; the tip element presented in Sec. 40 is useful for modeling the behavior near the tip of the cutoff wall. Liggett and Liu present density distributions in terms of real functions for incorporating singularities of various kinds, such as the square root singularity that occurs near the tip of the cutoff wall.

Equation (58.2) is generally applicable to thin features, even if they lie inside the domain; the feature is then connected to the boundary by line integrals that cancel in the boundary integral formulation (compare Figure 8.1). It may be applied also to model interzonal boundaries such as inhomogeneities; along the boundaries of the inhomogeneities only the line-doublets remain, and the same approach may be used as outlined in Sec. 35. As for the problem of a cutoff wall, the modeling of inhomogeneities by the use of line-doublets has advantages over the approach suggested by Liggett and Liu, and Hunt and Isaacs.

It is of interest to note that the boundary values $\Omega(\delta)$ in (58.2) still represent the boundary values of the complete complex potential, even though another potential is added to the contour integral. Correction of the boundary values for the combined influences of the additional complex potentials is unnecessary, provided that the sum of these additional functions is analytic at infinity. This observation follows from the derivation of (58.2), but will be clarified by the use of the theorems presented in Sec. 25 as follows.

Let the complex potential $\overset{2}{\Omega} = \overset{2}{\Omega}(z)$ represent the sum of the additional functions. Then $\Omega = \Omega(z)$ may be written as

$$\Omega = -\frac{1}{2\pi i}\oint_C \frac{\Omega(\delta)}{z-\delta}d\delta + \overset{2}{\Omega}(z) \tag{58.5}$$

We require that $\overset{2}{\Omega}(z)$ be both analytic and zero at infinity:

$$\overset{2}{\Omega} = \frac{a_1}{z} + \frac{a_2}{z^2} + \ldots \qquad (z \to \infty) \tag{58.6}$$

This condition implies that $\overset{2}{\Omega}(z)$ can not represent flow from or toward infinity. The complex potential $\overset{2}{\Omega}$, then, represents a function, analytic in $\mathcal{D}^-$ and continuous on $\mathcal{C}$ from the outside. Therefore, by a theorem of Sec. 25 (compare [25.85]), the integral

$$\Omega = -\frac{1}{2\pi i}\oint_C \frac{\overset{2}{\Omega}(\delta)}{z-\delta}d\delta \tag{58.7}$$

vanishes in $\mathcal{D}^+$. It thus follows that the formulation

$$\Omega = -\frac{1}{2\pi i}\oint_C \frac{\Omega(\delta) - \overset{2}{\Omega}(\delta)}{z-\delta}d\delta + \overset{2}{\Omega}(z) \tag{58.8}$$

is, for z in $\mathcal{D}^+$, equivalent to (58.5); it is not necessary to correct the boundary values $\Omega(\delta)$ for the influence of the function $\overset{2}{\Omega}(z)$, provided that $\overset{2}{\Omega}(z)$ satisfies (58.6).

If the function $\overset{2}{\Omega}(z)$ represents flow from or toward infinity, then this function may be separated into a part that meets (58.6), and a part $\overset{2}{\Omega}^*(z)$ that does not meet this condition. The latter function contributes to the contour integral, and its boundary values must be subtracted from $\Omega(\delta)$; in that case $\overset{2}{\Omega}(\delta)$ in the contour integral in (58.8) must be replaced by $\overset{2}{\Omega}^*(\delta)$.

Summarizing, the Cauchy integral method may be combined with analytic elements to solve general boundary value problems with features inside the flow domain discussed for regional aquifer modeling by the analytic element method in Chapter 6.

59 BOUNDARY INTEGRAL FORMULATION IN THREE DIMENSIONS

For the purpose of deriving a boundary integral formulation in three dimensions we make use of a Cartesian coordinate system x_i ($i = 1, 2, 3$) and denote the coordinates of a point on the boundary of the domain $\mathcal{D}$ as ξ_i ($i = 1, 2, 3$). We derive the boundary integral for the three-dimensional case by choosing the function G in (55.8) as

$$G = \frac{1}{\sqrt{(x_1 - \xi_1)^2 + (x_2 - \xi_2)^2 + (x_3 - \xi_3)^2}}$$
$$= \frac{1}{\sqrt{(x_i - \xi_i)(x_i - \xi_i)}} = \frac{1}{R} \quad \text{(sum on } i\text{)} \tag{59.1}$$

We decompose the boundary $\mathcal{B}$ in (55.8) in an integral over the boundary $\mathcal{S}$ of the domain and a small sphere s of radius R_0 around point $\mathcal{P}$ with coordinates x_i. We take the normal n_i again as pointing out of the domain $\mathcal{D}$ and obtain, noting that n_i points into the sphere s so that $\partial/\partial n$ equals $-\partial/\partial R$ along the boundary of s,

$$\int_{\mathcal{S}} \left[\Phi \frac{\partial}{\partial n} \left[\frac{1}{R} \right] - \frac{1}{R} \frac{\partial \Phi}{\partial n} \right] ds + \int_{-\pi/2}^{\pi/2} \int_0^{2\pi} \left[\frac{\Phi}{R_0^2} - \frac{1}{R_0} \left[\frac{\partial \Phi}{\partial R} \right]_{R=R_0} \right] R_0^2 \cos\theta \, d\varphi \, d\theta = 0 \tag{59.2}$$

Since Φ is harmonic and finite inside $\mathcal{D}$, the second term in the second integral vanishes in the limit for $R_0 \to 0$ and we obtain

$$\boxed{\Phi = -\frac{1}{4\pi} \int_{\mathcal{S}} \left[\Phi \frac{\partial}{\partial n} \left[\frac{1}{R} \right] - \frac{1}{R} \frac{\partial \Phi}{\partial n} \right] ds} \tag{59.3}$$

As for the two-dimensional case, the integral represents a distribution of doublets and sinks along the surface $\mathcal{S}$. We will not go into the details of evaluating the boundary integrals for three-dimensional problems. The interested reader is referred to Liggett and Liu [1983] who discuss procedures of numerical integration for three-dimensional problems, including some cases with free boundaries.

Chapter 9

Finite Difference and Finite Element Methods

Finite difference and finite element methods are presently the most common numerical techniques for modeling groundwater flow problems. These methods differ from those discussed thus far in that the potential is approximated by discretization throughout the flow domain. As a result, not only the boundary conditions are approximated, but also the differential equation itself. Characteristic for both methods is that the flow domain is bounded. This is true also for applications to regional flow in contrast to the analytic element method, where the aquifer system is modeled as being infinite in extent; boundary conditions are applied only along internal boundaries that are physically present in the aquifer system.

Advantages of both the finite difference and finite element methods are that the hydraulic conductivity can be easily varied throughout the aquifer system, that the formulations are suitable for modeling transient flow, and that they are comparatively straightforward. Both methods are discussed in detail in the literature on groundwater mechanics, e.g., Wang and Anderson [1982], Verruijt [1982], and Bear and Verruijt [1987]. These authors list computer programs applicable to a variety of problems of both steady-state and transient flow. Verruijt [1980] succeeded in modeling major problems by means of the finite element method using a microcomputer with only 32K bytes of memory. We will discuss both the finite difference and finite element methods in some detail. The interested reader is referred to the above references for further reading.

60 FINITE DIFFERENCE METHODS

Finite difference methods are based on a discretization of the flow domain into a mesh that is usually rectangular, where the potential or head is computed at the gridpoints by solving the differential equation in finite difference form throughout the mesh. There exist a variety of numerical techniques for solving the resulting system of linear equations, and the method may be applied with comparative ease to problems where the hydraulic conductivity varies from node to node. Our purpose is to explain only the basic principles involved in the method for a square mesh, and we will make restriction to cases where the hydraulic conductivity is constant.

Steady Flow

The differential equation for steady flow in an aquifer with homogeneous hydraulic conductivity is

$$\frac{\partial^2 \Phi}{\partial x^2} + \frac{\partial^2 \Phi}{\partial y^2} = -N(x,y) \tag{60.1}$$

where N represents a given rate of infiltration. The flow region is discretized by a square mesh as shown in Figure 9.1(a). The nodes are identified as (i,j) where i and j are integers such that the coordinates of node (i,j) are given by

$$x = (i-1)a \qquad y = (j-1)a \tag{60.2}$$

where a is the distance between grid points,

$$a = \Delta x = \Delta y \tag{60.3}$$

The node (i,j) with its four surrounding grid points is shown in Figure 9.1(b). Points $\mathcal{A}$, $\mathcal{B}$, $\mathcal{C}$, and $\mathcal{D}$ lie midway between nodes and the derivative $\partial \Phi / \partial x$ may be approximated at points $\mathcal{A}$ and $\mathcal{B}$ as

$$\left[\frac{\partial \Phi}{\partial x}\right]_\mathcal{A} = \frac{\Phi_{i,j} - \Phi_{i-1,j}}{a} \qquad \left[\frac{\partial \Phi}{\partial x}\right]_\mathcal{B} = \frac{\Phi_{i+1,j} - \Phi_{i,j}}{a} \tag{60.4}$$

An expression for the second derivative of Φ with respect to x at node (i,j) is obtained in a similar fashion,

$$\left[\frac{\partial^2 \Phi}{\partial x^2}\right]_{i,j} = \frac{\Phi_{i-1,j} + \Phi_{i+1,j} - 2\Phi_{i,j}}{a^2} \tag{60.5}$$

The derivative $\partial^2 \Phi / \partial y^2$ at node (i,j) is expressed similarly and we obtain the following expression for the Laplacian:

$$\left[\nabla^2 \Phi\right]_{i,j} = \frac{\Phi_{i-1,j} + \Phi_{i+1,j} + \Phi_{i,j-1} + \Phi_{i,j+1} - 4\Phi_{i,j}}{a^2} \tag{60.6}$$

Using this expression in the differential equation (60.1) and representing N at node i,j as $N_{i,j}$, we obtain

$$\Phi_{i-1,j} + \Phi_{i+1,j} + \Phi_{i,j-1} + \Phi_{i,j+1} - 4\Phi_{i,j} = -a^2 N_{i,j} \tag{60.7}$$

Sec. 60 Finite Difference Methods

We solve this equation for $\Phi_{i,j}$:

$$\Phi_{i,j} = \tfrac{1}{4}\left[\Phi_{i-1,j} + \Phi_{i+1,j} + \Phi_{i,j-1} + \Phi_{i,j+1} + a^2 N_{i,j}\right] \tag{60.8}$$

This equation is applied at each node and is used in conjunction with the boundary conditions to obtain values of the potential at all nodes as explained below.

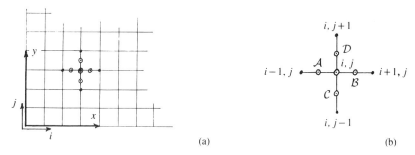

Figure 9.1 Discretization by a rectangular mesh.

Boundary Conditions in Terms of Φ

In the Dirichlet problem the potential is known along the entire boundary, and therefore at each boundary node. The equations (60.8) then need not be applied at the boundary nodes but only at the interior ones. The simplest method for solving the system of equations (60.8) is by the Jacobi method, where each value for $\Phi_{i,j}$ is first estimated at all interior nodes and values for successive iterations are obtained by applying (60.8) at each node using the values computed in the previous iteration:

$$\Phi_{i,j}^{(m+1)} = \tfrac{1}{4}\left[\Phi_{i-1,j}^{(m)} + \Phi_{i+1,j}^{(m)} + \Phi_{i,j-1}^{(m)} + \Phi_{i,j+1}^{(m)} + a^2 N_{i,j}\right] \tag{60.9}$$

where m is the iteration index: $m = 1$ for the first iteration.

A method that converges more rapidly is the Gauss-Seidel iteration. In this method we apply (60.8) at nodes $(2, 2), (3, 2), \ldots, (n_i - 1, 2)$ successively, where n_i is the number of nodes in the row $j = 2$. We then continue with row $j = 3$ in the same way, and proceed until all nodes have been used. In this way, two newly computed values are used in the expression for $\Phi_{i,j}^{(m+1)}$:

$$\boxed{\Phi_{i,j}^{(m+1)} = \tfrac{1}{4}\left[\Phi_{i-1,j}^{(m+1)} + \Phi_{i+1,j}^{(m)} + \Phi_{i,j-1}^{(m+1)} + \Phi_{i,j+1}^{(m)} + a^2 N_{i,j}\right]} \tag{60.10}$$

Boundary Conditions in Terms of $\partial\Phi/\partial n$; Mixed Boundary Conditions

In the Neumann problem the boundary conditions are specified in terms of $\partial\Phi/\partial n$. In order to approximate $\partial\Phi/\partial n$ at boundary nodes, the mesh is extended one row of nodal points beyond the boundary as illustrated in Figure 9.2. The values for either i or

j are zero for nodes (i,j) outside the boundary, which are called imaginary nodes. For the case illustrated in Figure 9.2, $\partial\Phi/\partial n$ is given along the boundary $x = 0$, $i = 1$ and $\partial\Phi/\partial n$ is approximated for node $1, j$ as

$$\left[\frac{\partial \Phi}{\partial n}\right]_{1,j} = \left[\frac{\partial \Phi}{\partial x}\right]_{1,j} = \frac{\Phi_{2,j} - \Phi_{0,j}}{2a} \tag{60.11}$$

so that

$$\Phi_{0,j} = \Phi_{2,j} - 2a\left[\frac{\partial \Phi}{\partial n}\right]_{1,j} \tag{60.12}$$

which is used to compute the value of $\nabla^2 \Phi$ at the boundary nodes; expression (60.10) becomes

$$\Phi_{1,j}^{(m+1)} = \tfrac{1}{4}\left\{2\Phi_{2,j}^{(m)} + \Phi_{1,j-1}^{(m+1)} + \Phi_{1,j+1}^{(m)} + a^2 N_{1,j} - 2a\left[\frac{\partial \Phi}{\partial n}\right]_{1,j}\right\} \tag{60.13}$$

Boundary values of $\partial\Phi/\partial n$ are thus applied by using (60.13) instead of (60.10) at the boundary nodes.

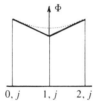 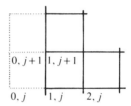

Figure 9.2 A boundary with $\partial\Phi/\partial n$ given.

Mixed boundary value problems are solved by extending the mesh beyond those boundaries where $\partial\Phi/\partial n$ is given, using equations of the form (60.13), while setting Φ equal to prescribed values at those nodes where the head is prescribed.

As a result of the discretization, it is a relatively simple matter to adapt the finite difference scheme to include such phenomena as leakage. The right-hand side of the equation then will involve a leakage term which is dependent upon the potential Φ, which merely results in a modification of the equation (60.10) for the potential at node i,j. Varying values of the hydraulic conductivity can be taken into account by formulating the problem in terms of a function F,

$$F = \Phi/k \tag{60.14}$$

that is independent of the hydraulic conductivity. The differential equation to be discretized then becomes

$$\frac{\partial}{\partial x}\left[k\frac{\partial F}{\partial x}\right] + \frac{\partial}{\partial y}\left[k\frac{\partial F}{\partial y}\right] = -N(x,y) \tag{60.15}$$

Sec. 60 Finite Difference Methods

Problems of transient flow may be solved by discretizing the applicable differential equation both in space and in time. The resulting scheme may be written either explicitly or implicitly in terms of the potential, as is discussed below.

Problem 60.1

Write a computer program using Gauss-Seidel iteration for solving the problem of flow through a strip shown in Figure 8.4.

Transient Flow

We will discuss application of the finite difference method to problems of transient flow in unconfined aquifers. The formulation for problems of transient flow in confined aquifers is simpler, and it is left to the reader to adapt the scheme discussed below to problems of confined transient flow.

The differential equation for shallow transient flow in an unconfined aquifer is given by (16.15),

$$\nabla^2 \Phi = S_p \frac{\partial \phi}{\partial t} - N \qquad (60.16)$$

There are two different methods of discretization of this equation with respect to time: the explicit formulation and the implicit one.

Explicit Finite Difference Formulation

In an explicit formulation, the discretization in time is as follows:

$$\left[\nabla^2 \Phi\right]_t = S_p \frac{\phi_{t+\Delta t} - \phi_t}{\Delta t} - N_t \qquad (60.17)$$

This discretization is called explicit because the Laplacian is computed at time t so that an explicit expression for the head at time $t + \Delta t$ is obtained as

$$\boxed{\phi_{t+\Delta t} = \phi_t + \frac{\Delta t}{S_p} \left[\nabla^2 \Phi + N\right]_t} \qquad (60.18)$$

A benefit of the scheme is that linearization of the differential equation is not necessary. The head at each node at time $t + \Delta t$ is obtained by computing $\nabla^2 \Phi$ by application of (60.6), in which the values of Φ at time t are to be used.

The explicit scheme has the advantages of flexibility and simplicity, but gives meaningful answers only if Δt is less than some critical value, Δt_{crit}, thus

$$\Delta t < \Delta t_{\text{crit}} \qquad (60.19)$$

The value of the critical time step may be found by trial and error: if Δt exceeds Δt_{crit} the solution will become unstable, giving results that are clearly in error. Alternatively, the value of Δt_{crit} may be estimated by using an expression derived below for the case that $N = 0$.

The Critical Time Step

Stability of the numerical solution requires that a small disturbance, caused for example by rounding off, damps out rather than amplifies. We consider the case that all values of Φ in the mesh are equal to some constant value Φ_0, except for node i,j where $\Phi = \Phi_0 - \delta\Phi$, and for the four neighboring nodes where $\Phi = \Phi_0 + \delta\Phi$; $\delta\Phi$ represents a disturbance. The expression (60.6) for the Laplacian then becomes:

$$[\nabla^2 \Phi]_{i,j} = \frac{4\Phi_0 + 4\delta\Phi - 4(\Phi_0 - \delta\Phi)}{a^2} = 8\frac{\delta\Phi}{a^2} \tag{60.20}$$

Substitution in (60.18) yields, with $N = 0$,

$$\phi_{t+\Delta t} = \phi_t + \frac{8\Delta t}{S_p}\frac{\delta\Phi}{a^2} \tag{60.21}$$

where $\phi_{t+\Delta t}$ and ϕ_t represent the head at node (i, j). The disturbance $\delta\Phi$ corresponds to a disturbance $\delta\phi$ in the head; $\phi_{t+\Delta t} - \phi_t$ must be less than $2\delta\phi$ in order that the disturbance decreases with time. Hence, the condition for stability is

$$\frac{8\Delta t}{S_p a^2}\delta\Phi < 2\delta\phi \tag{60.22}$$

With

$$\delta\Phi = \tfrac{1}{2}k[\phi_0 + 2\phi_0\delta\phi + (\delta\phi)^2] - \tfrac{1}{2}k\phi_0^2 = k\phi_0\delta\phi + \tfrac{1}{2}k(\delta\phi)^2 \tag{60.23}$$

(60.22) becomes

$$\frac{8\Delta t}{S_p a^2}[k\phi_0\delta\phi + \tfrac{1}{2}k(\delta\phi)^2] < 2\delta\phi \tag{60.24}$$

so that

$$\Delta t < \frac{S_p a^2}{4k(\phi_0 + \tfrac{1}{2}|\delta\phi|)} \tag{60.25}$$

A lower bound for the critical time step is obtained by taking for $\phi_0 + \tfrac{1}{2}|\delta\phi|$ the maximum value of ϕ that is expected to occur in the grid. Denoting this value as $\phi_{\max}$, we obtain

$$\Delta t_{\text{crit}} = \frac{S_p a^2}{4k\phi_{\max}} \tag{60.26}$$

where

$$a = \Delta x = \Delta y \tag{60.27}$$

Sec. 60 Finite Difference Methods

The reader may verify that for problems of one-dimensional flow the expression for the critical time step becomes

$$\Delta t_{\text{crit}} = \frac{S_p(\Delta x)^2}{2k\phi_{\max}} \tag{60.28}$$

It is good practice to keep the time step well below the critical value, e.g., $\Delta t = \frac{1}{2}\Delta t_{\text{crit}}$.

Problem 60.2

Adapt (60.26) to the case of shallow transient confined flow.

Problem 60.3

Write an explicit finite difference program to solve a problem of transient one-dimensional flow in a confined aquifer of width L with the following initial and boundary conditions:

$$\begin{array}{lll} t = 0 & 0 < x < L & \Phi = \Phi_0 \\ t > 0 & \left\{ \begin{array}{ll} x = 0 & \Phi = \Phi_1 < \Phi_0 \\ x = L & \dfrac{\partial \Phi}{\partial x} = 0 \end{array} \right\} \end{array} \tag{60.29}$$

Write a computer program to evaluate the following exact solution to this problem (see Carslaw and Jaeger [1959]):

$$\Phi = \Phi_1 + \frac{4}{\pi}(\Phi_0 - \Phi_1)\sum_{m=0}^{\infty} \frac{1}{2m+1} e^{-[\pi^2(2m+1)^2/4]\tau} \sin\left[\frac{\pi}{2}(2m+1)\frac{x}{L}\right] \tag{60.30}$$

where

$$\tau = \frac{k}{S_s}\frac{t}{L^2} \tag{60.31}$$

Choose $L = 100$ m, $H = 10$ m, $S_s = 2 * 10^{-4}$ m^{-1}, $k = 10^{-6}$ m/s, $\phi_0 = 15$ m and $\phi_1 = 12$ m, and compare the explicit finite difference solution with the exact one.

Implicit Finite Difference Formulation

In the implicit finite difference formulation we take the value of $\nabla^2 \Phi$ at some time between t and $t + \Delta t$. The resulting system of equations is implicit in terms of the unknowns and must be solved at each time step, for example, by Gauss-Seidel iteration. Because of the implicit nature of the formulation, the differential equation for shallow transient unconfined flow must be linearized. The linearized form of the differential equation is given by (16.18),

$$\nabla^2 \Phi = \frac{S_p}{k\bar{\phi}}\frac{\partial \Phi}{\partial t} - N \tag{60.32}$$

We represent the Laplacian as a weighted average between the approximations at times $t_0 + n\Delta t$ and $t_0 + (n+1)\Delta t$, where the time t_0 corresponds to the onset of transient flow, as follows:

$$\nabla^2 \Phi = \nu \nabla^2 \overset{n+1}{\Phi} + (1-\nu)\nabla^2 \overset{n}{\Phi} \qquad (0 \leq \nu \leq 1) \qquad (60.33)$$

where the superscripts refer to the time level. We introduce a function $\overset{n}{F}_{i,j}$ as

$$\overset{n}{F}_{i,j} = \tfrac{1}{4}\left[\overset{n}{\Phi}_{i-1,j} + \overset{n}{\Phi}_{i+1,j} + \overset{n}{\Phi}_{i,j-1} + \overset{n}{\Phi}_{i,j+1}\right] \qquad (60.34)$$

so that $\nabla^2 \Phi$ may be written as follows (see [60.33] and [60.6]),

$$\nabla^2 \Phi = \frac{4}{a^2}\left[\nu(\overset{n+1}{F}_{i,j} - \overset{n+1}{\Phi}_{i,j}) + (1-\nu)(\overset{n}{F}_{i,j} - \overset{n}{\Phi}_{i,j})\right] \qquad (60.35)$$

Discretization of the differential equation (60.32) gives

$$\nu(\overset{n+1}{F}_{i,j} - \overset{n+1}{\Phi}_{i,j}) + (1-\nu)(\overset{n}{F}_{i,j} - \overset{n}{\Phi}_{i,j}) = \frac{a^2 S_p}{4k\overset{n}{\phi}_{i,j}}\frac{(\overset{n+1}{\Phi}_{i,j} - \overset{n}{\Phi}_{i,j})}{\Delta t} - \frac{a^2}{4}\overset{n}{N}_{i,j} \qquad (60.36)$$

where $\overline{\phi}$ is taken as $\overset{n}{\phi}_{i,j}$. This equation is solved for the unknown value $\overset{n+1}{\Phi}_{i,j}$, which gives

$$\overset{n+1}{\Phi}_{i,j} = \frac{4k\overset{n}{\phi}_{i,j}\Delta t}{a^2 S_p + 4k\overset{n}{\phi}_{i,j}\nu\Delta t}\left[\nu \overset{n+1}{F}_{i,j} + (1-\nu)(\overset{n}{F}_{i,j} - \overset{n}{\Phi}_{i,j}) + \frac{a^2 S_p}{4k\overset{n}{\phi}_{i,j}\Delta t}\overset{n}{\Phi}_{i,j} + \frac{a^2}{4}\overset{n}{N}_{i,j}\right] \qquad (60.37)$$

Further improvement may be achieved by replacing the term $\overset{n}{N}_{ij}$ by a weighted average over the time step, using a formula of the form (60.33). Equations (60.37) may be solved by the use of Gauss-Seidel iteration. The implicit scheme reduces to an explicit one for $\nu = 0$ (compare [60.33]). For $\nu \neq 0$ the scheme is implicit; the finite difference formulation then is called a backward one, in contrast to the forward one for $\nu = 0$. For $\nu = 1$ the scheme is fully implicit; the method is called a central finite difference scheme for $\nu = \tfrac{1}{2}$.

The advantage of the implicit formulation is that the time step may be chosen much larger than for the explicit one. Disadvantages are a greater complexity, and that the Gauss-Seidel iteration must be applied at each time step. The explicit formulation is particularly attractive for three-dimensional problems; the formulation may be adapted to problems of three-dimensional flow simply by generating a three-dimensional mesh and changing the expression for $\nabla^2 \Phi$ in (60.18).

61 THE FINITE ELEMENT METHOD

The finite element method differs from finite difference methods in two respects. In the first place the domain is discretized into a mesh of finite elements of any shape, such as triangles. In the second place the differential equation is not solved directly but replaced by a variational formulation. There exist various variational formulations (see Zienkiewicz [1977] or Desai and Abel [1972]); we will follow the approach on the basis of a minimum principle here; the discussion closely follows Verruijt [1982].

We will first discuss the variational principle, apply the finite element method to an elementary flow problem, and finally discuss the finite element formulation for transient flow.

The Variational Principle

We write the differential equation for two-dimensional groundwater flow in a form that is commonly used in finite element formulations. Darcy's law may be written in terms of the discharge vector (Q_x, Q_y) as follows:

$$Q_x = -T\frac{\partial \phi}{\partial x}$$
$$Q_y = -T\frac{\partial \phi}{\partial y} \qquad (61.1)$$

where the transmissivity T may be a function of position. For shallow confined flow T equals kH, for shallow unconfined flow T equals $k\phi$, and for two-dimensional flow in the vertical plane T equals kB. We write the continuity equation in the form

$$\frac{\partial Q_x}{\partial x} + \frac{\partial Q_y}{\partial y} = N - \frac{\phi - \phi^*}{c} \qquad (61.2)$$

where the term $(\phi - \phi^*)/c$ is added to cover cases of flow in leaky aquifers (see Sec. 14). Combining (61.1) with (61.2) we obtain

$$\frac{\partial}{\partial x}\left[T\frac{\partial \phi}{\partial x}\right] + \frac{\partial}{\partial y}\left[T\frac{\partial \phi}{\partial y}\right] + N - \frac{\phi - \phi^*}{c} = 0 \qquad (61.3)$$

The differential equation is applied in a domain $\mathcal{D}$ with boundary $\mathcal{B}$, which is subdivided in segments $\mathcal{B}_1$ and $\mathcal{B}_2$ with boundary conditions of the form

$$\phi = f \qquad \text{(on } \mathcal{B}_1\text{)} \qquad (61.4)$$

and

$$-T\frac{\partial \phi}{\partial n} = g \qquad \text{(on } \mathcal{B}_2\text{)} \qquad (61.5)$$

where $\partial/\partial n$ represents differentiation in the direction of the outward normal to $\mathcal{B}_2$.

The variational principle that replaces the differential equation (61.3) is based on minimizing the following functional U (see Verruijt [1982]):

$$U(\phi) = \tfrac{1}{2} \iint_{\mathcal{D}} \left\{ T\left[\frac{\partial \phi}{\partial x}\right]^2 + T\left[\frac{\partial \phi}{\partial y}\right]^2 - 2N\phi + \frac{\phi^2 - 2\phi\phi^*}{c} \right\} dx\,dy + \int_{\mathcal{B}_2} g\phi\,ds \quad (61.6)$$

We will prove that the functional U has an absolute minimum if ϕ satisfies (61.3) with the boundary conditions (61.4) and (61.5). Let χ be a function that differs from ϕ by a variation λ,

$$\chi = \phi + \lambda \quad (61.7)$$

where λ is an arbitrary function of x and y that vanishes on $\mathcal{B}_1$,

$$\lambda = 0 \quad (\text{on } \mathcal{B}_1) \quad (61.8)$$

For the sake of brevity, we perform the analysis in indicial notation, replacing the coordinates x, y by x_i ($i = 1, 2$), and adopting the Einstein summation convention. The expression for $U(\chi)$ then becomes

$$U(\chi) = \tfrac{1}{2} \iint_{\mathcal{D}} \left[T\chi_{,i}\chi_{,i} - 2N\chi + \frac{\chi^2 - 2\chi\phi^*}{c} \right] dx_1\,dx_2 + \int_{\mathcal{B}_2} g\chi\,ds \quad (61.9)$$

where $\chi_{,i}$ represents the derivative of χ with respect to x_i, and the Einstein summation convention implies that

$$\chi_{,i}\chi_{,i} = \chi_{,1}\chi_{,1} + \chi_{,2}\chi_{,2} = \left[\frac{\partial \chi}{\partial x_1}\right]^2 + \left[\frac{\partial \chi}{\partial x_2}\right]^2 \quad (61.10)$$

Substitution of $\phi + \lambda$ for χ in (61.9) gives, with

$$\chi_{,i}\chi_{,i} = (\phi + \lambda)_{,i}(\phi + \lambda)_{,i} = \phi_{,i}\phi_{,i} + 2\phi_{,i}\lambda_{,i} + \lambda_{,i}\lambda_{,i} \quad (61.11)$$

the following expression for $U(\chi)$:

$$\begin{aligned} U(\chi) = \; & \tfrac{1}{2} \iint_{\mathcal{D}} \left[T\phi_{,i}\phi_{,i} - 2N\phi + \frac{\phi^2 - 2\phi\phi^*}{c} \right] dx_1\,dx_2 + \int_{\mathcal{B}_2} g\phi\,ds \\ & + \iint_{\mathcal{D}} \left[T\phi_{,i}\lambda_{,i} - N\lambda + \lambda\frac{\phi - \phi^*}{c} \right] dx_1\,dx_2 + \int_{\mathcal{B}_2} g\lambda\,ds \\ & + \tfrac{1}{2} \iint_{\mathcal{D}} \left[T\lambda_{,i}\lambda_{,i} + \frac{\lambda^2}{c} \right] dx_1\,dx_2 \end{aligned} \quad (61.12)$$

The first two terms represent $U(\phi)$; we use that $(T\phi_{,i}\lambda)_{,i} - (T\phi_{,i})_{,i}\lambda = T\phi_{,i}\lambda_{,i} + (T\phi_{,i})_{,i}\lambda - (T\phi_{,i})_{,i}\lambda = T\phi_{,i}\lambda_{,i}$ to write (61.12) as

$$U(\chi) = U(\phi) - \iint_{\mathcal{D}} \lambda \left[(T\phi_{,i})_{,i} + N - \frac{\phi - \phi^*}{c}\right] dx_1 dx_2$$

$$+ \iint_{\mathcal{D}} (T\phi_{,i}\lambda)_{,i} dx_1 dx_2 + \int_{\mathcal{B}_2} g\lambda ds$$

$$+ \tfrac{1}{2} \iint_{\mathcal{D}} \left[T\lambda_{,i}\lambda_{,i} + \frac{\lambda^2}{c}\right] dx_1 dx_2 \quad (61.13)$$

The first integral vanishes since ϕ satisfies the differential equation (61.3). The integrand in the last integral is positive since $T > 0$ and $1/c \geq 0$ so that this integral equals some positive number $|M(\lambda)|$,

$$|M(\lambda)| = \tfrac{1}{2} \iint_{\mathcal{D}} \left[T\lambda_{,i}\lambda_{,i} + \frac{\lambda^2}{c}\right] dx_1 dx_2 \quad (61.14)$$

We apply the divergence theorem (55.1) to the second integral in (61.13) with $T\lambda\phi_{,i}$ replacing Q_i and obtain

$$U(\chi) = U(\phi) + \int_{\mathcal{B}} T\lambda \frac{\partial \phi}{\partial n} ds + \int_{\mathcal{B}_2} g\lambda ds + |M(\lambda)| \quad (61.15)$$

The first integral vanishes along $\mathcal{B}_1$, by (61.8), and cancels with the second one along $\mathcal{B}_2$, by (61.5). What remains is

$$U(\chi) = U(\phi) + |M(\lambda)| \quad (61.16)$$

We observe from (61.14) that $M(\lambda)$ vanishes only if $\lambda = 0$, i.e., if $\chi = \phi$: the functional $U(\chi)$ reaches an absolute minimum when the function χ equals the solution to the differential equation with boundary conditions, as asserted.

The application of the variational principle consists of minimizing the functional U for a chosen class of functions that meet the boundary condition along $\mathcal{B}_1$ and contain a number of parameters. For example, if χ is a function of x and y and a set of n parameters ϕ_j,

$$\chi = \chi(x, y, \phi_1, \phi_2, \ldots, \phi_n) \quad (61.17)$$

then $U(\chi)$ will be a minimum if

$$\frac{\partial U}{\partial \phi_j} = 0 \quad (j = 1, 2, \ldots, n) \quad (61.18)$$

which represents n equations for the n unknowns ϕ_j. The accuracy of the solution is affected both by the choice of the function χ and by the number of parameters: That

(61.18) are fulfilled does not mean that the solution obtained is exact; it means that the best solution is found within the class of solutions considered.

The Finite Element Method for Steady Flow

A common choice for the function χ is obtained by discretization of the flow region in a mesh of n triangular elements, approximating the head as varying linearly inside each triangle, and taking the transmissivity T as being constant in each triangle. The integral $U(\chi)$ can then be written as a sum of integrals U_m over the n elements $\mathcal{D}_m$ ($m = 1, 2, \ldots, n$)

$$U = \sum_{m=1}^{n} U_m \qquad (61.19)$$

The head inside the triangle $\mathcal{D}_m$, $\overset{m}{\phi}$, may be represented as follows:

$$\overset{m}{\phi} = \overset{m}{a}_1 x + \overset{m}{a}_2 y + \overset{m}{a}_3 \qquad (61.20)$$

We have determined three similar coefficients a_j in Sec. 38 where the density distribution γ was taken to vary linearly between the values $\overset{e}{\gamma}_1$, $\overset{e}{\gamma}_2$, and $\overset{e}{\gamma}_3$ at the three nodes (ξ_j, η_j) ($j = 1, 2, 3$) of a triangle. The coefficients $\overset{m}{a}_j$ may be obtained from (38.23) through (38.26) by replacing ξ_j by $\overset{m}{x}_j$, η_j by $\overset{m}{y}_j$, and $\overset{e}{\gamma}_j$ by $\overset{m}{\phi}_j$, where $(\overset{m}{x}_j, \overset{m}{y}_j)$ represent the coordinates of, and $\overset{m}{\phi}_j$ the head at, the jth corner point of the mth element. It follows that the coefficients $\overset{m}{a}_j$ may be expressed as

$$\overset{m}{a}_1 = (\overset{m}{b}_1 \overset{m}{\phi}_1 + \overset{m}{b}_2 \overset{m}{\phi}_2 + \overset{m}{b}_3 \overset{m}{\phi}_3)/2A_m$$
$$\overset{m}{a}_2 = (\overset{m}{c}_1 \overset{m}{\phi}_1 + \overset{m}{c}_2 \overset{m}{\phi}_2 + \overset{m}{c}_3 \overset{m}{\phi}_3)/2A_m \qquad (61.21)$$
$$\overset{m}{a}_3 = (\overset{m}{d}_1 \overset{m}{\phi}_1 + \overset{m}{d}_2 \overset{m}{\phi}_2 + \overset{m}{d}_3 \overset{m}{\phi}_3)/2A_m$$

where A_m is the area of the triangle and where

$$\overset{m}{b}_1 = \overset{m}{y}_2 - \overset{m}{y}_3 \qquad \overset{m}{b}_2 = \overset{m}{y}_3 - \overset{m}{y}_1 \qquad \overset{m}{b}_3 = \overset{m}{y}_1 - \overset{m}{y}_2$$
$$\overset{m}{c}_1 = \overset{m}{x}_3 - \overset{m}{x}_2 \qquad \overset{m}{c}_2 = \overset{m}{x}_1 - \overset{m}{x}_3 \qquad \overset{m}{c}_3 = \overset{m}{x}_2 - \overset{m}{x}_1 \qquad (61.22)$$
$$\overset{m}{d}_1 = \overset{m}{x}_2 \overset{m}{y}_3 - \overset{m}{x}_3 \overset{m}{y}_2 \qquad \overset{m}{d}_2 = \overset{m}{x}_3 \overset{m}{y}_1 - \overset{m}{x}_1 \overset{m}{y}_3 \qquad \overset{m}{d}_3 = \overset{m}{x}_1 \overset{m}{y}_2 - \overset{m}{x}_2 \overset{m}{y}_1$$

We will consider the special case that the boundary $\mathcal{B}_2$ is a streamline so that $g = 0$ (see [61.5]) and the integral along $\mathcal{B}_2$ vanishes. We further set N and $1/c$ equal to zero so that the functional (61.6) becomes, representing the approximate solution as ϕ,

$$U = \tfrac{1}{2} \iint_{\mathcal{D}} \left\{ T \left[\frac{\partial \phi}{\partial x} \right]^2 + T \left[\frac{\partial \phi}{\partial y} \right]^2 \right\} dx\, dy \qquad (61.23)$$

Sec. 61   The Finite Element Method

The contribution of the mth element to U is

$$U_m = \tfrac{1}{2} T_m \iint_{\mathcal{D}} \left\{ \left[\frac{\partial \phi}{\partial x}\right]^2 + \left[\frac{\partial \phi}{\partial y}\right]^2 \right\} dx\, dy \qquad (61.24)$$

where T_m represents the transmissivity of the mth element, which is taken to be constant. Using (61.20) in (61.24) we obtain

$$U_m = \tfrac{1}{2} T_m \iint_{\mathcal{D}} (\overset{m}{a}{}_1^2 + \overset{m}{a}{}_2^2)\, dx\, dy = \tfrac{1}{2} T_m A_m (\overset{m}{a}{}_1^2 + \overset{m}{a}{}_2^2) \qquad (61.25)$$

Using (61.21), this may be written as

$$U_m = \sum_{i=1}^{3} \sum_{j=1}^{3} \frac{\tfrac{1}{2} T_m A_m}{4 A_m^2} \left[\overset{m}{b}_i \overset{m}{\phi}_i \overset{m}{b}_j \overset{m}{\phi}_j + \overset{m}{c}_i \overset{m}{\phi}_i \overset{m}{c}_j \overset{m}{\phi}_j \right] = \tfrac{1}{2} \sum_{i=1}^{3} \sum_{j=1}^{3} \overset{m}{P}_{ij} \overset{m}{\phi}_i \overset{m}{\phi}_j \qquad (61.26)$$

where

$$\overset{m}{P}_{ij} = \frac{T_m}{4 A_m} (\overset{m}{b}_i \overset{m}{b}_j + \overset{m}{c}_i \overset{m}{c}_j) \qquad (61.27)$$

An expression for the functional U in terms of the heads at the nodes is obtained by summation over all contributions U_m:

$$U = \tfrac{1}{2} \sum_{m=1}^{n} \sum_{i=1}^{3} \sum_{j=1}^{3} \overset{m}{P}_{ij} \overset{m}{\phi}_i \overset{m}{\phi}_j = \tfrac{1}{2} \sum_{i=1}^{M} \sum_{j=1}^{M} P_{ij} \phi_i \phi_j \qquad (61.28)$$

where M represents the number of nodal points and ϕ_i the head at node i. Application of (61.18) gives the following system of linear equations

$$\boxed{\sum_{j=1}^{M} P_{ij} \phi_j = 0 \qquad (i = 1, 2, 3, \ldots, n)} \qquad (61.29)$$

where use is made of the symmetry of P_{ij} (compare [61.27]).

The finite element method is quite suitable for numerical applications; the coefficients of the matrix P_{ij} can be generated from the geometry of the triangular mesh. The system of equations (61.29) may be solved by Gauss-Seidel iteration as discussed for the finite difference method in Sec. 60. For listings of finite element programs see, for example, Verruijt [1982], Wang and Anderson [1982], and Bear and Verruijt [1987].

Example 61.1: Flow Through a Strip

We will consider the trivial case of flow through a strip of uniform transmissivity T as an illustration of the method. The flow region is the rectangle with sides of 2 m and 1 m shown in Figure 9.3(a). The heads along sides 1-2 and 5-6 are ϕ_a and ϕ_b, and sides 2-4-6 and 1-3-5 are streamlines. The flow domain is subdivided in four triangles as shown in Figure 9.3(b). The first step is to express the constants $\overset{m}{a}_1$ and $\overset{m}{a}_2$ in terms of the heads at the nodal points. Application of (61.21) and (61.22) to the four triangles $\mathcal{D}_m$ gives, representing the head ϕ at node j as ϕ_j ($j = 1, 2, \ldots, 6$)

$$\overset{1}{a}_1 = \phi_4 - \phi_2 \quad \overset{1}{a}_2 = \phi_1 - \phi_2 \quad \overset{2}{a}_1 = \phi_3 - \phi_1 \quad \overset{2}{a}_2 = \phi_3 - \phi_4$$
$$\overset{3}{a}_1 = \phi_6 - \phi_4 \quad \overset{3}{a}_2 = \phi_3 - \phi_4 \quad \overset{4}{a}_1 = \phi_5 - \phi_3 \quad \overset{4}{a}_2 = \phi_5 - \phi_6 \tag{61.30}$$

The heads ϕ_1, ϕ_2, ϕ_5, and ϕ_6 are given

$$\phi_1 = \phi_2 = \phi_a \qquad \phi_5 = \phi_6 = \phi_b \tag{61.31}$$

so that the only two unknowns in the problem are ϕ_3 and ϕ_4. Differentiation of the functional U obtained from (61.19) and (61.25) with respect to ϕ_3 and ϕ_4 yields

$$T[(\phi_3 - \phi_1) + 2(\phi_3 - \phi_4) - (\phi_5 - \phi_3)] = T[4\phi_3 - \phi_1 - 2\phi_4 - \phi_5] = 0$$
$$T[(\phi_4 - \phi_2) - 2(\phi_3 - \phi_4) - (\phi_6 - \phi_4)] = T[4\phi_4 - \phi_2 - 2\phi_3 - \phi_6] = 0$$
$$\tag{61.32}$$

with the solution

$$\phi_3 = \phi_4 = \tfrac{1}{2}(\phi_a + \phi_b) \tag{61.33}$$

which corresponds to the exact solution for this case.

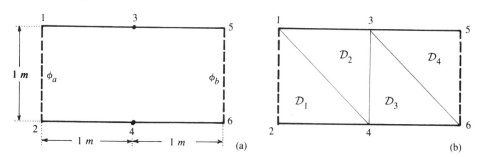

Figure 9.3 Flow in a strip.

Problems with Leakage and Infiltration

For problems with leakage and infiltration, i.e., $N \neq 0$ and $1/c \neq 0$, the integrands in the expressions for U_m will contain a term $-N\phi + \tfrac{1}{2}(\phi^2 - 2\phi\phi^*)/c$. In the finite element method it is straightforward to account for variation with position of the infiltration rate N, the resistance c, and the head ϕ^* in the adjoining aquifer. This is in contrast to the analytic approach (compare, e.g., Sec. 14). Both the infiltration rate N and the resistance

c are taken constant inside the elements and the head $\overset{m}{\phi^*}$ is interpolated linearly inside the triangle $\mathcal{D}_m$ between the values $\overset{m}{\phi_j^*}$ ($j = 1, 2, 3$) at the three corner points as follows:

$$\overset{m}{\phi^*} = \overset{m}{p_1}x + \overset{m}{p_2}y + \overset{m}{p_3} \qquad (61.34)$$

where $\overset{m}{p_j}$ ($j = 1, 2, 3$) are obtained from 61.21 by replacing $\overset{m}{a_j}$ by $\overset{m}{p_j}$ and $\overset{m}{\phi_j}$ by $\overset{m}{\phi_j^*}$. Following Verruijt [1982], we introduce local coordinate systems $(\overset{m}{\xi}, \overset{m}{\eta})$ for each triangle, with the origin at the centroid,

$$\begin{aligned}\overset{m}{\xi} &= x - \tfrac{1}{3}(\overset{m}{x_1} + \overset{m}{x_2} + \overset{m}{x_3}) \\ \overset{m}{\eta} &= y - \tfrac{1}{3}(\overset{m}{y_1} + \overset{m}{y_2} + \overset{m}{y_3})\end{aligned} \qquad (61.35)$$

An expression for U_m may be obtained following a procedure similar to the one used to derive (61.26). This expression is presented here without derivation:

$$U_m = \sum_{i=1}^{3}\sum_{j=1}^{3}\left[\tfrac{1}{2}\overset{m}{P_{ij}}\overset{m}{\phi_i}\overset{m}{\phi_j} - \frac{1}{c_m}\overset{m}{R_{ij}}\overset{m}{\phi_i}\overset{m}{\phi_j^*}\right] - \sum_{i=1}^{3}\overset{m}{Q_i}\overset{m}{\phi_i} \qquad (61.36)$$

where c_m is the resistance in element m, and

$$\overset{m}{P_{ij}} = \frac{T_m}{4A_m}\left[\overset{m}{b_i}\overset{m}{b_j} + \overset{m}{c_i}\overset{m}{c_j}\right] + \gamma_m \overset{m}{R_{ij}} \qquad (61.37)$$

$$\overset{m}{R_{ij}} = \frac{1}{4A_m}\left[\overset{m}{d_i}\overset{m}{d_j} + Z_{xx}\overset{m}{b_i}\overset{m}{b_j} + Z_{yy}\overset{m}{c_i}\overset{m}{c_j} + Z_{xy}(\overset{m}{b_i}\overset{m}{c_j} + \overset{m}{b_j}\overset{m}{c_i})\right] \qquad (61.38)$$

$$\overset{m}{Q_i} = \tfrac{1}{3}A_m N_m \qquad (61.39)$$

where N_m represents the infiltration rate in element m and the coefficients γ_m are expressed in terms of the resistance c_m as

$$\gamma_m = \frac{1}{c_m} \qquad (61.40)$$

The constants Z_{xx}, Z_{yy} and Z_{xy} are given by

$$\overset{m}{Z_{xx}} = \frac{1}{A_m}\iint_{\mathcal{D}_m}\overset{m}{\xi^2}d\overset{m}{\xi}d\overset{m}{\eta} = \tfrac{1}{12}(\overset{m}{\xi_1^2} + \overset{m}{\xi_2^2} + \overset{m}{\xi_3^2}) \qquad (61.41)$$

$$\overset{m}{Z_{yy}} = \frac{1}{A_m}\iint_{\mathcal{D}_m}\overset{m}{\eta^2}d\overset{m}{\xi}d\overset{m}{\eta} = \tfrac{1}{12}(\overset{m}{\eta_1^2} + \overset{m}{\eta_2^2} + \overset{m}{\eta_3^2}) \qquad (61.42)$$

$$\overset{m}{Z_{xy}} = \frac{1}{A_m}\iint_{\mathcal{D}_m}\overset{m}{\xi}\overset{m}{\eta}d\overset{m}{\xi}d\overset{m}{\eta} = \tfrac{1}{12}(\overset{m}{\xi_1}\overset{m}{\eta_1} + \overset{m}{\xi_2}\overset{m}{\eta_2} + \overset{m}{\xi_3}\overset{m}{\eta_3}) \qquad (61.43)$$

where $\overset{m}{\xi}_j$ and $\overset{m}{\eta}_j$ ($j = 1, 2, 3$) denote the local coordinates of the corner points of the triangle.

We minimize the functional U obtained from (61.19) and (61.36) and obtain a system of n linear equations in terms of the n heads ϕ_m, which may be written in the form (compare [61.29])

$$\sum_{j=1}^{n} P_{ij}\phi_j = Q_i + \frac{1}{c_m}\sum_{j=1}^{n} R_{ij}\phi_j^* \qquad (i = 1, 2, 3, \ldots, n) \tag{61.44}$$

where ϕ_j^* represents the head in the adjoining aquifer at node j.

The Finite Element Method for Transient Flow

The above finite element formulation may be readily modified to cover problems of transient flow. We adapt the differential equation (60.16) to the case of variable transmissivity and obtain, including the leakage term,

$$\frac{\partial}{\partial x}\left[T\frac{\partial \phi}{\partial x}\right] + \frac{\partial}{\partial y}\left[T\frac{\partial \phi}{\partial y}\right] + N - \frac{\phi - \phi^*}{c} = S_p\frac{\partial \phi}{\partial t} \tag{61.45}$$

Verruijt [1982] integrates this differential equation over a short time interval Δt and obtains

$$\frac{\partial}{\partial x}\left[T\frac{\partial \overline{\phi}}{\partial x}\right] + \frac{\partial}{\partial y}\left[T\frac{\partial \overline{\phi}}{\partial y}\right] + N - \frac{\overline{\phi} - \phi^*}{c} - \frac{S_p}{\Delta t}(\phi_+ - \phi_-) = 0 \tag{61.46}$$

where $\overline{\phi}$ represents the average value of ϕ over the time interval and ϕ_+ and ϕ_- are the values of ϕ at the end and the beginning of the time interval, respectively. We use an implicit formulation, as in Sec. 60 (compare [60.33]), so that $\overline{\phi}$ may be written in terms of ϕ_-, ϕ_+, and a coefficient ν as

$$\overline{\phi} = \nu\phi_+ + (1 - \nu)\phi_- \tag{61.47}$$

The interpolation corresponds to a forward difference for $\nu = 0$, to a backward difference for $\nu = 1$, and to a central difference for $\nu = 1/2$. Verruijt [1982] suggests a value somewhat smaller than 0.5. Using (61.47), (61.46) becomes

$$\frac{\partial}{\partial x}\left[T\frac{\partial \overline{\phi}}{\partial x}\right] + \frac{\partial}{\partial y}\left[T\frac{\partial \overline{\phi}}{\partial y}\right] + N - \frac{\overline{\phi} - \phi^*}{c} - \frac{S_p}{\nu\Delta t}[\overline{\phi} - \phi_-] = 0 \tag{61.48}$$

It follows that the transient effects are represented in (61.48) by a term that is similar in form to the leakage term. The equations derived for flow with leakage may be adapted to cases of transient flow by replacing ϕ by $\overline{\phi}$, ϕ^* by ϕ_-, and $1/c$ by β where

$$\beta = \frac{S_p}{\nu\Delta t} \tag{61.49}$$

Sec. 61   The Finite Element Method

If the flow is both leaky and transient, (61.36) becomes

$$U_m = \sum_{i=1}^{3}\sum_{j=1}^{3}\left[\tfrac{1}{2}\overset{m}{P}_{ij}\overline{\overset{m}{\phi}}_i\overline{\overset{m}{\phi}}_j - \frac{1}{c_m}\overset{m}{R}_{ij}\overline{\overset{m}{\phi}}_i\overset{m}{\phi}^*_j - \beta_m \overset{m}{R}_{ij}\overline{\overset{m}{\phi}}_i\overset{m}{\phi}_{j-}\right] - \sum_{i=1}^{3} Q_i \overline{\overset{m}{\phi}}_i \qquad (61.50)$$

where β_m is the value of β in element m, and $\overline{\overset{m}{\phi}}_j$ and $\overset{m}{\phi}_{j-}$ represent the average head and the head at the beginning of the time interval at corner point j of element m, respectively. The expressions (61.37) through (61.39) and (61.41) through (61.43) for the various constants in (61.50) remain valid, but (61.40) is replaced by

$$\gamma_m = \frac{1}{c_m} + \beta_m \qquad (61.51)$$

The system of equations (61.44) is modified to account for the additional term in (61.50) as follows:

$$\boxed{\sum_{j=1}^{n} P_{ij}\overline{\phi}_j = Q_i + \sum_{j=1}^{n} R_{ij}(\phi^*_j + \beta^* \phi_{j-})} \qquad (61.52)$$

The head ϕ_+ valid at the end of the time interval is obtained from (61.47) as

$$\phi_+ = \phi_- + \frac{1}{\nu}(\overline{\phi} - \phi_-) \qquad (61.53)$$

The above developments were carried out in terms of heads in order to facilitate the understanding of computer programs on the market today, most of which are written with the head as the dependent variable. A formulation in terms of potentials may have advantages over the one in terms of heads. In the first place, problems where both confined and unconfined flows occur can be handled elegantly (see Sec. 8); in the second place the programs can be readily adapted to various types of flow, e.g., interface flow (compare Chapter 2). A formulation in terms of potentials may therefore be preferred and will be discussed below.

Formulation in Terms of Potentials

We introduce a potential F for a vector with components U_x, U_y so that

$$\begin{aligned} U_x &= -\frac{\partial F}{\partial x} \\ U_y &= -\frac{\partial F}{\partial y} \end{aligned} \qquad (61.54)$$

where

$$U_x = \frac{Q_x}{k}$$
$$U_y = \frac{Q_y}{k}$$
(61.55)

The function F, defined as

$$F = \Phi/k$$
(61.56)

is independent of the hydraulic conductivity. Expressions for F applicable to the various types of flow discussed in Chapter 2 are obtained from the definitions of Φ by division by k; for example, F equals $\frac{1}{2}\phi^2$ for unconfined flow and $H\phi - \frac{1}{2}H^2$ for confined flow. Constants, such as $-\frac{1}{2}H^2$, may be added to F without affecting the governing differential equations as follows from (61.54): the derivatives with respect to x and y of these constants vanish. It is important to note, however, that the aquifer thickness H now must be kept constant. Discontinuities in H may be allowed, provided that the corresponding jumps in potential are properly accounted for. Note also that the varying thickness of the flow domain, as occurring in shallow unconfined flows and interface flows, is automatically accounted for in the definition of the potential F. With

$$\frac{\partial Q_x}{\partial x} + \frac{\partial Q_y}{\partial y} = \frac{\partial [kU_x]}{\partial x} + \frac{\partial [kU_y]}{\partial y} = -\frac{\partial}{\partial x}\left[k\frac{\partial F}{\partial x}\right] - \frac{\partial}{\partial y}\left[k\frac{\partial F}{\partial y}\right]$$
(61.57)

the differential equation (61.45) becomes

$$\frac{\partial}{\partial x}\left[k\frac{\partial F}{\partial x}\right] + \frac{\partial}{\partial y}\left[k\frac{\partial F}{\partial y}\right] + N - \frac{\phi - \phi^*}{c} = S_p\frac{\partial \phi}{\partial t}$$
(61.58)

For problems of horizontal confined flow the head ϕ is a linear function of F. In all other cases the relation between F and ϕ may be written as

$$\phi = \frac{F}{h}$$
(61.59)

where h is the thickness of the flow domain. This thickness may vary with position, but depends on the solution and is to be determined iteratively. The differential equation (61.58) may be written in terms of F and h as

$$\frac{\partial}{\partial x}\left[k\frac{\partial F}{\partial x}\right] + \frac{\partial}{\partial y}\left[k\frac{\partial F}{\partial y}\right] + N - \frac{F - F^*}{c^*} = S_p\frac{\partial (F/h)}{\partial t}$$
(61.60)

where

$$F^* = h\phi^* \qquad c^* = hc$$
(61.61)

If the flow is steady and $1/c = 0$, the finite element formulation in terms of potentials is obtained from the one given above in terms of head simply by replacing T

Sec. 61 The Finite Element Method

everywhere by k and Φ by F. For cases of leakage, the formulation remains valid, and F^* and c^* may either be kept constant using some average value for h, or be updated in the iterative process using newly computed values of h.

Transient flow in confined aquifers may be covered by replacing S_p in the equations by S_s with

$$S_s = \frac{S_p}{h} = \frac{S_p}{H} \qquad (61.62)$$

For problems of shallow unconfined transient flow, integration of the differential equation over a time interval Δt yields, taking h to be constant over the time step and equal to the value h_-, obtained at the end of the previous time step,

$$\frac{\partial}{\partial x}\left[k\frac{\partial \overline{F}}{\partial x}\right] + \frac{\partial}{\partial y}\left[k\frac{\partial \overline{F}}{\partial y}\right] + N - \frac{\overline{F} - F_-^*}{c_-^*} - \frac{S_{s-}}{\Delta t}[F_+ - F_-] = 0 \qquad (61.63)$$

where

$$F_-^* = \phi^* h_- \qquad c_-^* = c h_- \qquad S_{s-} = S_p/h_- \qquad (61.64)$$

It is seen by comparison of (61.63) with (61.46) that the finite element scheme in terms of potentials is obtained from the one in terms of heads by replacing T by k, c by c_-^*, ϕ^* by F_-^*, S_p by S_{s-}, $\overline{\phi}$ by $\overline{F}$, ϕ_+ by F_+, and ϕ_- by F_-. The difference between the two approaches, for the case of unconfined flow, is that in the formulation in terms of heads the transmissivity T is taken constant over each element, whereas in the present formulation the aquifer thickness varies as $\sqrt{2F}$ (F is a linear function) over the elements, but is kept constant over a time interval. Thus, the error is shifted from an error in aquifer thickness to an error in the computation of the terms for leakage and storage.

Chapter 10

Analogue Methods

Numerous physical processes exist that are governed by equations that can be written in the same form as Darcy's law and the continuity equations. Such processes are analogous to groundwater flow, and an analogue method consists of studying any one of these processes and interpreting the response to boundary and initial conditions in terms of groundwater flow. It is necessary for this interpretation to determine scale factors and to translate the physical constants into those involved in the groundwater flow equations, such as the intrinsic permeability, the viscosity, and the storage coefficient (see Karplus [1958], Bear [1972]). Examples of analogues are conservative force fields, electrokinetics and magnetism, conduction of heat in solids, aerodynamics (flow around air foils), deflections of a thin membrane under load, and slow viscous flow between parallel plates.

Knowledge of analogue physical processes may serve various purposes. In some cases the analogue is useful simply because numerous mathematical solutions have been generated that are directly applicable to groundwater flow. Many of the solutions to problems of heat flow given by Carslaw and Jaeger [1959] fall in this category; the first application of a solution to a problem of transient groundwater flow was obtained by Theis [1935] by interpreting the solution for an infinite linear heat source in terms of groundwater flow. Another example is the panel method commonly used for three-dimensional problems in aerodynamics, which may be adapted for use in three-dimensional groundwater flow (Hess and Smith [1966]). In other cases the analogue may serve to construct a visual model of groundwater flow as an aid in understanding the effect of modeling by using mathematical functions. The membrane analogue (De Josselin de Jong [1961]) is an example of this and is described in Sec. 62.

A third category of analogues consists of the ones that are amenable to physical experimentation; the measurements must then be translated in terms of groundwater flow. These physical experiments may be done either as illustrations for educational purposes, to solve field problems, to verify the validity of a mathematical formulation or, finally, to study a phenomenon in a controlled environment. The membrane analogue, besides being useful as a conceptual tool, has been used to verify a theory (De Josselin de Jong [1961]), but is not commonly used in groundwater mechanics. The electric analogue, discussed in Sec. 63, has been used in past years for modeling but currently is in use primarily for educational purposes. A modern version of the analogue, however, based on etching patterns of cracks on printed circuit boards, has been used recently for studying flow in fractured rock (Hudson and Lapointe [1980]). The viscous flow, or Hele Shaw, analogue is treated in Sec. 64 and is useful for educational purposes, for validating mathematical solutions, and to study complex phenomena such as two-phase flow. A variant of the Hele Shaw model, rather like a scaled down version of an aquifer, is sometimes useful for studying physical phenomena where the presence of a porous medium is required. Such scale models are discussed briefly in Sec. 65.

62 THE MEMBRANE ANALOGUE

The membrane analogue is based on the similarity between the normal displacement w of points on a membrane that is placed under tension by connecting it to some boundary (see Figure 10.1) and is loaded, for example, by applying pressure. This displacement is measured with respect to a reference plane parallel to the plane of the membrane prior to loading. For small displacements w fulfills Laplace's equation, so that contours of constant w correspond to equipotentials. The slope of the membrane measured with respect to the reference plane is proportional to the discharge vector. Thus, for example, the potential for flow toward a well in a circular island may be visualized by imagining a membrane attached under constant tension along a circular ring, parallel to the reference level. The effect of the well is generated by pressing a small ring downward at the position of the well. The resulting deflection of the membrane may easily be visualized and is a measure of the potential; the slope of the membrane is a measure of discharge. Application of a constant air pressure to one side of the membrane causes the governing differential equation for w to become the Poisson equation. The effect of infiltration on the above circular aquifer may thus be visualized as the deflection of the membrane due to a constant pressure underneath it. If the well is not present, the normal displacement w is controlled, for a given membrane, both by the pressure underneath it and the tension in the membrane; the former variable is proportional to the rate of infiltration and the latter one to the hydraulic conductivity: an increase in tension corresponds to an increase in k.

The membrane analogue described above serves well to create a visual image of the modeling of regional flow by the use of analytic elements. Imagine the infinite aquifer to be represented by a very large membrane, connected along its outer boundary to a ring. Setting the elevation of a selected point on the ring at a given elevation above the reference level corresponds to specifying the potential at a reference point in the analytic element model. Tilting the ring in space corresponds to specifying a uniform far-field. Regional infiltration may be visualized as applying a constant pressure underneath the entire membrane. Analytic elements for modeling finite canals may be visualized in the analogue by imagining the effect of pressing a thin rod into the membrane at a specified elevation, whereas drains are imagined in the analogue as rods that are attached to the membrane in such a way that they are parallel to the reference level but are free to move up or down. Area-sinks are visualized as zones of constant pressure applied above the membrane over the area of the area-sink. The line-sink may be visualized as a narrow area-sink with high pressure.

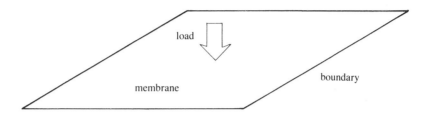

Figure 10.1 The membrane analogue.

The visual image is particularly useful for assessing the difference in effect of various analytic elements that may be used to approximate the same feature. For example, the difference between a line-sink of constant strength and a double-root element becomes quite clear. The constant pressure applied in the analogue to represent the line-sink is adjusted so as to produce a given elevation at the center of the element. This will cause a much more gradual variation in w near the tips of the element than the double-root element that corresponds to a thin rod and causes sharp depressions near its ends.

The effect of setting the condition at the reference point may now be explained visually. If all elements are discharge-specified, i.e., pressure-specified in the analogue, then a parallel raising or lowering of the ring will result in a uniform raising or lowering of the potential throughout the domain. This raising or lowering has no effect on the flow from infinity represented by the average slope of the membrane normal to the ring. The effect of the condition at the reference point is quite different if heads are specified along any of the elements, i.e., if the elevation of the membrane is fixed somewhere inside the ring. A raising or lowering of the ring then will result in an increase in the normal slope of the membrane along the ring, i.e., in an increase of the flow from or toward infinity.

Jumps in the stream function correspond in the analogue to discontinuities of slope, as occurring for example on either side of the rods referred to above. Jumps in the potential correspond to discontinuities in the elevation of the membrane.

Although useful as a visual tool, the membrane analogue is not a practical alternative to numerical modeling.

63 THE ELECTRIC ANALOGUE

The electric analogue is based on the similarity between Ohm's law for the movement of electrons through a conducting material and Darcy's law. Ohm's law may be written as

$$i = -\frac{1}{\rho}\frac{dV}{ds} \qquad (63.1)$$

where i is the current density in amperes per square meter, ρ the specific resistance of the material in ohm-meters, and V the electric potential in volts. The continuity equation is analogous to the property that electrons can neither be lost nor produced. The electrical analogue is thereby complete; $1/\rho$ represents the hydraulic conductivity, i the discharge vector, and V the piezometric head.

A simple instructional tool may be obtained by using a specially prepared type of paper with nearly uniform conductivity (commercially available under the name Teledeltos paper) in the simple experimental setup described below. Free edges of the paper correspond to impermeable boundaries, and boundary segments of constant potential may be implemented by clamping strips of copper with controlled voltage to the paper. A simple aquifer model may be constructed as shown in Figure 10.2. The two copper strips are attached to the two poles of a battery. Equipotentials may be contoured by the use of a probe, attached via an ammeter to a variable resistance. The voltage V_p applied at the probe may then be adjusted to the voltage V_0 applied on the negative electrode plus a value between zero and 100% of the difference in voltage. The ammeter will register zero when V_p equals the voltage at the point where the probe touches the paper. Tracing equipotentials thus is a matter of setting the variable resistance to the desired value and searching for points where the ammeter registers zero.

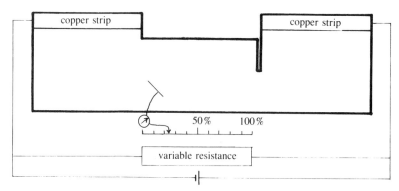

Figure 10.2 The electric analogue.

The analogue is also applicable to problems with a free boundary. The paper is first cut along an estimated position of this boundary and the voltage measured at preset elevations of it. The condition that the free surface is a streamline is identically satisfied, and the position of the free surface is adjusted by cutting until the drops in potential between points at elevations on fixed intervals are identical. It is obviously easier to cut paper than to reattach it; the initial estimate of the phreatic surface must be on the high side.

Resistance networks have been built in the past to simulate the flow of groundwater. Such networks are analogue computers that solve the differential equation in exactly the same fashion as is done digitally in the finite difference methods. The latter methods are cheaper and more flexible, and, along with other numerical techniques, have virtually replaced the analogue computer.

Problem 63.1

Explain how the total current through the model may be measured. Representing this measurement as I_{tot}, express the discharge flowing through the aquifer represented by the analogue in terms of I_{tot}, ρ, and ΔV.

Problem 63.2

Sketch an experimental setup that can be used to generate a plot of streamlines and specify the boundary conditions to be applied, using the quantity I_{tot} defined in Problem 63.1.

64 THE HELE SHAW ANALOGUE

The slow flow of a viscous fluid in the narrow space between two parallel plates, investigated by Hele Shaw (see Lamb [1932]), is analogous to groundwater flow, with the quantity

$$k = \frac{\rho g d^2}{12\mu} \tag{64.1}$$

representing the hydraulic conductivity. The constants in (64.1) are the specific density ρ of the fluid, the acceleration of gravity g, the dynamic viscosity μ, and the distance d between the parallel plates. The velocity corresponds to the specific discharge, and the pressure and head to the quantities of the same names.

The analogue is particularly useful for modeling problems with a free boundary, problems of transient flow, and problems of multiple fluid flow. The latter problems are of special interest because no exact solutions exist, and the behavior of the interface under transient conditions is not always fully understood (e.g., Curtis [1983]).

Hele Shaw experiments may be set up in two different ways. In the direct application of the Hele Shaw model, the entire space between the parallel plates is the flow domain, and the discontinuities in slot width at the reservoirs $\mathcal{A}$ and $\mathcal{B}$ are two boundary equipotentials (see Figure 10.3). The upper and lower boundaries are the phreatic surface and the model base. The problem thus modeled is the classical Dupuit problem of phreatic flow through a dam with vertical faces. Due to the completeness of the analogy, the boundary conditions along the phreatic surface and the seepage face at the downstream face of the dam are met identically in the analogue model.

A photograph of another direct application of the Hele Shaw model is shown in Figure 10.4. It represents the problem of interface flow from a horizontal equipotential

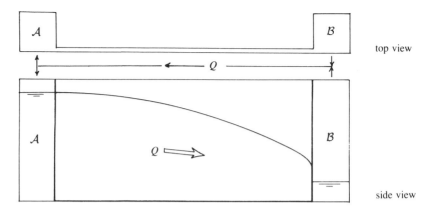

Figure 10.3 Direct application of the Hele Shaw model.

toward drains. The figure is a photograph of one of many experiments carried out as validations of the solution presented in Sec. 51. The drains were included by drilling small holes in the front plate, connected via hoses to reservoirs that were adjusted vertically in order to set the discharges of the drains. The streamlines shown in the photograph were produced by carefully injecting dye through needles into the slot at selected points. The hydraulic conductivity is proportional to the square of the distance between the plates which varies somewhat in the model and was computed as an average. The average distance between the plates was determined by dividing the volume of fluid that completely fills the model by the surface area of the plates. The results obtained by using the hydraulic conductivity computed with this average value for d corresponded exactly to the Hele Shaw experiment: the computed and photographed interfaces were indistinguishable (Strack [1972]). It may be noted that the sensitivity of the instability of the interface to the aquifer parameters was clearly observed in the experiments. The discharges of the drains had to be increased very slowly toward the critical value. Once the critical discharge was established, a stable-looking interface was observed to jump suddenly to an unstable form, sometimes triggered by a small disturbance.

An indirect use of the Hele Shaw model consists of setting the model up with a large slot width (e.g., 2 cm rather than 2 mm) and using filler plates that locally reduce the slot width to the value normally used in the analogue (e.g., 2 mm). In this way, phreatic flow through oddly shaped structures such as the dam shown in Figure 10.5 may be simulated. The effect of a clay core may be included by using a second filler plate to reduce the hydraulic conductivity to match that of the core. A drain at the toe of the dam is included by omitting the filler plate at the position of the drain. The impermeable base of the dam is represented by a filler plate that fills the gap completely.

It is not necessary to use Hele Shaw experiments to validate solutions obtained by the use of methods capable of producing accurate flow nets. The flow net can be checked visually for accuracy, giving sufficient information regarding the validity of the solution. The flow net produced for phreatic flow in an aquifer with a lamina shown in

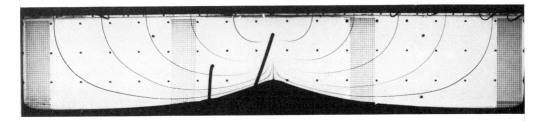

Figure 10.4 Hele Shaw experiment for interface flow.

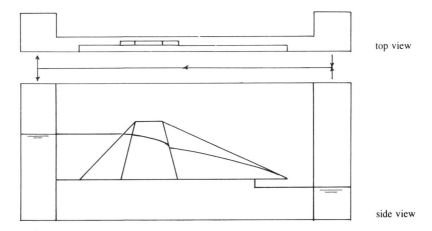

Figure 10.5 Indirect application of the Hele Shaw model.

Figure 7.29, Sec. 50, and the contour plots shown in Figures 6.31, 6.32, and 6.34 in Sec. 39 are examples of such visual validations.

65 SCALE MODELS; MEASURING HYDRODYNAMIC DISPERSION

Scale models are scaled-down versions of aquifers, with the medium consisting of a granular assembly. The most common version of a scale model is the sand box, which simply is a glass box filled with sand, and is used primarily for the purpose of illustration. A more useful scale model resembles the Hele Shaw model, but differs from it in that the space between the plates is filled with fine crushed glass. By selecting a fluid and a glass that have the same breaking index for the passage of light, a clear vision of the aquifer is obtained. Streamlines are produced by injecting dye. This kind of scale model is used if effects need be included that occur only in a porous medium. An example of such an effect is hydrodynamic dispersion (see De Josselin de Jong [1958],

Bear [1972]). Hydrodynamic dispersion is caused by the mixing occurring in the pore space as a result of diffusion, combined with the dispersing effect of the solid particles on the flow: the mixed fluid flowing in the pore space is divided each time that a solid particle is in its path. The magnitude of this dispersive effect is a function of the grain size, and should therefore be amplified in a scale model, where the relative grain size is usually larger than in an aquifer. An example of the use of a scale model is shown in Figure 10.6. The model was built by the Royal Dutch Shell in Rijswijk, the Netherlands, and used to verify the applicability of an exact solution to predict flow of oil-polluted groundwater in a shallow unconfined aquifer with rainfall (see CONCAWE [1977]). The oil is observed in practice to remain in an immobile quantity roughly on the phreatic surface (see Dietz [1972]). The contamination is carried into the aquifer by rainwater, which passes through the immobile oil, and carries small but measurable quantities of soluble components into the aquifer. A possible solution to this problem, examined at the time, consisted of flushing out the soluble components of oil by infiltrating water in the area above the immobile oil and recapturing it by means of a drain. The position of the drain so as to capture all contaminated water was determined by the use of a computer program based on an exact solution to the idealized problem; hydrodynamic dispersion was not taken into account. The streamline pattern shown in Figure 10.6(a) shows good agreement with the exact solution shown in Figure 10.6(b). It also shows that the effect of the hydrodynamic dispersion is small, even though the grain size in the model corresponds to very coarse gravel in the aquifer.

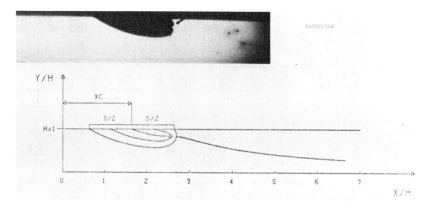

Figure 10.6 Prototype used to validate a model predicting flow of oil-contaminated water. (Source: CONCAWE [1977].)

Dispersive effects in aquifers become important when the inhomogeneities are much larger than particles. Dispersion in aquifers that are nearly homogeneous is small with respect to spreading caused by diverging streamlines (see Nelson [1978]).

Appendix A

Expansion of a Function Raised to a Power

Let F be a complex function of z, analytic at z_0, with a Taylor expansion of the form,

$$F = \sum_{j=0}^{\infty} \frac{F^{(j)}(z_0)}{j!}(z - z_0)^j \qquad (A.1)$$

Furthermore, $F(z_0)$ is unequal to zero,

$$F(z_0) \neq 0 \qquad (A.2)$$

The function

$$\overset{\alpha}{g}_0 = F^\alpha \qquad (A.3)$$

where α is a real number, then is analytic at z_0. We will outline a procedure for determining the first $n + 2$ coefficients in the expansion of F^α about z_0,

$$F^\alpha = \overset{\alpha}{g}_0(z_0) + \overset{\alpha}{g}'_0(z_0)(z - z_0) + \ldots + \frac{\overset{\alpha}{g}_0^{(j)}(z_0)}{j!}(z - z_0)^j + \ldots \qquad (A.4)$$

We introduce the function $\overset{\alpha}{g}_m$, $m = 1, 2, \ldots$, defined as the mth derivative of $\overset{\alpha}{g}_0$ with respect to F,

$$\overset{\alpha}{g}_m = \alpha(\alpha - 1)(\alpha - 2)\ldots(\alpha - m + 1)F^{\alpha - m} \quad \begin{Bmatrix} \alpha \neq k \\ \alpha = k \quad m \leq \alpha \end{Bmatrix} \qquad (k = 1, 2, \ldots)$$

(A.5)

$$\overset{\alpha}{g}_m = 0 \qquad\qquad \alpha = k \quad m > \alpha \qquad (k = 1, 2, \ldots)$$

We further introduce the function h as the derivative of F with respect to z:

$$h = \frac{dF}{dz} = F' \tag{A.6}$$

It follows that

$$\overset{\alpha}{g}_n{}' = \overset{\alpha}{g}_{n+1} h \tag{A.7}$$

We may apply Leibnitz's rule for differentiating the product of two functions to the function $\overset{\alpha}{g}_{n-m}{}'$,

$$\overset{\alpha}{g}_{n-m}{}' = \overset{\alpha}{g}_{n-m+1} h \tag{A.8}$$

as follows:

$$\frac{\overset{\alpha(j)}{g}_{n-m}}{(j-1)!} = \frac{\overset{\alpha(j-1)}{g}_{n-m+1}}{(j-1)!} h + \frac{\overset{\alpha(j-2)}{g}_{n-m+1}}{(j-2)!} \frac{h'}{1!} \cdots$$

$$+ \frac{\overset{\alpha(j-k)}{g}_{n-m+1}}{(j-k)!} \frac{h^{(k-1)}}{(k-1)!} + \ldots + \overset{\alpha}{g}_{n-m+1} \frac{h^{(j-1)}}{(j-1)!}$$

$$(m \le n, \quad j = 1, 2, \ldots, m+1) \tag{A.9}$$

Using this equation for $j = 1, 2, \ldots, m+1$ we obtain the first $m+1$ derivatives of $\overset{\alpha}{g}_{n-m}$ at $z = z_0$ expressed in terms of the first m derivatives of $\overset{\alpha}{g}_{n-m+1}$ and h at $z = z_0$. We apply (A.9) repeatedly, starting with $m = 1$ and $j = 1, m+1$, until $m = n$. This process yields the first $n+1$ derivatives of $\overset{\alpha}{g}_0$ at z_0. The procedure is covered in more detail below.

Equation (A.9) may be used to compute the first $n+2$ coefficients

$$a_j = \frac{1}{j!} \overset{\alpha(j)}{g}_0(z_0) \qquad (j = 0, 1, 2, \ldots, n) \tag{A.10}$$

in the expansion (A.4)

$$F^\alpha = \sum_{j=0}^{\infty} a_j (z - z_0)^j \tag{A.11}$$

of F^α about $z = z_0$ as follows. The coefficients in the expansion (A.1) of F about z_0 are assumed to be known; we introduce the constants b_j as

$$b_j = \frac{1}{j!} F^{(j)}(z_0) \tag{A.12}$$

It follows from (A.6) that

$$\frac{1}{j!} h^{(j)}(z_0) = \frac{1}{j!} F^{(j+1)}(z_0) = (j+1) b_{j+1} \tag{A.13}$$

App. A   Expansion of the Function F^α

The value of $\overset{\alpha}{g}_m$ at z_0 can be computed from (A.5):

$$\overset{\alpha}{g}_m(z_0) = \alpha(\alpha-1)(\alpha-2)\ldots(\alpha-m+1)\bigl[F(z_0)\bigr]^{\alpha-m} \tag{A.14}$$

or

$$\overset{\alpha}{g}_m(z_0) = \alpha(\alpha-1)(\alpha-2)\ldots(\alpha-m+1)b_0^{\alpha-m} \tag{A.15}$$

We apply (A.9) for $m = 1$ and $j = 1, 2$:

$$\overset{\alpha}{g}'_{n-1}(z_0) = \overset{\alpha}{g}_n(z_0)h(z_0) \tag{A.16}$$

$$\overset{\alpha}{g}''_{n-1}(z_0) = \overset{\alpha}{g}'_n(z_0)h(z_0) + \overset{\alpha}{g}_n(z_0)h'(z_0) \tag{A.17}$$

The terms at the right-hand sides of these equations can be computed using (A.15), (A.7), and (A.13). We next apply (A.9) for $m = 2$ and $j = 1, 2, 3$ with the values for $\overset{\alpha}{g}'_{n-1}(z_0)$ and $\overset{\alpha}{g}''_{n-1}(z_0)$ computed from (A.16) and (A.17):

$$\overset{\alpha}{g}'_{n-2}(z_0) = \overset{\alpha}{g}_{n-1}(z_0)h(z_0) \tag{A.18}$$

$$\overset{\alpha}{g}''_{n-2}(z_0) = \overset{\alpha}{g}'_{n-1}(z_0)h(z_0) + \overset{\alpha}{g}_{n-1}(z_0)h'(z_0) \tag{A.19}$$

$$\frac{1}{2!}\overset{\alpha(3)}{g}_{n-2}(z_0) = \frac{1}{2!}\overset{\alpha}{g}''_{n-1}(z_0)h(z_0) + \overset{\alpha}{g}'_{n-1}(z_0)h'(z_0) + \frac{1}{2!}\overset{\alpha}{g}_{n-1}(z_0)h''(z_0) \tag{A.20}$$

We apply (A.9) with these three values, with $m = 3$, and with $j = 1, 2, 3, 4$, which yields the first four derivatives of $\overset{\alpha}{g}_{n-3}$ at z_0. This process is continued until $m = n$; at this final stage (A.9) is applied for $j = 1, 2, \ldots, n+1$ to obtain the first $n+1$ derivatives of $\overset{\alpha}{g}_0$ at z_0. (A.9) yields $\overset{\alpha(j)}{g}_0/(j-1)!$; division by j yields the desired coefficient in the series (A.4).

The procedure is suitable for numerical implementation; it suffices to write a routine for computing the jth derivative of F at z_0, a routine to compute $\overset{\alpha}{g}_j$, and a routine that implements (A.9) in a double loop with j running from 1 to $m+1$ inside a loop where m runs from 1 to n. The listing of such a routine is provided on the diskette supplied with this book.

Appendix B

The Dilogarithm of Complex Argument

We define the principal branch of the function $\ln(1-z)$ as $\ln_p(1-z)$ with

$$-\pi < \arg_p(1-z) \leq \pi \tag{B.1}$$

so that any branch of $\ln(1-z)$ may be expressed in terms of the principal branch as

$$\ln(1-z) = \ln_p(1-z) + 2m\pi i \quad (m = 0, \pm 1, \pm 2, \dots) \tag{B.2}$$

We first consider the principal branch of the dilogarithm, defined as

$$\boxed{\operatorname{Li}_{2p}(z) = -\int_0^z \frac{\ln_p(1-t)}{t}\,dt} \tag{B.3}$$

For $|z| \leq 1$ the following series expansion exists for $\operatorname{Li}_{2p}(z)$ (see Lewin [1958]):

$$\boxed{\operatorname{Li}_{2p}(z) = \sum_{n=1}^{\infty} \frac{z^n}{n^2} \quad (|z| \leq 1)} \tag{B.4}$$

This series converges well and is suitable for numerical evaluation. For values of $|z| > 1$ a relation between $\operatorname{Li}_{2p}(1/z)$ and $\operatorname{Li}_{2p}(z)$ may be employed, which is derived below. It follows upon differentiation of (B.3) that

$$\frac{d}{dz}\operatorname{Li}_{2p}(z) = -\frac{\ln_p(1-z)}{z} \tag{B.5}$$

Replacing z by $1/z$ we obtain

$$\frac{d}{d(1/z)}\text{Li}_{2p}(1/z) = -\frac{\ln_p(1-1/z)}{1/z} \tag{B.6}$$

We apply the chain rule to obtain the derivative with respect to z,

$$\frac{d}{dz}\text{Li}_{2p}(1/z) = \frac{\ln_p(1-1/z)}{z} = \frac{\ln_p(z-1)}{z} - \frac{\ln_p(z)}{z} \tag{B.7}$$

We differentiate between the case that $\Im z \geq 0$ and the case that $\Im z \leq 0$. In the former case we have:

$$\Im z \geq 0 \qquad \ln_p(z-1) = \ln_p(1-z) + i\pi \tag{B.8}$$

and in the latter

$$\Im z \leq 0 \qquad \ln_p(z-1) = \ln_p(1-z) - i\pi \tag{B.9}$$

Defining a constant λ such that

$$\Im z \geq 0 \qquad \lambda = 1 \qquad \Im z \leq 0 \qquad \lambda = -1 \tag{B.10}$$

we may write (B.7) as

$$\frac{d}{dz}\text{Li}_{2p}(1/z) = \frac{\ln_p(1-z)}{z} - \frac{\ln_p(z)}{z} + \frac{i\lambda\pi}{z} \tag{B.11}$$

Integration of this expression gives, with (B.3):

$$\text{Li}_{2p}(1/z) = -\text{Li}_{2p}(z) - \tfrac{1}{2}\ln_p^2(z) + i\lambda\pi\ln z + \tfrac{\pi^2}{3} \tag{B.12}$$

where the constant of integration is obtained by setting z equal to one and by using the equation $\text{Li}_{2p}(1) = \pi^2/6$ (see Lewin [1958]).

This expression may be used to compute the dilogarithm for values of z with $|z| > 1$. In that case we write

$$\text{Li}_{2p}(z) = -\text{Li}_{2p}(1/z) - \tfrac{1}{2}\ln_p^2(z) + i\lambda\pi\ln z + \tfrac{\pi^2}{3} \tag{B.13}$$

where $1/|z| < 1$ so that (B.4) may be used to evaluate the dilogarithm. Equation (B.13) is due to Euler [1811].

The Many-Valued Dilogarithm

In some applications the principal branch of the dilogarithm cannot be used. We therefore express the dilogarithm for cases where (B.1) is not satisfied in terms of the principal branch. We choose the path of integration of the many-valued dilogarithm,

$$\text{Li}_2(z) = -\int_0^z \frac{\ln(1-t)}{t}dt \tag{B.14}$$

as follows. At the lower bound we take $\arg(1-t)$ equal to zero, and write

$$\text{Li}_2(z) = -\int_0^{1-\varepsilon} \frac{\ln_p(1-t)}{t}dt - \sum_{j=1}^{M} \oint_{\mathcal{B}_j} \frac{\ln(1-t)}{t}dt - \int_{1-\varepsilon}^{z} \frac{\ln(1-t)}{t}dt \qquad (B.15)$$

where m is defined by (B.2), and

$$M = |m| \qquad (B.16)$$

The contours $\mathcal{B}_j$ represent full circles of radius ε around $z = 1$, and the integration along these circles is taken in counterclockwise direction if $m > 0$ and in clockwise direction if $m < 0$. Upon each revolution, the argument of $1-t$ either increases or decreases by 2π, depending upon the sign of m. We continue this process until the argument of $\ln(1-t)$ in the last term in (B.15) equals that of $\ln(1-z)$ given by (B.2). The integrals along the contours $\mathcal{B}_j$ vanish in the limit for $\varepsilon \to 0$. This may be seen by substituting μ for $1-t$ in these integrals, and expanding $1/t = 1/(1-\mu)$ in a Taylor series about $\mu = 0$. We write $\ln(1-t)$ as $\ln_p(1-t) + 2m\pi i$ and take the limit for $\varepsilon = 0$:

$$\text{Li}_2(z) = -\int_0^1 \frac{\ln_p(1-t)}{t}dt - \int_1^z \frac{\ln_p(1-t)}{t}dt - \int_1^z \frac{2m\pi i}{t}dt \qquad (B.17)$$

By combining the first two terms and integrating the last one, we obtain an expression for the many-valued dilogarithm in terms of $\text{Li}_{2p}(z)$:

$$\boxed{\text{Li}_2(z) = \text{Li}_{2p}(z) - 2m\pi i \ln_p(z)} \qquad (B.18)$$

The principal branch $\ln_p(z)$ occurs in the last term because the difference in argument of t between the upper and lower bounds of the last integral in (B.17) equals $\Im \ln_p(z)$. It is possible to include the possibility that $\arg(t)$ deviates from the principal value at the origin by selecting a path that starts at $z = 0$, continues to $z = 1$, winds around $z = 1$, returns to $z = 0$, winds around the origin, returns to $z = 1$, and continues in this fashion until the desired increase in $\arg t$ is obtained. This is rarely necessary, however, and the additional term in (B.18) that is generated in this way (which is a constant) will not be presented.

Separation into Real and Imaginary Parts

Lewin [1958] gives the following expressions for the real and imaginary parts of $\text{Li}_{2p}(z)$:

$$\Re\text{Li}_{2p}(z) = -\tfrac{1}{2}\int_0^r \frac{\ln(1 - 2t\cos\theta + t^2)}{t}dt \qquad (B.19)$$

and

$$\Im\text{Li}_{2p}(z) = -\omega_p \ln r - \tfrac{1}{2}\text{Cl}_2(2\omega) + \tfrac{1}{2}\text{Cl}_2(2\omega - 2\theta) + \tfrac{1}{2}\text{Cl}_2(2\theta) \qquad (B.20)$$

where

$$z = re^{i\theta} \qquad \omega = \Im \ln(1-z) \qquad \omega_p = \Im \ln_p(1-z) \qquad (B.21)$$

The expression for the imaginary part is due to Kummer [1840] and involves Clausen's integral, defined as

$$\text{Cl}_2(\theta) = -\tfrac{1}{2}\int_0^\theta \ln(4\sin^2 \tfrac{\theta}{2})d\theta = \sum_{n=1}^\infty \frac{\sin(n\theta)}{n^2} \qquad (B.22)$$

It is not necessary to use the principal value for ω in Clausen's integrals because their period is 2π and the difference between ω_p and ω can only be a multiple of 2π.

Expressions for the real and imaginary parts of the many-valued dilogarithm are obtained by the use of (B.18), (B.19), and (B.20) as

$$\Re \text{Li}_2(z) = -\tfrac{1}{2}\int_0^r \frac{\ln(1-2t\cos\theta+t^2)}{t}dt + 2m\pi\theta \qquad (B.23)$$

and

$$\Im \text{Li}_2(z) = -(\omega_p + 2m\pi i)\ln r - \tfrac{1}{2}\text{Cl}_2(2\omega) + \tfrac{1}{2}\text{Cl}_2(2\omega - 2\theta) + \tfrac{1}{2}\text{Cl}_2(2\theta) \qquad (B.24)$$

It is of interest to note that the imaginary part of the dilogarithm is finite at $r = 0$ if $m = 0$, since ω_p is zero at $r = 0$; this imaginary part is not finite at $r = 0$, however, if $m \neq 0$. For such cases the dilogarithm has a logarithmic singularity at the origin.

The Boundary Condition Along the Phreatic Surface for the Problem Solved in Section 50

The phreatic surface corresponds to the negative real axis in the ζ plane. An expression for Φ on this segment is obtained from (50.15) as

$$\eta = 0 \qquad -\infty < \xi < 0 \qquad \Phi = -\frac{Q_a}{\pi}\ln|\xi| + \frac{\overset{l}{Q_b}}{\pi}\ln|1-\xi| + \Phi_0 \qquad (B.25)$$

The imaginary part of $z = z(\zeta)$ in (50.20) involves the imaginary part of the dilogarithm,

$$\Im \text{Li}_2(1-\zeta) = -\omega\ln|1-\zeta| - \tfrac{1}{2}\text{Cl}_2(2\omega) + \tfrac{1}{2}\text{Cl}_2(2\omega - 2\theta) + \tfrac{1}{2}\text{Cl}_2(2\theta) \qquad (B.26)$$

where

$$\omega = \arg(\zeta) \qquad \theta = \arg(1-\zeta) \qquad (B.27)$$

The ranges of definition of these angles are given by (50.23). Along the negative ξ axis, ω and θ are equal to π and 0, respectively, so that (B.26) becomes

$$\eta = 0 \qquad -\infty < \xi < 0 \qquad \Im \text{Li}_2(1-\zeta) = -\pi \ln|1-\xi| \qquad (B.28)$$

App. B   The Dilogarithm of Complex Argument   643

where it is used that $Cl_2(0) = Cl_2(2\pi) = 0$ (see [B.22]). Using this in (50.20) we obtain

$$\eta = 0 \quad -\infty < \xi \leq 0 \quad y = -\frac{Q_a}{\pi kB}\ln|\xi| + A_0\pi - H + \overset{l}{\frac{Q_b}{\pi kB}}\ln|1-\xi| + C_2 \quad \text{(B.29)}$$

where use is made of (50.23). Using (50.24) we obtain, after multiplication by kB,

$$\eta = 0 \quad -\infty < \xi \leq 0 \quad kBy = -\frac{Q_a}{\pi}\ln|\xi| + \overset{l}{\frac{Q_b}{\pi}}\ln|1-\xi| + \pi kBA_0 \quad \text{(B.30)}$$

This equals expression (B.25) for Φ, provided that

$$A_0 = \frac{\Phi_0}{\pi kB} \quad \text{(B.31)}$$

as asserted in (50.25).

Appendix C

Computer Programs SLW and SLWL

This appendix contains instructions for using the diskette provided with this book, as well as a listing of the computer program SLWL.

In order to make it possible to update the software contained on the diskette without the need for modifying this appendix, the instructions for running and modifying the programs are given in the file MANUAL.TXT contained on the diskette. Information on the minimum system configuration is given on the label of the diskette.

The file MANUAL.TXT is a standard text file and can be displayed on the screen. The manual will be printed on the standard printing device in response to the command PRINTMAN. The routines for free-format input and the plotting routines are contained on the diskette, but are not listed here. The listings of the remaining routines of SLWL are given below:

```
c       file sl.for
        PROGRAM SL
        IMPLICIT COMPLEX (C), LOGICAL (L)
        DIMENSION RMAT(100,100),RKNVEC(324),RSOLUT(100)
c       RMAT    array used for solving
c                   and for contouring: RMAT(imxsze,imxsze)
c       RKNVEC  used for solving:     RKNVEC(imxsze)
c                   for contouring: RKNVEC(1+(imxsze*imxsze-1)/31)
c       RSOLUT  used for solving:     RSOLUT(imxsze)
        IMXSZE=100
        CALL SETUP(IMXSZE,RMAT,RKNVEC,RSOLUT)
        STOP
        END
```

```
c      file slmn.for
c      Routine for processing input
c      RMAT    = matrix of maximum size IMXSZE*IMXSZE
c      RKNVEC  = vector with known coefficients; also used as
c                scratch array
c      RSOLUT  = vector with coefficients computed by SOLVE
       SUBROUTINE SETUP(IMXSZE,RMAT,RKNVEC,RSOLUT)
       IMPLICIT COMPLEX(C),LOGICAL(L),CHARACTER*1(A)
       DIMENSION RMAT(*),RKNVEC(*),RSOLUT(*)
       DIMENSION AWORD(90),AWOLS(10),AWOGR(15)
       INTEGER*2 IDVHDL,IWKING,ISTAT,IXYIN(2),IXYOUT(2),INK,IRBAND,
      .          VRQLOC
       COMMON /DWMN/ IDVHDL,IWKING(19)
c      AWORD, AWOLS, and AWOGR contain key-words.
c      Common block used for free-format input. ALINE=input line,
c      LERROR,LMISS are set true if an error is detected.
       CHARACTER*32 AFILE
       COMMON /CBMATC/ ILPNT(40),LMISS,LERROR,NPBAD,NCHAR
       COMMON /CBMATA/ ALINE(80),AFILE
c      Common block with grid data.
       COMMON /CBGRID/ RGRX1,RGRY1,RGRX2,RGRY2,RGRMAX,RGRMIN,
      .          NGRX,NGRY,RGRXG(150),RGRYG(150),RGRDX,RGRDY,
      .          IGRX1,IGRY1,IGRX2,IGRY2,RGRFAX,RGRFAY,NGRDTX,
      .          NGRDTY,RGREPS
c      Common block with aquifer data.
       COMMON /CBAQ/ NAQLAY,RAQK(20),RAQB(21),RAQT(20),RAQPHI(21),
      .          RAQTSM,NAQLMX
c      Common block with reference point data.
       COMMON /CBRF/ CRFZ,RRFFI,RRFCON
c      Common block with data for given functions.
       COMMON /CBGV/ RGVRN,CGVRZ,RGVRA,RGVRB,RGVRAL,
      .          RGVUQ0,RGVUAL,CGVMIA,
      .          NGVP,CGVPZC(20),RGVPRD(20),RGVPSG(20)
c      Common block with well data.
       COMMON /CBWL/ NWL,CWLZC(200),RWLQ(200),RWLRAD(200)
c      Common block with line-sink data.
       COMMON /CBLS/ NLS,NLSS,NLSF,
      .          CLSZS(200),CLSZE(200),CLSZC(200),RLSDIS(200),
      .          RLSFI(200),ILSPTS(200),ILSPTF(200)
c      Functions that can be contoured: RFPOT=potential,
c      RFPSI=stream function, RFHEAD=piezometric head.
       EXTERNAL RFPOT,RFPSI,RFHEAD
c      List of main keywords. The routine MATCH matches the word entered
c      in the input line to the words below. It returns the position in
```

```
c       the list of the keyword for which a match is found. Thus, if you
c       enter WELL then MATCH will return position number 4. The words are
c       separated by blanks and an exclamation mark denotes the end of the
c       list.
        DATA AWORD  /'A','Q','U','I',' ',
     .               'L','A','Y','E',' ',
     .               'R','E','F','E',' ',
     .               'R','A','I','N',' ',
     .               'U','N','I','F',' ',
     .               'P','O','N','D',' ',
     .               'W','E','L','L',' ',
     .               'L','I','N','E',' ',
     .               'S','O','L','V',' ',
     .               'W','I','N','D',' ',
     .               'G','R','I','D',' ',
     .               'P','L','O','T',' ',
     .               'L','A','Y','O',' ',
     .               'C','H','E','C',' ',
     .               'T','R','A','C',' ',
     .               'P','A','U','S',' ',
     .               'S','W','I','T',' ',
     .               'S','T','O','P','!'/
c       Keywords defining the kind of line-sink.
        DATA AWOLS  /'G','I','V','E',' ',
     .               'H','E','A','D','!'/
c       Keywords defining the function to be contoured
        DATA AWOGR  /'H','E','A','D',' ',
     .               'P','O','T','E',' ',
     .               'S','T','R','E','!'/
        DATA INK,IRBAND /2*0/
        WRITE(*,9003) IMXSZE
        OPEN(UNIT=7,FILE='CON')
        LERROR=.FALSE.
        LMISS=.FALSE.
 10     IF(LMISS) WRITE(*,9001)
        IF(LERROR.AND..NOT.LMISS) WRITE(*,9002)
        LERROR=.FALSE.
        LMISS=.FALSE.
        WRITE(*,9005)
        READ(5,9000) ALINE
        CALL TIDY
        CALL MATCH(AWORD,1,JUMP)
        IF(LMISS.OR.LERROR.OR.ILPNT(2).EQ.0) GOTO 10
        GOTO( 100, 200, 300, 400, 500, 600, 700, 800, 900, 1000,
```

```
               1100,1200,1300,1400,1500,1600,1700,8900),JUMP
c     aquifer data
 100  NAQLAY=IVAR(2)
      IF(NAQLAY.GT.NAQLMX) THEN
         WRITE(*,9010) NAQLMX
         GOTO 10
      ENDIF
      RAQB(1)=RVAR(3)
      IF(LMISS.OR.LERROR) GOTO 10
c     Prompt the user to enter the hydraulic conductivity and thickness
c     of each layer.
      ILAYER=1
 110  WRITE(*,9011) ILAYER
      READ(5,9000) ALINE
      CALL TIDY
      RAQK(ILAYER)=RVAR(1)
      RAQB(ILAYER+1)=RAQB(ILAYER)+RVAR(2)
      IF(LMISS.OR.LERROR) THEN
c     If an error occurred, prompt the user to enter the data for this
c     layer again.
         WRITE(*,9012) ILAYER
      ELSE
c     Otherwise, add 1 to the number of layers.
         ILAYER=ILAYER+1
      ENDIF
c     If the data set for the layers is not yet complete, read the
c     data for the next layer.
      IF(ILAYER.LE.NAQLAY) GOTO 110
c     Initialize the aquifer parameters.
      CALL AQINIT
      GOTO 10
c     Overwrite the data of layer nr. ILAYER.
 200  ILAYER=IVAR(2)
      RAQK(ILAYER)=RVAR(3)
      RAQB(ILAYER+1)=RAQB(ILAYER)+RVAR(4)
c     Initialize the aquifer parameters.
      CALL AQINIT
      GOTO 10
c     Reference point (the function cvar(i) returns the complex
c     variable entered on positions i and i+1 of the input line).
 300  CRFZ=CVAR(2)
      RRFFI=RVAR(4)
      GOTO 10
c     Infiltration from rainfall.
```

App. C Computer Programs SLW and SLWL

```
 400    RGVRN=RVAR(2)
        CGVRZ=CVAR(3)
        RGVRA=RVAR(5)*RVAR(5)
        RGVRB=RVAR(6)*RVAR(6)
        RGVRAL=RVAR(7)*3.1415926/180.
        GOTO 10
c       Uniform flow.
 500    RGVUQ0=RVAR(2)
        RGVUAL=RVAR(3)
        RARG=RGVUAL*3.1415926/180.
        CGVMIA=CEXP(CMPLX(0.,-RARG))
        GOTO 10
c       Ponds.
 600    NGVP=NGVP+1
        CGVPZC(NGVP)=CVAR(2)
        RGVPRD(NGVP)=RVAR(4)
        RGVPSG(NGVP)=RVAR(5)
        GOTO 10
c       Wells.
 700    NWL=NWL+1
        CWLZC(NWL)=CVAR(2)
        RWLQ(NWL)=RVAR(4)
        RWLRAD(NWL)=RVAR(5)
        IF (LERROR) NWL=NWL-1
        GOTO 10
c       Line-sinks.
 800    NLS=NLS+1
        CLSZS(NLS)=CVAR(3)
        CLSZE(NLS)=CVAR(5)
        RDUM=RVAR(7)
        IF (LERROR) THEN
          NLS=NLS-1
          GOTO 10
        ENDIF
        CLSZC(NLS)=.5*(CLSZS(NLS)+CLSZE(NLS))
        CALL MATCH(AWOLS,2,JUMP)
        IF (LERROR) THEN
          NLS=NLS-1
          GOTO 10
        ENDIF
        GOTO(710,720),JUMP
c       Line-sinks with given strength.
 710    NLSS=NLSS+1
        ILSPTS(NLSS)=NLS
```

```
              RLSDIS(NLS)=RDUM
          GOTO 10
c     Line-sinks with given head.
 720      NLSF=NLSF+1
              ILSPTF(NLSF)=NLS
              RLSFI(NLS)=RDUM
          GOTO 10
c    Solve.
 900      CALL SOLVE(IMXSZE,RMAT,RKNVEC,RSOLUT)
          GOTO 10
c    Window.
1000      RGRX1=RVAR(2)
          RGRY1=RVAR(3)
          RGRX2=RVAR(4)
          RGRY2=RVAR(5)
          GOTO 10
c    Grid.
1100      NGRX=IVAR(2)+1
          IF(LERROR) GOTO 10
          IF(NGRX.GE.IMXSZE) NGRX=IMXSZE-1
          IF(ILPNT(4).EQ.0) THEN
             JUMP=1
          ELSE
             CALL MATCH(AWOGR,3,JUMP)
          ENDIF
          CALL GRINIT
          GOTO(1110,1120,1130),JUMP
c     Piezometric heads.
1110      CALL FILL(RMAT,NGRX,IMXSZE,RFHEAD,LWRONG)
          IF(LWRONG) WRITE(*,9100)
          GOTO 10
c     Equipotentials.
1120      CALL FILL(RMAT,NGRX,IMXSZE,RFPOT,LWRONG)
          IF(LWRONG) WRITE(*,9100)
          GOTO 10
c     Streamlines.
1130      CALL FILL(RMAT,NGRX,IMXSZE,RFPSI,LWRONG)
          IF(LWRONG) WRITE(*,9100)
          GOTO 10
c    Contouring.
1200      IF(NGRX.EQ.0.OR.NGRY.EQ.0) THEN
             WRITE(*,9120)
             GOTO 10
          ENDIF
```

```
              WRITE(*,9121) RGRMIN,RGRMAX
              READ(5,9000) ALINE
              CALL TIDY
              RLEVEL=RVAR(1)
              RINCRM=RVAR(2)
              IF(LERROR) GOTO 10
              CALL DRAWIN
              CALL LAYOUT
              ISIZE=NGRX
 1210   CALL CONTUR
        .      (RMAT,ISIZE,NGRX,NGRY,IGRX1,IGRY1,RGRDX,RGRDY,RLEVEL,RKNVEC)
              RLEVEL=RLEVEL+RINCRM
              IF (RLEVEL.LE.RGRMAX) GOTO 1210
              READ(5,*)
              CALL DRAWOUT
              GOTO 10
c       layout.
 1300   CALL DRAWIN
              CALL LAYOUT
              READ(5,*)
              CALL DRAWOUT
              GOTO 10
c       check
 1400   CALL CHECK
              GOTO 10
c       tracing
 1500   RSTEP=RVAR(2)
              IF(LERROR) GOTO 10
              IF(NGRX.NE.0.AND.NGRY.NE.0) THEN
                WRITE(*,9121) RGRMIN,RGRMAX
                READ(5,9000) ALINE
                CALL TIDY
                RLEVEL=RVAR(1)
                RINCRM=RVAR(2)
                IF(LERROR) GOTO 10
                CALL DRAWIN
                CALL LAYOUT
                ISIZE=NGRX
 1510         CALL CONTUR
        .        (RMAT,ISIZE,NGRX,NGRY,IGRX1,IGRY1,RGRDX,RGRDY,RLEVEL,RKNVEC)
                RLEVEL=RLEVEL+RINCRM
                IF (RLEVEL.LT.RGRMAX) GOTO 1510
              ELSE
                CALL DRAWIN
```

```
          CALL LAYOUT
        ENDIF
c       Set the initial cursor position to the center of the window.
        IXYOUT(1)=(IGRX1+IGRX2)/2
        IXYOUT(2)=(IGRY1+IGRY2)/2
 1520   IXYIN(1)=IXYOUT(1)
        IXYIN(2)=IXYOUT(2)
c       Display the cursor and get the cursor position in VDI units
c       returned in IXYOUT, and the character entered from the keyboard.
        ISTAT=VRQLOC(IDVHDL,IXYIN,INK,IRBAND,IDVHDL,IXYOUT,ACHAR)
c       If either an r or an R was entered, leave graphics mode and
c       return to main menu.
        IF(ACHAR.EQ.'R'.OR.ACHAR.EQ.'r') THEN
        READ(5,*)
        CALL DRAWOUT
          GOTO 10
c       Otherwise translate the cursor position in window units and
c       display streamline.
        ELSE
          RX=RGRX1+FLOAT(IXYOUT(1)-IGRX1)/RGRFAX
          RY=RGRY1+FLOAT(IXYOUT(2)-IGRY1)/RGRFAY
          CZ=CMPLX(RX,RY)
          CALL TRACE(RSTEP,CZ)
          GOTO 1520
        ENDIF
        READ(5,*)
        CALL DRAWOUT
        GOTO 10
c       The call to PAUSE will cause the program to pause; given
c       sufficient memory, other programs such as an editor can
c       be executed. The program will continue upon pressing the
c       ENTER key.
 1600   PAUSE
        GOTO 10
c       The following statements make it possible to switch input
c       and output devices. The first argument of the command
c       SWITCH denotes the new input device (e.g. the name of a
c       file with input data prepared in advance). The command
c       SWITCH CON will cause the input to return to the keyboard.
c       The latter command must be the last command in an input file.
c       The second argument (e.g. a filename or the name of a device
c       such as PRN) will cause the output generated by the routine
c       CHECK to be directed to this file or device. The second
c       argument is optional. The function GETFN(IPAR) contained in
```

App. C   Computer Programs SLW and SLWL

```
c     the match package returns in the variable AFILE the filename
c     entered on position IPAR.
 1700 CALL GETFN(2)
      IF(LMISS.OR.LERROR) GOTO 10
      CLOSE(5)
      OPEN(UNIT=5,FILE=AFILE)
      IF(ILPNT(4).NE.0) THEN
        CALL GETFN(3)
        CLOSE(7)
        OPEN(UNIT=7,FILE=AFILE)
      ENDIF
      GOTO 10
 8900 RETURN
 9000 FORMAT(80A1)
 9001 FORMAT('++++ILLEGAL OR MISSING PARAMETERS IN MAIN+++'/)
 9002 FORMAT('++++ILLEGAL COMMAND IN MAIN+++'/)
 9003 FORMAT('+MAXIMUM NUMBER OF EQUATIONS IS ',I4/)
 9005 FORMAT(
     .'+<AQUIFER>(NLAYERS,BASE)'/
     .' <LAYER>(LAYER NR.,PERM,THICKNESS)'/
     .' <UNIFLOW>(Q,AL)'/
     .' <RAIN>(N,X,Y,A,B,AL)'/
     .' <POND>(X,Y,R,N)'/
     .' <WELL>(X,Y,Q,R)'/
     .' <LINESINK>(<GIVEN>/<HEAD>)(X1,Y1 X2,Y2)(DISCHARGE/HEAD)'/
     .' <REFERENCE>(X,Y,FI)'/
     .' <SOLVE>'/
     .' <WINDOW>(X1,Y1 X2,Y2)'/
     .' <GRID>(N)(<HEAD>/<POTENTIAL>/<STREAM-FUNCTION>)'/
     .' <PLOT>'/
     .' <LAYOUT>'/
     .' <TRACE>(STEP)once cursor appears, enter either <T> or <R>'/
     .' <CHECK>'/
     .' <SWITCH>(FILEIN)[FILEOUT]'/
     .' <PAUSE>'/
     .' <STOP>'/)
 9010 FORMAT('++++MAXIMUM NR. OF LAYERS (= ',I4,') EXCEEDED+++'/)
 9011 FORMAT(
     .'+ENTER HYDRAULIC CONDUCTIVITY AND THICKNESS FOR LAYER',I4/)
 9012 FORMAT(
     .'++++ILLEGAL OR MISSING PARAMETERS; RE-ENTER DATA FOR LAYER',I4/)
 9100 FORMAT(
     .'++++GRID TOO LARGE, PLEASE REDUCE NUMBER OF GRIDPOINTS+++')
 9120 FORMAT('++++GRID NOT FILLED+++'/)
```

```
 9121 FORMAT('+MIN. LEVEL= ',E14.6,' MAX. LEVEL=',E14.6/
     .       ' (MIN. LEVEL, INCREMENT)'/)
      END
c     Complex potential.
      COMPLEX FUNCTION COMEGA(CZ)
      IMPLICIT COMPLEX (C)
      COMEGA=COMGVN(CZ)+RFRFC(CZ)+CFLSU(CZ)
      RETURN
      END
c     Complex discharge function.
      COMPLEX FUNCTION CW(CZ)
      IMPLICIT COMPLEX (C)
      CW=CWGVG(CZ)+CWWLSM(CZ)+CWLSSM(CZ)
      RETURN
      END
c     Potential function.
      REAL FUNCTION RFPOT(CZ)
      IMPLICIT COMPLEX (C)
      RFPOT=REAL(COMEGA(CZ))
      RETURN
      END
c     Stream function.
      REAL FUNCTION RFPSI(CZ)
      IMPLICIT COMPLEX (C)
      RFPSI=AIMAG(COMEGA(CZ))
      RETURN
      END
c     Piezometric head.
      REAL FUNCTION RFHEAD(CZ)
      IMPLICIT COMPLEX (C)
      RPOT=RFPOT(CZ)
      RFHEAD=RFHEDP(RPOT)
      RETURN
      END
c     Contributions to the complex potential of all functions with
c     given coefficients.
      COMPLEX FUNCTION COMGVN(CZ)
      IMPLICIT COMPLEX (C)
      COMGVN=CFGVG(CZ)+CFWLG(CZ)+CFLSG(CZ)
      RETURN
      END
c     Layout.
      SUBROUTINE LAYOUT
      CALL GVLAY
```

```fortran
      CALL WLLAY
      CALL LSLAY
      RETURN
      END
c     Solving.
c     RMAT   = coefficient matrix.
c     RKNVEC = vector with known coefficients.
c     RSOLUT = vector with computed coefficients, returned by SOLVE.
      SUBROUTINE SOLVE(IMXSZE,RMAT,RKNVEC,RSOLUT)
      IMPLICIT COMPLEX (C), LOGICAL (L)
      DIMENSION RMAT(IMXSZE,*),RKNVEC(*),RSOLUT(*)
      COMMON /CBRF/ CRFZ,RRFFI,RRFCON
      COMMON /CBLS/ NLS,NLSS,NLSF,
     .              CLSZS(200),CLSZE(200),CLSZC(200),RLSDIS(200),
     .              RLSFI(200),ILSPTS(200),ILSPTF(200)
c     Set number of equations to 1 + number of line-sinks.
      NEQ=NLSF+1
c     Print error message if maximum number of equations is exceeded.
      IF(NEQ.GT.IMXSZE) THEN
         WRITE(*,9003) IMXSZE
         RETURN
      ENDIF
c     For columns 1 to NLSF = number of head specified line-sinks:
      IF(NLSF.GE.1) THEN
         DO 200 JCOL=1,NLSF
c        Make temporary storage for starting and end points.
         CZSLS=CLSZS(ILSPTF(JCOL))
         CZELS=CLSZE(ILSPTF(JCOL))
c        For each column, compute the coefficient of the strength of
c        the line-sink number JCOL. The coefficient is computed at
c        the center of line-sink number IROW.
         DO 100 IROW=1,NLSF
           RMAT(IROW,JCOL)=
     .          REAL(COMLS(CLSZC(ILSPTF(IROW)),CZSLS,CZELS))
  100    CONTINUE
c        Compute the coefficient of the strength of line-sink number
c        JCOL at the reference point.
         RMAT(NEQ,JCOL)=REAL(COMLS(CRFZ,CZSLS,CZELS))
  200    CONTINUE
      ENDIF
c     Store the coefficients of the constants, which are all 1.
      DO 300 IROW=1,NEQ
         RMAT(IROW,NEQ)=1.
  300 CONTINUE
```

```
c       Store the known vector in RKNVEC. The elements in this vector
c       are the potentials corresponding to the given heads at the
c       centers of the line-sinks and at the reference point minus
c       the effect of all given functions.
        IF(NLSF.GE.1) THEN
           DO 1000 IROW=1,NLSF
              RKNVEC(IROW)=RFPOTH(RLSFI(ILSPTF(IROW)))-
     .                REAL(COMGVN(CLSZC(ILSPTF(IROW))))
 1000      CONTINUE
        ENDIF
        RKNVEC(NEQ)=RFPOTH(RRFFI)-REAL(COMGVN(CRFZ))
c       Solve the system of equations.
        CALL SSOLVE(IMXSZE,RMAT,RKNVEC,RSOLUT,NEQ)
c       Store all computed strengths in the array.
        IF(NLSF.GE.1) THEN
           DO 2000 IROW=1,NLSF
              RLSDIS(ILSPTF(IROW))=RSOLUT(IROW)
 2000      CONTINUE
        ENDIF
c       Retrieve and store the constant.
        RRFCON=RSOLUT(NEQ)
        RETURN
 9003   FORMAT('+MAXIMUM NUMBER OF EQUATIONS , ',I4,' EXCEEDED;'/
     .         SOLVING ABORTED'/)
        END
        SUBROUTINE CHECK
        IMPLICIT COMPLEX(C),LOGICAL(L),CHARACTER*1(A)
        DIMENSION AWORD(60)
        CHARACTER* 32 AFILE
        COMMON /CBMATC/ ILPNT(40),LMISS,LERROR,NPBAD,NCHAR
        COMMON /CBMATA/ ALINE(80),AFILE
        COMMON /CBAQ/ NAQLAY,RAQK(20),RAQB(21),RAQT(20),RAQPHI(21),
     .                RAQTSM,NAQLMX
        COMMON /CBRF/ CRFZ,RRFFI,RRFCON
        COMMON /CBGV/ RGVRN,CGVRZ,RGVRA,RGVRB,RGVRAL,
     .                RGVUQ0,RGVUAL,CGVMIA,
     .                NGVP,CGVPZC(20),RGVPRD(20),RGVPSG(20)
        COMMON /CBWL/ NWL,CWLZC(200),RWLQ(200),RWLRAD(200)
        COMMON /CBLS/ NLS,NLSS,NLSF,
     .                CLSZS(200),CLSZE(200),CLSZC(200),RLSDIS(200),
     .                RLSFI(200),ILSPTS(200),ILSPTF(200)
c       List of keywords in check.
        DATA AWORD /'A','Q','U','I',' ',
     .              'R','E','F','E',' ',
```

App. C Computer Programs SLW and SLWL

```
                .           'R','A','I','N',' ',
                .           'U','N','I','F',' ',
                .           'P','O','N','D',' ',
                .           'W','E','L','L',' ',
                .           'L','I','N','E',' ',
                .           'C','O','N','T',' ',
                .           'H','E','A','D',' ',
                .           'D','I','S','C',' ',
                .           'O','M','E','G',' ',
                .           'R','E','T','U','!'/
          LERROR=.FALSE.
          LMISS=.FALSE.
   10     IF(LMISS) WRITE(*,9001)
          IF(LERROR.AND..NOT.LMISS) WRITE(*,9002)
          LERROR=.FALSE.
          LMISS=.FALSE.
          WRITE(*,9005)
          READ(5,9000) ALINE
          CALL TIDY
          CALL MATCH(AWORD,1,JUMP)
          IF(LMISS.OR.LERROR) GOTO 10
          GOTO( 100, 200, 300, 400, 500, 600, 700, 800, 900,1000,
                .      1100,8900),JUMP
    c     Aquifer data.
  100     WRITE(7,9010)
          DO 190 ILAYER=1,NAQLAY
             RTHICK=RAQB(ILAYER+1)-RAQB(ILAYER)
             WRITE(7,9011) ILAYER,RAQK(ILAYER),RAQB(ILAYER),RTHICK,
                .            RAQT(ILAYER)
  190     CONTINUE
          GOTO 10
    c     Reference point.
  200     WRITE(7,9020) CRFZ,RRFFI,RRFCON
          GOTO 10
    c     Infiltration from rainfall.
  300     WRITE(7,9030) RGVRN,CGVRZ,SQRT(RGVRA),SQRT(RGVRB),
                .            RGVRAL*180./3.1415926
          GOTO 10
    c     Uniform flow.
  400     RCOSA=REAL(CGVMIA)
          RSINA=-AIMAG(CGVMIA)
          WRITE(7,9040) RGVUQ0,RGVUAL,RCOSA,RSINA
          GOTO 10
    c     Ponds.
```

```
  500 IF(NGVP.EQ.0) THEN
         WRITE(*,9003)
         GOTO 10
      ENDIF
      WRITE(7,9050)
      WRITE(7,9051) (I,CGVPZC(I),RGVPRD(I),RGVPSG(I),I=1,NGVP)
      GOTO 10
c     Wells.
  600 IF(NWL.EQ.0) THEN
         WRITE(*,9003)
         GOTO 10
      ENDIF
      WRITE(7,9060)
      WRITE(7,9061) (I,CWLZC(I),RWLQ(I),RWLRAD(I),I=1,NWL)
      GOTO 10
c     Line-sinks.
  700 IF(NLS.EQ.0) THEN
         WRITE(*,9003)
         GOTO 10
      ENDIF
      IF(NLSS.GT.0) THEN
        WRITE(7,9070)
        WRITE(7,9071)
     .   (I,CLSZS(ILSPTS(I)),CLSZE(ILSPTS(I)),
     .   CABS(CLSZE(ILSPTS(I))-CLSZS(ILSPTS(I)))*RLSDIS(ILSPTS(I)),
     .   I=1,NLSS)
      ENDIF
      IF(NLSF.GT.0) THEN
        WRITE(7,9072)
        WRITE(7,9071)
     .   (I,CLSZS(ILSPTF(I)),CLSZE(ILSPTF(I)),
     .   CABS(CLSZE(ILSPTF(I))-CLSZS(ILSPTF(I)))*RLSDIS(ILSPTF(I)),
     .   I=1,NLSF)
      ENDIF
      GOTO 10
c     Control of specified heads of reference point and line-sinks with
c     unknown strength versus heads obtained from solution.
  800 WRITE(7,9080)  RRFFI,RFHEAD(CRFZ)
      WRITE(7,9081)
     .(RLSFI(ILSPTF(I)),RFHEAD(CLSZC(ILSPTF(I))),I=1,NLSF)
      GOTO 10
c     Head.
  900 CZ=CVAR(2)
      IF(LERROR) GOTO 10
```

```
        WRITE(7,9090) CZ,RFHEAD(CZ)
        GOTO 10
c       Discharge vector.
 1000   CZ=CVAR(2)
        IF(LERROR) GOTO 10
        WRITE(7,9100) CZ,CONJG(CW(CZ))
        GOTO 10
c       Complex potential.
 1100   CZ=CVAR(2)
        IF(LERROR) GOTO 10
        WRITE(7,9110) CZ,COMEGA(CZ)
        GOTO 10
 8900   RETURN
 9000   FORMAT(80A1)
 9001   FORMAT('++++ILLEGAL OR MISSING PARAMETERS IN CHECK+++'/)
 9002   FORMAT('++++ILLEGAL COMMAND IN CHECK+++'/)
 9003   FORMAT('+THERE ARE NO SUCH ELEMENTS'/)
 9005   FORMAT(
       .'+<AQUIFER><REFERENCE><RAIN><UNIFLO><POND><WELL><LINESINK>'/
       .' <CONTROL><HEAD>(X,Y)<DISCHARGE>(X,Y)<OMEGA>(X,Y)<RETURN>'/)
 9010   FORMAT(
       .'  LAYER  HYDR.CONDUCT.  BASE           THICKNESS     ',
       .'TRANSMISSIVITY')
 9011   FORMAT(I3,3X,4E14.6)
 9020   FORMAT(
       .'REFERENCE:    X              Y             HEAD          CONSTANT'/
       .'           ',4E14.6)
 9030   FORMAT('  RAIN           XCENTER       YCENTER'/
       .       ,3E14.6/
       .'                         A-AXIS        B-AXIS        ANGLE'/
       .'                       ',3E14.6)
 9040   FORMAT('  UNIFORM FLOW  ANGLE         COS(ANGLE)    SIN(ANGLE)'/
       .       4E14.6)
 9050   FORMAT('PONDS: NO XCENTER       YCENTER        RADIUS ',
       .'       INFILTRATION')
 9051   FORMAT('    ',I4,4E14.6)
 9060   FORMAT('WELLS: GIVEN: NO XCENTER       YCENTER ',
       .'      DISCHARGE    RADIUS')
 9061   FORMAT('              ',I4,4E14.6)
 9070   FORMAT('LINESINK: GIVEN'/
       .'  NO XSTART        YSTART        XEND          YEND ',
       .'         DISCHARGE')
 9071   FORMAT(I4,5E14.6)
 9072   FORMAT('LINESINK: HEAD'/
```

```
             .'  NO XSTART        YSTART         XEND          YEND ',
             .'          DISCHARGE')
        9080 FORMAT('          SPECIFIED HEAD   CALCULATED HEAD'/
             .         'REFERENCE:',E14.6,4X,E14.6/
             .         'LINESINK :')
        9081 FORMAT('             ',E14.6,4X,E14.6)
        9090 FORMAT('X,Y,HEAD    ',3E14.6)
        9100 FORMAT('X,Y,QX,QY   ',4E14.6)
        9110 FORMAT('X,Y,PHI,PSI ',4E14.6)
             END
```
c **file slaq.for**
c   Initialization of common block with aquifer data.
c   Set aquifer data to default values.
c   NAQLAY   = number of layers in the aquifer.
c   RAQK(I) = hydraulic conductivity of layer I.
c   RAQT(I) = transmissivity of layer I.
c   RAQB(I) = elevation of the base of layer I.
c   RAQPHI(I) = value of the potential corresponding to a head equal
c   to the elevation of the base of aquifer I.
c   RAQTSM   = transmissivity of the aquifer if it is entirely
c   saturated
c   NAQLMX   = maximum number of layers, given by the dimension
c   statement in the common block.
```
       BLOCK DATA BDAQ
       COMMON /CBAQ/ NAQLAY,RAQK(20),RAQB(21),RAQT(20),RAQPHI(21),
      .              RAQTSM,NAQLMX
       DATA NAQLAY /1/
       DATA RAQK(1),RAQB(1),RAQB(2),RAQT(1),RAQPHI(1),RAQPHI(2),RAQTSM
      .  / 1.    ,.0     ,1.     ,1.    ,.0       ,.5       ,1.   /
       DATA NAQLMX /20/
       END
```
c   Computation of RAQTSM, and (for all layers) of RAQT(I) and
c   of RAQPHI(I).
```
       SUBROUTINE AQINIT
       COMMON /CBAQ/ NAQLAY,RAQK(20),RAQB(21),RAQT(20),RAQPHI(21),
      .              RAQTSM,NAQLMX
       RAQTSM=.0
       DO 100 ILAYER=1,NAQLAY
         RAQT(ILAYER)=RAQK(ILAYER)*(RAQB(ILAYER+1)-RAQB(ILAYER))
         RAQTSM=RAQTSM+RAQT(ILAYER)
   100 CONTINUE
       DO 200 ILAYER=1,NAQLAY+1
         RAQPHI(ILAYER)=RFAQPH(ILAYER)
   200 CONTINUE
```

```
        RETURN
        END
c       Function for computing the value of RAQPHI(M) (see [11.21]).
        REAL FUNCTION RFAQPH(M)
        COMMON /CBAQ/ NAQLAY,RAQK(20),RAQB(21),RAQT(20),RAQPHI(21),
       .              RAQTSM,NAQLMX
        RFAQPH=.0
        IF(M.EQ.1) RETURN
        DO 100 J=1,M-1
          RFAQPH=RFAQPH+RAQT(J)*(RAQB(M)-RAQB(J))
       .         -.5*RAQK(J)*(RAQB(J+1)-RAQB(J))**2
  100   CONTINUE
        RETURN
        END
c       Function to convert head RFI to potential RFPOTH.
        REAL FUNCTION RFPOTH(RFI)
        COMMON /CBAQ/ NAQLAY,RAQK(20),RAQB(21),RAQT(20),RAQPHI(21),
       .              RAQTSM,NAQLMX
        RFPOTH=.0
c       Return 0 if the head is below the base of the aquifer.
        IF(RFI.LE.RAQB(1)) RETURN
c       Confined flow in all layers, apply (11.24).
        IF(RFI.GE.RAQB(NAQLAY+1)) THEN
          RFPOTH=RFAQPC(RFI,NAQLAY)
          RETURN
        ENDIF
c       Unconfined conditions occur; find the layer that contains the
c       water table, then compute the potential, using equation (11.31).
        DO 100 ILAYER=1,NAQLAY
          IF(RFI.GE.RAQB(ILAYER).AND.RFI.LT.RAQB(ILAYER+1)) THEN
            RFPOTH=.5*RAQK(ILAYER)*(RFI-RAQB(ILAYER))**2
            IF(ILAYER.GT.1) RFPOTH=RFPOTH+RFAQPC(RFI,ILAYER-1)
            RETURN
          ENDIF
  100   CONTINUE
        RETURN
        END
c       Function for computing the contributions of all fully saturated
c       layers to the potential.
        REAL FUNCTION RFAQPC(RFI,ILAYER)
        COMMON /CBAQ/ NAQLAY,RAQK(20),RAQB(21),RAQT(20),RAQPHI(21),
       .              RAQTSM,NAQLMX
c       First determine the sum of the transmissivities of all fully
c       saturated layers.
```

```
            RTJ=.0
            DO 100 J=1,ILAYER
              RTJ=RTJ+RAQT(J)
      100   CONTINUE
c     Next compute the contribution of all fully saturated layers
c     (the first 2 terms in [11.31]).
            RFAQPC=RTJ*(RFI-RAQB(ILAYER+1))+RAQPHI(ILAYER+1)
            RETURN
            END
c     Function to convert potential RPHI to head RFHEDP.
            REAL FUNCTION RFHEDP(RPOT)
            COMMON /CBAQ/ NAQLAY,RAQK(20),RAQB(21),RAQT(20),RAQPHI(21),
           .              RAQTSM,NAQLMX
            RFHEDP=RAQB(1)
c     Return elevation of base if the potential is <= 0.
            IF(RPOT.LE..0) RETURN
c     Confined flow in all layers; use (11.29).
            IF(RPOT.GE.RAQPHI(NAQLAY+1)) THEN
              RFHEDP=RAQB(NAQLAY+1)+(RPOT-RAQPHI(NAQLAY+1))/RAQTSM
              RETURN
            ENDIF
c     Unconfined conditions occur; first determine the sum of
c     the transmissivities of all fully saturated layers.
            DO 100 ILAYER=1,NAQLAY
c       find the layer that contains the water table, then compute the
c       head, using (11.34).
              IF(RPOT.GE.RAQPHI(ILAYER).AND.RPOT.LT.RAQPHI(ILAYER+1)) THEN
                RTILM1=.0
                DO 50 J=1,ILAYER-1
                  RTILM1=RTILM1+RAQT(J)
       50       CONTINUE
                RFHEDP=RAQB(ILAYER)-RTILM1/RAQK(ILAYER)
           .         +SQRT((RTILM1/RAQK(ILAYER))**2
           .         +2.*(RPOT-RAQPHI(ILAYER))/RAQK(ILAYER))
                RETURN
              ENDIF
      100   CONTINUE
            RETURN
            END
c     **file slgv.for**
c     Initialization of common block with data of rain, uniform flow
c     and ponds.
c     RGVRN  = infiltration from rainfall.
c     CGVRZ  = complex coordinate of center of elliptical
```

App. C Computer Programs SLW and SLWL

```
c               equipotentials generated by rainfall alone.
c     RGVRAS = square of a axis of ellipse.
c     RGVRBS = square of b axis of ellipse.
c     RGVRAL = angle between a axis of ellipse and x axis.
c     RGVUQ0 = discharge of uniform flow.
c     RGVUAL = angle in degrees between direction of uniform flow
c              and x axis.
c     CGVMIA = e(-i*angle of uniform flow in radians).
c     NGVP   = number of ponds.
c     CGVPZC = center of pond.
c     RGVPRD = radius of pond.
c     RGVPSG = infiltration rate of pond.
      BLOCK DATA BDGV
      IMPLICIT COMPLEX (C)
      COMMON /CBGV/ RGVRN,CGVRZ,RGVRAS,RGVRBS,RGVRAL,
     .              RGVUQ0,RGVUAL,CGVMIA,
     .              NGVP,CGVPZC(20),RGVPRD(20),RGVPSG(20)
      DATA RGVRN /0./
      DATA RGVUQ0,RGVUAL,CGVMIA /0.,0.,(1.,0.)/
      DATA NGVP /0/
      END
c     Contribution of rain uniform flow and ponds to the
c     complex potential.
      COMPLEX FUNCTION CFGVG(CZ)
      IMPLICIT COMPLEX(C),LOGICAL(L)
      COMMON /CBGV/ RGVRN,CGVRZ,RGVRAS,RGVRBS,RGVRAL,
     .              RGVUQ0,RGVUAL,CGVMIA,
     .              NGVP,CGVPZC(20),RGVPRD(20),RGVPSG(20)
      CFGVG=-RGVUQ0*CGVMIA*CZ
      IF(RGVRN.NE.0.) THEN
         RX=REAL(CZ-CGVRZ)
         RY=AIMAG(CZ-CGVRZ)
         RXBAR=RX*COS(RGVRAL)+RY*SIN(RGVRAL)
         RYBAR=RY*COS(RGVRAL)-RX*SIN(RGVRAL)
         CFGVG=CFGVG-.5*RGVRN*(RGVRBS*RXBAR*RXBAR+RGVRAS*RYBAR*RYBAR)/
     .                (RGVRAS+RGVRBS)
      ENDIF
      CFGVG=CFGVG+CFGVGP(CZ)
      RETURN
      END
c     Contribution of all ponds to the complex potential.
      COMPLEX FUNCTION CFGVGP(CZ)
      IMPLICIT COMPLEX(C),LOGICAL(L)
      COMMON /CBGV/ RGVRN,CGVRZ,RGVRAS,RGVRBS,RGVRAL,
```

```
                    .       RGVUQ0,RGVUAL,CGVMIA,
                    .       NGVP,CGVPZC(20),RGVPRD(20),RGVPSG(20)
            CFGVGP=(.0,.0)
            IF(NGVP.EQ.0) RETURN
            DO 100 IPOND=1,NGVP
              CRAD=CZ-CGVPZC(IPOND)
              RDIST=CABS(CRAD)
              RAD=RGVPRD(IPOND)
              IF(RDIST.GE.RAD) THEN
                COMPD=-.5*RAD*RAD*CLOG(CRAD/RAD)
              ELSE
                COMPD=CMPLX(.25*(RAD*RAD-RDIST*RDIST),.0)
              ENDIF
              CFGVGP=CFGVGP+RGVPSG(IPOND)*COMPD
        100 CONTINUE
            RETURN
            END
c     Contribution to the complex discharge (W=Qx-iQy) of rain
c     uniform flow and ponds.
            COMPLEX FUNCTION CWGVG(CZ)
            IMPLICIT COMPLEX(C),LOGICAL(L)
            COMMON /CBGV/ RGVRN,CGVRZ,RGVRAS,RGVRBS,RGVRAL,
                    .       RGVUQ0,RGVUAL,CGVMIA,
                    .       NGVP,CGVPZC(20),RGVPRD(20),RGVPSG(20)
            CWGVG=RGVUQ0*CGVMIA
            IF(RGVRN.NE.0.) THEN
              RX=REAL(CZ-CGVRZ)
              RY=AIMAG(CZ-CGVRZ)
              RCOSA=COS(RGVRAL)
              RSINA=SIN(RGVRAL)
              RXBAR=RX*RCOSA+RY*RSINA
              RYBAR=RY*RCOSA-RX*RSINA
              RBX=RGVRBS*RXBAR
              RAY=RGVRAS*RYBAR
              RQX=RGVRN*(RBX*RCOSA-RAY*RSINA)/(RGVRBS+RGVRAS)
              RQY=RGVRN*(RBX*RSINA+RAY*RCOSA)/(RGVRBS+RGVRAS)
              CWGVG=CWGVG+CMPLX(RQX,-RQY)
            ENDIF
            CWGVG=CWGVG+CWGVGP(CZ)
            RETURN
            END
c     Contribution of all ponds to the complex discharge.
            COMPLEX FUNCTION CWGVGP(CZ)
            IMPLICIT COMPLEX(C),LOGICAL(L)
```

```
      COMMON /CBGV/ RGVRN,CGVRZ,RGVRAS,RGVRBS,RGVRAL,
     .              RGVUQ0,RGVUAL,CGVMIA,
     .              NGVP,CGVPZC(20),RGVPRD(20),RGVPSG(20)
      CWGVGP=(.0,.0)
      IF(NGVP.EQ.0) RETURN
      DO 100 IPOND=1,NGVP
        CRAD=CZ-CGVPZC(IPOND)
        RDIST=CABS(CRAD)
        RRAD=RGVPRD(IPOND)
        IF(RDIST.GE.RRAD) THEN
          CWPD=.5*RRAD*RRAD/CRAD
        ELSE
          CWPD=.5*CONJG(CRAD)
        ENDIF
        CWGVGP=CWGVGP+RGVPSG(IPOND)*CWPD
  100 CONTINUE
      RETURN
      END
c     Layout of all ponds.
      SUBROUTINE GVLAY
      IMPLICIT LOGICAL(L),COMPLEX(C)
      COMMON /CBGV/ RGVRN,CGVRZ,RGVRAS,RGVRBS,RGVRAL,
     .              RGVUQ0,RGVUAL,CGVMIA,
     .              NGVP,CGVPZC(20),RGVPRD(20),RGVPSG(20)
      DO 100 IPOND=1,NGVP
        CALL DRAWC(CGVPZC(IPOND),RGVPRD(IPOND))
  100 CONTINUE
      RETURN
      END
c     file slrf.for
c     Initialization of common block for reference point.
c     CRFZ  = complex coordinate of reference point
c     RRFFI = head at reference point
c     RRFCON= constant in complex potential
      BLOCK DATA BDRF
      IMPLICIT COMPLEX (C)
      COMMON /CBRF/ CRFZ,RRFFI,RRFCON
      DATA RRFFI,CRFZ,RRFCON /.0,(.0,.0),.0/
      END
c     Function returning the constant in the complex potential :RRFCON
      REAL FUNCTION RFRFC(CZ)
      IMPLICIT COMPLEX (C)
      COMMON /CBRF/ CRFZ,RRFFI,RRFCON
      RFRFC=RRFCON
```

```
      RETURN
      END
c     file slwl.for
c     Initialization of common block with well data:
c     Set the number of wells to zero.
c     NWL    = number of wells
c     CWLZC  = complex coordinate of center of well
c     RWLQ   = discharge of well
c     RWLRAD = radius of well
      BLOCK DATA BDWL
      IMPLICIT COMPLEX (C),LOGICAL (L)
      COMMON /CBWL/ NWL,CWLZC(200),RWLQ(200),RWLRAD(200)
      DATA NWL /0/
      END
c     Contribution of all wells to the complex potential.
      COMPLEX FUNCTION CFWLG(CZ)
      IMPLICIT COMPLEX (C)
      COMMON /CBWL/ NWL,CWLZC(200),RWLQ(200),RWLRAD(200)
      CFWLG=(.0,.0)
      DO 100 IWL=1,NWL
         CFWLG=CFWLG+RWLQ(IWL)*CFWELL(CZ,IWL)
 100  CONTINUE
      RETURN
      END
c     Contribution of a single well of unit discharge to the complex
c     potential.
      COMPLEX FUNCTION CFWELL(CZ,IWL)
      IMPLICIT COMPLEX (C)
      COMMON /CBWL/ NWL,CWLZC(200),RWLQ(200),RWLRAD(200)
      DATA RPI /3.1415926/
      CRAD=CZ-CWLZC(IWL)
      RW=RWLRAD(IWL)
      IF(CABS(CRAD).GE.RW) THEN
         CFWELL=.5/RPI*CLOG(CRAD/RW)
      ELSE
         CFWELL=(.0,.0)
      ENDIF
      RETURN
      END
c     contribution of all wells to the complex discharge (=Qx-iQy).
      COMPLEX FUNCTION CWWLSM(CZ)
      IMPLICIT COMPLEX (C)
      COMMON /CBWL/ NWL,CWLZC(200),RWLQ(200),RWLRAD(200)
      CWWLSM = (0.,0.)
```

```
      DO 100 IWL=1,NWL
        CWWLSM=CWWLSM+RWLQ(IWL)*CWWELL(CZ,IWL)
100   CONTINUE
      RETURN
      END
c     Contribution of a single well of unit discharge to the complex
c     discharge (=Qx-iQy)
      COMPLEX FUNCTION CWWELL(CZ,IWL)
      IMPLICIT COMPLEX (C)
      COMMON /CBWL/ NWL,CWLZC(200),RWLQ(200),RWLRAD(200)
      DATA RPI /3.1415926/
      CRAD=CZ-CWLZC(IWL)
      RW=RWLRAD(IWL)
      IF(CABS(CRAD).GE.RW) THEN
        CWWELL=-1./(2.*RPI*CRAD)
      ELSE
        CWWELL=-1./(2.*RPI*CMPLX(RW,.0))
      ENDIF
      RETURN
      END
c     layout of all wells.
      SUBROUTINE WLLAY
      IMPLICIT LOGICAL (L),COMPLEX(C)
      COMMON /CBWL/ NWL,CWLZC(200),RWLQ(200),RWLRAD(200)
      DO 10 IWL=1,NWL
        CALL DRAWC(CWLZC(IWL),RWLRAD(IWL))
10    CONTINUE
      RETURN
      END
c     Logical function, set true when CZ is less than RTOL away
c     from a well.
      LOGICAL FUNCTION LFATWL(CZ,RTOL)
      IMPLICIT COMPLEX (C), LOGICAL(L)
      COMMON /CBWL/ NWL,CWLZC(200),RWLQ(200),RWLRAD(200)
      LFATWL=.FALSE.
      DO 100 IWL=1,NWL
        IF(RWLQ(IWL).LE..0) GOTO 100
        IF (CABS(CZ-CWLZC(IWL)).LT.(RTOL+RWLRAD(IWL))) LFATWL=.TRUE.
100   CONTINUE
      RETURN
      END
```

c   **file slls.for**
c   Variables contained in the common block BDLS.
c   NLS   = total number of line-sinks.

```
c      NLSS     = total number of line-sinks with given strength.
c      NLSF     = total number of line-sinks with given head.
c      CLSZS(I) = complex coordinate of the starting point of
c                 line-sink I.
c      CLSZC(I) = complex coordinate of the center of line-sink I.
c      CLSZE(I) = complex coordinate of the end point of line-sink I.
c      RLSDIS(I)= strength of line-sink I (discharge per unit length).
c      RLSFI(I) = head at the center of line-sink I.
c      ILSPTS(J)= pointer to the position of the discharge-specified
c                 line-sink number J in the arrays.
c      ILSPTF(J)= pointer to the position of the head-specified
c                 line-sink number J in the arrays.
c      Initialization of the common block with line-sink data:
c      Set the number of line-sinks equal to zero.
       BLOCK DATA BDLS
       IMPLICIT COMPLEX (C),LOGICAL (L)
       COMMON /CBLS/ NLS,NLSS,NLSF,
      .              CLSZS(200),CLSZE(200),CLSZC(200),RLSDIS(200),
      .              RLSFI(200),ILSPTS(200),ILSPTF(200)
       DATA NLS,NLSS,NLSF /3*0/
       END
c      Contibution of all strength-specified line-sinks to the
c      complex potential.
       COMPLEX FUNCTION CFLSG(CZ)
       IMPLICIT COMPLEX (C), LOGICAL (L)
       COMMON /CBLS/ NLS,NLSS,NLSF,
      .              CLSZS(200),CLSZE(200),CLSZC(200),RLSDIS(200),
      .              RLSFI(200),ILSPTS(200),ILSPTF(200)
       CFLSG=(.0,.0)
       DO 100 ILSS=1,NLSS
         IAD=ILSPTS(ILSS)
         CFLSG=CFLSG+RLSDIS(IAD)*COMLS(CZ,CLSZS(IAD),CLSZE(IAD))
 100   CONTINUE
       RETURN
       END
c      Contribution of all head-specified line-sinks to the complex
c      potential.
       COMPLEX FUNCTION CFLSU(CZ)
       IMPLICIT COMPLEX (C), LOGICAL (L)
       COMMON /CBLS/ NLS,NLSS,NLSF,
      .              CLSZS(200),CLSZE(200),CLSZC(200),RLSDIS(200),
      .              RLSFI(200),ILSPTS(200),ILSPTF(200)
       CFLSU=(.0,.0)
       DO 100 ILSF=1,NLSF
```

```
      IAD=ILSPTF(ILSF)
      CFLSU=CFLSU+RLSDIS(IAD)*COMLS(CZ,CLSZS(IAD),CLSZE(IAD))
100   CONTINUE
      RETURN
      END
c     Contribution of a single line-sink of unit strength to the
c     complex potential.
      COMPLEX FUNCTION COMLS(CZ,CZS,CZE)
      IMPLICIT COMPLEX (C), LOGICAL (L)
      DATA RPI /3.1415926/
      CBZ=(2.0*CZ-(CZS+CZE))/(CZE-CZS)
      COM1=(.0,.0)
      COM2=(.0,.0)
      IF(CABS(CBZ+1.).GT..0001) COM1=(CBZ+1.)*CLOG(CBZ+1.)
      IF(CABS(CBZ-1.).GT..0001) COM2=(CBZ-1.)*CLOG(CBZ-1.)
      COMLS=COM1-COM2+2.*CLOG(.5*(CZE-CZS))-2.
      COMLS=.25/RPI*CABS(CZE-CZS)*COMLS
      RETURN
      END
c     Contribution of all line-sinks to the complex discharge (=Qx-iQy).
      COMPLEX FUNCTION CWLSSM(CZ)
      IMPLICIT COMPLEX (C), LOGICAL (L)
      COMMON /CBLS/ NLS,NLSS,NLSF,
     .              CLSZS(200),CLSZE(200),CLSZC(200),RLSDIS(200),
     .              RLSFI(200),ILSPTS(200),ILSPTF(200)
      CWLSSM=(.0,.0)
      DO 100 ILS=1,NLS
         CWLSSM=CWLSSM+RLSDIS(ILS)*CWLS(CZ,ILS)
100   CONTINUE
      RETURN
      END
c     Complex discharge function due to a single line-sink of unit
c     strength.
      COMPLEX FUNCTION CWLS(CZ,ILS)
      IMPLICIT COMPLEX (C)
      COMMON /CBLS/ NLS,NLSS,NLSF,
     .              CLSZS(200),CLSZE(200),CLSZC(200),RLSDIS(200),
     .              RLSFI(200),ILSPTS(200),ILSPTF(200)
      DATA RPI /3.1415926/
      CZS=CLSZS(ILS)
      CZE=CLSZE(ILS)
      CBZD=1./(CZE-CZS)
      CBZ=(2.*CZ-(CZS+CZE))*CBZD
c     Avoid the end points of the line-sink because the discharge
```

```
c       function is singular there.
        IF(CABS(CBZ+1.).LT..0001) CBZ=-.9999
        IF(CABS(CBZ-1.).LT..0001) CBZ=.9999
        CWLS=.5/RPI*CABS(CZS-CZE)*CLOG((CBZ-1.)/(CBZ+1.))*CBZD
        RETURN
        END
c       Layout of all line-sinks.
        SUBROUTINE LSLAY
        IMPLICIT LOGICAL(L),COMPLEX(C)
        COMMON /CBLS/ NLS,NLSS,NLSF,
     .                CLSZS(200),CLSZE(200),CLSZC(200),RLSDIS(200),
     .                RLSFI(200),ILSPTS(200),ILSPTF(200)
        DO 10 I=1,NLS
          CALL DRAW(CLSZS(I),0)
          CALL DRAW(CLSZE(I),1)
   10   CONTINUE
        RETURN
        END
c       Logical function which is set true if CZ is less than RTOL away
c       from a line-sink of positive strength.
        LOGICAL FUNCTION LFATLS(CZ,RTOL)
        IMPLICIT COMPLEX (C), LOGICAL (L)
        COMMON /CBLS/ NLS,NLSS,NLSF,
     .                CLSZS(200),CLSZE(200),CLSZC(200),RLSDIS(200),
     .                RLSFI(200),ILSPTS(200),ILSPTF(200)
        LFATLS=.FALSE.
        DO 100 ILS=1,NLS
          IF(RLSDIS(ILS).LE.0) GOTO 100
          IF(CABS(CLSZS(ILS)-CZ).LT.RTOL
     .    .OR.CABS(CLSZE(ILS)-CZ).LT.RTOL) THEN
            LFATLS=.TRUE.
          ELSE
            CDZLS=CLSZE(ILS)-CLSZS(ILS)
            CCZ=(CZ+CZ-CLSZE(ILS)-CLSZS(ILS))/CDZLS
            IF(ABS(REAL(CCZ)).LT.1.
     .      .AND.ABS(AIMAG(CCZ))*CABS(CDZLS)*.5.LT.RTOL) LFATLS=.TRUE.
          ENDIF
  100   CONTINUE
        RETURN
        END
c       **file sltr.for**
c       Subroutine for plotting a streamline.
c       RSTEP  = step along the streamline ($\Delta s$).
c       CZ     = starting point of the streamline.
```

```
      SUBROUTINE TRACE(RSTEP,CZ)
      IMPLICIT COMPLEX(C),LOGICAL(L)
C     Tolerance used to determine whether a well or line-sink has
C     been reached.
      RTOL=1.1*RSTEP
      CZN=CZ
      CQN=CONJG(CW(CZN))
C     Move the pen to the first point of the streamline.
      CALL DRAW(CZN,0)
  100 CZO=CZN
      CQO=CQN
C     Proceed only if the point is not at a well, not at a line-sink,
C     and inside the window.
      IF(.NOT.LFATWL(CZO,RTOL).AND..NOT.LFATLS(CZO,RTOL)
     .  .AND.LFINWN(CZO)) THEN
C        Abort if a stagnation point is encountered.
         IF(CQO.EQ.(0.,0.)) THEN
            WRITE(*,*) 'DISCHARGE ZERO: STREAMLINE TERMINATED'
         ELSE
C           Determine the complex coordinate of the next point on the
C           discharge streamline, using (26.7).
            CZN=CZO+RSTEP*CQO/CABS(CQO)
C           Compute the complex discharge Qx+iQy at this point.
            CQN=CONJG(CW(CZN))
C           Estimate the average value of the complex discharge on the
C           interval.
            CQ=.5*(CQO+CQN)
C           Obtain the second approximation of the point on the
C           streamline using (26.8)
            CZN=CZO+RSTEP*CQ/CABS(CQ)
C           Plot the point.
            CALL DRAW(CZN,1)
C           Repeat the procedure for the next point.
            GOTO 100
         ENDIF
      ENDIF
      RETURN
      END
```

Appendix D

Additional Reading

The objective of this appendix is to suggest additional reading material that either might give a better understanding of the subject matter covered in this text, or contains other applications of some of the methods discussed in the present work. The list of publications consists primarily of textbooks; papers are included only if the material is not covered in books (to the author's knowledge) and is closely related to the subject matter covered in this text.

The reading material is divided in three categories:

1. General background material in geohydrology
2. General background material in applied mathematics
3. Material related to specific topics discussed in the various chapters and sections of this text

The author has not attempted to compile a complete list of references to publications in the field of groundwater mechanics. For such lists the reader is referred to other textbooks and publications, e.g., Polubarinova-Kochina [1962], Aravin and Numerov [1965], and Bear [1972].

General Reading Material

Texts that treat groundwater flow in the context of hydrology are Todd [1959], Davis and de Wiest [1966], Freeze and Cherry [1979], McWhorter and Sunada [1977], and De Marsily [1986]. A classical treatise that falls in this category is Hubbert [1940].

For general texts on the theory of groundwater flow and its applications, see Muskat [1937], Scheidegger [1960], Polubarinova-Kochina [1962], Harr [1962], Aravin and Numerov [1965], Bear [1972], Walton [1970], Bear [1979], Verruijt [1982], Hunt [1983], de Marsily [1986], and Bear and Verruijt [1987]. Huisman and Olsthoorn [1983] cover the subject of groundwater flow from the viewpoint of artificial recharge. De Wiest [1969] contains treatises by a number of authors on special topics in groundwater flow.

Handbooks on mathematics that the reader might find useful for reference are Korn and Korn [1968] and Spiegel [1968]. For mathematical tables and formulas see Abramowitz and Stegun [1965]. For definite and indefinite integrals, see Gröbner and Hofreiter [1965] or Gradshteyn and Ryzhik [1980].

For the methods of applied mathematics used in this textbook, the reader is referred to texts on mathematical methods in engineering and physics such as Morse and Feshbach [1953] and Wylie [1960].

For general numerical methods, see Carnahan et al. [1969].

Selected Topics

The references listed below give additional information on topics discussed in this text. Most of such information is already contained in the texts listed above; these texts will be mentioned again only if they provide additional information on specific topics. The references are listed for each chapter and section for which additional reading is recommended.

Chapter 2

Section 11   For additional reading on the applications of Girinski potentials, see Polubarinova-Kochina [1962] and Aravin and Numerov [1965].

Section 12   Bear [1972], [1979] contains numerous solutions to problems of shallow interface flow.

Section 13   A paper by Todd and Huisman [1959] contains an interesting method of solution for shallow one-dimensional flow in a coastal aquifer with a leaky layer. The approach is restricted to cases where the head in the main aquifer is a function of x only, but does not require linearization of the differential equation.

Section 15   Polubarinova-Kochina [1962] and Aravin and Numerov [1965] treat problems of steady shallow flow in aquifers with leaky layers in detail.

Section 16   For details on pumping test analysis, see Lohman [1972] and Kruzeman and De Ridder [1983]. The literature on transient flow is extensive; an overview of this literature will not be given here. For an introduction to transient flow with wells, see the treatise by Hantush [1964].

Section 17   Harr [1962] and Cedergren [1967] provide numerous solutions to problems of two-dimensional flow in the vertical plane.

Section 18   Only few solutions to problems of three-dimensional flow exist; some of these are found in Muskat [1937] and Polubarinova-Kochina [1962]. General potential theory in three dimensions is covered in Kellogg [1953].

Chapter 3

Nearly all of the texts on groundwater flow listed above contain at least some of the material covered in Chapter 3. For an introduction to complex variable methods, see Churchill [1960]. For a detailed treatise on singular Cauchy integrals, see Muskhelishvili [1958].

Chapter 4

For more information on contaminant movement in groundwater, see Bear [1972], Bear [1979], Hunt [1983], Javandel et al. [1984], Bear and Verruijt [1987], and Bear and Bachmat [1987]. A treatise contained in a series of four papers by Nelson [1978] focuses on the importance of contaminant spreading due to advective transport as compared to hydrodynamic dispersion. Anderson [1979] gives an overview of approaches to contaminant transport modeling.

Chapter 5

Possibly the most complete text on conformal mapping is Von Koppenfels and Stallmann [1959] (in German). Churchill [1960] gives an introduction to conformal mapping. For a comprehensive text on conformal mapping, written in the English language, see Nehari [1975]. The texts written by Polubarinova-Kochina [1962] and Aravin and Numerov [1965] contain much information on conformal mapping applications to groundwater flow problems. Texts on groundwater flow that contain either a chapter or a section on conformal mapping applications include Harr [1962], Bear [1972], Verruijt [1982], and Hunt [1983].

Chapter 6

Haitjema [1985] suggests a method for combining regional analytic element models with three-dimensional functions.

Chapter 7

The hodograph method is treated in detail in Polubarinova-Kochina [1962] and Aravin and Numerov [1965]. Harr [1962], Bear [1972], Verruijt [1982] and Hunt [1983] also cover the hodograph method. For problems involving many-valued hodographs see Strack [1972] and Strack [1973].

Approximate methods for solving analytically problems of phreatic flow toward horizontal drains are found in Kirkham [1958] and Dagan [1964]. Kirkham's approximation consists of neglecting the resistance to flow in the zone between a horizontal plane and the phreatic surface. Dagan suggested to transfer the boundary condition to a horizontal plane, assuming that the actual elevation of the phreatic surface is equal to the head along this plane. Both approximations have the advantage that they can be applied to problems that cannot be solved conveniently by the hodograph method.

De Josselin de Jong [1969, 1981] presents an approximate method for solving problems of transient interface flow by using distributed singularities along the interface.

Chapter 8

The boundary element method in terms of real variables as applied to groundwater flow problems is presented in detail in Liggett and Liu [1983]. A textbook entirely devoted to the complex variable boundary element method is Hromadka [1984]. The complex variable boundary integral equation method as applied to groundwater flow problems is discussed by Hunt and Isaacs [1981] for the direct formulation, and by Sidiropoulos et al. [1983] for the indirect formulation.

Chapter 9

Textbooks recommended for additional reading on the application of finite difference and finite element methods to groundwater flow problems are Wang and Anderson [1982], Verruijt [1982], and Bear and Verruijt [1987].

Chapter 10

Bear [1972] contains a detailed description of most analogue methods used in groundwater flow. Karplus [1958] covers analogue methods in general.

References

ABRAMOWITZ, M., and I. A. STEGUN. 1965. *Handbook of Mathematical Functions.* Dover Publications, New York.

ALLEN, E. E. 1954. Note 169. *MTAC*, 8, p. 240.

ANDERSON, M. P. 1979. Using models to simulate the movement of contaminants through groundwater flow systems. *Critical Reviews in Environmental Control*, 9(2), pp. 97-156.

ARAVIN, V. I., and S. N. NUMEROV. 1953. *Theory of Motion of Liquids and Gases in Undeformable Porous Media* (translated from Russian). Israel Program for Scientific Translation, Jerusalem.

ARAVIN, V. I., and S. N. NUMEROV. 1965. *Theory of Fluid Flow in Undeformable Porous Media.* Daniel Davey, New York.

BADON GHYBEN, W. 1888. Nota in verband met de voorgenomen putboring nabij Amsterdam. *Tijdschr. Kon. Inst. Ing. 1888-1889*, pp. 8-22.

BEAR, J. 1972. *Dynamics of Fluids in Porous Media.* American Elsevier, New York.

BEAR, J. 1979. *Hydraulics of Groundwater.* McGraw-Hill, New York.

BEAR, J., and Y. BACHMAT. 1987. Macroscopic modelling of transport phenomena in porous media, 1. The continuum approach and 2. Applications to mass momentum and energy transport. *Transport in Porous Media*, 1 (3).

BEAR, J., and A. VERRUIJT. 1987. *Modeling Groundwater Flow and Pollution.* Reidel Publishing Company, Dordrecht, The Netherlands.

BIOT, M. A. 1941. General theory of three-dimensional consolidation. *J. Appl. Phys.*, 12, pp. 155-164.

BISSHOPP, F. 1983. Numerical conformal mapping and analytic continuation. *Quarterly of Applied Mathematics*, April 1983, pp. 125-139.

BOUSSINESQ, J. 1904. Recherches théoriques sur l'écoulement des nappes d'eau infiltreés dans le sol. *Journal de Math. Pures et Appl.*, 10, pp. 363-394.

BROCK, R. R. 1976. Hydrodynamics of perched mounds. *J. Hyd. Div. ASCE.*, 102(HY8), pp. 1083-1099.

CARLSTON, C. W. 1963. An early American statement of the Badon Ghyben-Herzberg principle of static fresh-water-salt-water balance. *Am. Jour. Sci.*, 261,(1), pp. 89-91.

CARNAHAN, B., H. A. LUTHER, and J. O. WILKES. 1969. *Applied Numerical Methods.* Wiley, New York.

CARSLAW, H. S., and J. C. JAEGER. 1959. *Conduction of Heat in Solids*, 2nd ed. Oxford University Press, London.

CEDERGREN, H. R. 1967. *Seepage, Drainage, and Flow Nets.* John Wiley and Sons, New York.

CHARNY, I. A. 1951. A rigorous derivation of Dupuit's formula for unconfined seepage with a seepage surface. *Dokl. Akad. Nauk SSSR*, 79, pp. 937-940.

CHURCHILL, R. V. 1960. *Complex Variables and Applications*, 2nd ed. McGraw-Hill, New York.

CONCAWE. 1977. Strack model to establish optimum locations for a drain to collect oil polluted groundwater: an experimental and numerical evaluation. Report nr. 13/77, The Hague, The Netherlands.

COOPER, H. H., and C. E. JACOB. 1946. A generalized graphical method for evaluating formation constants and summarizing well field history. *Trans. Am. Geophys. Un.*, 27, pp. 526-534.

CURTIS, T. G., JR. 1983. *Simulation of Salt Water Intrusion by Analytic Elements.* Ph. D. Thesis, Dep. of Civ. and Min. Eng., Univ. of Minn., Minnesota.

DAGAN, G. 1964. Second order linearized theory of free-surface flow in porous media. *La Houille Blanche*, 8, pp. 901-909.

DARCY, H. 1856. *Les Fountaines Publiques de la Ville de Dijon.* Dalmont, Paris.

DAVIS, S. N., and R. J. M. DE WIEST. 1966. *Hydrogeology.* Wiley, New York.

DAVISON, B. B. 1936. On the steady two-dimensional motion of ground-water with a free surface. *Phil. Mag.*, 21(143), pp. 881-903.

DAVISON, B. B., and L. ROSENHEAD. 1940. Some cases of steady two-dimensional flow percolation of water through ground. *Proc. Roy. Soc. London*, A, 962, pp. 346-366.

DE JOSSELIN DE JONG, G. 1958. Longitudinal and transverse diffusion in granular deposits. *Trans. Am. Geophys. Un.*, 39, pp. 67-74.

DE JOSSELIN DE JONG, G. 1960. Singularity distributions for the analysis of multiple fluid flow through porous media. *J. Geophys. Res.*, 65, pp. 3739-3758.

DE JOSSELIN DE JONG, G. 1961. Moiré patterns of the membrane analogy for groundwater movement applied to multiple fluid flow. *J. Geophys. Res.*, 66, pp. 3625-3628.

DE JOSSELIN DE JONG, G. 1965. A many valued hodograph in an interface problem. *Water Resour. Res.*, 1(4), pp. 543-555.

DE JOSSELIN DE JONG, G. 1969. Generating functions in the theory of flow through porous media. Ch. 9 in *Flow Through Porous Media*, R. J. M. De Wiest, ed., Academic Press, New York.

DE JOSSELIN DE JONG, G. 1981. The simultaneous flow of fresh and salt water in aquifers of large horizontal extension determined by shear flow and vortex theory. In *Flow and Transport in Porous Media*, A. Verruijt and F. B. J. Barends, ed., pp. 75-82, Balkema, Rotterdam.

DE MARSILY, G. 1986. *Quantitative Hydrogeology.* Academic Press, Orlando, Florida.

DESAI, C. S., and J. F. ABEL. 1972. *Introduction to the Finite Element Method.* Van Nostrand Reinhold, New York.

DETOURNAY, C. 1985. *Applications of Boundary Elements to the Hodograph Method*. Ph. D. Thesis, Dep. of Civ. and Min. Eng., Univ. of Minn., Minnesota.

DE VOS, H. C. P. 1929. Enige beschouwingen omtrent de verweekingslijn in aarden dammen, *Waterstaatsingenieur*, 17, pp. 335-54.

DE WIEST, R. J. M., ed. 1969. *Flow Through Porous Media*. Academic Press, New York.

DIETZ, D. N. 1972. Behavior of components from spilled oil on their way through the soil. 47th Annual Meeting of the Society of Petroleum Engineers of AIME, San Antonio, Texas.

DU COMMUN, J. 1828. On the cause of fresh water springs, fountains, & c., *Am. Jour. Sci.*, 1st ser., 14, pp. 174-176.

DUPUIT, J. 1863. *Études Théoriques et Pratiques sur le Mouvement des Eaux dans les Canaux Decouverts et à Travers les Terrains Perméables*, 2nd ed. Dunod, Paris.

EULER, L. 1811. *Mémoires de l'Acad. de Saint-Pétersbourg*, 3, p. 38.

FITTS, C. R. 1985. *Modeling Aquifer Inhomogeneities with Analytic Elements*. M. S. Thesis, Dep. of Civ. and Min. Eng., Univ. of Minn., Minnesota.

FORCHHEIMER, P. 1886. Ueber die Ergiebigkeit von Brunnen-Anlagen und Sickerschlitzen, *Z. Architekt. Ing. Ver. Hannover*, 32, pp. 539-563.

FREEZE, R. A., and J. A. CHERRY. 1979. *Groundwater*. Prentice Hall, Englewood Cliffs, New Jersey.

GAIER, D. 1964. *Konstructive Methoden der Konformen Abbildung*. Springer-Verlag, Berlin.

GIRINSKI, N. K. 1946. Complex potential of flow with free surface in a stratum of relatively small thickness and $k = f(z)$, (in Russian). *Dokl. Akad. Nauk SSSR*, 51(5), pp. 337-338.

GIRINSKI, N. K. 1947. Complex potential of fresh groundwater flow in contact with brackish water, (in Russian). *Dokl. Akad. Nauk SSSR*, 58(4), pp. 559-561.

GLOVER, R. E. 1974. *Transient Ground Water Hydraulics*. Colorado State University, Fort Collins, Colorado.

GRADSHTEYN, I. S., and I. M. RYZHIK. 1980. *Table of Integrals, Series, and Products*. A. Jeffrey, ed., Academic Press, Orlando, Florida.

GRÖBNER, W., and N. HOFREITER. 1965. *Integraltafel, 1 and 2*, 4th ed. Springer-Verlag, Vienna.

HAITJEMA, H. M. 1977. Numerical application of vortices to multiple fluid flow in porous media. *Delft Progr. Rep.*, 2, pp. 237-248.

HAITJEMA, H. M. 1982. *Modeling Three-Dimensional Flow in Confined Aquifers Using Distributed Singularities*. Ph. D. Thesis, Dep. of Civ. and Min. Eng., Univ. of Minn., Minnesota.

HAITJEMA, H. M. 1985. Modeling three-dimensional flow in confined aquifers by superposition of both two- and three-dimensional analytic functions. *Water Resour. Res.*, 21(10), pp. 1557-1566.

HAITJEMA, H. M. 1987. Comparing a three-dimensional and a Dupuit-Forchheimer solution to a circular recharge area in a confined aquifer. *J. of Hydrol.*, 91, pp. 83-101.

HAITJEMA, H. M., and O. D. L. STRACK. 1985. An initial study of thermal energy storage in unconfined aquifers. Report for Battelle Memorial Institute, Pacific Northwest Laboratories, PNL-5818 UC-94e.

HAMEL, G. 1934. Ueber Grundwasser. *Z. Angew. Math. Mech.*, 14(3), pp. 129-157, Berlin.

HANTUSH, M. S. 1964. Hydraulics of wells. In *Advances in Hydroscience, 1*, V. T. Chow, ed., Academic Press, New York, pp. 281-442.

HARR, M. E. 1962. *Groundwater and Seepage*. McGraw-Hill, New York.

HASTINGS JR., C. 1955. *Approximations for digital computers.* Princeton University Press, Princeton, New Jersey.

HE, H. Y. 1987. *Groundwater Modeling of Leaky Walls.* M. S. Thesis, Dep. of Civ. and Min. Eng., Univ. of Minn., Minnesota.

HEITZMAN, G. 1977. *Analytical Modeling of Multi-Layered Groundwater Flow.* M.S. Thesis, Dep. of Civ. and Min. Eng., Univ. of Minn., Minnesota.

HELMHOLZ, C. 1868. Ueber discontinuirliche Flüssigkeitsbewegungen. *Berl. Monatsber.,* 4.

HENRY, H. R. 1959. Salt intrusion into freshwater aquifer. *J. Geophys. Res.,* 64(11), pp. 1911-1919.

HERZBERG, A. 1901. Die Wasserversorgung einiger Nordseebaden. *Z. Gasbeleucht. Wasserverzorg.,* 44, pp. 815-819, 824-844.

HESS, J. L., and A. M. O. SMITH. 1967. Calculation of potential flow about arbitrary bodies. In *Progress in Aeronautical Sciences,* 8, Pergamon, New York.

HROMADKA, T. V. 1984. *The Complex Variable Boundary Element Method.* Springer-Verlag, Berlin.

HUBBERT, M. K. 1940. The theory of ground-water motion. *J. Geol.,* 48, pp. 785-944.

HUDSON, J. A., and T. R. LAPOINTE. 1980. Printed circuit boards for studying rock mass permeability. *Int. Journ. Rock. Mech. and Min. Sci.,* 17, pp. 297-301.

HUISMAN L., and J. KEMPERMAN. 1951. Bemaling van spannings grondwater. *De Ingenieur,* 62(13).

HUISMAN, L., and T. N. OLSTHOORN. 1983. *Artificial Groundwater Recharge.* Pitman, London.

HUNT, B. 1983. *Mathematical Analysis of Groundwater Resources.* Butterworths, Cambridge.

HUNT, B., and L. T. ISAACS. 1981. Integral equation formulation for groundwater flow. *J. Hyd. Div. ASCE,* 107(HY10), pp. 1197-1209.

IRMAY, S. 1960. Calcul du rabattement des nappes aquiferes. VI^{emes} *Journées de l'Hydraulique,* 7, Question I, Nancy, France.

JACOB, C. E. 1940. The flow of water in an elastic artesian aquifer. *Trans. Am. Geophys. Un.,* 21, pp. 574-586.

JAVANDEL, I., C. DOUGHTY, and C-F. TSANG. 1984. *Groundwater Transport: Handbook of Mathematical Models.* American Geophysical Union, Washington, D. C.

KARPLUS, W. J. 1958. *Analog Simulation.* McGraw-Hill, New York.

KEIL, G. D. 1982. *A Dupuit Analysis for Leaky Unconfined Aquifers.* M.S. Thesis, Dep. of Civ. and Min. Eng., Univ. of Minn., Minnesota.

KELLOGG, O. D. 1953. *Foundations of Potential Theory.* Dover Publications, New York.

KIRCHHOFF, G. 1869. Zur Theorie freier Flüssigkeitsstrahlen. *Crelle,* 70, p. 416.

KIRKHAM, D. 1958. Seepage of steady rainfall through soil into drains. *Trans. Amer. Geophys. Un.,* 39, pp. 892-908.

KIRKHAM, D., and W. L. POWERS. 1971. *Advanced Soil Physics.* Wiley-Interscience, New York.

KOBER, H. 1952. *Dictionary of Conformal Representations.* Dover Publications, New York.

KORN, G. A., and T. M. KORN. 1968. *Mathematical Handbook for Scientists and Engineers,* 2nd ed. McGraw-Hill, New York.

KOZENY, J. 1931. *Wasserkraft und Wasserwirtshaft,* 26, pp. 28-31.

KRUZEMAN, G. P., and N. A. DE RIDDER. 1983. *Analysis and Evaluation of Pumping Test Data.* International Institute of Land Reclamation and Improvement, Wageningen, The Netherlands.

KUMMER, E. E. 1840. *J. Pure and Appl. Math. (Crelle)*, 21, pp. 74-90.

LAMB, H. 1932. *Hydrodynamics*, 6th ed. Cambridge University Press, Cambridge, England.

LAWRENTJEW, M. A., and B. W. SCHABAT. 1967. *Methoden der Komplexen Funktionentheorie*. VEB Deutscher Verlag der Wissenschaften, Berlin.

LEWIN, L. 1958. *Dilogarithms and Associated Functions*. Macdonald, London.

LIGGETT, J. A. 1977. Location of free surface in porous media. *J. Hyd. Div. ASCE*, 102(HY4), pp. 353-365.

LIGGETT, J. A., and P. L-F. LIU. 1983. *The Boundary Integral Equation Method for Porous Media Flow*. George Allen and Unwin, London.

LIST, 1964. The steady flow of precipitation to an infinite series of tile drains above an impervious layer. *J. Geophys. Res.*, 69(16), pp. 3371-3381.

LOHMAN, S. W. 1972. Ground-water hydraulics. *U.S. Geol. Survey Professional Paper 708*.

MCWHORTER, D. B., and D. K. SUNADA. 1977. Ground-water hydrology and hydraulics. *Water Resources Publications*, Fort Collins, Colorado.

MJATIEV, A. N. 1947. Pressure complex of underground water and wells, (in Russian). *Izv. Akad. Nauk SSSR Otd. Tekh. Nauk.*, 9.

MORSE P. M., and H. FESHBACH. 1953. *Methods of Theoretical Physics, I and II*. McGraw-Hill, New York.

MUALEM, Y., and J. BEAR. 1974. The shape of the interface in steady flow in a stratified aquifer. *Water Resour. Res.*, 10(6), pp. 1207-1215.

MUSKAT, M. 1937. *The Flow of Homogeneous Fluids Through Porous Media*. McGraw-Hill, Ann Arbor.

MUSKHELISHVILI, N. I. 1958. *Singular Integral Equations*. Wolters-Noordhoff, Groningen, The Netherlands.

NEHARI, Z. 1952. *Conformal Mapping*. McGraw-Hill, New York.

NEHARI, Z. 1975. *Conformal Mapping*. Dover Publications, New York.

NELSON, R. W. 1978. Evaluating the environmental consequences of groundwater contamination, 1, 2, 3, 4. *Water Resour. Res.*, 14(3), pp. 409-450.

POLUBARINOVA-KOCHINA, P. Y. 1940. Computations of seepage through an earthdam. *PMM*, 4(1).

POLUBARINOVA-KOCHINA, P. Y. 1952. *Theory of Groundwater Movement*, (in Russian). Gostekhizdat, Moscow.

POLUBARINOVA-KOCHINA, P. Y. 1962. *Theory of Groundwater Movement*, (translated by J. M. R. De Wiest). Princeton University Press, Princeton, New Jersey.

SCHEIDEGGER, A. E. 1960. *The Physics of Flow Through Porous Media*, 2nd ed. Univ. of Toronto Press, Toronto.

SIDIROPOULOS, E., C. TZIMOPOULOS, and P. TOLIKAS. 1983. Modelling of discontinuities through dipole distribution. *Appl. Math. Modelling*, 7, October, pp. 306-310.

SPIEGEL, M. R. 1968. *Mathematical Handbook of Formulas and Tables*. McGraw-Hill, New York.

STRACK, O. D. L. 1972. Some cases of interface flow towards drains. *J. Eng. Math.*, 6(2), pp. 175-191.

STRACK, O. D. L. 1973. *Many-valuedness Encountered in Groundwater Flow*. Ph.D. Thesis, Delft University of Technology, Delft, The Netherlands.

STRACK, O. D. L. 1976. A single potential solution for regional interface problems in coastal aquifers. *Water Resour. Res.*, 12(6), pp. 1165-1174.

STRACK, O. D. L. 1981(a). Flow in aquifers with clay lamina, 1. The comprehensive potential. *Water Resour. Res.*, 17(4), pp. 985-992.

STRACK, O. D. L. 1981(b). Flow in aquifers with clay lamina, 2. Exact solution. *Water Resour. Res.*, 17(4), pp. 993-1004.

STRACK, O. D. L. 1982(a). Analytic modeling of flow in a permeable fissured medium. Report for Battelle Memorial Institute, Pacific Northwest Laboratories, PNL-4005 UC-70.

STRACK, O. D. L. 1982(b). Boundary element formulation in terms of complex variables. In *Proceedings of the Fourth International Conference on Numerical Methods in Geomechanics*, Edmonton.

STRACK, O. D. L. 1984. Three-dimensional streamlines in Dupuit-Forchheimer models. *Water Resour. Res.*, 20(7), pp. 812-822.

STRACK, O. D. L. 1985. An application of determining streamlines in a Dupuit-Forchheimer model. *Hydrol. Science and Technol.; Short Papers*, 1(1), pp. 17-23.

STRACK, O. D. L., and M. ASGIAN. 1978. A new function for use in the hodograph method. *Water Resour. Res.*, 14(6), pp. 1045-1058.

STRACK, O. D. L., and T. G. CURTIS, JR. 1980. Outflow from a layered slope. *LGM Med.*, Part XXI(2), pp. 185-192.

STRACK, O. D. L., and H. M. HAITJEMA. 1981(a). Modeling double aquifer flow using a comprehensive potential and distributed singularities, 1, Solution for homogeneous permeabilities. *Water Resour. Res.*, 17(5), pp. 1535-1549.

STRACK, O. D. L., and H. M. HAITJEMA. 1981(b). Modeling double aquifer flow using a comprehensive potential and distributed singularities, 2, Solution for inhomogeneous permeabilities. *Water Resour. Res.*, 17(5), pp. 1551-1560.

STRACK, O. D. L., H. M. HAITJEMA, and J. MELNYK. 1980. Interactive modeling of the aquifers near the Tennessee-Tombigbee Waterway. In *Water and Related Land Resource Systems, IFAC*, pp. 261-269.

TERZAGHI, K. 1943. *Theoretical Soil Mechanics*. Chapman & Hall, London.

THEIS, C. V. 1935. The relation between lowering of the piezometric surface and the rate and duration of discharge of a well using ground water storage. *Trans. Am. Geophys. Un.*, 16th annual meeting, Pt. 2, pp. 519-524.

THIEM, G. 1906. *Hydrologische Methoden*. J. M. Gebhardt, Leipzig, Germany.

TODD, D. K. 1959. *Ground Water Hydrology*. Wiley, New York.

TODD, D. K., and L. HUISMAN. 1959. Ground water flow in the Netherlands coastal dunes. *J. Hyd. Div. ASCE*, 82(HY7), pp. 63-81.

VAN DEEMTER, J. J. 1949. Results of mathematical approach to some flow problems connected with drainage and irrigation. *Appl. Sci. Res.*, A, 2, pp. 333-353.

VAN DER VEER, P. 1976. Calculation method for two-dimensional groundwater flow. In *Delft Prog. Rep.*, 2, pp. 35-49.

VAN DER VEER, P. 1978. *Calculation Methods for Two-Dimensional Groundwater Flow*. Ph.D. Thesis, Delft University of Technology, Delft, The Netherlands.

VERRUIJT, A. 1970. *Theory of Groundwater Flow*. Macmillan, London.

VERRUIJT, A. 1972. Solution of transient groundwater flow problems by the finite element method. *Water Resour. Res.*, 8, pp. 725-727.

VERRUIJT, A. 1980. Finite element calculations on a micro-computer. *Int. J. Num. Meth. Engng.*, 15, pp. 1570-1574.

VERRUIJT, A. 1982. *Theory of Groundwater Flow*, 2nd ed. Macmillan, London.

VON KOPPENFELS, W., and F. STALLMANN. 1959. *Praxis der Konformen Abbildung.* Springer-Verlag, Berlin.

VREEDENBURGH, C. G. J. 1929. Over de vlakke stationnaire waterstroming door een homogene grondmassa, onder aanname van twee stelsels confocale parabolen als stroom- en potentiaallijnen. *T. H.*, Bandoeng, Ned.-Indië.

WALTON, W. C. 1970. *Groundwater Resource Evaluation.* McGraw-Hill, New York.

WANG, H., and M. P. ANDERSON. 1982. *Introduction to Groundwater Modeling.* Freeman, San Fransisco.

WASSYNG, A. 1982. Solving $Ax = b$: A method with reduced storage requirements. *SIAM J. Numer. Anal.*, 19(1), pp. 197-204.

WOODING, R. A. 1966. Groundwater flow over a sloping impermeable layer, 2, Exact solutions by conformal mapping. *J. Geophys. Res.*, 71(12), pp. 2903-2910.

WYLIE, C. R. 1960. *Advanced Engineering Mathematics*, 2nd ed. McGraw-Hill, New York.

YOUNGS, E. G. 1971. Seepage through unconfined aquifers with lower boundaries of any shape. *Water Resour. Res.*, 7(3), pp. 624-631.

ZHUKOVSKI, N. E. 1920. Seepage through dams, *Collected Works*, Vol. VII, Gostekhizdat.

ZIENKIEWICZ. O. C. 1977. *The Finite Element Method*, 3rd ed. McGraw-Hill, London.

List of Symbols

INTRODUCTION

The list of symbols is intended to assist the reader in finding the meaning of the symbols occurring in the formulae. Often the equation where a symbol is defined is given in the list. The list is not complete; symbols that are used incidentally are omitted.

There is a general list, which contains symbols that are used throughout the text, and there are lists of symbols grouped by chapter. Thus, to find a symbol, one might first check the general list, and then the chapter in which the symbol occurs.

The symbols are ordered alphabetically, with lower case letters preceding upper case letters: the order is a, A, b, B, c, C,... Greek letters follow the Latin letters; as an aid to the reader, a Greek alphabet follows. Many letters are used with superscripts, subscripts, exponents and indices. These items, which we will call appendices here for ease of reference, are assigned priorities as follows: index 4, subscript 3, superscript 2, and exponent 1. This may be displayed graphically as

$$\overset{2}{\underset{3}{A}}{\overset{1}{_4}}$$

The rule for ordering is that the symbols are ordered first for the appendix with the highest priority, then for the one with the next highest priority, and so on. Furthermore, it is understood that if an appendix of lower order is absent, the highest order character is to be substituted, and if an appendix of higher order is absent, the lowest order character is to be substituted. For ordering of appendices, the character 0 has the highest order (i.e., numerals precede letters), and the symbols $+$, $-$, and $*$ have the lowest order. Thus, for the purpose of ordering, $\overset{2}{A}$ is equivalent to $\overset{2}{A}{}^{0}_{*}{}_{*}$ and A_2 to $A^0_2{}_0{}^2$.

Greek and other characters

A	α	alpha
B	β	beta
Γ	γ	gamma
Δ	δ	delta
	∂	sign for partial derivative
E	ε	epsilon
Z	ζ	zeta
H	η	eta
Θ	θ, ϑ	theta
I	ι	iota
K	κ	kappa
Λ	λ	lambda
M	μ	mu
N	ν	nu
Ξ	ξ	xi
O	o	omicron
Π	π	pi
P	ρ	rho
Σ	σ	sigma
T	τ	tau
Υ	υ	upsilon
Φ	ϕ, φ	phi
X	χ	chi
Ψ	ψ	psi
Ω	ω	omega
∇		nabla

General

a_j	coefficient of the term with exponent j in a polynomial or expansion
b_j	elevation of the base of layer j; coefficient of the term with exponent j in a polynomial or expansion
B	width of aquifer in the direction normal to the plane of flow (in two-dimensional flow in the vertical plane)
C	constant
d	length or thickness

General List of Symbols

g	acceleration due to gravity
h	thickness of the flow domain in an aquifer
h_f	elevation of the phreatic surface above sea level
h_s	distance between sea level and the interface between fresh groundwater and salt groundwater at rest
$\overset{u}{h}$	thickness of the flow domain in the upper aquifer
H	thickness of a horizontal confined or semiconfined aquifer; elevation of a confining layer above the reference level (in interface flow)
H_j	thickness of layer j
H_s	elevation of the sea level above the reference level
$\overset{l}{H}$	thickness of the lower aquifer
$\overset{u}{H}$	thickness of the upper aquifer
i	purely imaginary complex number of modulus 1
$\Im$	imaginary part of
k	hydraulic conductivity
k_{ij}	(i is either x, y or z and j is either x, y or z) the components of the hydraulic conductivity tensor
k_j	hydraulic conductivity of layer j
$\overset{l}{k}$	hydraulic conductivity in the lower aquifer
$\overset{u}{k}$	hydraulic conductivity in the upper aquifer
L	length
L_j	length of element j
n	porosity
N	infiltration due to rainfall
O	order of
p	pressure
p_f	pressure in the fresh water
p_s	pressure in the salt water
q	specific discharge
q_i	(i is either x, y or z) the components of the specific discharge vector
q_r	radial component of the specific discharge vector
q_x	x component of the specific discharge vector
q_y	y component of the specific discharge vector
q_z	z component of the specific discharge vector

Q	discharge of a well
Q_j	discharge of well number j
Q_n	normal component of the discharge vector
Q_n^+	normal component of the discharge vector on the positive side of a contour, branch cut, or element
Q_n^-	normal component of the discharge vector on the negative side of a contour, branch cut, or element
Q_r	radial component of the discharge vector
Q_s	component of the discharge vector in the direction of flow
Q_x	x component of the discharge vector
Q_{x0}	x component of uniform flow
Q_{x0}^l	x component of uniform flow in the lower aquifer
Q_{x0}^u	x component of uniform flow in the upper aquifer
Q_y	y component of the discharge vector
Q_{y0}	y component of uniform flow
Q_{y0}^l	y component of uniform flow in the lower aquifer
Q_{y0}^u	y component of uniform flow in the upper aquifer
Q_θ	tangential component of the discharge vector
r	radial coordinate
r_j	local radial coordinate emanating from well number j
r_w	radius of a well
R	radius of a circle
$\Re$	real part of
s	arc length measured along a path line or streamline; drawdown
S_p	coefficient of phreatic storage
S_s	coefficient of specific storage
t	time
t_0	time of initiation of a process
T	transmissivity
T_j	transmissivity of layer j if fully saturated
v	seepage velocity
v_x	x component of the seepage velocity vector
v_y	y component of the seepage velocity vector
v_z	z component of the seepage velocity vector
W	discharge function $W = Q_x - iQ_y$
x	x coordinate (horizontal)

General List of Symbols

x_3	x_3 coordinate (vertical)
y	y coordinate (vertical for two-dimensional flow in the vertical plane, horizontal in all other cases)
z	z coordinate (vertical, Secs. 1 through 21); complex variable $z = x + iy$ (from Sec. 22 onward)
Δ	incremental distance
ε	small deviation or variation
θ	angle; tangential polar coordinate
θ_j	tangential coordinate of a polar coordinate system, centered at either well j or vortex j
κ	intrinsic permeability
μ	dynamic viscosity
ρ	density of water
ρ_f	density of fresh groundwater
ρ_s	density of salt groundwater
ϕ	hydraulic head (or piezometric head), used with indices 0, 1, or 2 to denote given values
ϕ_s	head in salt groundwater
ϕ_w	head at a well
$\overset{i}{\phi}$	head in aquifer i
$\overset{l}{\phi}$	head in the lower aquifer
$\overset{u}{\phi}$	head in the upper aquifer
ϕ^+	value of the head on the positive side of a contour, branch cut, or element
ϕ^-	value of the head on the negative side of a contour, branch cut, or element
Φ	discharge potential, used with indices 0, 1, or 2 to denote given values
Φ_w	potential at a well
$\overset{u}{\Phi}_w$	potential in the upper aquifer at a well
$\overset{l}{\Phi}_w$	potential in the lower aquifer at a well
$\overset{d}{\Phi}_e$	potential in the dual aquifer zone at the edge of a lamina
$\overset{l}{\Phi}_e$	potential in the lower aquifer at the edge of a lamina
$\overset{s}{\Phi}_e$	potential in the single aquifer zone at the edge of a lamina

$\overset{u}{\Phi}_c$	potential in the upper aquifer at the edge of a lamina
$\overset{d}{\Phi}$	potential in the dual aquifer zone
$\overset{j}{\Phi}$	potential in aquifer j
$\overset{l}{\Phi}$	potential in the lower aquifer
$\overset{s}{\Phi}$	potential in the single-aquifer zone
$\overset{u}{\Phi}$	potential in the upper aquifer
Φ^+	value of the potential on the positive side of a contour, branch cut, or element
Φ^-	value of the potential on the negative side of a contour, branch cut, or element
Ψ	stream function
Ψ^+	value of the stream function on the positive side of a contour, branch cut, or element
Ψ^-	value of the stream function on the negative side of a contour, branch cut, or element
Ω	$(= \Phi + i\Psi)$ complex potential
$\overset{l}{\Omega}$	complex potential defined in the lower aquifer
$\overset{u}{\Omega}$	complex potential defined in the upper aquifer
Ω^+	value of the complex potential on the positive side of a contour, branch cut, or element
Ω^-	value of the complex potential on the negative side of a contour, branch cut, or element
∇^2	Laplacian, $\nabla^2 = \frac{\partial^2}{\partial x^2} + \frac{\partial^2}{\partial y^2} + \frac{\partial^2}{\partial z^2}$

Chapter 1

A	area
q_x^*	component of the specific discharge vector in the x^* direction
q_y^*	component of the specific discharge vector in the y^* direction
Re	Reynolds number
x^*	Cartesian coordinate parallel to the major principal direction of the hydraulic conductivity tensor
y^*	Cartesian coordinate parallel to the minor principal direction of the hydraulic conductivity tensor
Z	elevation head

α	angle between the major principal direction of the hydraulic conductivity tensor and the x axis

Chapter 2

a	major axis of an elliptical boundary; constant in ditch function
$a_{i,j}$	constants defined in (15.4)
$\overset{i}{A}_j$	constants introduced in the text between equations (15.16) and (15.17)
$\overset{i}{A}$	constants defined in (15.14) through (15.16)
b	minor axis of an elliptical boundary; width of a ditch
c	resistance of a leaky layer, equal to its thickness divided by its hydraulic conductivity
C_c	constant involved in the definition of the potential for confined flow (see [8.5])
$\overset{l}{C}_c$	constant involved in the definition of the potential for confined flow in the lower aquifer (see [13.21] or [13.44])
C_{ci}	constant involved in the definition of the potential for confined interface flow (see [10.30])
$\overset{l}{C}_{ci}$	constant involved in the definition of the potential for confined interface flow in the lower aquifer (see [13.18])
C_{sc}	constant involved in the definition of the potential for semi-confined flow (see [14.2] and [14.9])
C_u	constant involved in the definition of the potential for unconfined flow (see [8.5] and [13.33])
$\overset{l}{C}_u$	constant involved in the definition of the potential for unconfined flow in the lower aquifer (see [13.20])
$\overset{u}{C}_u$	constant involved in the definition of the potential for unconfined flow in the upper aquifer (see [13.6])
C_{ui}	constant involved in the definition of the potential for unconfined interface flow (see [10.52] and [13.32])
$\overset{l}{C}_{ui}$	constant involved in the definition of the potential for unconfined interface flow in the lower aquifer (see [13.19])
$\overset{u}{C}_{ui}$	constant involved in the definition of the potential for unconfined interface flow in the upper aquifer (see [13.6])

$\overset{i}{C}$	constants introduced in (15.114), see also (15.124) through (15.127)
e_0	volume strain of the grain skeleton
e_w	volume strain of water
E_1	exponential integral or well function
F_1	influence function defined in (9.39)
F_2	influence function defined in (9.39)
F_r	influence function for rainfall defined in (9.44)
$\overset{i}{F}$	functions introduced in (15.51); functions introduced in (15.114)
G_0	influence function defined in (9.47), (9.52) and (9.54), compare Figure 2.29
G_d	influence function for a ditch, defined in equations (9.56) through (9.58)
G_p	influence function for a pond, defined in (9.70) and (9.71)
$\overset{i}{G}$	functions introduced in (15.51); functions introduced in (15.114)
H^*	thickness of a leaky layer
I_0	modified Bessel function of the first kind and order zero
I_1	modified Bessel function of the first kind and order one
k^*	hydraulic conductivity of a leaky layer
K_0	modified Bessel function of the second kind and order zero
K_1	modified Bessel function of the second kind and order one
N^*	rate of leakage; rate of decrease in phreatic storage
$\overset{i}{q}{}^*_z$	specific discharge through leaky layer i
$\overset{d}{Q}_n$	normal component of the discharge vector in the dual aquifer zone at the boundary of a lamina
$\overset{s}{Q}_n$	normal component of the discharge vector in the single aquifer zone at the boundary of a lamina
$\overset{d}{Q}_x$	x component of the discharge vector in the dual aquifer zone
$\overset{l}{Q}_x$	x component of the discharge vector in the lower aquifer
$\overset{s}{Q}_x$	x component of the discharge vector in the single aquifer zone
$\overset{u}{Q}_x$	x component of the discharge vector in the upper aquifer
$\overset{l}{Q}_{x0}$	x component of the uniform flow in the lower aquifer

Chap. 2 List of Symbols

$\overset{u}{Q}_{x0}$	x component of the uniform flow in the upper aquifer
$\overset{d}{Q}_{y}$	y component of the discharge vector in the dual aquifer zone
$\overset{l}{Q}_{y}$	y component of the discharge vector in the lower aquifer
$\overset{s}{Q}_{y}$	y component of the discharge vector in the single aquifer zone
$\overset{u}{Q}_{y}$	y component of the discharge vector in the upper aquifer
$\overset{l}{Q}_{y0}$	y component of the uniform flow in the lower aquifer
$\overset{u}{Q}_{y0}$	y component of the uniform flow in the upper aquifer
$\overset{j}{Q}$	discharge of a well in aquifer j
$\overset{n}{T}$	$\sum_{j=1}^{n} T_j$ (T_j is transmissivity of layer j, if fully saturated, see *General*)
u	dimensionless constant occurring in the Theis equation, given by (16.25)
V	volume
W, W_j	particular solutions to the Bessel equation, see (15.18)
x_j	x coordinate of the center of the jth well
x^*	Cartesian coordinate parallel to the major principal direction of the hydraulic conductivity tensor
y_j	y coordinate of the center of the jth well
y^*	Cartesian coordinate parallel to the minor principal direction of the hydraulic conductivity tensor
Z_b	elevation of the lower boundary of the flow domain in an aquifer
Z_t	elevation of the upper boundary of the flow domain in an aquifer
α	angle; constant in the definition of the potential Φ used in Sec. 10, see (10.5) and (10.8)
$\overset{i}{\alpha}_j$	constants defined in (15.17)
β	compressibility of water; constant in the definition of the potential Φ used in Sec. 10, see (10.5) and (10.8)
$\overset{i}{\gamma}$	constants, defined in (15.58)
λ	leakage factor ($\lambda = \sqrt{kHc}$) used in Secs. 14 and 15
λ_c	dimensionless constant defined in (10.38)
λ_u	dimensionless constant defined in (10.59)

μ	dimensionless constant defined in (10.38)
μ_j	constants, used in Sec. 15, see (15.18) and (15.19)
μ_j^i	constants used in Sec. 15, see text between (15.21) and (15.22)
ξ	dimensionless radial coordinate equal to r/λ, used in Sec. 14
ξ_1	distance from the origin to the left border of a ditch (see Sec. 9)
ξ_2	distance from the origin to the right border of a ditch (see Sec. 9)
ϕ_t	head at the tip of a saltwater tongue, see (10.28)
ϕ_t^l	head in the lower aquifer at the tip of a saltwater tongue, see text immediately preceding (13.60)
ϕ_t^u	head in the upper aquifer at the tip of a saltwater tongue, see (13.23)
ϕ_e	head at the edge of a lamina
ϕ_e^l	head in the lower aquifer at the edge of a lamina
ϕ_e^u	head in the upper aquifer at the edge of a lamina
ϕ_{e1}	head at edge number 1 of a lamina
ϕ_{e2}	head at edge number 2 of a lamina
ϕ^*	constant head above a semipermeable layer
Φ_m	constant value of the potential, see (11.21)
Φ_t	value of the potential at the tip of a saltwater tongue (see [10.35], [10.56] and [13.50])
Φ_t^u	value of the potential $\overset{u}{\Phi}$ at the tip of a saltwater tongue in the upper aquifer (see [13.68])
ω	constant defined in (15.12)
ω_j	particular value of ω, see (15.18)

Chapter 3

A	constant; strength of a vortex
b_j	real polynomial coefficient
p	distance of a well from the center of a circular equipotential; far-field correction function
p^*	distance of an image well from the center of a circular equipotential, see (20.3)

Chap. 3 List of Symbols

q	far-field correction function for a line-dipole
Q_0	flow drawn by a lake from infinity; magnitude of the discharge of uniform flow
Q_0^*	discharge of uniform flow
Q_η	normal component of the discharge vector at an element
Q_η^+	normal component of the discharge vector at the $+$ side of an element
Q_η^-	normal component of the discharge vector at the $-$ side of an element
Q_ξ	tangential component of the discharge vector at an element
Q_ξ^+	tangential component of the discharge vector at the $+$ side of an element
Q_ξ^-	tangential component of the discharge vector at the $-$ side of an element
s	strength of a dipole
s_x	x component of the strength of a dipole
s_y	y component of the strength of a dipole
W^*	discharge function, transformed such that its real part equals the tangential component of the discharge vector at an element
x_0	x coordinate of a dipole
x_c	x coordinate of the center of a circular equipotential
x_s	x coordinate of a stagnation point
x_v	x coordinate of a vortex
y_0	y coordinate of a dipole
y_c	y coordinate of the center of a circular equipotential
y_s	y coordinate of a stagnation point
y_v	y coordinate of a vortex
z_0	complex coordinate of either a dipole, or a reference point
$\overset{1}{z}_j$	complex coordinate of the starting point of the jth line-sink
$\overset{2}{z}_j$	complex coordinate of the end point of the jth line-sink
z_s	complex coordinate of a stagnation point
z_v	complex coordinate of a vortex
z_w	complex coordinate of a well
$\overset{1}{z}$	complex coordinate of the starting point of a line-sink
$\overset{2}{z}$	complex coordinate of the end point of a line-sink
$\overline{z}$	complex conjugate of z

Z	($= X + iY$) dimensionless complex variable chosen such that a line-element corresponds to the line segment $-1 \leq X \leq 1$, $Y = 0$ of the real axis in the Z plane
Z_j	($= X_j + iY_j$) dimensionless complex variable chosen such that the jth element corresponds to the line segment $-1 \leq X_j \leq 1$, $Y_j = 0$ of the real axis in the Z_j plane
α	angle between the x axis and an element; angle between the x axis and uniform flow
β	angle between the x axis and the orientation of either uniform flow or a dipole
β_m	coefficient in an expansion
δ	complex coordinate of a point on a line-element
ζ	($= \xi + i\eta$) local complex coordinate with the dimension of length (the ξ axis points in the direction of the element)
λ	strength of a line-doublet
λ_j	strength at the jth node of a string of line-doublets
Λ_{ls}	real part of the complex potential for a line-sink of unit strength
μ	strength of a line-dipole
μ_j	strength at the jth node of a string of line-dipoles
σ	strength of a line-sink
σ_j	strength of the jth line-sink
Φ_0	value of the potential at the reference point
Φ_g	potential due to all functions with given coefficients
Φ_j	value of the potential along the jth equipotential; value of the potential at the center of the jth line-sink
Ψ_j	value of the stream function along the jth streamline

Chapter 4

c	concentration of the mobile phase: mass of contaminant per unit volume of fluid
c_0	concentration at time t_0
c_s	concentration of the sorbed phase: mass of contaminant per unit volume of fluid
E	rate of change in elastic storage of an aquifer, integrated over the thickness (see [27.88])
K_d	distribution factor (see [28.17])
N_p	correction term (occurring when $S_p \neq n$) in the calculation of the elevation of path lines, see (27.95)

Chap. 4 List of Symbols

$\overset{b}{N}$	infiltration through the lower boundary of the flow domain
$\overset{t}{N}$	infiltration through the upper boundary of the flow domain
$\overset{b}{Q}_x$	x component of the discharge vector corresponding to the flow between a streamline and the lower boundary of the flow domain
$\overset{t}{Q}_x$	x component of the discharge vector corresponding to the flow between a streamline and the upper boundary of the flow domain
$\overset{b}{Q}_y$	y component of the discharge vector corresponding to the flow between a streamline and the lower boundary of the flow domain
$\overset{t}{Q}_y$	y component of the discharge vector corresponding to the flow between a streamline and the upper boundary of the flow domain
R	retardation factor
s_0	value of s at the starting point of a streamline or path line
$\overset{p}{s}_0$	value of s at the intersection of a discharge streamline with a pond
$\underset{1}{t}$	time at which a fluid particle is at $\underset{1}{x_i}$
$\overset{n}{\underset{2}{t}}$	n^th estimate of the time at which a fluid particle is at $\underset{2}{x_i}$
$\overset{m}{T}$	constant equal to the transmissivity of all layers from one through m if fully saturated (see [11.27]): $\overset{m}{T} = \sum_{j=1}^{m} T_j$
v_3^*	x_3 component of the vector of apparent velocity (see [28.9])
v_i	($i = 1, 2, 3$) components of the seepage velocity vector
v_x^*	x component of the vector of apparent velocity (see [28.9])
v_y^*	y component of the vector of apparent velocity (see [28.9])
v^*	magnitude of the vector of apparent velocity
x_i	($i = 1, 2, 3$) Cartesian coordinates
$\underset{1}{x_i}$	coordinates of a point on a streamline or path line
$\overset{n}{\underset{2}{x_i}}$	n^th approximation of the coordinates of the next point on a streamline or path line
x_s^*	constant defined by (27.54)
$\overset{i}{\underset{s}{x}}$	x coordinate of a stagnation point inside a pond

$\overset{o}{x}_s$	x coordinate of a stagnation point outside a pond
X_3	elevation of a point of a streamline or path line above the lower boundary of the flow domain (see [27.28], [27.29], [27.70], and [27.74])
$\overset{b}{X}_3$	value of x_3 at the lower boundary of the flow domain
$\overset{t}{X}_3$	value of x_3 at the upper boundary of the flow domain
$\overset{i}{y}_s$	y coordinate of a stagnation point inside a pond
$\overset{o}{y}_s$	y coordinate of a stagnation point outside a pond
$\overset{o}{z}_s$	complex coordinate of a stagnation point outside a pond
$\overset{z}{1}$	complex coordinate of a point on a streamline
$\overset{n}{z}_2$	n^{th} approximation of the complex coordinate of the next point on a streamline
ϵ	rate of change in elastic storage per unit volume
Δs	incremental distance along a streamline or path line
Δt	time increment corresponding to the travel time necessary for a fluid particle to move from the first to the second of two consecutive points computed on a streamline or path line
λ	decay rate factor
μ	function involved in the computation of the elevation of points on a streamline or path line, see (27.21), (27.28), and (27.67)
ρ_b	bulk density of solid material
τ	transmissivity of a vertical section, extending from the base of the flow domain to a point on a streamline or path line, see (27.66)

Chapter 5

a	major axis of an ellipse
A	major axis of the map of an ellipse onto the dimensionless Z plane
$|A|$	scale factor involved in the Schwarz-Christoffel transformation, and in the transformations with piecewise constant argument
b	minor axis of an ellipse
B	minor axis of the map of an ellipse onto the dimensionless Z plane; additive constant involved in the Schwarz-Christoffel transformation

Chap. 5 List of Symbols

f	distance between the center and the focus of an ellipse
k_ν	constant involved in the Schwarz-Christoffel integral, defined in (32.1) and in Figure 5.9
n	number of corner points of a polygon mapped by the use of the Schwarz-Christoffel transformation
L_{eq}	equivalent length (in partial penetration), see (32.100)
Q_0	discharge of uniform flow; discharge into the slot, or ditch, in Figure 5.17; inflow into a ditch assuming partial penetration
Q_0^*	constant related to uniform flows, see (31.57); inflow into a ditch, assuming full penetration
Q_1	flow toward point 1 (infinity) in Figure 5.17(a)
Q_5	flow from point 5 (infinity) in Figure 5.17(a)
Q^*	discharge flowing through an elliptic opening; discharge flowing from infinity into either an elliptic lake or a finite canal
z_ν	corner point of a polygon mapped by the use of the Schwarz-Christoffel transformation; point of a polygon, corresponding to a jump in argument of the transformation of piecewise constant argument
z_w	complex coordinate of a well
Z	dimensionless complex reference parameter (see [31.2], [32.75], or [33.29])
Z_w	complex coordinate of a well in the dimensionless Z plane
α	angle between the x axis and the major axis of an ellipse
α_ν	angle enclosed by two sides of a polygon at a point that lies either at the origin or at infinity
β	angle between the x axis and uniform flow
γ	distance between the center and the focus of an ellipse in the dimensionless Z plane; angle between the collectors of a collector well
γ_ν	constants defined in (33.36)
δ	complex coordinate of a well or drain in the upper half plane $\Im \zeta \geq 0$; pole in the transformation of piecewise constant argument
ζ	$(= \xi + i\eta)$ complex reference parameter ζ, defined in the upper half plane
η	imaginary part of the complex reference parameter ζ
λ	dimensionless parameter, for definitions see (31.48), (32.73), and Figure 5.15(b)
μ	dimensionless parameter, for definitions see (32.72) and Figure 5.17(b)

μ_ν	dimensionless parameter involved in the transformation of piecewise constant argument, equal to α_ν/π if $z_\nu = 0$, and to $-\alpha_\nu/\pi$ if $z_\nu = \infty$ (see [33.6])
ν	constant defined in (31.52)
ξ	real part of the complex reference parameter ζ
ξ_ν	coordinate of a point on the real axis of the ζ plane corresponding to the νth corner point of a polygon mapped onto the real axis
ξ_ν^*	coordinate of a point on the real axis of the ζ plane, where the argument of the transformation of piecewise constant argument jumps
σ	point in $\Im\zeta \geq 0$ where the transformation of piecewise constant argument is zero
τ	real part of χ; complex variable defined in the upper half plane ($= \tau_1 + i\tau_2$)
τ_1	real part of τ
τ_2	imaginary part of τ
φ	angle involved in the Schwarz-Christoffel transformation (see, e.g., [32.9])
φ^*	angle involved in the transformation of piecewise constant argument, see (33.5), (33.11), or (33.28)
Φ_0^l	potential at the reference point in the lower aquifer
Φ_1	potential near point 1 (at $z = L + iH$) in Figure 5.17
Φ_5	potential near point 5 (at $z = -L + iH$) in Figure 5.17
χ	($= \tau + i\omega$) dimensionless complex reference parameter, usually defined outside the unit circle
χ_0	real constant defined by (31.61)
χ_w	complex coordinate of a well in the dimensionless χ plane
ω	imaginary part of χ

Chapter 6

a	major axis of an elliptical equipotential generated by the potential V_{rn}; far-field correction function involved in the definition of the function F, see (35.18)
a_{1_m}	real part of a_m
a_{2_m}	imaginary part of a_m

Chap. 6 List of Symbols

a_j	components of the array of unknown constants, see (34.21); (for $j = 1, 2, 3$) constants associated with a second-order area-sink, defined by (38.23) through (38.25); real coefficients in one of the Laurent expansions involved in double-root elements (see [39.13]); real coefficients in the density distribution of circular analytic elements (see [41.15])
a_j^m	($m = 0, 1$) far-field correction coefficients involved in the function $\overset{e}{\underset{j}{\Omega}}$, see (37.67) and (37.68)
a_m	complex constants involved in analytic elements for closed boundaries, see (42.13)
a^l	constant involved in the linearized expression for the head in terms of potential for the lower aquifer, see (43.31) and (43.32)
a^u	constant involved in the linearized expression for the head in terms of potential for the upper aquifer, see (43.29) and (43.30)
A	area used in the continuity equation, see (34.19); constant defined in (35.9); area of an area-sink
A_{ij}	coefficient matrix, see (34.20)
A^*	($= 1/A$) reciprocal of the constant A defined in (35.9)
b	minor axis of elliptical equipotential generated by the potential V_{rn}; far-field correction function involved in the definition of the function G, (35.18)
b_m^1	real part of b_m
b_m^2	imaginary part of b_m
b_i	components of the known vector, see (34.20); real coefficients occurring in expression (41.23) for the far-field correction function for a circular analytic element
b_j^m	far-field expansion coefficients involved in the potential $\overset{e}{\underset{j}{\Omega}}$, see (37.65); real constants in the expansions about poles used in the double-root elements (see [39.18])
b_r^m	constants defined by (38.51)
b_m	complex constant involved in analytic elements for closed boundaries, see (42.18)
b^l	constant involved in the linearized expression for the head in terms of potential for the lower aquifer, see (43.31) and (43.32)

$\overset{u}{b}$	constant involved in the linearized expression for the head in terms of potential for the upper aquifer, see (43.29) and (43.30)
b^+	base elevation in domain $\mathcal{D}^+$
b^-	base elevation in domain $\mathcal{D}^-$
b^*	width of a slurry wall
B_0	complex constant involved in the expression for the function $\overset{2}{F}_m$ associated with analytic elements for closed boundaries, see (42.39)
B_1	complex constant involved in the expression for the function $\overset{2}{F}_m$ associated with analytic elements for closed boundaries, see (42.39)
B_m	real constants involved in the density distribution of a single-root dipole, see (40.19)
c	resistance of the bottom of a river or a lake; resistance of a leaky layer
$\overset{}{c}{}^m_j$	($m = 0, 1, 2$) complex constants defined by (37.60); real constants in the expansions about poles used in the double-root elements (see [39.18])
$\overset{r}{c}{}^{}_{m_j}$	($r = 1, 2, 3; m = 0, 1, 2, 3$) complex constants involved in the complex potential associated with a triangular second-order area-sink, see (38.46) through (38.49)
C_1	real constant involved in the function U
$\overset{u}{C}_1$	real constant involved in the function $\overset{u}{U}$
d_m	real constants involved in the far-field correction of a single-root dipole, see (40.19); complex coordinate of the map of the pole at $\chi = \delta_m$ from the χ plane onto the Z plane, see (42.34)
$d_{r\nu_j}$	constants defined by (38.54)
f	function of the head ϕ that is independent of the hydraulic conductivity (see [35.2]); function involved in the doublet strength, defined by (35.16)
f_j	function involved in double-root elements, see (39.12)
$\overset{1}{f}_m$	function involved in analytic elements for closed boundaries, see (42.17)
$\overset{2}{f}_m$	function involved in analytic elements for closed boundaries, see (42.32)

Chap. 6 List of Symbols

F	function involved in the line-doublet, defined by (35.18)
F_j	function involved in double-root elements, see (39.16)
$\overset{1}{F}_m$	function involved in analytic elements for closed boundaries, see (42.25)
$\overset{2}{F}_m$	function involved in analytic elements for closed boundaries, see (42.40)
$\overset{e}{F}_j$	complex function associated with the potential for an area-sink, see (37.82)
F^*	function associated with the second-order line-doublet, see (38.18)
g	function involved in the doublet strength, defined by (35.16)
$\overset{1}{g}_m$	function involved in analytic elements for closed boundaries, see (42.17)
$\overset{2}{g}_m$	function involved in analytic elements for closed boundaries, see (42.29)
G	function involved in the line-doublet, defined by (35.18)
G_j	function involved in double-root elements, see (39.17)
$\overset{1}{G}_m$	function involved in analytic elements for closed boundaries, see (42.25)
$\overset{2}{G}_m$	function involved in analytic elements for closed boundaries, see (42.33)
G_p	potential due to a circular pond of unit infiltration rate
$\overset{1}{h}_m$	complex function involved in analytic elements for closed boundaries, defined by (42.13)
H_b	elevation of a point in an aquifer just below a river bottom
$\overset{1}{H}_m$	complex function involved in analytic elements for closed boundaries, defined by (42.18)
H^+	aquifer thickness in domain $\mathcal{D}^+$
H^-	aquifer thickness in domain $\mathcal{D}^-$
k_1	hydraulic conductivity in domain $\mathcal{D}_1$
k_r	hydraulic conductivity of the rth inhomogeneity
k^+	hydraulic conductivity in domain $\mathcal{D}^+$
k^-	hydraulic conductivity in domain $\mathcal{D}^-$
k^*	hydraulic conductivity of a slurry wall
L	length of a second-order line-sink

L_j	length of the jth side of the polygon that bounds an area-sink
m_r	number of line-doublets in the boundary of the rth inhomogeneity
M	total number of equations due to all line-doublets used to model the boundaries of inhomogeneities
n_{ls}	number of line-sinks with given head
n_{ls}^*	number of line-sinks with given strength
n_o	number of line-sinks along the boundary of an opening in a layer separating two aquifers
n_p^*	number of circular ponds
n_w	number of wells with given head
n_w^*	number of wells with given discharge
$N_{p_j}^*$	rate of infiltration of the jth circular pond
N^*	leakage into an aquifer through the bottom of a river
p	far-field correction function for a line-dipole
$\overset{e}{p}_{r_j}$	far-field correction functions associated with a triangular second-order area-sink, see (38.57)
$\overset{e}{p}_j(\zeta)$	far-field correction function for $\overset{e}{\Omega}_j$, see (37.57)
p^*	far-field correction of the component Ω^* of a second-order line-doublet (see [38.12]); far-field correction of a single-root line-doublet (see [40.26])
$\overset{e}{P}_j(Z_j)$	far-field correction function for $\overset{e}{\Omega}_j$, see (37.47)
q^*	far-field correction function for a line-dipole, see (40.22)
Q_0	discharge of cross-flow
$\overset{e}{Q}_0$	discharge of an area-sink
Q_j	discharge of the jth well of given head
Q_j^*	discharge of the jth well of given discharge
$\overset{e}{Q}_n^+$	component of the discharge vector corresponding to the potential $\overset{e}{\Phi}$ normal to, and at the inside of, the boundary of an area-sink
$\overset{e}{Q}_n^-$	component of the discharge vector corresponding to the potential $\overset{e}{\Phi}$ normal to, and at the outside of, the boundary of an area-sink

Chap. 6 List of Symbols

$\overset{i}{Q}{}_n^+$	component of the discharge vector corresponding to the potential $\overset{i}{\Phi}$ normal to, and at the inside of, the boundary of an area-sink.
$\overset{i}{Q}{}_n^-$	component of the discharge vector corresponding to the potential $\overset{i}{\Phi}$ normal to, and at the outside of, the boundary of an area-sink.
$\overset{c}{Q}_s$	discharge flowing through a draining object (e.g., a crack), see (36.7)
$\overset{e}{Q}_x$	x component of the discharge vector corresponding to the potential $\overset{e}{\Phi}$
$\overset{i}{Q}_x$	x component of the discharge vector corresponding to the potential $\overset{i}{\Phi}$
$\overset{e}{Q}_y$	y component of the discharge vector corresponding to the potential $\overset{e}{\Phi}$
$\overset{i}{Q}_y$	y component of the discharge vector corresponding to the potential $\overset{i}{\Phi}$
$\overset{e}{Q}_\eta$	η component of the discharge vector corresponding to the potential $\overset{e}{\Phi}$
$\overset{i}{Q}_\eta$	η component of the discharge vector corresponding to the potential $\overset{i}{\Phi}$
$\overset{i}{Q}{}_\eta^*$	as $\overset{i}{Q}_\eta$, but associated with only the linear component of the strength of a second-order triangular area-sink
$\overset{e}{Q}_\xi$	ξ component of the discharge vector corresponding to the potential $\overset{e}{\Phi}$
$\overset{i}{Q}_\xi$	ξ component of the discharge vector corresponding to the potential $\overset{i}{\Phi}$
$\overset{i}{Q}{}_\xi^*$	as $\overset{i}{Q}_\xi$, but associated with only the linear component of the strength of a second-order triangular area-sink
R	arbitrary constant with the dimension of length; radius of a circular inhomogeneity; radius of a circular analytic element

R^*_{pj}	radius of the jth circular pond
U	sum of the potentials with unknown coefficients
U_{db}	sum of the potentials for line-doublets
U_{ls}	sum of the potentials for line-sinks with given strength
U_{wl}	sum of the potentials for wells with given head
$\overset{u}{U}$	sum of the potentials with unknown coefficients for the upper aquifer
V	sum of the potentials with given coefficients
V_{cf}	potential due to cross-flow
V_{ls}	sum of the potentials for line-sinks with given strength
V_{pd}	sum of the potentials for ponds
V_{rn}	potential due to infiltration from rainfall
V_{wl}	sum of the potentials for wells with given discharge
$\overset{u}{V}$	sum of the potentials with given coefficients for the upper aquifer
w_j	complex coordinate of the jth well of given head
w^*_j	complex coordinate of the jth well of given discharge
$\overset{e}{W}$	discharge function corresponding to the complex potential that has a real part equal to the potential for an area-sink
$\overset{e}{x}_0$	real part of $\overset{e}{z}_0$
x^*_{pj}	x coordinate of the center of the jth circular pond
$\overset{e}{y}_0$	imaginary part of $\overset{e}{z}_0$
y^*_{pj}	y coordinate of the center of the jth circular pond
z_0	complex coordinate of the reference point
$\overset{e}{z}_0$	complex coordinate of the origin of the local coordinate system, used in the potential for an area-sink, see (37.20)
z_c	complex coordinate of the center of a circular element
$\overset{1}{z}_j$	complex coordinate of the starting point of the jth line-sink with given head
$\overset{1}{z}{}^*_j$	complex coordinate of the starting point of the jth line-sink of given strength
$\overset{2}{z}_j$	complex coordinate of the end point of the jth line-sink with given head
$\overset{2}{z}{}^*_j$	complex coordinate of the end point of the jth line-sink of given strength

Chap. 6 List of Symbols

$\overset{c}{z}_j$	complex coordinate of the center of the jth line-sink
$\overset{e}{z}_j$	coordinate of the jth corner point of the polygon that bounds an area-sink
$\overset{1}{z}$	complex coordinate of the starting point of a line-sink, of a single-root element, or of a curvilinear element
$\overset{2}{z}$	complex coordinate of the end point of a line-sink, of a single-root element, or of a curvilinear element
Z	($= X + iY$) dimensionless complex reference parameter such that a line-sink corresponds to the slot $-1 \leq X \leq 1$ of the real axis; complex reference parameter such that a circular element corresponds to a segment of the unit circle in the Z plane
Z_m	($= X_m + iY_m$) dimensionless complex reference parameter such that the mth line element is mapped onto the slot $-1 \leq X_m \leq 1$ on the real axis; dimensionless complex parameter such that the mth side of the polygon that bounds an area-sink is mapped onto the aforementioned slot
$\underset{r}{Z}_m$	($= \underset{r}{X}_m + i\underset{r}{Y}_m$) dimensionless complex reference parameter such that the mth line-doublet of the boundary of the rth inhomogeneity is mapped onto the slot $-1 \leq \underset{r}{X}_m \leq 1$ on the real axis
α	orientation of cross-flow with respect to the x axis
α_j	orientation with respect to the x axis of the jth side of an area-sink; argument of the complex coordinate δ_j of the jth pole involved in an analytic element for a closed boundary
β	angle involved in the definitions of $\theta^+_{m-1,m+1}$ and $\theta^-_{m-1,m+1}$, see (35.31); angle involved in a circular analytic element, see Figure 6.37; angle between uniform flow and the x axis
$\underset{r}{\beta}_j$	complex constants involved in the expansions about poles associated with double-root elements, see (39.21)
β^m_j	complex coefficients in the expansion of χ^m as a function of Z, see (42.31)
β^*_j	complex coefficients in the expansion of the function $\chi = \chi(Z)$ used to map closed boundaries onto a circle, see (42.30)
γ	strength of an area sink, positive for flow out of the aquifer; angle between the chord of a circular element and the x axis
$\overset{e}{\gamma}_j$	($j = 1, 2, 3$) value at node j of the density distribution γ for a second-order triangular area-sink (see [38.23] through [38.25])

$\overset{l}{\gamma}$	strength of an area sink in the lower aquifer
$\overset{u}{\gamma}$	strength of an area sink in the upper aquifer,
δ	complex coordinate of a point on the circle on which a circular analytic element lies, see Figure 6.37
δ_j	complex coordinate of the jth node of a string of line-doublets; complex coordinate of the jth pole associated with a double-root element; complex coordinate of the jth pole of an analytic element for a closed boundary
δ_{r^j}	complex coordinate of the jth node of a string of line-doublets bounding the rth inhomogeneity
$\delta_{r^j}^-$	complex coordinate of the point at an infinitesimal distance to the outside of the jth node of a string of line-doublets bounding the rth inhomogeneity
ζ	($=\xi+i\eta$) local complex coordinate used in the expression for the potential for an area-sink, with the ξ axis parallel to the longest side of the boundary of the area-sink and the origin at $\overset{e}{z}_0$; complex reference parameter such that a circular analytic element corresponds to a segment of a straight line in the ζ plane
ζ_1	starting point of the map of a circular element onto the ζ plane
ζ_2	end point of the map of a circular element onto the ζ plane
ζ_j	complex coordinate in the ζ plane of the jth node of the polygon that bounds an area-sink
ζ_∞	map of the point $z=\infty$ onto the ζ plane
η	imaginary part of ζ
η_j	imaginary part of ζ_j
$\theta_{m-1,m+1}$	angle associated with two adjoining line-doublets, see (35.29)
$\theta_{m-1,m+1}^+$	angle associated with the $+$ side of two adjoining line-doublets, see Figure 6.6 and (35.30)
$\theta_{m-1,m+1}^-$	angle associated with the $-$ side of two adjoining line-doublets, see Figure 6.6 and (35.30)
λ	strength function of a line-doublet
λ_m	strength at the mth node of a string of line-doublets
λ_{r^m}	strength at the mth node of a string of line-doublets bounding the rth inhomogeneity
λ_m^c	strength at the center of the mth second-order line-doublet

$\overset{*}{\lambda}_m$	value at the center of the mth second-order line-doublet of the strength component that vanishes at the end points and has a maximum at the center of the element
$\overset{e}{\lambda}_j$	the function $\overset{e}{\lambda}$ along side j of the polygon that bounds an area-sink
$\overset{e}{\lambda}{}^*_j$	as $\overset{e}{\lambda}_j$, but associated only with the linear component of the strength of a second-order triangular area-sink
$\overset{e}{\lambda}$	jump in the potential $\overset{e}{\Phi}$ associated with an area sink, equal to $-\overset{i}{\Phi}$
$\overset{u}{\lambda}$	strength of doublets in the upper aquifer
λ^*	strength of a single-root line-doublet as a function of χ_1
Λ_{db}	Potential due to two adjoining line-doublets with a linearly varying strength that is zero at the two end nodes and one at the common node (see [35.35])
Λ_{ls}	potential due to a line-sink of unit strength, see (34.10)
Λ_{wl}	potential due to a single well of unit discharge, see (34.7)
μ_j	strength at the jth node of a string of line-dipoles
$\overset{e}{\mu}_j$	the value of $\overset{e}{\mu}$ at the jth node of the polygon that bounds an area-sink
$\overset{e}{\mu}{}^*_j$	as $\overset{e}{\mu}_j$, but associated only with the linear component of the strength of a second-order triangular area-sink
$\overset{e}{\mu}_j$	the function $\overset{e}{\mu}$ along the jth side of the polygon that bounds an area-sink
$\overset{e}{\mu}{}^*_j$	as $\overset{e}{\mu}_j$, but associated with only the linear component of the strength of a second-order triangular area-sink
$\overset{e}{\mu}$	jump in the stream function $\overset{e}{\Psi}$, which is equal to the imaginary part of the potential $\overset{e}{\Omega}$, associated with an area-sink
μ^*	density distribution of a single-root line-dipole as a function of χ_1, see (40.23)
$\overset{e}{\nu}_j$	dimensionless constant, equal to $\overset{e}{\mu}_j/\gamma$ (see [37.55])
ξ	real part of ζ
ξ_j	real part of ζ_j
ρ_m	modulus of the complex coordinate δ_m of the mth pole involved in an analytic element for a closed boundary
σ_j	strength of the jth line-sink of given head

$\overset{u}{\sigma}_j$	strength of the jth line-sink of given head in the upper aquifer
σ_j^*	strength of the jth line-sink of given strength
$\overset{e}{\sigma}_j$	the strength $\overset{e}{\sigma}$ along the jth side of the polygon that bounds an area-sink
$\overset{e}{\sigma}_j^*$	as $\overset{e}{\sigma}_j$, but now associated only with the linear component of the strength of a second-order triangular area-sink
$\overset{1}{\sigma}$	strength at the starting point of a second order line-sink
$\overset{2}{\sigma}$	strength at the end point of a second order line-sink
$\overset{e}{\sigma}$	line-sink strength associated with the potential for an area-sink, equal to $\overset{i}{Q}_n^+$
τ	real part of χ
τ_1	real part of χ_1
$\overset{m}{\tau}_j$	($m=1,2$) constants involved in the expression for the strength $\overset{e}{\mu}_j^*$, see (38.42) through (38.44)
$\tilde{\phi}(\Phi)$	function that converts potentials to heads
ϕ^*	head in a river
$\overset{1}{\Phi}_0$	potential at the reference point in the domain $\mathcal{D}_1$, see (35.10)
$\overset{e}{\Phi}_0$	as $\overset{e}{\Phi}$, but now associated only with the constant component of the strength distribution of a second-order triangular area-sink
$\overset{i}{\Phi}_0$	as $\overset{i}{\Phi}$, but now associated only with the constant component of the strength distribution of a second-order triangular area-sink
Φ_0^*	constant involved in the complex potential $\overset{1}{\Omega}$ used for analytic elements for closed boundaries (see [42.26])
Φ_1	potential for a single-root line-dipole
$\Phi_\mathcal{A}$	potential at an end point of a drain
Φ_{ν_c}	potential at the center of the νth line-sink
$\overset{u}{\Phi}_{\nu_o}$	value of the potential $\overset{u}{\Phi}$ at the center of the νth line-sink along the boundary of an opening in a separating layer
Φ_{ν_w}	potential at the boundary of the νth well
Φ_0	potential at the reference point
$\overset{u}{\Phi}_0$	potential in the upper aquifer at the reference point

Chap. 6 List of Symbols

$\overset{e}{\Phi}$	potential associated with an area sink, defined by (37.5) and (37.8)
$\overset{e}{\Phi}{}^{*}$	as $\overset{e}{\Phi}$, but now associated only with the linear component of the strength distribution of a second-order triangular area-sink
$\overset{i}{\Phi}$	potential associated with an area sink, defined by (37.5), (37.6), and (37.7)
$\overset{i}{\Phi}{}^{*}$	as $\overset{i}{\Phi}$, but now associated only with the linear component of the strength distribution of a second-order triangular area-sink
$\tilde{\Phi}(\phi)$	function that converts heads to potentials
χ	($= \tau + i\omega$) complex reference parameter defined outside the unit circle, and chosen such that either thin isolated features or closed boundaries of isolated features correspond to the unit circle
χ_1	($= \tau_1 + i\omega_1$) complex reference parameter chosen such that a tip element corresponds to the segment $-1 \leq \tau_1 \leq 1$ of the real axis
$\overset{e}{\chi}{}_r^j$	($r = 1, 2, 3$) functions involved in the complex density distribution for side j of a triangular second-order area-sink, see (38.53)
$\overset{e}{\chi}{}^j$	complex strength function involved in the complex potential $\overset{e}{\Omega}{}^j$, see (37.54)
χ^+	value of χ along the upper half of the unit circle
χ^-	value of χ along the lower half of the unit circle
Ψ_0	constant involved in analytic elements for closed boundaries, see (42.6) and (42.7)
Ψ_1	stream function for a single-root line-dipole
$\Psi_{\mathcal{A}}$	stream function at the end point of an impermeable wall
ω	imaginary part of χ
ω_1	imaginary part of χ_1
$\overset{e}{\Omega}{}_0$	as $\overset{e}{\Omega}$, but associated only with the constant component of the strength distribution of a second-order triangular area-sink
Ω_1	complex potential for a single-root line-dipole
Ω_{cf}	complex potential for cross-flow
$\overset{e}{\Omega}{}^j$	contribution to $\overset{e}{\Omega}$ of side j of the polygon that bounds an area-sink

$\overset{1}{\Omega}$ complex potential in the domain $\mathcal{D}_1$; complex potential involved in analytic elements for closed boundaries, see (42.5) through (42.10)

$\overset{2}{\Omega}$ complex potential in the domain $\mathcal{D}_2$; complex potential involved in analytic elements for closed boundaries, see (42.9) and (42.10)

$\overset{e}{\Omega}$ complex potential that has a real part equal to the harmonic part ($\overset{e}{\Phi}$) of the potential for an area-sink.

$\overset{e}{\Omega}{}^*$ as $\overset{e}{\Omega}$, but associated only with the linear component of the strength distribution of a second-order triangular area-sink

Ω^* complex potential due to two adjoining line-doublets (see [35.25]); component of the complex potential for a second-order line-doublet, chosen such that its strength vanishes at the end points and has a maximum at the center of the element

Chapter 7

a	real constant
$\lvert A \rvert$	real and positive constant
A_0	real constant, see (50.25)
$\lvert A_1 \rvert$	real and positive constant
$\lvert A_2 \rvert$	real and positive constant
C	complex constant
C_1	real part of C
$\underset{k}{C_1}$	($k = 1, 2$) real part of $\underset{k}{C}$
$\underset{z}{C_1}$	real constant defined by (51.45)
C_2	imaginary part of C
$\underset{k}{C_2}$	($k = 1, 2$) imaginary part of $\underset{k}{C}$
C_z	complex constant involved in the function $z = z(\zeta)$
$\underset{k}{C}$	($k = 1, 2$) complex constants
$\overset{j}{C}$	($j = 1, 2$) complex constants
$\overset{p}{d}$	constant equal to the negative of the x coordinate of a drain in the physical plane, see (54.34)
h	distance between a drain and a recharge drain
h_1	distance between a drain and a horizontal upper boundary

Chap. 7 List of Symbols

h_s	elevation of the point of intersection of a seepage face and a phreatic surface; distance between a stagnation point on the interface and the horizontal upper boundary, see Figure 7.34
$\overset{p}{h}$	distance between a drain and a recharge drain in the physical plane
$\overset{t}{h}$	distance between a drain and a recharge drain in the transformed domain
H_0	elevation of the phreatic surface at $x = -L$, see (49.12); distance between a horizontal boundary equipotential and an interface at infinity
k_1	major principal value of the hydraulic conductivity tensor
k_2	minor principal value of the hydraulic conductivity tensor
k_ν	constant involved in the Schwarz-Christoffel integral, defined in (32.1)
$\overset{1}{k_\nu}$	value of k_ν involved in the function that maps the upper half plane onto the domain in either the W^{-1} or the $\overset{1}{W}$ plane
$\overset{2}{k_\nu}$	value of k_ν involved in the function that maps the upper half plane onto the domain in either the Ω or the $\overset{1}{T}$ plane
k^*	factor with the dimension of a hydraulic conductivity, and equal to $k(\rho_s - \rho_f)/\rho_f$ (see [45.10])
K	constant with the dimension of hydraulic conductivity, defined by (54.18)
L	value of x for which the elevation of the phreatic surface is H_0, see (49.12)
L_b	distance fom the y axis to the plane where the heads are given, see Figure 7.27
Li_2	dilogarithm, see (50.22) and Appendix B
n	number of corner points involved in the Schwarz-Christoffel transformation
n_e	number of extraction drains
n_i	number of infiltration drains
$\overset{1}{n}$	number of corner points involved in the Schwarz-Christoffel transformation that maps the upper half plane onto the domain in either the W^{-1} or the $\overset{1}{W}$ plane

$\overset{2}{n}$	number of corner points involved in the Schwarz-Christoffel transformation that maps the upper half plane onto the domain in either the Ω or the $\overset{1}{T}$ plane
$\overset{c}{q}$	constant defined by (54.39)
$\overset{i}{q}$	constant defined by (54.39)
q^*	dimensionless discharge defined by (51.24)
Q	discharge to or from infinity; discharge of a horizontal drain
Q_a	discharge flowing from negative infinity for the problem shown in Figure 7.27
$\overset{l}{Q_b}$	discharge flowing to infinity in the lower aquifer, see Figure 7.27
$\overset{u}{Q_b}$	discharge flowing to infinity in the upper aquifer, see Figure 7.27
Q_j	discharge of the jth drain
$\overset{e}{Q_j}$	discharge of the jth extraction drain
$\overset{i}{Q_j}$	discharge of the jth infiltration drain
$\overset{e}{Q}$	discharge of the extraction drain
$\overset{i}{Q}$	discharge of the infiltration drain
R	reference function, see (47.1)
s	distance from the origin to the map of the stagnation point onto the negative real axis of the ζ plane
$\underset{j}{T}_1$	real part of $\underset{j}{T}$
$\underset{j}{T}_2$	imaginary part of $\underset{j}{T}$
$\underset{j}{T}$	($j = 1, 2$) auxiliary functions defined by (47.25)
w	specific discharge function, defined as W/B
$\underset{j}{W}$	($j = 1, 2$) transformation of the discharge function, see (46.22)
$\overset{p}{x}$	horizontal Cartesian coordinate in the physical plane
$\overset{p}{x}*$	Cartesian coordinate parallel to the major principal direction of the hydraulic conductivity tensor, see (54.2)
$\overset{t}{x}$	coordinate in the transformed domain chosen such that the $\overset{t}{x}$ axis corresponds to the $\overset{p}{x}$ axis in the physical plane
$\overset{t}{x}*$	coordinate in the transformed domain, see (54.3)

Chap. 7 List of Symbols

X	real part of the Zhukovski function Z
y_r	y coordinate of the reference level for the head
$\overset{p}{y}_r$	y coordinate of the reference level for the head
$\overset{p}{y}$	vertical Cartesian coordinate in the physical plane
$\overset{p}{y}{}^*$	Cartesian coordinate parallel to the minor principal direction of the hydraulic conductivity tensor, see (54.2)
$\overset{t}{y}$	coordinate in the transformed domain
$\overset{t}{y}{}^*$	coordinate in the transformed domain, see (54.3)
Y	elevation head; imaginary part of the Zhukovski function Z
$\overset{e}{z}_m$	complex coordinate of the mth extraction drain
$\overset{i}{z}_m$	complex coordinate of the mth infiltration drain
$\overset{t}{z}$	complex variable defined as $\overset{t}{x} + i\overset{t}{y}$
$z'(\zeta)$	derivative of the function $z = z(\zeta)$ with respect to ζ
Z	Zhukovski function, see (49.5)
α	slope of an embankment; argument of δ; angle between the major principal direction of the hydraulic conductivity tensor and the x axis
α_e	coordinate of the first point of a segment of the real axis of the ζ plane, that corresponds to a straight equipotential
α_j	value of $\arg z'(\xi)$ along a boundary segment of type j
α_r	coordinate of the first point of a segment of the real axis of the ζ plane, that corresponds to a phreatic surface with accretion
$\overset{t}{\alpha}$	angle between the $\overset{t}{x}{}^*$ and $\overset{t}{x}$ axes
β	real constant equal to $(k_1/k_2)^{1/2}$
β_e	coordinate of the second point of a segment of the real axis of the ζ plane, that corresponds to a straight equipotential
β_r	coordinate of the second point of a segment of the real axis of the ζ plane, that corresponds to a phreatic surface with accretion
γ	ratio of extraction rate to infiltration rate ($\overset{e}{Q}/\overset{i}{Q}$)
γ_j	j^{th} winding point
δ	complex coordinate of the map of a drain onto the ζ plane
δ_j	complex coordinate of the map of the jth drain onto the ζ plane
$\Delta\Phi_j$	jump in potential at the jth boundary vortex

ζ	complex reference parameter defined in the upper half plane
η	imaginary part of ζ
λ	real part of ω_0, see (49.14); parameter; distance of the first extraction drain above the real axis of the ζ plane ($\rho_1 = i\lambda$)
μ	coordinate in the ζ plane corresponding to the tip of a lamina in the z plane, see (50.26)
ν	angle defined in (54.24) and in Figure 7.45
ξ	real part of ζ
ξ_ν	coordinate of a point on the real axis of the ζ plane corresponding to the νth corner point of a polygon mapped onto the real axis
$\overset{1}{\xi}_\nu$	value of ξ_ν involved in the function that maps the upper half plane onto the domain in either the W^{-1} or the $\underset{1}{W}$ plane
$\overset{2}{\xi}_\nu$	value of ξ_ν involved in the function that maps the upper half plane onto the domain in either the Ω or the $\underset{1}{T}$ plane
ξ_ν^*	coordinate of a point on the real axis of the ζ plane, where the argument of the transformation of piecewise constant argument jumps
ρ_j	complex coordinate of the map of the jth extraction drain onto the ζ plane
σ	complex coordinate of the map of a stagnation point onto the ζ plane
σ^*	point in the ζ plane where W equals τ_1
τ_j	($j = 1, 2$) constant involved in the definition of $\underset{j}{W}$, see (46.22)
ϕ_a	head at $z = -L_a$
$\overset{l}{\phi}_b$	head in the lower aquifer at $z = L_b$
$\overset{u}{\phi}_b$	head in the upper aquifer at $z = L_b + iH$
φ	angle involved in the Schwarz-Christoffel transformation (see [32.6])
$\overset{j}{\varphi}$	($j = 1, 2$) angle involved in the Schwarz-Christoffel transformation
Φ_0	boundary value of the potential along a straight equipotential; constant
Ψ_0	boundary value of the stream function along a straight streamline
$\Psi'(\xi)$	derivative of the stream function with respect to ξ
ω	scaled discharge function, see (49.10)
ω_0	value of ω at $z = -L + iH_0$

Chap. 9 List of Symbols

$\Omega'(\zeta)$ — derivative of the function $\Omega = \Omega(\zeta)$ with respect to ζ

Chapter 8

G	Green's function
n_i	($i = 1, 2, 3$) components of the unit vector normal to the boundary
Q_i	($i = 1, 2$) Cartesian components of the discharge vector
r	distance between the boundary point with coordinates (ξ, η) and the point with coordinates (x, y), see (56.4)
s	arc length along a boundary contour
V	volume
x_i	($i = 1, 2, 3$) Cartesian coordinates
δ	complex coordinate of a point on the boundary
$\dfrac{\partial}{\partial n}$	derivative in a direction normal to, and out of, the boundary
η	y coordinate of a boundary point
λ	jump in potential, see (58.4)
ξ	x coordinate of a boundary point
Ω_j	($= \Phi_j + i\Psi_j$) value of Ω at boundary node j
$\overset{2}{\Omega}$	sum of additional functions, see (58.5)
$\overset{n}{\Omega}$	complex potential along the nth boundary segment

Chapter 9

a	distance between grid points in a finite difference grid
$\overset{m}{a}_j$	($j = 1, 2, 3$) constants involved in the expression for $\overset{m}{\phi}$, see (61.21)
A_m	area of the mth triangle in a finite element mesh
$\overset{m}{b}_j$	($j = 1, 2, 3$) constants defined in (61.22)
c	resistance of a semipermeable layer
$\overset{m}{c}_j$	($j = 1, 2, 3$) constants defined in (61.22)
c_m	resistance of a semipermeable layer in element m
$\overset{m}{d}_j$	($j = 1, 2, 3$) constants defined in (61.22)
F	function equal to the potential divided by the hydraulic conductivity, see (61.56)
$\overset{n}{F}_{i,j}$	constant defined by (60.34)

N_m	rate of infiltration for element m
$\overset{m}{p}_j$	constants involved in the expression of $\overset{m}{\phi^*}$, see (61.34)
P_{ij}	coefficient matrix defined by (61.28)
$\overset{m}{P}_{ij}$	coefficient matrix for element m, defined by either (61.27) or (61.37)
Q_i	constant defined in (61.44)
$\overset{m}{Q}_i$	constant defined by (61.39)
R_{ij}	matrix defined in (61.44)
$\overset{m}{R}_{ij}$	matrix defined by (61.38)
s	distance measured along the boundary
T	transmissivity
T_m	transmissivity inside element m
U	functional
U_m	contribution of element m to the functional
$\overset{m}{x}_k$	($k = 1, 2, 3$) x coordinate of corner point k of element m
$\overset{m}{y}_k$	($k = 1, 2, 3$) y coordinate of corner point k of element m
β	constant defined by (61.49)
γ_m	constant defined by either (61.40) or (61.51)
Δt	time step
Δt_{crit}	critical time step
$\overset{m}{\eta}$	local coordinate for element m, see (61.35)
λ	variation, see (61.7)
$\overset{m}{\xi}$	local coordinate for element m, see (61.35)
ν	interpolation coefficient defined by (61.47) (see also [60.33])
ϕ_i	head at node i
$\overset{m}{\phi}_i$	head at corner point i of triangle m
$\overset{m}{\phi^*_i}$	value of ϕ^* at node i of element m
$\overset{m}{\overline{\phi}_i}$	value of $\overline{\phi}$ at node i of element m
$\overline{\phi}_i$	value of $\overline{\phi}$ at node i
ϕ^*_i	value of ϕ^* at node i
$\phi^n_{i,j}$	head at node (i, j) and time level n
ϕ_t	head at time t
$\phi_{t+\Delta t}$	head at time $t + \Delta t$
ϕ_+	head at the end of a time interval

Chap. 10 List of Symbols

ϕ_-	head at the beginning of a time interval
$\overset{m}{\phi}$	head inside the mth triangle, see (61.20)
$\overset{m}{\phi^*}$	value of ϕ^* for element m
$\overline{\phi}$	average value of the head over a time interval, see (61.47)
ϕ^*	constant head at a point separated from the aquifer by a semipermeable layer
$\Phi_{i,j}$	potential at node (i,j)
$\overset{n}{\Phi}_{i,j}$	potential at node (i,j) and time level n
$\Phi_{i,j}^{(m)}$	potential at node (i,j) and at the mth iteration
$\overset{n}{\Phi}$	potential at time level n
χ	function that differs from Φ by a variation λ, see (61.7)

Chapter 10

d	width of the slot in a Hele Shaw analogue
i	density of electrical current
V	electric potential
ρ	specific resistance; density of fluid

Author Index

Author Index

A

Abel, J. F., 615
Abramowitz, M., 118, 169, 197, 198, 674
Allen, E. E., 198
Anderson, M. P., 3, 607, 619, 675, 676
Aravin, V. I., 2, 195, 514, 673, 674, 675
Asgian, M., 514, 532

B

Bachmat, Y., 675
Badon Ghyben, W., 97
Bear, J., 2, 3, 131, 149, 152, 159, 607, 619, 627, 634, 673, 674, 675, 676
Biot, M. A., 192
Bisshopp, F., 481
Boussinesq, J., 195
Brock, R. R., 131

C

Carlston, C. W., 97
Carnahan B., 314, 566, 674
Carslaw, H. S., 202, 613, 627
Cedergren, H. R., 674
Charny, I. A., 41
Cherry, J. A., 195, 200, 673
Churchill, R. V., 262, 352, 675
Cooper, H. H., 199
Cundall, P. A., 56
Curtis T. G., Jr., 533, 534, 535, 536, 539, 540, 631

D

Dagan, G., 675
Darcy, H., 5, 8
Davis, S. N., 200, 673
Davison, B. B., 513, 532
De Josselin de Jong, G., 423, 571, 627, 628, 633, 675
De Marsily, G., 673, 674
De Ridder, N. A., 674
Desai, C. S., 615
Detournay, C., 479, 481, 526
De Vos, H. C. P., 281
De Wiest, R. J. M., 200, 673, 674
Dietz, D. N., 634
Doughty, C., 342, 675
Du Commun, J., 97
Dupuit, J., 36

E

Euler, L., 640

F

Feshbach, H., 674
Fitts, C.R., 413, 414
Forchheimer, P., 36
Freeze, R. A., 195, 200, 673

G

Gaier, D., 481
Girinski, N. K., 2, 40, 48, 113, 119
Glover, R. E., 202
Gradshteyn, I. S., 674
Gröbner, W., 674

H

Haitjema, H. M., 2, 218, 283, 333, 412, 423, 451, 492, 500, 510, 675
Hamel, G., 513, 532
Hantush, M. S., 202, 674
Harr, M. E., 349, 514, 674, 675
Hastings Jr., C., 198
He, H. Y., 424, 474
Heitzman, G., 175
Helmholz, C., 513
Henry, H. R., 110
Herzberg, A., 97
Hess, J. L., 627
Hofreiter, N., 674
Hromadka, T. V., 676
Hubbert, M. K., 673
Hudson, J. A., 628
Huisman, L., 172, 469, 674
Hunt, B., 2, 589, 596, 602, 603, 674, 675, 676

I

Irmay, S., 283
Isaacs, L. T., 2, 589, 596, 602, 603, 676

J

Jacob, C. E., 193, 199
Jaeger, J. C., 202, 613, 627
Javandel, I., 342, 675

K

Karplus, W. J., 627, 676
Keil, G. D., 191, 192
Kellogg, O. D., 674

Kemperman, J., 172
Kirchhoff, G., 513
Kirkham, D., 2, 675
Kober, H., 352
Korn, G. A., 357, 590, 674
Korn, T. M., 357, 590, 674
Kozeny, J., 281
Kruseman, G. P., 674
Kummer, E. E., 642

L

Lamb, H., 631
Lapointe, T. R., 628
Lawretjew, M. A., 390
Lewin, L., 550, 639, 640, 641
Liggett, J. A., 2, 3, 283, 589, 594, 602, 605, 676
List, E. J., 532
Liu, P. L-F., 2, 3, 589, 594, 602, 605, 676
Lohman, S. W., 199, 674
Luther, A., 314, 566, 674

M

McWhorter, D. B., 673
Melnyk, J., 500
Mjatiev, A. N., 172
Morse, P. M., 674
Mualem, Y., 131, 149, 152, 159
Muskat, M., 2, 218, 349, 510, 514, 674
Muskhelishvili, N. I., 298, 300, 301, 675

N

Nehari, Z., 262, 352, 532, 675
Nelson, R. W., 634, 675
Numerov, S. N., 2, 195, 514, 673, 674, 675

O

Olsthoorn, T. N., 469, 674

P

Polubarinova-Kochina, P. Y., 2, 172, 349, 514, 532, 562, 673, 674, 675
Powers, W. L., 2

R

Rosenhead, L., 532
Ryzhik, I. M., 674

S

Schabat, B. W., 390
Scheidegger, A. E., 674
Sidiropoulos, E., 2, 589, 603, 676
Smith, A. M. O., 627
Spiegel, M. R., 487, 674
Stallmann, F., 262, 352, 375, 390, 468, 532, 675
Stegun, I. A., 118, 169, 197, 198, 674
Strack, O. D. L., 2, 36, 98, 127, 283, 319, 322, 330, 332, 333, 336, 370, 412, 415, 454, 468, 469, 470, 473, 492, 500, 510, 514, 532, 533, 534, 535, 536, 539, 540, 546, 547, 552, 553, 560, 568, 569, 589, 632, 675
Sunada, D. K., 673

T

Terzaghi, K., 193
Theis, C. V., 196, 627
Thiem, G., 26
Todd, D. K., 673, 674
Tolikas, P., 2, 589, 603, 676
Tsang, C-F., 342, 675
Tzimopoulos, C., 2, 589, 603, 676

V

Van Deemter, J. J., 530
Van der Veer, P., 2, 283, 596
Verruijt, A., 2, 3, 10, 12, 200, 208, 209, 388, 607, 615, 616, 619, 621, 622, 674, 675, 676
Von Koppenfels, W., 262, 352, 375, 390, 468, 532, 675
Vreedenburgh, C. G. J., 281

W

Walton, W. C., 674
Wang, H., 3, 607, 619, 676
Wassyng, A., 511
Wilkes, J. O., 314, 566, 674
Wooding, R. A., 532
Wylie, C. R., 674

Y

Youngs, E. G., 110, 113, 119

Z

Zhukovski, N. E., 513, 542
Zienkiewicz, O. C., 615

Subject Index

Subject Index

Page numbers refer to pages that contain either a description of the subject listed, a formula representing it, or a discussion relevant to it.

A

Abandoned well, 139, 142
Advection, 340
Advective transport, 340
Aerodynamics, 627
Analogue:
  electric, 630
  Hele-Shaw, 631
  membrane, 628
Analogue method, 627
Analytic element, 404
  circular, 476
  curvilinear, 474
  double-root, 462
  one-sided, 480
  two-sided, 480
Analytic element for closed boundaries, 480
Analytic element method, 403
Analytic function: general properties, 275
Anisotropic hydraulic conductivity, 13, 211, 579
Apparent path line, 343
Apparent velocity, 343
Aquiclude, 112, 502
Aquifer, coastal, 96, 143, 553
Aquifer, layered:
  contaminant transport in, 345
  streamlines in, 335, 336
Aquifer base elevation, discontinuity in the, 413
Aquifers, system of two, 492
Aquifer systems:
  confined leaky, 178
  regional, 404
  unconfined leaky, 185
Aquifer thickness, discontinuity in the, 413
Aquitard, 112
Area-sink, 425, 426, 454, 497
  of the first order, potential for a, 438
  of the second order, potential for a, 461
  bounded by a polygon, 429
Argument, 266
Auxiliary function, 513, 514

B

Backward finite difference, 614
Base elevation, discontinuity in the aquifer, 413
Basic auxiliary functions, 514
BASIC program, 135
Behavior at a node, 417
Bessel functions, 167
Bessel's equation, 167

BIEM, 589
Bilinear transformation, 352, 353
Boundaries, analytic elements for closed, 480
  confining, 20
Boundary:
  circular, 241
  conditions, 221, 514
  elliptical, 80
  integral, 592
  interzonal, 49, 102, 113, 114, 116
Boundary integral equation method, 589
Boundary types, 514
Boundary value problem, 221, 352, 513, 589
  mixed, 221, 610
Boussinesq equation, 195

C

Canal (finite, narrow), 363, 461
Cauchy integral, 299, 596
  singular, 298, 595
Cauchy's integral theorem, 300
Cauchy-Goursat integral theorem, 299
Cauchy-Riemann equations, 222
Central finite differences, 614
Characteristic, 343, 344
Circular boundary, 241
Circular element, 476
Circular equipotential, 241
Circular island, 241
Circular lake (fully penetrating), 244, 355
Circular pond, 87, 407
Clausen's integral, 642
Clay laminae, (see also Impermeable laminae), 126, 143, 533
Coastal aquifer, 96, 143, 553
Coefficient of specific storage, 196
Coefficient of volume compressibility, 194
Collector well, radial, 396
Complex conjugate, 265, 276
Complex discharge function, 278
Complex functions, 269
Complex potential, 278
Comprehensive discharge vector, 130
Comprehensive potential, 130
Computer program, 56, 89, 120, 135, 302, 318, 510, 645
Concentration, 341
  of the mobile phase, 341
  of the sorbed phase, 341
Conduction of heat in solids, 627
Conductivity, hydraulic, 8, 10, 411
Confined flow, horizontal, 20

Confined leaky systems, 178
Confining boundaries, 20
Conformal mapping, 349, 513
   direct and indirect applications of, 352
Conjugate, complex, 265, 276
Connection between aquifers, 139
Conservative force fields, 627
Constraint point, 404
Contaminant spreading, 142
Contaminant transport, 340
Continuity equation, 16, 39
   in a regional model, 409
Contouring (of stream function), 311
Contours:
   piezometric, 34
   potentiometric, 23
Crack, 422, 473
Creek, 283, 289, 290, 404, 405
Critical time step, 611, 612
Cross-flow, 407
Curve of symmetry, 355
Curvilinear elements, 474
Cusp (of an interface), 559
Cylindrical impermeable object, 253
Cylindrical inhomogeneity, 414

D

Dam, flow underneath a, 379
Darcy's law, 11, 12, 14
Datum, 7
Decay, 341, 342
De Moivre's formula, 267
Density, 8, 12
Digitizer, 445
Dilogarithm, 550, 639
   many-valued, 640
   principal branch, 639
   real and imaginary parts, 641, 642
Dipole, 247
   complex potential for a, 279
   double-root, 462
Direct application of conformal mapping, 352
Dirichlet problem, 221, 594, 609
Discharge, 8
   discontinuous, 287
   specific, 6
Discharge function, 278
Discharge potential, 23, 38, 98, 114
   for horizontal confined flow, 23, 49
   for shallow confined interface flow, 102
   for shallow unconfined interface flow, 109
   for shallow unconfined flow, 38, 49
   for vertically varying hydraulic conductivity, 114
Discharge streamline, 24, 323
Discharge vector, 22, 37

comprehensive, 130
Discharge well, 28
Discontinuity:
   in the aquifer base elevation, 413
   in the aquifer thickness, 413
   in the hydraulic conductivity, 412
   in the potential, 298, 412, 413, 421, 423, 466, 480
   in the stream function, 288, 295, 423, 463, 472, 473, 479
Discontinuous discharge, 287
Dispersion, 341
   hydrodynamic, 341, 633, 634
   macroscopic, 341
   microscopic, 341, 633
Distribution factor, 345
Ditch, 82, 461
   drainage, 461
   partially penetrating, 381
Divide, 77, 80
Dividing layer, 498
Double-root dipole, 462
Double-root doublet, 462, 466
Double-root elements, 462
   applications of, 466
Doublet, double-root, 462, 466
Drain, 425, 461, 553, 571
   narrow ideal, 425, 461
Drainage ditch, 425, 461
Draining object, 420, 422
Draw-down, 200
Dual-aquifer zone 127
Dupuit-Forchheimer assumption, 36, 319
Dynamic viscosity, 10

E

Eccentric well, 241
Edge, lamina, 130
Einstein summation convention, 590
Elastic storage, 195, 339
Electric analogue, 630
Electrokinetics, 627
Element (*see also* Analytic element):
   circular, 476
   curvilinear, 474
   tip, 469
Elements:
   double-root, 462
   finite, 615
Elevation head, 8
Elliptic impermeable object, 364
Elliptic inhomogeneity, 487
Elliptic lake (fully penetrating), 358
Elliptic opening, 365
Elliptical boundary, 80
Equipotential, 23, 515, 518

Subject Index

circular, 241
Equivalent length, 388
Euler's formula, 267
Euler's method, 314, 317
   modified, 314
Eutaw formation, 500
Expansion:
   Laurent, 357, 466, 478
   Taylor, 272, 273
Explicit finite differences, 611
Exponential integral, 197

F

Far-fields, 88
Finite canal, 363
Finite difference methods, 607
Finite differences:
   backward, 614
   central, 614
   explicit, 611
   implicit, 613
   forward, 614
Finite element method, 615
First-order area-sink, potential for the, 438
First-order line-dipole, 424
First-order line-doublet, 415
Flow:
   transient, 192, 509, 611, 622
   uniform, 57, 279
Flow underneath a dam, 379
FORTRAN, 56, 510
FORTRAN program, 56, 89, 120, 302, 318, 645
Forward finite differences, 614
Free boundary, 209, 513
Free surface, 209
Function:
   auxiliary, 513, 514
   basic auxiliary, 514
   complex, 269
   influence, 83, 415
Function U with unknown coefficients, 406, 408
Function V with given coefficients, 406
Functional, 616

G

Gauss-Seidel iteration, 609, 613, 619
Gaussian elimination, 511, 594
Generalized Schwarz-Christoffel transformation, 568
Ghyben-Herzberg equation, 97
Girinski potentials, 118
Gordo formation, 500
Gradient, hydraulic, 9, 12
Green's second identity, 589, 591

Grey Cloud Island, 443

H

Harmonic function, 219
Head, 7
   elevation, 8
   hydraulic, 7
   pressure, 8
Hele-Shaw analogue, 631
Hodograph, 518
Hodograph method, 513
Hodograph plane, 518
Horizontal confined flow, 20
Horner's rule, 198
Hydraulic conductivity, 8, 10, 411
   anisotropic, 13, 211, 579
   inhomogeneity of the, 412
   vertically varying, 112, 334
Hydraulic conductivity tensor, 15
Hydraulic gradient, 9, 12
Hydraulic head, 7
Hydrodynamic dispersion, 341, 633, 634

I

Images, method of, 27, 28, 555, 573
Image well, 31, 241
Imaginary axis, 263
Imaginary part, 263
Impermeable laminae (*see also* Clay laminae), 143, 533, 546
Impermeable layer (*see also* Aquiclude), 139
Impermeable object:
   cylindrical, 253
   elliptic, 364
Implicit finite differences, 613
Indirect application of conformal mapping, 352
Infiltration (or rainfall), 74, 111, 406, 425, 516, 521, 526, 582, 608, 620
Influence functions, 83, 415
Inhomogeneity, 411, 412, 413, 496
   cylindrical, 414
   elliptic, 487
Instability, 105, 558, 576
Interface, 96, 143, 517, 523, 527, 553
Interzonal boundary, 49, 102, 113, 114, 116
Intrinsic permeability, 10
Inversion, point of, 525, 528, 531, 532
Island, circular, 241
Iteration index, 609

J

Jacobi method, 609
Jump in potential, 298, 412, 413, 421, 423, 466, 480
Jump in stream function 288, 295, 423, 463, 472, 473, 479

L

Lake, 405, 425
   circular (fully penetrating), 244, 355
   elliptic (fully penetrating), 358
Lake bottom, 425
Lake (leakage), 425
Lamina, clay or impermeable, 126, 143, 533
Lamina edge, 130
Laplace transform, 197
Laplace's equation, 16, 23, 40, 50, 204, 513
Laplacian, 166
Laurent expansion, 357, 466, 478
Layer, leaky, 112, 160, 178, 185, 498
Layered aquifer:
   contaminant transport in, 345
   streamlines in, 335, 336
Leakage, 425, 620
   through a separating bed, 497
Leaky layers, 112, 160, 178, 185, 498
Leaky object, 420, 421
Leaky systems:
   confined, 178
   unconfined, 185
Leaky wall, 421
Leibnitz's rule, 636
Line-dipole, 291, 296, 424
   double-root, 462
   first-order, 424
   single-root, 470
Line-doublet, 297, 423, 424
   double-root, 466
   first-order, 415
   second-order, 451
   single-root, 474
Line-sink, 283, 296, 408
   first-order, 285
   second-order, 450
Liouville's theorem, 273
Local radial coordinate, 28
Lock, 255, 256
Logarithmic singularity, 227

M

Macroscopic dispersion, 341
Magnetism, 627
Many-valued dilogarithm, 640
Mass balance equation, 15

Material-time derivative, 344
Maximum modulus theorem, 219, 221
McShan formation, 500
Mean value theorem, 220
Membrane, 627, 628
Membrane analogue, 628
Method of images, 27, 28, 555, 573
Microscopic dispersion, 341, 633
Mixed boundary value problem, 221, 610
Mobile phase, concentration of, 341
Modulus, 266
Möbius transformation, 352
Moivre's formula, De, 267

N

Narrow canal, 461
Narrow ideal drain, 425, 461
Neumann problem, 221, 609
Newton-Raphson procedure (for streamlines), 313
Node:
   behavior at a, 417
   singularity at a, 423

O

Object, leaky, 420, 421
Ohm's law, 630
One-sided analytic element, 480
Opening, elliptic, 365
Outflow face, 515, 519

P

Partial penetration, 34
   error if full penetration assumed, 388
Partially penetrating ditch, 381
Path line, 311, 313, 316, 317
   apparent, 343
Path lines, pattern of, 313
Phreatic storage, 195
Phreatic storage coefficient, 195, 337
Phreatic surface, 36, 516, 520, 526, 581, 582
Piecewise constant argument, transformations with, 389
Piezometric contour, 34, 38
Piping, 532
Plug flow, 340, 341
Plume, 326
Point of inversion, 525, 528, 531, 532
Poisson equation, 76, 117, 132
Pole of order m, 357
Poles, 466, 467, 483, 569
Polygon (conformal mapping), 371, 374, 390, 392, 395

Polygon, area-sink bounded by a, 429
Pond, circular, 87, 407
Porosity, 6, 337
Positive sense (in tracing a contour), 351
Potential:
 complex, 278
 comprehensive, 130
 discharge, 23, 38, 98, 114
 for variable hydraulic conductivity, 623, 624
 Girinski, 118
 jump in, 298, 412, 413, 421, 423, 466, 480
Potentiometric contours, 23
Pressure, 8
Pressure head, 8
Principal branch of $\ln(1 - z)$, 639
Principal branch of the dilogarithm, 639
Pumping test, 199

R

Radial collector well, 396
Radial coordinate, local, 28
Radius of influence, 199
Rainfall (or infiltration), 74, 111, 406, 425, 516, 521, 526, 582, 608, 620
Real axis, 263
Real part, 263
Recharge well, 28
Reference function, 514, 525
Reference level, 7
Reference parameter, 352, 514
Reference point, 58, 290, 409, 629
Regional aquifer systems, 404
Residue condition, 375
Resistance network, 631
Resistance to flow, 8, 34, 36, 319
Retardation, 341
Retardation factor, 342, 345
Reynolds number, 10
Riemann surface, 468, 568, 569
River, 404, 425
River bottom, 425

S

Saltwater tongue, 102, 104, 109, 147
Saturated zone, 36
Scale model, 633
Schwarz-Christoffel integral for the exterior of a polygon, 375
Schwarz-Christoffel integral for the interior of a polygon, 372
Schwarz-Christoffel transformation, 369
 generalized, 569, 570
Schwarz Reflection Principle, 355

Second-order area-sink, 454
Second-order line-doublet, 451
Second-order line-sink, 450
Seepage face, 210, 515, 519
Seepage velocity, 6, 7
Semiconfined flow, 160
Semipermeable layer, 160
Sense, positive (in tracing a contour), 351
Series, Taylor, 272, 273
Shallow semiconfined flow, 160
Shallow unconfined flow, 36
Shiely Company, 443
Single-aquifer zone, 127
Single-root line-dipole, 470
Single-root line-doublet, 474
Single-valued function, 224
Singular Cauchy integrals, 298, 595
Singularity, logarithmic, 227
Singularity at node, 423
SLAEM, 445, 452, 510
Slow viscous flow, 627
Slurry wall, 420, 421, 461
SLW, 56, 89, 120, 645
SLWL, 302, 318, 645
Solute transport, 340
Sorbed phase, concentration of the, 341
Sorption, 340
Specific discharge, 6
Specific storage, coefficient of, 196
Stability, condition for, 576
Stagnation point, 23, 33, 45, 46, 105, 230, 234, 236, 253, 280, 327, 328, 547, 569
Storage, 192, 195, 339
Storage coefficient, specific, 196, 199
Storage equation, 192
Storativity, 199
Stream function, 222
 jump in the, 288, 295, 423, 463, 472, 473, 479
Streamline, 23, 311, 313, 322, 335, 336, 515, 519
 discharge, 24, 323
 for transient flow, 313
Stress-strain relation, 193
Summation convention, Einstein, 590
Superposition, 27, 79, 170, 201
Superposition in $\Im z \geq 0$, 563, 574
Superposition principle, 27
SYLENS, 500, 510
System of two aquifers, 492

T

Taylor expansion 272, 273
Teledeltos paper, 630
Tennessee-Tombigbee Waterway, 500
Tensor, hydraulic conductivity, 15
Theis solution, 197, 509

Thickness, discontinuity in the aquifer, 413
Thiem equation, 26
Thin isolated features, 461
Three-dimensional effects, 510
Time step, critical, 611, 612
Tip element, 469
Tip of saltwater tongue, 102, 104, 109, 147
Two-sided analytic element, 480
Tracing a contour in positive sense, 351
Transformation, bilinear, 352, 353
Transformations with piecewise constant argument, 389
Transient effects (in analytic element models), 509
Transient flow, 192, 509, 611, 622
Transmissivity, 114, 115, 199
Transport:
  advective, 340
  contaminant, 340
  solute, 340
Transpose elimination, 511
Two-dimensional flow in the vertical plane, 203
Two-sided analytic element, 480

U

Unconfined aquifer, shallow, 36
Unconfined flow, shallow, 36
Unconfined leaky systems, 185
Uniform flow, 57, 279
U.S. Army Corps of Engineers, Nashville District, 500
U.S. Bureau of Mines, 571
U.S. Geological Survey, 507
Unsaturated zone, 340, 582
Unstable interface, 105, 558
Unstable phreatic surface, 575

V

Variable infiltration, 425
Variational principle, 615, 616
Velocity, apparent, 343
Vertical embankment, 541
Vertical plane, two-dimensional flow in the, 203
Vertical stream surface, 329
Vertically varying hydraulic conductivity, 112, 334
Viscosity, dynamic, 10
Visual image of regional flow, 628
Volume compressibility, coefficient of, 194
Volume strain, 193
Vortex, 255
  complex potential for a, 279

W

Well, 25, 167, 176, 183, 191, 407, 408
  abandoned, 139, 142
  complex potential for a, 279
  discharge, 28
  eccentric, 241
  image, 31, 241
  radial collector, 396
  recharge, 28
Well between parallel rivers, 375
Well between two intersecting rivers, 377
Well function (transient), 197
Winding point, 569

Z

Zhukovski function, 542, 572